Hot-electron science and microscopic processes in plasmonics and catalysis

London, UK

18–20 February 2019

FARADAY DISCUSSIONS

Volume 214, 2019

ROYAL SOCIETY OF CHEMISTRY

The Faraday Division of the Royal Society of Chemistry, previously the Faraday Society, was founded in 1903 to promote the study of sciences lying between chemistry, physics and biology.

Editorial Staff

Executive Editor
Philippa Ross

Deputy Editor
Heather Montgomery

Editorial Production Manager
Claire Darby

Publishing Editors
Joseph Gibson, Kirsten Hall

Editorial Assistant
Aliya Anwar

Publishing Assistants
Robert Griffiths, David Bishop

Publisher
Jamie Humphrey

Faraday Discussions (Print ISSN 1359-6640, Electronic ISSN 1364-5498) is published 8 times a year by the Royal Society of Chemistry, Thomas Graham House, Science Park, Milton Road, Cambridge, UK CB4 0WF.

Volume 214 ISBN 13: 978-1-78801-670-4

2019 annual subscription price: print+electronic £1220 US $2148; electronic only £1162, US $2046. Customers in Canada will be subject to a surcharge to cover GST. Customers in the EU subscribing to the electronic version only will be charged VAT.

All orders, with cheques made payable to the Royal Society of Chemistry, should be sent to the Royal Society of Chemistry Order Department, Royal Society of Chemistry, Thomas Graham House, Science Park, Milton Road, Cambridge, CB4 0WF, UK Tel +44 (0)1223 432398; E-mail **orders@rsc.org**

If you take an institutional subscription to any Royal Society of Chemistry journal you are entitled to free, site-wide web access to that journal. You can arrange access via Internet Protocol (IP) address at **www.rsc.org/ip**

Customers should make payments by cheque in sterling payable on a UK clearing bank or in US dollars payable on a US clearing bank.

Printed in the UK

MIX Paper from responsible sources FSC® C013604

ROYAL SOCIETY OF CHEMISTRY | Celebrating IYPT 2019

Faraday Discussions are unique international discussion meetings that focus on rapidly developing areas of chemistry and its interfaces with other scientific disciplines.

Hot-electron science and microscopic processes in plasmonics and catalysis

Faraday Discussions

www.rsc.org/faraday_d

A General Discussion on Hot-electron science and microscopic processes in plasmonics and catalysis was held in London, UK on the 18th, 19th and 20th of February 2019.

RSC Publishing is a not-for-profit publisher and a division of the Royal Society of Chemistry. Any surplus made is used to support charitable activities aimed at advancing the chemical sciences. Full details are available from www.rsc.org

CONTENTS

ISSN 1359-6640; ISBN 978-1-78801-670-4

Cover
Front cover image courtesy of Stefan Maier, Imperial College London.

INTRODUCTORY LECTURE

PAPERS AND DISCUSSIONS

Faraday Discussions

EDITORIAL

The 300th Faraday Discussion

Susan Weatherby, *[a] John Seddon [b] and Claire Vallance [c]

DOI: 10.1039/c9fd90015b

This *Faraday Discussion* volume on *Hot-electron science and microscopic processes in plasmonics and catalysis* is a record of the 300th Discussion meeting, held at Burlington House, London from 18th to 20th February 2019. Given Michael Faraday's contributions to electromagnetism, and his studies on the optical properties of colloidal metallic nanoparticles from the 1850s, plasmonics was a very fitting topic for this anniversary Discussion.

In addition to the participants at the 300th meeting, a number of guests who have been instrumental to the organisation of the Discussions over the past ten years attended the 300th meeting celebrations on the first day. These included many past and current members of the Faraday Division Council and the Faraday Standing Committee on Conferences (Fig. 1). The celebrations commenced with an overview talk on the history and significance of the Discussions given by Claire Vallance, President of the Faraday Division Council, and John Seddon, Chair of the Faraday Standing Committee on Conferences (FSCC). The regular Discussion poster session and wine reception that followed was upgraded to a celebratory glass of bubbly and celebratory anniversary cake, which was enjoyed by all participants.

The Faraday Discussions have an interesting history, which began in 1902, when a group of British physical chemists decided to set up a new society. The Faraday Society was founded in 1903, with the aim of "promoting the study of electrochemistry, electrometallurgy, chemical physics, metallurgy, and kindred subjects". There was some grumbling from famous scientists of the day at the prospect of a new society and the inevitable new journal, on the grounds that there were already too many journals to keep up with.

The Faraday Society originally held what were called Ordinary Meetings, at which a wide variety of scientific topics were discussed. Right from the start, the focus was on discussion, and all papers were circulated beforehand. In 1906, the

[a]*Royal Society of Chemistry, Thomas Graham House, Science Park, Milton Road, Cambridge CB4 OWF, UK. E-mail: weatherbys@rsc.org*

[b]*Chemistry Department, Molecular Sciences Research Hub, Imperial College London, Wood Lane, London W12 0BZ, UK*

[c]*Department of Chemistry, University of Oxford, Chemistry Research Laboratory, 12 Mansfield Rd, Oxford OX1 3TA, UK*

Fig. 1 Celebrating the 300th Discussion: Wendy Niu, Andy Mount, Eleanor Campbell, Philippa Ross, Sandra Macaskill, John Seddon, Claire Vallance, Susan Weatherby and Dwayne Heard.

Council proposed holding scientific meetings for which all papers presented were on a common theme. This marked the true start of the Faraday Discussions.

The first Faraday Discussion was held on 29th January 1907, on the topic of Osmotic Pressure, with Professor H. E. Armstrong FRS in the Chair. Indeed, this was the topic for which the very first Nobel Prize in Chemistry had been awarded, to Jacobus H. van't Hoff, in 1901.

Since 1910 there has been at least one Discussion each year, even through two world wars. Nowadays there are approximately eight Discussions a year, on a wide variety of topics suggested by the scientific community. While the first meetings focused on topics that would have been of direct interest to Michael Faraday himself, over the years the scope of the meetings has broadened, and they now often cover topics at the interfaces of chemistry, biology, and physics.

The concerns of those who grumbled about the setting up of the Faraday Society turned out to be very well justified: alongside the meetings, the Faraday Society did indeed start its own journal. *Transactions of the Faraday Society* was established in 1905 to publish the Society's Discussions. In 1947 this was split into two journals, *Faraday Transactions*, and the *Faraday Discussions*. The numbering of the current *Faraday Discussions* volume starts from 1947 with volume 1, hence this 300th meeting is numbered as the 214th volume.

A number of traditions have been introduced during the history of the Discussions. One is the Spiers Memorial Award, presented to an Introductory Lecturer in recognition of their outstanding contribution to the field of the Discussion. The award commemorates F. S. Spiers, the first secretary of the Faraday Society. One of the 2019 joint winners of this award, Naomi Halas, was present at the 300th Discussion. Another tradition is the Loving Cup ceremony, the Loving cup was purchased by the Faraday Society in 1950, to commemorate G. S. W. Marlow, Secretary of the Faraday Society 1926–1948. The cup is used at Faraday Discussion conference dinners, where a toast to the memory of Marlow is given, and to the memory of Angela and Tony Fish. Angela Fish worked for the Royal Society of Chemistry from 1968–1995 and was most closely associated with the organisation of the Faraday Discussion meetings. She left a legacy to the Royal Society of Chemistry in both her name and that of her husband Tony. The 300th meeting was no exception to this unique tradition, and all delegates enjoyed the Loving Cup ceremony, if not necessarily all partaking of the port therein.

From a Discussion in 1935 that is recognised as the birth of polymer physics, to a Discussion on atmospheric chemistry in 1995 taking place just when the discipline was expanding, Faraday Discussions have always been at the cutting edge of chemistry, attended by world-leading scientists, and have paved the way for significant scientific advances. The discussion-driven format of the meetings is a unique feature. Many attendees have commented on the fact that the longer discussion time really allows for open debates, where young researchers can interact with world-leading scientists in their field. The 300th meeting had its fair share of robust and insightful discussion, as you will find later in this volume and as shown in Fig. 2.

Since their inception, more than 40 000 delegates have attended Faraday Discussions. During the last decade, the Royal Society of Chemistry has increasingly organised Discussion meetings outside the UK, in countries where there is substantial high-quality research. In recent years, Discussions have been held in China, India, and South Africa, and we expect this trend to continue.

Of course, the Discussions would not be possible without the many volunteers and Royal Society of Chemistry staff members who contribute to their organisation. On the occasion of this 300th meeting, we would like to formally thank all those who have made the Discussions what they are today. To all the past members of the

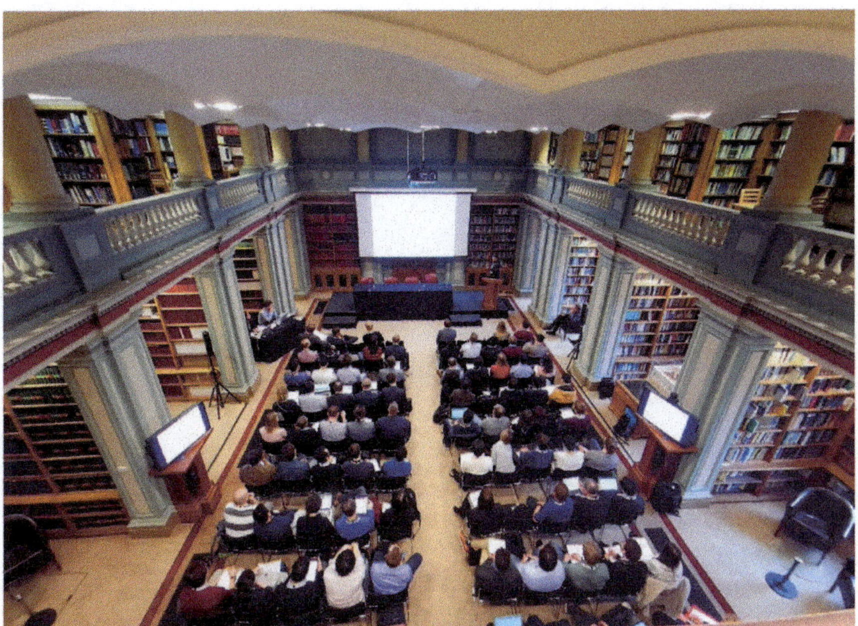

Fig. 2 Discussion in progress at the 300th meeting.

FSCC, the members of each Discussion scientific committee, the authors/speakers and the participants themselves – thank you. To all the Royal Society of Chemistry staff – thank you. Around 50 staff each year are currently involved in some aspect of their organisation and many more across the years. All play a role in making *Faraday Discussions* the unique and world-leading series that they are.

A list of upcoming Discussions can be found on the Royal Society of Chemistry website (http://rsc.li/faraday-discussion). Over the coming two years, topics such as quantum effects in complex systems, peptide–membrane interactions, and the chemistry of 2D materials will be covered. The FSCC is always looking for new proposals in exciting and rapidly developing areas of physical sciences. More information and contact details are available on the Royal Society of Chemistry website (http://www.rsc.org/faraday_d).

Susan Weatherby, Programme Manager, Royal Society of Chemistry

Claire Vallance, President of Faraday Division and University of Oxford

John M. Seddon, Chair, Faraday Standing Committee on Conferences and Imperial College London

Faraday Discussions

PAPER

Spiers Memorial Lecture

Introductory lecture: Hot-electron science and microscopic processes in plasmonics and catalysis

N. J. Halas

Received 3rd January 2019, Accepted 31st January 2019
DOI: 10.1039/c9fd00001a

In these introductory remarks we discuss the generation of nonequilibrium electrons in metals, their properties, and how they can be utilized in two emerging applications: for extending the capabilities of photodetection, and for photocatalysis, lowering the barriers of chemical reactions. Because direct illumination of noble/coinage metal nanoparticles results in the excitation of their localized surface plasmons, these nanostructures can serve as active optical antennas, central to the goal of efficient hot electron generation to drive these processes. Currently, noble/coinage metal nanoparticles are being supplemented by earth-abundant, sustainable alternatives. Herein, we discuss how active optical antennas can expand the wavelength accessibility and alter the properties of traditional photoconductive detectors in new ways. We also examine how active optical antennas, when combined with conventional catalytic nanoparticles in an integrated manner, can convert catalysts into photocatalysts to change chemical product specificities and even alter chemical reaction pathways.

Introduction

The topic of this meeting, the 300th Faraday Discussion, lies at the confluence of electricity, chemistry and nanoscience, a region of interest that owes a great deal to Michael Faraday himself. Metals are appreciated for their ability to conduct electricity, a phenomenon that Faraday took advantage of with his invention of the electrical motor (alternator) in 1831, which represented the first reliable method for generating electrical current. Prior to that, as a chemist working with Humphry Davy, his interest in how chemical reactions could induce electrical currents provided much of the foundations of electrochemistry and the design of electrical batteries. Even metal nanoparticles and their interactions with light captured Faraday's attention; he was the first scientist to realize that the optical properties of gold were dramatically transformed from reflection into the absorption of green light, once the metal was reduced in dimension to size-scales

Department of Electrical and Computer Engineering, Department of Physics and Astronomy, Department of Chemistry, Laboratory for Nanophotonics, Smalley-Curl Institute, and Department of Materials Science and NanoEngineering, Rice University, 6100 Main Street, Houston, Texas 77005, USA

far smaller than optical wavelengths. Our modern understanding of electrons that follow Fermi–Dirac statistics has spawned interest in phenomena associated with electron populations driven out of equilibrium in metals, where optical excitation can give rise to electrons with energies as high as the energy of the photon that excited them. While photons with energies greater than the work function of a metal result in electrons that are photoemitted from the metal into vacuum, the regime in which electrons in a metal are excited but have insufficient energy for photoemission provides its own distinct phenomena. Here, visible light can excite electrons at a metal–semiconductor interface that can traverse an interfacial barrier, or that can tunnel into an unoccupied energy level of a molecule adsorbed at the metal surface. To study and ultimately exploit this regime of electron behavior, however, requires that nonequilibrium electrons are excited efficiently. While in bulk metals the direct optical excitation of hot electrons is an inefficient process, the excitation of surface plasmons can result in the generation of hot electrons through plasmon decay. In metallic structures of subwavelength dimensions or in periodic arrays, surface plasmons can be easily excited by direct optical illumination.[1] In other words, the direct optical excitation of surface plasmons by illuminating specific types of metallic structures provides an efficient mechanism for the generation of nonequilibrium, or "hot", electrons (and holes).

The development of metal nanoparticles and nanostructures specifically for the direct optical excitation of surface plasmons—the coherent oscillation of electrons within the metal—has been the topic of intense research activity during the past two decades and, as an extremely active and dynamic field, has been reviewed extensively.[2–16] Much of the earlier research on this topic focused on the dependence of the plasmonic properties of nanostructures on their shape, size and dielectric environment, primarily for sensing applications.[17–21] The intense near-field that can be generated in the junction between two metallic nanostructures in close proximity to each other when appropriately illuminated has been shown to give rise to surface-enhanced Raman spectroscopic enhancements. This has also been the topic of multiple reviews, as well as a recent Faraday Discussion.[22–28] The near-field enhancements of plasmonic structures have also been shown to benefit device properties, in particular enhancing photovoltaic efficiencies.[29–32] While the focus of this field on hot electron science has been a more recent development,[33,34] in many ways it has the potential to eclipse earlier advancements in terms of importance. Because there are already several useful and comprehensive reviews on hot carriers in plasmonics,[35,36] in this introductory article we will focus specifically on two topics within hot electron science: hot electron photodetection and hot electron photocatalysis. Our introduction begins with a theoretical analysis of hot carrier generation in metallic nanoparticles induced by plasmon decay.[37] Here, through a straightforward theoretical model, we examine some of the essential dependencies of hot carrier generation, such as the dependence of hot carrier generation rate and the energies of the hot carriers produced on the absorption spectrum of a particle, particle size, and carrier lifetime. We then turn to hot electron-based photodetection, a mechanism that drives a subset of the growing number of plasmonic photodetection devices of various types.[38] Hot electron-based photodetection brings new light-detection mechanisms to photocurrent-based photodetector devices, making it possible to detect light below the semiconductor bandgap.[39,40] The combination of plasmonics and electronic photodetection allows for spectral and polarization selectivity in light detection, which offers the opportunity to design specialized photodetection strategies that would by conventional approaches be far more complex.[39,41] Hot electron-based photodetection can also provide insight into the properties of the hot carrier generation process itself, providing clues as to how enhancing this process could result in higher-performance photodetectors.[42] On the topic of plasmon-based photodetection, extending device designs to new materials such as graphene creates new benchmarks for photodetector device performance.[31,43] Hot-carrier photocatalysis provides new approaches for the design of catalysts for chemical reactions, and presents new possibilities for converting thermally-driven catalytic processes into light-driven processes with substantially lower operating temperatures. We also discuss the importance of incorporating new materials, in particular catalytic metals whose surface reactivities but poor optical absorption properties complement those of noble/coinage metals alone, and how plasmonic and catalytic metals can be combined into "antenna-reactor" complexes. Finally, since the nonradiative decay of a surface plasmon results in both hot electron generation and photothermal heating, we discuss how to clearly distinguish between the two processes in plasmonic photocatalytic reactions.

Properties of plasmon-induced hot carriers in metal nanoparticles

By applying a straightforward theoretical model, Manjavacas and Nordlander have elucidated several essential characteristics of hot carriers resulting from the plasmon-resonant excitation of a metal nanoparticle.[37] Their model describes a spherical Ag nanoparticle with a diameter ranging from 5 to 25 nm (quasistatic limit) as free carriers in a finite spherical potential well, considering only the conduction electrons (Fig. 1A). External illumination results in the excitation of a plasmon, described as a classical field. Hot carriers are generated by the action of the external and plasmon-induced electric fields, which act upon the conduction electrons, inducing transitions from initial states below the Fermi level to final states above it. The probability of hot electron generation is calculated using Fermi's golden rule. Within this picture, the particle size and carrier lifetime are both varied to better understand the hot carrier properties. Carrier lifetimes (not to be confused with plasmon lifetimes of $\sim$10 fs) are limited by electron–electron, electron–surface and electron–phonon scattering, and range between $0.05 < \tau < 1$ ps. Although this model does not capture the fully quantum nature of the plasmon response of the metal,[44] several important aspects of hot electron generation rates and energies are well described with this model. First, the number of hot electrons generated by illumination per unit time and volume as a function of wavelength establishes that hot electron generation directly follows the absorption spectrum of the nanoparticle (Fig. 1B and C). This means that the direct excitation of hot electron–hole pairs, which is present also for nonresonant excitation, is a many orders of magnitude less efficient process for hot carrier generation than plasmon decay. This result was also compared directly with equivalent calculations performed using density functional theory (DFT) (Fig. 1C). The striking similarity of the two theoretical approaches verified that this model

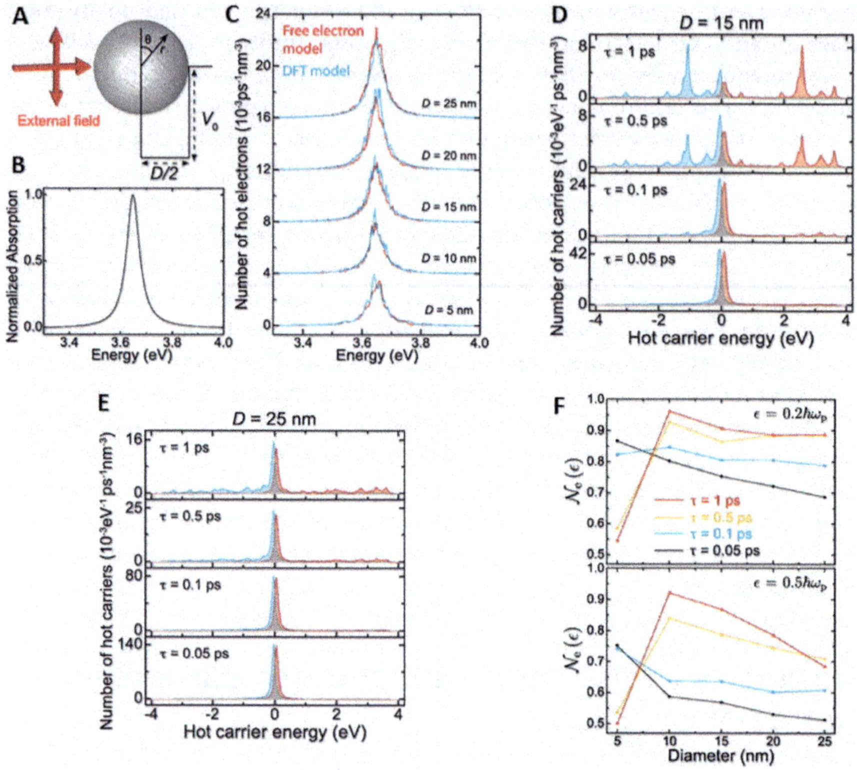

Fig. 1 (A) Schematic of a theoretical model for plasmon-induced hot carrier generation. (B) Calculated optical absorption of a Ag nanoparticle. (C) Number of hot electrons generated per unit time and volume as a function of wavelength. (D) Hot carrier number and energy distribution for a resonantly illuminated Ag nanoparticle with a diameter of 15 nm, for carrier lifetimes ranging from 0.05 to 1 ps. Red: electrons; blue: holes. (E) Hot carrier number and energy distribution for a resonantly illuminated Ag nanoparticle with a diameter of 25 nm, for the same range of carrier lifetimes as in (D). (F) The efficiency of hot electron generation due to resonant illumination of a Ag nanoparticle, as a function of particle diameter and carrier lifetime. The decrease in hot electron generation efficiency, which is clearly indicated for both increasing particle diameter and decreasing carrier lifetime, is clearly reversed for nanoparticle diameters in the 5 nm range. Reprinted (adapted) with permission from ref. 37. Copyright 2014 American Chemical Society.

was useful for the regime being studied, and also that many-body effects such as exchange–correlation have only a minor impact on hot carrier generation in the regime being considered (high plasmon quantum number; classical excitation), and so can be neglected. The energy distribution of hot electrons and holes was then calculated for a range of carrier lifetimes and nanoparticle diameters (Fig. 1D and E). This calculation presents several striking results. It is clearly apparent that the carrier lifetime has a direct influence on the energy of the hot carriers that can be generated by illumination (this model applies to illumination times much longer than the carrier lifetimes). Short carrier lifetimes result in a higher number, but lower energy, of "cooler" hot carriers. The particle diameter has a similar influence, with smaller nanoparticles yielding more energetic carriers, a result that reflects several experimental plasmonic photocatalysis studies where smaller nanoparticles appeared to generate transient negative ion states of adsorbate molecules more efficiently than larger nanoparticles.[45] However, when we look at the total generation rate independent of energy distribution, we find that it is largest for larger nanoparticles and shorter carrier lifetimes (Fig. 1D and E, y axes). Hot carrier generation efficiency was also defined as the number of hot electrons generated per excited plasmon that have an energy larger than a certain threshold (Fig. 1F). This generation efficiency was calculated as a function of nanoparticle diameter, for the range of hot carrier lifetimes being considered. Here we see that the hot carrier generation efficiency decreases with increasing nanoparticle diameter for sizes of 10 nm and larger, and also decreases with shorter carrier lifetimes. For nanoparticles smaller than 10 nm, these trends do not hold, which is most likely because of the discrete nature of the energy levels that begins to become apparent for this smaller size regime.

While this study yields insight into the properties of plasmon-induced hot electrons, the essential point of this analysis is that the time-dependent E field within the nanoparticle during plasmon-resonant optical excitation is directly responsible for hot carrier generation. This is the essential starting point for the design of effective structures and devices that use hot electrons for various applications. This is a unique aspect of plasmonic behavior that differs from the properties of bulk metals, since bulk metals screen external fields extremely well. In the sections that follow, we will examine the importance of this property in two applications: hot electron-based photodetection and hot carrier plasmonic photocatalysis.

Hot electron-based photodetection

A hot electron-based photodetector is created when a plasmonic structure is deposited on a semiconductor such that a Schottky barrier is formed at the interface (Fig. 2A).[39,40] When the plasmon energy is larger than the Schottky barrier height, hot electrons are injected into the conduction band of the semiconductor. The plasmonic nanostructures must be electrically connected to avoid charging and to complete the electrical circuit, which in this case is accomplished by covering the front face of the active area of the device with a conductive transparent oxide (indium tin oxide, ITO) layer (Fig. 2B). The initial device consisted of a square array of nanorods of equivalent length (Fig. 2C) patterned by e-beam lithography onto an n-type silicon substrate with a titanium adhesion layer, followed by deposition of the ITO layer. The device can be operated without a bias voltage, although applying a bias voltage greatly improves its responsivity. There are several interesting and unique properties of this type of photodetector. Firstly, this photocurrent generation mechanism is not limited by the bandgap of the semiconductor, thus enabling photocurrent generation at longer wavelengths, where the photon energy is insufficient to create electron–hole pairs in the semiconductor itself (Fig. 2D). Secondly, the resonance of the plasmonic nanostructure used for hot carrier generation defines the spectral response of the photodetector. Therefore this type of photocurrent generation mechanism is inherently spectrally selective, which enables the possibility of hot electron photodetector arrays of various resonances serving as a fully integrated spectrometer-detector. For plasmonic nanostructures with polarization sensitivity, this characteristic is also imparted onto the photodetector. By

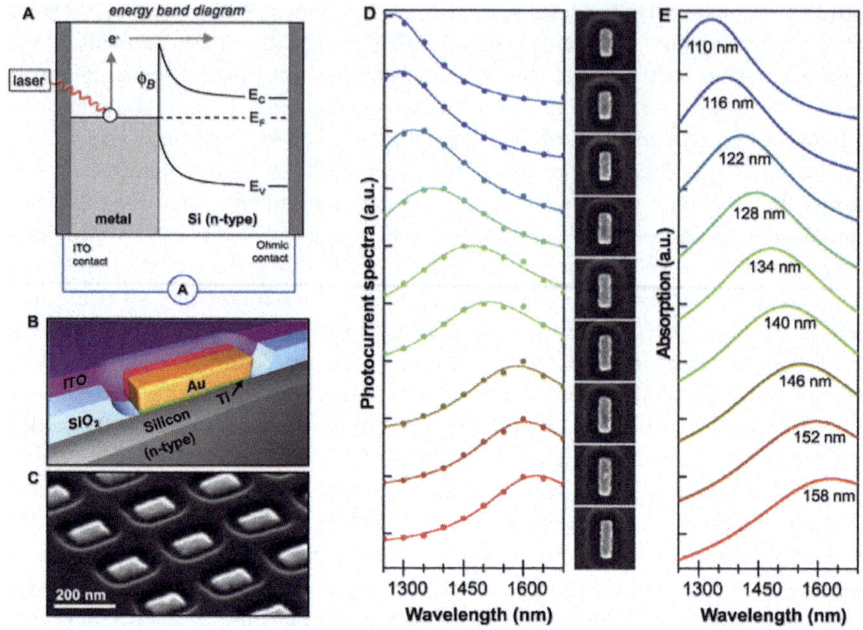

Fig. 2 (A) Band diagram for plasmonically driven internal photoemission over a nanoantenna-semiconductor Schottky barrier (fB). (B) Representation of a single Au resonant antenna on an n-type silicon substrate. (C) Scanning electron micrograph of a representative device array prior to ITO coating, imaged at a 65° tilt angle. (D) Photocurrent spectra of nine devices with arrays of nanorods of various lengths (shown). (E) Corresponding calculated absorption spectra. From ref. 39. Reprinted with permission from AAAS.

systematically modifying the aspect ratio of the nanorod components of the array, the spectral response of the detector can be tuned (Fig. 2D). As mentioned in the earlier theoretical discussion on the spectral response of hot electron generation, we see that the photocurrent spectral response here follows the optical absorption of the nanorod arrays used in each device (Fig. 2E).

By modifying the photodetector device design, one can directly investigate and quantitatively compare the properties of hot electrons induced by plasmon excitation relative to those induced by absorption in the metal itself.[42] As we saw in the theoretical study discussed earlier, hot carrier generation is dependent upon the E-field inside the metal, which is strongly enhanced when illuminated on resonance. Generally, it is localized to regions of large field enhancements on more complex structures. Hot carriers can also be generated due to light absorption in the metal, limited by its optical absorption depth at that wavelength. In Au, carrier generation by direct photoexcitation results from the excitation of d-band electrons into the conduction band.

In Fig. 3 we show a hot-electron photodetector that consists of a Au grating that extends from a Au film. The grating structure is fabricated onto a TiO_2 substrate, a wide-bandgap n-type semiconductor. The wide bandgap of the semiconductor allows the preferential collection of electrons. Surface plasmons are launched by illumination onto the grating structure[41] (their resonant wavelength is ~675 nm); the film "pad" allows for good electrical continuity between the plasmonic grating

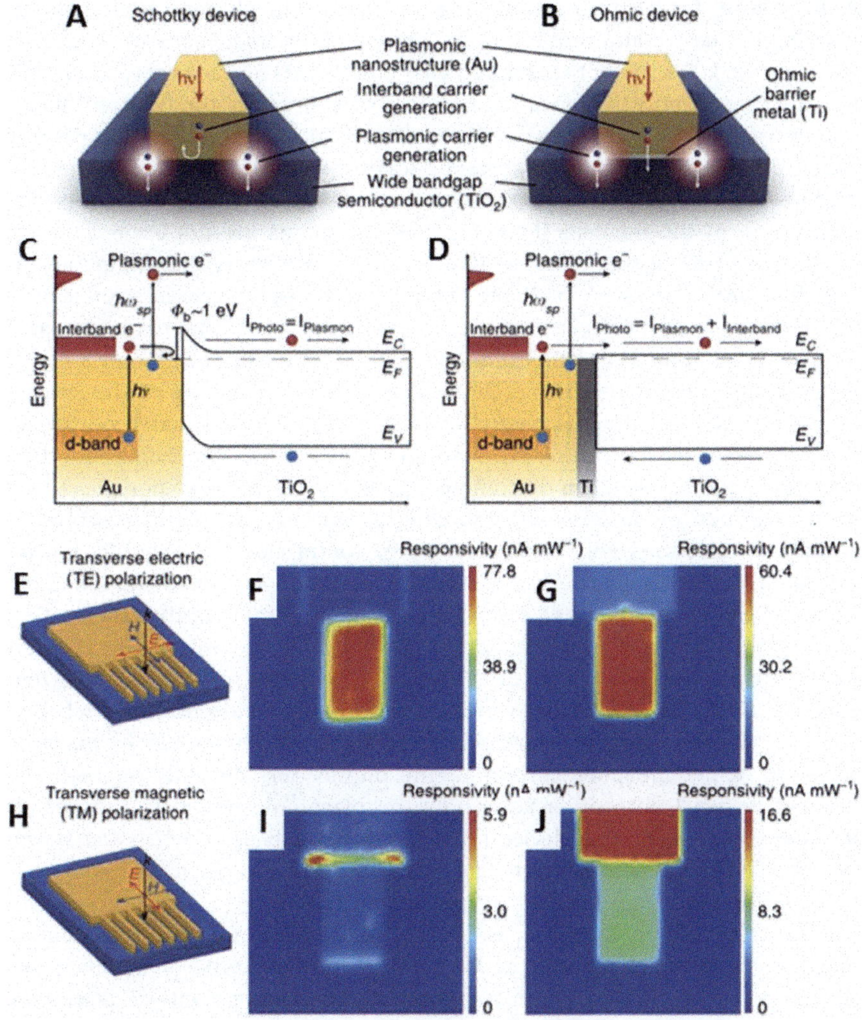

Fig. 3 Schematic of hot-carrier generation and collection over a Schottky (A) or ohmic (B) barrier. Band diagram schematics of (C) a Au–TiO_2 Schottky device and (D) a Au–Ti–TiO_2 ohmic device. (E) Schematic of TE excitations used to generate photocurrent maps. The laser wavelength is tuned to the resonance of the plasmonic nanowires. Photocurrent maps of the (F) Schottky and (G) ohmic devices using TE-polarized light. (H) Schematic of TM-polarized light excitation. Photocurrent maps of the (I) Schottky and (J) ohmic devices. Reproduced from ref. 42 with permission from Springer Nature, copyright 2015.

structure and the semiconductor substrate. When the Au grating structure is fabricated directly onto TiO_2, it forms a Schottky barrier that is measured to be, on average, 1.07 eV (Fig. 3A). This compares quite well with the literature value of 1 eV. When an intervening Ti layer is deposited on the TiO_2 prior to Au deposition and fabrication, the barrier is ohmic instead (Fig. 3B). The band diagrams for the Schottky and ohmic devices are shown in Fig. 3C and D, respectively. For the Schottky device, the photocurrent must be due to hot electrons, while all photocurrent generated will be detectable in the ohmic device. Resonant illumination of the grating for TE polarization (Fig. 3E) results in plasmon excitation. By scanning the device and measuring the photocurrent response, a photocurrent map is created (Fig. 3F for the Schottky case and Fig. 3G for the ohmic case). This map indicates from which location on the device structure the detected photocurrent originates. For the case of plasmon excitation, we observe that there is a relatively strong photocurrent signal originating from the region of the grating structure, for both the Schottky and ohmic cases (Fig. 3F and G). By rotating the polarization of the incident light (Fig. 3H), plasmons are not excited on the grating structure. In this case, the photocurrent maps show that the photocurrent, substantially less for both the Schottky and ohmic devices, originates in different regions of the device. For the Schottky case, virtually no photocurrent is observed for illumination either of the grating or the Au film, with only a minor response observed on the edges of the structure where the film meets the grating. For the ohmic case, the observed photocurrent originates primarily on the Au film, with a smaller contribution from the grating structure. These data show that plasmon excitation results in much hotter electrons (Fig. 3F) than non-plasmonic direct excitation of carriers, with the latter excitation mechanism resulting in virtually no electrons with sufficient energy to traverse the Schottky barrier (Fig. 3I). In the ohmic, barrierless case, we can see that plasmon excitation of the grating gives rise to more than seven times as many carriers in the grating itself (Fig. 3G) than in the case where optical absorption is the only mechanism for photocurrent generation (Fig. 3J). Theoretical analysis of the photocurrent response of this detector shows explicitly that the photocurrent in the Schottky case depends only on the near-field inside the metal generated by plasmon excitation, while the photocurrent in the ohmic case depends on both the inherent material absorption as well as the plasmonic near-field induced by plasmon excitation.[42] This device provides clear evidence of what was shown theoretically,[37] *i.e.*, that the internal near-field generated by plasmon excitation is the quantity that determines hot carrier generation.

Since initial demonstrations, hot-carrier photodetectors have been demonstrated using 2D materials such as MoS_2, graphene[12] and other materials as the current collector. Greater sophistication in the development of plasmonic antennas with Fano resonances[43,46] and perfect absorber geometries[38] has also been implemented in order to increase device responsivities. The combination of these effects with other current-enhancing properties in semiconductors, most notably free-carrier absorption, has also been demonstrated.[47] The further development of hot electron-based photodetectors should ultimately lead to performances that are comparable to those of current commercial systems, which should result in new spectrally sensitive detection capabilities in compact and lightweight device geometries.

Plasmon-induced hot electron-based photocatalysis

Emerging primarily from earlier studies of femtochemistry on well-defined index faces of metallic crystals, the transition to metallic nanoparticle-based photocatalysis has enabled the direct optical excitation of nanoparticle plasmons, thus substantially boosting hot electron generation. The transition from index face studies using high-power lasers to studies of metal nanoparticles was motivated by the desire to develop photocatalysts for reactions that are of direct industrial interest. The chemical industry is one of the highest consumers of energy worldwide, and lowering this energy consumption level has long been a primary motivating factor for catalyst development. On metallic nanoparticles, processes that are driven by hot electrons can dominate surface chemistry, change reaction outcomes relative to thermal processes, and lower reaction barriers such that reactions can proceed at substantially lower temperatures than under conventional conditions.[48] Pioneering work in the area of hot electron photocatalysis includes liquid phase water splitting,[49–51] H_2 production from alcohol,[52,53] gas phase oxidation reactions[54,55] and hydrocarbon conversion.[56]

In plasmonic photocatalysis, plasmon excitation leads to hot electron–hole pairs; the hot electron cloud extends from the nanoparticle surface due to the shape of the surface potential (Fig. 4A–C).[45] For the case of plasmonic H_2 dissociation illustrated here, if a H_2 molecule is present, the electron can transfer from the metal to the LUMO level of the adsorbed species on or near the metal surface, thus creating a transient negative ion, $H_2^{\delta-}$ (Fig. 4D and E). In general, the detailed nature of this charge transfer process in specific systems has been studied and debated in the literature. However, its role as an important mechanism in plasmonic photocatalysis is widely accepted. The additional Coulomb repulsion in the molecule increases the H–H bond distance and lowers the barrier to dissociation relative to the case of the neutral molecule (Fig. 4E). Calculations

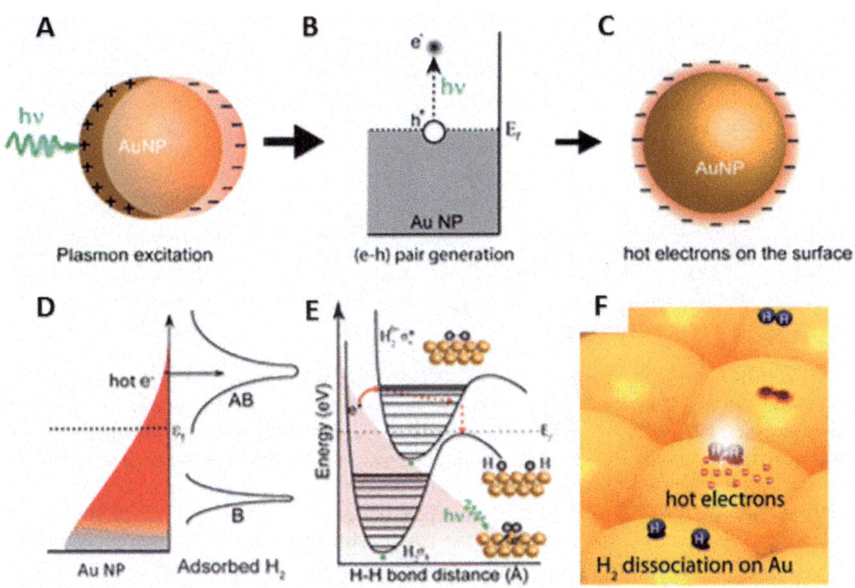

Fig. 4 (A) Schematic of plasmon excitation, (B) e–h pair generation and (C) hot electrons available at the nanoparticle surface. (D) Schematic of the Fermi–Dirac distribution of hot electrons permitting hot electron transfer into the antibonding state of H_2. (E) Proposed mechanism of the hot-electron induced dissociation of H_2 at the Au nanoparticle surface. (F) Schematic of H_2 dissociation mediated by hot electron transfer and the formation of a transient negative ion state, $H_2^{\delta-}$, followed by dissociation. Reprinted (adapted) with permission from ref. 45. Copyright 2013 American Chemical Society.

on an isolated hydrogen molecule show that H_2^- has a metastable state located about 1.7 eV above the ground state of the neutral molecule.[45,57] The bond dissociation energy for this state is 2 eV lower than for the neutral case (2.6 eV *vs.* 4.6 eV). The process occurs at room temperature with visible light at a wavelength that corresponds to excitation of the localized surface plasmon of the nanoparticle.

The experimental system for H_2 dissociation is shown in Fig. 5. Very small Au nanoparticles were dispersed in an oxide matrix (TiO_2). The size dependence of the Au nanoparticles proved to be quite important, and for nanoparticles larger than 10 nm in diameter the photocatalytic activity was reduced relative to the smaller nanoparticles. Smaller size Au nanoparticles likely have slower electron thermalization rates due to reduced electron–phonon interactions. The role of the TiO_2 matrix is to reduce the diffusion of the gas molecules on the surface of the Au, and does not directly participate in the reaction. It was subsequently

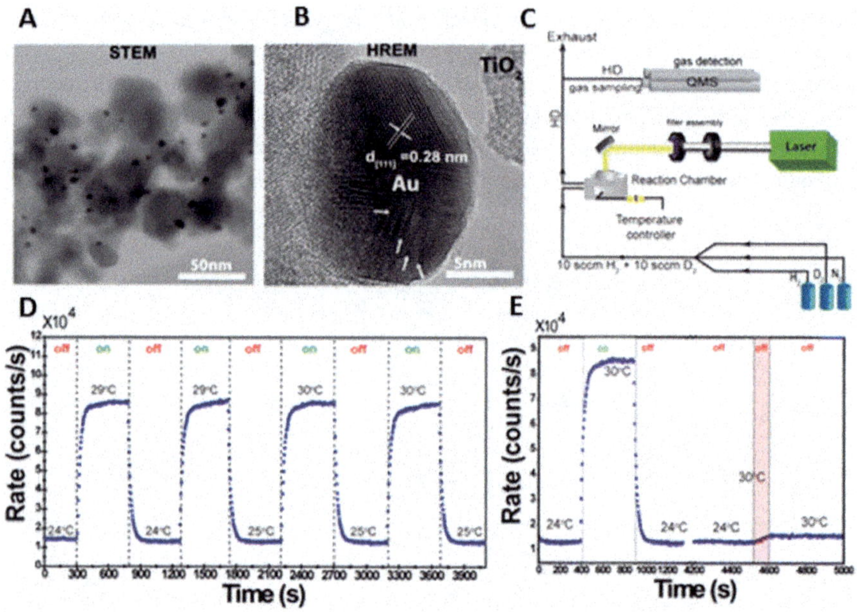

Fig. 5 Plasmon-induced dissociation of H_2/D_2 using a 1.75 wt% Au/TiO_2 photocatalyst at room temperature (297 K). (A) Scanning transmission electron micrograph showing the distribution of 1.75 wt% Au NPs (dark particles) over a TiO_2 support (gray). (B) High-resolution transmission electron micrograph of a single Au NP supported on a TiO_2 matrix, showing darker contrast for the Au NP in comparison to the TiO_2 support. Characteristic lattice fringes of Au (111) and (200) are visible at ~2.8 Å and 2.3 Å, respectively. (C) Schematic of the overall experimental setup. (D) Real time monitoring of the rate with laser excitation (on, 2.41 W cm^{-2}) and without laser excitation (off). During 8 min of laser irradiation on the photocatalyst, the temperature increases by 5 K, as shown in the figure, from 297 to 303 K. The dotted vertical lines indicate the laser on/off times. (E) Distinction between the photocatalytic rate (only laser on, 2.41 W cm^{-2}) and thermal heating rate (laser off, heating on). The shaded gray area displays the photocatalytic rate due to the laser. The shaded red area displays the rate of HD formation due to heating of the photocatalyst sample from room temperature (297 K) to 303 K without laser illumination. Reprinted (adapted) with permission from ref. 45. Copyright 2013 American Chemical Society.

discovered, by performing the same experiment using a SiO_2 matrix, that TiO_2 can actually compete as an electron scavenger with the H_2 adsorbate molecules.[57] H_2 dissociation is detected by using both H_2 and D_2 reactants and looking for the presence of HD using a mass spectrometer (Fig. 5C). The HD signal corresponds to illumination of the plasmonic photocatalyst (Fig. 5D). The small amount of photothermal heating, roughly 5–6 degrees Celsius, does not contribute significantly to HD formation (Fig. 5E). This very simple chemical reaction with no side products can serve as a benchmark for plasmonic photocatalyst design, and allows us to learn in greater detail about the role of the metal nanoparticle in mediating hot electron processes. For example, the same experiment was performed on aluminum nanocrystals (Al NCs), which were found to be photocatalytically reactive relative to Au nanoparticles for the case of H_2 dissociation.[58] Here, Al was chosen as the plasmonic metal because it is the most earth-abundant plasmonic metal, and also because of its highly attractive properties in its nanostructure form.[59] In the case of Al NCs acting as the plasmonic photocatalyst for H_2 dissociation, charge transfer was already found to be substantial for ground state H_2, thus suggesting that the electron affinity level of the H_2 molecule lies very close to the Fermi level of the Al NC. Because the affinity level is close to the Al Fermi level, it overlaps with the photo-induced hot electron distribution. The reaction barrier for hydrogen dissociation is further lowered after the injection of a hot electron from the Al nanoparticle into the antibonding orbital. For Al, the energy of maximum HD generation corresponds to its interband transition, not its plasmon resonance.

Antenna–reactor hybrid nanoparticles for hot-electron photocatalysis

Plasmon-based photocatalysis was initially demonstrated with noble/coinage metals, with their strong plasmon-derived optical near field and resultant hot carrier generation. Conventional heterogeneous catalysis has developed using weakly light-absorbing metals such as Pd, Pt, Ru and Rh to lower the activation energy for chemical reactions under conditions that typically require high temperatures and pressures. Combining the optical properties of plasmonic metals with the chemical reactivities of catalytic ones would open the door to new types of photocatalytic processes, essentially converting catalysts into photocatalysts. This can be accomplished by developing hybrid systems where a plasmonic nanoparticle acts as an "antenna" for a nearby catalytic "reactor" nanoparticle. Other variations of this concept, such as reactive surface alloys on a plasmonic nanoparticle, can be developed based on the same principle.

The initial realization of the antenna–reactor principle is shown in Fig. 6A.[60] The idea is based on the concept originally discussed in Fig. 1, *i.e.* that the photoexcited plasmon decays by generating hot electrons. Here, the plasmon mode of the antenna itself is photoexcited, and if the reactor nanoparticle is within the antenna particle's near field, it will also be photoexcited even though it may have no inherent optical absorption itself at the plasmon-resonant wavelength of the antenna particle. It is important to realize that the photoexcitation of the reactor particle by the antenna particle does not require the two to be

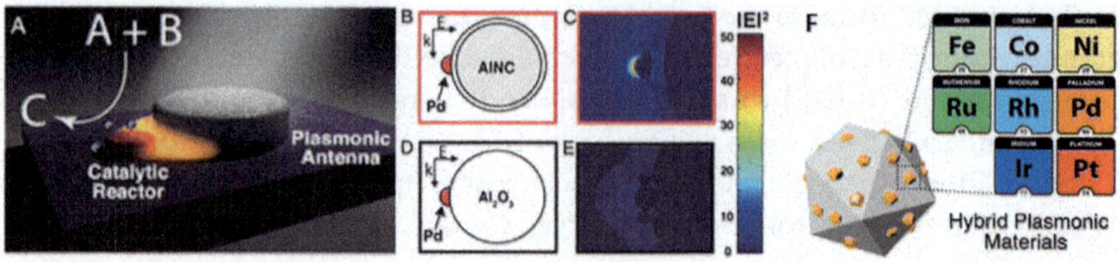

Fig. 6 (A) The antenna–reactor concept, which shows a plasmonic nanoparticle being illuminated and the near-field being generated, with the catalytic reactor nanoparticle located within the fringing field of the antenna nanoparticle. (B) A prototype antenna–reactor is illustrated consisting of an aluminum nanocrystal antenna with a small Pd reactor nanoparticle. (C) The local field enhancement of (B) shows a strong field on the Pd surface at the resonant plasmon wavelength of the Al NC. (D) For a Pd reactor on an aluminum oxide nanoparticle, no light absorption, and therefore no near field at the Pd surface, is visible. (F) The modularity of the antenna–reactor concept is illustrated, whereby a range of different types of catalytic reactor nanoparticles can be adsorbed onto the Al NC surface for enhancing specific reaction rates and outcomes. Parts A–E are reproduced from ref. 60. Part F is reprinted (adapted) with permission from ref. 61. Copyright 2017 American Chemical Society.

conductively coupled; there does not need to be charge transfer between the antenna and reactor nanoparticles. In fact, if the two particles were in conductive contact with each other, the reactor would likely dampen the plasmon resonance of the antenna particle. The case illustrated in Fig. 6A involves two nanoparticles deposited adjacent to each other on a planar substrate, for example, where the only (non-mechanical) coupling between the two nanoparticles is through the antenna particle's fringing field. Physically, the fringing field of the antenna particle induces a similar electron oscillation in the reactor particle as in the antenna particle itself; we can think of this as a "forced plasmon" in the reactor driven solely by the antenna particle under photoexcitation. In contrast to the direct excitation of hot electron–hole pairs which is a single-particle excitation process, this forced plasmon is a collective but strongly damped excitation which can only decay nonradiatively. Just as the plasmon mode of the antenna particle decays by the generation of electron–hole pairs, the same process should occur in the "forced plasmon" of the reactor particle. In Fig. 6B, the geometry of an Al nanoantenna particle (approximated as a sphere) with a 2–4 nm thick native oxide coating, on which a Pd hemispherical reactor particle is deposited, is shown. Under resonant illumination of the antenna particle, the local $|E|^2$ at the reactor particle surface is calculated (Fig. 6C). This geometry is compared to the case of a Pd hemispherical reactor particle deposited instead on a nanosphere consisting entirely of Al_2O_3; in this case there is no field enhancement at the Pd surface (Fig. 6D and E). A schematic illustration of the spherical antenna-reactor geometry, envisioned as a modular photocatalyst design, is depicted in Fig. 6F. Here, the Al NC antenna, whose plasmon mode can be tuned in energy by varying the particle size, can be decorated with a variety of different catalytic metals as reactor particles.[61]

In light of our previous discussion regarding H_2 dissociation on plasmonic Au and Al antenna particles, it is interesting to see how this reaction proceeds on an antenna–reactor particle. As illustrated in Fig. 6, the Al NC antenna–Pd reactor particle was both chemically synthesized[60] and also fabricated using a planar plasmonic heterodimer geometry.[62] A head-to-head comparison of H_2 dissociation on pristine Al NC nanoparticles and on chemically synthesized Al NC–Pd antenna–reactor particles is shown in Fig. 7. The wavelength dependence (Fig. 7A) clearly indicates that the reaction proceeds differently on the two different plasmonic photocatalysts. What is not indicated in Fig. 7A is that the HD yield for the antenna–reactor particles is almost two orders of magnitude higher than for the pristine Al NCs.[60] For the antenna–reactor complex, the maximum reactivity occurs when the dipolar plasmon of the antenna particle is excited, which is in stark contract to the pristine Al NC case, where maximum reactivity occurs at wavelengths consistent with the interband transition of Al. From these data it is clear that the reaction pathways on the two photocatalysts are very different. It is well established that dissociative adsorption of H_2 occurs in an essentially barrierless manner on Pd.[63,64] In contrast to the hot electron transfer process for H_2 dissociation on pristine Au and Al nanoparticle surfaces, here the hot electrons generated in the Pd by forced plasmon decay are responsible for inducing the desorption of H and D, which quickly recombine. This is a charge neutral process; once the chemical reaction is completed, the hot electrons recombine with the holes in the Pd islands. It is also evident in Fig. 7A that HD production on the Al NC–Pd antenna–reactor nanoparticles follows the optical absorption of the nanoparticle complex very closely, which is in agreement with our earlier-discussed theoretical analysis of hot electron generation in plasmonic nanoparticles (Fig. 1B and C). In Fig. 7B the laser power dependence of the HD generation rate is shown for the two wavelengths corresponding to the dipolar antenna–reactor plasmon mode and the interband transition at 800 nm, respectively. The supralinear power dependence observed here is consistent with other hot electron-induced photocatalytic processes.[65]

In addition to studying the H_2 dissociation reaction on Al–Pd antenna–reactor particles, the hydrogenation of acetylene was studied, comparing the photocatalytic reaction to the thermocatalytic reaction on the same antenna–reactor complex (Fig. 7C). In this reaction, which is quite important industrially, both ethylene and ethane are the reaction products, and ethylene selectivity is highly desired. What is observed is a high selectivity of the photocatalytic

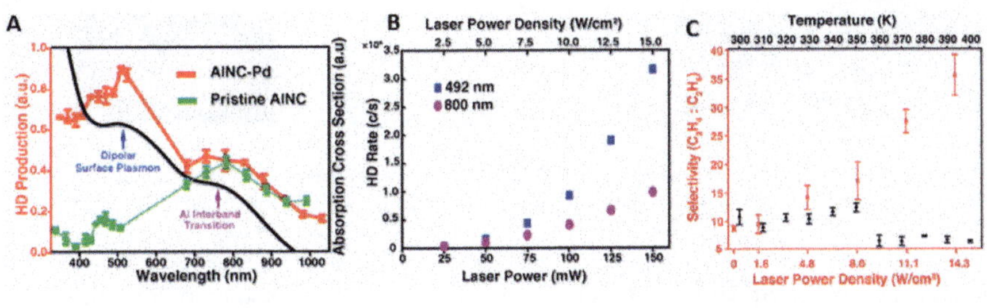

Fig. 7 (A) Wavelength dependence of the H_2 photocatalytic dissociation reaction performed using Al NCs (green) and Al NC–Pd antenna–reactor complexes (red). (B) Power dependence of H_2 dissociation on Al–Pd antenna–reactor particles. (C) Acetylene hydrogenation reaction on Al NC–Pd antenna–reactor particles; here we examine the ethylene/ethane selectivity, comparing the photocatalyzed reaction (red) and the thermally catalyzed reaction (same complexes, reaction run thermally with no photoexcitation, black). Reproduced from ref. 60.

process for ethylene synthesis relative to that of the thermal process. Ethylene/ethane ratios as high as 37 : 1 were observed for the photocatalysis case, whereas for the thermal case ethylene/ethane ratios never increased above 10 : 1. It has been postulated that the selectivity enhancement of photo-hydrogenation is likely due to the availability of dissociated H_2 in this case, again, induced by hot electron desorption on the reactor nanoparticles. In a theoretical analysis it was found that the Pd–H antibonding interactions were energetically just above the Fermi level, thus easily populated by hot electrons. Therefore a hot-electron-initiated destabilization of the Pd–H bond, leading to desorption, is a plausible mechanism. The facile hot-electron-induced desorption of H_2 from the surface limits the hydrogenation of acetylene, promoting ethylene over ethane as the reaction product.

To additionally demonstrate the antenna–reactor photocatalysis concept (Fig. 6A), a large-area sample of Al–Pd plasmonic heterodimers was fabricated (Fig. 8) using the method of hole-mask lithography (Fig. 8A–C). The heterodimers were photocatalytically active, and since their density on the sample surface was known, the number of HD molecules produced per laser pulse, per photocatalytic antenna–reactor, could be determined as a function of incident laser intensity (Fig. 8D). As plasmonic heterodimers, the fabricated structures with well-defined gaps showed a polarization dependence for their photocatalytic activity (Fig. 8E), consistent with the near-field enhancement induced in the Pd reactor disk of the

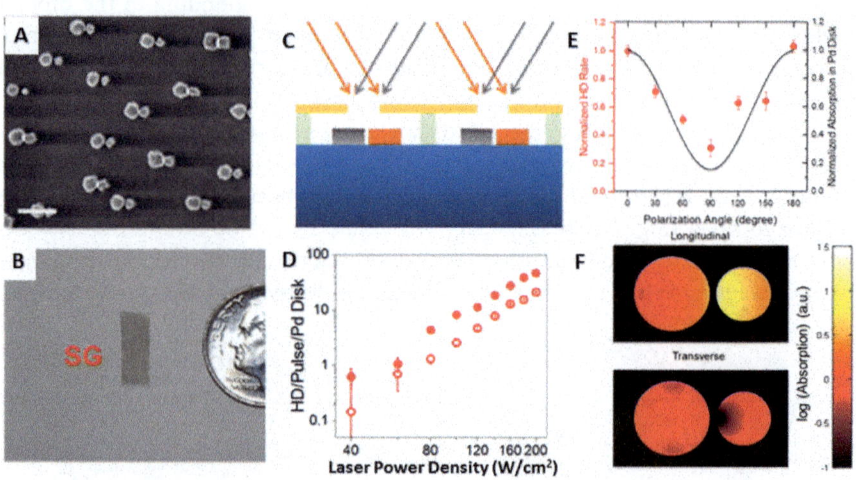

Fig. 8 (A) Scanning Electron Microscope (SEM) image of a sample of Al–Pd heterodimers, fabricated over a large area (B) suitable for photocatalysis. A fabrication method known as hole-mask lithography (C) was used to fabricate heterodimers with precisely defined nanogaps over such a large area. (D) By performing the H_2 dissociation reaction on a fabricated substrate where the density of antenna–reactors is known, the production rate corresponding to the number of HD molecules produced per laser pulse per antenna–reactor structure was determined as a function of the incident laser power density (W cm^{-2}). (E) The Al–Pd heterodimers also showed a polarization sensitivity to their photocatalytic response. (F) Theoretical analysis using the finite-difference time-domain method shows the absorption induced by the Al nanodisk on the Pd nanodisk for longitudinal (top) and transverse (bottom) incident light polarization. Reprinted (adapted) with permission from ref. 62. Copyright 2016 American Chemical Society.

heterodimer, which varied strongly as a function of polarization angle (Fig. 8F). These types of nanofabricated antenna–reactors may be useful for antenna and reactor material combinations that are difficult to fabricate using chemical synthesis approaches.

Quantifying hot electron and thermal contributions to plasmonic photocatalysis

It is important to remember that the resonant illumination of a metal nanoparticle results in both hot electron generation and photothermal heating[6,66,67] due to nonradiative energy transfer processes, such as electron–phonon coupling in the metal. Both processes can simultaneously play a critical role in driving chemical reactions (resonant energy transfer and radiative processes[33] may also occur but are not the focus of this article). To understand which fundamental processes are occurring in plasmonic photocatalysis, it is important to monitor the reaction temperature at the photocatalyst surface in order to elucidate the relative role of electron- and phonon-driven processes in the overall chemical reaction.

In a recent study of ammonia decomposition using a plasmonic photocatalyst,[68] the relationship between thermal and electronic excitations was addressed by introducing the concept of a light-dependent activation barrier, in order to account for the effect of light illumination on both electronic and thermal excitations within the same unified framework (Fig. 9). In this study an antenna–reactor particle of a new geometry was developed, namely a Cu nanoparticle with a Cu–Ru surface alloy. The surface stoichiometry was such that the presence of Ru on the surface did not appreciably compromise or dampen the plasmonic response of the Cu nanoparticle. The Cu–Ru antenna–reactor nanoparticles were dispersed within a MgO support matrix (Fig. 9A). Ru was chosen as the reactor site because of its affinity for the reversible binding of nitrogen-containing molecules, such as ammonia. The Cu–Ru antenna–reactor nanoparticle was found to be substantially more photocatalytically active than either Cu or Ru nanoparticles alone, and substantially more photocatalytically than thermocatalytically active (Fig. 9B). The ammonia decomposition reaction was monitored by measuring the H_2 product generation rate under both thermocatalytic and photocatalytic conditions (Fig. 9C). Here, it is important to note that the photocatalytic reaction can produce the same rate of ammonia decomposition as the thermocatalytic reaction but at a substantially lower temperature.[69] In both cases the surface temperature was monitored using an infrared-detecting thermometer with temperatures that were found to be in excellent agreement with theoretical simulations. Illumination of the photocatalyst also increases its surface temperature due to photothermal heating; the question is how can we quantify these two contributions? By measuring the Arrhenius dependence of the photocatalytic reaction on the surface temperature of the catalyst, under conditions of illumination alone and combined with carefully controlled thermal heating and monitoring, the reaction barrier was found to be reduced under illumination conditions relating to both wavelength and intensity. Expressing this behavior in the form of a light-induced reaction barrier allows one to clearly quantify the photonic *versus* thermal contributions. For example, if this reaction

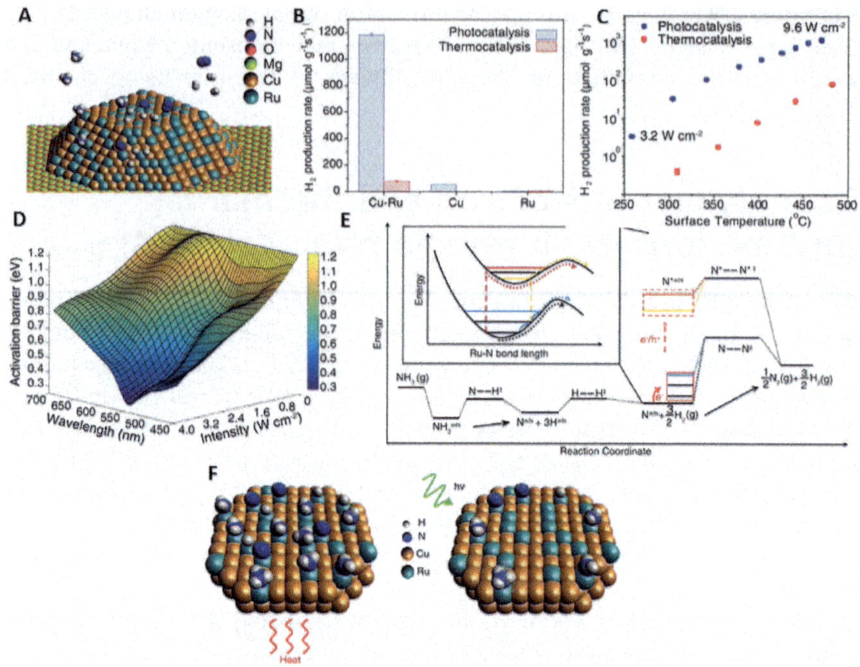

Fig. 9 (A) Schematic of the Cu–Ru antenna–reactor photocatalyst on its MgO support. (B) The photocatalytic activity of the antenna–reactor towards ammonia decomposition was found to be far greater than its thermocatalytic activity, and was in turn found to be more active than either a pristine Cu or Ru nanoparticle. (C) Both the photocatalysis and thermocatalysis reaction rates increased with increasing surface temperature. (D) Arrhenius plots of the apparent reaction barriers for illumination wavelengths and surface temperatures (controlled using a temperature controller) were measured for a range of illumination conditions, varying both the laser power and wavelength. From this data, a light-dependent decrease in the reaction barrier can be defined, where the reaction barrier in the dark (1.2 eV, right axis) is reduced as a function of both wavelength and light intensity, to a value as low as 0.2 eV at an illumination wavelength of 550 nm. (E) This reduction in the activation barrier changes the rate-limiting step of the reaction, from the associative desorption of N_2 to the N–H bond breaking step. (F) The physical mechanism for the change in the rate-limiting step is the hot electron-induced desorption of N_2 from the Ru reactive sites (right), thus freeing up those sites for further adsorption and decomposition of NH_3 adsorbate molecules. The lack of a desorption mechanism in the thermal case (left) makes N_2 desorption the rate-limiting step. From ref. 68. Reprinted with permission from AAAS.

was exclusively photothermal in nature with no hot electron contributions, the slope of the Arrhenius plot would not change with illumination intensity or wavelength. Any change in the slope due to illumination intensity shows that processes other than purely thermal ones are also contributing to the chemical reaction. For ammonia decomposition, this light-reduced reaction barrier is plotted as a function of both laser power and wavelength (Fig. 9D). The reaction barrier for the purely thermocatalytic reaction is shown at the right of this contour plot, corresponding to zero illumination intensity. The reaction barrier upon illumination is reduced, with both a wavelength and laser power dependence. The barrier reduction is at its maximum at the illumination wavelength of 550 nm, which corresponds to the spectral peak of the plasmon resonance of the Cu–Ru antenna–reactor particle. For ammonia decomposition with this antenna–reactor photocatalyst, the reaction barrier is reduced from 1.2 eV to 0.2 eV at the optimal wavelength (Fig. 9D). This large reduction in the reaction barrier changes the rate-limiting step in this well-studied reaction; the reaction coordinate for ammonia decomposition is shown in Fig. 9E. Under thermocatalytic conditions, the rate-limiting step for this reaction is the associative desorption of N_2 from the Ru reactive sites. For the photocatalyzed reaction, the hot electrons generated by plasmon decay contribute to the desorption of N_2 from the photocatalyst surface (Fig. 9F), similar to the photoinduced desorption discussed earlier in the case of acetylene hydrogenation and enhanced ethylene/ethane specificity (Fig. 7C).[60]

Outlook

The exciting results shown to date in the case of hot electron photodetection could potentially alter the way that photodetectors are envisioned, and in the nascent field of plasmonic photocatalysis could lead to a revolution in chemistry and in the chemical industry. Conventional chemical processes that require high temperatures and pressures also require substantial capital investments and benefit from an economy of scale; these practices have given rise to the large chemical processing plants that are the industry standard. Can these new photon-driven ways of doing chemistry be translated into industrial processes that rely on photons—either sunlight or inexpensive, high efficiency photon sources that are currently available—for energy? The translation of this work into industrial practice will require a complete change in both education and mindset for chemical engineers, and may result in the establishment of a new field of "photocatalysis engineering" in order to fully translate these new ideas into real-world applications.

Conflicts of interest

There are no conflicts to declare.

Acknowledgements

I would like to thank the many students, postdocs and collaborators whom I have worked with on these topics in the past years. I would also like to thank the funding agencies that have supported this work: The Air Force Office of Scientific Research, the Office of Naval Research, the Army Research Office, the Defense Threat Reduction Agency, the National Science Foundation (all in the USA), and the Welch Foundation for their ongoing and generous support. In particular, I would like to thank my main collaborator in these efforts, Peter Nordlander, for his critical reading of this manuscript.

References

1 C. F. Bohren, How can a particle absorb more than the light incident on it, *Am. J. Phys.*, 1983, **51**, 323–327, DOI: 10.1119/1.13262.

2 N. C. Lindquist, P. Nagpal, K. M. McPeak, D. J. Norris and S. H. Oh, Engineering metallic nanostructures for plasmonics and nanophotonics, *Rep. Prog. Phys.*, 2012, **75**, 036501, DOI: 10.1088/0034-4885/75/3/036501.
3 S. A. Maier and H. A. Atwater, Plasmonics: Localization and guiding of electromagnetic energy in metal/dielectric structures, *J. Appl. Phys.*, 2005, **98**, 01101, DOI: 10.1063/1.1951057.
4 M. Pelton, J. Aizpurua and G. Bryant, Metal-nanoparticle plasmonics, *Laser Photonics Rev.*, 2008, **2**, 136–159, DOI: 10.1002/lpor.200810003.
5 M. I. Stockman, Nanoplasmonics: past, present, and glimpse into future, *Opt. Express*, 2011, **19**, 22029–22106, DOI: 10.1364/oe.19.022029.
6 G. Baffou and R. Quidant, Thermo-plasmonics: using metallic nanostructures as nano-sources of heat, *Laser Photonics Rev.*, 2013, **7**, 171–187, DOI: 10.1002/lpor.201200003.
7 P. Biagioni, J. S. Huang and B. Hecht, Nanoantennas for visible and infrared radiation, *Rep. Prog. Phys.*, 2012, **75**, 024402, DOI: 10.1088/0034-4885/75/2/024402.
8 H. J. Chen, L. Shao, Q. Li and J. F. Wang, Gold nanorods and their plasmonic properties, *Chem. Soc. Rev.*, 2013, **42**, 2679–2724, DOI: 10.1039/c2cs35367a.
9 S. Eustis and M. A. El-Sayed, Why gold nanoparticles are more precious than pretty gold: Noble metal surface plasmon resonance and its enhancement of the radiative and nonradiative properties of nanocrystals of different shapes, *Chem. Soc. Rev.*, 2006, **35**, 209–217, DOI: 10.1039/b514191e.
10 X. H. Huang, S. Neretina and M. A. El-Sayed, Gold Nanorods: From Synthesis and Properties to Biological and Biomedical Applications, *Adv. Mater.*, 2009, **21**, 4880–4910, DOI: 10.1002/adma.200802789.
11 D. K. Gramotnev and S. I. Bozhevolnyi, Plasmonics beyond the diffraction limit, *Nat. Photonics*, 2010, **4**, 83–91, DOI: 10.1038/nphoton.2009.282.
12 Q. L. Bao and K. P. Loh, Graphene Photonics, Plasmonics, and Broadband Optoelectronic Devices, *ACS Nano*, 2012, **6**, 3677–3694, DOI: 10.1021/nn300989g.
13 J. M. Dieterich and E. A. Carter, Quantum solutions for a sustainable energy future, *Nat. Rev. Chem.*, 2017, **1**, DOI: 10.1038/s41570-017-0032.
14 N. J. Halas, S. Lal, W. S. Chang, S. Link and P. Nordlander, Plasmons in Strongly Coupled Metallic Nanostructures, *Chem. Rev.*, 2011, **111**, 3913–3961, DOI: 10.1021/cr200061k.
15 S. Lal, S. Link and N. J. Halas, Nano-optics from sensing to waveguiding, *Nat. Photonics*, 2007, **1**, 641–648, DOI: 10.1038/nphoton.2007.223.
16 N. I. Zheludev and Y. S. Kivshar, From metamaterials to metadevices, *Nat. Mater.*, 2012, **11**, 917–924, DOI: 10.1038/nmat3431.
17 K. A. Willets and R. P. Van Duyne, Localized surface plasmon resonance spectroscopy and sensing, *Annu. Rev. Phys. Chem.*, 2007, **58**, 267–297, DOI: 10.1146/annurev.physchem.58.032806.104607.
18 J. N. Anker, *et al.*, Biosensing with plasmonic nanosensors, *Nat. Mater.*, 2008, **7**, 442–453, DOI: 10.1038/nmat2162.
19 J. Homola, Present and future of surface plasmon resonance biosensors, *Anal. Bioanal. Chem.*, 2003, **377**, 528–539, DOI: 10.1007/s00216-003-2101-0.
20 E. Hutter and J. H. Fendler, Exploitation of localized surface plasmon resonance, *Adv. Mater.*, 2004, **16**, 1685–1706, DOI: 10.1002/adma.200400271.

21 F. Vollmer and S. Arnold, Whispering-gallery-mode biosensing: label-free detection down to single molecules, *Nat. Methods*, 2008, **5**, 591–596, DOI: 10.1038/nmeth.1221.

22 H. Aitchison, *et al.*, Analytical SERS: general discussion, *Faraday Discuss.*, 2017, **205**, 561–600, DOI: 10.1039/c7fd90096a.

23 J. Aizpurua, *et al.*, Ultrasensitive and towards single molecule SERS: general discussion, *Faraday Discuss.*, 2017, **205**, 291–330, DOI: 10.1039/c7fd90088k.

24 D. Cialla, *et al.*, Surface-enhanced Raman spectroscopy (SERS): progress and trends, *Anal. Bioanal. Chem.*, 2012, **403**, 27–54, DOI: 10.1007/s00216-011-5631-x.

25 D. Graham, *et al.*, Theory of SERS enhancement: general discussion, *Faraday Discuss.*, 2017, **205**, 173–211, DOI: 10.1039/c7fd90095c.

26 A. I. Henry, T. W. Ueltschi, M. O. McAnally and R. P. Van Duyne, Spiers Memorial Lecture Surface-enhanced Raman spectroscopy: from single particle/molecule spectroscopy to angstrom-scale spatial resolution and femtosecond time resolution, *Faraday Discuss.*, 2017, **205**, 9–30, DOI: 10.1039/c7fd00181a.

27 P. L. Stiles, J. A. Dieringer, N. C. Shah and R. R. Van Duyne, Surface-Enhanced Raman Spectroscopy, *Annu. Rev. Anal. Chem.*, 2008, **1**, 601–626, DOI: 10.1146/annurev.anchem.1.031207.112814.

28 G. A. Baker and D. S. Moore, Progress in plasmonic engineering of surface-enhanced Raman-scattering substrates toward ultra-trace analysis, *Anal. Bioanal. Chem.*, 2005, **382**, 1751–1770, DOI: 10.1007/s00216-005-3353-7.

29 H. A. Atwater and A. Polman, Plasmonics for improved photovoltaic devices, *Nat. Mater.*, 2010, **9**, 205–213, DOI: 10.1038/nmat2629.

30 S. Pillai and M. A. Green, Plasmonics for photovoltaic applications, *Sol. Energy Mater. Sol. Cells*, 2010, **94**, 1481–1486, DOI: 10.1016/j.solmat.2010.02.046.

31 Q. Q. Gan, F. J. Bartoli and Z. H. Kafafi, Plasmonic-Enhanced Organic Photovoltaics: Breaking the 10% Efficiency Barrier, *Adv. Mater.*, 2013, **25**, 2385–2396, DOI: 10.1002/adma.201203323.

32 C. F. Guo, T. Y. Sun, F. Cao, Q. Liu and Z. F. Ren, Metallic nanostructures for light trapping in energy-harvesting devices, *Light: Sci. Appl.*, 2014, **3**, e161, DOI: 10.1038/lsa.2014.42.

33 Y.-Y. Cai, *et al.*, Photoluminescence of Gold Nanorods: Purcell Effect Enhanced Emission from Hot Carriers, *ACS Nano*, 2018, **12**, 976–985.

34 J. G. Liu, H. Zhang, S. Link and P. Nordlander, Relaxation of Plasmon-Induced Hot Carriers, *ACS Photonics*, 2018, **5**, 2584–2595.

35 C. Clavero, Plasmon-induced hot-electron generation at nanoparticle/metal-oxide interfaces for photovoltaic and photocatalytic devices, *Nat. Photonics*, 2014, **8**, 95–103, DOI: 10.1038/nphoton.2013.238.

36 M. L. Brongersma, N. J. Halas and P. Nordlander, Plasmon-induced hot carrier science and technology, *Nat. Nanotechnol.*, 2015, **10**, 25–34, DOI: 10.1038/nnano.2014.311.

37 A. Manjavacas, J. G. Liu, V. Kulkarni and P. Nordlander, Plasmon-Induced Hot Carriers in Metallic Nanoparticles, *ACS Nano*, 2014, **8**, 7630–7638, DOI: 10.1021/nn502445f.

38 A. Dorodnyy, *et al.*, Plasmonic Photodetectors, *IEEE J. Sel. Top. Quantum Electron.*, 2018, **24**, 4600313, DOI: 10.1109/jstqe.2018.2840339.

39 M. W. Knight, H. Sobhani, P. Nordlander and N. J. Halas, Photodetection with Active Optical Antennas, *Science*, 2011, **332**, 702–704, DOI: 10.1126/science.1203056.

40 M. W. Knight, *et al.*, Embedding Plasmonic Nanostructure Diodes Enhances Hot Electron Emission, *Nano Lett.*, 2013, **13**, 1687–1692, DOI: 10.1021/nl400196z.

41 A. Sobhani, *et al.*, Narrowband photodetection in the near-infrared with a plasmon-induced hot electron device, *Nat. Commun.*, 2013, **4**, 1643, DOI: 10.1038/ncomms2642.

42 B. Y. Zheng, *et al.*, Distinguishing between plasmon-induced and photoexcited carriers in a device geometry, *Nat. Commun.*, 2015, **6**, 7797, DOI: 10.1038/ncomms8797.

43 Z. Y. Fang, *et al.*, Graphene-Antenna Sandwich Photodetector, *Nano Lett.*, 2012, **12**, 3808–3813, DOI: 10.1021/nl301774e.

44 J. Zuloaga, E. Prodan and P. Nordlander, Quantum Plasmonics: Optical Properties and Tunability of Metallic Nanorods, *ACS Nano*, 2010, **4**, 5269–5276.

45 S. Mukherjee, *et al.*, Hot Electrons Do the Impossible: Plasmon-Induced Dissociation of H-2 on Au, *Nano Lett.*, 2013, **13**, 240–247, DOI: 10.1021/nl303940z.

46 J. A. Fan, *et al.*, Self-Assembled Plasmonic Nanoparticle Clusters, *Science*, 2010, **328**, 1135–1138, DOI: 10.1126/science.1187949.

47 M. Tanzid, *et al.*, Combining Plasmonic Hot Carrier Generation with Free Carrier Absorption for High-Performance Near-Infrared Silicon-Based Photodetection, *ACS Photonics*, 2018, **5**, 3472–3477, DOI: 10.1021/acsphotonics.8b00623.

48 E. Cortes, Activating plasmonic chemistry, *Science*, 2018, **362**, 28–29, DOI: 10.1126/science.aav1133.

49 J. Lee, S. Mubeen, X. L. Ji, G. D. Stucky and M. Moskovits, Plasmonic Photoanodes for Solar Water Splitting with Visible Light, *Nano Lett.*, 2012, **12**, 5014–5019, DOI: 10.1021/nl302796f.

50 C. G. Silva, R. Juarez, T. Marino, R. Molinari and H. Garcia, Influence of Excitation Wavelength (UV or Visible Light) on the Photocatalytic Activity of Titania Containing Gold Nanoparticles for the Generation of Hydrogen or Oxygen from Water, *J. Am. Chem. Soc.*, 2011, **133**, 595–602, DOI: 10.1021/ja1086358.

51 D. B. Ingram and S. Linic, Water Splitting on Composite Plasmonic-Metal/Semiconductor Photoelectrodes: Evidence for Selective Plasmon-Induced Formation of Charge Carriers near the Semiconductor Surface, *J. Am. Chem. Soc.*, 2011, **133**, 5202–5205, DOI: 10.1021/ja200086g.

52 M. Murdoch, *et al.*, The effect of gold loading and particle size on photocatalytic hydrogen production from ethanol over Au/TiO_2 nanoparticles, *Nat. Chem.*, 2011, **3**, 489–492, DOI: 10.1038/nchem.1048.

53 Z. W. Seh, *et al.*, Janus Au-TiO_2 Photocatalysts with Strong Localization of Plasmonic Near-Fields for Efficient Visible-Light Hydrogen Generation, *Adv. Mater.*, 2012, **24**, 2310–2314, DOI: 10.1002/adma.201104241.

54 P. Christopher, H. L. Xin and S. Linic, Visible-light-enhanced catalytic oxidation reactions on plasmonic silver nanostructures, *Nat. Chem.*, 2011, **3**, 467–472, DOI: 10.1038/nchem.1032.

55 W. H. Hung, M. Aykol, D. Valley, W. B. Hou and S. B. Cronin, Plasmon Resonant Enhancement of Carbon Monoxide Catalysis, *Nano Lett.*, 2010, **10**, 1314–1318, DOI: 10.1021/nl9041214.
56 W. B. Hou, *et al.*, Photocatalytic Conversion of CO_2 to Hydrocarbon Fuels *via* Plasmon-Enhanced Absorption and Metallic Interband Transitions, *ACS Catal.*, 2011, **1**, 929–936, DOI: 10.1021/cs2001434.
57 S. Mukherjee, *et al.*, Hot-Electron-Induced Dissociation of H-2 on Gold Nanoparticles Supported on SiO_2, *J. Am. Chem. Soc.*, 2014, **136**, 64–67, DOI: 10.1021/ja411017b.
58 L. Zhou, *et al.*, Aluminum Nanocrystals as a Plasmonic Photocatalyst for Hydrogen Dissociation, *Nano Lett.*, 2016, **16**, 1478–1484, DOI: 10.1021/acs.nanolett.5b05149.
59 M. J. McClain, *et al.*, Aluminum Nanocrystals, *Nano Lett.*, 2015, **15**, 2751–2755, DOI: 10.1021/acs.nanolett.5b00614.
60 D. F. Swearer, *et al.*, Heterometallic antenna-reactor complexes for photocatalysis, *Proc. Natl. Acad. Sci. U. S. A.*, 2016, **113**, 8916–8920, DOI: 10.1073/pnas.1609769113.
61 D. F. Swearer, *et al.*, Transition-Metal Decorated Aluminum Nanocrystals, *ACS Nano*, 2017, **11**, 10281–10288, DOI: 10.1021/acsnano.7604960.
62 C. Zhang, *et al.*, Al-Pd Nanodisk Heterodimers as Antenna-Reactor Photocatalysts, *Nano Lett.*, 2016, **16**, 6677–6682, DOI: 10.1021/acs.nanolett.6b03582.
63 W. Dong and J. Hafner, H_2 dissociative adsorption on Pd(111), *Phys. Rev. B: Condens. Matter Mater. Phys.*, 1997, **56**, 15396–15403.
64 T. Mitsui, M. K. Rose, E. Fomin, D. F. Ogletree and M. Salmeron, Dissociative hydrogen adsorption on Pd requires aggregates of three or more vacancies, *Nat. Catal.*, 2003, **422**, 705–707.
65 S. Linic, U. Aslam, C. Boerigter and M. Morabito, Photochemical transformations on plasmonic metal nanoparticles, *Nat. Mater.*, 2015, **14**, 567–576.
66 L. R. Hirsch, *et al.*, Nanoshell-mediated near-infrared thermal therapy of tumors under magnetic resonance guidance, *Proc. Natl. Acad. Sci. U. S. A.*, 2003, **100**, 13549–13554, DOI: 10.1073/pnas.2232479100.
67 N. J. Hogan, *et al.*, Nanoparticles Heat through Light Localization, *Nano Lett.*, 2014, **14**, 4640–4645, DOI: 10.1021/nl5016975.
68 L. A. Zhou, *et al.*, Quantifying hot carrier and thermal contributions in plasmonic photocatalysis, *Science*, 2018, **362**, 69–72, DOI: 10.1126/science.aat6967.
69 J. M. P. Martirez and E. A. Carter, Prediction of a low-temperature N-2 dissociation catalyst exploiting near-IR-to-visible light nanoplasmonics, *Sci. Adv.*, 2017, **3**, eeao4710, DOI: 10.1126/sciadv.aao4710.

Faraday Discussions

PAPER

Hot carriers generated by plasmons: where are they generated and where do they go from there?

Jacob B. Khurgin

Received 21st November 2018, Accepted 5th December 2018
DOI: 10.1039/c8fd00200b

A physically transparent unified theory of optically- and plasmon-induced hot carrier generation in metals is developed with all of the relevant mechanisms included. Analytical expressions that estimate the carrier generation rates and their locations, energies and directions of motion are obtained. Among the four mechanisms considered: interband absorption, phonon and defect assisted absorption, electron–electron scattering assisted absorption and surface-collision assisted absorption (Landau damping), it is the last one that generates hot carriers, which are most useful for practical applications in photodetection and photocatalysis.

Introduction

Recent developments in plasmonics[1–3] have resulted in renewed interest in the issue of absorption in metals. The fact that metal absorption is an important limiting factor in every imaginable application of plasmonics is universally known,[4] and, in the absence of efficient means for reducing loss, the emphasis has been shifting towards such applications where absorption in metals is not the main factor (typically sensors[5]). More recently, attention has been drawn to an altogether different class of applications for which absorption is a positive rather than deleterious factor, *e.g.* plasmonic Schottky photo-detectors[6–8] and light harvesting.[9,10] In these applications the incoming photons excite surface plasmons, either propagating or localized, and the rapid decay of plasmons causes the excitation of hot carriers which, once they reach the metal surface, can produce a photo-current in photodetectors[6–8] or initiate a chemical reaction *via* photocatalysis.[11] In order to understand the operation and evaluate the performance of these plasmon-assisted schemes, it is important to know not only the rate at which plasmons decay but also the distribution of energies and momentum of the hot carriers excited in the process.

As a general example, consider a plasmonic Schottky detector,[6–8] such as that shown in Fig. 1. In this detector, the decay of a surface plasmon with energy $\hbar\omega$

Johns Hopkins University, Baltimore, MD 21218, USA. E-mail: jakek@jhu.edu

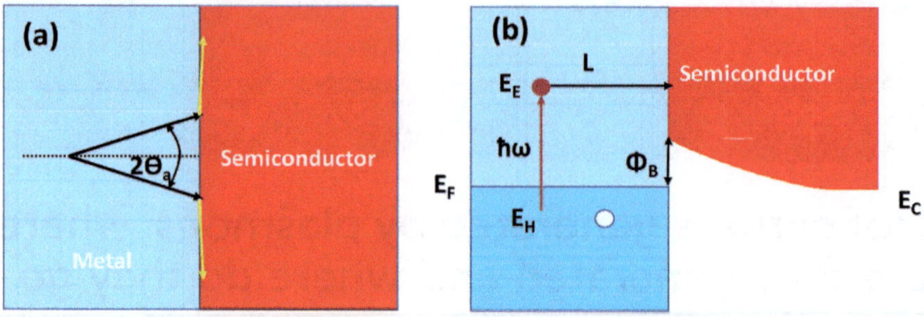

Fig. 1 Hot carrier Schottky barrier metal semiconductor photodetector: (a) spatial picture and (b) energy band diagram.

creates a hot electron above the Fermi level with kinetic energy E_E and a hole with energy E_H. If the electron has energy above the Schottky barrier, Φ_B, it has a chance to be injected into the semiconductor and contribute to the photocurrent. For that the electron needs to travel an unimpeded distance, L, to the interface, and more significantly it should also have a sufficiently small in-plane component of its momentum so it can enter the semiconductor rather than being reflected in a manner similar to the total internal reflection observed in optics. This fact has to do with the enormous momentum mismatch between the metal, where the carriers have a high kinetic energy that is commensurate with the Fermi energy, E_F, *i.e.* 5–6 eV for Au or Ag, and the semiconductor, where the kinetic energy of the carriers is only $E_E - \Phi_B$, or 1 eV at best. In order to avoid total reflection, the electrons must propagate inside a cone subtended by an "acceptance angle"[12]

$$\Theta_a = \sin^{-1}\left(\sqrt{\frac{m_s}{m_0}\frac{E_E - \Phi_B}{E_F}}\right) \tag{1}$$

where m_s and m_0 are the effective masses for the semiconductor and metal, respectively. The angle is typically on the scale of 10–20 degrees, leading to injection efficiencies that are often less than 1% for carriers propagating omnidirectionally. For this reason, it is important to investigate the energies as well as the spatial and angular distributions of the hot carriers engendered by the decay of surface plasmons.

While there has been a significant amount of work performed in this respect recently,[13–18] no clear consensus has been achieved and no unified picture of the process has been presented. Four different mechanisms of plasmon decay have been identified: interband absorption, phonon (or defect) assisted absorption, electron–electron scattering assisted absorption and interband absorption, and each of these has been treated using a different formalism, ranging from a simple phenomenological treatment to a full quantum one. Some of the processes, such as electron–electron scattering assisted absorption, have been totally neglected which, as we show below, is not justified in the visible and even IR range. On the other hand, interband absorption was considered as a promising source of hot carriers, while in reality most of the energy in this process goes to the holes residing in the d-band and having a very low velocity, which prevents them from reaching the interface and performing any useful function. Furthermore, while most of the studies have focused on the energy distribution of the hot carriers, little attention[15] has been paid to their angular and spatial distributions which, as we have just explained, are of paramount importance to hot-carrier enabled devices. First and foremost, practically all of the current work has been performed numerically, which may be a convenient method but does not reveal the intricate physics of hot carrier generation.

In this work we develop a simple analytical model of hot carrier generation based on the Fermi golden rule, which allows us to ascertain the energy, spatial and angular distributions of hot carriers generated by different absorption mechanisms. This allows us to make estimates of the efficiency of injection of the photo-generated carriers and to suggest ways towards its improvement.

Background: Drude theory of free carrier absorption

Currently, in order to estimate the FCA (free carrier absorption) of a metal, one uses standard expressions for the AC conductivity using Drude formulae for the dielectric constant of the metal

$$\varepsilon(\omega) = \varepsilon_b - \frac{Ne^2}{\varepsilon_0 m(\omega^2 + j\omega\gamma)}, \tag{2}$$

where ε_∞ is the relative permittivity due to bound carriers (inter-band transitions), m is the effective mass, and the scattering rate, γ, is usually introduced phenomenologically. The power dissipation rate is then

$$\frac{dU}{dt} = -\omega\varepsilon_0 \mathrm{Im}[\varepsilon(\omega)]\frac{E^2}{2} = \frac{Ne^2\gamma}{m\omega^2}\frac{E^2}{2} = -\varepsilon_0\frac{\omega_p{}^2}{\omega^2}\gamma\frac{E^2}{2}. \tag{3}$$

It should be noted here that the total energy density in the metal is

$$U = \frac{1}{4}\varepsilon_0 \mathrm{Re}\left(\omega\frac{\partial\varepsilon}{\partial\omega}\right) + \frac{1}{4}\mu_0 H^2 = \frac{1}{4}\varepsilon_0\varepsilon_\infty E^2 + \frac{1}{4}\mu_0 H^2 + \frac{1}{4}\varepsilon_0\frac{\omega_p{}^2}{\omega^2}E^2, \tag{4}$$

where the last term is the density of kinetic energy of the carriers, therefore the power dissipation is

$$\frac{dU}{dt} = -2\gamma U_K, \tag{5}$$

which of course makes perfect sense since the energy is lost only when the carriers are in motion. The scattering rate is usually written as a sum of the different scattering processes, namely scattering on phonons, on various lattice imperfections, and due to electron–electron interactions. However, it remains questionable how one can apply a theory developed for low frequencies, typically for radio-frequencies all the way to visible light and even UV. At low frequencies all of the processes take place in the vicinity of the Fermi surface, so one needs to know only the material properties near the Fermi surface, while at higher frequencies almost all of the Brillouin zone starts to play a role. This fact leads to some interesting observations; some materials that are good low-loss conductors in the RF region, such as Cu, may suffer from higher losses in the optical region than Al for example, which is not such a great conductor. Clearly, the scattering rate, γ, is a strong function of frequency. Therefore, in this work we establish this dependence and with it we re-derive the Drude formula for the frequency dependence of the dielectric constant.

Is there a difference between the absorption of a photon and a plasmon?

Before proceeding, it is important to establish once and for all what the difference is between the processes of hot carrier generation as a result of photon or plasmon decay. In order to understand this, it is important to realize that what is normally referred to as a "plasmon" is in reality a (surface) plasmon polariton, or SPP, *i.e.* a coupled mode (or a quasiparticle) in which the energy is split between the energy of the electro-magnetic field, $\frac{1}{4}\varepsilon_0 E^2 + \frac{1}{4}\mu_0 H^2$, and the matter. In metals, at frequencies significantly less than the plasma frequency, the energy of the matter is mostly the kinetic energy of the collective motion of free carriers, $\frac{1}{4}\varepsilon_0({\omega_p}^2/\omega^2)E^2$, as well as a contribution from the potential energy of the collective motion of bound carriers $\frac{1}{4}\varepsilon_0(\varepsilon_\infty - 1)E^2$. In dielectrics, what is normally referred to as a "photon" is in fact also a polariton, and the energy is split between the energy of the electro-magnetic field and the energy of the bound carriers. The only difference between plasmons and photons in dielectrics is that in plasmons a significant part of the energy is contained as the kinetic energy of free carriers, while in photons the dielectric energy is contained in the potential energy of bound carriers and in the magnetic field.[19] The distribution of energies is shown in Fig. 2. The interaction between polaritons and single electron–hole pair

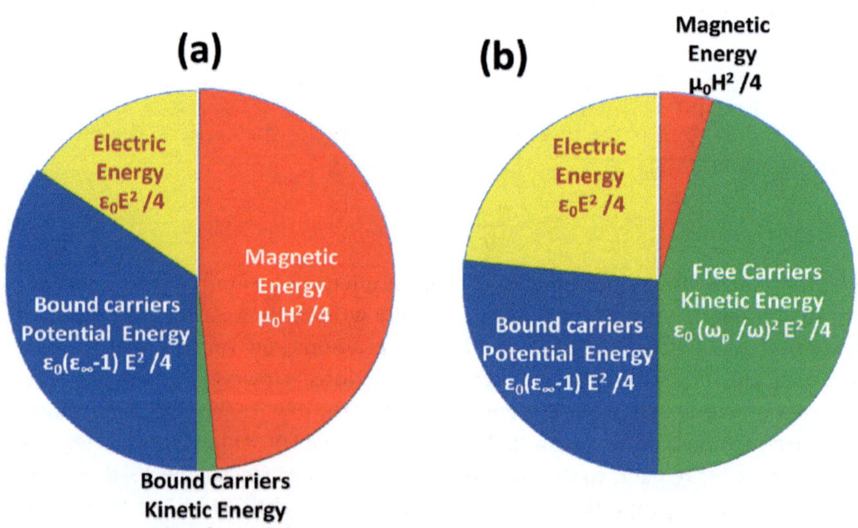

Fig. 2 Energy breakdown for (a) a photon polariton in a dielectric and (b) a plasmon polariton in a metal. The only difference is in the relative contributions of the magnetic and kinetic energies, and only the electric field determines the interactions between polaritons and single particle excitations in the matter.

excitations in metals or dielectrics is still characterized by the same Hamiltonian, $e\mathbf{r}\cdot\mathbf{E}$(or $e\mathbf{p}\cdot\mathbf{A}/m$ in the momentum gauge), *i.e.* only the electric field of the polariton interacts with the matter.

When it comes to a comparison between the absorption of localized SPPs and the incoming photons in a metal, once again there is no physical difference between the two processes; in fact, if a localized SPP can be considered a bound state, the photon propagating in the vicinity of a metal can be considered as simply an un-bound state. Of course, the actual loss will depend on the electric field distribution in the states, and the absorption will differ as a result, but the difference is quantitative and not qualitative. However, different absorption mechanisms will dominate in the bulk of the metal and near its surface, and their relative strengths will affect the properties of the photo-generated hot carriers, whether they are generated by SPPs or by incoming photons penetrating the metal. In the following sections we consider these mechanisms one by one.

Free carrier absorption arising from phonon and defect scattering

This is the most ubiquitous mechanism responsible for intra-band free carrier absorption, and can be understood phenomenologically as simple friction accompanying the motion of free carriers in the electromagnetic field. For this reason, it is often referred to as "classical"[18] or "resistive",[16] and somehow not leading to hot carrier generation. This is strictly speaking not correct as, of course, hot carriers are generated in this process that can be perfectly well described by a quantum mechanical formalism, as shown below.

Consider a simple parabolic conduction band of a metal, such as that shown in Fig. 3. When a photon (SPP) with energy $\hbar\omega \gg k_BT$ is inside the metal, the momentum conservation rule does not allow a direct absorption of it. The momentum mismatch can be provided by either a phonon or an impurity

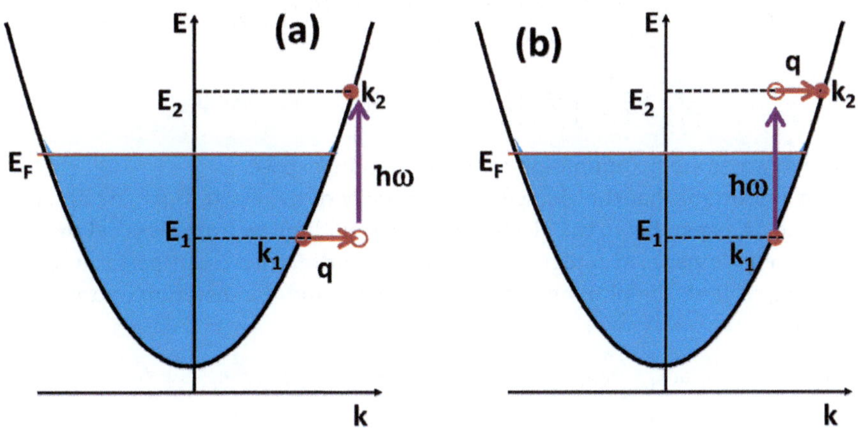

Fig. 3 Two processes involved in phonon-assisted absorption in a metal: (a) phonon scattering into the virtual state is followed by photon (SPP) absorption into the real state, k_2, and (b) photon (SPP) absorption into the virtual state is followed by scattering into the real state, k_2.

(imperfection). Here we consider a phonon with energy $\hbar\omega_p \ll \hbar\omega$ and wavevector $\boldsymbol{q}$ that provides momentum matching between the initial state of the electron with energy and momentum $(E_1,\boldsymbol{k}_1)$ and the final state $(E_2 = E_1 + \hbar\omega, \boldsymbol{k}_2 = \boldsymbol{k}_1 \pm \boldsymbol{q})$. The $\pm$ sign corresponds to phonon emission and absorption, respectively. Imperfections can be treated essentially as superpositions of phonons with vanishingly low frequencies and different wavevectors. The interaction between the electromagnetic field and the electron is described by the Hamiltonian $H_{ef} = e\boldsymbol{A}\cdot\boldsymbol{v}$, where $\boldsymbol{A} = \boldsymbol{E}/\omega$ is the vector potential and $\boldsymbol{v} = \hbar\boldsymbol{k}/m$ (assuming a parabolic band) is the electron velocity. The electron–phonon interaction is described by the Hamiltonian

$$H_q = D\sqrt{\frac{2\hbar\omega_p(n_p + 1/2 \pm 1/2)}{\rho V v_s^2}}, \tag{6}$$

where D is a deformation potential, ρ is the density, v_s is the sound velocity, and $n_p(T)$ is the temperature-dependent number of phonons with a given frequency. The phonons involved typically have wavevector q on the order of the Fermi wavevector. Hence their frequency is on the order of the Debye frequency of a few THz, and we can introduce the average parameter $\langle n_p \rangle$ which describes both absorption and emission.

What is important to consider is that the process of phonon assisted absorption can occur *via* two alternative sequences (or pathways). In the first pathway, shown in Fig. 3a, at first an electron scatters from $\boldsymbol{k}_1$ to a virtual state $(E_1,\boldsymbol{k}_2)$ with the help of a phonon, and the energy is not conserved, with an energy deficit of $-\hbar\omega$. Following that, the photon gets absorbed and the electron goes into the final state $(E_2,\boldsymbol{k}_2)$. In the second pathway, shown in Fig. 3b, the electron first absorbs the photon and ends up in a virtual state $(E_2,\boldsymbol{k}_1)$, which now has excess energy $(+\hbar\omega)$. This is followed by a phonon-assisted transition into the final state $(E_2,\boldsymbol{k}_2)$. The amplitudes of both pathways will interfere, hence, according to the Fermi golden rule, one can write for the absorption rate of a photon by an electron in the initial state $(E_1,\boldsymbol{k}_1)$ with spin conservation

$$\begin{aligned} R_{k_1} &= \frac{2\pi}{\hbar}\sum_{k_2}\left[\frac{1}{2}\frac{(H_q)(e\boldsymbol{A}\cdot\boldsymbol{v}_2)}{\hbar\omega} - \frac{1}{2}\frac{(e\boldsymbol{A}\cdot\boldsymbol{v}_1)(H_q)}{\hbar\omega}\right]^2 \delta(E_2 - E_1 - \hbar\omega) \\ &= \frac{2\pi e^2}{\hbar^3\omega^4}V\frac{1}{8\pi^3}\int H_q^{\,2}\left[\frac{E}{2}(v_2 - v_1)\right]^2 \delta(E_2 - E_1 - \hbar\omega)\mathrm{d}k_2 \end{aligned} \tag{7}$$

in which we go from summation to integration in k-space.

It is important that the change in the velocity of the electron, *i.e.* its acceleration vector, is parallel to the electric field, as would follow from classical theory. Next, when we perform integration in polar coordinates we can perform angle averaging of both the phonon scattering strength and the direction and magnitude of $\boldsymbol{v}_2 - \boldsymbol{v}_1$

$$\begin{aligned} \left\langle (v_{2,z} - v_{1,z})^2 \right\rangle &= v_F^{\,2}\left\langle (\cos\theta_2 - \cos\theta_1)^2 \right\rangle_{\theta_1,\theta_2} \approx v_F^{\,2} \times \left(\left\langle \cos^2\theta_1 \right\rangle_{\theta_1} + \left\langle \cos^2\theta_2 \right\rangle_{\theta_2}\right) \\ &= \frac{2}{3}v_F^{\,2}, \end{aligned} \tag{8}$$

where v_F is the Fermi velocity, and obtain

$$R_{k_1} = \frac{2\pi e^2}{\hbar^3 \omega^4} V \frac{1}{8\pi^3} \frac{E^2}{6} v_F^{\ 2} \int H_q^{\ 2} \delta(E_2 - E_1 - \hbar\omega) \mathrm{d}\boldsymbol{k}_2 \approx \frac{\pi}{3} \frac{e^2 E^2 v_F^{\ 2}}{\hbar^3 \omega^4} V H_q^{\ 2} \rho_{E_1+\hbar\omega}, \tag{9}$$

where $\rho_{E_1+\hbar\omega} = (1/8\pi^3)\int \delta(E_2 - E_1 - \hbar\omega)\mathrm{d}\boldsymbol{k}_2$ is the density of the final states. We can now introduce the average electron-phonon scattering time as

$$\tau_{ep}^{\ -1}(\omega) = \langle (2\pi/\hbar) H_q^{\ 2} V \rho_{E_1+\hbar\omega} \rangle_{E_1} \tag{10}$$

and then obtain for the rate of absorption from the given energy level, E_1,

$$R_{E_1} = \frac{e^2 v_F^{\ 2} \tau_{ep}^{\ -1} E^2}{6\hbar^2 \omega^4} \tag{11}$$

Next we calculate the rate of hot carrier generation from all levels, *i.e.*

$$\frac{\mathrm{d}N_e}{\mathrm{d}t} = \int_{E_F - \hbar\omega}^{E_F} R_{E_1} \rho(E_1) \mathrm{d}E_1 \approx \hbar\omega R_{E_1} \rho(E_F) = \frac{e^2 v_F^{\ 2} \tau_{ep}^{\ -1}(\omega) E^2}{6\hbar\omega^3} \rho(E_F). \tag{12}$$

The energy relaxation rate is then

$$\frac{\mathrm{d}U_\omega}{\mathrm{d}t} = -\hbar\omega \frac{\mathrm{d}N_e}{\mathrm{d}t} \approx \frac{e^2 v_F^{\ 2} \tau_{ep}^{\ -1}}{6\omega^2} \rho(E_F) E^2 = \frac{e^2 v_F^{\ 2}}{3\omega^2} \rho(E_F) \tau_{ep}^{\ -1} \frac{E^2}{2}. \tag{13}$$

To make (13) look like (3) we multiply and divide by the square of the plasma frequency

$$\frac{\mathrm{d}U_\omega}{\mathrm{d}t} = -\frac{e^2 v_F^{\ 2}}{3\varepsilon_0 \omega_p^{\ 2}} \rho(E_F) \varepsilon_0 \frac{\omega_p^{\ 2}}{\omega^2} \tau_{ep}^{\ -1} \frac{E^2}{2} = -\gamma_{ph} \varepsilon_0 \frac{\omega_p^{\ 2}}{\omega^2} \frac{E^2}{2}. \tag{14}$$

where

$$\gamma_{ph} = \frac{e^2 v_F^{\ 2}}{3\varepsilon_0 \omega_p^{\ 2}} \rho(E_F) \tau_{ep}^{\ -1}. \tag{15}$$

However, for the spherical band, the AC conductivity can be found by integrating over the Fermi surface as

$$\sigma(\omega) = \mathrm{i}\frac{1}{3} \frac{e^2 v_F^{\ 2} \rho(E_F)}{\omega} = \mathrm{i}\varepsilon_0 \frac{\omega_p^{\ 2}}{\omega}. \tag{16}$$

Therefore, the plasma frequency is

$$\omega_p^{\ 2} = \frac{1}{3} \frac{e^2 v_F^{\ 2} \rho(E_F)}{\varepsilon_0} \tag{17}$$

and (15) becomes simply

$$\gamma_{ph}(\omega) = \tau_{ep}^{\ -1}(\omega), \tag{18}$$

which is of course an intuitive result. However, it should be noted that this result is frequency dependent as the scattering is averaged over the range of energies from the Fermi level up to $E_F + \hbar\omega$, and thus can be somewhat different from the scattering rate at the Fermi level that enters the expression for conductivity at low frequencies. One can make a rough estimate of (10), assuming a deformation potential of $D \sim 10$ eV, a density of states of $\rho_{E_F} \approx \frac{3}{2} N_e/E_F$, where the electron density is 6×10^{22} cm^{-3}, a mass density of $\rho \sim 20$ g cm^{-3}, a speed of sound of $v_s = 3.2 \times 10^3$ m s^{-1}, and a mean phonon energy of $\langle \hbar\omega_p \rangle \sim 10$ meV that is equal to the Debye energy. One obtains for gold at room temperature, $\gamma_{ph} = 6 \times 10^{13}$ s^{-1}. Recent calculations[20] have estimated room temperature values of $\gamma_{ph} = 3 \times 10^{13}$ s^{-1} for silver and $\gamma_{ph} = 1 \times 10^{14}$ s^{-1} for gold.

Next, it is important to estimate the directionality of the photo-excited hot carriers. What we need to do is ascertain the photo-excitation dependence on the angle θ_2 while averaging over the angle θ_1, or

$$R(\theta_2) \sim \frac{\left\langle (v_{2,z} - v_{1,z})^2 \right\rangle_{v_1}}{\left\langle (v_{2,z} - v_{1,z})^2 \right\rangle_{v_1,v_2}} \sim \frac{\left\langle (\cos\theta_2 - \cos\theta_1)^2 \right\rangle_{\theta_1}}{\left\langle (\cos\theta_2 - \cos\theta_1)^2 \right\rangle_{\theta_1,\theta_2}} = \frac{\cos^2\theta_2 + \frac{1}{3}}{\frac{2}{3}}$$
$$= \frac{1}{2}\left(3\cos^2\theta_2 + 1\right). \tag{19}$$

Therefore, the distribution is not uniform and is skewed along the direction of the electric field, which for many plasmonic structures is directed normal to the surface. We shall consider the impact of this directionality later on (Fig. 7).

Absorption assisted by electron–electron scattering

This is the process that is often overlooked in studies of hot carrier generation in plasmonics. This is partially due to the fact that it is a second order process involving two electrons. In the field of photonics, most closely related to plasmonics, higher order processes involving multiple photons are indeed often disregarded since they are quite weak and manifest themselves only at high powers, as is known to anyone familiar with nonlinear optics.[21] However, the situation could not be more different in condensed matter physics where multi-phonon and multi-electron interactions play a key role in many important phenomena, such as heat transport.[22] The difference between optics and condensed matter physics is of course in the density of states that is at least 10 orders of magnitude higher in the solid state than in optics. The large number of states available for scattering makes multi-particle processes important, and the electron–electron scattering-assisted absorption of photons is just one of these processes. At low frequencies these processes are not important since the phase space available for scattering is within a narrow layer, $\hbar\omega$, from the Fermi surface, but at higher photon frequencies this phase space increases proportionally to $\hbar\omega$, and, since there are two electrons involved in scattering, the probability of these processes occurring increases as ω^2. This fact has been observed in multiple experiments dating back to the 1970s and 1980s,[23–25] and theoretical calculations[26] performed on the basis of Landau's Fermi liquid theory[27] have shown that the electron–electron scattering rate at high frequencies can indeed be evaluated as

$$\gamma_{\mathrm{ee},\omega}(\omega,T) \sim \gamma_{\mathrm{ee},0}(T)(\hbar\omega/2\pi k_{\mathrm{B}}T)^2, \quad (20)$$

where for gold $\gamma_{\mathrm{ee},0} \sim 2 \times 10^{11}\ \mathrm{s}^{-1}$ at room temperature.[23] At an energy of 1 eV the rate of electron–electron scattering involved in photon absorption is $3 \times 10^{13}\ \mathrm{s}^{-1}$, while calculations performed more recently[20] have predicted somewhat higher rates of $7 \times 10^{13}\ \mathrm{s}^{-1}$, which is commensurate with phonon-assisted absorption, γ_{ph}.

A rigorous theoretical description of the electron–electron scattering process is quite involved, and should follow the full treatment of a Fermi liquid,[28] so in this section we outline the physics of the process and perform a simplified and physically transparent order-of-magnitude estimation of absorption aided by electron–electron scattering.

There are four different pathways, all shown in Fig. 4, that correspond to the absorption of a photon, $\hbar\omega$, and the simultaneous transition of two electrons (of different spins) from the initial state $(E_1,\boldsymbol{k}_1;E_3,\boldsymbol{k}_3)$ to the final state $(E_2,\boldsymbol{k}_2;E_4,\boldsymbol{k}_4)$. Since this is elastic scattering the energy is conserved: $E_1 + E_3 + \hbar\omega = E_2 + E_4$. The momentum is also conserved: $\boldsymbol{k}_1 + \boldsymbol{k}_3 - \boldsymbol{k}_4 - \boldsymbol{k}_2 = K\boldsymbol{G}$, where $\boldsymbol{G}$ is one of the inverse lattice basis wavevectors. Two electrons are scattered by the screened Coulomb potential

$$H_{\mathrm{ee,K}} = f_{\mathrm{K}} \frac{e^2}{\varepsilon_0 V\left(\Delta k^2 + {k_{\mathrm{s}}}^2\right)}, \quad (21)$$

where k_{s} is the screening wavevector that for most metals is close to the Fermi wavevector, k_{F}, and $\Delta\boldsymbol{k} = \boldsymbol{k}_3 - \boldsymbol{k}_1$ is the momentum transfer between the two electrons. Since $\Delta k < k_{\mathrm{F}} \sim k_{\mathrm{s}}$, in order to make an order-of-magnitude estimate one can assume

$$H_{\mathrm{ee,K}} \approx f_{\mathrm{K}} \frac{e^2}{\varepsilon_0 V {k_{\mathrm{F}}}^2}. \quad (22)$$

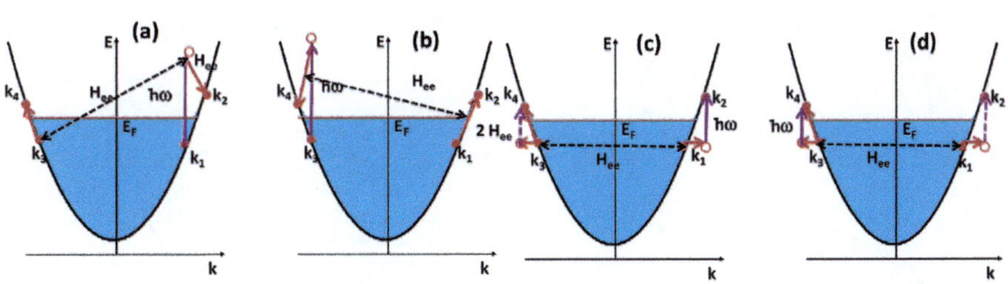

Fig. 4 Absorption assisted by electron–electron scattering. Four different pathways resulting in the generation of two electron–hole pairs of different spins are shown. (a) A photon excites the electron on the right and transfers the system into a virtual state $(E_2,\boldsymbol{k}_1;E_3,\boldsymbol{k}_3)$ with excess energy, $\hbar\omega$; electron–electron scattering then moves it into the final state $(E_2,\boldsymbol{k}_2;E_4,\boldsymbol{k}_4)$. (b) A photon excites the electron on the left, transferring the system into a virtual state $(E_1,\boldsymbol{k}_1;E_4,\boldsymbol{k}_3)$ with excess energy, $\hbar\omega$; electron–electron scattering then moves it into the final state $(E_2,\boldsymbol{k}_2;E_4,\boldsymbol{k}_4)$. (c) Electron–electron scattering transfers the system into a virtual state $(E_1,\boldsymbol{k}_2;E_3,\boldsymbol{k}_4)$; a photon then interacts with the electron on the right and moves the system into its final state $(E_2,\boldsymbol{k}_2;E_4,\boldsymbol{k}_4)$. (d) Electron–electron scattering transfers the system into a virtual state $(E_1,\boldsymbol{k}_2;E_3,\boldsymbol{k}_4)$; a photon then interacts with the electron on the left and moves the system into its final state $(E_2,\boldsymbol{k}_2;E_4,\boldsymbol{k}_4)$.

When $K = 0$ the process is normal (N), otherwise it is an Umklapp (U) process[22] and the matrix element of the Hamiltonian reported in (21) is reduced by the overlap integral $f_K = \int u^2(\boldsymbol{r})e^{jK\boldsymbol{G}\cdot\boldsymbol{r}}d\boldsymbol{r}$, where $u(\boldsymbol{r})$ is the periodic part of the Bloch function and the integral is taken over one unit cell. Obviously $f_0 = 1$. Let us now consider the four sequences shown in Fig. 4, demonstrating that there are four different sequences for absorption involving two electrons of different spins.

The interference of the four pathways shown in Fig. 4 results in the following probability for the absorption of a phonon or exciton by an electron in state $\mathbf{k}_1$:

$$\begin{aligned} R_{k_1} &= \frac{2\pi}{\hbar}V^3\sum_{k_3}\sum_{k_4}\sum_{k_2}\left[\frac{(e\boldsymbol{A}\cdot v_1)(H_{\text{ee,K}})}{-\hbar\omega} + \frac{(e\boldsymbol{A}\cdot v_3)(H_{\text{ee,K}})}{-\hbar\omega} + \frac{(H_{\text{ee,K}})(e\boldsymbol{A}\cdot v_2)}{\hbar\omega}\right.\\ &\quad \left. + \frac{(H_{\text{ee,k}})(e\boldsymbol{A}\cdot v_4)}{\hbar\omega}\right]^2 \times\delta(\boldsymbol{k}_3 + \boldsymbol{k}_1 - \boldsymbol{k}_4 - \boldsymbol{k}_2 - K\boldsymbol{G})\delta(E_1 + E_3 + \hbar\omega - E_2 - E_4)\\ &= \frac{2\pi e^2}{\hbar^3\omega^4}V^3\sum_{k_3}\sum_{k_4}\sum_{k_2}H_{\text{ee,K}}{}^2[\boldsymbol{E}(\boldsymbol{v}_3 + \boldsymbol{v}_1 - \boldsymbol{v}_4 - \boldsymbol{v}_2)]^2\delta(\boldsymbol{k}_3 + \boldsymbol{k}_1 - \boldsymbol{k}_4 - \boldsymbol{k}_2 - K\boldsymbol{G})\\ &\quad \times\delta(E_1 + E_3 + \hbar\omega - E_2 - E_4). \end{aligned} \tag{23}$$

We now see that the parabolic band N-process with $K = 0$ results in no absorption as the sum of velocities is equal to zero. This is precisely the classical result that follows from momentum conservation – elastic electron scattering conserves the total momentum of electrons and, since velocity is proportional to momentum, the total velocity and current are conserved, thus indicating zero resistance and hence zero energy dissipation. Therefore, one should only consider U processes, realistically only with the order $K = 1$ so that one of the generated hot

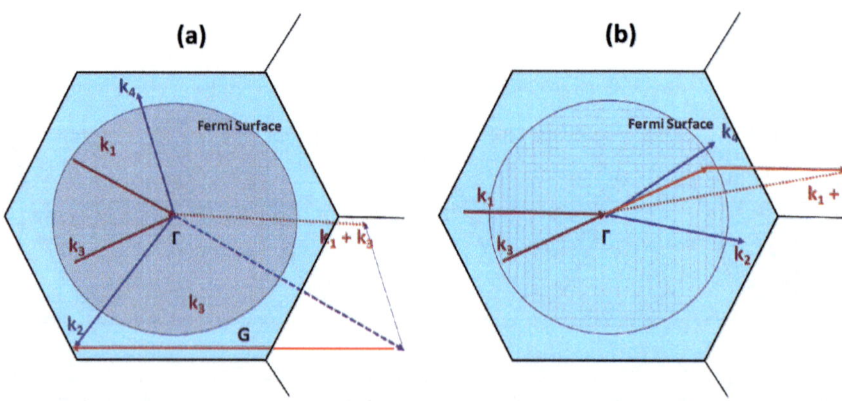

Fig. 5 (a) Photon (SPP) absorption with energy $\hbar\omega$ assisted by Umklapp scattering of the electrons. As the photon (SPP) gets annihilated the electrons k_1,k_3 below the Fermi surface get scattered into the states k_2,k_4 above the Fermi surface, with momentum conservation preserved by the lattice vector $\boldsymbol{G}$. (b) Thermalization of a hot electron. A hot electron, k_1, with an energy $\hbar\omega$ above the Fermi level, and a cold electron, k_3, scatter into the states k_2,k_4 above the Fermi surface, each with an energy less than $\hbar\omega$. Both normal and Umklapp processes are possible. The similarity between the two processes indicates that the rate of e–e scattering assisted absorption is commensurate with the rate of hot electron thermalization.

carriers (k_2) ends up in the adjacent Brillouin zone, as shown in Fig. 5a. Obviously velocity is no longer conserved and the energy becomes dissipated. Of course, another possibility is that the second electron, k_4, ends up in the adjacent Brillouin zone.

Therefore, the rate of absorption from a particular state, k_1, is obtained by integration over all of the possible initial and final states of the second electron, while the state k_4 is determined by the momentum conservation

$$R_{\boldsymbol{k}_1} = \frac{2\pi e^6 E^2 (\hat{\boldsymbol{e}}\cdot\boldsymbol{G})^2}{{k_F}^4 {\varepsilon_0}^2 \hbar m^2 \omega^4}\left(\frac{1}{4\pi^3}\right)^2 \int_{\boldsymbol{k}_3}\int_{\boldsymbol{k}_2}\left[A_U(\boldsymbol{k}_2){f_1}^2(\boldsymbol{k}_2) + A_U(\boldsymbol{k}_4){f_1}^2(\boldsymbol{k}_4)\right]\delta(E_1 + E_3 + \hbar\omega - E_2 - E_4)\mathrm{d}\boldsymbol{k}_2\mathrm{d}\boldsymbol{k}_3 \tag{24}$$

where $\hat{\boldsymbol{e}}$ is a unit vector denoting polarization. Of course, integration assumes that the energies of the initial states 1 and 3 are below the Fermi surface while the energies of the final states 2 and 4 are above it. Let us now elaborate on the nature of the coefficients $A_U(\boldsymbol{k})$ and ${f_1}^2(\boldsymbol{k})$. The first coefficient $A_U(\boldsymbol{k}_2)(A_U(\boldsymbol{k}_4) = 1$ when the final wavevector $\boldsymbol{k}_2(\boldsymbol{k}_4)$ is outside the first Brillouin zone and is equal to zero otherwise, while the overlap integral

$$f_1(\boldsymbol{k}) = \int u^*_{k-g}(\boldsymbol{r})\mathrm{e}^{\mathrm{i}\boldsymbol{G}\cdot\boldsymbol{r}} u_{\boldsymbol{k}}(\boldsymbol{r})\mathrm{d}\boldsymbol{r} \tag{25}$$

is essentially the first order Fourier component of the periodic Bloch function $u_{\mathbf{k}}(\boldsymbol{r})$, which for nearly free electrons in noble metals obviously becomes large only in the vicinity of the Brillouin zone boundary. Hence the whole term in the square bracket is substantial when one of the final wavevectors exceeds the Brillouin zone boundary by a relatively small value.

Next, following example (10) we introduce the effective electron–electron Umklapp scattering rate as

$$\tau_{ee,U}^{-1}(\omega) = \frac{2\pi e^4}{\hbar {\varepsilon_0}^2 {k_F}^4}\left(\frac{1}{4\pi^3}\right)^2 \int_{\boldsymbol{k}_3}\int_{\boldsymbol{k}_2}\left[A_U(\boldsymbol{k}_2){f_1}^2(\boldsymbol{k}_2) + A_U(\boldsymbol{k}_4){f_1}^2(\boldsymbol{k}_4)\right]\delta(E_1 + E_3 + \hbar\omega - E_2 - E_4)\mathrm{d}\boldsymbol{k}_2\mathrm{d}\boldsymbol{k}_3 \approx F_U(\hbar\omega)\times\tau_{ee}^{-1}(\omega), \tag{26}$$

where

$$\tau_{ee}^{-1}(\omega) = \frac{2\pi e^4}{\hbar {\varepsilon_0}^2 {k_F}^4}\left(\frac{1}{4\pi^3}\right)^2 \int_{\boldsymbol{k}_3}\int_{\boldsymbol{k}_2}\delta([E_F + \hbar\omega] + E_3 - E_2 - E_4)\mathrm{d}\boldsymbol{k}_2\mathrm{d}\boldsymbol{k}_3 \tag{27}$$

is the electron–electron relaxation (thermalization) rate of the hot carriers having an energy $\hbar\omega$ above the Fermi level (as shown in Fig. 5b), for which a large amount of experimental data has been accumulated,[29–33] and

$$F_U(\hbar\omega) = \langle A_U(\boldsymbol{k}_2){f_1}^2(\boldsymbol{k}_2) + A_U(\boldsymbol{k}_4){f_1}^2(\boldsymbol{k}_4)\rangle_{\boldsymbol{k}_2,\boldsymbol{k}_4} \tag{28}$$

is the relative strength of the Umklapp process. It is important to note that, in addition to the scattering of two different electrons with different spins there is also a contribution from the scattering of two identical electrons, which is smaller due to exchange. This interaction has been considered in ref. 34 and found to be of the same order but somewhat less (order of 1/2) than the interaction of carriers with different spins. As long as exchange interactions are factored into the expression (27) for electron thermalization they will also be factored into the expression for photon (plasmon) decay obtained from (24):

$$R_{E_1} = \frac{e^2 G^2}{3m^2\omega^4} F_U(\hbar\omega)\tau_{ee}{}^{-1}(\hbar\omega)E^2, \tag{29}$$

where the factor (1/3) has been obtained by averaging over the directions of the reciprocal lattice, ***G***. Next, following all the steps taken for the phonon-assisted absorption shown in (12)–(14), we obtain

$$\frac{dU_\omega}{dt} = -\hbar^2\omega^2 R_{E_1}\rho(E_F) \approx \frac{\hbar^2 e^2 G^2}{3m^2\omega^2}\rho(E_F)F_U(\hbar\omega)\tau_{ee}{}^{-1}(\hbar\omega)E^2. \tag{30}$$

Then, using (17), we transform this expression into

$$\frac{dU_\omega}{dt} = 2\frac{\hbar^2 G^2}{m^2 v_F{}^2} F_U(\hbar\omega)\tau_{ee}{}^{-1}(\hbar\omega)\varepsilon_0 \frac{\omega_p{}^2}{\omega^2}\frac{E^2}{2} \tag{31}$$

Now the term $2\hbar^2 G^2/m^2 v_F{}^2$ is commensurate with the ratio of the energy at the boundary of the BZ to the Fermi energy, and is on the order of unity for a metal, hence it can also be incorporated into the "Umklapp factor" $F_U(\hbar\omega)$. In accordance with this, one finally obtains an expression for the effective electron–electron scattering that can be entered into the Drude formula as

$$\gamma_{ee}(\omega) = F_U(\hbar\omega)\tau_{ee}{}^{-1}(\hbar\omega). \tag{32}$$

The term $\tau_{ee}{}^{-1}(\hbar\omega)$ has been studied in numerous theoretical[35–38] and experimental[29–33] works, which have all shown that for energies 1 eV above the Fermi level it is on the scale of 10^{14} s^{-1}, and so here we should ascertain whether this order of magnitude follows from our theoretical model.

Following Ziman's derivation of AC resistivity, reported in ref. 39, we obtain an expression for the integral reported in (27) as roughly

$$I = \int_{k_3}\int_{k_2} \delta([E_F + \hbar\omega] + E_3 - E_2 - E_4)d\boldsymbol{k}_2 d\boldsymbol{k}_3 \approx \frac{\pi^4 k_F{}^6}{192}\frac{1}{E_F}\left(\frac{\hbar\omega}{E_F}\right)^2 \tag{33}$$

and then substitute it into (27) to obtain

$$\tau_{ee}{}^{-1}(\omega) = \frac{\pi k_F{}^2}{96\hbar}\left(\frac{e^2}{4\pi\varepsilon_0}\right)^2 \frac{1}{E_F}\left(\frac{\hbar\omega}{E_F}\right)^2. \tag{34}$$

This expression can be simplified by introducing the Bohr radius

$$a_0 = \frac{4\pi\varepsilon_0\hbar^2}{e^2 m_0} = 0.53 \text{ A} \tag{35}$$

to obtain, assuming $k_F \sim a_0{}^{-1}$,

$$\tau_{ee}{}^{-1}(\omega) = \frac{\pi k_F{}^2}{96\hbar}\left(\frac{\hbar^2 k_F}{m_0 a_0}\right)^2 \frac{1}{E_F}\left(\frac{\hbar\omega}{E_F}\right)^2 \approx \frac{\pi}{24}\frac{E_F}{\hbar}\left(\frac{\hbar\omega}{E_F}\right)^2. \tag{36}$$

Using the relation between the plasma frequency and Fermi energy, $\omega_p \approx \frac{4}{\sqrt{3\pi}}\frac{E_F}{\hbar}$, one finally obtains

$$\tau_{ee}^{-1}(\omega) \approx \frac{\pi^{3/2}\sqrt{3}}{96}\omega_p\left(\frac{\hbar\omega}{E_F}\right)^2. \tag{37}$$

This expression is close to the one reported by Pines in ref. 28 where the numerical factor equals $\pi^2\sqrt{3}/128$. For gold, one obtains τ_{ee}^{-1} values ranging from 4.5×10^{13} s^{-1} at 1 eV above the Fermi level to 1.8×10^{14} s^{-1} at 2 eV above it. The Umklapp factor $F_U(\hbar\omega)$ is not as simple to ascertain, but various calculations[34,39] have estimated it to be on the scale of 30 to 40%, from which it follows that the Umklapp electron–electron scattering rate, $\gamma_{ee}(\omega)$, reaches 10^{13} s^{-1} at photon (SPP) energies less than 1 eV, and becomes the dominant factor in absorption at energies exceeding 2 eV, which is in agreement with experimental data.[23] Therefore, neglecting this process is erroneous. Since two electron–hole pairs are excited by the absorption of a single photon (SPP), the average hot carrier energy is only $\hbar\omega/4$.

Absorption *via* Landau damping (a.k.a. surface collision assisted absorption)

Another way that momentum conservation in the transition between two states with different wavevectors can be restored is simply by using the "recoil" that occurs when the electron is reflected from the surface and the momentum is transferred between the electron and the wall (*i.e.* the entire metal body). The phenomenological theory of this process was first considered by Kreibig and Vollmer,[40] who simply introduced a "surface collision scattering rate" $\gamma_s \sim v_F/a$, where v_F is the Fermi velocity and a is the characteristic dimension of the metallic object. According to this phenomenological treatment, the additional damping is caused by the limited physical dimension of the system and is the result of the restriction of the mean free path of electrons. It was later shown that this process is related to the Landau damping of electrons.[19,41–44] Landau damping[45] is the process of energy transfer between free electrons and electromagnetic waves occurring when their velocities are matched, for which the electromagnetic wave must have a larger (by a factor of 100 or so) wavevector than the free space wavevector. While no propagating SPP can have such a large wavevector along its direction of propagation, lateral confinement of the field means that large wavevectors are present in the spatial Fourier spectrum of the field. Here we estimate the rate of hot carrier generation *via* Landau damping and their properties, namely the energy, directionality and spatial positions, with some not entirely expected results.

The matrix element of the Hamiltonian for the transition between two states separated by energy $\hbar\omega$ can be estimated using either the A.p or E.r Hamiltonian as

$$H_{12}^{E\cdot r} = -\frac{e}{j2\Delta kL}\int E(x)e^{+j\Delta kx}dx = -\frac{e}{j2\Delta kL}\tilde{E}(\Delta k), \tag{38}$$

which is essentially a Fourier transform of the electric field shape. Here (since the wavevectors in the y and z directions are the same), $\Delta k = \omega/v_{Fx}$ and L is the quantization length. Next, let us evaluate the Fourier transform. First, let us assume that the surface is really sharp and we can represent the electric field, shown in Fig. 6a, as a product

$$E(x) = H(x)F(x) \tag{39}$$

of a step function $H(x)$ whose Fourier transform shown in Fig. 6b is

$$\tilde{H}(k) = \pi\delta(k) - \frac{j}{k}, \tag{40}$$

and the shape function $F(x)$, represented in Fig. 5a by a dashed line, whose spatial extent is significantly broader than $1/\Delta k \sim v_F \cos\theta/\omega < 1\ nm$. Therefore its spatial spectrum $\tilde{F}(k)$, also shown in Fig. 5b, is much narrower than Δk.

The Fourier transform of the field can then be determined to be a convolution of two Fourier transforms:

$$\tilde{E}(\Delta k) = \frac{1}{2\pi}\int_{-\infty}^{\infty}\tilde{H}(q)\tilde{F}(\Delta k - q)\mathrm{d}q \approx \frac{1}{2}\tilde{F}(\Delta k) - \frac{j}{2\pi\Delta k}\int_{-\infty}^{\infty}\tilde{F}(q)\mathrm{d}q \approx \frac{j}{\Delta k}E(0) \tag{41}$$

In other words, the result is independent of the exact shape of the mode and depends only on the value of the field at the surface. Clearly, in the absence of a sharp feature in the electric field profile, the value of $\tilde{E}(\Delta k)$ would be far less and the Landau damping would become insignificant. Therefore we obtain

$$|H_{12}|^2 = \frac{e^2E^2(0)}{4\Delta k^4L^2} = \frac{e^2E^2(0){v_F}^4}{4\omega^4L^2}\cos^4\theta. \tag{42}$$

In the next step we calculate the absorption rate from a given state with the wavevector k_1

$$R_{k_1} = (2\pi/\hbar){H_{12}}^2V\rho_{E_1+\hbar\omega}, \tag{43}$$

where $\rho_x(E_F) = (2\pi\hbar v_F|\cos\theta|)^{-1}L$ is the one-dimensional density of the final states, evaluated under a consideration that neither the spin nor direction of

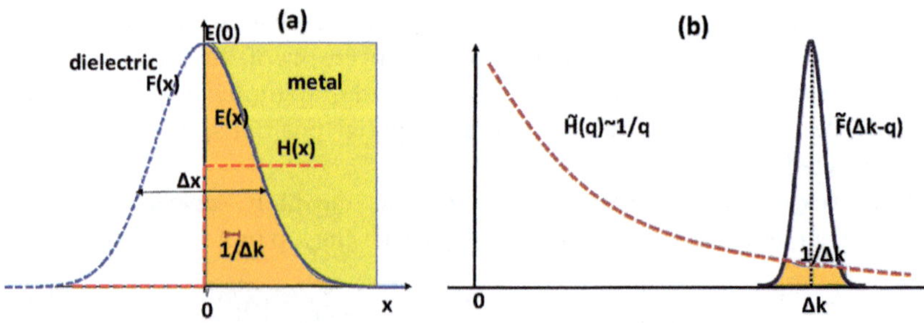

Fig. 6 (a) Electric field $E(x)$ (shaded) at the sharp interface between the metal and dielectric. (b) Fourier transform of the field $\tilde{F}(q)$ and its convolution with a step function $\tilde{H}(q)$ (shaded).

propagation change as the transition takes place. For the change of direction the required wavevector would be $2k_F \gg \Delta k$ and therefore this process can be neglected. Thus we obtain

$$R_{k_1} = \frac{1}{\hbar^2}\frac{e^2E^2(0)v_F{}^3}{4\omega^4 L}\left|\cos^3\theta\right|. \tag{44}$$

Averaging over the directions $\langle|\cos^3\theta|\rangle_\theta = 1/4$ we obtain

$$R_{E_1} = \frac{e^2E^2(0)v_F{}^3}{4\hbar^2\omega^4 L}. \tag{45}$$

The total rate of absorption then requires integration over all of the states within $\hbar\omega$ from the Fermi level and volume. Integration over the volume can be split over the integration of the quantization length L (which may be different for different transverse coordinates y and z) and then surface integration over y and z. What one obtains as a result is the total power transferred from the plasmons (electric field) to the single electron–hole excitations

$$\int\frac{dU}{dt}dV = -\frac{e^2v_F{}^3}{16\omega^2}\rho(E_F)\int E_\perp{}^2dS, \tag{46}$$

where the integral on the right hand side is taken over the surface and $E_\perp$ is the component of the electric field that is normal to the surface. Using (17) we obtain

$$\int\frac{dU}{dt}dV = -\frac{3}{16}\frac{\omega_p{}^2}{\omega^2}v_F\int\varepsilon_0 E_\perp{}^2dS. \tag{47}$$

At the same time, the integration of (3) over the volume of the metal formally yields

$$\frac{dU}{dt}dV = -\frac{1}{2}\frac{\omega_p{}^2}{\omega^2}\int\gamma(r)\varepsilon_0 E(r)^2dV = -\frac{\gamma_{\text{eff}}}{2}\frac{\omega_p{}^2}{\omega^2}\int_{\text{metal}}\varepsilon_0 E(r)^2dV. \tag{48}$$

Therefore, we may introduce the effective Landau damping (or surface collision) rate as

$$\gamma_s = \frac{3}{8}v_F/d_{\text{eff}}, \tag{49}$$

where the effective surface to volume ratio is

$$d_{\text{eff}}{}^{-1} = \frac{\int\varepsilon_0 E_\perp{}^2(r)dS}{\int_{\text{metal}}\varepsilon_0 E(r)^2dV}. \tag{50}$$

A similar expression has been previously obtained in ref. 46 and 47. The whole Landau damping rate can then be interpreted as simply the inverse of the time that the average electron propagating along the direction of the electric field takes to reach the surface. For an SPP propagating along the interface plane between the metal and dielectric, one obtains the previously reported result $\gamma_s = \frac{3}{4}qv_F$, where q is the decay constant inside the metal, while for a spherical nanoparticle of diameter a the result is $\gamma_s = \frac{3}{4}v_F/a$, which is very close to the very intuitive interpretation given by Kreibig.[40]

Directionality of excited carriers

We can now compare the directions of propagation of the carriers excited by three different mechanisms. For Landau damping it follows from (44) that the normalized distribution of photogenerated hot carriers is $R(\theta_2) \sim |\cos^3\theta_2|/\pi$. For phonon and defect assisted absorption, according to (19) the angular distribution of hot carriers is $R(\theta_2) \sim (3\cos^2\theta_2 + 1)/8\pi$. For electron–electron scattering assisted absorption, the distribution of hot carriers is omnidirectional ($R(\theta_2) \sim 1/4\pi$). These distributions are shown in Fig. 7a. The impact of this difference in directionality is shown in Fig. 7b, where the injection efficiency, $I.E.(\Theta_a) = \int_0^{\Theta_a} R(\theta_2)\mathrm{d}\theta_2$, is shown as a function of the acceptance angle Θ_a defined in (1). One can see that the injection efficiency of the carriers generated by Landau damping is four times higher than that of omnidirectional carriers engendered by e–e scattering and two times higher than that of carriers generated with the assistance of phonons and imperfections. This factor of 4 is important as it may be one of the factors that explain why the experimentally observed rate of hot carrier photoinjection[48,49] sometimes significantly exceeds the rate predicted by a simple approach that does not take into account the directionality of the hot carriers.[50]

Where do the hot carriers get excited?

Previous calculations have shown us that Landau damping originates from the momentum conservation violation at the abrupt boundary. It is also related to the

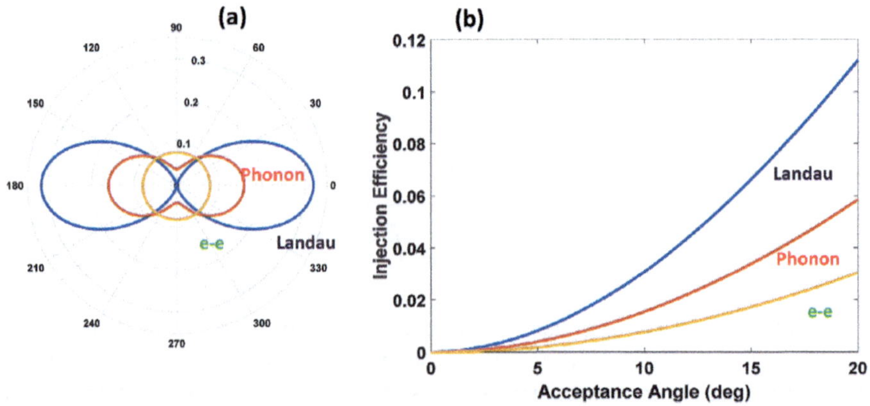

Fig. 7 (a) Angular distributions of hot carriers photoexcited by different processes. (b) Injection efficiency of the hot carriers.

phenomenon of non-locality,[51–54] *i.e.* the dependence of the dielectric function of the free electron gas on the wavevector, or spatial dispersion that can be described by Lindhard's formula[55]

$$\varepsilon(\omega,k) = \varepsilon_\mathrm{b} + \frac{3{\omega_\mathrm{p}}^2}{k^2{v_\mathrm{F}}^2}\left[1 - \frac{\omega}{2kv_\mathrm{F}}\ln\frac{\omega + kv_\mathrm{F}}{\omega - kv_\mathrm{F}}\right]. \tag{51}$$

As one can see, when $|k|^3 \geq \omega/v_\mathrm{F} \equiv \Delta k$, Landau damping becomes possible and one can write an expression for the imaginary part of the dielectric function as

$$\varepsilon_\mathrm{i}(\omega, |k| > \Delta k) = 3\pi{\omega_\mathrm{p}}^2\omega\big/2k^3{v_\mathrm{F}}^3 = \frac{3}{2}\pi\frac{{\omega_\mathrm{p}}^2\omega}{k^3{v_\mathrm{F}}^3}. \tag{52}$$

This spatial dispersion is shown in Fig. 8.

For each value of the wavevector $k^3 \geq \Delta k$, the spectral power density of the field according to (41) is

$$\left|\tilde{E}(k)\right|^2 \approx \frac{|E(0)|^2}{k^2}. \tag{53}$$

Only the longitudinal component of the field $\tilde{E}_\|(\boldsymbol{k}) = \tilde{E}(\boldsymbol{k})\cdot\boldsymbol{k}/k$ gets absorbed; hence, one can then evaluate the effective dielectric constant, $\varepsilon_\mathrm{eff,i}(\omega)$, as a function of the size and shape of an SPP mode by evaluating the overlap with $\varepsilon_\mathrm{i}(\omega,k)$, as seen in Fig. 8:

$$\varepsilon_\mathrm{eff,i} = \frac{3\pi{\omega_\mathrm{p}}^2\omega}{2{v_\mathrm{F}}^3}\int_{k>\Delta k}^{\infty} k^{-3}\left|\tilde{E}_\|(k)\right|^2\mathrm{d}^3k\bigg/\int_{k>\Delta k}^{\infty}\left|\tilde{E}(k)\right|^2\mathrm{d}^3k. \tag{54}$$

Now, splitting integration first along the direction of the electric field and then in the normal plane and using Parseval's theorem we obtain

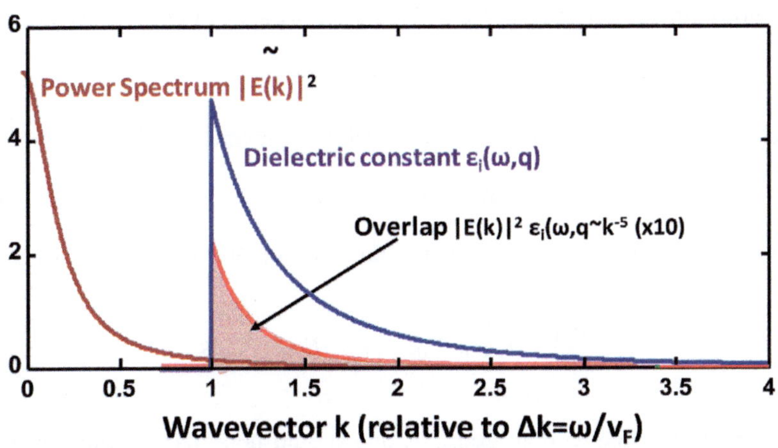

Fig. 8 Calculating the effective dielectric constant (imaginary part) and damping rate as an overlap of the spatial spectral power density of the field and the spatial spectrum of the imaginary part of the dielectric constant.

$$\varepsilon_{\text{eff,i}} = \frac{3\pi\omega_{\text{p}}^{2}\omega}{2v_{\text{F}}^{3}} \oint |E_{\perp}(0,r_{\perp})|^{2} \int_{|k|>\Delta k}^{\infty} k^{-5}\text{d}k\text{d}S \Big/ 2\pi \int_{\text{metal}} E(r)^{2}\text{d}V$$
$$= \frac{3}{8}\frac{\omega_{\text{p}}^{2}}{\omega^{3}} v_{\text{F}} \frac{\int E_{\perp}^{2}(r)\text{d}S}{\int_{\text{metal}} E(r)^{2}\text{d}V}. \tag{55}$$

It then immediately follows from (2) that $\varepsilon_{\text{eff,i}} = \omega_p^2 \gamma_s/\omega^3$ and therefore $\gamma_{\text{s}} = \frac{3}{8} v_{\text{F}}/d_{\text{eff}}$, *i.e.* precisely eqn (49).

The analysis of Landau damping performed so far allows us to estimate the total rate of absorption, but by its nature the integral in (55) precludes us from determining exactly where the absorption takes place. A naïve way of thinking would lead us to conclude that since the electron wavefunction is spread out along the entire metal volume, so is the absorption, which is absurd. In fact, the absorption involves the excitation of a wave packet near the interface, because that is where all the free electron wavefunctions get locked in phase by the boundary conditions. Since the energies of the photo-excited electrons are spread over $\hbar\omega$, their wavevectors are spread over the range $\Delta k = \omega/v_F$, hence one should expect that the absorption should be localized within roughly $\Delta L \sim 2\pi/\Delta k$ from the boundary.

To find the spatial distribution of the absorption we need to modify the expression for power dissipation (3) for the case of a non-local medium whose dielectric constant is

$$\varepsilon(r - r_1) = \frac{1}{2\pi} \int \varepsilon(k) \text{e}^{\text{j}k(r-r_1)} \text{d}k. \tag{56}$$

Power dissipation is then

$$\frac{\text{d}U(\boldsymbol{r})}{\text{d}t} = \frac{1}{2} \boldsymbol{E}(\boldsymbol{r}) \cdot \boldsymbol{J}_{\text{r}}(\boldsymbol{r}) = \frac{1}{2} \omega \boldsymbol{E}(\boldsymbol{r}) \cdot \boldsymbol{P}_{\text{i}}(\boldsymbol{r}), \tag{57}$$

where $\boldsymbol{J}_{\text{r}}$ is the in-phase (real) part of the current density and $P_{\text{i}} = \omega^{-1} \boldsymbol{J}_i$ is the quadrature (imaginary) part of the polarization, which can be expressed as

$$\boldsymbol{P}_{\text{i}}(\boldsymbol{r}) = \varepsilon_0 \int \varepsilon_{\text{i}}(\boldsymbol{r} - \boldsymbol{r}_1) \boldsymbol{E}(\boldsymbol{r}_1) \text{d}\boldsymbol{r}_1, \tag{58}$$

or, taking a Fourier transform of this convolution integral:

$$\tilde{P}_{\text{i}}(k) = 2\pi\varepsilon_0 \varepsilon_{\text{i}}(k) \tilde{E}_{\|}(k). \tag{59}$$

Finally, by taking the inverse Fourier transform of (59) and substituting it into (57) we obtain

$$\frac{\text{d}U(\boldsymbol{r})}{\text{d}t} = \frac{1}{2} \varepsilon_0 \omega \boldsymbol{E}(\boldsymbol{r}) \cdot \int \varepsilon_{\text{i}}(k) \tilde{\boldsymbol{E}}_{\|}(\boldsymbol{k}) \text{e}^{\text{j}\boldsymbol{k}\cdot\boldsymbol{r}} \text{d}\boldsymbol{k}. \tag{60}$$

Before proceeding with an example, one should note that the expression for the dielectric constant (51) does not include the bulk losses ($\gamma_{\text{b}}(\omega) = \gamma_{\text{ph}}(\omega) + \gamma_{\text{ee}}(\omega) + \ldots$) caused by scattering on phonons, defects and other electrons as well as

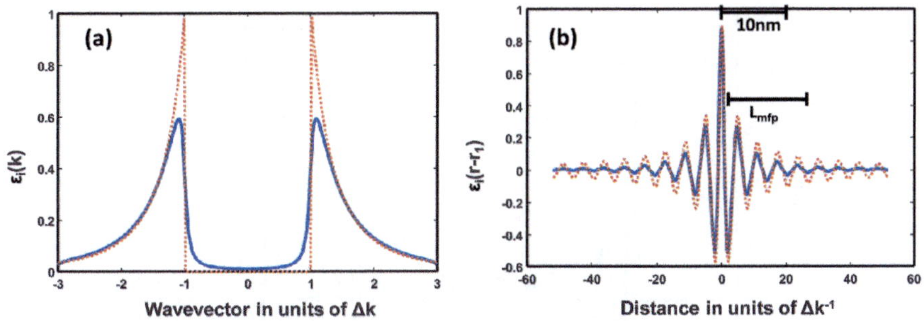

Fig. 9 (a) Imaginary part of the metal dielectric constant according to Lindhard's formula (52): dashed curve, and modified for bulk scattering according to (61): solid line. (b) Dielectric response of the metal in real space according to Lindhard's formula (52): dashed curve, and modified for bulk scattering according to (61): solid line.

residual inter-band absorption. A simple way to include them is to introduce the uncertainty of the wavevector commensurate with the inverse of the mean free path, *i.e.* $\delta k = {L_{\rm mfp}}^{-1} = 4\gamma_{\rm b}/3v_{\rm F}$. The reason for the inclusion of a factor of 4/3 will be made clear below.

A modified imaginary part of the dielectric constant incorporating the bulk absorption is then the convolution

$$\varepsilon'_i(\omega, k) = \int \frac{\pi^{-1}\delta k}{(k - k_1)^2 + \delta k^2} \varepsilon_i'(\omega, k_1)\mathrm{d}k_1, \tag{61}$$

as shown in Fig. 9a for the case of $\gamma_{\rm b} = 10^{14}$ s^{-1} and $\omega = 2.5 \times 10^{15}$ s^{-1} ($\lambda = 754$ nm), meaning that $\delta k/\Delta k = 4\gamma_{\rm b}/3\omega = 0.053$. As one can see, two obvious changes occur when bulk scattering is included. First of all, for small wavevectors ($k \ll \Delta k$) one recovers precisely the Drude expression for the bulk dielectric constant (imaginary part) $\varepsilon_i'(\omega, 0) = {\omega_{\rm p}}^2\gamma_{\rm b}/\omega^3$ (which is why a factor of 4/3 has been included in the uncertainty of momentum). Secondly, the sharp step of $\varepsilon(k = \pm\Delta k)$ has been smoothed by δk. From looking at the properties of the Fourier transform it follows now that in real space the dielectric response will decay over the length δk^{-1}, *i.e.* roughly one mean free path. The response in real space is shown in Fig. 9b, which shows rapid oscillations with a period of $\Delta L = 2\pi/\Delta k$ and which decays nearly completely over $L_{\rm mfp}$.

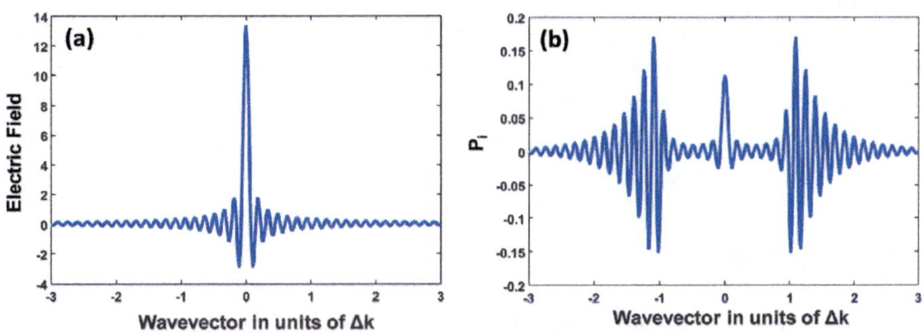

Fig. 10 (a) Spatial spectrum of the electric field in the 40 nm Au plate. (b) Spatial spectrum of the imaginary part of polarization (in phase current).

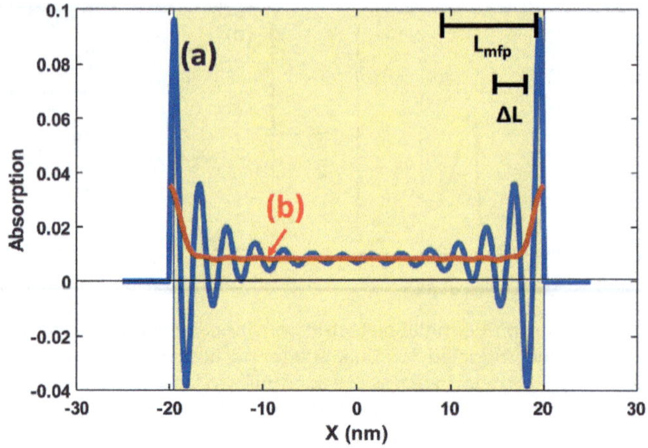

Fig. 11 (a) Spatial dependence of the energy dissipation rate in a 40 nm thick Au plate. (b) Same as in (a) but with the oscillations smoothed out over each period.

Let us now consider a simple example of a thin metal plate of thickness $d =$ 40 nm that is entirely penetrated by the electric field normal to its surface, which results in the sinc-like spatial spectrum of $\tilde{\boldsymbol{E}}_{\|}(\boldsymbol{k})$ shown in Fig. 10a and the spectrum of the imaginary (quadrature) part of polarization (in phase current) (59) that is shown in Fig. 10b. The peak at $k = 0$ represents bulk absorption and the oscillating wings at $|k| > \Delta k$ represent the effect of surface collisions (Landau damping).

Finally, in Fig. 11 we show the rate of energy dissipation (60), where curve (a) clearly shows a strong oscillatory response with a period of $\Delta L = 2\pi/\Delta k = 3$ nm. These oscillations are due to the singularity of the dielectric constant of the metal at $k = \pm\Delta k$ and are conceptually similar to the Friedel oscillations of the dielectric response at zero frequency. As expected, the oscillations decay away from the surfaces over one mean free path of about 12 nm. In addition to the oscillatory response one can also see a constant response of bulk absorption. The physical meaning of oscillation is simply that electron–hole pairs created at the positive peaks move into the valleys where the absorption is then negative. However, the overall absorption is obviously positive. In fact, if we average the absorption over a period of oscillation (curve b) while keeping the integral under the curve constant, one can see that the Landau damping takes place entirely within $\Delta L = 2\pi/\Delta k = v_F T_{opt}$. This result simply means that all of the carriers that can reach the surface within one optical period, T_{opt}, can participate in the Landau damping. One can also easily observe from Fig. 11 that the ratio of Landau-damping-caused free carrier absorption to bulk absorption is about 0.3, which is of course simply the ratio of $\gamma_a/\gamma_b \sim L_{mfp}/d$.

Conclusions

In this work I have developed a simple quantum theory of free carrier absorption in metals and determined the rates of hot carrier generation, and their energies, spatial positions and angular distributions for different mechanisms. Below I present a short summary and some advice, which the reader can choose to either follow or disregard.

Phonon and defect assisted absorption occurs throughout the entire volume of the SPP (or photon) mode in the metal. The rate of damping caused by this process is on the scale of 10^{13} to 10^{14} s^{-1}. The energy of the SPP is on average split evenly between the hole and electron, thus each carrier on average has an energy of $\hbar\omega/2$. The hot electrons have their momentum preferentially directed along the electric field but the directionality is weak (factor of 2). Since only a fraction of hot electrons reaches the surface, this process is moderately useful for applications such as photodetection and photocatalysis.

Electron–electron scattering assisted absorption should not be overlooked. This process is strongly dependent on the SPP (photon) energy and in the visible range its rate is comparable to that of phonon-assisted SPP damping. Since two pairs of carriers are generated, each carrier on average has an energy of only $\hbar\omega/4$. Furthermore, the carriers are omnidirectional. Therefore, from the applications point of view this process is deleterious and may cause a reduction in efficiency at shorter wavelengths.

Interband absorption, which arises in the visible region for gold and in the UV region for silver, has not been considered in this work, since it is obvious that this process does not generate useful carriers. Most of the SPP (photon) energy is consumed by promoting the electron from the d-shell to the Fermi level – therefore the electron is generated not far above the Fermi level and typically does not have enough kinetic energy to overcome the barrier. The hole in the d-shell has a very low velocity and cannot reach the surface at all. For these reasons, hot electron devices probably cannot operate with any degree of efficiency at short wavelengths.

Finally, **Landau damping** (or surface collision assisted absorption) is by far the best mechanism for generating hot carriers. The average energy of the hot carriers is $\hbar\omega/2$ and their angular distribution is relatively directional (factor of 4). Most importantly, the carriers are all generated right at the surface and thus are capable of being emitted, provided they have enough kinetic energy in the forward direction. The rate of Landau damping surpasses the rate of phonon and defect assisted damping when the mode penetration (or the entire nanoparticle size) becomes less than the mean free path.

Stripped of all the superfluous veneer, the recommendation is then very simple: to achieve the rates of hot carrier injection that are borderline practical, it would be prudent to use metal structures with dimensions that are less than the mean free path, so that Landau damping would dominate the absorption process as long as the wavelength is long enough to avoid deleterious interband absorption and absorption due to electron–electron scattering.

With any luck, this accessible, comprehensive and physically transparent treatment of all the processes of hot carrier generation by plasmons will be found to be, at least to some extent, useful to the developers of practical photonic and plasmonic devices.

Conflicts of interest

There are no conflicts to declare.

Acknowledgements

The author is grateful to NSF (DMR-1507749) for funding, and, more than ever, to his longtime colleague and supporter Prof. P. Noir for their thought-provoking discussions that made this modest contribution to the ongoing hot carrier explorations possible.

References

1 M. I. Stockman, Nanoplasmonics: past, present, and glimpse into future, *Opt. Express*, 2011, **19**(22), 22029–22106.
2 H. A. Atwater and A. Polman, Plasmonics for improved photovoltaic devices (vol. 9, pg 205, 2010), *Nat. Mater.*, 2010, **9**(10), 865.
3 D. K. Gramotnev and S. I. Bozhevolnyi, Plasmonics beyond the diffraction limit, *Nat. Photonics*, 2010, **4**(2), 83–91.
4 J. B. Khurgin, How to deal with the loss in plasmonics and metamaterials, *Nat. Nanotechnol.*, 2015, **10**(1), 2–6.
5 M. E. Stewart, *et al.*, Nanostructured plasmonic sensors, *Chem. Rev.*, 2008, **108**(2), 494–521.
6 I. Goykhman, *et al.*, Locally Oxidized Silicon Surface-Plasmon Schottky Detector for Telecom Regime, *Nano Lett.*, 2011, **11**(6), 2219–2224.
7 M. W. Knight, *et al.*, Photodetection with Active Optical Antennas, *Science*, 2011, **332**(6030), 702–704.
8 E. Panchenko, *et al.*, Plasmonic Metasurface-Enabled Differential Photodetectors for Broadband Optical Polarization Characterization, *ACS Photonics*, 2016, **3**(10), 1833–1839.
9 M. Moskovits, The case for plasmon-derived hot carrier devices, *Nat. Nanotechnol.*, 2015, **10**(1), 6–8.
10 M. L. Brongersma, N. J. Halas and P. Nordlander, Plasmon-induced hot carrier science and technology, *Nat. Nanotechnol.*, 2015, **10**(1), 25–34.
11 S. Linic, P. Christopher and D. B. Ingram, Plasmonic-metal nanostructures for efficient conversion of solar to chemical energy, *Nat. Mater.*, 2011, **10**(12), 911–921.
12 M. Grajower, U. Levy and J. B. Khurgin, The Role of Surface Roughness in Plasmonic-Assisted Internal Photoemission Schottky Photodetectors, *ACS Photonics*, 2018, **5**(10), 4030–4036.
13 R. Sundararaman, *et al.*, Theoretical predictions for hot-carrier generation from surface plasmon decay, *Nat. Commun.*, 2014, **5**, 5788.
14 C. Clavero, Plasmon-induced hot-electron generation at nanoparticle/metal-oxide interfaces for photovoltaic and photocatalytic devices, *Nat. Photonics*, 2014, **8**(2), 95–103.
15 A. Manjavacas, *et al.*, Plasmon-Induced Hot Carriers in Metallic Nanoparticles, *ACS Nano*, 2014, **8**(8), 7630–7638.
16 A. M. Brown, *et al.*, Nonradiative Plasmon Decay and Hot Carrier Dynamics: Effects of Phonons, Surfaces, and Geometry, *ACS Nano*, 2016, **10**(1), 957–966.
17 E. Blandre, *et al.*, Limit of efficiency of generation of hot electrons in metals and their injection inside a semiconductor using a semiclassical approach, *ACS Photonics*, 2018, **5**(9), 3613–3620.

18 G. V. Hartland, *et al.*, What's so Hot about Electrons in Metal Nanoparticles?, *ACS Energy Lett.*, 2017, **2**(7), 1641–1653.

19 J. B. Khurgin, Relative merits of phononics *vs.* plasmonics: the energy balance approach, *Nanophotonics*, 2018, **7**(1), 305–316.

20 R. Citroni, F. Di Paolo and A. Di Carlo, Replacing Noble Metals with Alternative Metals in MID-IR Frequency: A Theoretical Approach, *Nanoinnovation 2017*, vol. 2018, p. 1990.

21 Y. R. Shen, *The principles of nonlinear optics*, New York, J. Wiley. xii, 1984, p. 563.

22 N. W. Ashcroft and N. D. Mermin, *Solid state physics*, New York, Holt. xxi, 1976, p. 826.

23 M. L. Theye, Investigation of the optical properties of Au by means of thin semitransparent films, *Phys. Rev. B: Solid State*, 1970, **2**(8), 3060–3078.

24 F. Abelès, *Optical properties of solids*, Amsterdam, New York, North-Holland Pub. Co.; American Elsevier, 1972, p. 1034.

25 G. P. Motulevich, Optical Properties of Polyvalent Non-Transition Metals, *Phys.–Usp.*, 1969, **12**(1), 80–104.

26 R. N. Gurzhi, Mutual Electron Correlations in Metal Optics, *Sov. Phys.–JETP*, 1959, **8**(4), 673–675.

27 L. D. Landau, E. M. Lifshits, and L. P. Pitaevskiĭ, *Statistical physics. Course of theoretical physics*, Oxford; New York, Pergamon Press, 1980.

28 D. Pines and P. Nozières, *The theory of quantum liquids*, New York, W.A. Benjamin. v, 1966.

29 H. Petek and S. Ogawa, Femtosecond time-resolved two-photon photoemission studies of electron dynamics in metals, *Prog. Surf. Sci.*, 1997, **56**(4), 239–310.

30 C. A. Schmuttenmaer, *et al.*, Femtosecond Studies of Carrier Relaxation Processes at Single-Crystal Metal-Surfaces, *Laser Techniques for Surface Science*, 1994, vol. 2125, pp. 98–106.

31 W. S. Fann, *et al.*, Electron Thermalization in Gold, *Phys. Rev. B: Condens. Matter Mater. Phys.*, 1992, **46**(20), 13592–13595.

32 W. S. Fann, *et al.*, Direct Measurement of Nonequilibrium Electron-Energy Distributions in Subpicosecond Laser-Heated Gold-Films, *Phys. Rev. Lett.*, 1992, **68**(18), 2834–2837.

33 M. Bauer, A. Marienfeld and M. Aeschlimann, Hot electron lifetimes in metals probed by time-resolved two-photon photoemission, *Prog. Surf. Sci.*, 2015, **90**(3), 319–376.

34 W. E. Lawrence and J. W. Wilkins, Electron-Electron Scattering in Transport Coefficients of Simple Metals, *Phys. Rev. B*, 1973, **7**(6), 2317–2332.

35 A. V. Lugovskoy and I. Bray, Ultrafast electron dynamics in metals under laser irradiation, *Phys. Rev. B: Condens. Matter Mater. Phys.*, 1999, **60**(5), 3279–3288.

36 B. Rethfeld, *et al.*, Ultrafast dynamics of nonequilibrium electrons in metals under femtosecond laser irradiation, *Phys. Rev. B: Condens. Matter Mater. Phys.*, 2002, **65**(21), 214303.

37 B. Y. Mueller and B. Rethfeld, Relaxation dynamics in laser-excited metals under nonequilibrium conditions, *Phys. Rev. B: Condens. Matter Mater. Phys.*, 2013, **87**(3), 035139.

38 A. H. Macdonald and D. J. W. Geldart, Electron-Electron Scattering and the Thermal Resistivity of Simple Metals, *J. Phys. F: Met. Phys.*, 1980, **10**(4), 677–692.

39 J. M. Ziman, *Electrons and phonons : the theory of transport phenomena in solids, Oxford classic texts in the physical sciences*, Oxford New York, Clarendon Press ; Oxford University Press. xiv, 2001, p. 554.

40 U. Kreibig and M. Vollmer, *Optical properties of metal clusters. Springer series in materials science*, Berlin, New York, Springer. xx, 1995, p. 532.

41 C. Yannouleas and R. A. Broglia, Landau Damping and Wall Dissipation in Large Metal-Clusters, *Ann. Phys.*, 1992, **217**(1), 105–141.

42 A. V. Uskov, *et al.*, Broadening of Plasmonic Resonance Due to Electron Collisions with Nanoparticle Boundary: a Quantum Mechanical Consideration, *Plasmonics*, 2014, **9**(1), 185–192.

43 J. B. Khurgin, Ultimate limit of field confinement by surface plasmon polaritons, *Faraday Discuss.*, 2015, **178**, 109–122.

44 Z. Yuan and S. W. Gao, Landau damping and lifetime oscillation of surface plasmons in metallic thin films studied in a jellium slab model, *Surf. Sci.*, 2008, **602**(2), 460–464.

45 G. D. Mahan, Physics of solids and liquids, *Many-particle physics*, New York, Kluwer Academic/Plenum Publishers. xii, *3rd edn*, 2000, p. 785.

46 T. V. Shahbazyan, Landau damping of surface plasmons in metal nanostructures, *Phys. Rev. B*, 2016, **94**(23), 235431.

47 T. V. Shahbazyan, Surface-Assisted Carrier Excitation in Plasmonic Nanostructures, *Plasmonics*, 2018, **13**(3), 757–761.

48 D. C. Ratchford, *et al.*, Quantification of Efficient Plasmonic Hot-Electron Injection in Gold Nanoparticle TiO2 Films, *Nano Lett.*, 2017, **17**(10), 6047–6055.

49 I. Goykhman, *et al.*, Waveguide based compact silicon Schottky photodetector with enhanced responsivity in the telecom spectral band, *Opt. Express*, 2012, **20**(27), 28594–28602.

50 R. H. Fowler, The analysis of photoelectric sensitivity curves for clean metals at various temperatures, *Phys. Rev.*, 1931, **38**(1), 45–56.

51 F. J. G. de Abajo, Nonlocal Effects in the Plasmons of Strongly Interacting Nanoparticles, Dimers, and Waveguides, *J. Phys. Chem. C*, 2008, **112**(46), 17983–17987.

52 N. A. Mortensen, *et al.*, A generalized non-local optical response theory for plasmonic nanostructures, *Nat. Commun.*, 2014, **5**, 3809.

53 S. Raza, *et al.*, Nonlocal optical response in metallic nanostructures, *J. Phys.: Condens. Matter*, 2015, **27**(18), 183204.

54 J. Aizpurua and A. Rivacoba, Nonlocal effects in the plasmons of nanowires and nanocavities excited by fast electron beams, *Phys. Rev. B: Condens. Matter Mater. Phys.*, 2008, **78**(3), 035404.

55 J. Lindhard, On the Properties of a Gas of Charged Particles, *Mat.-Fys. Medd.–K. Dan. Vidensk. Selsk.*, 1954, **28**(8), 1–57.

Faraday Discussions

PAPER

Impact of chemical interface damping on surface plasmon dephasing

Andrew J. Therrien,[a] Matthew J. Kale, [b] Lin Yuan,[ef] Chao Zhang, [cf] Naomi J. Halas[cdefg] and Phillip Christopher *[a]

Received 10th October 2018, Accepted 26th October 2018

DOI: 10.1039/c8fd00151k

The excellent light harvesting ability of plasmonic nanoparticles makes them promising materials for a variety of technologies that rely on the conversion of photons to energetic charge carriers. In such applications, including photovoltaics and photocatalysis, the excitation of surface plasmons must induce charge transfer across the metal–adsorbate or metal–semiconductor interface. However, there is currently a lack of molecular level understanding of how the presence of a chemical interface impacts surface plasmon dephasing pathways. Here, we report an approach to quantitatively measure the influence of molecular adsorption on the spectral shape and intensity of the extinction, scattering, and absorption cross-sections for nanostructured plasmonic surfaces. This is demonstrated for the case of thiophenol adsorption on lithographically patterned gold nanodisk arrays. The results show that the formation of a chemical interface between thiophenol and Au causes surface plasmons to decay more prominently through photon absorption rather than photon scattering, as compared to the bare metal. We propose that this effect is a result of the introduction of adsorbate-induced allowable electronic transitions at the interface, which facilitate surface plasmon dephasing *via* photon absorption. The results suggest that designed chemical interfaces with well-defined electronic structures may enable engineering of hot electron distributions, which could be important for understanding and controlling plasmon-mediated photocatalysis and, more generally, hot carrier transfer processes.

[a]*Department of Chemical Engineering, University of California, Santa Barbara, USA. E-mail: pchristopher@ucsb.edu*

[b]*Department of Chemical Engineering and Materials Science, University of Minnesota, 421 Washington Ave. SE, Minneapolis, Minnesota 55455, USA*

[c]*Department of Electrical and Computer Engineering, Rice University, 6100 Main Street, Houston, Texas 77005, USA*

[d]*Department of Physics and Astronomy, Rice University, 6100 Main Street, Houston, Texas 77005, USA*

[e]*Department of Chemistry, Rice University, 6100 Main Street, Houston, Texas 77005, USA*

[f]*Laboratory for Nanophotonics, Smalley-Curl Institute, Rice University, 6100 Main Street, Houston, Texas 77005, USA*

[g]*Department of Materials Science and NanoEngineering, Rice University, 6100 Main Street, Houston, Texas 77005, USA*

Introduction

Plasmonic materials offer a promising platform for use in a variety of technologies including photovoltaics, photocatalysis, sensing, and imaging, due primarily to the large photon extinction cross-sections and localized electromagnetic field enhancements when illuminated by a resonant photon flux.[1–5] In applications that rely on charge transfer, particularly photovoltaics and photocatalysis, localized surface plasmon excitation must lead to charge transfer at a chemical interface between the plasmonic particle and an adsorbate or semiconductor. The formation of chemical interfaces on plasmonic nanostructures is known to influence the optical properties of the material. However, there is a lack of understanding of how the surface plasmon dephasing pathways are influenced, which has important implications for photovoltaics and photocatalysis.

A surface plasmon is a coherent oscillation of electrons that only lasts for a short time, on the order of femtoseconds.[6,7] The energy distribution of charge carriers during surface plasmon excitation results in the formation of a nonthermal distribution of low-energy (<0.5 eV from E_F) excited Drude electrons.[8] After excitation, surface plasmons can dephase by either radiative scattering by re-emission of a photon, or by electron–hole pair formation induced by the absorption of the photon. In optical measurements, radiative emission is observed as photon scattering, while electron–hole pair formation is observed as photon absorption, with the sum of the two defined as the extinction. Electron–hole pair excitation occurs as a result of surface plasmon decay through surface scattering, often referred to as Landau damping, and generates hot charge carriers with a difference in electron and hole energy equal to the photon energy. Since the Drude electrons responsible for the plasmon oscillation represent a bulk effect and the high-energy charge carriers formed by Landau damping represent a surface scattering effect, the carrier distribution is particle size dependent.[9] The non-thermalized hot carrier distribution produced during Landau damping quickly transfers energy to low energy carriers through electron–electron scattering, resulting in electron thermalization on the sub-picosecond time scale. After thermalization, charge carriers couple to phonon modes of the metal and ultimately lead to heat dissipation to the surrounding environment, which occurs on the order of tens to hundreds of picoseconds.[1,10,11] For bare plasmonic nanoparticles, the dominant dephasing pathway is dictated by the size of the nanoparticle, with relatively larger particles exhibiting predominately scattering and smaller particles favoring absorption, although the exact nanoparticle size dependence of dephasing pathways is also material dependent.[12]

In conjunction with photon scattering and photon absorption, the addition of a chemical interface to the metal, in the form of adsorbates or semiconductors, adds a new pathway for surface plasmon dephasing, called chemical interface damping (CID). The three possible surface plasmon dephasing pathways, and their implications to photocatalysis, are shown schematically in Fig. 1. For simplicity, the schematic representation shows an idealized homogeneous density of states with respect to energy for the photon absorption pathways. The initial electron distributions are depicted as having some density of low energy electrons and holes surrounding the Fermi energy to represent the Drude electrons present from the surface plasmon oscillation.[8] In the case of CID, the

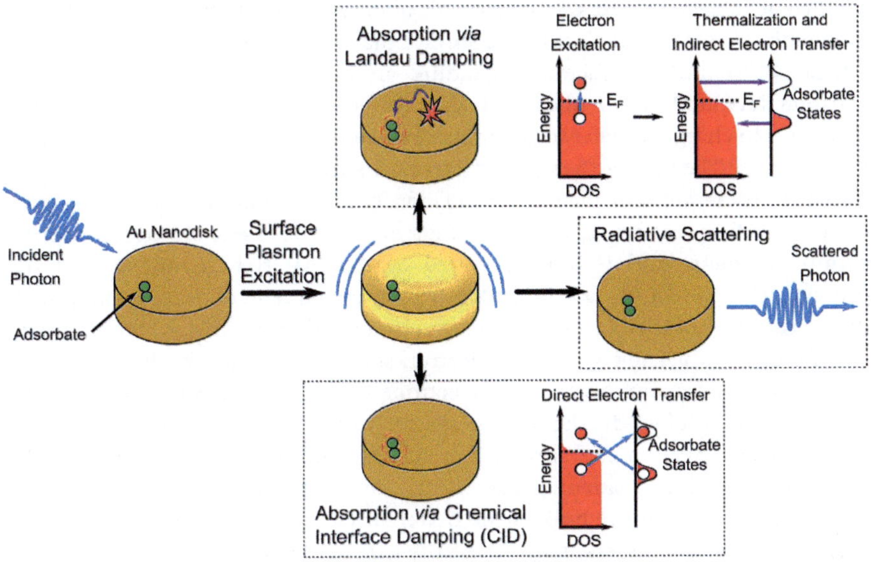

Fig. 1 Schematic representation of surface plasmon excitation and three possible dephasing pathways in the presence of a surface bound adsorbate. Absorption pathways are shown with corresponding band diagrams depicting the involved electronic transitions, where E_F denotes the Fermi energy and DOS is the density of states.

surface plasmon dephases *via* photon absorption through carrier excitation directly to, or from, electronic states of the adsorbate.[13–17] After promotion of an electron to an acceptor state of the adsorbate, or removal of an electron from an adsorbate donor state, reactions can occur by transient ion processes that enable bond-selective surface chemistry.[18]

It has been proposed that charge transfer required for photocatalysis can be optimized by using small nanoparticles, due to the high density of hot charge carriers that result from the primary light extinction pathway of photon absorption.[11,19] This strategy may be particularly suitable for processes that require charge transfer at adsorbate states near the Fermi energy, as there is a relatively long-lived high concentration of electron density just above the Fermi energy during electron thermalization.[20] However, indirect charge transfer to adsorbate states has been shown to be relatively slow,[21] and therefore a significant amount of charge transfer from hot charge carriers in the metal to adsorbate states following Landau damping is unlikely.[11] On the other hand, direct charge transfer *via* CID has been implicated to play a predominant role in photocatalysis by wavelength dependent studies.[22] In fact, multi-photon photoemission measurements find that charge transfer occurs extremely quickly (<10 fs) when a chemical interface is present, which is too fast to be attributed to indirect electron transfer.[14,23–26] Although maximizing hot carrier production has been hypothesized to be important for photocatalysis, the potential importance of CID has not been directly established.

The optical properties of plasmonic materials are inherently tied to the surface plasmon dephasing pathways described above. With respect to a bare metal nanoparticle, experimental signatures of CID include a decrease in the extinction coefficient, a red-shift in the surface plasmon resonance energy, and spectral broadening.[27–36] The red-shift in the surface plasmon resonance energy has been attributed to multiple factors, including an increase in the local dielectric constant surrounding the nanoparticle, a decrease in the electron density in the metal due to charge transfer to adsorbates, and a resonance shift due to increased damping.[32–39] The linewidth is a function of the surface plasmon coherence time,[7] for which there are four factors to consider: bulk damping, radiation damping, electron-surface scattering, and CID.[40] Bulk damping is due to electron scattering within the metal and is well described by the material's complex dielectric function; therefore it is material and frequency dependent. Radiation damping is due to coupling of the plasmon oscillation to the radiation field, which increases as a function of particle volume.[41] Electron-surface scattering leads to damping due to the larger mean free path of electrons than the dimensions of the nanoparticle, and therefore the damping is inversely proportional to the effective path length of electrons to the surface.[42] It has recently been shown that surface plasmon linewidth broadening due to CID also depends on the effective path length of electrons in the bulk of the nanoparticle to reach the surface.[34,36] This observed correlation offers insight into the nature of CID, however, such careful measurements of the homogeneous surface plasmon linewidth employed single particle spectroscopy,[40] which has only been executed in scattering mode for the analysis of CID.[32,34,36] Thus, there is essentially no understanding of how CID affects the relative absorption and scattering cross-sections with respect to bare plasmonic nanoparticles.

Despite many promising and successful demonstrations of plasmonic photovoltaics and photocatalysts, it is not clear what characteristic of the plasmonic materials, beyond large extinction cross-sections, is responsible for controlling the efficiency of photocatalytic processes. The primary candidates are the degree of electromagnetic field enhancement, energy of hot charge carriers, concentration of hot charge carriers relating to the photon absorption efficiency, and CID. Determination of the critical design parameters for optimization of plasmonic materials for such applications will require a detailed understanding of each of the mentioned properties. In this study we develop insights into CID using a newly designed approach to measure the optical properties of nanostructured plasmonic surfaces. This approach allows quantitative measurements of the photon absorption and scattering cross-sections of plasmonic surfaces using a reflective sample housed in an ultra-high vacuum (UHV) chamber coupled to an integrating sphere UV-vis-NIR spectrophotometer. This approach is demonstrated through the controlled deposition of thiophenol onto well-defined Au nanodisk arrays, allowing for quantification of the influence CID has on the absorption and scattering plasmon dephasing pathways. This approach has the potential to provide deeper insights into the dominant surface plasmon dephasing pathways that contribute to plasmon-mediated photocatalysis and other charge transfer processes.

Methods

Sample fabrication

The Au nanodisk array used here as a model plasmonic surface was fabricated on microscope cover glass (Fisherbrand) by utilizing a hole-mask colloidal lithography process.[43] A schematic of the approach is shown in Fig. 2. First, a layer of poly(methyl methacrylate) (PMMA, MircoChem, 950 PMMA A4) was spin-coated on a glass substrate (2000 rpm for 1 min), then baked on a hot plate at 180 °C for 10 min. The sample was then cleaned under Ar–O_2 plasma for 8 seconds (model 1020 plasma cleaner, Fischione Instruments) to make the surface of the PMMA more hydrophilic. An aqueous solution of 0.2 wt% poly(-diallyldimethylammonium chloride) (PDDA) was drop-casted onto the PMMA and soaked for 5 min. Deionized water was then used to rinse the surface, which was subsequently dried by N_2 gas flow. The addition of PDDA serves to functionalize the PMMA surface with a positive charge. An aqueous solution of 0.02 wt% 100 nm polystyrene (PS) beads (Sulfate Latex Beads, 8% w/v, 0.1 μm, Thermo Fisher Scientific) was drop-casted onto the sample and left to sit for 10 min. Then the solution was directly removed using dry N_2 gas flow. The interplay of coulombic attraction between the PS beads and the substrate and the coulombic repulsion between PS beads keeps the PS dispersed across the entire PMMA surface. Next, a 20 nm Au layer was deposited followed by tape peeling to remove the PS beads from the surface, forming the hole-mask. Ar–O_2 plasma etching was employed again here for 13.5 minutes to etch the PMMA layer and expose the coverslip substrate. 50 nm of Au was then directly deposited onto the substrate. Lift off of the PMMA and the hole-mask was achieved by soaking in acetone followed by rinsing with isopropanol and N_2 gas flow to clean the surface. After Au nanodisk fabrication, in-line pass-through UV-vis-NIR measurements were performed to measure the extinction. Finally, a mirrored backing of 200 nm of Al was deposited onto the substrate.

Optical measurements

Optical measurements were performed on reflective samples prepared by the nanofabrication procedure described above and were housed in an ultra-high vacuum (UHV) chamber with a base pressure < 1×10^{-8} torr. Prior to introduction into the UHV chamber, samples were rinsed with ethanol and cleaned by O_2 plasma followed by H_2 flame annealing to remove any contamination. An integrating sphere with a 3 inch light tube (Labsphere) was employed for light collection, which was pressed flush against a quartz window re-entrant into the UHV chamber, as diagrammed in Fig. 3A. A Xe light source (Avantes, AvaLight-XE) and a CCD detector (Avantes, AvaSpec-ULS-RS-TEC) were used for UV-vis-NIR measurements. Recorded spectra were averaged 25 times using a 14 s

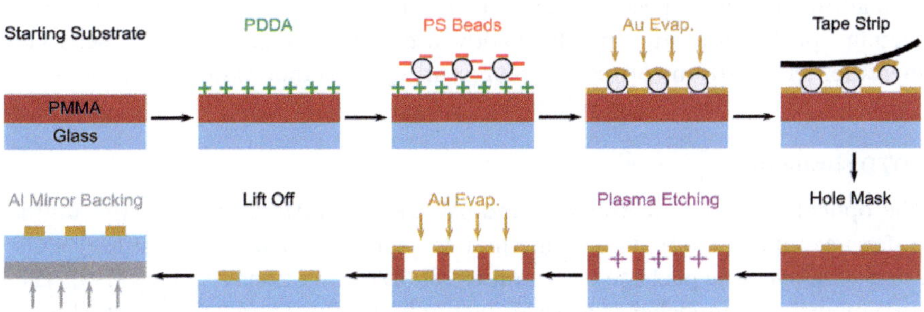

Fig. 2 Schematic of hole-mask lithography for sample fabrication.

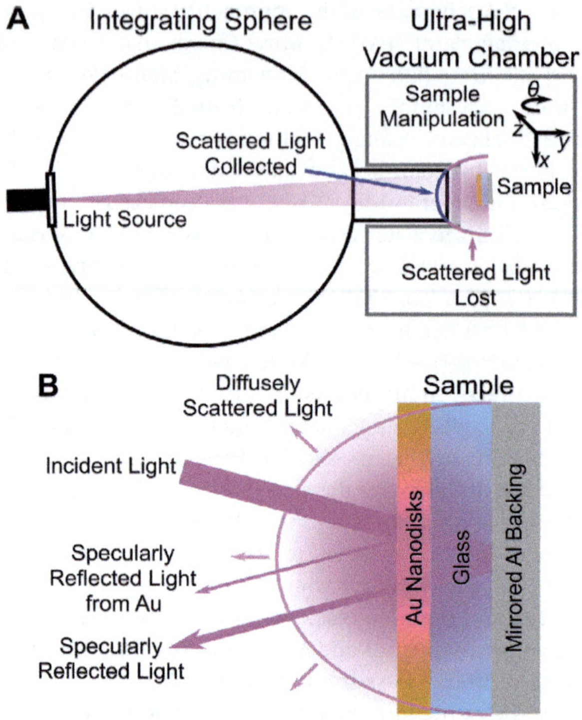

Fig. 3 (A) Schematic of UV-vis-NIR measurements using an integrating sphere and reflective sample. (B) Diagram of light interaction with the sample.

integration time taken over 1300 light source pulses. Experimental spectra are reported in units of $1 - T$, where T is the optical transmission, and spectra were collected with a bare microscope cover glass backed by 200 nm of Al as a reference. Spectra were smoothed using a Savitzky–Golay filter, fitting 2nd order polynomials over ~18 nm windows. Spectra were converted to units of optical cross-section by dividing $1 - T$ by the Au nanodisk density of $6.2 \pm 0.2\ \mu m^{-2}$. The inherent noise in the spectra before smoothing is on the order of ~200 nm^2 (in units of optical cross-section) and the spectrum step size is ~0.6 nm. High precision leak valves allowed for controlled exposure to thiophenol (analytical standard, Sigma-Aldrich), which was purified by freeze-pump-thaw cycles prior to deposition. The purity of thiophenol exposure in UHV was confirmed with a quadrupole mass spectrometer (Hiden). Reported values of optical cross-section, peak position, and linewidth are the average of three separate measurements with the error bars representing the standard deviation.

FDTD simulations

The optical properties of the Au nanodisk were simulated by using the finite-difference-time-domain (FDTD) method for solving Maxwell's equations. This study employed the package from Lumerical FDTD Solutions (Version 8.19). The Au nanodisk was modeled as a circle 100 nm in diameter and 50 nm in thickness and the "Au (Gold) – CRC" refractive index was used from the material database provided by Lumerical. Total-field/scattered-field light source conditions were used in all simulations. The incident light was injected toward the $+z$ direction, and polarized along the x-axis. A uniform mesh was used with a mesh step of $dx = dy = dz = 0.5$ nm. The absorption and scattering were calculated within the total-field/scattered-field formalism.

Scanning electron microscopy (SEM)

Scanning electron microscopy (SEM) was performed using an FEI Sirion Field Emission SEM. Samples were sputter coated with PdAu prior to imaging to ensure conductivity. Imaging conditions of a 5 kV electron beam at a working distance of 5 mm were used.

Results and discussion

UV-vis-NIR measurements

We developed an approach to measure the photon absorption and scattering cross-sections of nanostructured surfaces using a reflective sample housed in an UHV chamber. As shown in Fig. 3A, an integrating sphere was used for light collection with a light tube that was re-entrant into the UHV chamber. Because the mirrored sample is positioned outside of the integrating sphere (in the UHV chamber interfaced by the light tube) only a fraction of the diffusely scattered light by the nanostructured surface is collected in the UV-vis-NIR measurement, while the remainder is scattered into the UHV chamber. Since the light source is not polarized and, in the case of the Au nanodisk array examined here, the Au particle density of the sample is spatially homogeneous, the light is scattered isotropically. Movement of the sample by a manipulator within the UHV chamber to different distances from the light tube inlet to the integrating sphere causes a different fraction of scattered light to be collected by the UV-vis-NIR measurements. However, the absorption contribution to all spectra is the same at all positions, as the specularly reflected light off the mirrored backing is completely collected and subject to intensity loss by photon absorption. By manipulating the distance of the sample from the inlet of the light tube, the amount of light diffusely scattered by the sample and collected by the integrating sphere varies and thus can be separated from the absorbed light. The process of light interacting with the sample in the UHV chamber is schematically shown in Fig. 3B. Also noted is that in this design light is extinct twice, as it passes through the Au nanodisk array twice, which is accounted for in the data analysis.

Optical properties of bare Au nanodisks

SEM imaging of the Au nanodisk array was carried out to evaluate the sample homogeneity and particle density for quantification of extinction, scattering, and absorption cross-sections. A representative SEM image is given in Fig. 4A, which shows that the Au nanodisks were deposited with a density of 6.2 ± 0.2 μm^{-2}. The diameter of the Au nanodisks was also measured, and a histogram of the particle size is shown in Fig. 4B. The particle size histogram matches well with the 100 nm targeted diameter, and the population was fit to a Gaussian distribution with a diameter size distribution of 102 ± 16 nm.

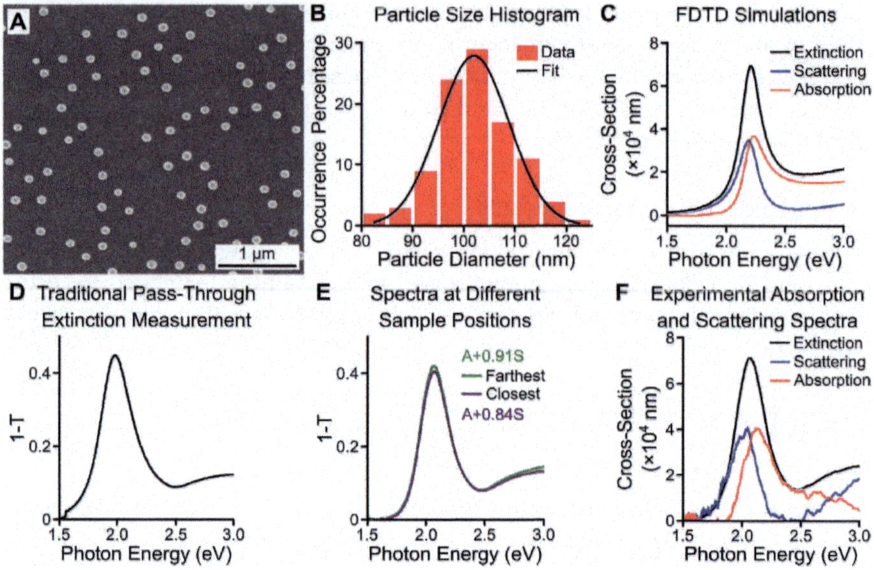

Fig. 4 (A) A representative SEM image of a 100 nm diameter Au nanodisk sample. (B) Histogram of Au nanodisk size measured by SEM. (C) FDTD simulation for a 100 nm diameter by 50 nm thick Au disk. (D) Traditional pass-through UV-vis-NIR measurement of sample extinction. (E) Integrating sphere measurements at different sample positions. (F) Experimental deconvolution of scattering and absorption from total extinction.

Using the described methodology for UV-vis-NIR measurements, a reference for the absorption and scattering contributions to extinction is required. FDTD simulations were performed and used as a reference for the experimental measurements, as shown in Fig. 4C. The simulations were used only to calibrate the relative fraction of scattered light that was collected by the integrating sphere as a function of sample position. The use of FDTD simulations as a reference is validated by previous studies that show that theory agrees well with experimentally measured scattering and absorption cross-sections of plasmonic nanostructures.[12] Transmission UV-vis-NIR measurements made before the addition of the mirrored backing of 200 nm of Al were performed to calibrate the total light extinction, as shown in Fig. 4D.

As previously discussed, the amount of diffusely scattered light by the Au nanodisk array collected in the measurement is a function of the sample distance from the integrating sphere inlet. The absorption and scattering contributions to the UV-vis-NIR extinction spectrum at a given sample position can be written as $A + xS$, where A is the absorption, S is the total scattering, and x is the fraction of diffusely scattered light not collected by the integrating sphere. The largest amount of diffusely scattered light loss (x is greatest) occurs when the sample position is farthest from the integrating sphere inlet, making it the most similar to an extinction measurement. When the sample is closer to the integrating sphere, more diffusely scattered light is collected (x is smaller). Thus, the difference in the spectra at different sample positions is proportional to photon scattering by the nanostructures. The spectra closest to and farthest from the integrating sphere for the Au nanodisk array are shown in Fig. 4E, along with the scattering and absorption contributions. The amount of diffusely scattered light by the sample that was not collected in the measurement, x, at a given sample position was determined in the following way. The scattering spectrum for a given sample was assumed to be the difference in the measured spectra at two different sample positions divided by the change in scattered light collected by the detector, Δx. The absorption spectrum was then calculated as the measured spectrum at a given position minus x at that position multiplied by the scattering spectrum. This definition yielded an absorption to scattering ratio that was used with x as a fitting parameter to match the absorption to scattering ratio predicted from FDTD simulations.

In our experimental system the difference in the amount of diffusely scattered light collected in the measurement at the closest and further positions from the light tube, Δx, was determined to be 0.07 (or 7%). Therefore, the scattering spectrum from the Au nanodisk array was calculated by taking the difference between the closest and farthest spectra and dividing by Δx. To produce the absorbance spectrum, the scattering spectrum was multiplied by x at a given sample position and then subtracted from either of the raw UV-vis-NIR spectra shown in Fig. 4E, both giving identical results. Then the sum of the scattering and absorption spectra yielded extinction. This extinction spectrum was calibrated to the transmission extinction measurement (Fig. 4D) to enable calculation of optical cross-sections. Spectra were converted to units of cross-section by dividing $1 - T$ by the Au nanodisk density of $6.2 \pm 0.2\ \mu m^{-2}$. From these measurements the total extinction, absorption, and scattering of the Au nanodisks were determined and are shown in Fig. 4F. The extinction cross-section was measured to be $7.3 \times 10^4\ nm^2$ at the peak maximum of 1.98 eV. The geometric cross-section for 102 nm Au disks is $8.2 \times 10^3\ nm^2$, which yields an extinction efficiency of 8.9, defined as the ratio of optical to geometric cross-sections. Note that the total extinction is red-shifted from the FDTD simulations (Fig. 4C) by 0.09 eV due to the refractive index of the glass support.

The shape and relative positions of the scattering and absorption spectra match well with the FDTD simulation (Fig. 4C), which validates the experimental approach. We performed analogous measurements with Ag nanodisks, which showed similar agreement between the experimental and simulated scattering and absorption spectra in both relative position and peak shape. By this analysis, the scattering and absorption cross-sections of the Au nanodisks were determined to be $3.9 \times 10^4\ nm^2$ and $4.0 \times 10^4\ nm^2$ at their peak maximums, respectively. The results are in excellent agreement with previous measurements of similarly sized Au nanodisks,[12] and also in very good agreement with the FDTD simulation (Fig. 4C), which predicts scattering and absorption cross-sections of $3.5 \times 10^4\ nm^2$ and $3.7 \times 10^4\ nm^2$, respectively. The measured optical cross-sections correspond to a scattering efficiency of 4.5 and absorption efficiency of 4.8 for the bare Au nanodisks.

Chemical interface damping induced by thiophenol adsorption

Thiophenol was adsorbed onto the plasmonic Au nanodisks to study the influence of CID on the extinction, absorption and scattering cross-sections of the Au nanodisks. Single-crystal studies have shown that thiophenol adsorbs dissociatively to Au(111), breaking the S–H bond.[44–46] Thiophenol readily forms ordered structures on Au, with the phenyl ring of the adsorbed thiolate tilted off the surface normal.[44–46] To saturate the Au nanodisks with thiophenol, the sample was exposed to increasing amounts of gas-phase thiophenol in the UHV chamber until no further spectral changes were observed, as shown in Fig. 5A. This was achieved after total exposure to thiophenol of approximately 300 Langmuirs (1 Langmuir = 1×10^{-6} torr s). After addition of thiophenol, UV-vis-NIR measurements were made to determine the extinction, scattering, and absorption cross-sections, as shown in Fig. 5B and C for the bare Au nanodisks and after adsorption of a saturated layer of thiophenol, respectively. By comparing the optical cross-sections of the bare and thiol covered Au nanodisks it is visually clear that the ratio of absorption to scattering cross-section increased after thiophenol adsorption. The scattering and absorption cross-sections of the thiophenol saturated Au nanodisks were determined to be 3.4×10^4 nm^2 and 4.1×10^4 nm^2 at their peak maximum, respectively.

The benefit of our approach is that the effects of CID on both scattering and absorption cross-sections can be quantitatively measured and compared. The changes in optical cross-section, surface plasmon resonance energy, and FWHM were quantified and are shown in Fig. 5D. Reported values are the average of three separate measurements with the error bars representing the standard deviation. The total light extinction and scattering spectra show a decreased cross-section, surface plasmon energy red-shift, and increase in the FWHM after thiophenol

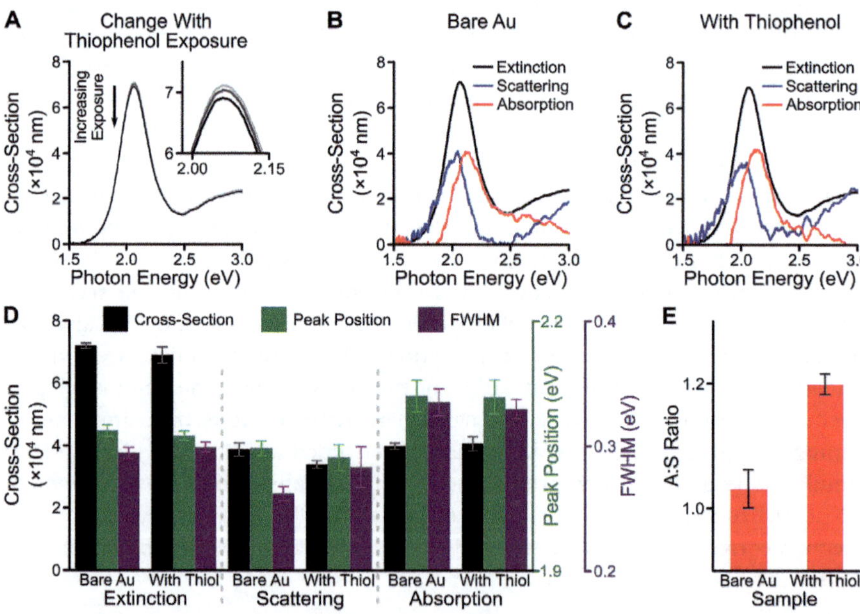

Fig. 5 (A) Series of extinction measurements showing spectral changes with increasing thiophenol exposure. The measured scattering and absorption contribution to extinction for (B) bare Au nanodisks and (C) after saturation adsorption of thiophenol. (D) Quantification of changes to the optical cross-sections, peak position, and full-width-at-half-maximum (FWHM) for bare Au and thiophenol saturated samples. (E) Absorption to scattering ratio before and after thiol adsorption. Reported values are the average of three separate measurements with the error bars representing the standard deviation.

adsorption, as characteristic of CID. However, the absorption spectra remained relatively unchanged after thiophenol adsorption within error. Thiophenol adsorption had a relatively small impact on the surface plasmon peak broadening, resulting in an increase of ~0.01 eV for the total extinction spectrum. The surface plasmon linewidth of Au is known to increase with increasing plasmon resonance energy due to bulk damping and the onset of interband transitions.[41,47] However, the bulk damping induced surface plasmon frequency red-shift is small, <0.01 eV, and therefore any influence on the FWHM due to bulk damping is negligible. The broadening observed here is consistent with previous measurements that correlated the influence of alkanethiol adsorption on the scattering spectra of various sizes of Au nanorods and the effective path length of electrons to the particle surface.[34] For our Au nanodisks the effective path length of electrons to the surface is about 50 nm, which would predict a surface plasmon linewidth broadening from alkanethiol adsorption of about 0.01 eV, consistent with our results.[34]

It is clear from the measurements reported here that the photon scattering dephasing pathway of surface plasmons is more significantly influenced by thiophenol adsorption than the photon absorption dephasing pathway. This can be interpreted as thiol adsorption inducing an increase in the ratio of surface plasmon dephasing by absorption compared to scattering, as shown in Fig. 5D. The bare Au nanodisks had an absorption to scattering cross-section ratio of 1.03, but after the addition of a chemical interface *via* thiophenol adsorption that ratio increases to 1.20. While this may seem like a small effect, this is induced by a very small amount of thiophenol with respect to the amount of Au in the nanodisks. Assuming a thiophenol packing of 0.71 nmol cm^{-2} on Au,[44] the thiophenol to Au ratio in the nanodisks is $\sim 2 \times 10^{-8}$, showing that modification of only a small fraction of Au within the nanoparticle can significantly influence the surface plasmon dephasing process. The very small amount of thiophenol with respect to Au indicates that the enhancement in the relative photon absorption is localized at the chemical interface. Based on previous work it would be expected that using smaller particles would magnify the effect further.[34]

Adsorption of thiophenol adds new electronic states at the surface of the Au nanodisks that are characterized by the hybridization of S and Au states. We propose that the new electronic states introduced by the chemical interface form new allowable interband electron transitions,[48] similar to d-band to s,p-band transitions in non-plasmonic metals.[49] In this way, the addition of a chemical interface makes the plasmonic material behave more like transition metals that do not have a full d-band. This is evidenced by an increase in the absorption to scattering ratio and a decrease in the overall extinction cross-section, which is more similar to non-coinage metals.[12,49] This mechanism is also consistent with the recent demonstration that Pt-decorated Ag nanocubes have a higher absorption to scattering ratio than pure Ag nanocubes.[50] Although in multimetallic nanostructures it can be difficult to determine the optical contributions of the different components, there is similarity in that the presence of new allowable interband transitions increases the absorption to scattering ratio in plasmonic materials. In the system presented here of thiophenol bound to Au, the influence of CID toward the behavior of plasmonic materials to be more like other transition metals is not substantial enough to significantly impact the excellent light harvesting capability of the plasmonic nanoparticles. However, in more strongly coupled systems this effect may be so dramatic as to completely change the nature of the surface plasmon excitation.[11,24]

We hypothesize that the presence of new electronic states at the surface of plasmonic materials influences the electron distribution following surface plasmon dephasing; the distribution should be different for bare metals compared to metals dressed with chemical interfaces. The energy distribution of charge carriers upon initial surface plasmon dephasing should reflect the energy profile of allowable electronic transitions, specifically s,p-intraband transitions and metal–adsorbate interband transitions. This mechanism is consistent with ultrafast multi-electron photoemission studies showing that the hot electron distribution at chemical interfaces is sensitive to the energy of electron acceptor states.[25,26,51] Ultrafast transient absorption spectroscopy measurements also show that changing the surface chemistry of plasmonic nanoparticles can increase the hot electron lifetime.[52] Our results show that CID enhances the photon absorption to scattering ratio during surface plasmon dephasing, and suggest that photon absorption is localized at the chemical interface. Based on the direct involvement of adsorbate states in the CID plasmon dephasing pathway, it is hypothesized that CID should be a more efficient and energy specific charge transfer process for charge transfer at chemical interfaces compared to indirect carrier transfer, and therefore may play a more dominant role in photocatalysis and photovoltaics.

Conclusions

The photon scattering and absorption cross-sections of Au nanodisks were characterized before and after thiophenol adsorption to provide fundamental insight into the influence of a chemical interface on surface plasmon damping and dephasing. Thiophenol adsorption induced a decrease in the extinction coefficient, red-shift in the plasmon resonance energy, and spectral broadening. In particular, the measured broadening is quantifiably consistent with previous studies of the influence of alkanethiol adsorption on photon scattering by Au nanostructures. However, CID had almost no impact on the photon absorption plasmon dephasing pathway. The greater influence of CID on the photon scattering surface plasmon dephasing pathway relative to photon absorption leads to an increase in the absorption to scattering ratio. Our physical interpretation of the results is that the new electronic states associated with the adsorbed thiophenol present new surface plasmon dephasing pathways by interband electron transitions at the chemical interface. It is reasonable to expect the energy distribution of charge carriers to reflect the energy profile of allowable electronic transitions, and therefore should reflect the energy of the adsorbate states given the CID surface plasmon dephasing pathway of direct interband electron transfer. It is hypothesized that direct charge transfer to adsorbate states characteristic of the CID surface plasmon dephasing pathway should be a more efficient and bond selective mechanism of photocatalysis as compared to the indirect charge transfer pathway.

Conflicts of interest

There are no conflicts to declare.

Acknowledgements

P. C. and N. J. H. acknowledge funding from the Air Force Office of Scientific Research MURI Grant FA9550-15-10022. PC acknowledges the Army Research Office Grants W911NF-17-1-0340 and W911NF-15-1-0321 for support in construction of the apparatus used to make measurements in these studies.

References

1 K. Watanabe, D. Menzel, N. Nilius and H.-J. Freund, *Chem. Rev.*, 2006, **106**, 4301–4320.
2 N. J. Halas, S. Lal, W. S. Chang, S. Link and P. Nordlander, *Chem. Rev.*, 2011, **111**, 3913–3961.
3 G. V. Hartland, *Chem. Rev.*, 2011, **111**, 3858–3887.
4 P. Christopher and M. Moskovits, *Annu. Rev. Phys. Chem.*, 2017, **68**, 379–398.
5 Y. Zhang, S. He, W. Guo, Y. Hu, J. Huang, J. R. Mulcahy and W. D. Wei, *Chem. Rev.*, 2018, **118**, 2927–2954.
6 U. Kreibig, M. Gartz and A. Hilger, *Ber. Bunsenges. Phys. Chem.*, 1997, **101**, 1593–1604.
7 S. Link and M. A. El-Sayed, *J. Phys. Chem. B*, 1999, **103**, 4212–4217.
8 A. O. Govorov and H. Zhang, *J. Phys. Chem. C*, 2015, **119**, 6181–6194.
9 L. V. Besteiro and A. O. Govorov, *J. Phys. Chem. C*, 2016, **120**, 19329–19339.
10 S. Link and M. A. El-Sayed, *J. Phys. Chem. B*, 1999, **103**, 8410–8426.
11 G. V. Hartland, L. V. Besteiro, P. Johns and A. O. Govorov, *ACS Energy Lett.*, 2017, **2**, 1641–1653.
12 C. Langhammer, B. Kasemo and I. Zorić, *J. Chem. Phys.*, 2007, **126**, 194702.
13 B. N. J. Persson, *Surf. Sci.*, 1993, **281**, 153–162.
14 H. Petek, *J. Chem. Phys.*, 2012, **137**, 091704.
15 S. Linic, P. Christopher, H. Xin and A. Marimuthu, *Acc. Chem. Res.*, 2013, **46**, 1890–1899.
16 M. J. Kale, T. Avanesian and P. Christopher, *ACS Catal.*, 2014, **4**, 116–128.
17 M. J. Kale and P. Christopher, *Science*, 2015, **349**, 587–588.
18 D. Menzel, *J. Chem. Phys.*, 2012, **137**, 091702.
19 Q. Wei, S. Wu and Y. Sun, *Adv. Mater.*, 2018, 1802082.
20 T. Avanesian and P. Christopher, *J. Phys. Chem. C*, 2014, **118**, 28017–28031.
21 A. Listorti, B. O'Regan and J. R. Durrant, *Chem. Mater.*, 2011, **23**, 3381–3399.
22 C. Boerigter, R. Campana, M. Morabito and S. Linic, *Nat. Commun.*, 2016, **7**, 10545.
23 D. Menzel, K. H. Kim, D. Mulugeta and K. Watanabe, *J. Vac. Sci. Technol., A*, 2013, **31**, 050817.
24 K. Wu, J. Chen, J. R. McBride and T. Lian, *Science*, 2015, **349**, 632–635.
25 S. Tan, A. Argondizzo, J. Ren, L. Liu, J. Zhao and H. Petek, *Nat. Photonics*, 2017, **11**, 806–812.
26 S. Tan, Y. Dai, S. Zhang, L. Liu, J. Zhao and H. Petek, *Phys. Rev. Lett.*, 2018, **120**, 126801.
27 H. Hövel, S. Fritz, A. Hilger, U. Kreibig and M. Vollmer, *Phys. Rev. B: Condens. Matter Mater. Phys.*, 1993, **48**, 18178–18188.
28 A. Iline, M. Simon, F. Stietz and F. Träger, *Surf. Sci.*, 1999, **436**, 51–62.

29 C. Voisin, N. Del Fatti, D. Christofilos and F. Vallée, *J. Phys. Chem. B*, 2001, **105**, 2264–2280.

30 M. D. Malinsky, K. L. Kelly, G. C. Schatz and R. P. Van Duyne, *J. Am. Chem. Soc.*, 2001, **123**, 1471–1482.

31 C. Hendrich, J. Bosbach, F. Stietz, F. Hubenthal, T. Vartanyan and F. Träger, *Appl. Phys. B*, 2003, **76**, 869–875.

32 P. Zijlstra, P. M. R. Paulo, K. Yu, Q. H. Xu and M. Orrit, *Angew. Chem., Int. Ed.*, 2012, **51**, 8352–8355.

33 J. Wei, N. Schaeffer and M.-P. Pileni, *J. Phys. Chem. B*, 2014, **118**, 14070–14075.

34 B. Foerster, A. Joplin, K. Kaefer, S. Celiksoy, S. Link and C. Sönnichsen, *ACS Nano*, 2017, **11**, 2886–2893.

35 T. Lünskens, A. Von Weber, M. Jakob, T. Lelaidier, A. Kartouzian and U. Heiz, *J. Phys. Chem. C*, 2017, **121**, 9331–9336.

36 S. W. Moon, P. V. Tsalu and J. W. Ha, *Phys. Chem. Chem. Phys.*, 2018, **20**, 22197–22202.

37 K. L. Kelly, E. Coronado, L. L. Zhao and G. C. Schatz, *J. Phys. Chem. B*, 2003, **107**, 668–677.

38 M. M. Miller and A. A. Lazarides, *J. Opt. A: Pure Appl. Opt.*, 2006, **8**, S239–S249.

39 C. Langhammer, M. Schwind, B. Kasemo and I. Zoric, *Nano Lett.*, 2008, **8**, 1461–1471.

40 J. Olson, S. Dominguez-Medina, A. Hoggard, L.-Y. Wang, W.-S. Chang and S. Link, *Chem. Soc. Rev.*, 2015, **44**, 40–57.

41 C. Sönnichsen, T. Franzl, T. Wilk, G. von Plessen, J. Feldmann, O. Wilson and P. Mulvaney, *Phys. Rev. Lett.*, 2002, **88**, 77402.

42 V. Juvé, M. F. Cardinal, A. Lombardi, A. Crut, P. Maioli, J. Pérez-Juste, L. M. Liz-Marzán, N. Del Fatti and F. Vallée, *Nano Lett.*, 2013, **13**, 2234–2240.

43 H. Fredriksson, Y. Alaverdyan, A. Dmitriev, C. Langhammer, D. S. Sutherland, M. Zäch and B. Kasemo, *Adv. Mater.*, 2007, **19**, 4297–4302.

44 L.-J. Wan, M. Terashima, H. Noda and M. Osawa, *J. Phys. Chem. B*, 2000, **104**, 3563–3569.

45 J. Nara, S. Higai, Y. Morikawa and T. Ohno, *J. Chem. Phys.*, 2004, **120**, 6705–6711.

46 P. Maksymovych and J. T. Yates, *J. Am. Chem. Soc.*, 2008, **130**, 7518–7519.

47 A. Derkachova, K. Kolwas and I. Demchenko, *Plasmonics*, 2016, **11**, 941–951.

48 M. J. Kale, T. Avanesian, H. Xin, J. Yan and P. Christopher, *Nano Lett.*, 2014, **14**, 5405–5412.

49 I. Zorić, M. Zäch, B. Kasemo and C. Langhammer, *ACS Nano*, 2011, **5**, 2535–2546.

50 U. Aslam, S. Chavez and S. Linic, *Nat. Nanotechnol.*, 2017, **12**, 1000–1005.

51 J. Stähler, M. Mehlhorn, U. Bovensiepen, M. Meyer, D. O. Kusmierek, K. Morgenstern and M. Wolf, *Phys. Rev. Lett.*, 2007, **98**, 206105.

52 K. O. Aruda, M. Tagliazucchi, C. M. Sweeney, D. C. Hannah, G. C. Schatz and E. A. Weiss, *Proc. Natl. Acad. Sci. U. S. A.*, 2013, **110**, 1–6.

Faraday Discussions

PAPER

Monitoring plasmonic hot-carrier chemical reactions at the single particle level

Sabrina Simoncelli, [*ab] Evangelina L. Pensa, [a] Thomas Brick,[a] Julian Gargiulo, [a] Alberto Lauri, [a] Javier Cambiasso, [a] Yi Li, [ac] Stefan A. Maier[ac] and Emiliano Cortés [*ac]

Received 30th September 2018, Accepted 24th October 2018
DOI: 10.1039/c8fd00138c

Plasmon excitation in metal nanoparticles triggers the generation of highly energetic charge carriers that, when properly manipulated and exploited, can mediate chemical reactions. Single-particle techniques are key to unearthing the underlying mechanisms of hot-carrier generation, transport and injection, as well as to disentangling the role of the temperature increase and the enhanced near-field at the nanoparticle–molecule interface. Gaining nanoscopic insight into these processes and their interplay could aid in the rational design of plasmonic photocatalysts. Here, we present three different approaches to monitor hot-carrier reactivity at the single-particle level. We use a combination of dark-field microscopy and photoelectrochemistry to track a hot-hole driven reaction on a single Au nanoparticle. We image hot-electron reactivity with sub-particle spatial resolution using nanoscopy techniques. Finally, we push the limits by looking for a hot-electron induced chemical reaction that generates a fluorescent product, which should enable imaging plasmonic photocatalysis at the single-particle and single-molecule levels.

Introduction

Recent years have seen tremendous progress in the design of plasmonic antennas that can amplify and confine optical fields at the nanoscale.[1] These optical antennas, fabricated with metallic nanostructures, convert free-space light into localized fields by coupling electronic-optical oscillations at the metal interface. The magnitude and spatial distribution of these high-field regions are very sensitive to the size, geometry and composition of the nanostructure.[2] Indeed, these characteristics make them powerful tools for controlling external radiation

[a]*The Blackett Laboratory, Department of Physics, Imperial College London, London SW7 2AZ, UK. E-mail: s.simoncelli@imperial.ac.uk*

[b]*Department of Physics, Randall Division of Cell and Molecular Biophysics, King's College London, London SE1 1UL, UK*

[c]*Chair in Hybrid Nanosystems, Nanoinstitute Munich, Faculty of Physics, Ludwig-Maximilians-Universität München, 80799 München, Germany. E-mail: Emiliano.Cortes@lmu.de*

down to the sub-wavelength regime which, during the past two decades, has been exploited to modify the optical properties of nearby emitters, thus revolutionizing fields as diverse as bio-sensing and homeland security.

In addition to the opportunities presented by these optical antennas for controlling the photophysical processes of molecules, recent developments have also emerged *via* the combination of plasmonics and chemistry.[3] The optical excitation of surface plasmons in metallic nanostructures leads not only to strongly enhanced light scattering, but also to enhanced light absorption. The non-radiative relaxation of the plasmon, which results in the generation of energetic charge-carriers (electron–hole pairs), can serve as an efficient source of energy for triggering chemical reactions.[4,5] Unlike semiconductors, in which the properties of the carriers are determined by the band gap, plasmonic nanostructures allow for versatile manipulation. The tunability of nanostructure shape, size, composition and interface offers a high degree of control for tailoring the energy, spatial distribution and lifetime of the plasmon-derived hot-carriers. Fine control over these processes could assist in adapting the reactivity of plasmonic nanoparticles (NPs), thus providing exciting opportunities for photocatalysis.[5] In this way, some of the pioneering studies in this area have demonstrated the possibility to plasmonically induce the water splitting reaction,[6] the dissociation of H_2 on Au NPs[7] or the epoxidation of ethylene and the oxidation of CO and NH_3 on Ag NPs,[8] amongst other examples.[9–12]

Despite the growing interest and advances in the design of plasmonic nanoantennas that can be exploited to activate chemical transformations, many challenges still remain regarding their use as efficient hot-carrier injectors. The overall power conversion ultimately depends on the efficient extraction of the hot-carriers before thermalization and the subsequent regeneration of the surface charge density, either by restoring lost electrons or holes.[12,13] In order to design efficient extraction schemes, much remains to be learnt about the energy and momentum distribution of the generated hot-carriers, the complex energy-transfer processes and the crucial role of the interface.[14–16] Furthermore, it is essential to disentangle the role of hot carriers from other phenomena linked to plasmon-excitation and decay that can also contribute to chemical transformations, such as increased temperature and near-field enhancement.[17,18]

Insights into the mechanistic details of the reactivity of hot-carriers have recently been unearthed by combining optical information on the plasmonic nanomaterials with ensemble kinetic analysis of the chemical transformations, either in the gas- or condensed-phase.[19–21] While these ensemble schemes are a powerful source for helping to start uncover the complex interplay of processes controlling hot-carrier reactivity, they fail to address the role of the structural inhomogeneities of nanomaterials in their overall efficiency.[22] Gaining nanoscopic insight at the single particle level and under *in operando* conditions is critical for guiding the efficient design and fabrication of reactive plasmonic platforms, *i.e.* for deciding the best shape, size, energy input, crystalline structure, reactive sites and material configuration (for multi-metallic materials or particle–substrate combinations), among many other examples.[23–26] However, pushing the current bulk-inspired methods down to the single-particle level presents the challenge of tracking chemical transformations of only a few thousand molecules.[23,27,28]

Here we present three different approaches for tracking and quantifying the reactivity of hot-carriers on single plasmonic nanoantennas. First, we explore a plasmonic hot-hole driven reaction under *in operando* conditions using a combination of dark-field microscopy and photoelectrochemistry. Second, we show that single-molecule based super-resolution schemes can be used to map hot-carrier reactivity beyond the diffraction limit of light, which we tested by visualizing the hot-electron driven desorption of thiol anchor groups from the surfaces of single Au nanostructures. Finally, we demonstrate the hot-electron driven reduction of Resazurin into Resorufin using Ag NPs. This fluorogenic reaction can open up new paths for revealing the nanoscopic aspects of hot-carrier driven reactions at the single-particle and single-molecule levels.

Results

Hot-hole reactivity at the single particle level

We combined electrochemistry and dark-field microscopy to study the role of a photoexcited Au NP in charge transfer processes, as schematized in Fig. 1a.

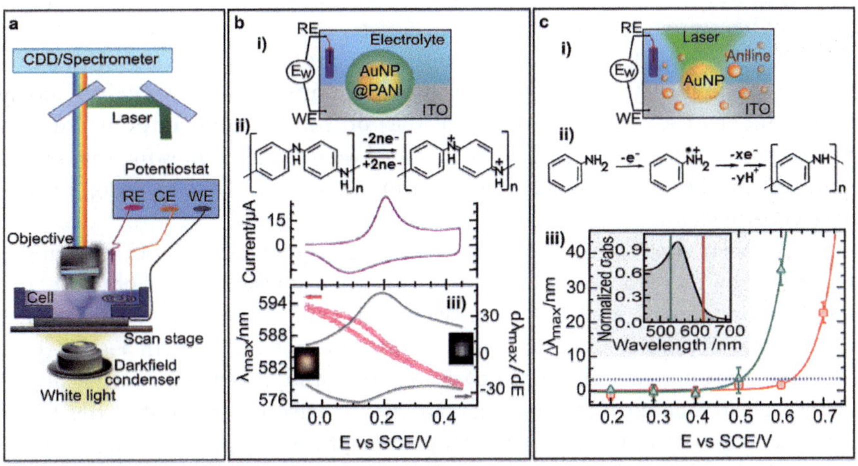

Fig. 1 Optoelectrochemical measurements of single Au NPs. (a) Scheme of the opto-electrochemical setup. (b) Monitoring the oxidation/reduction of the Au NP capping layer (polyaniline, PANI) at the single-particle level: (i) composition of the WE: PANI capped Au NPs (Au NPs@PANI) on ITO; (ii) cyclic voltammogram of PANI on polycrystalline Au. The upper inset image depicts the redox reaction; (iii) maximum scattering wavelength (λ_{max}) of a single Au NP@PANI as a function of E (pink dots). The insets show dark-field images of a single Au NP@PANI at the two limit potentials. Through derivation of the spectroelectrochemical data it is possible to recover the redox profiles for PANI oxidation and reduction at the single particle level (grey solid lines). In all cases the applied potential was set between −0.05 to 0.45 V (*vs.* SCE) at 0.05 V s^{-1}. (c) Plasmon-assisted electropolymerization of aniline: (i) scheme of the WE: Au NPs drop-casted on ITO. The reaction is conducted in aniline solution and under CW laser illumination; (ii) simplified aniline polymerization mechanism highlighting that radical formation is the rate-limiting step; (iii) red-shift in the λ_{max} ($\Delta\lambda_{max}$) after the Au NPs were illuminated for 30 s at a given potential E with either 532 nm (green triangles) or 633 nm (red squares) CW lasers. The irradiance for each laser wavelength was chosen such that the power absorbed by the Au NPs was the same.

Electrochemical control was performed using a potentiostat and a three-electrode cell configuration using 0.5 M H_2SO_4 aqueous as supporting electrolyte. The saturated calomel electrode (SCE) and Pt coil served as the reference electrode (RE) and counter electrode (CE), respectively. The working electrode (WE) consisted of 80 nm Au NPs drop-casted on ITO. The three-electrode electrochemical cell was mounted on a dark-field microscope with a water immersion objective. This optoelectrochemical setup allowed for the tuning of the WE electron's energy as well as the illumination and spectroscopic characterization of single nanoparticles.[29]

As a first example of the capabilities of our setup, we monitored the changes in the optical properties of a single Au NP when the electro-oxidation and electroreduction of its capping layer took place. As shown in Fig. 1b(i), the WE consists of Au NPs capped with polyaniline (PANI) and drop-casted on a ITO substrate. Fig. 1b(ii) shows that the redox state of PANI can be switched by the applied potential, as depicted by the cyclic voltammogram curve of the system and the inset scheme of the reaction. As the potential (E) increases linearly from −0.05 to 0.45 V, the Au NPs drain electrons from the PANI layer, thus oxidizing it. As a result of the charge transfer process the current increases, peaking at 0.2 V. It should be noted that the final state (at 0.45 V) exhibits a higher current value than the initial state (at −0.05 V), which is characteristic of a more conductive PANI layer.[30] When reversing the scan from 0.45 V to −0.05 V, the Au NPs inject electrons into the PANI layer and a reduction peak is observed at 0.1 V. It must be noted that the measured current is an ensemble property corresponding to the total number of Au NPs (acting as the WE) on the ITO substrate. In order to monitor the process at the single-particle level we turned to following the optical changes (*i.e.* the scattering spectra) of a single Au NP whilst carrying out the electrochemical scan.

It has been shown that the dielectric function of PANI drastically changes depending on its redox state, being lower for the oxidized state.[31] Therefore, the optical properties of Au NPs coated with PANI are expected to change when performing a cyclic voltammetry scan similar to that reported in Fig. 1b(ii).[32] Following the approach developed by the Landes group,[33] we performed a single-particle electrochemical experiment, but in this case we monitored the redox changes in the PANI capping layer through the optical response of a single Au NP. We used dark-field microscopy to localize and track the changes at the single particle level. Dark-field scattering spectra of a single PANI-capped Au NP (PANI@Au NP) were recorded as the potential was cyclically scanned between −0.05 and 0.45 V. Fig. 1b(iii) shows that the Au NP maximum scattering wavelength (λ_{max}) blue-shifts when the potential is increased and red-shifts when the potential is decreased (see the pink curve). The non-linear response and the hysteresis observed for λ_{max} *vs.* E – when comparing the direct and reverse electrochemical scans – reveal a non-reversible redox behaviour for the PANI layer of this Au NP (*i.e.* the PANI oxidation and reduction peaks are centered at different potentials). This can be extracted further when calculating the derivative of the scattering peak against the applied potential ($d\lambda_{max}/dE$), as shown in the grey curves of Fig. 1b(iii). In this way, it is possible to determine the electrochemical response of a single Au NP from the optical changes in its scattering spectra.[33] This is further reinforced by the similarity between the electrochemically-measured cyclic voltammogram for many Au NPs (Fig. 1b(ii)) and the optically-extracted redox behaviour of a single Au NP (Fig. 1b(iii)).

Taking advantage of the single-particle sensitivity and the precise electrochemical control of the WE electron's energy, we finally studied the electro-oxidation of aniline on an optically excited Au NP. This configuration could assist in revealing the quanta of energy supplied by the illuminated Au NP in order to drive the oxidation process; we should be able to monitor changes in the potential required to drive the reaction if there is any photocatalytic contribution from the Au NP. To this end, Au NPs were drop-casted on ITO and immersed in an aniline solution (Fig. 1c(i)). The potential was step-increased by 0.1 V from 0.2 to 0.7 V, as an oxidation reaction needs to take place as shown in the diagram of Fig. 1c(ii).

For each applied potential, the Au NP was excited for 30 seconds of illumination with a diffraction-limited CW laser coupled through a water immersion objective (see Fig. 1a). Two laser lines were employed: 532 nm and 633 nm. This allowed us to explore two nearly extreme situations: the maximum and minimum of the absorption spectrum for the 80 nm Au NP (see the inset spectrum in Fig. 1c(iii)). The laser irradiances were set at 2.2 mW cm^{-2} and 9.3 mW cm^{-2} for 532 nm and 633 nm, respectively. These values guarantee that the absorbed power, and therefore the increase in temperature, is the same for both wavelengths. Fig. 1c(iii) shows the shift in the λ_{max} ($\Delta\lambda_{max}$) as a function of the applied potential after single Au NPs were subjected to laser illumination. The green triangles correspond to a Au NP illuminated at 532 nm while the red squares correspond to 633 nm illumination. In both cases, after a certain potential is reached, a drastic red-shift in the scattering peak is observed. These changes in the scattering spectra are due to PANI formation around the Au NP. Interestingly, the polymerization of aniline starts at different potentials for both irradiation wavelengths. When irradiated at 532 nm, a red-shift of $\Delta\lambda_{max} = 3.4$ nm is observed at 0.50 V, while at 633 nm irradiation the same red-shift is reached at 0.62 V (see the dotted line in Fig. 1c(iii)). This result is even more surprising when considering that the absorbed power is the same for both wavelengths and thus the increase in temperature is the same. Aniline being a transparent molecule in the visible regime also rules out any possible near-field contribution or excitation.[18] Hence, these results suggest that the electrochemical reaction is plasmon-catalyzed by the generated hot-holes on the Au NP and that the energy of the available holes depends on the excitation wavelength of light.

These results show that photoelectrochemical measurements can serve as a way to estimate the energy and efficiency of plasmon generated hot-carriers in single particles, thus revealing unknown aspects of the light-induced electron transfer mechanism. We then went on to explore the spatial reactivity of single plasmonic photocatalysts.

Hot-electron reactivity-mapping at the single-particle level

In recent years, single-molecule based super-resolution fluorescence microscopy techniques have emerged as a versatile method for mapping molecular targets with nanometer precision (*i.e.* beyond Abbe's diffraction limit of light).[34] By separating the fluorescence emission of individual fluorophores in time using photo-conversion, photo-switching or transient binding approaches, these techniques can provide high resolution information thanks to the subsequent determination of the single fluorophore's position.[35] Borrowing these ideas, we developed a nanoscopic approach that enables the imaging of reactive sites in plasmonic photocatalysts.

Fig. 2a illustrates the concept of our imaging approach, named metallic DNA-PAINT (m-PAINT).[36,37] DNA-PAINT (DNA-Points Accumulation for Imaging in Nanoscale Topography),[38] is a recently developed localization based super-resolution nanoscopy technique that uses single-stranded fluorescent-labelled DNA probes (imager strands) which transiently bind and quickly dissociate from their complementary DNA targets (docking strands) to reconstruct super-resolved images. In m-PAINT, we combined this technique with single-particle femtosecond illumination to map the plasmon-induced hot-electron active sites on the surfaces of metallic nanomaterials.[36,37]

First, we chemically coupled DNA targets (docking strands) to the surfaces of the metallic nanostructures through a binder ligand. Mindful design of the binder ligand is of paramount importance to both: (i) achieve highly-resolved images and (ii) study different hot-electron reactions (*i.e.* by changing the functional anchor group of the binder or by incorporating an electron-acceptor molecule in it). In order to optimize the imaging capabilities of our technique, we set the docking strand height between the fluorophores and the metallic surface to ~7 nm, which diminishes the spatial fluorescence quenching effects on the metals.[39] In our case, this was achieved by introducing a hydrophilic PEG unit in the form of hexaethylene glycol between the docking strand and a mercaptohexyl linker. Additionally, we used a backfilling molecule (6-mercapto-1-hexanol, MCH) to ensure a more compact monolayer and therefore prevent unspecific

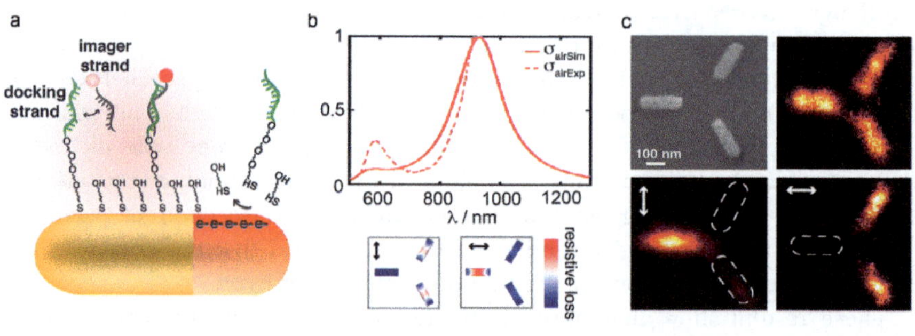

Fig. 2 Chemical-desorption mapping of thiolated-DNA molecules upon plasmonic hot-electron transfer. (a) Schematic of metallic DNA-PAINT. Docking strands (thiolated-DNA) are immobilized onto the Au surface of the plasmonic element providing transient binding sites for the complementary fluorescent-labelled DNA strands (imager strands). Upon plasmon excitation and hot-electron transfer, some docking strands are desorbed from the surface. (b) Optical performance of the Au nanorod trimer antenna. Each nanorod has a dimension of 180 × 55 nm^2. Top: Calculated (solid line) and experimental (dashed line) scattering cross-sections. Bottom: Near-field maps of the absorbed power in water (or resistive loss maps) for vertical or horizontal polarizations at 950 nm excitation. (c) Scanning electron microscope image of the plasmonic structure. Super-resolution m-PAINT images for the non-irradiated sample and for the vertical/horizontal polarization excitation with 220 fs laser illumination at 950 nm wavelength. Dashed curves indicate the borderline of the Au elements where thiol desorption has taken place (*i.e.* there is no-binding of the imager strand).

binding of DNA strands on the surface of the metallic nanostructure.[40] Our experimental approach allowed the imaging of single Au nanorod trimer structures with nanometric resolution (Fig. 2c, top right), perfectly agreeing with the obtained scanning electron microscope (SEM) images (Fig. 2c, top left).

Next, we studied the hot-electron driven cleavage of Au–sulfur bonds using a Au nanorod trimer system. Fig. 2b (top) shows the scattering spectrum of the system in air. A plasmonic dipolar resonance is present at 910 nm, corresponding to the resonance of each individual rod. At this wavelength, the scattering spectrum of the system is independent of the excitation polarization (data not shown). However, the spatial distribution of the non-radiative decay that governs the generation of hot electrons can largely differ between sub-elements depending on light polarization.[36] Fig. 2b (bottom) depicts the near-field power absorption maps (also named as resistive loss maps in the literature) for the two polarizations. We illuminated the system using a linearly polarized 950 nm femtosecond laser. By spinning the polarization of incident light, it is possible to selectively excite either the horizontal rod or the two diagonal rods. Fig. 2c (bottom) shows m-PAINT images of the trimers after irradiation. The dark areas of the antennas indicate the places where photocleavage has occurred, while the bright areas indicate the remaining binding domains after irradiation. The resemblance between the m-PAINT maps and simulated absorption maps for both polarizations proves that the spatial distribution of energetic electrons responsible for triggering the Au–S bond cleavage can be imaged with sub-diffraction resolution.

Hot-electron reaction for tracking *in operando* single-photocatalyst activity

This section aims to extend the study of hot-carrier reactions from the single photocatalyst to the single molecule level as they happen *in situ*. Single-molecule studies of single nanocatalysts have recently been implemented using fluorogenic reactions,[28,41] where a non-fluorescent reactant is converted into a fluorescent product. One of the most employed of such reactions is the reduction of the non-fluorescent Resazurin (Rz) molecule when reacting with hydroxylamine on the surfaces of Au nanoparticles (heterogeneous catalysis) to produce the highly fluorescent Resorufin (Rs) molecule,[28] as schematized in Fig. 3a. The high sensitivity of fluorescence allows for the detection and quantification of single turn-over events, which appear as spikes in the collected fluorescence signal, as shown in the inset of Fig. 3a. Following this approach, kinetics studies of several single heterogeneous nanocatalysts have been studied, revealing reaction pathways,[28] size[41] and facet dependence,[42] sub-particle distribution of active sites,[34] or even activation energies.[43]

In order to extend this technique to the study of hot-carrier driven chemical reactions at the single photocatalyst and single molecule levels, it is crucial to identify a fluorogenic reaction driven by hot-carriers. Here, we show that Ag spherical nanoparticles (Ag NPs) illuminated at their plasmon resonance can produce hot-electrons capable of reducing Rz into Rs without the use of hydroxylamine or any other reducing agent, as schematized in Fig. 3b.

We carried out ensemble-level fluorescence measurements to study the photocatalysis of Rz to Rs in the presence of plasmonic 60 nm Ag NPs. Fig. 3c shows the absorption spectra of these three components. Due to excitation of the plasmon resonance, the Ag NPs present an absorption maximum at 430 nm. We

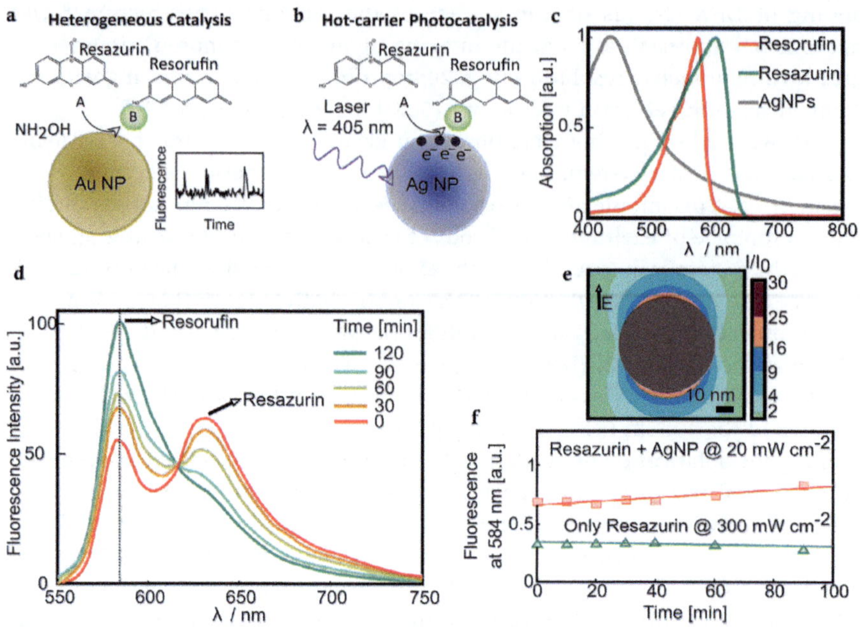

Fig. 3 (a) Example of a heterogeneous catalysis experiment. A weakly fluorescent molecule, **A**, is converted into a highly fluorescent one, **B**, on the nanocatalyst surface. Each turn-over event is detected as a spike in the fluorescence signal. (b) Working principle of hot-carrier driven photocatalysis. A plasmonic Ag NP is resonantly illuminated with a continuous wave laser at 405 nm, generating highly energetic electrons. These electrons reduce the weakly fluorescent Resazurin molecule into the fluorescent Resorufin molecule. (c) Normalized absorption spectra of 60 nm Ag nanoparticles (grey), Resorufin (red) and Resazurin (green). (d) Photoconversion of an initial 5 μM Resorufin solution (red) over increasing irradiation times. The spectra were normalized to the iso-emissive point at 613 nm. The vertical dashed line indicates the Rs emission peak at 584 nm used in (f). (e) Map of the intensity enhancement produced by a 60 nm Ag sphere illuminated at 405 nm. The polarization of the beam is indicated with an arrow. (f) Fluorescence intensity at the Resazurin peak (584 nm) *vs.* time for a solution containing 5 μM Resazurin and 20 pM Ag NPs illuminated at 20 mW cm^{-2} (pink line), and for a solution of 5 μM Resazurin without Ag NPs and with 15 times more power (300 mW cm^{-2}) (green line). The curves have been vertically shifted for comparison.

illuminated a cuvette containing 20 pM Ag NPs and 5 μM Rs with a 405 nm CW laser at a power density of 300 mW cm^{-2}. At this wavelength, the direct absorption of light by Rs or Rz is very low. Fig. 3d illustrates the variation in the fluorescence spectra for increasing irradiation times. Before irradiation (red curve), an emission peak at 634 nm corresponding to the weakly fluorescent Rz molecule could be observed. In addition, a small number of highly fluorescent Rs molecules contribute to an emission peak at 585 nm. Dye conversion can be followed through the increase and accompanying decrease in the fluorescence emission peaks of Rs and Rz, respectively. After 150 minutes, almost the entire solution is converted, with the final spectrum resembling that of Rs. Pertinent controls performed in the absence of Ag NPs or without illumination showed no significant conversion. These experiments effectively demonstrate that it is possible to photocatalyse the reduction of Rs into Rz when exciting the plasmon resonance of the Ag NPs. However, plasmonic NPs can catalyse chemical transformations due to enhanced fields, thermal effects, or by the generated carriers.[5] In order to prove that this reaction can be used as a hot-electron probe, thermal and field induced conversion should be ruled out.

The temperature increase ΔT of an illuminated single Ag NP can be estimated using the equation $\Delta T = \frac{\sigma_{abs} I}{4\pi k a}$, where σ_{abs} corresponds to the absorption cross section, I the irradiance, k the thermal conductivity of the immersion media and a the NP radius.[44] Applying this function using values of $\sigma_{abs} = 1.2 \times 10^{-14}$ m^2, $k =$ 0.59 W K m^{-1} and a = 30 nm, we envisage a negligible temperature increase on the order of just 10^{-4} K. The above assumes an isolated nanoparticle, but recent studies have shown that collective effects for highly concentrated solutions can lead to a marked increase in the temperature of the solution, which is some orders of magnitude greater than that expected for isolated NPs.[45] In order to rule out such macroscopic temperature rises a thermocouple probe was inserted into a solution of 20 pM Ag nanoparticles, and the solution was illuminated with a 405 nm laser at the maximum power intensity used in our experiments. No temperature increase was observed, meaning that the temperature increase was smaller than the sensitivity of the thermocouple (0.1 K).

Next, we investigated whether the conversion reaction can be attributed solely to the enhancement in the electric field around the NPs upon resonant illumination. Fig. 3e depicts finite difference time domain (FDTD) calculations of field intensity enhancements around a 60 nm diameter Ag NP suspended in water upon illumination at 405 nm. Within a shell of 10 nm around the nanoparticle surface, the average field intensity over all three planes is 14.9 times greater than that of the incident wave. We can then rule out the notion that it is solely a field induced effect through a comparison of the dye conversion with and without nanoparticles, using a 15 times higher power density in the latter case (see Fig. 3f). This way, both systems exhibit the same magnitude of field intensity. Fig. 3f shows the fluorescence intensity at the emission peak of Rs as a function of time. In the absence of Ag NPs and with the use of a 15 times higher beam intensity (300 mW cm^{-2}), no clear increase in the fluorescence peak intensity is observed. For a lower beam intensity (20 mW cm^{-2}) in the presence of Ag NPs, a slight but discernible increase in the peak intensity over time is observed, which is indicative of the conversion of Rz to Rs. The lower magnitude of this conversion relative to that illustrated in Fig. 3d is to be expected given the low laser power involved.

The results presented here allow us to claim that the reduction of Rz to Rs is driven by hot-carriers induced upon the resonant excitation of plasmons in Ag nanoparticles. This result paves the way for the investigation of hot-carrier reactions at the single molecule level.

Conclusions

The rational design of plasmonic photocatalysts requires a deeper understanding of the mechanisms contributing to plasmon-induced chemical reactions. Single-particle and single-molecule studies were carried out in order to decouple the ensemble effects, thus facilitating the interpretation of the experimental results and offering easier routes to improve the performance of the photocatalyst under *in operando* conditions. This has been exemplified in the present contribution through a study of the energy or spatial distribution of plasmonic hot-carriers in single plasmonic particles coupled to different reactive interfaces. Finally, we have demonstrated the reaction conditions that are required in order to drive the hot-electron conversion of a fluorogenic molecule, which could serve as a reactive probe for future single-molecule studies of a diverse number of super-resolution schemes.

Experimental

Sample preparation for optoelectrochemical measurements

All reagents were of high-purity and were used as received. For aqueous solutions, ultra-pure water was employed (18.2 MΩ cm, MilliQ®). Au NPs (80 nm in diameter, capped with CTAB, OD solution: 0.05) were synthesized and purified according to the method reported by Yoon J. H. *et al.*[46] Au NPs@PANI were obtained *via* the chemical synthesis of PANI layers (19 nm in thickness) around the Au NPs, following the protocol described by Xing S. *et al.*[47]

Sample preparation for hot-electron reactivity mapping

Gold trimer nanorods with dimensions of 180 × 55 nm^2 were fabricated on borosilicate glass using electron beam lithography (E-line, Raith GmbH). The substrate was first coated with a positive-tone poly(methyl methacrylate) (950 K A4) resist and then baked for 3 min at 180 °C. The nanostructures were then defined by exposure to an electron beam, followed by a development procedure. Chromium and gold thermal evaporation at 1.5 Å s^{-1} (Amod, Angstrom Engineering Inc.) and a standard lift-off process at 70 °C completed the fabrication process. For DNA functionalization the samples were first left for 1.5 min in a UVO cleaner/oxidizer (Femto, Diener Electronic GmbH) and mounted onto a bottomless self-adhesive slide (sticky-Slide VI 0.4, Ibidi) in order to form a flow chamber with an inner volume of ~30 μL. DNA functionalization was carried out *via* the overnight incubation at RT of a 1 μM solution of the single stranded DNA docking strand (ThiolC6-SpacerC18-ATGAGTTAATT, biomers.net GmbH) diluted in 10 mM Tris–HCl, 1 M NaCl, pH = 7 (buffer A). Next, the sample was incubated in a 1 mM 6-mercapto-1-hexanol (MCH, Sigma Aldrich) solution (30 minutes, RT) prior to a high-volume wash with buffer A. Finally, the sample was left in a solution of 10 mM Tris–HCl, 50 mM NaCl, pH = 7 (buffer B) or 5 mM Tris–HCl, 10 mM $MgCl_2$, 1 mM EDTA, pH = 8 (buffer C) for use in femtosecond pulsed laser irradiation or fluorescence imaging, respectively, after washing with buffer B.

For the scanning electron microscopy imaging, a 20 nm thin film of Espacer 300Z was coated onto the sample before DNA functionalization at 2000 rpm for 60 s, followed by a 60 s, 90 °C bake.

Dark-field microscopy and spectroscopy

Optical measurements for the optoelectrochemical configuration were carried out using a Witec microscope. For dark-field imaging, the sample was illuminated from the bottom with a white light source (OSL2, Thorlabs, United States) through an oil-immersion dark-field ultra-condenser (NA = 1.2–1.4, Zeiss, Germany). Scattered light was collected using a water immersion objective (NA = 1.0, ×63 magnification, Zeiss, Germany) and directed to a colour CCD camera. Scattering spectra were recorded before and after each Au NP was illuminated. Single-particle dark-field scattering spectroscopy measurements of the gold nanorod trimers in air were carried out using an upright Nikon Eclipse Ti-U microscope (Nikon Instruments) equipped with a dark-field objective (Nikon LU Plan ELWD 100× NA 0.80), which is used to both provide the incident illumination and to collect the scattered light. For illumination, a 100 W halogen lamp (Nikon) was used. The scattered light was measured using a diffractive grating and a charge-coupled device camera cooled to 70 K. The plotted spectra were corrected for the system wavelength response (by measuring the cross section of a perfect reference white rough scatter) as well as for the dark and background counts.

Electrochemistry

Electrochemical measurements were carried out at room temperature (24.0 ± 0.2 °C) in a three-electrode configuration cell and with a CHI760C potentiostat (CH Instruments, United States). A 0.5 M H_2SO_4 (99.99%, Sigma-Aldrich) aqueous solution was used as the supporting electrolyte. Aniline (99%, Sigma Aldrich) was dissolved in the supporting electrolyte until a final concentration of 0.2 M was reached. The working electrode consists of 80 nm Au nanoparticles drop-casted on ITO (Nanoscribe, Germany). A Pt coil (99.99%, Goodfellow, UK) and a saturated calomel electrode (SCE) served as the counter and reference electrodes, respectively. Before every experiment the Pt coil was cleaned using a butane flame annealing process followed by a quenching procedure in ultrapure water.

Femtosecond pulsed-laser irradiation

For the femtosecond pulsed-laser irradiation of the individual gold trimer nanorods a pulsed Yb:KGW PHAROS laser system was used as the pump of an ORPHEUS collinear optical parametric amplifier with a LYRA wavelength extension option (Light Conversion Ltd., pulse duration of 180 fs, repetition rate of 100 kHz). The excitation beam ($\lambda = 950$ nm) was reflected by a short-pass dichroic mirror (Thorlabs DMSP805) and focused onto a $\sim$1 μm^2 spot on the sample with a 40× (CFI S Plan Fluor ELWD 40×, NA 0.60, Nikon) air objective. The beam polarization was adjusted using a half-wave plate and a polarizer, and the power fluence was set to 1.7 mJ cm^{-2}. The sample was fixed to an XYZ piezo-scanner stage (Nano-Drive, Mad City Laboratories), which allows for automatic single nanoantenna irradiation (10 seconds per antenna). Alignment of the antennas with the excitation beam was made possible by inspecting the sample with white light illumination and detecting the sample using a CCD camera (QICAM Fast 1394, QImaging).

Fluorescence microscopy imaging

Fluorescence imaging was carried out using an inverted Nikon Eclipse Ti-U microscope (Nikon Instruments). The excitation beam at 640 nm was provided by a 20 mW CW laser (LDH, Picoquant) spectrally filtered with a clean-up filter (BrightLine HC 636/8, Semrock). A single-edge dichroic beam splitter (Semrock BrightLine FF 649) was used to reflect the excitation light into the 100×, 1.49 NA oil immersion Apo DIC N2 TIRF objective. The fluorescence light was spectrally filtered using an emission filter (ET705/72m, Chroma Technology) and collected onto an electron-multiplying charge-coupled device (EMCCD) camera (Evolve 512 Delta, Photometrics) operated at 100 ms per frame. For the DNA-PAINT imaging, a solution of the imager strand (Atto655-TTAACTCATG-3′, biomers.net GmbH) diluted to 5 nM in buffer C was added to the sample immediately before starting the measurements. 20 000 frames per super-resolution image were recorded at an excitation intensity of 450 W cm^{-2}.

Super-resolution data processing and image analysis

m-PAINT images were reconstructed from the raw data based on the subsequent localization of single molecules *via* Gaussian fitting using the rainSTORM Matlab application. Prior to single-molecule localization, stage drift correction and background subtraction of the photoluminescence (static, autofluorescence emission) from the gold trimer nanorods were carried out *via* the utilisation of an in-house Matlab routine. Images of the individual gold trimer nanorods were reconstructed from ~700 to 3000 localizations. Particle averaging of the gold trimer nanorods was carried out using a translation registration algorithm (Matlab built-in function). Average m-PAINT images were usually reconstructed by overlaying super-resolution images of 30 to 36 individual gold nano-antennas that had been exposed to the same chemical/optical conditions.

Resazurin photoconversion

Absorption spectra were recorded using a UV-Vis Agilent Technologies Cary 60 spectrometer. Fluorescence spectra were recorded using an Agilent Technologies Cary Eclipse Spectrophotometer, exciting at 520 nm and monitoring from 530–800 nm.

A 10 mm optical path fused quartz cuvette was illuminated using a continuous wave Coherent Cube laser expanded through a telescope to a beam waist of 1 cm, such that the entire solution was illuminated. After a determined irradiation time the cuvette was removed from the beam path and measured in the spectrometer.

Numerical simulations

Numerical calculations of the absorption and scattering cross-section spectra and the near-field intensity distributions were carried out using both Lumerical FDTD 2017a and Comsol Multiphysics v4.4 software.[36]

Conflicts of interest

There are no conflicts to declare.

Acknowledgements

This work has been supported by the EPSRC through the Reactive Plasmonics Programme (EP/M013812/1), the Royal Society and the Lee-Lucas Chair in Physics. S. S. acknowledges the financial support received from the Human Frontier Science Program Organization through a postdoctoral fellowship. J. G. acknowledges the financial support received from the European Commission through a Marie Skłodowska-Curie fellowship. Y. L. acknowledges the funding received from the European Union's Framework Programme for Research and

Innovation Horizon 2020 (2014-2020) under the Marie Skłodowska-Curie Grant Agreement No. 754388 and from LMU Munich's Institutional Strategy LMU excellent within the framework of the German Excellence Initiative (No. ZUK22). E. C. acknowledges the financial support received from the European Commission through the ERC Starting Grant CATALIGHT (802989).

References

1 N. J. Halas, S. Lal, W.-S. Chang, S. Link and P. Nordlander, Plasmons in Strongly Coupled Metallic Nanostructures, *Chem. Rev.*, 2011, **111**, 3913–3961.

2 V. Giannini, A. I. Fernández-Domínguez, S. C. Heck and S. A. Maier, Plasmonic Nanoantennas: Fundamentals and Their Use in Controlling the Radiative Properties of Nanoemitters, *Chem. Rev.*, 2011, **111**, 3888–3912.

3 S. Linic, P. Christopher and D. B. Ingram, Plasmonic-metal nanostructures for efficient conversion of solar to chemical energy, *Nat. Mater.*, 2011, **10**, 911.

4 M. L. Brongersma, N. J. Halas and P. Nordlander, Plasmon-induced hot carrier science and technology, *Nat. Nanotechnol.*, 2015, **10**, 25.

5 E. Cortés, Efficiency and Bond Selectivity in Plasmon-Induced Photochemistry, *Adv. Opt. Mater.*, 2017, **5**, 1700191.

6 S. Mubeen, *et al.*, An autonomous photosynthetic device in which all charge carriers derive from surface plasmons, *Nat. Nanotechnol.*, 2013, **8**, 247.

7 S. Mukherjee, *et al.*, Hot Electrons Do the Impossible: Plasmon-Induced Dissociation of H_2 on Au, *Nano Lett.*, 2013, **13**, 240–247.

8 P. Christopher, H. Xin and S. Linic, Visible-light-enhanced catalytic oxidation reactions on plasmonic silver nanostructures, *Nat. Chem.*, 2011, **3**, 467.

9 T. Oshikiri, K. Ueno and H. Misawa, Selective Dinitrogen Conversion to Ammonia Using Water and Visible Light through Plasmon-induced Charge Separation, *Angew. Chem., Int. Ed.*, 2016, **55**, 3942–3946.

10 X. Zhang, *et al.*, Product selectivity in plasmonic photocatalysis for carbon dioxide hydrogenation, *Nat. Commun.*, 2017, **8**, 14542.

11 H. Robatjazi, *et al.*, Plasmon-induced selective carbon dioxide conversion on earth-abundant aluminum-cuprous oxide antenna-reactor nanoparticles, *Nat. Commun.*, 2017, **8**, 27.

12 W. Xie and S. Schlücker, Hot electron-induced reduction of small molecules on photorecycling metal surfaces, *Nat. Commun.*, 2015, **6**, 7570.

13 Y. Kim, D. Dumett Torres and P. K. Jain, Activation Energies of Plasmonic Catalysts, *Nano Lett.*, 2016, **16**, 3399–3407.

14 G. Tagliabue, *et al.*, Quantifying the role of surface plasmon excitation and hot carrier transport in plasmonic devices, *Nat. Commun.*, 2018, **9**, 3394.

15 M. J. Kale and P. Christopher, Plasmons at the interface, *Science*, 2015, **349**, 587–588.

16 S. Chavez, U. Aslam and S. Linic, Design Principles for Directing Energy and Energetic Charge Flow in Multicomponent Plasmonic Nanostructures, *ACS Energy Lett.*, 2018, **3**, 1590–1596.

17 E. Kazuma, J. Jung, H. Ueba, M. Trenary and Y. Kim, Real-space and real-time observation of a plasmon-induced chemical reaction of a single molecule, *Science*, 2018, **360**, 521–526.

18 I. L. Violi, J. Gargiulo, C. von Bilderling, E. Cortés and F. D. Stefani, Light-Induced Polarization-Directed Growth of Optically Printed Gold Nanoparticles, *Nano Lett.*, 2016, **16**, 6529–6533.

19 C. Boerigter, R. Campana, M. Morabito and S. Linic, Evidence and implications of direct charge excitation as the dominant mechanism in plasmon-mediated photocatalysis, *Nat. Commun.*, 2016, **7**, 10545.

20 K. Li, *et al.*, Balancing Near-Field Enhancement, Absorption, and Scattering for Effective Antenna–Reactor Plasmonic Photocatalysis, *Nano Lett.*, 2017, **17**, 3710–3717.

21 K. Wu, J. Chen, J. R. McBride and T. Lian, Efficient hot-electron transfer by a plasmon-induced interfacial charge-transfer transition, *Science*, 2015, **349**, 632–635.

22 E. Cortés, *et al.*, Plasmonic hot electron transport drives nano-localized chemistry, *Nat. Commun.*, 2017, **8**, 14880.

23 C.-Y. Wu, *et al.*, High-spatial-resolution mapping of catalytic reactions on single particles, *Nature*, 2017, **541**, 511.

24 C. Zhan, *et al.*, From plasmon-enhanced molecular spectroscopy to plasmon-mediated chemical reactions, *Nat. Rev. Chem.*, 2018, **2**, 216–230.

25 L. Liu and A. Corma, Metal Catalysts for Heterogeneous Catalysis: From Single Atoms to Nanoclusters and Nanoparticles, *Chem. Rev.*, 2018, **118**, 4981–5079.

26 C. Novo, A. M. Funston and P. Mulvaney, Direct observation of chemical reactions on single gold nanocrystals using surface plasmon spectroscopy, *Nat. Nanotechnol.*, 2008, **3**, 598.

27 J. B. Sambur and P. Chen, Approaches to Single-Nanoparticle Catalysis, *Annu. Rev. Phys. Chem.*, 2014, **65**, 395–422.

28 W. Xu, J. S. Kong, Y.-T. E. Yeh and P. Chen, Single-molecule nanocatalysis reveals heterogeneous reaction pathways and catalytic dynamics, *Nat. Mater.*, 2008, **7**, 992.

29 W. Wang, Imaging the chemical activity of single nanoparticles with optical microscopy, *Chem. Soc. Rev.*, 2018, **47**, 2485–2508.

30 D. Nicolas-Debarnot and F. Poncin-Epaillard, Polyaniline as a new sensitive layer for gas sensors, *Anal. Chim. Acta*, 2003, **475**, 1–15.

31 Y. Leroux, *et al.*, Tunable Electrochemical Switch of the Optical Properties of Metallic Nanoparticles, *ACS Nano*, 2008, **2**, 728–732.

32 T. Ding, J. Mertens, A. Lombardi, O. A. Scherman and J. J. Baumberg, Light-Directed Tuning of Plasmon Resonances *via* Plasmon-Induced Polymerization Using Hot Electrons, *ACS Photonics*, 2017, **4**, 1453–1458.

33 C. P. Byers, *et al.*, Single-Particle Spectroscopy Reveals Heterogeneity in Electrochemical Tuning of the Localized Surface Plasmon, *J. Phys. Chem. B*, 2014, **118**, 14047–14055.

34 K. A. Willets, A. J. Wilson, V. Sundaresan and P. B. Joshi, Super-Resolution Imaging and Plasmonics, *Chem. Rev.*, 2017, **117**, 7538–7582.

35 T. Chen, *et al.*, Optical Super-Resolution Imaging of Surface Reactions, *Chem. Rev.*, 2017, **117**, 7510–7537.

36 S. Simoncelli, Y. Li, E. Cortés and S. A. Maier, Nanoscale Control of Molecular Self-Assembly Induced by Plasmonic Hot-Electron Dynamics, *ACS Nano*, 2018, **12**, 2184–2192.

37 S. Simoncelli, Y. Li, E. Cortés and S. A. Maier, Imaging Plasmon Hybridization of Fano Resonances *via* Hot-Electron-Mediated Absorption Mapping, *Nano Lett.*, 2018, **18**, 3400–3406.

38 R. Jungmann, *et al.*, Multiplexed 3D cellular super-resolution imaging with DNA-PAINT and Exchange-PAINT, *Nat. Methods*, 2014, **11**, 313.

39 P. Anger, P. Bharadwaj and L. Novotny, Enhancement and Quenching of Single-Molecule Fluorescence, *Phys. Rev. Lett.*, 2006, **96**, 113002.

40 T. M. Herne and M. J. Tarlov, Characterization of DNA Probes Immobilized on Gold Surfaces, *J. Am. Chem. Soc.*, 1997, **119**, 8916–8920.

41 X. Zhou, W. Xu, G. Liu, D. Panda and P. Chen, Size-Dependent Catalytic Activity and Dynamics of Gold Nanoparticles at the Single-Molecule Level, *J. Am. Chem. Soc.*, 2010, **132**, 138–146.

42 T. Chen, *et al.*, Single-Molecule Nanocatalysis Reveals Facet-Dependent Catalytic Kinetics and Dynamics of Palladium Nanoparticles, *ACS Catal.*, 2017, **7**, 2967–2972.

43 T. Chen, Y. Zhang and W. Xu, Single-Molecule Nanocatalysis Reveals Catalytic Activation Energy of Single Nanocatalysts, *J. Am. Chem. Soc.*, 2016, **138**, 12414–12421.

44 G. Baffou, R. Quidant and F. J. García de Abajo, Nanoscale Control of Optical Heating in Complex Plasmonic Systems, *ACS Nano*, 2010, **4**, 709–716.

45 G. Baffou, R. Quidant and C. Girard, Thermoplasmonics modeling: A Green's function approach, *Phys. Rev. B*, 2010, **82**, 165424.

46 J. H. Yoon, F. Selbach, L. Langolf and S. Schlücker, Ideal Dimers of Gold Nanospheres for Precision Plasmonics: Synthesis and Characterization at the Single-Particle Level for Identification of Higher Order Modes, *Small*, 2018, **14**, 1702754.

47 S. Xing, *et al.*, Highly controlled core/shell structures: tunable conductive polymer shells on gold nanoparticles and nanochains, *J. Mater. Chem.*, 2009, **19**, 3286–3291.

Faraday Discussions

PAPER

Electron-induced molecular dissociation at a surface leads to reactive collisions at selected impact parameters

Kelvin Anggara, [a] Lydie Leung,[a] Matthew J. Timm,[a] Zhixin Hu[b] and John C. Polanyi *[a]

Received 27th September 2018, Accepted 29th October 2018
DOI: 10.1039/c8fd00137e

Crossed molecular beams of gases have provided definitive information concerning the dynamics of chemical reactions. The results have, however, of necessity been averaged over collisions with impact parameters ranging from zero to infinity, thus obscuring the effect of this important variable. Here we employ a method through which impact parameter averaging is suppressed in a surface reaction. We aim a highly collimated reactive 'projectile' molecule along a surface at a stationary adsorbed 'target' molecule, with both the projectile and target being observed by Scanning Tunnelling Microscopy (STM). The projectile was CF_2 recoiling from electron-induced bond-breaking in chemisorbed CF_3 on Cu(110) at 4.6 K. The collimation of the resulting CF_2 'surface-molecular-beam' restricted it to a lateral spread of $\pm 1°$ as a consequence of its interaction with the Cu rows below. This collimation was modelled by molecular dynamics simulations. In the experiments the recoiling CF_2 projectile was aimed, successively, at impact parameters of $b = 0$ and $+3.6$ Å at a chemisorbed second CF_2, $b = +1.8$ Å at a chemisorbed I-atom, or $b = -4.0$, $b = -0.4$ and $b = +3.2$ Å at a chemisorbed vinyl radical. The pattern of reactive and non-reactive scattering was determined by STM. These collimated surface-molecular-beams have the potential to aim molecular projectiles with selected impact parameters at the many target species identifiable at a surface by STM.

1. Introduction

A pivotal event in the development of experimental methods for studying inelastic molecular collisions, especially reactive collisions, was the introduction of the method of crossed molecular beams.[1,2] In that work gaseous beams were collimated by slits, so as to be aimed at one another. The low molecular-density ensured that the observed velocities and angular distributions in the products

[a]Lash Miller Chemical Laboratories, Department of Chemistry and Institute of Optical Sciences, University of Toronto, 80 St. George Street, Toronto, Ontario, M5H 3H6, Canada. E-mail: john.polanyi@utoronto.ca
[b]Center for Joint Quantum Studies and Department of Physics, Tianjin University, Tianjin, 300350, China

were the outcome of a single collision, thus revealing the fundamental event of a chemical reaction.

Though the gaseous beams in the molecular beam studies were aimed at one another, the individual molecules were not, so the collisions being examined were averaged over all collision impact-parameters. Selection of the impact parameter or 'collision miss-distance', being as central in guiding chemical reactions as in playing billiards, has been characterised by D. R. Herschbach as being beyond reach, thus constituting the 'forbidden fruit' of reaction dynamics.[3] Restriction of the impact parameter in a bimolecular collision has been achieved in earlier work by inducing a reaction within the confines of a van der Waals complex.[4] Moreover, the effect of variation in the impact parameter has been shown to evidence itself in the magnitude and planarity of rotation in the products of a crossed-beam reaction.[5] Surface Aligned Reactions (SARs)[6–16] and time-resolved SARs[8,17], in which a surface-aligned 'projectile' reagent formed photolytically or by electron-induced dissociation recoils so as to miss the centre of mass of a co-adsorbed 'target' molecule by a controlled distance, have afforded the possibility to restrict the collision impact parameter. Prior to the present study, an efficient and reproducible abstraction reaction between directed recoiling phenylene and an adjacent I-atom bound to a copper surface was observed.[15]

The present approach has furnished the missing element required for the more general selection of the impact parameter.[18] The missing element provides the means to obtain a high degree of alignment of the motion of the 'projectile' species, so that it collided at a selected distance, b (Å), from the centre of mass of a target, an atom or molecule remote from the point of origin of the projectile. This alignment is achieved here by exploiting the collimating effect of adsorbate–substrate interactions with the row of atoms below, present in the crystalline substrate – this constituting a nanoscale replacement of the macroscopic slits used to collimate gaseous beams. The resulting high degree of collimation of the CF_2 projectiles employed here renders them into a 'surface-molecular-beam', directed along a copper row of Cu(110). At the nanoscale it is shown to be possible to accurately direct the recoiling species along the substrate so as to control the miss-distance (*i.e.* impact parameter) in the molecular collision. Different impact parameters, b, were obtained in this work by employing different adsorption sites for the parent molecule of the projectile, so as to change the miss-distance relative to the centre of mass of the target. With a CF_2 projectile approaching targets successively along different rows of Cu-atoms, CF_2 moved from a known impact parameter on one side of the target, through on-target collisions ($b \approx 0$), to an impact parameter on the other side of the target. The magnitude of the impact parameter (miss-distance) was determined in each case to within 1%.

2. Methods

2.1 Experimental

The experiments were carried out in a low-temperature UHV-STM (Omicron) with a base pressure of $< 3.0 \times 10^{-11}$ mbar. The Cu(110) surface was cleaned *via* repeated cycles of Ar^+ sputtering (0.6 keV) and annealing at 800 K, until less than 1% of contaminants was observed by STM. The STM images were taken in the constant-current mode at 4.6 K at the sample scanning voltages stated. Iodotrifluoromethane (CF_3I; SynQuest, 99%) and vinyl iodide (C_2H_3I; Oakwood

Chemical, 95%) were deposited *via* a capillary tube aimed towards the Cu sample, whose temperature reached a maximum of 14 K after deposition. Vinyl iodide was outgassed by several freeze-pump-thaw cycles, prior to deposition. The C–I bond was observed to be broken upon the adsorption of CF_3I at 14 K to give chemisorbed trifluoromethyl (CF_3) and an I-atom on the Cu surface. The thermal dissociation of vinyl iodide, C_2H_3I, was induced by annealing the substrate to 80 K by removing the sample from the STM stage with a wobble stick for 1.12 min.[19] The difluorocarbene (CF_2) projectile was produced *via* the electron-induced dissociation of CF_3. Selective C–F bond-breaking of the C–F bond along a Cu-row was carried out by placing the STM tip over the middle of the CF_3-feature for several seconds at a bias voltage of +1.3 V, with the feedback loop disabled. The distances and directions of the product moieties were determined *via* analysis of the STM images using the WSxM software.[20]

2.2 Theory

Density Functional Theory (DFT) calculations implemented in the Vienna Ab initio Simulation Package (VASP)[21,22] were carried out in order to model the experimental results on the SciNet supercomputer.[23] The pathways for the reactions were obtained *via* the nudged elastic band (NEB) climbing-image method,[24] and from molecular dynamics (MD) trajectories. The calculations utilized the projector augmented wave (PAW) method[25,26] and the generalized-gradient approximation (GGA) with the Perdew–Burke–Ernzerhof (PBE) functional.[27] Grimme's semi-empirical dispersion correction (DFT-D3) was added to all calculations in order to correct for Van der Waals interactions.[28] The Cu(110) surface was modelled by employing periodic boundary conditions on a slab composed of five Cu atom layers, together with the adsorbates, and a 17 Å vacuum layer. All atoms were allowed to move except for the lowest two Cu atom layers. A single gamma point was used for the relaxed geometries. The system was relaxed until the forces on each atom were less than 0.01 eV $Å^{-1}$. For the NEB calculations on a (4 × 6) supercell, consisting of 120 Cu atoms, the system was relaxed until the forces orthogonal to the nudged band were less than 0.02 eV $Å^{-1}$. MD calculations were carried out on a (3 × 8) supercell (all 120 Cu-atoms). STM-image simulations were generated using the Tersoff–Hamann approximation[29] and were visualized using the Hive software.[30,31]

The dynamics of the electron-induced reaction of CF_3 were computed using an Impulsive Two-State (I2S) model that is described in detail elsewhere.[14–16,32–35] Briefly, the system at equilibrium in the ground state is promoted directly to an ionic state obtained by moving one electron from the F-atom core to its valence shell, thus generating the ionic configuration [He] $2s^2$ $2p^6$. The excited system was evolved for a short period of time (t^*=13 fs, giving the observed mean distance of recoil of CF_2 from CF_3 in electron-induced dissociation), during which the molecular bond extended away from its initial equilibrium position. At t^*, the system was returned to the ground potential energy surface (PES) to simulate the loss of the transferred electron, retaining the geometry and velocities acquired in the ionic state. MD trajectories were obtained by solving the equations of motion until the system reached its terminus on the ground PES. The MD calculations were carried out under microcanonical conditions, which conserved the number of atoms (N), the volume (V) and the total energy. The initial state was initialized by a set of random velocities on each atom taken from the Maxwell–Boltzmann distribution at 4.7 K. A time step of 0.5 fs was used for the MD calculations on both states, in order to give a negligible total energy drift (<4 meV ps^{-1}). The MD trajectories for the CF_2 + CF_2 and CF_2 + vinyl encounters were obtained by introducing a relaxed stationary CF_2 or vinyl target into the path of the CF_2 projectile at a selected impact parameter.

3. Results and discussion

3.1 CF_2 projectile

The 'surface-molecular-beam' employed here was obtained from the electron-induced dissociation of trifluoromethyl, CF_3, adsorbed on Cu(110) at 4.6 K, as

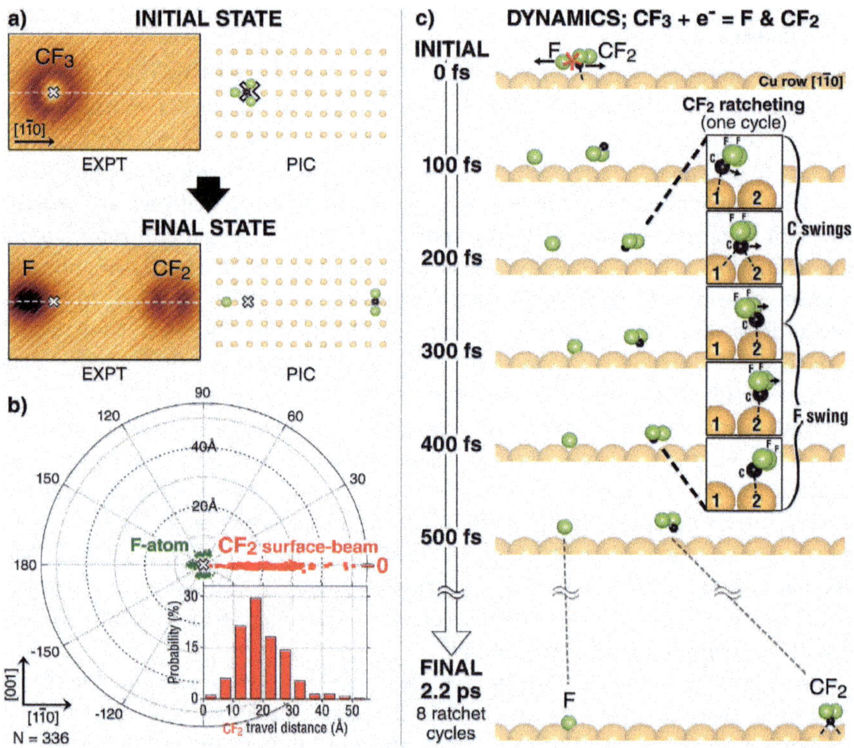

Fig. 1 Generation of the CF_2 surface-molecular-beam from the electron-induced dissociation of CF_3 on Cu(110) at 4.6 K. (a) STM images (I = 0.3 nA, V = −0.1 V) and schematics of the initial state and final state of the CF_3 reaction, showing a chemisorbed CF_2 and a F-atom. The white crosses in panels (a) and (b) indicate the position of the CF_3 feature prior to electron-induced dissociation. The dashed lines in the STM images mark a Cu-row along [1$\bar{1}$0] on the surface. (b) Distance and angle distributions given as a polar plot for the CF_2 products (red squares), constituting the surface-molecular-beam, and for the F-atom products (green squares). The inset in (b) shows the distance distribution of the CF_2 product along the [1$\bar{1}$0] direction as a histogram of travel distances (bin size: 5 Å). (c) Time-evolution of the CF_3 electron-induced reaction obtained *via* the Impulsive Two-State model (the red cross shows the locus of repulsion). The inset in (c) shows the dynamics of the 'ratcheting' of CF_2 along the Cu-row. The black dashed lines linking the adsorbate to the surface at 0 and 2.2 ps indicate C–Cu bonds, initially and finally.

shown in Fig. 1. The chemisorbed CF_3 was imaged as an asymmetric protrusion surrounded by a depression (Fig. 1a). Its apparent height was measured to be 0.07 ± 0.04 Å. DFT calculations showed that the CF_3 was bound to an atop Cu atom *via* its carbon atom with an adsorption energy of 2.53 eV, with the C–F bond along the Cu-row tilting towards the surface, thereby increasing the angle away from the surface of the other two C–F bonds. The interaction of the F-atom directed along the Cu-row with that row resulted in an increased C–F bond length of ~5% (0.07 Å bond extension), in comparison with the other two C–F bonds (equilibrium separation of 1.38 Å). Single C–F bond-dissociation exclusively of the stretched C–F bond in chemisorbed CF_3 was carried out by adding hot electrons from the STM tip with energies above +0.9 eV. Two products from this electron-induced dissociation were identified as being a difluorocarbene CF_2 radical recoiling along the $[1\bar{1}0]$ Cu-row, and an F-atom recoiling approximately in the opposite direction, as shown in Fig. 1a. The F-atom was imaged as a dark depression (with a depth of 0.36 ± 0.04 Å) observed near the CF_3 reagent position, whereas CF_2, which recoiled long distances along the row, was imaged as a depression elongated along the [001] direction (measured depth of 0.23 ± 0.04 Å). The experimental angle and distance distributions for both products, CF_2 and F, are shown in Fig. 1b. From the CF_3 reagent as a point of origin, the F-atom product was observed to have travelled 3.8 ± 0.7 Å on average, whereas the average distance of travel of CF_2 was much larger (19.7 ± 7.8 Å), being found at distances up to 52.6 Å. The resultant unidirectional array of CF_2's recoiling from electron-induced dissociation comprised the 'surface-molecular-beam'.

According to our images and DFT calculations, both products, F and CF_2, adsorbed on short-bridge sites, with computed adsorption energies of 5.01 eV for F and only 1.89 eV for CF_2, thus accounting for its higher mobility. As indicated, the F-atom and CF_2 angular distributions showed that these moieties, markedly different in appearance, recoiled in opposite directions. In the case of the F-atom, five angles relative to the Cu-row direction were observed, corresponding to an angular spread: ±71°, ±110° and 179° (±10° for all angles measured relative to the $[1\bar{1}0]$ direction, see Fig. 1b). In contrast, and significantly, for all of the many cases examined in the experiments ($N = 336$), CF_2 was observed at a single angle, 0°, directly along the Cu-row direction, with a very narrow angular spread of ±1°. This highly-collimated motion of the CF_2 radical gave the appearance of a molecular-beam traversing the surface along a $[1\bar{1}0]$ Cu-row. As a consequence of this high degree of directionality, the miss-distance or 'impact parameter' of the projectile relative to the centre of mass of a stationary target molecule in a collision was defined and measurable.

Details on the unidirectional travel of CF_2 in the 'surface-molecular-beam' were elucidated *via* molecular dynamics (MD) simulations, as shown in the inset in Fig. 1c. The electron-induced reaction of CF_3 demonstrated a measured yield of 8.8×10^{-12} at +1.3 V and was shown to proceed *via* electronic excitation involving a single electron.[18] This electronic excitation and the subsequent dynamics were described by an Impulsive Two-State (I2S) model, in which the added electronic charge was placed on the F-atom located along the $[1\bar{1}0]$ Cu-row. This single-electron excitation is supported by the observed extension of a single C–F bond directed along the Cu-row, and by the calculated partial density of states (published in ref. 18) which showed a hybrid state with C–F σ^* anti-bonding character at 1.3 eV above the Fermi level. The MD results indicated that the two products, the F-atom and CF_2, recoiled along the direction of the severed C–F bond in opposite directions, in accordance with previous studies of electron-induced carbon-halogen bond-breaking in aromatic molecules adsorbed on Cu(110).[15,32,34,35] In the case of the F-atom, the recoil distance was 1.3 Å (half a unit cell) from the initial position of the CF_3 reagent. This small displacement was due to the strong interaction between the F-atom and the Cu surface (see above). Further MD calculations showed that a misalignment of the parent CF_3 adsorption geometry by less than 7° from the Cu-row direction could account for the observed angular spread of the F-atom product. For CF_2, the long average travel distance that was computed, 21.6 Å (8.5 unit cells), is accounted for by the MD results as being due to a 'ratcheting motion', detailed in the inset of Fig. 1c. The repulsion acquired during the electronic excitation induced the CF_2 to ratchet along the $[1\bar{1}0]$ Cu-row as follows; first the C-atom interacted with a Cu-atom (#1) and then swung towards the adjacent Cu-atom (#2) on the row to form a C–Cu bond, thus maximizing the bonding of CF_2 to the Cu surface. The energy gained from the latter C–Cu bond formation generated momentum that caused the heavy F-atoms to swing around the chemisorbed C-atom, along the same direction (designated as forward). This internal energy transfer between the C-atom and the pair of F-atoms is more efficient than the energy dissipation to the surface, thus permitting CF_2 to remain 'hot' and travel long distances along the Cu-row. The collimation of the CF_2 recoil, due to the p-orbital of the C-atom directed along the Cu-row, caused CF_2 to bind successively to the close-packed Cu-atoms of the $[1\bar{1}0]$ row.

3.2 CF_2 target and I target

Collisions of the collimated CF_2 projectile were studied, separately, with two different stationary targets: CF_2 and I located on or adjacent to the CF_2 recoil path. The impact parameter, b, was given by the distance, measured perpendicular to the CF_2 travel direction, to the centre of mass of the target species. The results obtained with known impact parameters for the collisions between the CF_2 projectile and the two stationary targets, CF_2 and I, are summarised in Fig. 2.[18] In the case of the CF_2 target, two impact parameters were investigated, $b = 0$ and 3.6 Å. For the $CF_2 + CF_2$ collision at $b = 0$ Å, and, importantly, only at $b = 0$ Å, a new product was observed as shown in Fig. 2a. This product was identified as tetrafluoroethylene, C_2F_4, chemisorbed along the Cu-row by two C–Cu covalent bonds, with a measured height of 0.29 ± 0.04 Å. This new species was observed for all of the 29 reactive cases examined.

MD simulations showed C–C bond formation in C_2F_4 to be due to the collision at zero impact parameter between the C-atom of the CF_2 projectile and the C-atom of the CF_2 target. In the experiments, a further hot-electron pulse at +1.3 V was found to cause the chemisorbed C_2F_4 to diffuse bodily along the $[1\bar{1}0]$ direction.

Significantly, when the impact parameter between the CF_2 projectile and CF_2 target was increased to 3.6 Å, no associative or other reaction was observed (see Fig. 2b). Instead the collision was observed experimentally to result in an inelastic encounter transferring sufficient energy to the CF_2 target to cause it to diffuse by a substantial distance, almost 6 unit cells (13.5 Å), along the Cu-row. From analysis of the MD results, the interaction between the CF_2 projectile and CF_2 target at $b = 3.6$ Å was shown to constitute a strong F–F collision.

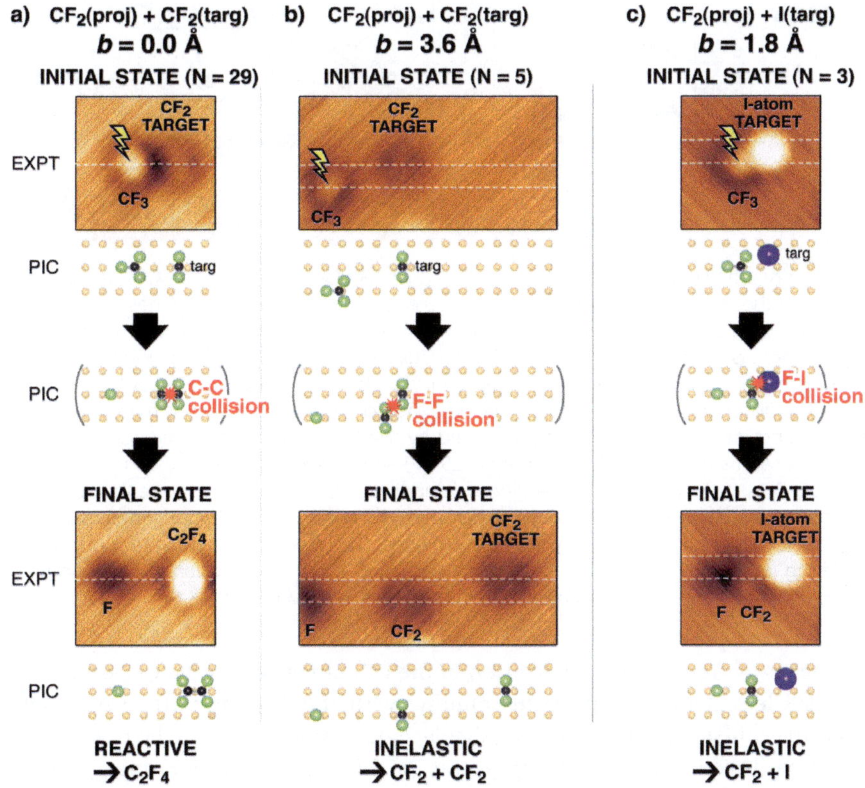

Fig. 2 Bimolecular collisions at selected impact parameters, b, for CF_2 (projectile) + CF_2 (target), and CF_2 (projectile) + I (target). In all cases, the bimolecular collision was initiated by the generation of a CF_2 surface-molecular-beam aimed towards the stationary target. (a) STM images ($I = 0.3$ nA, $V = -0.1$ V) and schematics of the initial and final states for $CF_2 + CF_2$ at $b = 0$ Å, giving a reactive collision that forms a chemisorbed tetrafluoroethylene (C_2F_4). (b) Images for $CF_2 + CF_2$ and schematics at $b = 3.6$ Å, giving an inelastic collision (displaced CF_2 target). (c) Images for CF_2 + I and schematics at $b = 1.8$ Å, giving an inelastic collision (displaced I-atom). The pictorial representations (PIC) highlight the loci of repulsion in the encounters with a red ×. In all of the panels the white dashed lines indicate Cu-rows along [1$\bar{1}$0] on the surface.

Inelastic scattering was also readily observable between the CF_2 projectile and an I-atom, for which b was set at 1.8 Å (Fig. 2c). The nature of the CF_2 ratcheting ensured that the collision occurred between the F-atom of the CF_2 projectile and the target I-atom. The resulting forward recoil of the I-atom was apparent (shown in the images of Fig. 2c) as a one unit cell (2.6 Å) displacement of the iodine. This experimental finding showed that the CF_2 projectile plus chemisorbed I-atom collision energy directed along [1$\bar{1}$0] was sufficient to overcome the I-atom diffusion barrier calculated as 0.1 eV.

The effect of the collision energy for the CF_2 projectile + CF_2 target collision at $b = 0$ Å was examined by varying the travel distance of the CF_2 projectile prior to collision with the CF_2 target. Two reaction pathways were observed, as shown in Fig. 3b, with one at a high collision energy and one at a low collision energy, with each demonstrating a characteristic displacement of the C_2F_4 chemisorbed

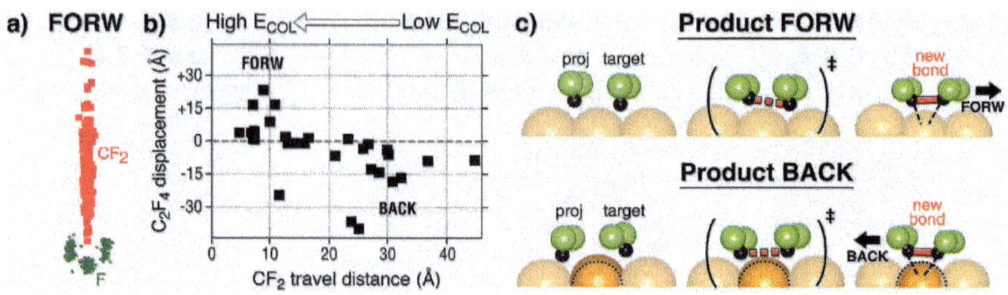

Fig. 3 Forward and backward scattering of the molecular product of the CF_2 association reaction at the selected impact parameter, $b = 0$ Å. (a) The surface beam defines the forward direction (Forw). (b) At an impact parameter of $b = 0$ Å, the measured C_2F_4 product displacement is given as a function of CF_2 travel distance prior to its collision with the target (hence a high collision-energy, E_{COL}, is seen on the left of (b), and a low E_{COL} is seen on the right). The observed C_2F_4 displacement, being either forward or backward, is measured from the initial position of the CF_2 target, indicated by the horizontal dashed line. (c) Molecular dynamics results for the observed forward (1.5 eV) and backward (1.2 eV) scattered reaction product, C_2F_4. The heavy dashed red line in (c) indicates C–C attraction, while the solid red line indicates a newly-formed C–C bond between the projectile and target (additionally, C-to-surface bonding is indicated by light broken lines). The results in (c) show that backward scattering involves an underlying Cu atom raised ~0.5 Å from its equilibrium position (the unperturbed position being indicated in the figure by a dashed circle).

product away from the initial position of the CF_2 target. When the CF_2 projectile travelled less than 11 Å prior to colliding with the CF_2 target (hence a high projectile collision energy), the C_2F_4 displacement was observed to be forward, in the same direction as the CF_2 projectile. At this higher collision energy, every encounter at $b = 0$ Å led to the formation of C_2F_4. For a travel distance of the CF_2 projectile of greater than 11 Å (reduced projectile collision energy), the C_2F_4 product of collision with a CF_2 target was observed to be scattered backward, in the opposite direction to the CF_2 projectile. At this lower collision energy, the collision efficiency to form C_2F_4 was lower, with the result that, averaged over all collision energies, the efficiency of converting a projectile plus target to a molecular product in the CF_2 + CF_2 collisions was 85% (*i.e.* 29 out of the total of 34 collisions studied).[18] The observed shift from forward to backward scattering of the product is attributable to the change in the projectile collision energy.

Evidence of changing dynamics with a change in the collision energy was also provided by previously reported MD calculations.[18] These showed that by reducing the collision energy between the CF_2 projectile and CF_2 target from 1.5 to 1.2 eV, the formation of C_2F_4 involved more complex interactions with the surface, as shown in Fig. 3c. In the MD results for the higher energy forward pathway, C–C bond formation transferred the translational energy of the CF_2 projectile to the C_2F_4 product, carrying it in the forward direction. At the lower collision energy of 1.2 eV that gave backward scattering, the motion of the Cu atoms of the slab was found by MD to result in the lifting of a Cu atom located midway between the projectile and the target. This raised Cu atom catalysed the C–C bond formation to form C_2F_4, as visualised in the lower panel of Fig. 3c.

3.3 Vinyl target

The collision between a CF_2 projectile and a third adsorbate, chemisorbed vinyl, was also studied experimentally. This third adsorbate was chosen since it is an intermediate in Fischer–Tropsch polymerization.[36,37] An STM image of the chemisorbed vinyl ($CH{=}CH_2$) on the Cu surface is shown in Fig. 4a. The vinyl (Vi) appears in the STM image as a bright asymmetric feature with a measured height of 0.28 ± 0.04 Å, accompanied by two adjacent depressions. The STM simulation shown in Fig. 4b, based on the computed adsorption geometry of vinyl, is in agreement with the STM image. The chemisorbed vinyl is strongly bound with a computed adsorption energy of 3.25 eV. Fig. 4c shows a C-atom of the vinyl to be bound to two adjacent Cu atoms along a row, with the CH_2 group at the opposite end of the vinyl tilted away from the surface plane. From analysis of the image and the adsorption geometry, the brighter side of the vinyl in Fig. 4a and b was ascribed to the raised CH_2 part of $CH{=}CH_2$, and the darker side to the chemisorbed C-atom at the opposite end of the vinyl, with the two dark depressions indicating binding to the surface.

Three different impact parameters, b = −4.0 Å, −0.4 Å and +3.2 Å, were selected for a study of the collision between the CF_2 projectile and a stationary vinyl target (see Fig. 5). These three impact parameters were obtained by electron-induced dissociation of the CF_3 chemisorbed (a) on Cu-row I, (b) on Cu-row II and (c) on Cu-row III.

The impact parameter of −4.0 Å (Fig. 5a), was obtained by aiming the CF_2 projectile along Cu-row I at the chemisorbed (darker) side of the vinyl target (designated as negative). For this impact parameter, the experimental outcome in all 17 cases was a final-state with a CF_2 projectile coming to a halt closely adjacent to the chemisorbed end of the vinyl target (2.6 Å away), without a detectable shift in the vinyl.

Changing to an impact parameter of approximately zero, the observed outcome of the collision between CF_2 and vinyl was markedly different, as shown in Fig. 5b. In all 60 cases, where the CF_2 projectile was aimed along Cu-row II corresponding to $b = -0.4$ Å, the outcome was a bright oval feature not previously seen with a height of 0.65 ± 0.04 Å. We attribute this new feature to a product molecule formed in the head-on collision. The feature invariably exhibited a slight depression on the side from which CF_2 approached, thus suggesting the covalent bonding of the new molecule to the Cu surface. The newly-formed molecule was always displaced forward along the Cu-row, in the direction of the CF_2 motion. We attribute this new molecular feature to a reaction forming chemisorbed 1,1-difluoroallyl (CF_2CHCH_2), as evidenced by the good agreement

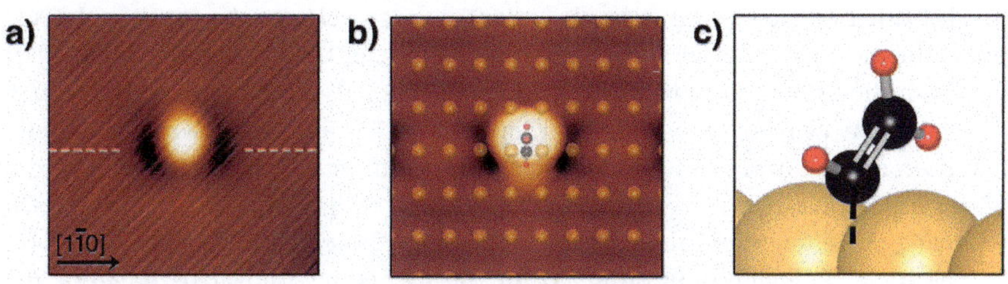

Fig. 4 Adsorption geometry of a chemisorbed vinyl (Vi). (a) STM image of a chemisorbed Vi (I = 0.1 nA, V = −0.1 V). The white dashed line indicates a [1$\bar{1}$0] Cu-row. (b) STM simulation image of a chemisorbed Vi on Cu(110). (c) Side-view of the computed structure of a Vi chemisorbed on a short-bridge site of Cu(110). The black dashed line in (c) indicates C–Cu interaction.

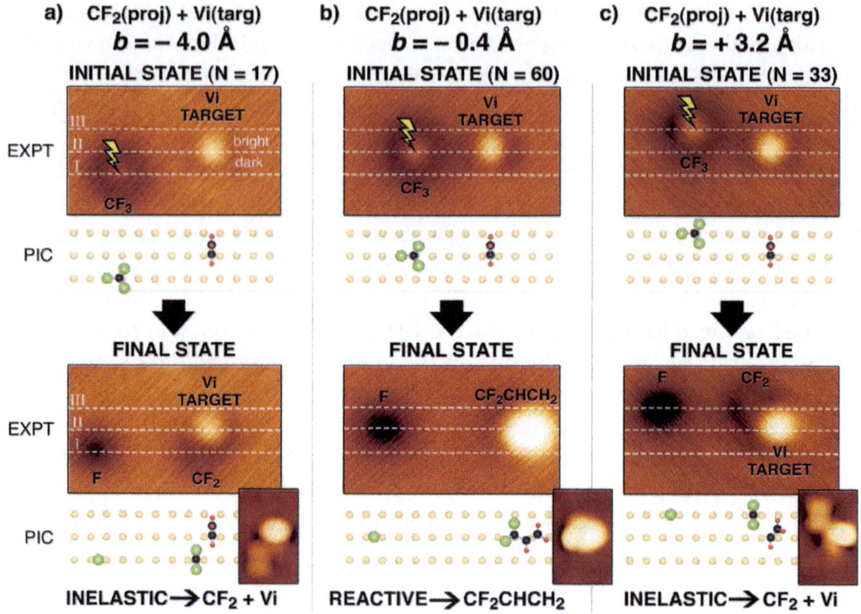

Fig. 5 Bimolecular collisions at selected impact parameters, b, for CF_2 (projectile) + Vi (target). The collision was initiated by the generation of a CF_2 surface-molecular-beam aimed towards the stationary Vi target. (a) STM images (I = 0.1 nA, V = −0.1 V) and schematics of the initial and final states for CF_2 + Vi at b = −4.0 Å, giving an inelastic collision at the chemisorbed end of the Vi target, with the impact parameter designated as negative. (b) Images (I = 0.1 nA, V = −0.1 V) for CF_2 + Vi at b = −0.4 Å, giving a reactive collision that forms a chemisorbed 1,1-difluoroallyl (CF_2CHCH_2) reaction product. (c) Images (I = 0.1 nA, V = −0.1 V) for b = +3.2 Å, giving an inelastic collision at the CH_2 end of the Vi target, with the impact parameter designated as positive. In all of the panels the horizontal white dashed lines indicate three adjacent Cu-rows, I, II and III, on the surface, with the Vi target chemisorbed on row II (brighter end above, darker end below). The insets in all of the final states show simulated STM images of the products generated from the respective CF_2 + Vi collisions at the designated impact parameters.

with the simulated STM image of this product molecule (Fig. 5b, final state). Further evidence of the existence of a stable chemisorbed product came from the observation of its bodily diffusion along the Cu-row when irradiated with electrons ranging from 0.3 to 2.4 eV.

The third impact parameter, b = +3.2 Å, was selected by aiming the CF_2 projectile along Cu-row III at the bright (CH_2) side of the vinyl target (Fig. 5c). In the 33 studied cases of inelastic collision at b = +3.2 Å, the CF_2 projectile came to a halt 2.6 Å away from the vinyl target, but at the CH_2 end of the vinyl, at which it had been aimed. In this b = +3.2 Å inelastic encounter with the raised CH_2 group of the vinyl, the target molecule was invariably tilted away from the colliding CF_2 following the encounter, also appearing brighter (height: 0.32 ± 0.04 Å).

The inelastic encounters observed for the two larger impact parameters, b = −4.0 and +3.2 Å, showed that in both cases the CF_2 projectile failed to transfer sufficient energy to form the reaction product CF_2CHCH_2, or to carry the vinyl over its 0.55 eV diffusion barrier. Instead the CF_2 projectile scattered backward (opposite to its direction of approach), coming to a halt 2.6 Å away from the undiffused vinyl.

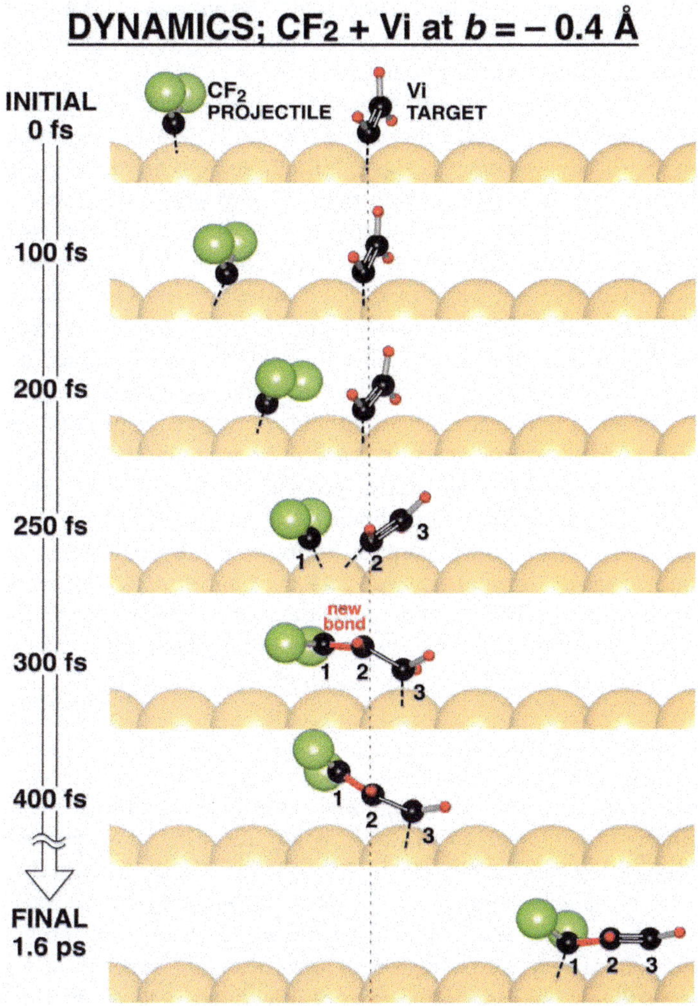

Fig. 6 Computed dynamics for a reactive collision between a CF_2 projectile and a Vi target, at $b = -0.4$ Å. The initial state of the trajectory ($t = 0$ fs) was obtained by placing a stationary Vi target along the path of a ratcheting CF_2 from Fig. 1c at 179 fs. The Vi target was placed 3.5 unit cells (8.92 Å) away from the CF_3 position. The fine vertical dashed line corresponds to the initial position of the vinyl target. C_1 corresponds to the C-atom of the CF_2 projectile. C_2 and C_3 correspond to the C-atoms of the CH and CH_2 parts of Vi, respectively. The black dashed lines indicate C–Cu bonds. The red solid line indicates a newly-formed C–C bond between the projectile and target at this low impact parameter.

For the CF_2 projectile plus vinyl target aimed to give an impact parameter of $b = -0.4$ Å, a qualitative correlation was established between the CF_2 projectile energy and the displacement of the 1,1-difluoroallyl product. A high collision energy, associated with short CF_2 recoil distances (<5.0 Å), resulted in a substantial forward product displacement of the 1,1-difluoroallyl product by 14.0 ± 4.0 Å on average. For lower collision energies obtained for the CF_2 projectile that recoiled >22.0 Å, and hence had 'cooled' translationally, the average forward scattered distance of the product was observed to be only 2.6 ± 1.2 Å. The forward scattering at a higher collision energy was attributed to the retention of the projectile momentum in the molecular product.

The dynamics computed *via* MD simulation for the reactive outcome between the CF_2 projectile and the vinyl target at the selected $b = -0.4$ Å impact parameter are shown for successive times in Fig. 6. Initially at $t = 0$ fs, the CF_2 projectile is shown 2.5 unit cells (6.4 Å) away from the stationary vinyl target. After one and a half ratcheting cycles, by 200 fs, CF_2 had reached its closest distance of approach, 2.6 Å, to the vinyl target. By 250 fs, the colliding CF_2 had tilted the vinyl away from the surface, thus suggesting that it was now bound to only a single Cu-atom. Fig. 6 shows CF_2 in collision with the neighbouring chemisorbed vinyl, binding to the same Cu-atom as the vinyl. By 300 fs, a new C–C bond had formed between CF_2 and vinyl. Judging from the C–C bond separations, the new C_1–C_2 bond was formed by reducing the C_2–C_3 double bond to a single bond, as shown. The product of this addition reaction, 1,1-difluoroallyl, was partially lifted away from the Cu surface at 300 fs, with only C_3 of the vinyl bound to the surface. In its final configuration at 1.6 ps, 1,1-difluoroallyl was computed to have migrated 2.5 unit cells (6.4 Å) away from the initial position of the vinyl target, giving an adsorption energy in its final state of 2.23 eV, bound at C_1. Satisfyingly, this computed migration of the 1,1-difluoroallyl product to a location well beyond the vinyl, was clearly observable in the STM images of the final state of the CF_2 + vinyl reaction at $b = -0.4$ Å.

4. Conclusions

In the experiments described herein, hot electrons were used to induce selective bond-breaking in chemisorbed CF_3 on Cu(110), resulting, through interactions with the underlying copper row, in a highly-collimated surface-beam of CF_2. This directed beam made it possible to aim the CF_2 projectiles at selected impact parameters at three stationary targets: chemisorbed CF_2, I-atoms and vinyl. For the impact parameters in the region of zero, the CF_2 projectile + CF_2 target collisions resulted in an association reaction, yielding a C_2F_4 product in 100% of the collisions at our higher collision energy that gave the forward scattered product. Similarly, for the CF_2 projectile + vinyl, collisions at close to zero impact parameter ($b = -0.4$ Å) resulted in 100% of the collisions giving rise to a reaction forming forward scattered chemisorbed 1,1-difluoroallyl as a product.

For larger impact parameters, the observed outcome was an inelastic encounter with molecular scattering but without a chemical reaction. Specifically for the CF_2 projectile colliding with a CF_2 target at $b = 3.6$ Å, or with an I-atom at $b = 1.8$ Å, the target was found to have recoiled along the direction of approach of the projectile (recoiling by 13.5 and 2.6 Å, respectively), despite the barrier to diffusion (0.3 eV in the case of the CF_2 target, 0.1 eV in the case of the I-atom).

Two cases of large impact-parameter collision were examined for a CF_2 projectile colliding with a vinyl target. In the first case, designated as a negative impact parameter, the collision was with the darker chemisorbed end of the vinyl target at $b = -4.0$ Å, while in the second case, designated as a positive impact parameter, the collision was with the raised CH_2 end of the vinyl, at $b = +3.2$ Å. In both cases, the CF_2 projectile was observed to come to rest adjacent to the vinyl, but in the former case at the depressed chemisorbed end of the vinyl at which it had been aimed, and in the latter case at the raised CH_2 aiming point. When the projectile was aimed at the CH_2 ($b = +3.2$ Å), the effect of the inelastic collision was additionally found to move the CH_2 moiety forward in the direction of the impact.

The molecular dynamics results were examined using an Impulsive Two-State model introduced earlier. In the model, electron-induced dissociation occurred due to excitation to a short-lived anionic state. The dynamics for $e^- + CF_3$ exhibited short-range scattering for the strongly surface-bound F-atom, and long-range recoil for the more weakly bound CF_2. The long-range recoil of the CF_2 was shown by MD to occur exclusively along the closely-spaced Cu-atoms of the $[1\bar{1}0]$ rows of Cu(110). Directed translational recoil of the CF_2 projectile combined with a rocking motion of this biradical to 'ratchet' it, by successive binding to adjacent Cu-atoms along the row. Since this motion was restricted to the row, it gave rise to a 'surface-molecular-beam'. CF_2 lost translational energy as it travelled along the beam direction.

Both the experimental and MD results demonstrated forward scattering of the reaction products, C_2F_4 or CF_2CHCH_2, at higher collision energies (corresponding to short projectile recoil). Following the convention established in gas-phase molecular beam experiments, we defined 'forward' as a continuation of the direction of the accelerated reagent, CF_2 (the 'projectile').

The molecular dynamics results showed that at a low projectile collision energy, the interaction between the projectile and substrate was of sufficient duration to permit surface reconstruction whereby a single Cu atom rose up so as to catalyse the association reaction between adjacent CF_2 moieties.

The observation here of inelastic non-reactive scattering in the interval of impact parameters selected to be in the range 2–4 Å afforded a means of probing long-range intermolecular forces. Since the inelastic scattering differed according to the side of the target that was impacted, the experiments revealed the asymmetry in these long-range forces.

The method employed here for selecting the impact parameter depends on the presence of a directed surface-molecular-beam. This, in turn, depends on the existence of anisotropy in the adsorbate–substrate interactions. Anisotropy is likely to be present in cases in which an adsorbate (whether neutral or charged) is bound to a crystalline substrate. Selection of the impact parameter can then be achieved by situating a target molecule at a site in the path of the projectile, or displaced from that path to one side or the other. It appears likely that these requirements can be met for atomic and molecular recoil across many crystals, thus permitting studies of the effect of impact parameter on reactive and non-reactive collisions at surfaces.

Conflicts of interest

There are no conflicts of interest to declare.

Acknowledgements

This work was funded in part by the Natural Sciences and Engineering Research Council of Canada (NSERC) and the University of Toronto NSERC General Research Fund. Computations were carried out on the General Purposed Cluster

(GPC) and Niagara at SciNet HPC Consortium. SciNet is funded by the Canada Foundation for Innovation under the auspices of Compute Canada, the Government of Ontario, Ontario Research Fund-Research Excellence and the University of Toronto. K. A. thanks the Connaught International Scholarship for Doctoral Students for its financial support.

References

1 D. R. Herschbach, *Angew. Chem., Int. Ed.*, 1987, **26**, 1221–1243.
2 Y. T. Lee, *Angew. Chem., Int. Ed.*, 1987, **26**, 939–951.
3 D. Herschbach, *Eur. Phys. J. D*, 2006, **38**, 3–13.
4 S. Buelow, G. Radhakrishnan, J. Catanzarite and C. Wittig, *J. Chem. Phys.*, 1985, **83**, 444–445.
5 K. S. Kalogerakis and R. N. Zare, *J. Chem. Phys.*, 1996, **104**, 7947–7964.
6 E. B. D. Bourdon, J. P. Cowin, I. Harrison, J. C. Polanyi, J. Segner, C. D. Stanners and P. A. Young, *J. Phys. Chem.*, 1984, **88**, 6100–6103.
7 E. B. D. Bourdon, P. Das, I. Harrison, J. C. Polanyi, J. Segner, C. D. Stanners, R. J. Williams and P. A. Young, *Faraday Discuss. Chem. Soc.*, 1986, **82**, 343–358.
8 J. C. Polanyi and A. H. Zewail, *Acc. Chem. Res.*, 1995, **28**, 119–132.
9 C. E. Tripa and J. T. Yates Jr, *Nature*, 1999, **398**, 591–593.
10 L. J. Lauhon and W. Ho, *Faraday Discuss.*, 2000, **117**, 249–255.
11 S.-W. Hla, L. Bartels, G. Meyer and K.-H. Rieder, *Phys. Rev. Lett.*, 2000, **85**, 2777–2780.
12 P. Maksymovych, D. C. Sorescu, K. D. Jordan and J. T. Yates Jr, *Science*, 2008, **322**, 1664–1667.
13 T. Kumagai, A. Shiotari, H. Okuyama, S. Hatta, T. Aruga, I. Hamada, T. Frederiksen and H. Ueba, *Nat. Mater.*, 2011, **11**, 167–172.
14 Z. Ning and J. C. Polanyi, *J. Chem. Phys.*, 2012, **137**, 091706.
15 A. Eisenstein, L. Leung, T. Lim, Z. Ning and J. C. Polanyi, *Faraday Discuss.*, 2012, **157**, 337–353.
16 Z. Ning and J. C. Polanyi, *Angew. Chem., Int. Ed.*, 2013, **52**, 320–324.
17 M. E. Vaida and T. M. Bernhardt in *Ultrafast Phenomena in Molecular Sciences*, ed. R. de Nalda and L. Bañares, Springer International Publishing, Switzerland, 2014, pp. 231–261.
18 K. Anggara, L. Leung, M. J. Timm, Z. Hu and J. C. Polanyi, *Sci. Adv.*, 2018, **4**, eaau2821.
19 D. B. Dougherty, J. Lee and J. T. Yates Jr, *J. Phys. Chem. B*, 2006, **110**, 20077–20080.
20 I. Horcas, R. Fernández, J. M. Gómez-Rodríguez, J. Colchero, J. Gómez-Herrero and A. M. Baro, *Rev. Sci. Instrum.*, 2007, **78**, 013705.
21 G. Kresse and J. Hafner, *Phys. Rev. B: Condens. Matter Mater. Phys.*, 1993, **47**, 558–561.
22 G. Kresse and J. Furthmüller, *Phys. Rev. B: Condens. Matter Mater. Phys.*, 1996, **54**, 11169–11186.
23 C. Loken, D. Gruner, L. Groer, R. Peltier, N. Bunn, M. Craig, T. Henriques, J. Dempsey, C.-H. Yu, J. Chen, L. J. Dursi, J. Chong, S. Northrup, J. Pinto, N. Knecht and R. van Zon, *J. Phys.: Conf. Ser.*, 2010, **256**, 012026.
24 G. Henkelman, B. P. Uberuaga and H. Jónsson, *J. Chem. Phys.*, 2000, **113**, 9901–9904.

25 P. E. Blöchl, *Phys. Rev. B: Condens. Matter Mater. Phys.*, 1994, **50**, 17953–17979.
26 G. Kresse and D. Joubert, *Phys. Rev. B: Condens. Matter Mater. Phys.*, 1999, **59**, 1758–1775.
27 J. P. Perdew, K. Burke and M. Ernzerhof, *Phys. Rev. Lett.*, 1996, **77**, 3865–3868.
28 S. Grimme, J. Antony, S. Ehrlich and H. Krieg, *J. Chem. Phys.*, 2010, **132**, 154104.
29 J. Tersoff and D. R. Hamann, *Phys. Rev. B: Condens. Matter Mater. Phys.*, 1985, **31**, 805–813.
30 D. Vanpoucke, *HIVE STM-program» The Delocalized Physicist*, http://dannyvanpoucke.be/hive-stm-en/.
31 D. E. P. Vanpoucke and G. Brocks, *Phys. Rev. B: Condens. Matter Mater. Phys.*, 2008, **77**, 241308.
32 L. Leung, T. Lim, Z. Ning and J. C. Polanyi, *J. Am. Chem. Soc.*, 2012, **134**, 9320–9326.
33 A. Chatterjee, F. Cheng, L. Leung, M. Luo, Z. Ning and J. C. Polanyi, *J. Phys. Chem. C*, 2014, **118**, 25525–25533.
34 L. Leung, T. Lim, Z. Ning, J. C. Polanyi, W. Ji and C.-G. Wang, *J. Phys. Chem. C*, 2015, **119**, 26038–26045.
35 K. Anggara, K. Huang, L. Leung, A. Chatterjee, F. Cheng and J. C. Polanyi, *Nat. Commun.*, 2016, **7**, 13690.
36 P. M. Maitlis, H. C. Long, R. Quyoum, M. L. Turner and Z.-Q. Wang, *Chem. Commun.*, 1996, **1**, 1–8.
37 P. M. Maitlis and V. Zanotti, *Chem. Commun.*, 2009, **13**, 1619–1634.

Faraday Discussions

PAPER

Hot electron effects during reactive scattering of H_2 from Ag(111): assessing the sensitivity to initial conditions, coupling magnitude, and electronic temperature

Reinhard J. Maurer, †*[a] Yaolong Zhang,†[b] Hua Guo [c] and Bin Jiang *[b]

Received 1st October 2018, Accepted 30th October 2018
DOI: 10.1039/c8fd00140e

Low-lying electronic excitations in metals, so-called hot electrons, efficiently mediate molecule-metal energy transfer and contribute to energy loss during molecular reactions at surfaces. They furthermore play an important role in plasmon-driven chemistry. Electronic friction represents a simple and effective concept to model hot electron-induced energy loss under ambient conditions. Different methods exist that vary in their description of magnitude, coordinate and directional dependence of friction during reactive molecular scattering at metal surfaces. Using molecular dynamics simulations with electronic friction, we systematically study the effect of hot electrons on measurable state-to-state scattering probabilities of molecular hydrogen from a (111) surface of silver. We assess the ability of *ab initio* electronic friction methods to accurately describe hot electron-mediated energy loss as a function of initial reaction conditions and electronic temperature. We furthermore find that dynamic scattering results and the ensuing energy loss are highly sensitive to the magnitude of electronic friction. Therefore, existing approximate models of electronic friction, which exhibit inherent uncertainties with respect to the magnitude of electronic friction, may not be applicable for a quantitative prediction of plasmon driven hot electron effects in their current state. We outline a development direction to potentially overcome these limitations.

1 Introduction

Plasmonic science[1] and hot electron chemistry[2–4] are based on the idea of controlling the flow of energy between electronic excitations in metallic surfaces

[a]Department of Chemistry, Centre for Scientific Computing, University of Warwick, Gibbet Hill Road, Coventry, CV4 7AL, UK. E-mail: R.Maurer@warwick.ac.uk

[b]Hefei National Laboratory for Physical Science at the Microscale, Department of Chemical Physics, University of Science and Technology of China, Hefei, Anhui 230026, China. E-mail: bjiangch@ustc.edu.cn

[c]Department of Chemistry and Chemical Biology, University of New Mexico, Albuquerque, New Mexico 87131, USA

† These two authors contributed equally to this work.

or nanostructures,[5] so-called hot electrons, and molecular adsorbate degrees of freedom, such as molecular vibrations. A high level of control of this molecule–surface energy flow can facilitate chemical bond breaking events with light by funneling energy into the reaction coordinate or impair a reaction by efficiently dissipating energy away from a reaction centre.[6,7] The underlying mechanism for this energy exchange is based on the efficient coupling of molecular adsorbate motion and metallic electrons due to the vanishing band-gap in metallic systems.[8,9] The consequence of this coupling is a deviation from thermal chemistry as described by the Born–Oppenheimer approximation and a failure of existing methods to describe chemical reaction dynamics.[10,11]

The proposed technological applications of hot electron effects in plasmonic sensors,[12] plasmon-enhanced catalysts,[13] and locally-enhanced cancer radiation treatment,[14] are driven by exciting fundamental experiments that provide quantitative evidence for hot electron effects. These include molecular beam scattering experiments that explicitly measure the state-to-state scattering probability and the energy loss profile of a gas–surface reaction,[15] but also the measurement of electric currents,[11,16] so-called chemicurrents, that occur due to hot electrons that are generated upon adsorption, desorption, or chemical transformation of molecules.

The ability to computationally predict the molecular details that underlie hot electron-mediated processes and the intricate coupling of electrons and atomic motion is essential for the design of new devices and catalysts based on plasmonics. As such, it is imperative to develop first-principles computational simulation methods that correctly capture how molecular adsorbate motion generates hot electrons or how hot electrons affect measurable dynamical properties of chemical reactions. The intrinsic dimensionality of systems involving many hundreds of atoms and thousands of electrons make exact nonadiabatic quantum dynamics simulations close to unfeasible and therefore ask for efficient approximate mixed quantum-classical dynamics methods.[17,18] Such methods are typically based on stochastic surface hopping models based on chemical master equations, where hot electron effects on the reaction dynamics are modelled as explicit electronic transitions.[19,20] An alternative is the molecular dynamics with electronic friction (MDEF) method,[21–23] where the classical motion of nuclei follows a single potential energy surface and hot electron effects are described as electronic friction forces that act on the adsorbate nuclei. These forces are governed by the so-called electronic friction tensor,[24] which arises from nonadiabatic coupling and the resulting equations of motion follow a stochastic Langevin model. The electronic friction tensor is in general frequency-dependent and anisotropic but is often approximated to be frequency independent and spatially isotropic, *i.e.*, nonadiabatic relaxation rates are equal in all spatial directions for rigid atomic and molecular motion. The most common is the Local Density Friction Approximation (LDFA),[25,26] where nonadiabatic coupling is expressed as a simple function of the metallic electron density within which the adsorbate atom is embedded. This model is spatially isotropic for rigid molecular motion, but does account for some anisotropic effects along intramolecular degrees of freedom.[26,27] Due to its simplicity and ease of computation, LDFA-based MDEF is a particularly appealing approach that, in the past, has been applied to describe hot electron-mediated dynamics in ambient[28] and laser-driven conditions.[29,30] However, especially for molecular scattering and light-driven plasmonics, the underlying neglect of frequency dependence and the role of molecular anisotropy of electronic friction forces and their effects on measurable observables have been the reason for exciting debates in recent literature.[24,31–33]

In recent work, we have achieved a description of electronic friction that captures the directional and the explicit vibrational mode dependence and tensorial nature of electronic friction using an efficient implementation of first-order time-dependent perturbation theory (TDPT) based on Density Functional Theory (DFT).[24,31] On the examples of vibrational energy loss of metal-adsorbed diatomics[31] and the dissociative chemisorption of molecular hydrogen on Ag(111),[34] we were able to show that the overall magnitude of friction is, on average, similar to the LDFA prediction, but the directional and vibrational mode dependences and their effect on the dynamic energy loss are different. Before us, Spiering and Meyer have reached a similar conclusion for reactive scattering of H_2 on Cu(111),[35] where mode-dependent electronic friction forces as described by TDPT provide a measurable signature of energy loss during state-to-state scattering of vibrationally excited molecules. We have recently developed a general method to construct an analytical representation of the electronic friction tensor as a function of the adsorbate coordinates that enabled us to simulate dynamic scattering probabilities and to discuss the subtle differences in electronic friction that give rise to the observation that hot electron effects are much more pronounced during reactive scattering of H_2 on Cu(111) than on Ag(111).[36]

In this work, we use this analytical representation of electronic friction for H_2 on Ag(111) further to systematically assess the validity and robustness of the current frequency-independent MDEF method based on TDPT electronic friction. We do this by studying the sensitivity of dynamic observables such as the vibrational excitation/de-excitation probability during molecular scattering and the integrated energy loss along a specific degree of freedom as a function of various initial conditions and the type of electronic friction description. The former includes the effects of the initial and final vibrational excited states and the electronic temperature of the substrate. The latter includes a comparison between LDFA and TDPT and a discussion of their limitations and sensitivity to intrinsic numerical parameters. We find that the energy loss predicted by LDFA and TDPT shows different trends with respect to the parameters we studied and that dynamic observables are particularly sensitive to the absolute magnitude of tensorial friction. The intrinsic sensitivity of friction to numerical parameters such as the broadening width will represent a particular challenge for quantitative predictions of hot electron mediated reaction dynamics at high electronic temperatures that correspond to light-driven conditions. We therefore stress the importance of overcoming this limitation introduced by assuming frequency independence of the friction tensor. This will enable further study of the applicability of electronic friction approaches for light-driven hot electron chemistry.

2 Theory

2.1 Electronic friction based on time-dependent perturbation theory

In MDEF simulations, the nuclear dynamics are defined *via* a Langevin equation (LE):[21]

$$M\ddot{R}_i = -\frac{\partial V}{\partial R_i} - \sum_j \mathbf{\Lambda}_{ij}\dot{R}_j + \mathcal{R}_i. \quad (1)$$

Herein, the three force components are the conservative force due to the potential energy V, the electronic friction force given as a product between the electronic friction tensor $\mathbf{\Lambda}$ and the velocity of the atoms, and a temperature- and friction-dependent random force term that ensures detailed balance. The second and third terms describe the interaction of the adsorbate degrees of freedom with the bath degrees of freedom that represent the metal electrons. The two major assumptions to arrive at an LE are that the electron-nuclear coupling is weak and that it occurs instantaneously, *i.e.*, coupling due to electron–hole pair (EHP) excitations has no memory of previous times and no dependence on the perturbing frequency (Markov approximation). The first assumption is problematic in cases where molecule–metal electron transfer significantly changes the underlying potential energy surfaces, such as in the case of nitrous oxide on Au(111).[37,38] The second approximation is equivalent to assuming that the electron-nuclear coupling is constant regardless of nuclear velocities or vibrational frequencies. The result is that, within 1st order TDPT, we evaluate the electronic friction tensor using Fermi's golden rule:[23,31]

$$\mathbf{\Lambda}_{ij} = 2\pi\hbar \sum_{k,\nu,\nu'>\nu} \left\langle \psi_{k\nu} \left| \frac{\partial}{\partial R_i} \right| \psi_{k\nu'} \right\rangle \left\langle \psi_{k\nu'} \left| \frac{\partial}{\partial R_j} \right| \psi_{k\nu} \right\rangle \cdot \left(\varepsilon_{k\nu'} - \varepsilon_{k\nu}\right) \cdot \delta\left(\varepsilon_{k\nu'} - \varepsilon_{k\nu}\right). \quad (2)$$

Herein, the factor 2 accounts for spin multiplicity in the case of non-spin-polarised calculations and we have not included occupation factors that arise from finite temperature state populations. As our DFT calculations generate a finite number of states at discrete points $\boldsymbol{k}$ in momentum space, we need to interpolate to describe the friction tensor elements, which correspond to relaxation rates due to electron-nuclear coupling along the (mixed) Cartesian directions. This can be done in two ways, as follows. By choosing a broadening large enough to achieve convergence and small enough not to affect relaxation rates one approaches the zero-frequency limit, which is the formally correct limit within 1st order TDPT. It was recently shown that this level of description struggles to capture vibrational energy loss for dense molecular overlayers[33] and it does not correspond to the correct limit for gas–surface scattering with incidence energies above several hundred meV. As detailed in our previous work[34] and by Spiering and Meyer,[35] we choose the second approach, namely to replace the delta function with a normalized Gaussian function of a finite width (0.6 eV). With such a broadening, we effectively model the inclusion of higher energy EHPs. It has recently been pointed out that such broadening introduces contributions that are not rigorously contained in a zero-frequency 1st order approximation[32] and leads to relaxation rates that depend on the choice of broadening. This dependence remains a weak spot in this theory and has been discussed in detail in our previous work,[31] yet, *in lieu* of a feasible non-Markovian MDEF approach,[39] this pragmatic choice delivers a meaningful way to calculate relaxation rates beyond the LDFA[24,31,40,41] that correctly capture the directional and electronic structure dependence of the friction tensor. All elements of the friction tensor are calculated as a function of the adsorbate atom position using the all-electron, local atomic orbital code FHI-aims[42] and the Perdew–Burke–Ernzerhof (PBE) functional.[43] Our computational settings regarding model set-up, basis set, and Brillouin zone sampling have been detailed in a previous publication.[34]

2.2 Neural network representation of the friction tensor

2.2.1 Symmetry mapping scheme. To solve eqn (1) efficiently, one has to evaluate the potential energy and friction tensor at each trajectory step analytically. Neural network (NN) based representations of the PES[44] and the embedding electron density[45] (directly associated with the LDFA friction coefficient) which fully comply with the surface symmetry have been described previously. We thus focus here on the more complex representation of the anisotropic friction tensor computed by TDPT. Because the TDPT-based friction tensor is directionally-dependent on molecular coordinates, the symmetry equivalent molecular configurations that share the same potential energy and LDFA friction coefficient would have totally different but symmetrically-correlated friction tensor elements. This feature immediately renders the extension of any standard methodology for PES construction difficult for representing the friction tensor as a function of molecular geometry. Specifically, we discuss the case of a diatomic molecule (H_2) on a rigid metal surface, in which six coordinates are involved. These six coordinates can either be the Cartesian coordinates (x_1, y_1, z_1, x_2, y_2, z_2) or internal coordinates (X, Y, Z, r, θ, φ). In the latter, the first three specify the center of mass of the molecule while the rest denote the internuclear distance and the polar and azimuthal angles to the surface normal.

Spiering and Meyer have very recently reported an approximate NN-based representation of the 6 × 6 friction tensor for the H_2 + Cu(111) system.[35] They first described individual friction coefficients in terms of four symmetry independent internal coordinates, as illustrated in Fig. 1, namely X, Y, Z, and r with angular coordinates fixed (*e.g.*, $\theta_0 = \varphi_0 = 90°$). Since the tensorial friction ($\mathbf{\Lambda}$) is a symmetric matrix, only the upper triangular elements $\mathbf{\Lambda}_{ij}(i \leq j)$ are to be determined in practice. This treatment avoids the complex symmetry involved in the full six-dimensional representation. Once $\mathbf{\Lambda}(X, Y, Z, r)$ with $\theta_0 = \varphi_0 = 90°$ is obtained, the friction

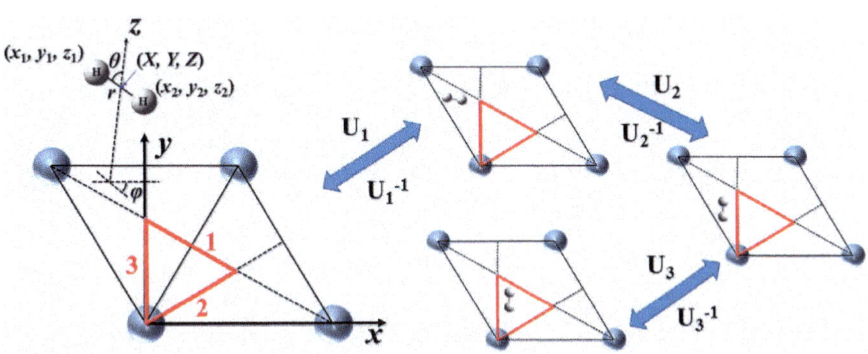

Fig. 1 Cartesian (x_1, y_1, z_1, x_2, y_2, z_2) and internal (X, Y, Z, r, θ, φ) coordinates to describe H_2 interacting with a rigid Ag(111) surface represented by a unit cell, with the symmetry unique triangle marked in red (left panel). Transformation matrices (U_1, U_2, U_3) and their inverses that move a molecular configuration outside the triangle into it and back to the original position, corresponding to the three reflection lines (1, 2, 3) that enclose the triangle.

tensor at an arbitrary configuration depending on all six internal coordinates can be approximately generated by the following transformation,[35]

$$\mathbf{\Lambda}(X, Y, Z, r, \theta, \varphi) \approx \mathbf{T}(\theta, \varphi)\mathbf{\Lambda}(X, Y, Z, r)\mathbf{T}^{-1}(\theta, \varphi), \tag{3}$$

where $\mathbf{T}(\theta, \varphi)$ is the transformation matrix that rotates configuration (θ, φ) to the corresponding reference geometry with $\theta_0 = \varphi_0 = 90°$. Although this decomposition of the full 6 × 6 friction tensor certainly depends on the choice of the reference orientation, Spiering and Meyer have shown that the dependence is relatively weak near the minimum energy path (MEP).

In a previous publication[36] and in this present work, we propose an alternative strategy. We choose to fit the elements of the full 6 × 6 matrix form of the electronic friction tensor in Cartesian coordinates inside an irreducible triangle in the (111) surface unit cell (marked in red in Fig. 1). In this triangle, each friction tensor element is symmetry unique with a one-to-one mapping to a molecular geometry, thus posing no problems for fitting. In practice, since we have collected many data scattered inside and out the unit cell, we first move those data distributed outside this triangle into it. This can be done by first translating a data point outside the unit cell into this region, which does not change the friction tensor, followed by a series of reflections when necessary.[46] For each reflection, the new Cartesian coordinate vector $\boldsymbol{q}'$ and friction tensor $\mathbf{\Lambda}'$ can be obtained by,

$$\boldsymbol{q}' = \mathbf{U}\mathbf{q}, \tag{4}$$

$$\mathbf{\Lambda}' = \mathbf{U}^{\mathrm{T}}\mathbf{\Lambda}\mathbf{U}, \tag{5}$$

where $\mathbf{q}$ and $\mathbf{\Lambda}$ are the original coordinate vector and friction tensor and $\mathbf{U}$ is the transformation matrix. For Ag(111), $\mathbf{U}_1$, $\mathbf{U}_2$ and $\mathbf{U}_3$ are given below and illustrated in Fig. 1,

$$\mathbf{U}_1 = \begin{bmatrix} 0.5 & -\sqrt{3}/2 & 0 & 0 & 0 & 0 \\ -\sqrt{3}/2 & -0.5 & 0 & 0 & 0 & 0 \\ 0 & 0 & 1 & 0 & 0 & 0 \\ 0 & 0 & 0 & 0.5 & -\sqrt{3}/2 & 0 \\ 0 & 0 & 0 & -\sqrt{3}/2 & -0.5 & 0 \\ 0 & 0 & 0 & 0 & 0 & 1 \end{bmatrix}, \tag{6}$$

$$\mathbf{U}_2 = \begin{bmatrix} 0.5 & \sqrt{3}/2 & 0 & 0 & 0 & 0 \\ \sqrt{3}/2 & -0.5 & 0 & 0 & 0 & 0 \\ 0 & 0 & 1 & 0 & 0 & 0 \\ 0 & 0 & 0 & 0.5 & \sqrt{3}/2 & 0 \\ 0 & 0 & 0 & \sqrt{3}/2 & -0.5 & 0 \\ 0 & 0 & 0 & 0 & 0 & 1 \end{bmatrix}, \tag{7}$$

$$\mathbf{U}_3 = \begin{bmatrix} -1 & 0 & 0 & 0 & 0 & 0 \\ 0 & 1 & 0 & 0 & 0 & 0 \\ 0 & 0 & 1 & 0 & 0 & 0 \\ 0 & 0 & 0 & -1 & 0 & 0 \\ 0 & 0 & 0 & 0 & 1 & 0 \\ 0 & 0 & 0 & 0 & 0 & 1 \end{bmatrix}. \tag{8}$$

In practice, the number of reflections operated on a given point may be less than three and, in the case that the molecule is already situated inside the triangle, no transformation is necessary. It should be noted that this treatment does not account for exchange symmetry of the two H atoms. We solve this issue by ordering the positions of the two hydrogen atoms when necessary (*i.e.* with $z_1 \leq z_2$) so that the related matrix elements in the friction tensor are exchanged accordingly. Similar mapping schemes have been successfully applied in constructing PESs for gas–surface systems using either NNs[47–49] or the modified Shepard interpolation.[46] One drawback of these PESs is that the gradients of PESs (*i.e.* forces) may not be continuous at the boundary, resulting in non-conservation of total energy when running classical trajectories. However, this is not a problem for the friction tensor as we only need values of friction coefficients but no gradients in eqn (1). Similar schemes can be devised for other facets of the metal surface.

2.2.2 **Fitting and evaluation.** After the mapping procedure, a single NN is utilized to describe the one-to-one mapping between the molecular geometry and each individual element of the friction tensor in the triangle. Six Cartesian coordinates of the molecule are set as the input layer of the NN, which pass geometric information to two hidden layers with 20 and 40 neurons *via* hyperbolic tangent activation functions. A hybrid algorithm combining the extreme learning machine and the Levenberg–Marquardt (ELM–LM) algorithm was used to train the NN.[50,51] The data set consists of ~1200 points sampled by previous AIMD trajectories[34] and ~2600 additional points used for constructing the PES[44] that cover the dynamically relevant region. Since the friction coefficients are very small far above the surface where the electron density vanishes, we exclude data points in that region. Instead, a switching function was imposed to smoothly dampen the friction coefficients to zero when the molecule–surface interaction is negligible ($Z \rightarrow \infty$, in Å),

$$f(Z) = \begin{cases} 1 & Z < 3.4 \\ 0.5 \times (\cos((Z - 3.4) \times \pi/0.6) + 1) & 3.4 \leq Z \leq 4.0 \\ 0 & Z > 4.0 \end{cases} \tag{9}$$

One should note that the NN fitting of each friction coefficient is essentially the same as that of the PES in previous works and very robust.[47–49] However, for our purpose, to evaluate a friction tensor for any point outside this symmetry-unique region, we need to first find the symmetry identical geometry of that point in the irreducible triangle by the transformation described above, then calculate the NN interpolated friction tensor there, followed by inversely transforming the obtained full 6 × 6 friction tensor back to that corresponding to the original geometry.

2.3 Molecular dynamics simulations with tensorial friction

The molecular dynamics (MD) and MDEF trajectories have been computed with various initial conditions sampled in a quasi-classical way as described in our previous publication.[36] The initial molecular center is chosen to be far above the Ag(111) surface ($Z = 8.0$ Å) with random lateral positions in the surface unit cell. The molecular internal coordinates and conjugate momenta are sampled semiclassically for given vibrational and rotational quantum numbers v and j.[52] The molecular orientation is randomly selected in the polar and azimuthal angles for most cases. For those m_j specific states, however, the molecular orientation is chosen randomly on a plane perpendicular to the angular momentum vector $\vec{j}$, which forms an angle of $\arccos(mj/\sqrt{j(j+1)})$ with the space-fixed Z axis (surface normal here). For the scattered molecule, the vibrational action number is determined by Einstein–Brillouin–Keller (EBK) semi-classical quantization and rotational quantum number j by the quantum mechanical expression for rotational angular momentum.[53] As in ref. 35, the standard histogram binning is used to obtain the state distributions with integer quantum numbers.

Given an initial condition, eqn (1) is propagated using third-order Beeman's algorithm with a time step of 0.1 fs. The MD and MDEF with LDFA simulations are easily implemented as friction coefficients only depend on the value of the electron density of the clean substrate.[25,28] The tensorial nature of the TDPT based electronic friction makes the equations of all coordinates coupled. In practice, we need to transform the coordinates and velocities into the friction eigenspace and apply the friction eigenvalues as damping coefficients onto the transformed velocities, and then transform the products back to the Cartesian space. This requires a diagonalization of the friction tensor at each time step. In addition, the random force term on the right-hand side of the Langevin equation, given a finite surface temperature (T), is described as a Gaussian white noise of zero mean and a standard deviation of $\sqrt{2\eta_i kT/\Delta t}$, where η_i is the ith friction eigenvalue (or simply the ith friction coefficient in LDFA) and k is the Boltzmann constant. However, we neglect surface temperature in most simulations because the contribution of the random force is typically small, except where we discuss temperature effects explicitly.

3 Results and discussion

3.1 Validation of the NN interpolation of the friction tensor

Based on our previously published analysis,[54] we have further examined the quality of our NN representation of the friction tensor. Fig. 2a shows the root mean square errors (RMSEs) averaged over all data for the 21 independent friction coefficients. The overall RMSEs uniformly range from $\sim$0.01 ps^{-1} to $\sim$0.03 ps^{-1} without any outliers, which are generally comparable or smaller than those reported for the H_2 + Cu(111) system.[35] In Fig. 2b, we compare several large components in the TDPT friction tensor with the NN interpolated values along the minimum energy path (MEP) for H_2 dissociative chemisorption, corresponding to the route for H_2 dissociation over the bridge site with the H–H bond parallel to the surface and stretching to two hollow sites. Clearly, there are strong couplings between the vibration (r) and translation (Z) directions, and between parallel motion (X/Y) and rotation (θ). Although the coupling terms can be both negative and positive, their absolute values increase as the molecule approaches the transition state. It has been previously shown that our NN diagonal tensor elements agree well with TDPT ones,[36] and we also find here excellent agreement between NN and TDPT off-diagonal friction coefficients. As a further validation, Fig. 2c compares two-dimensional electronic friction contour plots, as a function

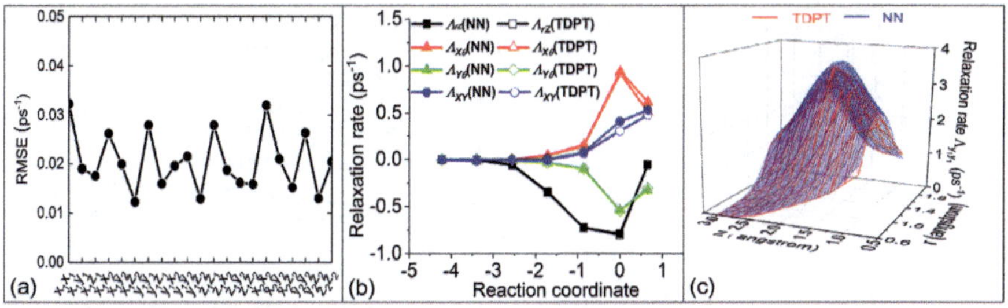

Fig. 2 (a) Root Mean Square Errors (RMSEs) of Neural Network (NN) fits for the 21 independent friction tensor elements with TDPT. (b) Comparison of several off-diagonal elements in the friction tensor with regard to internal coordinates obtained by NNs and TDPT along the minimum energy path (MEP). (c) Comparison of the original TDPT and NN interpolated $\Lambda_{y_1y_1}$ as a function of Z and r with the H_2 molecule dissociating with parallel orientation from the bridge site to two hollow sites.

of Z and r along the MEP, generated from NN interpolation and direct DFT calculations, respectively. Our NN representation faithfully captures the multidimensional topography of the friction tensor element, even though these data points were actually not included in the training set. These results indicate that the NN representation of the friction tensor is sufficiently accurate for subsequent MDEF simulations.

3.2 The physical differences between LDFA and TDPT

In our previous work,[36] we have shown different behaviors of the electronic friction in the H_2 + Cu(111) and H_2 + Ag(111) systems, which can be traced back to the differences in the underlying electronic structure of the two metals. Within the framework of the independent atom LDFA method, the friction coefficient depends only on the local electron density (ρ) at the position of the adsorbate atoms, whereas TDPT accounts for the response of the electronic wave functions (and therefore the nonadiabatic density response) of the combined molecule–surface system to the adsorbate motion. In Fig. 3a, we plot the clean surface density ρ varying with the position of one of the H atoms in the molecule along the MEP (the other H atom moves almost symmetrically to the opposite and hence feels nearly the same electron density). Because the transition state structures for H_2 dissociation on Cu(111) and Ag(111) appear at different distances from the surface, the comparison of ρ is best made as a function of the relative height with respect to that of the hydrogen atom at the respective transition state. Surprisingly, the embedded electron densities for H_2 scattering on Ag and Cu are very similar, resulting in very close values in LDFA friction coefficients. In the case of Fermi's golden rule, electronic friction does not depend on the density, but on the local density of states around the Fermi level and on the nonadiabatic coupling matrix elements that encode the difference in response along different directions, which sensitively depend on the underlying electronic structure and the nature of the metal. This level of difference is position and direction dependent and can be directly seen in Fig. 3b. In the following we will study how these different descriptions translate into measurable dynamic observables and how they depend on system parameters.

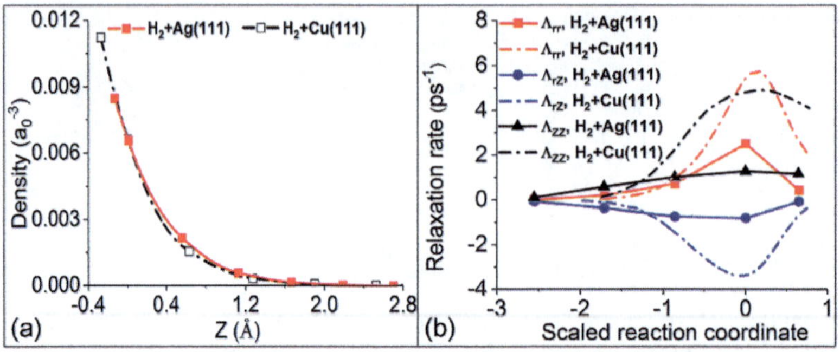

Fig. 3 (a) Comparison of the clean metal substrate electron density evaluated at the position of an H atom moving along the MEPs for dissociation on Cu(111) and Ag(111), as a function of the relative height with respect to the molecular center at the transition state (*i.e.* $Z_{TS} = 0$). (b) Comparison of the mode-specific electronic friction computed by TDPT along the MEPs for H_2 dissociation on Cu(111) and Ag(111).

3.3 The sensitivity of the vibrational deexcitation probability on the magnitude of friction

In our previous results on the vibrational deexcitation probabilities from H_2 ($\nu = 2, j = 0$) to H_2 ($\nu = 1, j = 0$) (*i.e.* $P_{2\to1}$) during H_2 scattering on Ag(111), we found that both TDPT and LDFA yield comparable deexcitation probabilities that are slightly elevated compared to the adiabatic MD case (see Fig. 4a). This is in contrast to findings for H_2 on Cu(111), where particularly the TDPT friction has been shown to lead to a significantly elevated deexcitation probability at intermediate translational incidence energies.[35] This suggests that dynamic observables are highly sensitive to the absolute magnitude of electronic friction. To test this sensitivity, we monitor the $P_{2\to1}$ varying with the electronic friction multiplied with a scaling factor for H_2 + Ag(111). We scale all tensor elements in Cartesian coordinates by the same factor. As shown in Fig. 4a, rather unsurprisingly, $P_{2\to1}$ increases for both TDPT and LDFA with increasing magnitude of electronic friction. When the strength of the TDPT friction tensor is doubled, corresponding to a similar magnitude as was found for the H_2 + Cu(111) system, the $P_{2\to1}$ with TDPT becomes much more significantly increased and sharply peaks at a translational energy (E_i) of 0.3 eV. At this specific E_i, we obtain a $P_{2\to1}$ with TDPT that is roughly 4 (3) times larger than the adiabatic (LDFA) results, in general accord with the observation in the H_2 + Cu(111) system. On the other hand, the doubly-increased LDFA friction coefficients also promote vibrational deexcitation at higher energies, yielding a greater enhancement of $P_{2\to1}$ at translational energies higher than 0.4 eV. The different dependence of $P_{2\to1}$ on translational energy with TDPT and LDFA can be understood by the corresponding mode-specific magnitude of electronic friction and nonadiabatic energy loss. As discussed in our previous work,[36] the LDFA electronic friction is dominated by the component along the translational coordinate, leading to almost linearly increased energy dissipation with increasing translational energy. This behavior becomes more remarkable when the strength of the friction coefficient is doubled, as shown in Fig. 4b, giving rise to the greater enhancement at high translational energies for LDFA. On the other hand, the fact that the energy loss

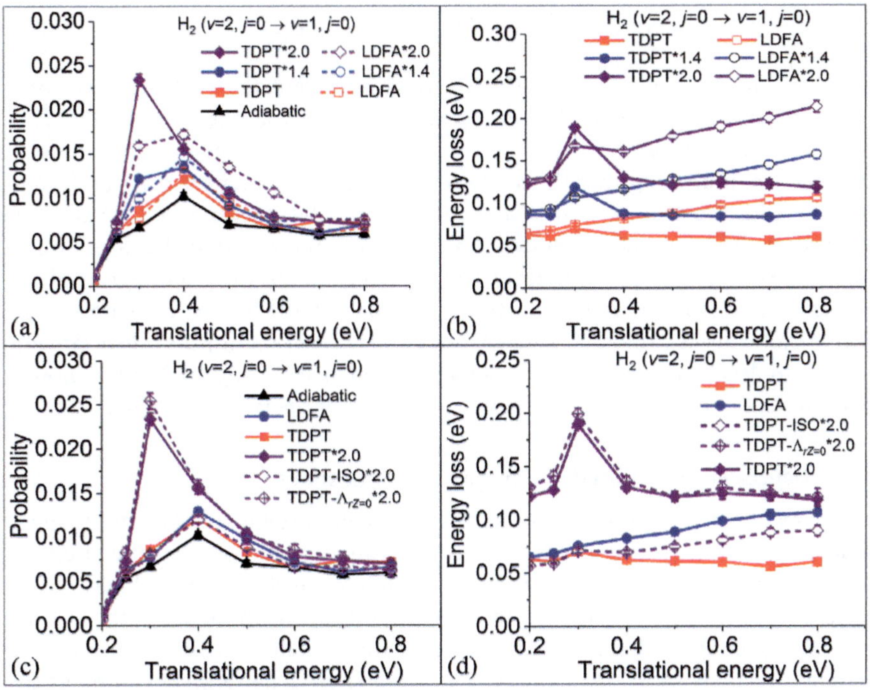

Fig. 4 (a) Vibrational deexcitation probabilities from H_2 ($v = 2, j = 0$) to H_2 ($v = 1, j = 0$) as a function of translational energy, with adiabatic dynamics (triangles), with electronic friction by LDFA (open symbols) and TDPT (solid symbols) models, and with 1.4 (circles) and 2.0 (diamonds) times larger friction coefficients. (b) Mean nonadiabatic energy losses for various simulations in panel (a). (c) On the basis of (a) but adding new data computed by removing anisotropy (TDPT-ISO*2) and the Z–r coupling term (TDPT-$\mathbf{\Lambda}_{rZ=0}$*2.0) in the TDPT friction tensor. (d) Mean nonadiabatic energy losses for various simulations in panel (b).

due to the TDPT electronic friction shows a peak at 0.3 eV is consistent with the same peak in the $P_{2\rightarrow1}$ distribution. This can be related to a balanced effect between the longer interaction time and smaller molecular velocity at this incidence energy.

Furthermore, it is interesting to analyse whether the nonadiabatic effects stem from the anisotropy along different directions or the mode-coupling between these directions in the friction tensor. In Fig. 4c, we compare the $P_{2\rightarrow1}$ with TDPT electronic friction where the off-diagonal coupling between the intra-atomic coordinate and the molecule-adsorbate coordinate (TDPT) is included and where it is artificially set to zero (TDPT-$\mathbf{\Lambda}_{rZ=0}$). To make the comparison clearer, a factor of two is also multiplied in both cases. Although the magnitude of $\mathbf{\Lambda}_{rZ}$ is not small as displayed in Fig. 2b, interestingly, the results with zero $\mathbf{\Lambda}_{rZ}$ are quite close to the original ones in the entire energy range, implying the negligible effect of the mode-coupling between the translational and vibrational DOFs in this case. In addition, to study the importance of directional anisotropy, we average the three diagonal components of the TDPT friction tensor (twice in magnitude) in the Cartesian space for each atom, then assign the same friction coefficient for each atomic Cartesian coordinate in the following way,

$$\Lambda_{x_ix_i}^{\mathrm{ISO}} = \Lambda_{y_iy_i}^{\mathrm{ISO}} = \Lambda_{z_iz_i}^{\mathrm{ISO}} = \frac{1}{3}\left(\Lambda_{x_ix_i} + \Lambda_{y_iy_i} + \Lambda_{z_iz_i}\right), \quad i = 1,\ 2, \tag{10}$$

and zero all off-diagonal terms (TDPT-ISO). This converts the TDPT friction tensor to a form comparable with LDFA. It is found in Fig. 4c that the calculated $P_{2\to1}$ in this condition becomes drastically decreased, but very similar to the unscaled LDFA ones, indicating that the anisotropy of the friction tensor plays an important role in this process. This is consistent with what Meyer and Spiering have found for the H_2 + Cu(111) system[35] and hints to the fact that the larger translational energy loss as predicted by LDFA really stems from its neglect of the directional anisotropy of the electronic friction tensor. It should be noted that while statically isotropic, the LDFA (much as TDPT) description does provide a certain level of dynamic directional dependence during molecular motion within MDEF due to the directional dependence of the velocity profile.

3.4 Dependence on initial conditions and final states

Our previous analyses suggested that the velocity profile and potential energy landscape also play a significant role in determining the nonadiabatic effects in the molecular scattering process.[36] One can of course change the velocity profile by choosing various initial conditions. We first study the influence of the angular velocity by performing MDEF simulations with initially different rotational and orientational states at the same ($\nu = 2$) vibrational state. Fig. 5a compares the vibrational deexcitation probabilities for H_2 ($\nu = 2, j = 5, m_j = 0/5$) to H_2 ($\nu = 1, j = 5$) with and without electronic friction. These two initial states correspond to the cartwheel ($m_j = 0$) and helicopter ($m_j = j$) orientations, respectively, which would impact the surface with similar vibrational and translational velocities but distinctly different angular velocities for a given incidence energy. Compared to previous results for the rotational ground state, the nonadiabatic energy dissipation is more important for these two states, resulting in up to an ~80% increase of the vibrational deexcitation probability at low translational energies. Both friction methods provide similar results with small EHP-induced enhancement for the cartwheel motion and larger effects for the helicopter motion. The latter can be understood in terms of an increased residence time of the molecule and rovibrational coupling during the scattering process. Our findings show that, despite some rotational state dependence of friction, there is little change of the nonadiabatic effects on $P_{2\to1}$ among different product rotational states (partly shown in ref. 36), suggesting that the angular velocities only play a minor role in such cases.

Following this line of thought, we show the excitation probability ($P_{2\to3}$) from H_2 ($\nu = 2, j = 0$) to H_2 ($\nu = 3, j = 0$) in Fig. 5b and deexcitation probability ($P_{3\to2}$) from H_2 ($\nu = 3, j = 2$) to H_2 ($\nu = 2, j = 2$) in Fig. 5c. Both scattering processes involve a higher vibrational state which naturally increases the molecular velocity with respect to the intramolecular stretching mode. We see much stronger nonadiabatic effects in both cases, leading to the significant decrease of $P_{2\to3}$ and increase of $P_{3\to2}$ due to electronic friction at low translational energies. In such cases, the translational velocity is low enough for the molecule to remain close to the surface for an extended time, and the vibrational velocity is high enough to cause electronic excitations. While the results with TDPT and LDFA are very similar with respect to $P_{2\to3}$, the former become qualitatively different from the

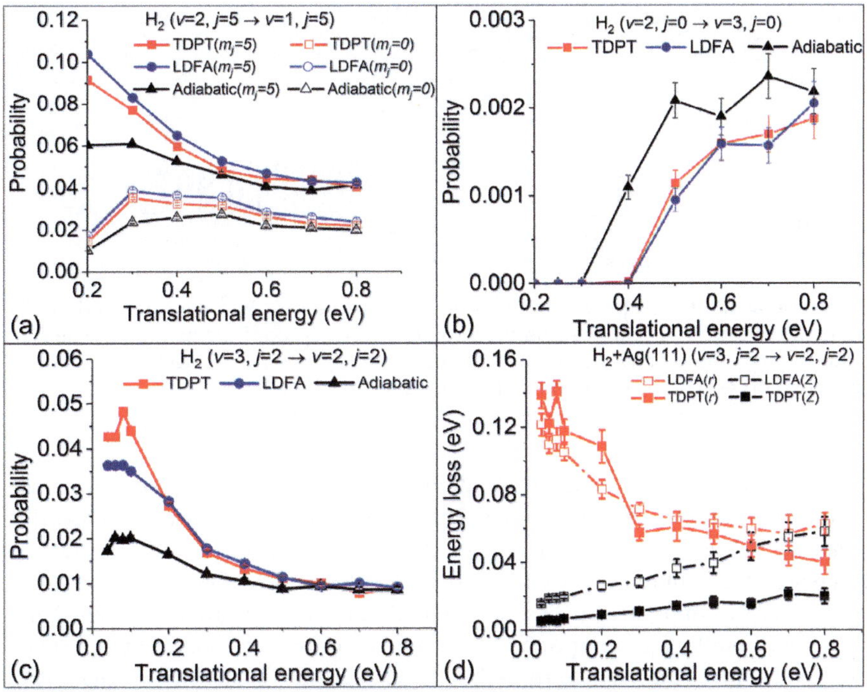

Fig. 5 (a) Vibrational state-to-state deexcitation probabilities from H_2 ($v = 2, j = 5$) to H_2 ($v = 1, j = 5$) for rotationally excited initial conditions as predicted with MD, MDEF(LDFA), and MDEF(TDPT). (b) Vibrational excitation probabilities upon scattering. (c) Vibrational deexcitation probability from H_2 ($v = 3, j = 2$) to H_2 ($v = 2, j = 2$). (d) Energy loss during vibrational deexcitation from H_2 ($v = 3, j = 2$) to H_2 ($v = 2, j = 2$).

latter in $P_{3\to2}$ below $E_i = 0.2$ eV. This interesting behavior is reflected by the nonadiabatic energy loss given in Fig. 5d, where the TDPT induced energy losses along the molecular vibration (r direction) are more significant than the LDFA ones in that range. Such a large vibrational energy loss with TDPT at low energy seems to be surprising given the similar magnitude of $\mathbf{\Lambda}_{rr}$ with TDPT and LDFA. We therefore monitor the velocity and friction coefficient profiles in the Z and r directions as a function of time for an exemplary trajectory at $E_i = 0.1$ eV in Fig. 6a and b. It appears that $\mathbf{\Lambda}_{rr}$ with TDPT is remarkably higher than those with LDFA in the strong interaction region, implying that the trajectory significantly departs from the MEP. Unlike our previous results (Fig. 4d in ref. 36) for H_2 ($\nu = 2 \to \nu = 1$), the H_2 ($\nu = 3$) molecules do not necessarily follow the MEP. As a result, the multidimensional topography of the friction tensor determines the total nonadiabatic energy loss and thus the scattering probability. This reflects the importance of correctly capturing this tensorial topology.

3.5 Electronic temperature

Within the electronic friction picture, plasmon driven or laser-heated hot electron effects are typically simulated *via* an increase of the electronic temperature that defines the heat of the electronic bath and the strength of the random force contribution. This can be done in the form of a constant elevated temperature or

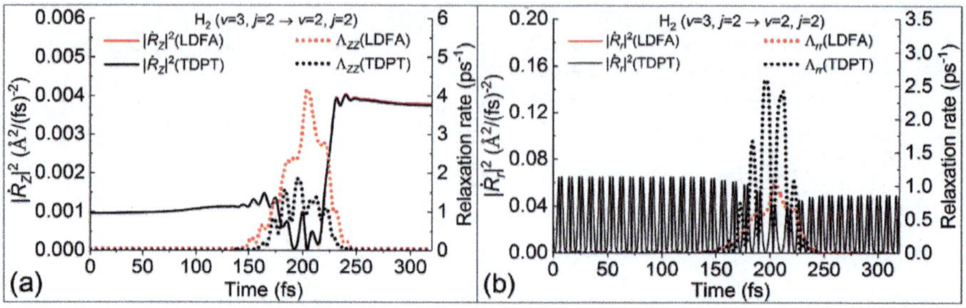

Fig. 6 Variation of the square of velocity (solid lines) in the Z ($|\dot{R}_Z|^2$, panel (a)) and r ($|\dot{R}_r|^2$, panel (b)) directions along with the corresponding friction coefficients (dotted lines), as a function of time for a representative trajectory for H_2 ($v = 3, j = 2$) to H_2 ($v = 2, j = 2$) scattering on Ag(111), at a translational energy of 0.1 eV.

by simulating a time-dependent temperature profile that is coupled with a two-temperature model of the coupled lattice-electron dynamics of the metallic substrate.[29,30] Electronic temperatures that arise from laser-heating a surface can reach transient temperatures up to several thousands of Kelvin. It should be noted that this approach will assume that all hot electrons are thermalised and the weak coupling assumption will neglect any possible resonant electron-nuclear coupling effects due to high-lying excitations. Furthermore, the use of a temperature ramp is not strictly valid within an equilibrium Langevin theory and corresponding implications on detailed balance need to be carefully considered. In previous work, we studied the effect of including electronic temperature in eqn (2) and found that over a range of up to 6000 K, electronic relaxation rates within our currently employed level of theory depend very weakly on temperature.[31] We therefore describe the electronic friction tensor as temperature independent. However, within the dynamic formalism of eqn (1) temperature still enters the Langevin expression and it is interesting to study how dynamic state-to-state scattering is affected by varying the electronic bath temperature. Whereas previous sections studied scattering at low temperature conditions and neglected the random force, Fig. 7 shows the dependence of vibrational deexcitation as a function of different electronic temperatures (0 K, 1000 K, and 6000 K). We find that, while overall deexcitation probabilities do not

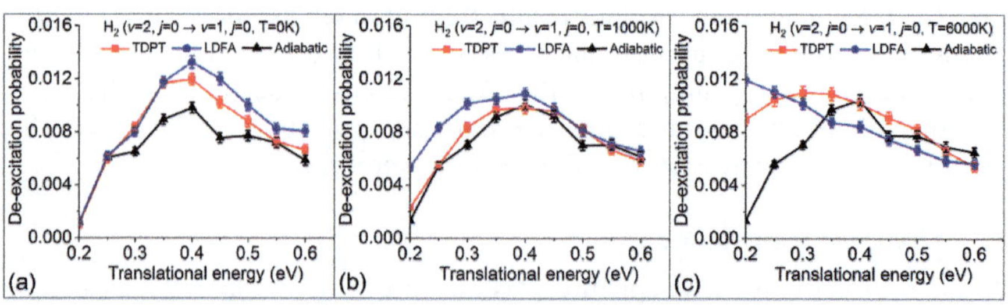

Fig. 7 Vibrational deexcitation probabilities from H_2 ($v = 2, j = 0$) to H_2 ($v = 1, j = 0$) simulated by MD, MDEF(LDFA), and MDEF(TDPT), with electronic temperatures of 0 K (a), 1000 K (b), and 6000 K (c).

significantly change for both friction models, higher temperatures lead to higher integrated energy loss and increased deexcitation probabilities for very low translational energies. For the LDFA method, this effect becomes relevant at lower temperatures than for the TDPT method. This can be related to the higher average friction magnitude of the LDFA method. The increase of deexcitation probability at low translational energies can be directly understood by the increased residence time of molecules close to the surface and the increased exposure to random force perturbations. From our results, it appears that state-to-state scattering is not strongly affected by plasmonic heating. This will, however, be different for rates of reactions such as recombinative desorption and dissociation reactions with high barriers and physisorbed precursor states where surface residence times are much larger.

4 Conclusions

We have presented extensive MDEF simulations within the Markov approximation for the state-to-state scattering of molecular hydrogen from a Ag(111) surface. These simulations are enabled by the construction of neural-network based potential energy surfaces and an electronic friction tensor with data calculated from density functional theory and 1st order TDPT. We find that nonadiabatic dynamic state-to-state scattering is strongly dependent on the absolute magnitude of friction with respect to the relevant internal coordinate, whereas coupling between internal coordinates does not appear to play a significant role. Different vibrational and rotational states are affected very differently by electronic friction, however, the two different friction models did not yield drastically different results. The rotational state dependence hereby is mostly due to the changed molecular orientation and particular trajectory route with respect to the surface during impingement that boost vibrational or translational energy loss. We have furthermore shown that the inclusion of elevated electronic temperatures as they appear in plasmon-driven hot electron chemistry dominantly affects reaction probabilities at low incidence energy, where surface residence times are larger than at high incidence energies. In particular the latter finding suggests that plasmon-driven MDEF studies with electronic friction directly computed from DFT orbitals are possible and could provide a basis for further systematic improvement beyond the current state-of-the-art represented by LDFA.[29,30] As guided by recent work of Novko *et al.*,[33] a word of warning on the Markov and zero-frequency approximation is in order. Our methodology is numerically robust but suffers from the choice of a broadening parameter that, despite a weak dependence in the chosen regime, does introduce some variation of frictional magnitude. Our results have shown that this dependence can strongly affect the dynamical scattering results on a level that would, at best, allow a qualitative prediction of reaction rate trends for hot electron chemistry. We conclude from this that, to investigate the ability of MDEF to act as a predictive tool for plasmonic science, the Markov approximation needs to be overcome. This endeavor is currently under way.

Conflicts of interest

There are no conflicts to declare.

Acknowledgements

YZ and BJ acknowledge support by the National Key R&D Program of China (2017YFA0303500), the National Natural Science Foundation of China (91645202, 21722306, and 21573203), the Anhui Initiative in Quantum Information Technologies, and computational resources offered by the Supercomputing Center of USTC and Advanced Materials High-Performance Company. RJM acknowledges computational resources provided by the University of Warwick Scientific Computing Research Technology Platform, and the EPSRC-funded HPC Midlands Plus centre (EP/P020232/1). HG is supported by the U.S. National Science Foundation (Grant No. CHE-1462109).

References

1 M. L. Brongersma, N. J. Halas and P. Nordlander, *Nat. Nanotechnol.*, 2015, **10**, 25–34.
2 W. Xie and S. Schlücker, *Nat. Commun.*, 2015, **6**, 7570.
3 J. Y. Park, S. W. Lee, C. Lee and H. Lee, *Catal. Lett.*, 2017, **147**, 1851–1860.
4 J. Y. Park, S. M. Kim, H. Lee and I. I. Nedrygailov, *Acc. Chem. Res.*, 2015, **48**, 2475–2483.
5 S. D. Forno, L. Ranno and J. Lischner, *J. Phys. Chem. C*, 2018, **122**, 8517–8527.
6 E. Cortés, W. Xie, J. Cambiasso, A. S. Jermyn, R. Sundararaman, P. Narang, S. Schlücker and S. A. Maier, *Nat. Commun.*, 2017, **8**, 14880.
7 S. Mukherjee, F. Libisch, N. Large, O. Neumann, L. V. Brown, J. Cheng, J. B. Lassiter, E. A. Carter, P. Nordlander and N. J. Halas, *Nano Lett.*, 2013, **13**, 240–247.
8 A. M. Wodtke, D. Matsiev and D. J. Auerbach, *Prog. Surf. Sci.*, 2008, **83**, 167–214.
9 S. P. Rittmeyer, V. J. Bukas and K. Reuter, *Adv. Phys.: X*, 2018, **3**, 1381574.
10 A. M. Wodtke, J. C. Tully and D. J. Auerbach, *Int. Rev. Phys. Chem.*, 2004, **23**, 513–539.
11 A. Kandratsenka, H. Jiang, Y. Dorenkamp, S. M. Janke, M. Kammler, A. M. Wodtke and O. Bünermann, *Proc. Natl. Acad. Sci. U. S. A.*, 2018, **115**, 680–684.
12 D. Sil, K. D. Gilroy, A. Niaux, A. Boulesbaa, S. Neretina and E. Borguet, *ACS Nano*, 2014, **8**, 7755–7762.
13 P. Christopher, H. Xin and S. Linic, *Nat. Chem.*, 2011, **3**, 467–472.
14 X. Huang, P. K. Jain, I. H. El-Sayed and M. A. El-Sayed, *Nanomedicine*, 2007, **2**, 681–693.
15 O. Bunermann, H. Jiang, Y. Dorenkamp, A. Kandratsenka, S. M. Janke, D. J. Auerbach and A. M. Wodtke, *Science*, 2015, **350**, 1346–1349.
16 H. Nienhaus, H. S. Bergh, B. Gergen, A. Majumdar, W. H. Weinberg and E. W. McFarland, *Phys. Rev. Lett.*, 1999, **82**, 446–449.
17 J. C. Tully, *J. Chem. Phys.*, 2012, **137**, 22A301.
18 Y. Zhang, T. R. Nelson, S. Tretiak, H. Guo and G. C. Schatz, *ACS Nano*, 2018, **12**, 8415–8422.
19 N. Shenvi, S. Roy and J. C. Tully, *J. Chem. Phys.*, 2009, **130**, 174107.
20 H. Guo, P. Saalfrank and T. Seideman, *Prog. Surf. Sci.*, 1999, **62**, 239–303.
21 M. Head Gordon and J. C. Tully, *J. Chem. Phys.*, 1995, **103**, 10137–10145.
22 W. Dou and J. E. Subotnik, *J. Chem. Phys.*, 2018, **148**, 230901.

23 B. Hellsing and M. Persson, *Phys. Scr.*, 1984, **29**, 360–371.

24 M. Askerka, R. J. Maurer, V. S. Batista and J. C. Tully, *Phys. Rev. Lett.*, 2016, **116**, 217601.

25 J. Juaristi, M. Alducin, R. Muiño, H. Busnengo and A. Salin, *Phys. Rev. Lett.*, 2008, **100**, 116102.

26 D. Novko, M. Blanco-Rey, M. Alducin and J. I. Juaristi, *Phys. Rev. B*, 2016, **93**, 245435.

27 S. P. Rittmeyer, J. Meyer and K. Reuter, *Phys. Rev. Lett.*, 2017, **119**, 176808.

28 M. Blanco-Rey, J. I. Juaristi, R. Díez Muiño, H. F. Busnengo, G. J. Kroes and M. Alducin, *Phys. Rev. Lett.*, 2014, **112**, 103203.

29 J. I. Juaristi, M. Alducin and P. Saalfrank, *Phys. Rev. B*, 2017, **95**, 125439.

30 I. Lončarić, M. Alducin, P. Saalfrank and J. I. Juaristi, *Phys. Rev. B*, 2016, **93**, 014301.

31 R. J. Maurer, M. Askerka, V. S. Batista and J. C. Tully, *Phys. Rev. B*, 2016, **94**, 115432.

32 D. Novko, M. Alducin, M. Blanco-Rey and J. I. Juaristi, *Phys. Rev. B*, 2016, **94**, 224306.

33 D. Novko, M. Alducin and J. I. Juaristi, *Phys. Rev. Lett.*, 2018, **120**, 156804.

34 R. J. Maurer, B. Jiang, H. Guo and J. C. Tully, *Phys. Rev. Lett.*, 2017, **118**, 256001.

35 P. Spiering and J. Meyer, *J. Phys. Chem. Lett.*, 2018, **9**, 1803–1808.

36 Y. Zhang, R. J. Maurer, H. Guo and B. Jiang, *Chem. Sci.*, 2019, **10**, 1089–1097.

37 S. Roy, N. A. Shenvi and J. C. Tully, *J. Chem. Phys.*, 2009, **130**, 174716.

38 N. Shenvi, S. Roy and J. C. Tully, *Science*, 2009, **326**, 829–832.

39 T. Olsen and J. Schiøtz, *J. Chem. Phys.*, 2010, **133**, 134109.

40 M. Forsblom and M. Persson, *J. Chem. Phys.*, 2007, **127**, 154303.

41 A. C. Luntz and M. Persson, *J. Chem. Phys.*, 2005, **123**, 074704.

42 V. Blum, R. Gehrke, F. Hanke, P. Havu, V. Havu, X. Ren, K. Reuter and M. Scheffler, *Comput. Phys. Commun.*, 2009, **180**, 2175–2196.

43 J. P. Perdew, K. Burke and M. Ernzerhof, *Phys. Rev. Lett.*, 1996, **77**, 3865–3868.

44 B. Jiang and H. Guo, *Phys. Chem. Chem. Phys.*, 2014, **16**, 24704–24715.

45 B. Jiang, M. Alducin and H. Guo, *J. Phys. Chem. Lett.*, 2016, **7**, 327–331.

46 C. Crespos, M. A. Collins, E. Pijper and G. J. Kroes, *J. Chem. Phys.*, 2004, **120**, 2392–2404.

47 T. Liu, B. Fu and D. H. Zhang, *Sci. China: Chem.*, 2014, **57**, 147–155.

48 T. Liu, Z. Zhang, B. Fu, X. Yang and D. H. Zhang, *Chem. Sci.*, 2016, **7**, 1840–1845.

49 X. Shen, J. Chen, Z. Zhang, K. Shao and D. H. Zhang, *J. Chem. Phys.*, 2015, **143**, 144701.

50 Y. Zhang, X. Zhou and B. Jiang, *Chin. J. Chem. Phys.*, 2017, **30**, 727–734.

51 M. T. Hagan and M. B. Menhaj, *IEEE Trans. Neural Network.*, 1994, **5**, 989–993.

52 X. Hu, W. L. Hase and T. Pirraglia, *J. Comput. Chem.*, 1991, **12**, 1014–1024.

53 M. C. Gutzwiller, *Chaos in classical and quantum mechanics*, Springer, New York, 1990.

54 P. Shi, Y. Yang, B. Ao, P. Zhang and X. Wang, *J. Phys. Chem. C*, 2014, **118**, 26634–26640.

Faraday Discussions

DISCUSSIONS

Dynamics of hot electron generation in metallic nanostructures: general discussion

Javier Aizpurua, Michael Ashfold, Francesca Baletto, Jeremy Baumberg, Phillip Christopher, Emiliano Cortés, Bart de Nijs, Yuri Diaz Fernandez, Julian Gargiulo, Sylwester Gawinkowski, Naomi Halas, Ruben Hamans, Bartlomiej Jankiewicz, Jacob Khurgin, Priyank Vijaya Kumar, Jie Liu, Stefan Maier, Reinhard J. Maurer, A Mount, Niclas Sven Mueller, Rupert Oulton, Matteo Parente, Jeong Y. Park, John Polanyi, Jhon Quiroz, Sebastian Rejman, Sebastian Schlücker, Zachary Schultz, Yonatan Sivan, Giulia Tagliabue, Madasamy Thangamuthu, Laura Torrente-Murciano, Xiaofei Xiao, Anatoly Zayats and Chao Zhan

DOI: 10.1039/C9FD90011J

Yonatan Sivan opened discussion of the introductory lecture by Naomi J. Halas: I have 3 comments, each addressing one of the key papers you presented.

(1) My first comment is regarding the theory paper.[1] In that paper, the authors calculated and plotted the electron excitation rate due to photon absorption, but claimed that it is actually the distribution itself. This is in principle not correct. Secondly, the calculation involved only the electron system, such that the energy flow into the phonons and environment was not correctly accounted for; specifically, this approach does not allow one to adjust the rate of heat flow from the phonons to the environment according to the particle size and shape. This leads, as we have shown in our paper (DOI: 10.1039/c8fd00147b), to quantitative errors in the electron distribution close to the Fermi energy. However, as we have also shown in our paper, a quantitative prediction for the distribution itself can effectively be obtained by multiplying the electron excitation rate by the e–e collision time. However, not only did the authors not do that, but they also did not plot the results on a logarithmic scale. Hence, it is impossible to know what the high energy electron population looks like, even qualitatively.

(2) As we showed during the poster session, my own presentation and in a paper we have submitted for publication (available on ArXiv[4]), the temperature measurement in the paper by Mukherjee *et al.*[2] was carried out far from where the reaction took place, thus underestimating the temperature rise and, therefore, the thermal contribution to the reaction. This means that the control experiment performed in that work was unfortunately meaningless. Furthermore, an Arrhenius fit to the data shows that the authors' claim that the activation energy in the

presence of Au nanoparticles is 1.8eV is incorrect; it also shows that all the data can be fit perfectly with the correct, intensity-independent activation-energy measured in the dark and with a correct evaluation of the reaction temperature.

(3) Finally, as we also showed during the poster session, my own presentation and in an accepted paper,[5] for the temperature measurement in the paper by Zhou *et al.*,[3] the thermal camera seemed to have been operated under the default settings, such that the temperatures were again underestimated. As we explained,[4,5] even if the temperatures were measured correctly, any slight difference between the pure thermal control (*i.e.* the data obtained in the dark) and the thermal contribution to the reaction in the photocatalysis experiment would be incorrectly interpreted as coming from the non-thermal (*i.e.* 'hot') electrons. In practice, not only are the spatial distributions of the temperature in these experiments quite different,[4,5] but also the exponential sensitivity of the reaction to the temperature amplifies these differences dramatically. Consequently, the conclusion drawn by the authors about the dominance of 'hot' electron effects over regular thermal effects is not supported by the obtained data.

1 A. Manjavacas, J. G. Liu, V. Kulkarni and P. Nordlander, *ACS Nano*, 2014, **8**, 7630–7638.
2 S. Mukherjee, F. Libisch, N. Large, O. Neumann, L. V. Brown, J. Cheng, J. Britt Lassiter, E. A. Carter, P. Nordlander and N. J. Halas, *Nano Lett.*, 2013, **13**, 240–247.
3 L. Zhou, D. F. Swearer, C. Zhang, H. Robatjazi, H. Zhao, L. Henderson, L. Dong, P. Christopher, E. A. Carter, P. Nordlander and N. J. Halas, *Science*, 2018, **362**, 69–72.
4 Y. Sivan, I. W. Un and Y. Dubi, 2019, arXiv:1902.03169 [physics.chem-ph].
5 Y. Sivan, J. Baraban, I. W. Un and Y. Dubi, Comment on "Quantifying hot carrier and thermal contributions in plasmonic photocatalysis", *Science*, 2019, **364**, eaaw9367.

Reinhard J. Maurer enquired: In your talk, you mentioned that plasmon-driven ammonia decomposition benefits from a reduced rate of ammonia desorption in comparison to thermal decomposition. I wonder how this relates to photo-induced ammonia desorption on copper, as studied by Gerhard Ertl and co-workers?[1] This is a process that is also understood to be driven by hot electrons in the metal and will contribute to desorption in the case of plasmonic catalysis.

1 T. Hertel, M. Wolf and G. Ertl, *J. Chem. Phys.*, 1995, **102**, 3414.

Naomi Halas replied: Yes, this work is very likely quite similar in that the process relies on desorption induced by electron transitions. However there are substantial differences in that one cannot generate hot electron populations directly on a bulk metal substrate sample as one can by optical illumination of subwavelength metal structures.

Priyank Vijaya Kumar asked: In comparison to an aluminium catalyst, does an aluminium–palladium catalyst produce higher HD reaction rates in your experiments? Do you have an estimate for this value?

Naomi Halas responded: Yes it most certainly does, by about 2 orders of magnitude. That is not clear in our presentation of the data because we focused instead on the difference in wavelength dependence of the HD yield. However, it isn't even close, because H_2 and D_2 dissociate spontaneously on Pd and the hot electrons in that system desorb the dissociated species; this was confirmed by E. A. Carter and her group.

Stefan Maier queried: What are the main design criteria for plasmonic catalysts? If you want to design an efficient plasmonic catalyst for a given reaction where do you start?

Naomi Halas answered: If one takes the antenna–reactor strategy, which is modular, then one has many options regarding reactor sites for binding reactants to the complex, as well as antenna tunability to address binding energies for bond dissociation. The weak link here is theory; there are only a few groups that address these problems in adsorbate levels and binding energies quantitatively.

Chao Zhan remarked: In the demonstration of antenna–reactor complexes, Pd is used as the reactor. However, as we know, Pd itself has a plasmonic effect. Can it be used to prove that it is the nanoantenna effect? Maybe it is required to extend the 'reactor' to another metal or metal oxide that does not have SP properties. Confirmation is also needed that there is no charge transfer between the antenna and reactor.

Naomi Halas replied: Yes, Pd has a weak plasmonic response but it does certainly have one. In the geometry we used, the antenna NP was much larger than the Pd NPs. The antenna NP already has a larger cross section and, with a larger size relative to the Pd, its optical properties dominate those of Pd. Here we take advantage of the ability of Pd for the barrierless dissociation of H_2 on its surface. We have also extended this general concept to other types of reactors, such as metal oxide layers. In the case of Al NPs, they have a 2–4 nm thick oxide layer and therefore it is highly improbable that there is any charge transfer between the antenna and the reactors.

Jhon Quiroz communicated: My question regards the antenna–reactor photocatalyst design. If I understood well the first example concerning the Al–Pd antenna reactor, both the plasmonic and catalytic components are not in direct contact. However, in the second example concerning the Cu–Ru photocatalyst intimate contact is suggested, producing an alloy. Could you please shed light on which design is more efficient or promising?

Naomi Halas communicated in reply: That is a very good question. They both have their merits. Because the reactive sites of Pd and Ru are different and because we have not tried them out on the same reaction we do not have a direct comparison. Since the reactor sites are different for the two complexes I don't think there is a way to compare them. However, if you were to synthesize both types of antenna–reactor complexes with the same reactor sites, one could make a direct comparison. Still, the properties of the specific reaction one ran would be very important in distinguishing the two A–R types from one another and assessing how one may perform better than the other. One thing is clear however, and that is that the Al-based A–R complex has a tunable plasmon resonance and the Cu plasmon is not nearly as tunable due to the more complex electronic structure of Cu and how this effects the plasmon resonant energy in this material.

Javier Aizpurua opened discussion of the paper by Jacob B. Khurgin: Your paper is very nice, and very helpful for obtaining the numbers and orders of magnitude of the rates of decay of hot electrons due to the different mechanisms. Most of the estimations are made for highly symmetric and homogenous bulks and surfaces, where you provide a nice description of each scattering mechanism, for example of a slab or a spherical particle. However, when one comes to realistic surfaces, these can show bumps, rough areas, or even atomic-scale protrusions, which can cause surface scattering processes to become more complicated. What happens when we have less-symmetric systems and angstrom-scale nano-features? How can we develop similar descriptions for systems of a few nanometers in size, or beyond these symmetric systems, in order to find out what the hot electrons do in those more extreme systems?

Jacob Khurgin responded: Thank you for this great question. I think that this theory stands for structures that have features larger than $1/\Delta k$ or v_{Fermi}/ω, which corresponds to a couple of nanometers. You can simply find the field and the emission will more or less follow the strength of the field normal to the surface. However, for smaller features this all starts breaking down and so probably the only correct way to handle this is to use a full atomistic model. A comparative study of two models would be great! However the general conclusion—that smaller features help—will stand; in the end all we need is to violate (relax) momentum conservation by all means possible.[1]

1 M. Grajower, U. Levy and J. B Khurgin, *ACS Photonics*, 2018, **5**, 4030–4036.

Jeremy Baumberg remarked: Your paper describes a very nice picture, and is very helpful. Since it is semi-classical, I wonder whether you have considered what happens in surfaces with sharp surface asperities such as vertices, edges or bumps (adatoms). It would seem that you can cast the theory to look at electron focussing (I suspect they will be beamed normal to the surface). Your theory also suggests that these are the places where hot electron emission would be strongest, since the field penetration is strongest. This would suggest that this semi-classical picture for catalysis also implies that there are active sites at step-edges, adatoms and vertices. This is an intriguing conclusion. Your comments about this would be very interesting, and could lay out a programme for tackling this.

Jacob Khurgin answered: Thank you for this intriguing question. I would not call my theory "semi-classical" as the only "classical" aspect is that the electric field is not quantized. Quantizing the field would not reveal anything new for applications in detection and catalysis. You are correct about the beaming—it does take place. In the text I derive an expression in which I show that the hot carriers inside the metal are generated proportional to some function (depending on the mechanism) of the cosine of the angle (Fig. 7a in DOI: 10.1039/c8fd00200b). Therefore, if you have a protruding (convex) feature on the surface I expect the carriers to spread out as the field tends to be normal to the surface. However, a concave feature will indeed cause focusing. Regarding the second half of your question, my belief is that the field is strongly affected by the presence of adatoms; in general any sharp feature further violates momentum conservation and increases hot electron emission.

Phillip Christopher commented: Very nice descriptions exist of how the electronic states of atoms at metal surfaces and adsorbates hybridize upon bond formation from a quantum mechanical perspective. The works of J. Norskov and many others have demonstrated these pictures over the past 2 decades based on density functional theory calculations.[1,2]

1 B. Hammer and J. K. Norskov, *Nature*, 1995, **376**, 238–240.
2 J. K. Norskov, F. Abild-Pedersen, F. Studt and T. Bligaard, *Proc. Natl. Acad. Sci. U. S. A.*, 2011, **108**, 937–943.

Jeremy Baumberg responded: Indeed, there are many works on the electronic states of adsorbates at surfaces, but my comment is about doing this for real systems where there is solvation, ionic double layers which respond to localised charges, Au atom movement *etc.* These are not typically included in such calculations, and certainly not in the systematic manner that is needed for the work presented here.

Julian Gargiulo asked: Where would "resistive losses" fit in the presented framework where absorption is led by carrier generation? Is there any energy transfer mechanism from the plasmon to the lattice without carrier generation?

Jacob Khurgin replied: "Resistive losses" as described in simple Drude theory at high (optical) frequencies should be described correctly with both energy and momentum conservation and not using phenomenological damping constants. This is exactly what is done in my work (see DOI: 10.1039/c8fd00200b), in which I show how phonon and impurity assisted transitions become Drude-like at low frequencies. It is important to understand that an intraband transition can only take place from an occupied state to an empty state and that the energy must be conserved. Using Drude theory for high frequencies without thinking about it is absolutely wrong. For example, Drude theory does not include electron–electron scattering contributions; this is correct at low frequencies but absolutely wrong at optical frequencies, as has been shown in numerous experiments dating back to the 1970s. In summary, there is no separate resistive contribution—it is already taken into account in my theory. Including it would lead to double counting, which has in fact been done in the literature.

Jie Liu queried: From a theoretical point of view, what is the difference between nanoparticles at room temperature and at higher temperatures (*e.g.* 1000 degrees)? Will plasmonic absorption change at all?

Jacob Khurgin answered: Yes, it will change with temperature, but only one contribution to the total phonon assisted absorption will increase. Since this absorption dominates at low optical frequencies, the relative change will be very large proportional to $(kT)^n$, where n can be as high as 4. However, when $\hbar\omega \gg kT$, the change will not be that large (maybe a factor of 2–3).

Niclas Sven Mueller remarked: Your study is mostly concerned with single nanoparticles. Can I also apply your theory to nanoparticle oligomers with plasmonic hot spots? If so, how would your advice and summary on the last page of your paper change? Would you still suggest using nanoparticles that are smaller than the effective mean free path?

Jacob Khurgin responded: Thank you for your question. Yes, I believe that the general conclusion will stand—you want to generate hot carriers at the surface and not deep inside. The theory is applicable to oligomers; of course having hot spots increases photoemission. In the end the total emission is increased with the surface to volume ratio (equation 50 in the text). Having hot spots will cause a change in the spatial distribution of the emitted carriers, which may be very important if you manage to have the molecules attached there.

Anatoly Zayats commented: Absorption of complex nanostructures will be different as, for example, an oligomer may have both bright and dark modes, thus light may be trapped more efficiently in the dark modes.

Jacob Khurgin replied: Yes, I agree. However my work is not really concerned with how much light gets absorbed, but more with what kinds of carriers are excited (in terms of energy and momentum) and what happens to them. The more light gets trapped the more carriers are excited, but their distribution does not change. In general, whether let's say 50% or 100% of light gets absorbed matters very little when considering that of all the photo-excited carriers, less than 0.1% will ever succeed in performing any useful function and not end up as just heat.

Sylwester Gawinkowski asked: In real situations, we very rarely have flat walls. How does the roughness of the metal–semiconductor interface affect the directionality of hot electrons?

Jacob Khurgin answered: Yes, I agree. In general the electrons will tend to move more or less normal to the walls locally. For more about roughness please refer to my previous work.[1]

1 M. Grajower, U. Levy and J. B Khurgin, *ACS Photonics*, 2018, **5**, 4030–4036.

Reinhard J. Maurer queried: In your paper, you conclude that phonon-assisted hot-electron generation is only moderately useful for photocatalysis. At the interface, the electronic structure will not be purely metallic due to the existence of adsorbates. Multi-phonon decay, multi-phonon-electron-coupling-induced decay and coupling of phonons and electrons with adsorbate vibrations have been shown before to be important energy dissipation channels. How do these different surface-dominated absorption processes fit into your analysis?

Jacob Khurgin responded: Thank you for your question. First and foremost, as I mentioned in my talk I am not really an expert on photocatalysis and so my conclusions are more relevant for hot electron emission, *i.e.* for detection and photovoltaics. Having said this, all that I meant was that phonon-assisted hot carrier generation occurs everywhere in the material and that this reduces the fraction of hot carriers reaching the surface. I agree entirely that hybridization between the metal and absorbate plays a very important role. However, once hybridization takes place there is no need for phonons to achieve the absorption of photons/plasmons as the momentum conservation impeding intraband absorption in the metal is broken. Therefore, there is no need for phonons to absorb the light. When it comes to decay, phonons do of course play a role, which I have never denied.

Jeremy Baumberg asked: Is your optimal hot electron system then not dense small nanoparticle aggregates, but perhaps instead a gold foam in order to allow the diffusion of reactants and products, and to use all of the gold as surface electrons with optimal penetration and absorption of light?

Jacob Khurgin replied: You are absolutely correct. However, one must be careful to avoid a situation whereby the outer layers of the gold foam screen the field so that it does not penetrate the inner layers.

Matteo Parente opened discussion of Phillip Christopher's paper: You depict the CID mechanism as a direct electron transfer from the metallic nanoparticle to an acceptor state of a molecule adsorbed on its surface. In other publications, the CID mechanism is depicted as an electron excitation from the HOMO to the LUMO of the molecule adsorbed on the surface of the metallic nanoparticle. In this latter case, the CID would look more like a near field effect. Could you explain why you prefer the first way of depicting it?

Phillip Christopher answered: The cartoon depiction we presented is too simplified. My picture of this situation is essentially a combination of the two ideas you describe. When a molecule chemisorbs to a metal surface, states of the molecule and metal will hybridize. The picture I have of CID is near field enhancement of electronic transitions involving these hybridized metal–molecule states. In these situations it isn't quite right to think of these transitions as a metal-to-molecule electron transfer, as seen in MLCT for organometallic complexes, because the states involved in electronic transitions have both metal and molecule character. However, it is possible that the states are localized more on either the metal or molecule and this probably plays a role in dictating how the excitation induces nuclear motion in the system.

Chao Zhan addressed Phillip Christopher and Sebastian Schlücker :In SERS, when the molecule interacts strongly with the plasmonic nanostructure, non-EM effects including CE/CT will contribute to the total enhancement. This has been widely investigated and can be separated into three types. Type I is a result of a charge transfer between the probe molecule and the surface. Type II involves the formation of surface complex which can be considered as new molecules that have new electronic transitions in resonance with the incident light, resulting in a resonant Raman process. Type III is photon-induced charge transfer from the plasmonic structure to the molecule, as has been observed in some electrochemical Raman spectroscopy experiments. However, usually there are no chemical reactions. Chemical interface damping does not mean a plasmon-mediated chemical reaction. For PMCR, excited carriers should have longer retention times for the reactants to take part in the reaction. Nevertheless, the excited carriers in the CT of PERS quickly decay back to the metal or the 'surface complex' on a femtosecond timescale. Additionally, in order to prove the existence of chemical interface damping, looking at only the absorption, scattering or extinction is not enough, and so more experiments are needed. For example, we can use the electrochemical method to regulate the Fermi level of the plasmonic nanostructure in order to prove its relation to the molecular energy level.

Phillip Christopher replied: I agree. Those interested in plasmon-induced catalysts can learn a lot from the many results and theoretical constructs existing in the SERS literature. The idea of using electrochemical approaches to learn more about these processes is a great one. It should be done.

Sebastian Schlücker responded: I agree that electrochemistry is an excellent way of controlling the Fermi level of a plasmonic nanostructure. In addition, we should also aim to carry out wavelength-dependent studies, as far as they are experimentally accessible (at least for a few fixed CW laser wavelengths or ideally a tunable CW laser source). A third important aspect in my opinion is to work on single nanostructures, *i.e.* single nanoparticles or single nanoantennas with defined structures.

Giulia Tagliabue queried: I notice that the absorption peak in your system is above 2eV, hence in a spectral region where interband transitions in Au are very strong. We also know that often molecule adsorption occurs through interaction with the d-band orbitals. Do you think that the two aspects are related? Can this give us some insight into the energetics of the process or any design guidelines? What would you expect to see in the case of a metal without a d-band, for example aluminum?

Phillip Christopher answered: This is an interesting point. It is important to note that similar measurements of chemical interface damping have been made for Au nanostructures with plasmon resonances of much lower energy (away from the interband threshold). Therefore, I am not sure whether the two aspects are related. I do think that the situation should look different for adsorbates on Al. The hybridized states created from molecule adsorption on Al will look quite different than those on Au/Ag/Cu, due to the involvement of the d-band, as you mentioned. This presumably would influence the probability of available electronic transitions associated with adsorbates being involved in plasmon decay processes.

Laura Torrente-Murciano remarked: Your paper nicely demonstrates that the adsorption of thiophenol on the surfaces of Au nanodisks reduces their extinction coefficient and affects the plasmon resonance energy. Based on this, would you expect a similar effect from the substrate upon which the plasmonic particles are supported? It seems to me that this is a generally disregarded point for understanding (and comparing) systems, especially in catalysis.

As a follow up question, do you think that it would be possible to use this effect in a beneficial manner?

Phillip Christopher responded: Absolutely there is an influence of the substrate. In some of the original work on CID, the influence of a SiO_2 matrix on plasmonic structures was examined, where the extent of CID is more significant than often seen with molecules. In addition, it has been shown that interfaces between plasmonic structures and semiconductors (not so different to many substrates used to support metal particles) can induce significant plasmon damping. For example, in Tim Lian's work[1] the strong interfacial damping induced by a semiconductor substrate has been shown to enable very efficient plasmon-induced electron transfer. Furthermore, Naomi Halas' work[2] has highlighted the importance of substrate composition for promoting plasmon-driven surface chemistry. As we learn more about these interfacial processes, we should be able to choose substrates to facilitate plasmon-mediated hot carrier transfer processes.

1 K. Wu, J. Chen, J. R. McBride and T. Lian, *Science*, 2015, **349**, 632–635.
2 S. Mukherjee, L. Zhou, A. M. Goodman, N. Large, C. Ayala-Orozco, Y, Zhang, P. Nordlander and N. Halas, *J. Am. Chem. Soc.*, 2014, **136**, 64–67.

Sebastian Schlücker asked: In contrast to the indirect mechanism, the direct mechanism depends, among other factors, on strong molecule–metal interactions including orbital coupling *etc.* Keeping this in mind, how universally applicable do you think the direct mechanism is with respect to establishing general design principles for plasmonic catalysis?

Phillip Christopher replied: In most catalytic processes on metal surfaces, reactants, intermediates or products will adsorb with energies of at least ~0.5 eV. This is at the low end for adsorbates on optimal catalytic surfaces, but already requires significant hybridization between metal and molecule electronic states. Therefore, from this perspective it is not unreasonable to think that plasmon excitation of catalytic materials will often involve plasmon decay by transitions involving electronic states associated with adsorbates.

Madasamy Thangamuthu queried: What is the reason for using thiophenol? Is it to introduce covalent bonding with gold, which would correspond to the chemisorption of the adsorbate on the metal surface? If so, this should also work with L-cysteine, which has sulfur that can easily form a self-assembled monolayer on gold.

Why is an ultra high vacuum needed to carry out integrating sphere experiments? Would that also work under normal conditions with other thiol molecules on the gold surface?

A Xe light source is used in your study for the integrating sphere experiments, but it is claimed that the observed mechanism is consistent with the ultrafast studies reported in the literature. Both clearly have different working conditions and hence is it reliable to compare these?

Phillip Christopher answered: Thiophenol was used as a model system where we knew that we would be able to see the influence of adsorption on the optical properties of the plasmonic particles. The goal moving forward is to focus on more catalytically relevant molecules, *e.g.* alcohols, and to use the vacuum chamber in typical surface science experiments to follow how plasmon excitation induces chemical transformations and changes in the optical properties of the system under well defined conditions. The comment that the conclusions drawn from low intensity Xe probing and ultrafast studies seem to be consistent is from a mechanistic interpretation, not similarity in the measurements. Our measurements suggest that electronic states hybridized at the metal–molecule interface induce plasmon dephasing. This is consistent with inferences from ultrafast studies that show almost instantaneous charge transfer to interfacial species upon plasmon decay.

Yuri Diaz Fernandez remarked: I appreciate that the net variations that you observed after thiophenol deposition are small even with complete coverage of the surface. However, did you observe any dependence of the different spectroscopic contributions with the surface coverage as you cycled your deposition steps?

Phillip Christopher responded: Yes we did. Systematic damping of the surface plasmon resonance peak was observed as a function of coverage mostly following a Langmuir-type adsorption process. This can be seen in the inset of Figure 5A in our paper (DOI: 10.1039/c8fd00151k). We did not deconvolute the absorption and scattering contributions as a function of thiophenol coverage, but expect that the effect would be similar to the complete coverage case where the scattering spectrum was damped/broadened/shifted more significantly than the absorption spectrum.

Reinhard J. Maurer commented: In your work, you clearly distinguished between Landau damping (plasmon decay which leads to indirect electron transfer to the adsorbate) and chemical interface damping (plasmon decay which leads to direct electron transfer). The latter is an effect due to the hybridisation of localised adsorbate and delocalised surface states. I wonder how clearly these two conceptual processes can be separated in reality. In the case of strong hybridisation, the adsorbate state will hybridise with metallic states over a wide range of energies. This is definitely true for thiolate linkers. Therefore, only a small part of the probability distribution of an individual (hybridised) eigenstate will be localised at the adsorbate. It also becomes increasingly difficult to distinguish between adsorbate and substrate states. Whilst there is clearly an adsorbate-induced change in the hot-electron profile upon plasmon decay, it is not absolutely evident to me that this will also mean that more electrons will be directly transferred to electronic states localised at the adsorbate.

Phillip Christopher replied: I don't disagree that the situation is quite complicated. There are nice demonstrations[1] that the linewidth broadening (decreased lifetime) of plasmon resonances can be related to the formation of bonds between the metal surface and adsorbates or solid-state materials. This suggests that the electronic states associated with these interfacial bound species, or their influence on the metal electronic states, provide more effective pathways for plasmon decay. Therefore, while this is supportive of the idea that direct plasmon dephasing is driven by electronic transitions involving states associated with adsorbed species, there are other possible explanations.

1 B. Foerster, A. Joplin, K. Kaefer, S. Celiksoy, S. Link and C. Sönnichsen, *ACS Nano*, 2017, **11**, 2886–2893.

Bartlomiej Jankiewicz communicated: Have you carried out for comparison studies on the chemical interface damping induced by the adsorption of thiophenol derivatives with either electron withdrawing (EWG) or electron donating (EDG) groups? If so, did the substitution of thiophenol affect the measured scattering and absorption contribution to extinction? If not, do you expect to see any effect of substitution with other types of groups? What will happen if alkanethiol is used instead of thiophenol?

Phillip Christopher communicated in response: We have been carrying out these measurements. While it is too early to provide all of the details, we do see an influence of the R group substitution on thiophenol on the overall magnitude of the plasmon damping. This suggests that regulating the electronic structure of the metal–molecule electronic states controls how effectively these states participate in plasmon decay processes. It is also important to consider the packing density of adsorbed species, which should be different in the case of alkanethiols and thiophenol.

Bart de Nijs communicated: In your paper you mention that the surface damping by the molecules will increase if smaller metal structures are used as the number of molecules increases with respect to gold. Could you comment on how this effect relates to field localisation as well? For example, bringing two such metal objects very close together will not change the amount of gold but the field will be highly localised in the hot-spot between the disks. Would this effect then be the same or also be increased?

Phillip Christopher communicated in reply: This is a great point. The answer should be yes. The field localization at hot spots has been shown to enable enhanced hot carrier production due to surface scattering.[1] With the idea that hybridized metal–molecule states are likely to be involved in surface scattering when adsorbates on the metal surface are present, it is reasonable to consider that molecules in hot spots will be quite efficient at mediating the decay of surface plasmons through chemical interface damping.

1 L. V. Besteiro and A. O. Govorov, *J. Phys. Chem. C*, 2015, **120**, 19329–19339.

Jeong Y. Park opened discussion of the paper by Emiliano Cortés: In your technique, hot electrons are generated on metallic nanoparticles and then transferred to the molecule. These hot electrons cause the reduction of molecules into the fluorescent resorufin molecule, which can be detected *via* fluorescence measurements. Can we quantify the hot electron flux transferred to the molecules? If so, can we determine the efficiency of hot electron transfer, *i.e.* the ratio of the number of reduced molecules to the number of transferred hot electrons?

Emiliano Cortés answered: Efficiencies can be estimated, after agreeing upon a definition of efficiency. It is challenging to count or estimate the number of carriers, but it is possible to estimate the number of absorbed photons and converted molecules. We could then estimate the efficiency defined as the number of converted molecules per absorbed photon. In the solution experiments presented in our paper, the number of absorbed photons is equal to the irradiance of the laser multiplied by the absorption cross section and divided by the energy of one photon. The number of converted molecules can be estimated from the increments in the fluorescent signal. On the other hand, experiments at the single molecule level enable the counting of reduction events. This can be translated into reduction events (or transferred electrons)/absorbed photons. This could serve as an estimation of the number of 'useful' carriers, as long as the limiting regent is the carriers and not the molecules.

Ruben Hamans queried: In your simulations of the resistive losses, I see that the losses are typically higher in the middle of the particle than at the edges. However, your DNA-PAINT maps look very binary in that a particle is either fully covered with DNA strands or completely lacking DNA strands. Do you think that the DNA-PAINT method can eventually be used to gain a more quantitative understanding of the resistive losses in a single plasmonic particle?

Emiliano Cortés responded: We have previously shown that by performing a power-dependent mapping of the structure it is possible to resolve the initial steps of thiol-desorption in the Au structures (please see Figure 3g in our previous study).[1] As shown in that figure, the thiol-desorption reaction is initiated in the middle part of the rod, matching with the resistive losses map of the structure. It is also possible to quantify the number of Au–S bond cleavages from that power-dependence study.

1 S. Simoncelli, Y. Li, E. Cortés and S. Maier, *ACS Nano*, 2018, **12**, 2184–2192.

Zachary Schultz asked: There are some reports that the Fermi level changes in nanoparticles as a function of nanoparticle size. Your single particle measurements should be able to detect these changes. Have you looked into this effect and do you expect to see any changes?

Emiliano Cortés replied: This is an interesting point. We have not yet carried out size-dependence measurements in order to answer this question. Changes in the Fermi level are expected as a result of changes in the external applied potential, the size of the Au NP and the chemical composition at its interface.[1] However, in our experiments, negligible changes in the scattering spectra were observed with the applied potential unless polymerization took place, consequently changing the refractive index around the AuNP drastically (see Figures 1d (ii/iii) and S4 in a previous study by Pensa et al.).[2] This points towards dark-field microscopy not being the most suitable detection method for evaluating changes in the Fermi level alone, unless these changes trigger a process that modifies the local refractive index around the nanoparticle. The combination of electrochemistry and another optical detection method might be better suited for this purpose.

1 M. D. Scanlon, P. Peljo, M. A. Méndez, E. Smirnov and H. H. Girault, *Chem. Sci.*, 2015, **6**, 2705–2720.
2 E. Pensa, J. Gargiulo, A. Lauri, S. Schlücker, E. Cortés and S. A. Maier, *Nano Lett.*, 2019, **19**, 1867–1874.

A Mount enquired: What is the synthesis protocol and resulting thickness of the capping layer in Fig. 1b?

There appears to be an error in the redox equation (Fig. 1b(ii)), which doesn't have charge balance. I presume that the reduced PANI form should be shown schematically as [R–NH–] in order to be consistent with the product shown in Fig. 1c. In any case, isn't the actual redox reaction in this experiment more complicated, as 1e and 2e oxidation is known to occur in PANI? What complication(s) would this additional complexity introduce into your mechanisms? It is also well established for bulk PANI that there is asymmetry in the PANI CV. Indeed the CV in Fig 1b(ii) looks very similar to that of bulk PANI modified electrodes. The asymmetry in the redox peak is not assigned to irreversible electron transfer but a change in phase between the conducting, solvated, charged and hydrophilic oxidised form and the hydrophobic and compact semiconducting or insulating reduced form. In this case, as a change in the redox potential of PANI would be expected to reflect this change in form, would the delta lambda effect not be more likely to correlate with charge (redox composition) than with potential E?

Emiliano Cortés answered: Fig. 1b shows the optoelectrochemical response of a single gold nanoparticle covered with PANI. The aim of this experiment was to show that by using the setup depicted in Fig. 1a it is possible to follow the redox process at the single particle level. Details regarding the synthesis and characterization of AuNP@PANI are included in the experimental section of the paper (see 'Sample preparation for optoelectrochemical measurements' in DOI: 10.1039/c8fd00138c). The synthesis was carried out *via* chemical oxidative polymerization following the protocol described by Xing *et al.*[1] Briefly, oxidation of the aniline monomer was achieved in a strong acidic medium by using ammonium persulfate as a redox initiator at room temperature. The PANI thickness was estimated to be *ca.* 19 nm.

In the above context, we agree that the redox reaction associated with PANI is more complicated than the simple electron transfer that is usually sketched (*cf.* Fig. 1b(i)). As mentioned here, the charge transfer process has been proposed to occur in parallel to other phenomena such as the exchange of counter ions or solvent molecules and elastic deformation in the polymer film, among other examples.[2] However, all these phenomena contribute to the change in the dielectric constant of PANI, which is well known to be different for the two PANI forms studied here (fully reduced 'leucoemeraldine' and the partially oxidized 'emeraldine').[3] This change in the dielectric constant, which modifies the refractive index around the plasmonic nanoparticle, is what we are monitoring optically. Then, for the experiment depicted in Fig. 1b, the shifts in the scattering maximum wavelength ($\Delta\lambda_{max}$) are a consequence of the applied potential. Further experiments and assumptions must be carried out and made in order to satisfactorily correlate $\Delta\lambda_{max}$ with the exact redox composition of the polymer, *i.e.* the charge of the polymer along the applied potential window must be unambiguously determined.

On the other hand, Fig. 1c corresponds to plasmon-assisted aniline electropolymerization on single gold nanoparticles. Here, we study the influence of plasmon excitation on the onset potential that is needed to trigger the polymerization of aniline on single Au nanoparticles, *i.e.* we are focused on the initial step of the reaction shown in Fig. 1c(ii). Then, the composition and rate of polymerization would only affect the growing-slope of the curve $\Delta\lambda_{max}$ – E. In this case we have verified that very thin layers of PANI (less than 5 nm) can be detected optically due to the large difference in the refractive index between water (*i.e.* prior to polymerization) and PANI around the nanoparticle.

1 S. Xing, L. H. Tan, M. Yang, M. Pan, Y. Lv, Q. Tang, Y. Yang and H. Chen, *J. Mater. Chem.*, 2009, **19**, 3286–3291.
2 E. M. Andrade, F. V. Molina, M. I. Florit and D. Posadas, *Electrochem. Solid-State Lett.*, 2000, **3**, 504–507.
3 A. Baba, S. Tian, F. Stefani, C. Xia, Z. Wang, R. C. Advincula, D. Johannsmann and W. Knoll, *J. Electroanal. Chem.*, 2004, **562**, 95–103.

Laura Torrente-Murciano queried: This is a fundamental question directly related to this paper but also to a number of papers related to plasmonic catalysis discussed during the meeting.

If I had to compare plasmonic-driven catalysis with conventional photocatalysis (*e.g.* TiO2), in the latter case there are reduction and oxidation half-reactions that consume electrons and holes, respectively. However, on many occasions in the case of hot-electron driven catalysis, like in Fig. 3 of this paper (hot electron reduction of resazurin), only electrons are consumed. Where are these electrons coming from or, in other words, where are the holes going to?

Emiliano Cortés responded: In every reaction involving electron or hole transfer there must be a counter reaction fulfilling the redox cycle. In many cases the solvent could play that role, however, that might limit the rate of the reaction. This is a fundamental point to address the optimization of efficiencies in systems involving a charge transfer process. It is important to play with electron or hole scavengers in order to disentangle this point (see for example a previous study by Rao *et al.*)[1] Another possibility would be to split the redox reactions, as these occur in any electrochemical experiment (see our example in Fig. 1).

1 V. G. Rao, U. Aslam and S. Linic, *J. Am. Chem. Soc.*, 2019, **141**, 643–647.

Phillip Christopher replied: In many of the plasmon-mediated processes discussed in the literature, redox chemistry is driven that looks a lot like semiconductor photocatalysis. There can also be plasmon enhanced electrocatalytic processes, where a connection to an electrical circuit maintains charge neutrality. However, there is also another proposed process where an electron (or hole) transiently transfers to an adsorbed species. The transient transfer of charge induces forces on the molecule, which can result in the deposition of vibrational energy into a potential energy surface and induced chemistry. This sort of inelastic transient electron transfer is well-known in the surface science field as the Menzel Gomer Redhead (MGR) model, which was proposed in the mid 1960s for chemistry on metal surfaces induced by electron bombardment.

Yuri Diaz Fernandez asked: For the example of single particle spectroscopy detection of the photo-electrochemical polymerisation of PANI:

a) Did you observe non-specific electrochemical polymerisation on the ITO substrate?

b) How did you deconvolute the particle–surface polymerisation from the ITO polymerisation, considering that the effective refractive index and the plasmonic response will result from the combination of both?

Emiliano Cortés answered: As shown in Figure 1d (ii/iii) in Pensa *et al.*,[1] non-specific polymerization on ITO was not observed for illuminated Au nanoparticles below the electrochemical polymerization threshold E_{onset} (E_{W} = 0.68 V). Only when the polymerization is electrochemically controlled, *i.e.* Ew ≥ 0.68 V, does the reaction take place over the whole electrode (ITO and Au NPs) and non-specific polymerization can be observed by dark-field microscopy.

As mentioned above, polymerization was only observed on Au NPs that were illuminated, meaning that PANI/ITO is only generated at the interface of AuNP/ITO. We have simulated this situation, as described in the supporting information of Pensa *et al.*[1] Briefly, 80 nm Au NPs surrounded by different thicknesses of PANI layer are placed on an ITO surface. The substrate is immersed in water. Numerical simulation shows a linear-like dependence of the maximum scattering wavelength on the thickness (see Figure S1). This trend is in line with experimental results where the PANI layer is increased around the Au NP by increasing either the illumination time or power (see Figure S5 in Pensa *et al.*).[1]

1 E. Pensa, J. Gargiulo, A. Lauri, S. Schlücker, E. Cortés and S. A. Maier, *Nano Lett.*, 2019, **19**, 1867–1874.

Yonatan Sivan enquired: Regarding your temperature control experiments, do you observe a signature unique to high energy non-thermal electrons and, if so, can you say how much energy goes into raising the temperature and how much goes into creating these non-thermal electrons?

Emiliano Cortés responded: We do not have an experiment where we can distinguish between high-energy non-thermal carriers and thermalized carriers. However, we have found that the effective energy contribution of the carrier to the photocatalytic effect (*i.e.* the energy that a carrier can transfer to a nearby molecule) depends on the excitation wavelength. This dependency has been measured at constant lattice temperature and at the single particle level.[1] If the temperature of the electrons was thermalized with the lattice (under our experimental conditions), our results would indicate that non-thermal electrons are driving the process. Regarding your last question, it is challenging to measure the steady state energy distribution between the lattice, thermal carriers and non-thermal carriers, and we have not done this yet.

1 E. Pensa, J. Gargiulo, A. Lauri, S. Schlücker, E. Cortés and S. A. Maier, *Nano Lett.*, 2019, **19**, 1867–1874.

Javier Aizpurua opened a general discussion of the papers by Jacob Khurgin, Phillip Christopher and Emiliano Cortés, and addressed Phillip Christopher: I am wondering what else we could do to access the dynamics of chemical reactions. Most of the current experimental approaches used by this community rely on optical spectroscopy to trace chemical reactions, but sometimes all we can detect is a small shift and/or broadening of an optical marking peak. This, sometimes, and after lots of nice control experiments and assumptions (for example an assumption of direct transfer), drives the main conclusions of the dynamics. One can detect a small percentage of change in the absorbance for instance, such as in your paper, and one can conclude with slight uncertainty regarding quenching. What can we do on top of this? What can we do to go beyond spectroscopy to address the dynamics? Can we simplify the systems and carry out real-time experiments in some way? What are the current efforts along these lines? This, of course, does not mean that the current information obtained *via* optical spectroscopy is not valuable or meritorious, as you have shown in your paper.

Phillip Christopher replied: There is work being carried out on very well defined systems that attempts to watch chemical transformations simultaneously with temporal and spatial resolution. This is a very tough task. Work with microscopy often does not have the required temporal resolution, while time-resolved work often does not have the required spatial resolution. Understanding the dynamics associated with processes like plasmon-mediated chemistry will likely require some insights to be gained from multiple experimental and theoretical approaches.

Naomi Halas asked: How can theorists move beyond the theory they are currently pursuing to also address the energies of adsorbates on surfaces, and to answer more of the questions we have regarding the processes and mechanisms of plasmonic photocatalysis? For accurate adsorbate energies, does this not involve extending theoretical efforts to include strategies such as embedded wavefunction approaches?

Javier Aizpurua answered: The theoretical challenge to address the energies involved in plasmonic photocatalysis, as well as the associated dynamics, is as demanding, if not more, than the experimental challenge. *Ab initio* efforts, together with molecular dynamic simulations, have been able to address some aspects of the energies of simple adsorbates, even in a dynamical way. However, those studies are restricted to 'good' metallic surfaces. In those studies, the density of states produced by plasmonic decay is still missing. To reconcile these two aspects would be a tremendous challenge, and I agree that theory is in need of a substantial methodological improvement to properly address all of the dynamical aspects of plasmon-induced catalysis.

Javier Aizpurua commented further: Although I encourage experimentalists to try to access the dynamics of reactions with more direct experimental techniques, I realize, and agree with Naomi Halas, that theory is in need of the same or even more development to address the challenges of plasmonically driven chemistry. In proper theoretical approaches, we would need to combine quantum chemistry with condensed matter physics on different timescales and for large systems, which is very challenging, and so this would also require a similar effort in theory.

Jeremy Baumberg returned to the discussion of the paper by Phillip Christopher: You use a thiol SAM and show that the optical properties change, which you ascribe to some sort of molecular hybridisation with gold. However it is known that at least 30% of Au atoms under such a thiol SAM are plucked out into adatom sites or staples between the S atoms. In this case, the roughness and surface electron states of the Au change drastically, and so we would already expect changes in the optical properties. Is this not what you see? Other techniques for probing the surface also have a tendency to move around these adatoms, so it is not clear whether other techniques, other than perhaps low temperature STM, are helpful for understanding this. Are there other techniques you would advocate that are gentle enough?

Phillip Christopher responded: Unravelling the roles of adsorbate-induced metal surface reconstruction and metal–molecule hybridized electronic states on plasmon decay and interfacial energy transfer is a very tough task, but could certainly teach us something. Both mechanisms make good sense as possible sources of enhanced surface scattering of electrons and plasmon decay. This could potentially be interrogated *via* variable temperature UV-Vis spectroscopy, where one might expect that at extremely low temperatures a metal surface would be kinetically trapped from reconstruction upon molecular adsorption, while reconstruction could occur at higher temperatures. However, the metal–molecule hybridization should also change upon surface reconstruction. In order to tease out these details, theory would likely be most effective once we can simultaneously describe plasmons and electronic structure with sufficient accuracy in a single calculation.

Sebastian Schlücker commented: The analysis of molecular adsorbates on plasmonic nanostructures requires experimental techniques to provide molecular/chemical specificity and surface selectivity. Such techniques are available, including SERS and vibrational SFG. However, theory is lacking behind in this respect. We therefore need strong support from our colleagues in theory who are able to reliably predict the properties of molecular adsorbates, in particular experimental observables such as vibrational spectra (surface-enhanced IR/Raman/fluorescence).

Emiliano Cortés replied: I agree with this comment and we should keep in mind that the excited states (vibrational and electronic) may play an important role in these transformations. We should be able to predict new reaction pathways if we know the energy landscape of the adsorbed molecules.

Jacob Khurgin also replied: I agree entirely. I am not sure that I can do much in this field but hopefully younger scientists can.

Rupert Oulton returned to the discussion of the paper by Phillip Christopher by communicating: Jacob Khurgin talked about the large density of states in metals, and Emiliano Cortés spoke of the relatively small number of molecules in comparison. Your paper even highlights this discrepancy. What is thus the intuitive picture for chemically induced damping?

There was also some discussion of the surface modification of gold by the molecule and electronic states of the molecule. Would it not be straightforward to rule out the role of electronic states by saturating them under high intensity illumination?

Phillip Christopher communicated in reply: The Landau damping process of plasmon decay and the generation of hot carriers that Jacob Khurgin discussed occurs due to the scattering of electrons off of the plasmonic nanostructure surface. In this case, the electronic transitions involved in the generation of the initial, nascent hot carriers will involve electronic states at the metal surfaces. The addition of adsorbates will specifically influence metal electronic states at the surface and further introduce new states due to hybridization. Thus, even though the number of molecules is small compared to the number of metal atoms, they are located at the surface where hot carrier formation initially occurs and will further introduce relatively localized electronic states that have a high probability of being excited during surface scattering.

Your idea of saturating these states through high intensity excitation is interesting. I would be worried that this would induce chemical transformations in the molecules, which would make analysis tough. Furthermore, the lifetimes of these excited states are very short (~few fs).

Emiliano Cortés provided a general comment: I think we should agree, sooner rather than later, on a set of basic principles for the field of hot-electron induced science. Otherwise we are at risk of reporting efficiencies that have been calculated in many different ways (which would make the data useless for benchmarking), or of reporting plasmonic photocatalysts that demonstrate no real impact or improvement over existing materials or technologies. Selectivity is a much more standardized concept, and it is clearer as to how to report it (% of each generated product). However, for efficiencies I think that we are already facing a problem. Taking into account all of the processes involved in a plasmonic photocatalyst, perhaps the ratio of (generated product(s) per gram of photocatalyst)/(total number of incident photons) could be a possible, rather simpler, metric that could be used to calculate the overall performance. There may be other options, but I think we should agree on this topic.

Jeremy Baumberg opened discussion of the paper by John C. Polanyi: You are making nice 'beams' on an atomic surface, but I am worried about constraining the molecular directions just along the lines of copper atoms. Does this not constrain the output directions of the products and so limit the type of reaction that you can do? Is this not like doing reactions on a chess board? How will this change the chemistry that is achievable? Can the molecules adapt and give you something useful, particularly if there are two products trying to head in different directions?

John Polanyi answered: We should have stressed that we are carrying out our reactions on an unmodified surface of a copper crystal, on which we and others have seen varied dynamics depending on the reagents. The surface beams are not due to irregularities in the surface. They are due to what I call 'funny walks'. We have previously reported a preference for CH_2, recoiling in the electron-induced reaction of CH_2I_2 on the same surface, Cu(110), to 'walk' on its residual bonds between the rows of Cu atoms.[1] In the present work we have another biradical, CF_2, walking (we call it 'ratcheting') directly along the close-packed rows of Cu atoms. Molecular dynamics showed that the reason for these specialized motions, in both cases, was a combination of linear recoil with a rocking motion. Linear momentum is essential to the overall motion. The rocking has a more subtle role, assisting in detaching the 'legs' of this walker successively from the surface as it proceeds along its path. What we have shown here is that this novel directional walking can be put to use in selecting the impact parameter in subsequent reactive encounters in which the walking 'projectile' strikes a fixed chemisorbed 'target'. The impact parameter is variable, and is determinable from the STM image. We can do something 'useful' (to use your term) by shedding light, as in Fig. 6 (DOI: 10.1039/c8fd00137e), on the dynamics of a reaction related to Fischer–Tropsch polymerization. It is true that we only do something 'useful' largely by accident.

1 A. Chatterjee, F. Cheng, L. Leung, M. Luo, Z. Ning and J. C. Polanyi, *J. Phys. Chem. C*, 2014, **118**, 25525–25533.

Michael Ashfold asked: What limits the range of collision energies that can be investigated with the 'surface-molecular-beam' method? The outcomes demonstrated in your paper involve the projectile and target species combining, or experiencing a non-reactive 'near-miss'. Can you envisage studying a sufficiently exothermic reaction by this method that might yield two products and, if so, how might their recoil be constrained by registration with the underlying array of metal atoms?

John Polanyi responded: The upper limit of the collision energy in the case of the CF_2 'projectile' is the energy released in the anti-bonding anionic state, as CF_2 recoils from the strongly surface-bound F-atom. This has been computed to be approximately 1.2 eV (see Fig. S2 of our previous study).[1] This is, therefore, our maximum possible collision energy. The CF_2 cools thereafter as it travels increasing distances along a copper row towards a target. One can see the effect of this cooling in the experimental results of Fig. 3b for CF_2 directed at zero impact parameter at a stationary chemisorbed CF_2 target. In the case that the CF_2 projectile travel distance was only ~10 Å, the reaction was found to give forward scattered C_2F_4 (this reaction product being scattered on the surface forward along the continuation of the direction of the incoming beam). If, however, the travel distance was increased to 30 Å, the cooled CF_2 was observed to give product C_2F_4 scattered largely backwards (back towards the origin of the beam). This is revealing of the molecular dynamics.

Returning to your question, this collision energy should suffice for barrier-crossing in some exothermic reactions. Preliminary results from our laboratory have indeed, as you suggest, shown evidence of exchange reactions at a selected impact parameter yielding two new products. It is a mistake to think that the product recoil in the present work is being artificially constrained; the reagents are coming together on an unmodified clean copper crystal. Reactions on such surfaces are known to provide a rich variety of outcomes.

1 K. Anggara, L. Leung, M. J. Timm, Z. Hu and J. C. Polanyi, *Sci. Adv.*, 2018, **4**, eaau2821.

Naomi Halas queried: The animation you showed in your presentation appeared to simulate bonds between copper and carbon forming and breaking on the surface. Is that what is really going on here? Why does the adsorbate molecule move across the surface?

John Polanyi replied: The animation that I showed did indeed simulate the 'ratcheting' recoil of CF_2, moving along a close-packed row of Cu atoms on Cu(110), and involving successive bonding to the copper atoms of the row. The clue to this binding came from our observation of the molecular dynamics of the successive shortening of the C–Cu separation as the travelling carbon end of CF_2 came atop a copper atom (single-bond from an sp^3 hybridised carbon) or subsequently between a pair of copper atoms (a pair of bonds from an sp^2 hybridised carbon). You have asked (sensibly) why CF_2 is moving along the row. This is because it was formed, following electron attachment, from an anti-bonding state of the CF_3^- anion in which $F.CF_2$ repulsion (the dot indicates the locus of the repulsion) leads to F bonding strongly to the surface and the CF_2 (in its ratcheting motion, discussed above) recoiling a substantial distance along the close-packed row of Cu atoms. The CF_2 recoil travel direction is opposite to a slightly (but visibly) stretched C–F bond directed along the Cu-row, which is the bond broken in electron-induced dissociation.

Reinhard J. Maurer enquired: How many electrons are typically transferred in a pulse? Can surface-molecular-beams be introduced with positive and negative bias voltage?

John Polanyi answered: We have shown *via* a study of reaction rate *versus* current (see Fig. S8 of our previous study)[1] that the electron-induced dissociation of CF_3 adsorbed on copper is a single-electron process. We typically used a current of a few nanoamps for several seconds to induce a reactive event. This translated into a yield of $\sim 10^{-11}$ reactive events per electron, at a bias voltage of +1.3 V, as reported here. Our experiments were restricted to positive surface bias since (as described in Section 2.2) we are able to compute approximate energy states for the CF_3 anion at the surface.

1 K. Anggara, L. Leung, M. J. Timm, Z. Hu and J. C. Polanyi, *Sci. Adv.*, 2018, **4**, eaau2821.

Phillip Christopher asked: In studies of molecular beams interacting with metal surfaces, there is often a question of whether energy dissipation from exothermic chemical events is dissipated through adiabatic or non-adiabatic mechanisms. In your demonstration of molecular beams on surfaces, do you have any thoughts as to whether the projectiles dissipate energy to the surface through adiabatic or non-adiabatic mechanisms?

John Polanyi responded: Our impulsive two-state (I2S) model (whose antecedents lie in the work of Menzel, Gomer and Redhead, the famous MGR) is able to describe the varied dynamics of electron-induced product recoil, as observed experimentally by STM. We reference examples in our paper. This model takes no account of non-adiabaticity. The CF_2 loses its energy to surface vibrations, *i.e.* to phonon excitation. Despite the neglect of non-adiabaticity, the dynamics (average recoil distances and recoil directions) resemble those observed for molecules engaged in long-range excursions and reactions at the surfaces studied. Of course, in reality, non-adiabaticity may nonetheless be involved, as your question suggests.

Bart de Nijs communicated: You mention in your article that you look for an anisotropic shape to identify the CF_3 molecule on the (110) surface. Do you have the resolution to determine the orientation of the molecule before breaking the molecular bond and launching the CF_2, or will the CF_2 occasionally be launched in the opposite direction?

John Polanyi communicated in reply: You are attentive. When our STM tip was working well we could indeed detect asymmetry in the image of the chemisorbed CF_3. This was indicative of the small extension in bond length of the C–F that pointed along a Cu-row, as compared with the other two C–F bonds. Electron irradiation at +1.3 eV (positive surface; threshold energy for reaction ~0.9eV) resulted in a CF_2 'projectile' recoiling preferentially away from the extended-bond end of the parent CF_3 molecule.

Jacob Khurgin opened discussion of Reinhard J. Maurer's paper: The question is regarding which hot carriers are responsible for the catalysis; they can be 'primary' hot carriers excited by the decay of plasmons or they can be 'secondary' carriers that have experienced multiple electron–electron collisions and then thermalized with electron temperature T_e. I presume that since T_e is proportional to the incoming light intensity and the effect (catalysis or just electron emission) depends on the presence of hot carriers in the tail of the distribution, one would observe some kind of nonlinear effect/intensity dependence in the experiments if the secondary thermalized carriers were responsible for this effect. Has this been observed?

Reinhard J. Maurer answered: This is indeed an open question of debate. In our approach, we only capture thermalized electrons at elevated T_e. There are several instances in the literature in which a time-dependent change in T_e can correctly capture the measurable effects of light illumination, *e.g.* in light-driven desorption. However, there is also some experimental evidence from molecular beam scattering that suggests that 'primary' transient molecular excitations dominate molecule–metal energy transfer.

Phillip Christopher remarked: There are some very nice experimental results from the 1980s and 1990s, for example the work of Gerhard Ertl,[1] Wilson Ho[2] and David King,[3] that demonstrated power law dependences of reaction yield on photon flux for pulsed laser driven chemical processes at metal surfaces. Further observations of a transition from linear photon intensity dependence to 'super-linear' dependence have also been reported in similar studies. Two pulse correlation experiments have been carried out to demonstrate that in the superlinear regime the reactions occur on timescales that are much too fast (<1 ps after the pulse) to be explained by thermal (phonon mediated) processes.

1 M. Bonn, S. Funk, C. Hess, D. N. Denzler, C. Stampfl, M. Scheffler, M. Wolf and G. Ertl, *Science*, 1999, **285**, 1042–1045.
2 F. M. Zimmermann and W. Ho, *Surf. Sci. Rep.*, 1995, **22**, 127–247.
3 S. A. Buntin, L. J. Richter, R. R. Cavanagh and D. S. King, *Phys. Rev. Lett.*, 1988, **61**, 1321–1324.

Jacob Khurgin commented: I think you mean lattice temperature. Can they still be 'secondary electrons' after a few e–e collisions?

John Polanyi enquired: As you well know, there has been much interest over the years in the dynamics of collisional vibrational ladder-climbing, which, in order to lead to molecular dissociation, must compete with its converse, collisional vibrational de-activation. Your molecular dynamics studies of the hot-electron mediated dissociation of H_2 on Ag(111) should teach us new things about how this balance of opposing processes can result in molecular dissociation. Would you care to comment?

Reinhard J. Maurer responded: On the basis of our current results, we are not in a position to confidently comment on this matter. However, we find that electronic friction due to hot-electron effects is larger for vibrational motion than it is for rotational or translational molecular motion of the adsorbate. The result is that hot electrons selectively promote molecule–metal energy transfer into and out of the intramolecular stretch of H_2. In the case of a hot-electron-heated metal surface, this points towards a qualitative trend of increasing the probability of dissociation.

Javier Aizpurua asked: You have developed a very nice description of molecular scattering on metallic surfaces which combines molecular dynamics with *ab initio* inputs. In this combined description, you assume an initial trajectory of the molecules; how much do the results depend on the initial conditions of the molecule (orientation and velocity, for instance)? Regarding the lack of memory, what are your plans to improve the current description based on the Markovian approximation? Finally, you treat a perfectly flat surface, which is a very useful geometry, but how could you incorporate more complex surfaces or nano-antennas? In other words, how could we transfer this to incorporate localised plasmons in your interactions with the molecule?

Reinhard J. Maurer replied: We find that the initial conditions play an important role in determining the extent of hot-electron-driven molecule–metal energy transfer and the subsequent measurable reaction probabilities (see, for example, Fig. 3 in DOI: 10.1039/c8fd00140e). Our plan is to improve our current description to include memory effects by explicitly integrating over electronic friction at past times using a generalised Langevin dynamics approach. We believe that this will enable us to employ electronic friction-based approaches for hot-electron chemistry in cases where thermalised hot-electron distributions are the main cause of molecule–metal energy transfer. While we are free to use any periodic surface structure in our simulations, we are currently not able to capture the light-matter interaction that gives rise to plasmon resonances directly. Therefore, simulating nanopatterned surfaces as such might not correctly capture the ensuing plasmonic effects. To capture the effects of plasmon scattering on molecular states, we would certainly have to model non-thermalised electronic excitations giving rise to transient molecule–metal charge transfer. This is clearly beyond the capabilities of electronic friction theory.

Xiaofei Xiao remarked: Thank you for giving your presentation. Current numerical methods can address relatively simple simulations. However, simulations become harder as structures become more complex. Some problems, such as many-body systems, cannot be solved due to various reasons, such as insufficient memory. Machine learning may be a solution. I saw that you have already used machine learning in your project. My questions are:

(1) What are the advantages and disadvantages of using machine learning in physics?

(2) What kinds of problems do you think this method are suitable for?

(3) What is the future or outlook of this method in physics, especially in plasmonics and catalysis?

Reinhard J. Maurer answered: We have used machine learning in the form of feed-forward neural-networks to generate interpolations of total energies and electronic friction tensors from density functional theory (DFT) as a function of the molecular degrees of freedom. In doing so, we were able to retain the predictive capabilities of DFT while performing simulations of tens of thousands of reaction events—an absolute necessity to determine statistically meaningful predictions of measurable reaction probabilities. In chemical reaction dynamics, machine learning has become a powerful tool for extending the time- and length-scales of electronic structure methods. I believe that in the near future we will see many more applications to this end that will enable previously impossible applications of *ab initio* methods, such as the simulation of reaction dynamics in the context of plasmonics and catalysis. It is important to note that, for machine learning to produce reliable representations of *ab initio* methods, reliable and consistent training data of high-quality needs to be generated, which is highly non-trivial.

John Polanyi asked: In your study of the reactive scattering of H_2 at a metal, you considered the effect on energy transfer to the H_2 of molecular cartwheeling. I recall a previous Faraday Discussion (No. 62, on potential energy surfaces) during which it was noted that enhanced rotation would lead to a regime in which the time spent in the preferred geometry for energy transfer or a reaction was too short to permit the process to occur. Is this outside the range of j that you examined?

Reinhard J. Maurer responded: In this work, we studied initial conditions with rotational quanta up to $j = 5$. We find that the vibrational de-excitation probability ($\nu = 2$ to $\nu = 1$) increases slightly for $j = 5$, in comparison to $j = 0$. While the residence time on the surface is indeed shorter for higher rotational states, the molecular velocity is also larger, which, together with the electronic friction tensor, defines the nonadiabatic forces that act on the impinging molecule. The latter is the main reason why nonadiabatic molecule-to-surface energy dissipation is larger for higher rotational states. However, it should be noted that we find significant differences between different orientations. Fig. 3(a) in our paper shows the de-excitation probability for ($\nu = 2, j = 5, m_j = 0/5$) to ($\nu = 1, j = 5$) scattering, where $m_j = 0$ and $m_j = 5$ correspond to the cartwheel and helicopter orientations, respectively. The de-excitation probability and ensuing energy loss are much larger in the helicopter motion due to the increased surface residence time and the ability of the molecule to penetrate regions of the potential energy surface in which larger electronic friction forces act on the molecule.

Francesca Baletto queried: Adsorption might alter the metallic substrate too. Is there any specific reason for keeping the Ag-substrate fixed?

Are there any limitations to the use of MDEF for depicting the adsorption of small molecules such as, O_2, H_2O, H_2, OH, CO and CO_2 on different metallic (mono- and bi-metallic) substrates?

Reinhard J. Maurer replied: Principally, our methodology is not restricted to a particular adsorbate size or to a frozen surface. Using on-the-fly *ab initio* calculations, we could use any choice of system. The main limitation is the computational cost associated with the evaluation of electronic friction from density functional theory at each time step. This is why we have used a neural-network-based interpolation to construct a 6-dimensional (adsorbate degrees of freedom) tensor landscape of electronic friction based on DFT data, allowing us to perform many hundreds of thousands of simulations. We are currently developing an interpolation representation of tensorial friction that will allow us to go beyond the frozen substrate approximation.

Sebastian Rejman returned to discussion of the paper by John Polanyi: Why does the CF_2 travel somewhat further on the surface than the F-atom? At least from a consideration of conservation of momentum, it should be the other way around, as the CF_2 is a lot heavier.

John Polanyi answered: The alert group doing the experiments were as amazed as you are; you are right about the expectation from conservation of momentum. Why does the light F-atom barely recoil (4 Å on average), whereas CF_2 recoils a long distance (20 Å on average)?

My co-workers' subsequent calculation of the binding energies of the two fragments to the underlying copper explained the short recoil of F; its binding energy to the surface was large, 5.01eV, in comparison with 1.89eV for CF_2 (both computed at a short-bridge site, between a pair of Cu atoms along a Cu-row).

That still left a mystery as to why the CF_2 travelled so far. Molecular dynamics gave the answer. It involved the walking (we call it 'ratcheting') of the divalent CF_2 radical from one Cu atom to the next, directly along a copper row. We had previously seen something similar for a different diradical, methylene CH_2, recoiling this time from an I-atom to which it had been bound and walking in this case between copper rows.[1] Perhaps such 'funny walks' will turn out to be common, especially for recoiling polyvalent radicals which have two or more 'legs' on which to walk.

1 A. Chatterjee, F. Cheng, L. Leung, M. Luo, Z. Ning and J. C. Polanyi, *J. Phys. Chem. C*, 2014, **118**, 25525–25533.

Conflicts of interest

There are no conflicts to declare.

Faraday Discussions

PAPER

Dynamics of electron-emission currents in plasmonic gaps induced by strong fields

Garikoitz Aguirregabiria,[ab] Dana-Codruta Marinica,[c] Markus Ludwig,[d] Daniele Brida,[de] Alfred Leitenstorfer,[d] Javier Aizpurua *[ab] and Andrey G. Borisov *[c]

Received 24th October 2018, Accepted 15th November 2018
DOI: 10.1039/c8fd00158h

The dynamics of ultrafast electron currents triggered by femtosecond laser pulse irradiation of narrow gaps in a plasmonic dimer is studied using quantum mechanical Time-Dependent Density Functional Theory (TDDFT). The electrons are injected into the gap due to the optical field emission from the surfaces of the metal nanoparticles across the junction. Further evolution of the electron currents in the gap is governed by the locally enhanced electric fields. The combination of TDDFT and classical modelling of the electron trajectories allows us to study the quiver motion of the electrons in the gap region as a function of the Carrier Envelope Phase (CEP) of the incident pulse. In particular, we demonstrate the role of the quiver motion in establishing the CEP-sensitive net electric transport between nanoparticles.

1 Introduction

Electron emission from metallic surfaces is a physical process which exploits the exchange of energy between incident photons and outgoing electrons.[1,2] The current trends in nanotechnology are able to design complex morphologies of metallic nanostructures which can act as effective optical nanoantennas trapping and enhancing light of specific wavelengths in their proximity.[3–5] This is achieved by means of the collective excitation of the electronic charge density at the interfaces of the metallic nanostructures, so-called surface plasmons.[6,7] Plasmonic nanoparticles are thus a very appealing structure for the generation of strong-field emitted electrons due to the electric-field enhancement and

[a]*Centro de Física de Materiales CSIC-UPV/EHU, Paseo Manuel de Lardizabal 5, 20018, Donostia-San Sebastián, Spain. E-mail: aizpurua@ehu.eus*

[b]*Donostia International Physics Center, Paseo Manuel de Lardizabal 4, 20018, Donostia-San Sebastián, Spain*

[c]*Institut des Sciences Moléculaires d'Orsay – UMR 8214, CNRS-Université Paris Sud, Batiment 520, 91405 Orsay Cedex, France. E-mail: andrei.borissov@u-psud.fr*

[d]*Department of Physics, Center for Applied Photonics, University of Konstanz, D-78457 Konstanz, Germany*

[e]*Physics & Materials Science Research Unit, University of Luxembourg, 162a Avenue de la Faïencerie, L-1511, Luxembourg*

localization produced near the particle surface.[8–15] Among others, metallic tips have been commonly used in the generation of photoemitted electrons, for instance in the guns of electron microscopes.[16]

More recently, plasmonic gaps, formed in a metal–insulator–metal junction, have been shown to host very intense bonding-gap plasmons which produce electromagnetic hot spots commonly exploited in surface-enhanced spectroscopy.[17–22] The plasmonic gap has also been the object of intense research regarding the interplay between optical and transport properties, as the two metallic particles forming the gap can be externally biased, and the current across the gap can thus be accessed while illuminating the structure with light.[22–27] Last, but not least, the ability to shape light into ultra-short single-cycle pulses with varying Carrier Envelope Phase (CEP) provides an additional degree of control on the photoelectrons emitted.[9,13,24] When those electrons are emitted in a plasmonic gap as a result of the incidence of an ultra-short laser pulse, a complex dynamical behaviour of the electrons can be foreseen which depends on the particular conditions of the emission, such as energy of the photons, duration of the pulse, local field enhancement, or size of the gap.[24] Differently from a metallic tip, characterized by electron emission into free-space, the plasmonic gap offers the possibility of achieving electron transport between nanoparticles controlled at the optical cycle time-scale thus granting mutual coherence between radiation and electron current. A theoretical scheme capable of tracing the dynamics of these ultra-short electron bursts in plasmonic gaps can be extremely useful when it comes to the design and control of the optoelectronic properties of ultrafast electron currents.

This discussion precisely addresses the challenge of describing the complex dynamics of photoemitted currents in plasmonic gaps under ultra-short illumination. A simple model of the structure will help to provide an understanding of the interplay between the carrier envelope phase of a single pulse and the motion of the electron bursts in the gap. We describe our model system, the methodological approach to the dynamics of the electrons, and the results in the following sections.

2 Plasmonic system

As a prototype of a plasmonic gap system we consider a dimer of parallel gold cylinders of diameter $D = 10$ nm as sketched in Fig. 1. The identical cylinders are infinite along the z-axis and the distance between the two nanoparticle surfaces at the closest point is $d_{\text{gap}} = 3$ nm as measured along the x-axis. The centre of the gap is located at ($x = 0$, $y = 0$). This structure is illuminated by a single-cycle femtosecond (fs) Gaussian pulse polarized along the x-axis. The electric field of the pulse is given by:

$$E(t) = \tilde{E}\cos(\omega t + \phi)\mathrm{e}^{-t^2/\tau^2}, \tag{1}$$

where the amplitude of the electric field is $\tilde{E} = 3.7$ V nm^{-1}, ω is the carrier frequency, ϕ is the CEP, and $\tau = (2\pi/\omega)/\sqrt{2\log(2)}$ is the duration of the pulse. Fig. 1 shows two of the pulse CEPs used in the calculation: in blue $\phi = 0$ and in red $\phi = \pi/2$. We use carrier frequencies ω in the near-infrared, thus far from the bonding dipolar plasmon resonance of the dimer at Ω_{p}.[25,28] With the present

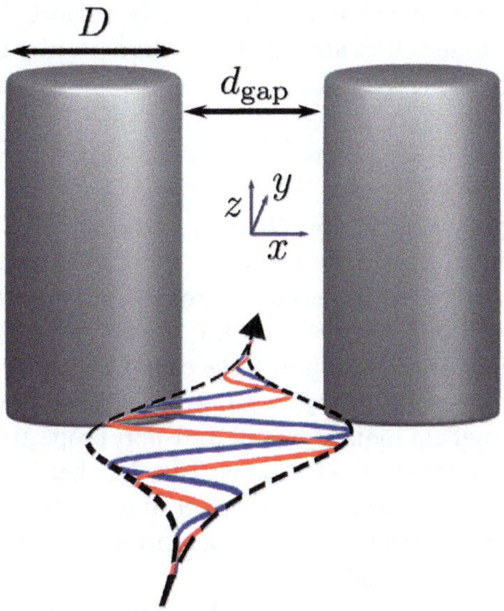

Fig. 1 Dimer of parallel gold cylinders of diameter $D = 10$ nm. The cylinders are separated by a gap of $d_{gap} = 3$ nm. The dimer is excited by a sub-cycle laser pulse with a Gaussian envelope characterized by a carrier frequency ω and a carrier envelope phase ϕ. Two examples of the laser electric field are shown for $\phi = 0$ (blue line) and $\phi = \pi/2$ (red line). We study the dynamics of the optical-field emitted electrons in the plasmonic gap and their dependence on ϕ and ω.

choice of model to describe the electronic structure of the gold cylinders, as discussed below, $\Omega_p = 5.5$ eV. Due to the off-resonance excitation of the system the time-dynamics of the local fields in the gap are similar to those of the incident pulse (sub-cycle). Since the local field in the gap triggers the electron current across the junction it is expected that the net electronic transport will be sensitive to the CEP of the external pulse. Contrary to this situation, for the resonant excitation the dynamics of the field in the junction would be determined by the damping rate of the corresponding plasmon mode. As a result, except for an extremely broad plasmon resonance, the local ac field in the gap would generally last a long time after the end of the external transient. In such a resonant situation, the averaging by many-cycle oscillations would lead to an almost zero net current across the junction and very low sensitivity to the CEP of the incident pulse.

Under these off-resonance illumination conditions, with the corresponding field enhancement in the gap, and the range of frequencies used, the Keldish parameter[29] (γ) obtained in this work is in the range of $\gamma \approx 0.48$–1.90, which is usually considered within the optical-field emission regime.[2,8,30,31]

3 Methodology

To treat the electron dynamics of the system under strong-field illumination we use Time-Dependent Density Functional Theory (TDDFT) based on the Kohn–Sham (KS) scheme.[32–34] The Au cylinders are described with the so-called

Stabilized Jellium Model (SJM).[35] Within the SJM, the valence electrons are treated explicitly, while the atomic cores are represented as a positive background charge density

$$n^+ = \left(\frac{4\pi}{3} r_s^{\ 3}\right)^{-1}, \tag{2}$$

where r_s is the Wigner–Seitz radius (here $r_s = 3.02a_0$, 1 $a_0 = 0.053$ nm). An attractive pseudopotential (V_{st}, see below) is used so that the empirical work function of Au[36], $\Phi = 5.5$ eV, is retrieved with the ground-state Density Functional Theory (DFT)[32] calculations. Even though the use of the SJM neglects atomistic details and effects linked with localized d-electron bands, it allows us to trace the dynamics of valence electrons in the metal, which are involved in the screening, photoemission, strong-field ionization and transport properties in the plasmonic dimer. The jellium model of metal nanoparticles has been successfully used in the context of plasmonics allowing the theoretical prediction of quantum effects later confirmed experimentally[23,37,38] as well as corroborated by TDDFT studies of strong-field emission from metal nanotips.[39]

The time evolution of the KS orbitals is obtained from the 2D time-dependent KS equations (all equations are written in atomic units, a.u., unless otherwise indicated):

$$\left[-\frac{1}{2}\nabla^2 + V_{\text{eff}}[n](x,y,t)\right]\Psi_{\text{k}}(x,y,t) = i\frac{\partial}{\partial t}\Psi_{\text{k}}(x,y,t), \tag{3}$$

where Ψ_{k} are the KS orbitals, $\nabla = \hat{x}\frac{\partial}{\partial x} + \hat{y}\frac{\partial}{\partial y}$, where $\hat{x}(\hat{y})$ is the unit length vector in the direction of the $x(y)$ axis. The effective KS potential $V_{\text{eff}}[n](x, y, t)$ is given by:

$$V_{\text{eff}}[n](x, y, t) = V_{\text{H}}[n](x, y, t) + V_{\text{xc}}[n](x, y, t) + V_{\text{st}}(x, y, t) + V_{\text{ext}}(x, t). \tag{4}$$

The Hartree potential, $V_{\text{H}}[n](x, y, t)$, accounts for the interaction with the charge density of the system. Due to the subwavelength scale of the relevant dimensions of the cylindrical dimer, the non-retarded approximation is considered in the calculations. The exchange–correlation potential, $V_{\text{xc}}[n](x, y, t)$, effectively incorporates the exchange and correlation effects of the many-electron interacting system. Here we use the functional of O. Gunnarsson and B. I. Lundqvist[40] within the Adiabatic Local Density Approximation (ALDA).[34,41] The stabilizing potential, $V_{\text{st}}(x, y, t)$, corresponds to an attractive constant potential in the region of space filled by the nanoparticles, allowing the correct work function of gold to be obtained as already discussed above. Finally, $V_{\text{ext}}(x, t) = xE(t)$ is the interaction potential between an electron and the external electromagnetic field.

The Fourier-grid Hamiltonian approach[42–45] with wavefunction representation on a grid of equidistant points in x- and y-coordinates, and the Split-Operator technique[43,46,47] are used to solve eqn (3) by short time-step propagation. The initial conditions are $\Psi_{\text{k}}(x, y, t = 0) = \Psi_{\text{k}}^{0}(x, y)$, where Ψ_{k}^{0} correspond to the ground state KS orbitals retrieved from self-consistent DFT[32] calculation of the system in the absence of the external electromagnetic field.

The time-dependent electronic density and electron current density, $n(x, y, t)$ and $\vec{j}(x, y, t)$ respectively, are computed at each time step as follows:

$$n(x,y,t) = 2\sum_{\mathrm{k}\in \mathrm{occ}} \chi_{\mathrm{k}} |\Psi_{\mathrm{k}}(x,y,t)|^2, \tag{5}$$

$$\vec{j}(x,y,t) = 2\sum_{\mathrm{k}\in \mathrm{occ}} \chi_{\mathrm{k}} \mathrm{Re}\left[\Psi_{\mathrm{k}}^{*}(x,y,t)\ \hat{\mathbf{p}}\ \Psi_{\mathrm{k}}(x,y,t)\right], \tag{6}$$

where $\hat{\mathbf{p}} = -i\nabla$, the sum runs over all the occupied KS orbitals, the factor 2 accounts for the spin degeneracy, and $\chi_{\mathrm{k}} = \frac{1}{\pi}\sqrt{2(E_{\mathrm{F}} - E_{\mathrm{k}})}$ accounts for the number of electronic states associated with the z-motion. E_{F} stands for the Fermi level of the nanostructures and E_{k} for the ground-state energy of the orbital k.

4 Results

To study the dynamics of strong-field emitted electrons in the gap, we illuminate the system with the field transient given by eqn (1) mimicking a single-cycle optical pulse, and at each time step we compute the electron current across the mid-plane of the gap,

$$J(x=0,t) = \int_{-\infty}^{\infty} \hat{x}\,\vec{j}(x=0,y,t)\mathrm{d}y. \tag{7}$$

Fig. 2 shows the result of such analysis for the case of a pulse carrier frequency $\omega = 0.67$ eV, where the electric field at the centre of the gap, E_{gap}, is plotted in blue and the current density across the mid-plane of the gap, $J(x = 0, t)$, is plotted in red. E_{gap} comprises both the electric field of the incident pulse and the induced electric field due to the polarisation of the nanowires. Prior to the central peak of the pulse at $t = 0$, the electromagnetic field in the junction is not strong enough to produce a significant electron current through the middle of the gap. Upon arrival, the central peak produces the emission of electrons from the surface of the cylinder on the right hand side in Fig. 1 and therefore a negative electronic current is observed at $t \approx 1$ fs. Around $t \approx 1.5$ fs, the field in the gap changes polarity and a second smaller current density peak can be observed, which flows in the opposite direction to the main one. This weaker current density is formed

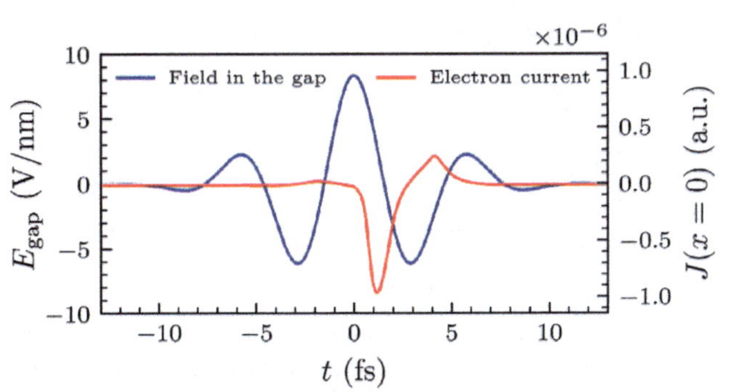

Fig. 2 TDDFT results for the dimer illuminated by a pulse of carrier frequency $\omega = 0.67$ eV and CEP $\phi = 0$. Blue line: electric field at the centre of the gap, E_{gap}. Red line: electron current across the mid-plane of the gap, $J(x = 0, t)$.

from two contributions. The first contribution is due to the cylinder situated on the left hand side in Fig. 1. The second contribution, as we will show below, is originating from electrons which were emitted from the right cylinder but did not reach the opposite nanoparticle before E_{gap} changed polarity, thus inducing electron quiver motion.

Integrating in time the current in eqn (7), one can obtain the number of electrons transferred across the gap. The analysis of the net electron transport as a function of CEP is shown in Fig. 3 for different carrier frequencies of the incident pulse. We first focus our attention on the lowest frequencies, $\omega = 0.32$ eV (solid grey line) and $\omega = 0.48$ eV (solid pink line). For these frequencies, the results in Fig. 3 show a behaviour consistent with the direct propagation of the emitted electrons across the junction without any effect of the quiver motion.[10–12] Indeed, for a CEP $\phi = \pi/2$ or $\phi = 3\pi/2$, the local field in the gap, E_{gap}, is antisymmetric (sine-like) with respect to the centre of the Gaussian envelope and thus, the field shows two identical peaks of opposite polarity (see the red line in the sketch of Fig. 1). In such a situation, the amount of electrons transferred in both directions is equal, they compensate each other, and the net transport is zero. The maximum electron transfer occurs for $\phi = 0$ or $\phi = \pi$. The incident electromagnetic pulse in this case has a symmetric electric field profile, with the main positive or negative peak centred at $t = 0$ (see Fig. 2). As the optical-field emission process is highly non-linear,[30,31] the electron current in the junction is dominated by the ultra-short burst of electrons emitted at the half-period with strongest field, as nicely observed in Fig. 2. Under the assumption that all the electrons arrive to the opposite surface, $\phi = 0$ or $\phi = \pi$ correspond to the

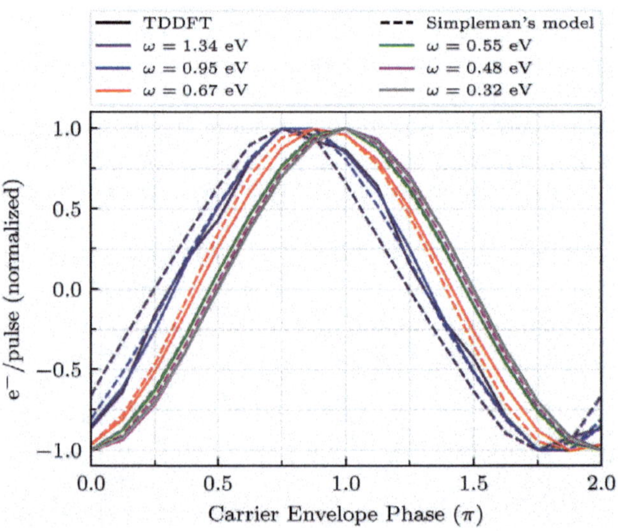

Fig. 3 Normalized electron transfer per pulse as a function of the CEP ϕ. Different colours indicate the various carrier frequencies considered. Solid lines: results of the TDDFT calculations. Dashed lines: results obtained using Simpleman's model for electron emission and transport. The quiver amplitude associated to each of the frequencies ($\omega = 0.32$ eV, 0.48 eV, 0.55 eV, 0.67 eV, 0.95 eV and 1.34 eV) is $X_q = 6.33$ nm, 2.82 nm, 2.12 nm, 1.42 nm, 0.72 nm and 0.37 nm, respectively.

maximum number of electrons transferred in positive ($\phi = \pi$) or negative ($\phi = 0$) directions of the x axis.

However, when the illumination frequency is increased, the Carrier Envelope Phase (CEP) at which the maximum electron transfer is produced shifts away from the expected $\phi = 0$ and $\phi = \pi$ values. In contrast to single nanoparticles where the evanescent near-field decays rapidly as the distance to the nanoparticle's surface is increased and the quiver motion might be quenched,[10] the field in small plasmonic gaps is almost constant across the full distance separating the nanoparticles. Therefore, the quiver motion of the emitted electrons needs to be carefully considered. This quiver motion of the electrons in the gap is responsible for the shift of the value of the CEP corresponding to the maximum net electron current shown in Fig. 3, as we analyse below.

Classical analysis

We employ the classical Simpleman's model (SMM) to study the quiver motion of the photoemitted electrons in the plasmonic gap, and the role that the quiver motion plays in the electron transfer between the nanoparticles. The SMM has been first developed to describe the optical field electron emission from atomic species,[48–50] and later successfully applied to electron emission from surfaces of metallic tips.[30,31,39] To qualitatively capture the main physical effects, while keeping the analysis of the results simple, we neglect the 2D aspect of the problem and only consider the electron motion along the x-axis. The optical field electron emission is treated as a two step process. First we use the field in the gap, E_{gap}, obtained from the TDDFT calculations to compute the instantaneous Fowler–Nordheim tunneling current,[51] J_{FN}, at each time step within the pulse,

$$J_{\text{FN}} \propto -E_{\text{gap}}(t_e)\left|E_{\text{gap}}(t_e)\right|\exp\left(-\frac{b\Phi^{\frac{3}{2}}}{\left|E_{\text{gap}}(t_e)\right|}\right), \tag{8}$$

where $b = 6.83$ V (nm^{-1} eV$^{-3/2}$), $\Phi = 5.5$ eV is the work function of gold, and $E_{\text{gap}}(t_e)$ is the electric field in the gap at the time instant of emission defined as t_e. The second step consists of calculating the classical trajectory on the x-axis of an electron emitted at time t_e and subjected to the time-dependent electric field E_{gap} (assuming that E_{gap} is homogeneous in the gap region):

$$\dot{v}_x = -E_{\text{gap}}. \tag{9}$$

The initial condition $v_x(t_e) = 0$ corresponds to a tunneling electron emerging from the potential barrier at the metal/vacuum interface into the classically allowed region in the gap. Using the relationship between the electric field and the vector potential (A), $E = -\mathrm{d}A/\mathrm{d}t$, the velocity of the classical electron at any time $t > t_e$ is:

$$v_x(t) = A_{\text{gap}}(t) - A_{\text{gap}}(t_e). \tag{10}$$

This trajectory is computed until the electron reaches the surface of one of the two cylinders. If the electron crosses the gap, the trajectory is considered for the calculation of the net charge transfer assuming that it contributes with a weight given by eqn (8). The trajectory is discarded if, because of the quiver motion, the electron returns back to the initial nanoparticle. We thus neglect the possible, albeit small, contribution of the electron back-scattering.[39]

Analysing eqn (10), a third scenario is possible. If an electron is emitted at a time t_e such that E_{gap} is a local maximum, the associated vector potential is $A_{gap}(t_e) = 0$. Furthermore, the finite duration of the illumination transient ensures that $A_{gap}(t \to \infty) = 0$. Therefore, electrons emitted within the condition $A_{gap}(t_e) = 0$ might end up in the middle of the gap with velocity $v_x(t \to \infty) = 0$. We will call these trajectories as "trapped" trajectories.

The electron transfer across the plasmonic gap calculated within the SMM as a function of the CEP and frequency of the incident pulse is shown with dashed lines in Fig. 3. The agreement with the TDDFT calculations (solid lines) is remarkable. The Simpleman's model allows the identification of the quiver amplitude X_q as a key parameter controlling the electron transfer. The quiver amplitude, X_q, is the amplitude of the oscillatory movement of an electron in a homogeneous harmonic electric field, $E(t) = E_0 \cos(\omega t)$, of frequency ω and amplitude strength E_0:

$$X_q = \frac{E_0}{\omega^2}. \tag{11}$$

Taking E_0 as the maximum of the total electric field generated in the gap, $E_0 = [E_{gap}]_{max} \approx 8.45$ V nm^{-1}, the carrier frequencies of the incident electromagnetic pulse $\omega = 0.32$ eV, 0.48 eV, 0.55 eV, 0.67 eV, 0.95 eV and 1.34 eV result in the corresponding quiver amplitudes $X_q = 6.33$ nm, 2.82 nm, 2.12 nm, 1.42 nm, 0.72 nm and 0.37 nm, respectively. As follows from the results presented in Fig. 3, the maximum electric transport is produced at CEP $\phi = 0$ or $\phi = \pi$ only for the cases where $X_q \gtrsim d_{gap}$.

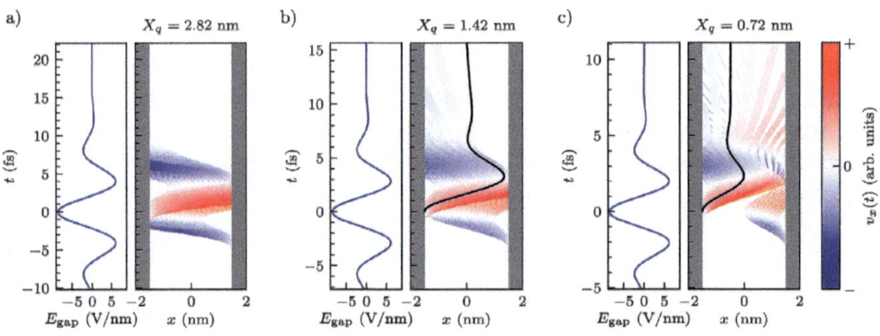

Fig. 4 Classical trajectories obtained using the Simpleman's model for carrier frequencies of the incident pulse (a) $\omega = 0.48$ eV, (b) $\omega = 0.67$ eV and (c) $\omega = 0.95$ eV. In all cases the CEP $\phi = \pi$. The electric field at the centre of the gap, E_{gap}, (horizontal axis) as a function of time (vertical axis) is shown to the left of each panel. The classical trajectories of electrons emitted along the x-axis are shown to the right of each panel. The position x is shown on the horizontal axis and time on the vertical axis. The x-component of the velocity of the electrons in each trajectory is colour-coded with positive velocity in red and negative in blue. The velocity in each panel is normalized by the maximum velocity in the x-direction to the range [−1, 1]. The weight of each trajectory as given by eqn (8) is represented by the width of the corresponding trajectory line. Wider lines imply a larger weight and thus a larger amount of charge moving along in such trajectory. The black lines in (b) and (c) mark two examples of "trapped" trajectories.

Fig. 4 shows the electric field in the gap (left line-plot in each panel) and the color-map formed by classical electron trajectories (right side of each panel) for three different carrier frequencies selected in such a way that the respective quiver amplitudes belong to the two different regimes, *i.e.*, $X_q \gtrsim d_{gap}$ and $X_q < d_{gap}$. In the three cases the CEP is $\phi = \pi$. The velocity of the electron in each trajectory is colour-coded, having positive velocity ($v_x > 0$) marked in red and negative velocity ($v_x < 0$) in blue. The width of each line is linked with the tunneling current in eqn (8), which indicates that a larger amount of charge is moving along wider lines. In the three examples quiver motion is observed depending on the moment t_e when the electrons are emitted. In Fig. 4a, $X_q = 2.82$ nm and thus, most of the electron trajectories reach the opposite nanoparticle before the ballistic motion is inverted. In particular it can be observed how almost all the trajectories involving a large number of emitted electrons (wider lines) directly reach the opposite nanoparticle, corroborating that the maximum net electron transfer occurs for $\phi = \pi$ (see the pink line in Fig. 3).

Fig. 4b and c show the situation where $X_q < d_{gap}$. In both cases the quiver motion is more obvious and relevant as X_q decreases. An increasing number of trajectories quiver back because the polarity of the field changes before they reach the opposite side of the gap. It is particularly relevant that in these two cases many of the trajectories involving a large number of emitted electrons (wide lines) are not able to cross the gap and thus the maximum electron transfer no longer occurs for the incident pulse with CEP $\phi = \pi$ which is the value displayed in this figure. This effect is consistent with the situation observed for larger illumination frequencies displayed with blue and red lines in Fig. 3. The black lines in Fig. 4b and c follow two examples of "trapped" trajectories where, as already mentioned, the final velocity is $v_x \approx 0$. The electrons moving along "trapped" trajectories are thus stopped in the middle of the gap upon the termination of the electromagnetic pulse. In practice, these electrons will be absorbed by the substrate or deviated by the stray fields without contributing to the coherent charge transport. As follows from the discussion above, the classical SMM analysis turns out to be a powerful and intuitive tool to trace the dynamics of ultrafast electron currents in plasmonic gaps driven by light.

5 Discussion and conclusion

Using a combination of quantum calculations (TDDFT) and classical modelling (Simpleman's model) we have been able to address the complex dynamics of strong-field emission currents in plasmonic gaps. In contrast with the reported results using metallic tips,[10,12] we observed that the homogeneous field distribution characteristic of small plasmonic gaps produces a situation where the quiver motion of electrons has a substantial impact on the electron current. The ratio between the electron quiver amplitude, X_q, and the gap distance separating the nanoparticles, d_{gap}, is a key parameter to establish the absolute Carrier Envelope Phase (CEP) which produces the maximum electron transport between nanoparticles. These results reveal the importance of the CEP of an incident pulse to control the mutual coherence between electron and photon dynamics in specific metallic junctions.

Conflicts of interest

There are no conflicts of interest to declare.

Acknowledgements

G. A. acknowledges project PI2017-30 of the Departamento de Educación, Política Lingüística y Cultura of the Basque government and, G. A. and J. A. acknowledge funding from project FIS2016-80174-P of the Ministry of Economy, Industry and Competitiveness MINEICO.

Notes and references

1 P. Dombi, S. E. Irvine, P. Rácz, M. Lenner, N. Kroó, G. Farkas, A. Mitrofanov, A. Baltuška, T. Fuji, F. Krausz and A. Y. Elezzabi, *Opt. Express*, 2010, **18**, 24206–24212.

2 S. V. Yalunin, M. Gulde and C. Ropers, *Phys. Rev. B: Condens. Matter Mater. Phys.*, 2011, **84**, 195426.

3 J. Aizpurua, G. W. Bryant, L. J. Richter, F. G. De Abajo, B. K. Kelley and T. Mallouk, *Phys. Rev. B: Condens. Matter Mater. Phys.*, 2005, **71**, 235420.

4 M. Pelton, J. Aizpurua and G. Bryant, *Laser Photonics Rev.*, 2008, **2**, 136–159.

5 L. Novotny and N. Van Hulst, *Nat. Photonics*, 2011, **5**, 83–90.

6 S. A. Maier, *Plasmonics: fundamentals and applications*, Springer Science & Business Media, 2007.

7 L. Novotny and B. Hecht, *Principles of nano-optics*, Cambridge university press, 2012.

8 R. Bormann, M. Gulde, A. Weismann, S. V. Yalunin and C. Ropers, *Phys. Rev. Lett.*, 2010, **105**, 147601.

9 M. Krüger, M. Schenk and P. Hommelhoff, *Nature*, 2011, **475**, 78.

10 G. Herink, D. R. Solli, M. Gulde and C. Ropers, *Nature*, 2012, **483**, 190.

11 D. J. Park, B. Piglosiewicz, S. Schmidt, H. Kollmann, M. Mascheck and C. Lienau, *Phys. Rev. Lett.*, 2012, **109**, 244803.

12 P. Dombi, A. Hörl, P. Rácz, I. Márton, A. Trügler, J. R. Krenn and U. Hohenester, *Nano Lett.*, 2013, **13**, 674–678.

13 B. Piglosiewicz, S. Schmidt, D. J. Park, J. Vogelsang, P. Groß, C. Manzoni, P. Farinello, G. Cerullo and C. Lienau, *Nat. Photonics*, 2014, **8**, 37.

14 M. Lehr, B. Foerster, M. Schmitt, K. Krüger, C. Sönnichsen, G. Schönhense and H.-J. Elmers, *Nano Lett.*, 2017, **17**, 6606–6612.

15 W. P. Putnam, R. G. Hobbs, P. D. Keathley, K. K. Berggren and F. X. Kärtner, *Nat. Phys.*, 2017, **13**, 335.

16 A. V. Crewe, D. N. Eggenberger, J. Wall and L. M. Welter, *Rev. Sci. Instrum.*, 1968, **39**, 576–583.

17 P. Nordlander, C. Oubre, E. Prodan, K. Li and M. I. Stockman, *Nano Lett.*, 2004, **4**, 899–903.

18 P. Mühlschlegel, H.-J. Eisler, O. J. F. Martin, B. Hecht and D. W. Pohl, *Science*, 2005, **308**, 1607–1609.

19 I. Romero, J. Aizpurua, G. W. Bryant and F. J. G. De Abajo, *Opt. Express*, 2006, **14**, 9988–9999.

20 H. Xu, E. J. Bjerneld, M. Käll and L. Börjesson, *Phys. Rev. Lett.*, 1999, **83**, 4357.

21 S. S. Acimovic, M. P. Kreuzer, M. U. González and R. Quidant, *ACS Nano*, 2009, **3**, 1231–1237.
22 M. Liao, S. Jiang, C. Hu, R. Zhang, Y. Kuang, J. Zhu, Y. Zhang and Z. Dong, *Nano Lett.*, 2016, **16**, 4040–4046.
23 K. J. Savage, M. M. Hawkeye, R. Esteban, A. G. Borisov, J. Aizpurua and J. J. Baumberg, *Nature*, 2012, **491**, 574–577.
24 T. Rybka, M. Ludwig, M. F. Schmalz, V. Knittel, D. Brida and A. Leitenstorfer, *Nat. Photonics*, 2016, **10**, 667.
25 D. C. Marinica, M. Zapata, P. Nordlander, A. K. Kazansky, P. M. Echenique, J. Aizpurua and A. G. Borisov, *Sci. Adv.*, 2015, **1**, e1501095.
26 D. R. Ward, F. Hüser, F. Pauly, J. C. Cuevas and D. Natelson, *Nat. Nanotechnol.*, 2010, **5**, 732.
27 T. L. Cocker, V. Jelic, M. Gupta, S. J. Molesky, J. A. J. Burgess, G. De Los Reyes, L. V. Titova, Y. Y. Tsui, M. R. Freeman and F. A. Hegmann, *Nat. Photonics*, 2013, **7**, 620.
28 G. Aguirregabiria, D. C. Marinica, R. Esteban, A. K. Kazansky, J. Aizpurua and A. G. Borisov, *Phys. Rev. B*, 2018, **97**, 115430.
29 L. Keldysh, *et al.*, *J. Exp. Theor. Phys.*, 1965, **20**, 1307–1314.
30 M. Krüger, M. Schenk, M. Förster and P. Hommelhoff, *J. Phys. B: At., Mol. Opt. Phys.*, 2012, **45**, 074006.
31 M. Krüger, C. Lemell, G. Wachter, J. Burgdörfer and P. Hommelhoff, *J. Phys. B: At., Mol. Opt. Phys.*, 2018, **51**, 172001.
32 W. Kohn and L. J. Sham, *Phys. Rev.*, 1965, **140**, A1133.
33 E. Runge and E. K. U. Gross, *Phys. Rev. Lett.*, 1984, **52**, 997.
34 M. A. Marques and E. Gross, *Annu. Rev. Phys. Chem.*, 2004, **55**, 427–455.
35 J. P. Perdew, H. Q. Tran and E. D. Smith, *Phys. Rev. B: Condens. Matter Mater. Phys.*, 1990, **42**, 11627.
36 E. V. Chulkov, V. M. Silkin and P. M. Echenique, *Surf. Sci.*, 1999, **437**, 330–352.
37 R. Esteban, A. G. Borisov, P. Nordlander and J. Aizpurua, *Nat. Commun.*, 2012, **3**, 825.
38 J. A. Scholl, A. García-Etxarri, A. L. Koh and J. A. Dionne, *Nano Lett.*, 2013, **13**, 564–569.
39 M. Krüger, M. Schenk, P. Hommelhoff, G. Wachter, C. Lemell and J. Burgdörfer, *New J. Phys.*, 2012, **14**, 085019.
40 O. Gunnarsson and B. Lundqvist, *Phys. Rev. B: Condens. Matter Mater. Phys.*, 1976, **13**, 4274.
41 K. Burke, J. Werschnik and E. Gross, *J. Chem. Phys.*, 2005, **123**, 062206.
42 D. Kosloff and R. Kosloff, *J. Comput. Phys.*, 1983, **52**, 35–53.
43 R. Kosloff, *J. Phys. Chem.*, 1988, **92**, 2087–2100.
44 C. C. Marston and G. G. Balint-Kurti, *J. Chem. Phys.*, 1989, **91**, 3571–3576.
45 R. Kosloff, *Dynamics of molecules and chemical reactions*, 1996, pp. 185–230.
46 M. D. Feit, J. A. Fleck and A. Steiger, *J. Comput. Phys.*, 1982, **47**, 412–433.
47 M. D. Feit and J. A. Fleck Jr, *J. Chem. Phys.*, 1983, **78**, 301–308.
48 P. B. Corkum, *Phys. Rev. Lett.*, 1993, **71**, 1994.
49 G. G. Paulus, W. Becker, W. Nicklich and H. Walther, *J. Phys. B: At., Mol. Opt. Phys.*, 1994, **27**, L703.
50 G. Paulus, W. Becker and H. Walther, *Phys. Rev. A: At., Mol., Opt. Phys.*, 1995, **52**, 4043.
51 R. G. Forbes, *Appl. Phys. Lett.*, 2006, **89**, 113122.

Faraday Discussions

PAPER

Direct optical excitation of dark plasmons for hot electron generation

Niclas S. Mueller,[a] Bruno G. M. Vieira,[ab] Dominik Höing,[c] Florian Schulz,[c] Eduardo B. Barros,[c] Holger Lange[c] and Stephanie Reich*[a]

Received 9th October 2018, Accepted 12th November 2018
DOI: 10.1039/c8fd00149a

An ideal plasmonic system for hot-electron generation allows the optical excitation of plasmons, limits radiation losses, exhibits strong non-radiative electron damping, and is made from scalable and cost-effective materials. Here we demonstrate the optical excitation of dark interlayer plasmons in bilayers of colloidal gold nanoparticles. This excitation is created by an antiparallel orientation of the dipole moments in the nanoparticle layers; it is expected to exhibit strongly reduced radiative damping. Despite the vanishing dipole moment, an incoming electromagnetic wave that is propagating normal to the surface will excite the dark mode due to field retardation. We observe a strong peak in the absorption spectrum of a colloidal gold bilayer (nanoparticle diameter = 46 nm); this peak is absent for a nanoparticle monolayer. The full width at half maximum of the dark mode is 230 meV for an ideal nanoparticle crystal and 320 meV for the structure produced by self-assembly out of solution. The position and width of the dark plasmon are efficiently tailored by the interparticle distance within the layer, nanoparticle size and layer number. We present time-resolved pump and probe experiments of hot-electron generation by bright and dark bilayer nanoparticle modes.

Introduction

Plasmons are an excellent source of excited electrons.[1–3] The excess energy of electrons may be used for driving catalytic reactions, in imaging applications and in photothermal heat generation, amongst other examples.[2,4–9] Maximizing the yield of hot electrons requires careful engineering of the plasmonic decay channels so that dephasing is preferred over energy loss. In particular, relaxation channels that distribute a large amount of energy, such as photon emission or Auger-type processes, should be minimized.

[a]*Department of Physics, Freie Universität Berlin, Arnimallee 14, 14195 Berlin, Germany. E-mail: sreich@zedat.fu-berlin.de*
[b]*Departamento de Física, Universidade Federal do Ceará, Fortaleza, Brazil*
[c]*Institute of Physical Chemistry, University of Hamburg, Martin-Luther-King-Platz 6, 20146 Hamburg, Germany*

Plasmon decay

$$\Gamma = \Gamma_\mathrm{b} + \Gamma_\mathrm{s} + \Gamma_\mathrm{r} \quad (1)$$

occurs *via* intrinsic (Drude) damping, Γ_b, which includes all of the damping processes of the bulk metal, surface-mediated (Laundau) damping, Γ_s, and radiative damping, Γ_r, by photon scattering.[1,10–13] The desired relaxation channel for hot electron generation is Γ_s, where the surface breaks the conservation of crystal momentum.[1,14] This mainly leads to dephasing with little energy loss, thus producing highly excited electrons.

One approach for optimising Γ_s is to reduce the size of the metal nanoparticles below the electron mean free path.[1,15–18] Since light scattering by metal nanoparticles scales with the volume squared, a small diameter also reduces Γ_r, making surface scattering a sizable contribution for particle diameters below 10 nm.[11,12,15] Bulk damping in gold is $\Gamma_\mathrm{b} \approx$ 100–200 meV below 2.1 eV;[19] Γ_s will add 20–30 meV for an effective particle size of 10 nm.[17,18] However, the small volumes also reduce the cross section for optical absorption.[12] Since optical excitation is the most viable and economic route for plasmon excitations, a low absorption cross section reduces the total rate of hot electron generation per nanoparticle, even if the conversion of plasmons into hot electrons is highly efficient.

Another idea to increase plasmon relaxation *via* surface scattering is through the formation of hot spots in nanoparticle agglomerates and plasmonic oligomers.[14,20–22] If two (or more) particles are in close contact, their plasmonic near fields interact, thus creating volumes of particularly high field enhancement (hot spots).[12] Hot spots increase the rate of hot electron generation, because damping by surface scattering depends on the magnitude of the electric field close to the metal surface.[1,21,23,24] A second effect is that the tight focusing of the field in a hot spot lifts the translational symmetry and thus the conservation of (quasi-) momentum.[1,14] The drawback of the second approach is the increase in radiative damping, Γ_r. Optically active, "bright" plasmons decay rapidly by emitting photons into the far field.[10,12] Since the radiative decay channel is highly efficient for coupled nanoparticles, the plasmon lifetime is limited to a few fs and energy is lost to the radiative processes.[25]

Dark plasmons are plasmonic excitations with a vanishing net dipole moment, *e.g.*, higher-order multipole modes in metal nanoparticles such as quadrupoles.[26,27] Assemblies of nanoparticles possess dark modes that arise by combining the dipolar plasmons of the individual particles into a collective mode with a vanishing dipole moment.[28–30] Because dark plasmons have no net dipole moment, they typically do not interact with far field radiation; light is neither absorbed nor emitted/scattered.[31] The absence of light emission is an advantage for hot-electron generation, because the entire excess energy is channelled into non-radiative processes, *i.e.* the creation of excited electrons. The absence of light absorption, on the other hand, is a major disadvantage because it prevents the optical excitation of dark modes. This disadvantage may be overcome, *e.g.* by exploiting Fano resonances for excitation in plasmonic oligomers, exciting the plasmon *via* a magnetic field instead of an electric field in metal–insulator–metal slabs, or exciting nanoparticle assemblies with cylindrical vector beams.[30,32–34] These routes, however, require the nanofabrication of plasmonic nanostructures *via* electron-beam lithography and tailored beams of structured light.

Here we demonstrate the excitation of dark modes and the creation of hot electrons using linearly polarized light and scalable, cost-effective plasmonic surfaces. The surfaces are created *via* the self-assembly of gold nanoparticles into colloidal mono- and bi-layers. Above a characteristic diameter (30–50 nm depending on the surrounding medium), light absorption is altered by retardation, as we show through our symmetry analysis and numerical calculations. Light is absorbed by bright dipole modes and a dark mode in the bilayer that arises through the combination of the dipolar modes of the two layers into an excitation with a vanishing net dipole moment. We experimentally demonstrate dark plasmon absorption in gold bilayers. The hot-electron dynamics subsequent to plasmon excitation are analyzed *via* time-resolved pump-probe spectroscopy.

Methods

Synthesis of colloidal gold nanoparticle layers

The synthesis and characterization of gold mono- to few-layers is described in detail in ref. 35 and 36. To summarize, quasispherical citrate-stabilized gold nanoparticles with a diameter of 40 nm or 46 nm were synthesized based on the seeded-growth protocol. The particles were functionalized with thiolated polystyrene and mixed (6 mL) with polystyrene (0.16 mM) and oleylamine (1 mM) in toluene (6 mL) as well as ethanol (6 mL). After rapid stirring (30 min) and phase separation, the organic phase was extracted, concentrated and purified by centrifugation, resulting in a final gold nanoparticle concentration of 2.1 nM. The nanoparticle films were prepared *via* self-assembly at the liquid–liquid interface. For the transient absorption measurements, the nanoparticle films were deposited on quartz substrates. The gold nanoparticle solution in toluene was pipetted onto diethylene glycol in a Teflon well. The well was covered with a quartz slide and left undisturbed at room temperature until the solvent had evaporated and a gold nanoparticle film had formed.[35,36] For their characterization by transmission electron microscopy (TEM) and microabsorbance measurements, the films were carefully skimmed off with a carbon-coated copper grid. The probability of bilayer formation was increased by repeating the procedure. The grids were dried afterwards on filter paper for at least 24 hours.

Microabsorbance spectroscopy

Optical absorbance and reflectance spectra of the gold nanoparticle layers were recorded using a home-built microabsorbance spectrometer. For a detailed description of the optical setup, please see ref. 36. A supercontinuum laser (Fianium) was used as a broadband light source. The light was linearly polarized, the spectral range was reduced to the measurement range through a combination of long- and short-pass filters, and the total laser power was reduced to below 100 μW. An inverted microscope was used to focus the light on the sample and find the region of interest (see Fig. 4c below). The light was focused on the substrate either by a 100× objective with a numerical aperture (NA) of 0.9 or by a 20× objective with an NA of 0.25. The transmitted light was collected by objectives with a similar numerical aperture and detected by a fibre spectrometer (Avantes). As a reference for the transmittance, T, we recorded the transmission through the substrate without nanoparticle layers. The reflected light was separated from the incoming light by a beam splitter and guided into the spectrometer *via* an optical fibre. A silver mirror was used as a reference to obtain the reflectance, R. The absorbance spectrum was calculated as $1 - T - R$. To determine the plasmon peak position and full width at half maximum (FWHM) we first subtracted the contribution from the gold interband transitions from the measured absorption spectrum. The absorption by the plasmons was fitted with two Lorentzian peaks (one peak for the monolayer that has only a bright mode).

Transient absorption measurements

For the transient absorption experiments, a commercial amplified Ti–sapphire laser system (Spitfire Ace, 800 nm, 1 kHz, 35 fs; Spectra Physics) was employed as the main excitation source. Its output was split into two parts. One seeded an optical parametric amplifier (TOPAS-Prime; Light Conversion) that provides a tunable excitation wavelength over the range of 290–1600 nm. The other output was used to generate the probe pulses. A commercial TA setup (Helios; Ultrafast Systems) was used to record the experiment. The quartz slides with the gold layers were mounted in a non-collinear geometry. The diameter of the focused probe beam was around 100 μm, whereas that of the pump beam was around 500 μm. Each differential spectrum was referenced by subtracting the ground-state spectrum from the pumped spectrum, using a 500 Hz chopper. The pump intensity was kept constant at 100 μJ cm^{-2} (200 nJ per pulse).

Finite-difference time-domain simulations

Simulated spectra were obtained *via* the commercial software package, Lumerical FDTD Solutions. The layers were implemented by defining a unit cell and using periodic boundary conditions. Mesh-override regions of 0.25 nm were used for greater accuracy. The nanoparticle layers were illuminated by a plane-wave source. One power monitor was included behind the layers to capture the transmittance, T, and another was included behind the source to gather the reflectance, R. The absorbance, A, was later calculated as $A = 1 - T - R$. The gold dielectric functions were implemented by fitting the experimental data reported by Johnson and Christy.[37] The nanoparticle layers were placed on top of a 10 nm thick amorphous carbon layer, as in the experiments, for the simulations shown in Fig. 5–7. The amorphous carbon was implemented by using the experimental data of Larruquert *et al.*[38] A gold slab with a thickness of 200 nm was added for the simulation of the layers over a gold substrate, as shown in Fig. 9. Convergence tests were carried out by varying the mesh sizes in order to ensure the accuracy of the simulation results.

Results and discussion

Optical selection rules for dark interlayer modes

In order to understand the optical absorption by nanoparticle layers, we carried out a symmetry-based analysis of colloidal bilayers or crystals in a hexagonal close packed (hcp) structure.[39,40] The hcp structure belongs to the D_{6h} point group; it has two nanoparticles in a hexagonal unit cell, one at the origin and the other at (2/3, 1/3, 1/2), as shown in Fig. 1a. We will restrict our analysis to plasmon modes

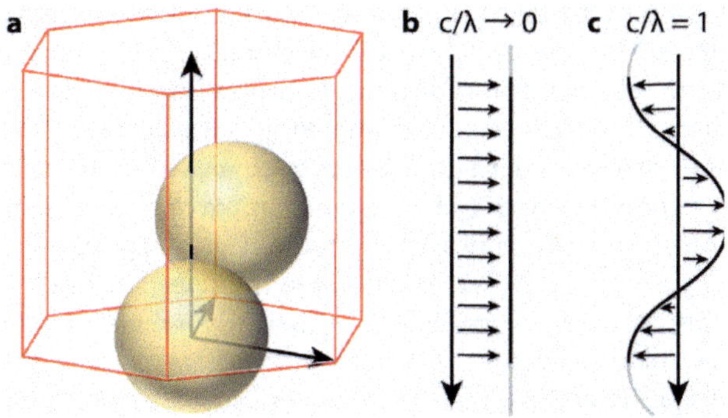

Fig. 1 (a) Unit cell of the hcp crystal with two nanoparticles in the unit cell. Electric field of the incoming light, assuming normal incidence under the assumption of (b) infinite wavelength and (c) a wavelength that matches the translational periodicity of the crystal shown in (a).

that arise as combinations of the dipole excitation in the nanoparticles; this argumentation extends straightforwardly to multipolar excitations.

The three-fold degenerate dipolar plasmons of the individual nanoparticles induce four dipole-derived eigenmodes in the bilayer[39]

$$\Gamma_{\mathrm{pl}}^{\mathrm{dip}} = \mathrm{A_{2u}} \oplus \mathrm{B_{2g}} \oplus \mathrm{E_{1u}} \oplus \mathrm{E_{2g}}. \tag{2}$$

Two of these modes are non-degenerate, whilst two are two-fold degenerate. The symmetry-adapted eigenvectors are obtained by projection operators, which are shown in Fig. 2. Out of the representations detailed in eqn (2), the A_{2u} and E_{1u}

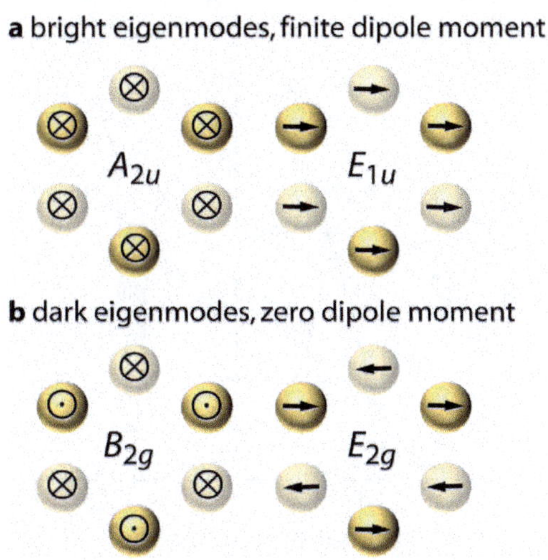

Fig. 2 Dipole-related plasmon eigenmodes of an hcp-stacked nanoparticle bilayer with a (a) finite and (b) vanishing dipole moment. The irreducible representation of each mode is shown. The dark (bright) disks represent the nanoparticles in the top (bottom) layer. The size of the nanoparticles is not to scale.

representations are normally considered as optically active and have finite dipole moments, whereas B_{2g} and E_{2g} are considered dark (dipole forbidden). Within group theory, the selection rules for optical absorption are derived by analysing the transformation properties of the incoming electric field.[39] The common approximations are to consider only the dipole expansion of the electric field and to assume the light wavelength to be large in comparison to the extension of the unit cell (Fig. 1b). Under this approximation, the electromagnetic field transforms like a vector, $\Gamma_{vec} = A_{2u} \oplus E_{1u}$, thus leading to the selection rules listed above.

The assumption of infinite wavelength is justified for molecules and crystalline solids, because their characteristic lengths (1 nm) are much smaller than the wavelength of light (100–1000 nm). However, in the case of colloidal layers and crystals, the much larger extension of the unit cell (10–100 nm) makes retardation effects dominant. To examine the other limiting case, we consider the wavelength of the electromagnetic field to be identical to the translational symmetry of a colloidal crystal (Fig. 1c). For normal incidence, as shown in Fig. 1c, the field transforms according to E_{2g}. It will be able to excite the in-plane antiparallel dark mode shown on the right of Fig. 2b.

Symmetry analysis predicted the optical selection rules of the nanoparticle bilayers to depend on the ratio between the nanoparticle diameter, d, and the wavelength, λ, of the incoming light. For realistic nanoparticles, d will always be much smaller than λ, but for at least some time during its period of oscillation, the electric field will be anti-parallel in the top and bottom layers. We expect this to activate absorption by the E_{2g} dark mode.

To study the effect of retardation on the optical spectra, we simulated a gold nanoparticle bilayer in a vacuum with hcp stacking, a diameter, d, of 20 nm, and

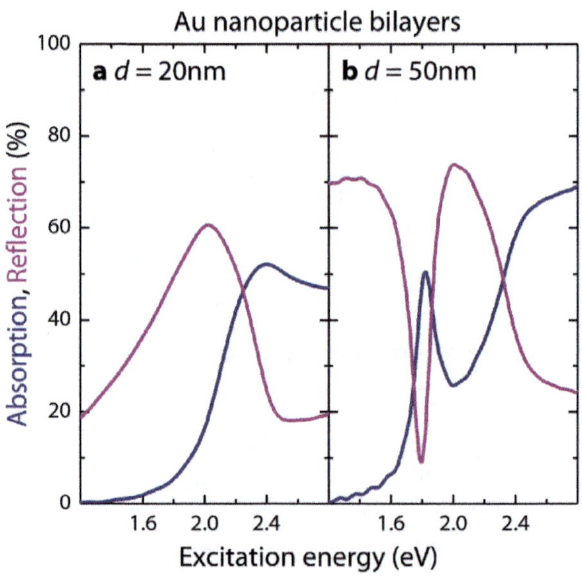

Fig. 3 Simulation of the absorption (blue) and reflection (magenta) of hcp-stacked gold nanoparticle bilayers with particle diameters of (a) 20 nm and (b) 50 nm. The absorption peak at 1.82 eV is caused by the dark plasmons. Gaps of 2 nm and a background medium with an index of $n = 1$ were assumed.

a gap size, g, of 2 nm.[41] The absorption spectrum (Fig. 3a) is dominated by the step function in the absorption coefficient due to interband transitions.[42] The plasmon peak at 2.24 eV overlaps with the interband transitions. It is strongly damped (FWHM = γ = 450 meV) and is hardly visible in the spectrum. Likewise, the reflection is broad and unstructured. After increasing the nanoparticle diameter to 50 nm (Fig. 3b), the simulation predicts a strong (50%) absorption peak at 1.82 eV. It is accompanied by a drop in reflectance to almost zero, thus clearly identifying it as an electromagnetic resonance. We analysed the induced currents in the upper and lower layer in order to assign the plasmon eigenmode.[36] Both currents point within the plane and are antiparallel. The peak at 1.82 eV thus corresponds to the E_{2g} eigenmode shown in Fig. 2b.

The FWHM of the E_{2g} mode ($\gamma(E_{2g})$ = 140 meV) is much narrower than the width of the bright E_{1u} mode ($\gamma(E_{1u})$ = 710 meV) which is observed at $E(E_{1u})$ = 2.07 eV for the 50 nm spheres. The bulk damping, Γ_b, of the gold nanoparticles is 110 meV at $E(E_{2g})$ = 1.82 eV and 200 meV at $E(E_{1u})$ = 2.07 eV.[19] This implies an additional damping of only 30 meV for the dark E_{2g} plasmon mode, but 510 meV for the bright E_{1u} mode. The additional broadening is caused by radiative damping where a large fraction of the absorbed energy is lost in the form of photons. The dark modes of the colloidal bilayers are potential excitations for minimizing radiation loss and maximizing the decay of plasmons into hot carriers for harvesting losses.

Absorption by dark interlayer plasmons

The excitation of dark modes through field retardation was predicted for nanoparticle diameters of $\gtrsim$50 nm. We recently developed an advanced synthesis technique for d = 46 nm gold particles and their assembly into colloidal hcp structures (see the methods section in ref. 35). Transmission electron microscopy (TEM) images show hexagonal packing due to self-assembly after drying the solution on a TEM grid with a 10 nm amorphous carbon film (Fig. 4a).[36] Parts of the sample were areas with bi-, tri- and few-layers (Fig. 4b) that often extended

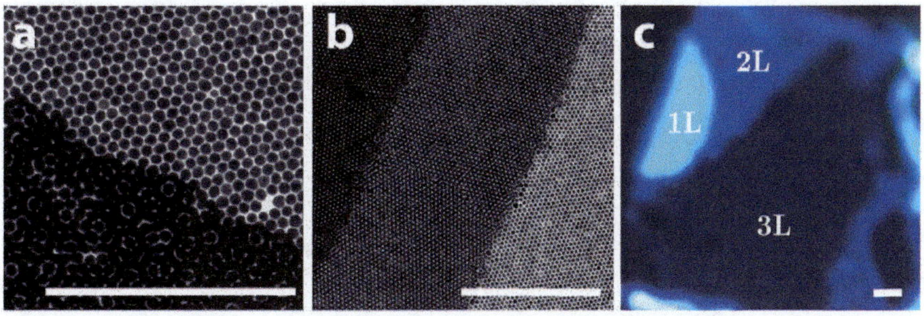

Fig. 4 (a) TEM image of a colloidal gold monolayer (top) and bilayer (bottom). The monolayer demonstrates a well-ordered structure with little defects (missing particles and smaller particles). The Moiré pattern in the bilayer indicates a finite angle of rotation between the two layers. (b) Survey TEM image of a monolayer (right), bilayer (middle) and trilayer (left). (c) The number of layers may also be determined *via* analysis of the change in contrast in the optical microscopy images, as shown for a sample spot with a monolayer (1L), bilayer (2L) and trilayer (3L). The scale bars correspond to 1 μm.

over many μm^2 (Fig. 4c).[35] A change in the optical contrast with changing layer number was also clearly visible *via* optical microscopy measurements (Fig. 4c). It allows the identification of the number of layers during the absorption and reflection measurements.

Fig. 5 shows the measured and simulated absorption spectra of a gold nanoparticle monolayer. It features a weak plasmon peak in the background of the interband transitions similar to the bilayer calculations of the 20 nm particles presented above. For the bilayers (Fig. 6), the absorption spectra are dominated by the E_{2g} peak at $E(E_{2g}) = 1.52$ eV. The peak is at a lower frequency than in Fig. 3, because the nanoparticles are stabilized by polystyrene and oleylamine molecules, thus giving rise to a background refractive index of $n = 1.4$. We recorded the spectra of various bilayers and consistently obtained almost identical spectra (see Fig. 6b and c for two spots measured with a 100× objective). This shows that the system is insensitive towards crystal imperfections such as missing nanoparticles, grain boundaries in the hexagonal packing, and the stacking of the two layers.[36] This robust behaviour is understandable when considering the macroscopic nature of the plasmon mode, whose main feature is the antiparallel dipole orientation in the two layers. The excitation of dark plasmons due to retardation is not limited to bilayers, but can be expected for any multilayer arrangement of plasmonic nanoparticles. In general, a sample with m gold nanoparticle layers will have $(m - 1)$ absorption peaks by dark plasmons in the visible and near infrared when excited with linearly polarized light.[36]

We evaluated the FWHM of the E_{1u} and E_{2g} plasmons for the gold monolayer and bilayer (see Table 1). The energetic position of the plasmon shows excellent

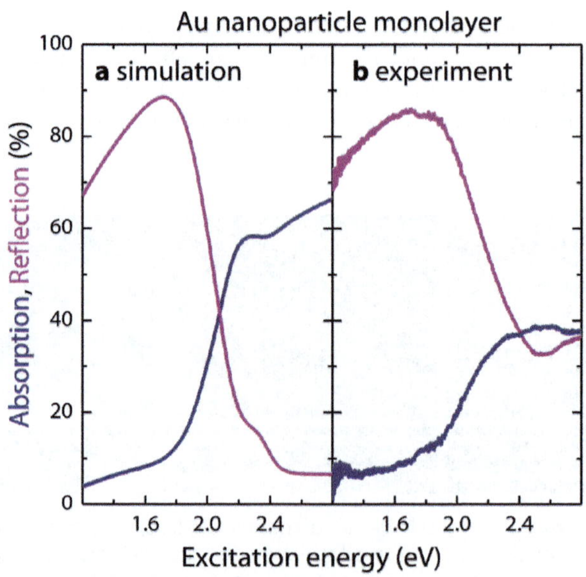

Fig. 5 (a) Simulated and (b) measured absorption (blue) and reflection (magenta) spectra of a gold monolayer with $d = 46$ nm and $g = 2$ nm. The simulation used a background dielectric constant with $n = 1.4$ and assumed hcp stacking. The nanoparticle layer was placed on top of a 10 nm thick amorphous carbon film.

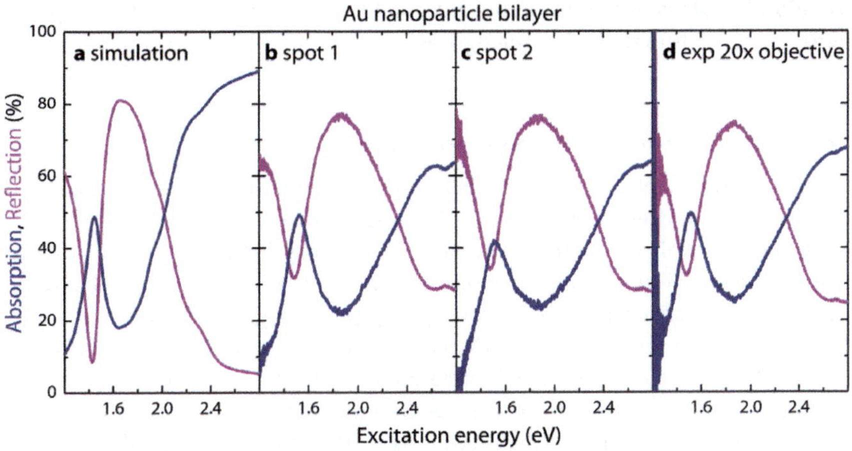

Fig. 6 Absorption (blue) and reflection (magenta) spectra of gold nanoparticle bilayers. (a) Simulation, (b and c) two spots measured with a 100× objective, and (d) a spot measured with a 20× objective. The peaks at 1.52 eV are caused by the dark E_{2g} plasmons. The parameters are identical to those reported for Fig. 5.

agreement between the experimental and simulation results for the gold monolayer. For the bilayer, the position of the dark mode was underestimated and the position of the bright mode overestimated in the simulations. The widths of the plasmon peaks for the monolayer and bilayer are consistently 100 meV broader in the experimental results than those predicted by the simulations. The contributions to this additional width arise from surface scattering and inhomogeneous broadening due to size and shape variations.[17,18,35]

Variations in the particle diameter (estimated as 46 ± 3 nm from analysis of the TEM images) lead to variations in the interparticle gap widths ($g = 2 \pm 1$ nm). The distance between the nanoparticles strongly affects the position of the E_{2g} mode, thus explaining the larger linewidth.[41] We find that the spectra measured with a 100× and 20× objective exhibit identical peak positions and FWHM values (Table 1 and Fig. 6b–d).[36] Inhomogeneous broadening must therefore happen at a length of well below 1 μm. On the other hand, the position of the bright mode is hardly affected by variation in the particle size and interparticle distance.[41] The additional broadening of both modes, therefore, rather points at least partially to an intrinsic mechanism such as surface scattering enhanced by hot spot

Table 1 Energy, E, and damping, γ, of the dark (E_{2g}) and bright (E_{1u}) modes of the monolayer and bilayer

	$E(E_{2g})$ in eV	$\gamma(E_{2g})$ in meV	$E(E_{1u})$ in eV	$\gamma(E_{1u})$ in meV
Monolayer, 100×	—	—	2.18 ± 0.05	510 ± 50
Monolayer, simulation	—	—	2.16	400
Bilayer, 100×	1.524 ± 0.005	325 ± 15	2.04 ± 0.03	630 ± 20
Bilayer, 20×	1.52 ± 0.01	330 ± 30	2.07 ± 0.05	620 ± 50
Bilayer, simulation	1.44	230	2.14	510

formation.[1,14] Measurements of the near-field resonance *via* scanning optical microscopy as a function of wavelength will help in clarifying this point.

Time-resolved transient absorption spectroscopy

To measure the electron dynamics in the gold nanoparticle layers, we utilized time-resolved transient absorption (TTA) spectroscopy. In these experiments, a probe pulse monitors the transient change in absorption in the visible (around the E_{1u} energy) region in order to follow the temperature increase in the electronic system after the absorption of a strong pump pulse. Fig. 7a displays exemplary TTA spectra with excitation at 1.27 eV, *i.e.* away from any pronounced resonance of the gold layers. The spectra are dominated by the response of the bilayer ($d = 40$ nm). The sample consisted of bi- and penta-layer domains, but no visible light was transmitted through the pentalayers over the spectral range of the probe pulse. We observe a temperature increase of the electron gas after plasmon excitation and dephasing; this results in a broadening of the E_{1u} absorption,[43] which generates a pronounced bleach with two positive sidebands in the differential spectrum[44] (Fig. 7a). The coupled plasmon modes at the long-wavelength tail behave differently, thus resulting in asymmetry of the positive sidebands. The intensity of the main bleach can be related to the electron temperature, but it is sensitive to the experimental conditions.

The dynamics of the main bleach can be separated into three phases with different timescales: contrast increase after plasmon dephasing and electron–electron scattering (10 fs), electron–phonon coupling (1 ps), and cooling of the nanoparticle (10 ps)[45] (see Fig. 7b). The data were evaluated with a bi-exponential fit to obtain the characteristic time constants. Due to the linear heat capacity and constant electron–phonon coupling, the observed electron–phonon coupling time constant, t_1, can be directly related to the initial excess temperature in the electron gas.[43,44] This quantity is less sensitive to experimental variations than the bleach intensity.

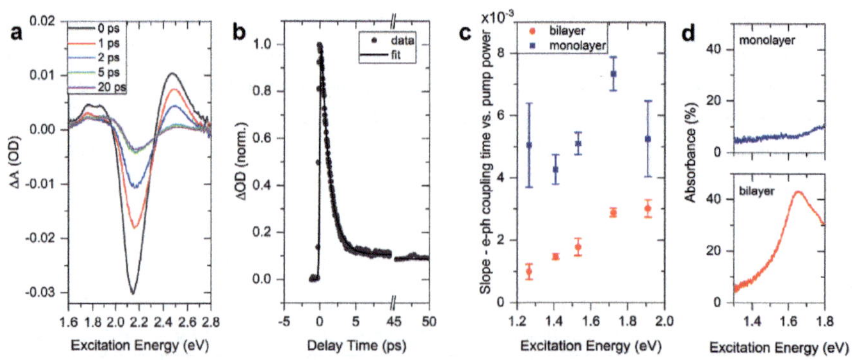

Fig. 7 Time-resolved transient absorption (TTA) spectroscopy of gold nanoparticle layers ($d = 40$ nm, $g = 2$ nm). (a) TTA spectra with excitation at 1.27 eV for selected pump-probe delays. (b) Normalized intensity of the main bleach shown in (a). (c) Slope of the electron–phonon coupling constant, t_1, *versus* pump power for different pump energies measured for a monolayer (blue) and a bi- and few-layer (red). (d) Absorbance spectra of the monolayer (top, blue) and the bi- and few-layer (bottom, red).

As the initial temperature scales linearly with the photon flux, the slope of the electron–phonon coupling time (initial temperature) with photon input allows us to assess the efficiency of the hot electron generation (Fig. 7c). We carried out a general screening of the excitation-dependence between 1.2 and 2.0 eV. The monolayer shows no dependence on the excitation energy, which is in agreement with its structure-less absorption spectrum (Fig. 7d). In the bilayer we observe an increase in the hot-electron generation efficiency at 1.65 eV, which corresponds to the E_{2g} resonance for this bilayer with $d = 40$ nm, as shown in Fig. 7d. The exact dependence and origin of the increase of the slope of the electron–phonon coupling constant is the subject of ongoing studies.

Optimization of nanostructure design for hot electron generation

Finally, we will show that the nanostructure design can be further optimized to increase the hot electron yield by dark interlayer plasmon excitation. The surface mediated damping, Γ_s, which is relevant for hot-electron generation, increases with decreasing nanoparticle size.[1,14,17,18] The quantum yield of hot electrons in single metal nanoparticles decreases with increasing particle size, but remains relatively large for diameters up to 30 nm. In Fig. 8a we show by simulation how the absorbance spectra of gold nanoparticle bilayers evolve with decreasing particle diameter. The narrow E_{2g} absorbance peak remains well resolved down to $d = 20$ nm. This is because the background refractive index ($n = 1.4$) reduces the incident wavelength and increases field retardation. With decreasing particle diameter, radiative damping is further suppressed thus reducing the width of the dark mode. The dark mode can be characterized by the quality factor $Q = E(E_{2g})/\gamma(E_{2g})$, which is associated with the field enhancement.[10,19] The highest quality factor ($Q = 22.5$) is expected for a nanoparticle bilayer with $d = 25$ nm (see the disks in Fig. 8b). This is much larger than the quality factor ($Q < 8$) of isolated spherical gold nanoparticles of similar size;[10] in fact, Q reaches its limit for gold nanostructures that is imposed by Γ_b.[19] For $d = 10$ nm, the width of the

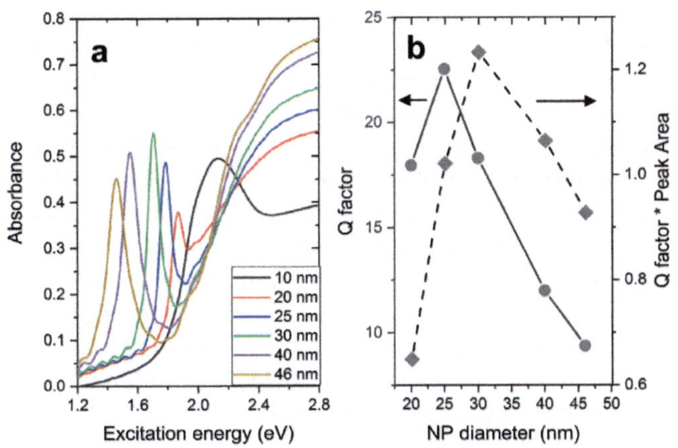

Fig. 8 Excitation of dark interlayer plasmons in gold nanoparticle bilayers with different particle sizes. (a) Simulated optical absorbance of gold nanoparticle bilayers with different particle diameters. Interparticle gaps of 2 nm and a background index of $n = 1.4$ were assumed. (b) Quality factor, $Q = E(E_{2g})/\gamma(E_{2g})$, and its product with the peak area, $A(E_{2g})$, as a function of nanoparticle diameter.

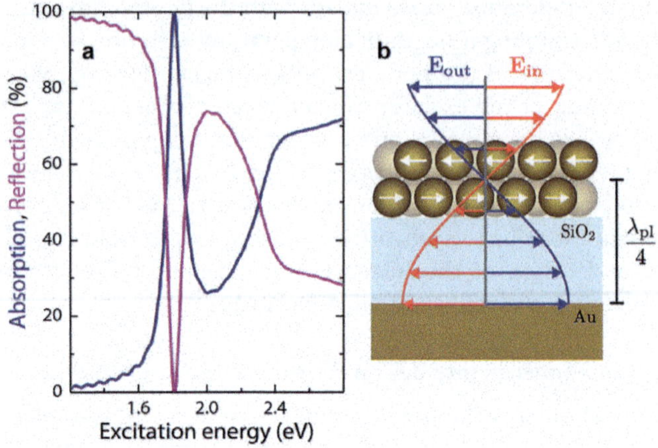

Fig. 9 Increasing the dark mode absorption to 100% by destructive interference above a gold mirror. (a) Simulated absorbance (blue) and reflectance (magenta) spectra of a 46 nm gold nanoparticle bilayer placed on top of a gold substrate with a 102 nm thick SiO_2 layer. Interparticle gaps of 2 nm and a background index of $n = 1.4$ were assumed. (b) Explanation of the increase in absorbance. The transmitted and reflected light is cancelled out by destructive interference when the optical distance between the bilayer center and gold surface is a quarter of the dark plasmon wavelength.

absorption peak strongly increases. This is partially due to the strong damping by the onset of the interband transitions, but an in-depth analysis of the simulations is necessary to understand this behaviour.

The second important parameter for the generation rate of hot electrons is the cross section for optical absorption. Considering both the quality factor and absorption (see the diamonds in Fig. 8b), we identify 30 nm as the most promising nanoparticle diameter for the generation of hot electrons in colloidal bilayers. Measuring the steady-state and time-resolved absorption of bilayers with varying nanoparticle diameter will be a way to verify our prediction.

Finally, we demonstrate that the optical absorption cross section of the dark plasmons can be significantly increased when placed on top of a reflective surface.[46,47] We simulate the optical absorbance of a gold nanoparticle bilayer on top of a gold substrate with a 102 nm thick SiO_2 layer. For a SiO_2 thickness that puts the inversion center of the bilayer at $\lambda_{pl}/4$ over the gold surface (λ_{pl} is the plasmon wavelength), the absorption reaches 100% (Fig. 9a). This striking increase in comparison with Fig. 6a is explained by a consideration of the destructive optical interference of the transmitted and reflected light, as illustrated in Fig. 9b. The colloidal gold surface may be optimized for perfect absorption by nanoparticles with different diameters by adjusting the thickness of the SiO_2 layer to the dark plasmon wavelength. The perfect optical absorption in combination with the suppressed radiative damping will enable the efficient generation of hot electrons for photocatalysis and plasmon-mediated chemical reactions.

Conclusions

In conclusion, we studied the optical excitation and decay of dark modes as a source of hot electrons. Dark plasmons may be excited in colloidal gold bilayers through the use of linearly polarized light under normal incidence. This activation of optical absorption is explained by field retardation and the comparatively large size of the colloidal gold nanoparticles. We showed that the damping of dark modes is strongly reduced in comparison to that of bright plasmons, because radiative damping is absent. The generation of plasmonic hotspots in hexagonal close packed crystals instead favours the decay by surface effects exciting hot electrons. We discussed how to maximize the rate of hot electron emission by increasing the absorption efficiency of the dark plasmons to 100% and reducing the nanoparticle size to make surface effects dominant. Colloidal multilayers can be prepared over larger areas and hold potential as large-scale and low-cost substrates for hot-electron harvesting.

Conflicts of interest

There are no conflicts of interest to declare.

Acknowledgements

This work was supported by the European Research Council under grant DarkS-ERS (772108). N. S. M. acknowledges Deutsche Telekom Stiftung for their financial support. B. G. M. V. acknowledges the Coordenação de Aperfeiçoamento de Pessoal de Nivel Superior (CAPES) for their financial support under the program PDSE (Grant No. 88881.134611/2016-01) and Dahlem Research School (DRS). E. B. B. acknowledges the financial support from CNPq. F. S. was supported by the DFG *via* the project SCHU 3019/2-1. H. L. acknowledges the German Research Foundation (DFG) for their financial support through the Cluster of Excellence CUI.

References

1 G. V. Hartland, L. V. Besteiro, P. Johns and A. O. Govorov, *ACS Energy Lett.*, 2017, **2**, 1641–1653.
2 M. L. Brongersma, N. J. Halas and P. Nordlander, *Nat. Nanotechnol.*, 2015, **10**, 25.
3 S. Lal, S. E. Clare and N. J. Halas, *Acc. Chem. Res.*, 2008, **41**, 1842–1851.
4 S. Linic, P. Christopher and D. B. Ingram, *Nat. Mater.*, 2011, **10**, 911.
5 C. Clavero, *Nat. Photonics*, 2014, **8**, 95.
6 S. Linic, U. Aslam, C. Boerigter and M. Morabito, *Nat. Mater.*, 2015, **14**, 567.
7 X. Huang, I. H. El-Sayed, W. Qian and M. A. El-Sayed, *J. Am. Chem. Soc.*, 2006, **128**, 2115–2120.
8 A. M. Schwartzberg and J. Z. Zhang, *J. Phys. Chem. C*, 2008, **112**, 10323–10337.
9 P. Narang, R. Sundararaman and A. Atwater Harry, *Nanophotonics*, 2016, 596.
10 C. Sönnichsen, T. Franzl, T. Wilk, G. von Plessen, J. Feldmann, O. Wilson and P. Mulvaney, *Phys. Rev. Lett.*, 2002, **88**, 077402.
11 A. Crut, P. Maioli, N. Del Fatti and F. Vallée, *Chem. Soc. Rev.*, 2014, **43**, 3921–3956.
12 S. A. Maier, *Plasmonics: Fundamentals and Applications*, Springer US, 2010.
13 M. Liu, M. Pelton and P. Guyot-Sionnest, *Phys. Rev. B: Condens. Matter Mater. Phys.*, 2009, **79**, 035418.
14 L. V. Besteiro and A. O. Govorov, *J. Phys. Chem. C*, 2016, **120**, 19329–19339.
15 J. A. Scholl, A. L. Koh and J. A. Dionne, *Nature*, 2012, **483**, 421.

16 S. Link and M. A. El-Sayed, *J. Phys. Chem. B*, 1999, **103**, 8410–8426.
17 C. Novo, D. Gomez, J. Perez-Juste, Z. Zhang, H. Petrova, M. Reismann, P. Mulvaney and G. V. Hartland, *Phys. Chem. Chem. Phys.*, 2006, **8**, 3540–3546.
18 S. Berciaud, L. Cognet, P. Tamarat and B. Lounis, *Nano Lett.*, 2005, **5**, 515–518.
19 F. Wang and Y. R. Shen, *Phys. Rev. Lett.*, 2006, **97**, 206806.
20 P. Christopher, H. Xin, A. Marimuthu and S. Linic, *Nat. Mater.*, 2012, **11**, 1044.
21 A. Sousa-Castillo, M. Comesaña-Hermo, B. Rodríguez-González, M. Pérez-Lorenzo, Z. Wang, X.-T. Kong, A. O. Govorov and M. A. Correa-Duarte, *J. Phys. Chem. C*, 2016, **120**, 11690–11699.
22 H. Harutyunyan, A. B. F. Martinson, D. Rosenmann, L. K. Khorashad, L. V. Besteiro, A. O. Govorov and G. P. Wiederrecht, *Nat. Nanotechnol.*, 2015, **10**, 770.
23 H. Zhang and A. O. Govorov, *J. Phys. Chem. C*, 2014, **118**, 7606–7614.
24 X.-T. Kong, Z. Wang and A. O. Govorov, *Adv. Opt. Mater.*, 2016, **5**, 1600594.
25 O. L. Muskens, V. Giannini, J. A. Sánchez-Gil and J. Gómez Rivas, *Opt. Express*, 2007, **15**, 17736–17746.
26 C. F. Bohren and D. R. Huffman, *Absorption and scattering of light by small particles*, Wiley, 1983.
27 M. Liu, T.-W. Lee, S. K. Gray, P. Guyot-Sionnest and M. Pelton, *Phys. Rev. Lett.*, 2009, **102**, 107401.
28 P. Nordlander, C. Oubre, E. Prodan, K. Li and M. I. Stockman, *Nano Lett.*, 2004, **4**, 899–903.
29 J. A. Fan, C. Wu, K. Bao, J. Bao, R. Bardhan, N. J. Halas, V. N. Manoharan, P. Nordlander, G. Shvets and F. Capasso, *Science*, 2010, **328**, 1135.
30 J. Sancho-Parramon and S. Bosch, *ACS Nano*, 2012, **6**, 8415–8423.
31 M.-W. Chu, V. Myroshnychenko, C. H. Chen, J.-P. Deng, C.-Y. Mou and F. J. García de Abajo, *Nano Lett.*, 2009, **9**, 399–404.
32 B. Luk'yanchuk, N. I. Zheludev, S. A. Maier, N. J. Halas, P. Nordlander, H. Giessen and C. T. Chong, *Nat. Mater.*, 2010, **9**, 707.
33 T. Pakizeh, M. S. Abrishamian, N. Granpayeh, A. Dmitriev and M. Käll, *Opt. Express*, 2006, **14**, 8240–8246.
34 R. Verre, Z. J. Yang, T. Shegai and M. Käll, *Nano Lett.*, 2015, **15**, 1952–1958.
35 F. Schulz, S. Tober and H. Lange, *Langmuir*, 2017, **33**, 14437–14444.
36 N. S. Mueller, B. G. d. M. Vieira, F. Schulz, P. Kusch, V. Oddone, E. B. Barros, H. Lange and S. Reich, *ACS Photonics*, 2018, **5**, 3962.
37 P. B. Johnson and R. W. Christy, *Phys. Rev. B: Condens. Matter Mater. Phys.*, 1972, **6**, 4370–4379.
38 J. I. Larruquert, L. V. Rodríguez-de Marcos, J. A. Méndez, P. J. Martin and A. Bendavid, *Opt. Express*, 2013, **21**, 27537–27549.
39 T. Inui, Y. Tanabe and Y. Onodera, *Group theory and its applications in physics*, Springer-Verlag, 1990.
40 D. W. Brandl, N. A. Mirin and P. Nordlander, *J. Phys. Chem. B*, 2006, **110**, 12302–12310.
41 B. G. d. M. Vieira, N. S. Mueller, E. B. Barros and S. Reich, 2018, in preparation.
42 P. Anatoliy, P. Gero von and K. Uwe, *J. Phys. D: Appl. Phys.*, 2004, **37**, 3133.
43 G. V. Hartland, *Chem. Rev.*, 2011, **111**, 3858–3887.
44 A. M. Brown, R. Sundararaman, P. Narang, W. A. Goddard and H. A. Atwater, *Phys. Rev. B*, 2016, **94**, 075120.

45 E. Minutella, F. Schulz and H. Lange, *J. Phys. Chem. Lett.*, 2017, **8**, 4925–4929.
46 A. Moreau, C. Ciracì, J. J. Mock, R. T. Hill, Q. Wang, B. J. Wiley, A. Chilkoti and D. R. Smith, *Nature*, 2012, **492**, 86.
47 C. Ng, J. J. Cadusch, S. Dligatch, A. Roberts, T. J. Davis, P. Mulvaney and D. E. Gomez, *ACS Nano*, 2016, **10**, 4704–4711.

Faraday Discussions

PAPER

Carrier dynamics and spin–valley–layer effects in bilayer transition metal dichalcogenides

Christopher J. Ciccarino,[ab] Chitraleema Chakraborty, [ac] Dirk R. Englund[c] and Prineha Narang *[a]

Received 26th October 2018, Accepted 26th November 2018
DOI: 10.1039/c8fd00159f

Transition metal dichalcogenides are an interesting class of low dimensional materials in mono- and few-layer form with diverse applications in valleytronic, optoelectronic and quantum devices. Therefore, the general nature of the band-edges and the interplay with valley dynamics is important from a fundamental and technological standpoint. Bilayers introduce interlayer coupling effects which can have a significant impact on the valley polarization. The combined effect of spin–orbit and interlayer coupling can strongly modify the band structure, phonon interactions and overall carrier dynamics in the material. Here we use first-principles calculations of electron–electron and electron–phonon interactions to investigate bilayer MoS_2 and WSe_2 in both the AA′ and AB stacking configurations. We find that in addition to spin–orbit coupling, interlayer interactions present in the two configurations significantly alter the near-band-edge dynamics. Scattering lifetimes and dynamic behavior are highly material-dependent, despite the similarities and typical trends in TMDCs. Additionally, we capture significant differences in dynamics for the AA′ and AB stacking configurations, with lifetime values differing by up to an order of magnitude between them for MoS_2. Further, we evaluate the valley polarization times and find that maximum lifetimes at room temperature are of the scale of 1 picosecond for WSe_2 in the AB orientation. These results present a pathway to understanding complex heterostructure configurations and 'magic angle' physics in TMDCs.

I. Introduction

Transition metal dichalcogenides (TMDCs) have been extensively studied in monolayer and few-layer forms for a variety of both fundamental and technological reasons. In the monolayer, TMDCs have been shown to be of optical

[a]*John A. Paulson School of Engineering and Applied Sciences, Harvard University, Cambridge, MA, USA. E-mail: prineha@seas.harvard.edu*

[b]*Department of Chemistry and Chemical Biology, Harvard University, Cambridge, MA, USA*

[c]*Department of Electrical Engineering and Computer Science, Massachusetts Institute of Technology, Cambridge, MA, USA*

interest due to their efficient emission properties as direct band-gap semiconductors.[1–3] Additionally, spin-selective excitation is possible using circularly polarized light,[4] enabling controlled valley population. Without an inversion center, spin and valley pseudospin can couple, leading to the spin–valley locking phenomena useful in valleytronic applications such as information processing.[5–7]

Bilayer TMDCs are compelling in a host of unique ways;[8] multilayer stacks of TMDCs interact with neighboring layers *via* van der Waals interactions. These weak interactions allow for the possibility of fabricating multilayer stacks with controllable precision in orientation and ordering.[9–11] Unique material phenomena have already been found using this stacking degree of freedom, including superconductivity in graphene[12] and band-gap tuning in heterostructure TMDCs,[13–15] for example. In general, control of interlayer twist angles can introduce long range Moiré patterns, effectively acting as a long range potential modulator. Moiré effects have been previously studied in bilayer MoS_2, which introduced novel phonon modes associated with long-range periodicity.[16] Theoretical studies have also investigated the effects of strain and its relation to interlayer coupling using tight-binding methods.[17–19] Interlayer coupling effects have shown modulation of photoluminescence[20] and more recently have been demonstrated to introduce relaxation pathways for excited carriers in heterostructures.[21] More fundamentally, band-edge states have been shown to be sensitive to interlayer coupling effects.[22] TMDC heterostructures have also been demonstrated to have ultra long valley polarization times in experiment.[23]

These interesting experimental demonstrations motivate our work towards a thorough, fundamental understanding of the physics in multilayer TMDC systems. In the bilayer, TMDCs are indirect band-gap materials, similar in this way to their bulk behavior.[24] Despite this, these systems retain interest from the valleytronic community for a number of reasons, including the fact that their inversion symmetry can be tuned by electric fields,[25] and have overall been more susceptible to electric field manipulation than monolayers.[26] Additionally, strong valley polarization has been observed in bilayer WS_2[27] due to spin–orbit splitting. Understanding valleytronic applicability in these systems requires a detailed understanding of the band structure and phonon properties of the material.

Overall, the dynamics of bilayer TMDCs present a substantial theoretical and computational challenge relative to their monolayer counterparts. The effects of interlayer coupling must be considered, which strongly influences the band edge states. Additionally, stacking orientation is paramount to understanding the symmetry of the system, which can severely affect valley dynamics through spin–orbit coupling. The experimentally observed AA′ (also known as the 2H phase) and AB (also known as the 3R phase) orientations of bilayer TMDCs[28,29] introduce differences in interlayer coupling and symmetries, which can induce unique material behavior. In the bilayer, spin and valley pseudospin degrees of freedom can additionally couple to the layer index, denoted as layer pseudospin, such that all three of these quantum numbers describe the K point valleys of the system.[30] Here, opposite spins from opposite layers are coupled at each of the split valence bands, such that transitions must consider interlayer hopping as well as spin flipping. Additionally, different stacking configurations can lead to shifts in phonon frequencies, spin–orbit splitting and the magnitude of van der Waals interactions.[31] All of these effects create a complicated system compared to the monolayer case and require new theoretical and computational implementations.

In this work, we present a novel *ab initio* approach for electron–electron and electron–phonon interactions to study bilayer MoS_2 and WSe_2 systems in both the AA′ and AB stacking orientations. We compare our self-consistent spin–orbit calculated dynamics with spin-degenerate calculations in order to understand the impact of spin–orbit coupling on dynamics in these systems. Importantly, we find that in both materials, a detailed understanding of spin–orbit splitting, interlayer coupling and symmetry effects is required in order to understand the different dynamics captured for each of the material configurations. Further, we find that spin–orbit effects are more relevant in WSe_2 due to its larger spin-splitting, while the smaller splitting in MoS_2 has a much weaker effect on dynamics. This is in contrast with the uniform TMDC trends seen in monolayers.[32] Additionally, we calculate hot carrier profiles and find that despite the indirect nature of all the systems considered, direct transitions dominate the calculated results, in qualitatively excellent agreement with the optical properties observed experimentally. These results provide a pathway for future studies in TMDC dynamics, including heterostructures, twisted bilayer 'magic angle' physics, and larger multilayer stacks.

II. Methods

All calculations were performed using JDFTx,[33] based on density-functional theory (DFT). We model the bilayers using PBEsol[34] pseudopotentials for a plane-wave basis set with a 25 hartree energy cutoff on a $9 \times 9 \times 1$ reciprocal space grid. To describe the bilayer effectively, we employ the Grimme D2 correction for interlayer coupling[35] as well as Coulomb truncation[36] in the out-of-plane direction, such that periodic images of the bilayer are neglected. Phonons are calculated using the frozen phonon approach on a $3 \times 3 \times 1$ supercell. In relativistic calculations, spin–orbit coupling is self-consistently included, distinct from previous work in these materials. Each cell was relaxed and the a lattice constant found to agree with experiment to within 1%. The interlayer spacing of the AB-stacked MoS_2 (WSe_2) bilayer corresponds well with the out-of-plane lattice constant of 12.27 Å (12.98 Å), which is within 1% of the bulk lattice value in each material. The AA′ configuration showed a 10% reduction in the interlayer spacing for both MoS_2 and WSe_2.

In order to effectively capture the dynamical properties of the system, a dense sampling of **k** points in the Brillouin zone (BZ) is required. Conventional DFT approaches to generating these meshes are inadequate given the computational cost. Instead, we employ a Wannier function scheme which describes the Bloch space Kohn–Sham states on an equivalent real space grid.[37] We can then transform back to reciprocal space on a finely-spaced grid of **k** points. This enables efficient, high-fidelity BZ sampling, crucial for our calculations. In the calculations presented, Wannier interpolation allows us to sample $\sim 10^6$ **k** points in the 2D BZ of the bilayer. We apply these interpolation methods in calculations which both omit and include spin–orbit coupling, where the latter employs spinorial trial functions to explicitly include spin. Trial centers are made up of transition metal d orbitals and chalcogen p orbitals for capturing bands nearest to the band gap.

III. Results and discussion

A. Electronic structure

It is useful to first consider the monolayer band structure before considering the bilayer. In the monolayer, spin–orbit coupling alone accounts for significant splitting of the valence band at the K point and the conduction band at the Q point. In conjunction with the lack of an inversion center, it is relatively easy to understand the origin of the splitting in these systems. In the bilayer case, especially when considering the multiple possible stacking configurations, corresponding symmetry and interlayer coupling effects can introduce complications in addition to spin–orbit coupling. Symmetry plays a pivotal role in understanding how spin–orbit splitting impacts the band structure, while interlayer coupling has already been shown to most strongly affect bands near the band edges.[22] Together these interactions promise to have important effects on the scattering possibilities of a given system.

The calculated spin–orbit split band structures of each of the bilayer orientations are plotted in Fig. 1. Despite the weak van der Waals forces governing interlayer interactions, the states near the band edge are clearly affected by interlayer coupling to some degree.[22,38] These states are composed of mostly d orbitals from the transition metal atom, where the K point of the valence band and the Q point of the conduction band are both comprised of the $d_{x^2-y^2}$ and d_{xy} orbitals which introduce significant splitting.[39] In the AA′ case, a combination of

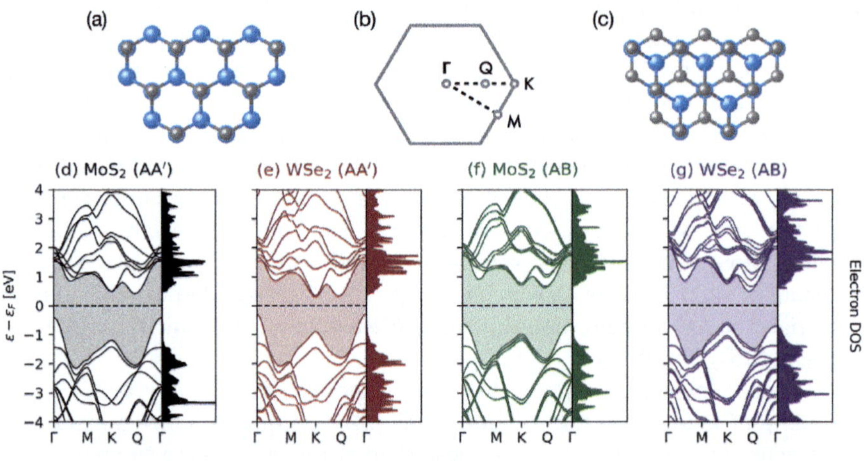

Fig. 1 Bilayer TMDC structure, orientation and electronic structure. Panels (a) and (c) show the stacking arrangement of the two TMDC layers in the out-of-plane direction for the AA′ and AB orientations, respectively. The band structures in panels (d)–(g) are plotted along the high-symmetry lines defined by the two dimensional Brillouin zone depicted in panel (b). The conduction band edge states in all cases are at the K and Q points, whereas the valence band edges are at the K and Γ points. All four of these points are sensitive to interlayer coupling, and the Q of the conduction band and K of the valence band are split further by a combination of spin–orbit and interlayer coupling. In the AA′ orientation the primary splitting is a result of interlayer coupling, while in the AB orientation it is the spin–orbit effects which introduce large splitting values. The smaller overall band gap of the AA′ orientation is suggestive of stronger interlayer coupling, as the predicted interlayer spacing for these layers is 10% smaller than the spacing in the AB configuration.

the out-of-plane interlayer coupling and intralayer spin–orbit splitting introduces sizable splits in the observed valence band edge at the K point (up to 550 meV in WSe_2). However, only two energetically distinct valleys are observed, which is a direct result of inversion symmetry. This configuration represents the spin–valley–layer locking state, where each valley is composed of opposite spin states for opposite layers of the TMDC,[30] which together can be indexed by the spin, valley pseudospin and layer pseudospin index. In the AB configuration, four non-degenerate valleys emerge due to the lack of an inversion center. Interestingly, the stacking orientation also impacts the interlayer spacing, suggesting orientation-dependent van der Waals interactions. For both MoS_2 and WSe_2, the interlayer spacing decreased by ~10% for the AA′ configurations relative to the AB configuration. This, as previously investigated and confirmed in our results, leads to blue-shifted phonon modes in addition to impacting the interlayer coupling effects on the electronic structure.[31]

The confluence of spin–orbit and interlayer coupling effects on the splitting in the band structure is distinct for the AA′ and AB stacking configurations. In the AA′ stacking, the splitting of the valence band at the K point is substantial, even in the spin-degenerate calculations. This suggests that splitting of the K valley is dominated by interlayer coupling effects. When spin–orbit effects are included, these splits then increase further, but there is no additional splitting of the two bands into four: this is forbidden in the inversion-symmetric cell. However, in the AB configuration, the splitting from interlayer coupling alone is considerably smaller. Instead, the observed splittings seen in Fig. 1(f and g) are primarily due to spin–orbit coupling effects. This correlates well with the larger interlayer spacing found in the AB-stacked cells, which indicates weaker interlayer interactions. The spin–orbit splitting mechanism which dominates the splitting in the AB orientation is important when understanding possible electronic transitions.

B. Lifetimes

In determining the total scattering rate for carriers, we consider Coulomb scattering between two electrons (e–e) as well as electron–phonon (e–ph) scattering. Together, these processes contribute to the total scattering rate, given by Matthiessen's rule:

$$\tau_{\mathbf{k}n}{}^{-1} = (\tau_{\mathbf{k}n}^{\text{e-e}})^{-1} + (\tau_{\mathbf{k}n}^{\text{e-ph}})^{-1}, \tag{1}$$

which is indexed by a carrier in band n and wave-vector $\mathbf{k}$ in the two-dimensional Brillouin zone. Each of these processes are modeled using a diagrammatic framework, where we compute the contribution of each process to the electronic self-energy $\Sigma_{\mathbf{k}n}$ as described formally in a field theoretic approach. Thus, in general the lifetime is given by $\tau_{\mathbf{k}n}^{\text{e-e(e-ph)}} = \hbar/2\Sigma_{\mathbf{k}n}^{\text{e-e(e-ph)}}$. The Coulomb-mediated electron scattering is therefore written as:[40]

$$\left(\tau_{\mathbf{k}n}^{\text{e-e}}\right)^{-1} = \frac{2\pi}{\hbar}\int_{\text{BZ}}\frac{\text{d}\mathbf{k}'}{(2\pi)^2}\sum_{n'}\sum_{\mathbf{GG}'}\tilde{\rho}_{\mathbf{k}'n',\mathbf{k}n}(\mathbf{G})\tilde{\rho}^{*}_{\mathbf{k}'n',\mathbf{k}n}\left(\mathbf{G}'\right)\times\frac{1}{\pi}\,\text{Im}\,W_{\mathbf{GG}'}\left(\mathbf{k}'-\mathbf{k},\varepsilon_{\mathbf{k}n}-\varepsilon_{\mathbf{k}'n'}\right). \tag{2}$$

Here we employ the single-particle electronic density matrices $\tilde{\rho}_{\mathbf{k}'n',\mathbf{k}n}(\mathbf{G})$ and capture the imaginary part of their interaction *via* the dynamically screened Coulomb interaction, $W_{\mathbf{GG}'}(\mathbf{k}' - \mathbf{k},\omega)$, which is evaluated within the random phase approximation. This calculation is performed in the plane-wave basis of reciprocal lattice vectors **G** and **G**′, and involves a sum over a full second set of electronic states ($\mathbf{k}'n'$).

The electron–phonon lifetime is determined using Fermi's Golden rule:

$$\left(\tau_{\mathbf{k}n}^{\mathrm{e-ph}}\right)^{-1} = \frac{2\pi}{\hbar}\sum_{n'\alpha\pm}\int_{\mathrm{BZ}}\frac{\Omega\mathrm{d}\mathbf{k}'}{(2\pi)^2}\delta\left(\varepsilon_{\mathbf{k}'n'} - \varepsilon_{\mathbf{k}n} \mp \hbar\omega_{\mathbf{q}\alpha}\right) \times \left[n_{\mathbf{q}\alpha} + \frac{1}{2} \mp \left(\frac{1}{2} - f_{\mathbf{k}'n'}\right)\right]\left|g_{\mathbf{k}'n',\mathbf{k}n}^{\mathbf{q}\alpha}\right|^2. \tag{3}$$

In this expression, final electronic bands n' and phonon polarizations α are summed over while integration is performed over the entire 2D BZ. The δ function imposes energy conservation for all scattering processes involving initial (final) electronic states $\mathbf{k}n$ ($\mathbf{k}'n'$) with corresponding energies $\varepsilon_{\mathbf{k}n}$ ($\varepsilon_{\mathbf{k}'n'}$) and a phonon in

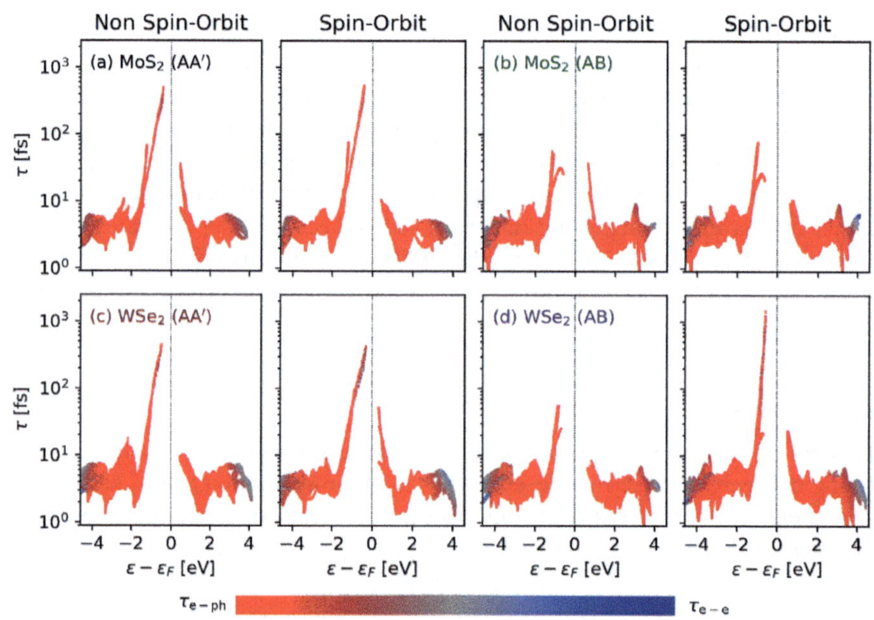

Fig. 2 (a–d) Calculated scattering lifetimes in each of the bilayer TMDCs studied. The lifetimes in the AA′ and AB configurations are relatively similar except at the band edges. The hole lifetimes of both TMDCs are not significantly affected by spin–orbit splitting in the AA′ configuration, as the effect of spin–orbit coupling is to increase the already-split valence band edge, which is induced by interlayer coupling. Electron lifetimes are increased in WSe_2 with spin and decreased in MoS_2 with spin because of the relative energies of the Q and K valleys. Spin–orbit coupling pushes the Q point to phonon-accessible energies relative to the K point in MoS_2, while it pushes the Q valley away from the K valley in the case of WSe_2. In the AB configuration, hole lifetimes of WSe_2 are found to significantly increase as a result of spin–orbit coupling, while hole lifetimes in MoS_2 are predicted to be relatively unaffected. Similar to the AA′ configuration, electron lifetime trends are governed by the relative energies of the Q and K valleys. The relative contribution of electron–electron and electron–phonon interactions to the lifetime is indicated by the color bar. All scattering observed is dominated by electron–phonon processes.

state $\mathbf{q}\alpha$ with energy $\hbar\omega_{\mathbf{q}\alpha}$. Diagrammatically this represents a three-vertex. The $\mp$ accounts for both phonon emission and absorption. These scattering events are weighted by the phononic (Bose) and electronic (Fermi) occupation factors described by $n_{\mathbf{q}\alpha}$ and $f_{\mathbf{k}n}$, respectively. Finally, the electron–phonon matrix element which describes the coupling of these two electronic states with the phonon mode is denoted by $g^{\mathbf{q}\alpha}_{\mathbf{k}'n',\mathbf{k}n}$. We note that in order to satisfy conservation of crystal momentum, $\mathbf{q} = \mathbf{k}' - \mathbf{k}$ for all terms which contribute to the above expression. We interpolate our DFT-based couplings using Wannier functions to generate an ultra-fine mesh of matrix elements for converged integration.

After solving for each lifetime individually, the total scattering lifetimes are determined from eqn (1) and presented in Fig. 2. As seen in the figure, the scattering within 4 eV of the Fermi energy is dominated by electron–phonon interactions. The AA′ stacking configuration shows that spin–orbit coupling has a negligible impact on the hole lifetime at the valence band edge. This is because the band structures even in the spin-degenerate calculations are sufficiently split by interlayer coupling,[22] with splitting values of 165 meV for MoS_2 and 240 meV for WSe_2. We note that the maximum phonon energy for these systems is 60 meV, thus scattering between each of these valleys is not accessible, enabling long-lived states. Introducing spin–orbit splitting indeed increases the splitting values by about 60 meV (310 meV) in MoS_2 (WSe_2), but this does not change the dynamics any further. Meanwhile excited electrons at the conduction band edge are slightly affected by spin–orbit coupling, however, the effect is opposite for each TMDC. In MoS_2, lifetimes decrease with spin–orbit coupling, as this shifts the Q point towards K for more energetically accessible K–Q scattering, while the opposite occurs in WSe_2: the energy separation between the Q and K valleys becomes larger.

The results in the AB configuration are equally interesting and material-dependent. Hole dynamics in MoS_2 are hardly affected by spin–orbit coupling, while electron lifetimes are again decreased for the same Q-related reason as in the AA′ scenario. Meanwhile, hole lifetimes in WSe_2 are significantly enhanced when spin–orbit coupling is introduced, similar to the behavior observed in monolayer systems. In this case, the larger splitting induced by WSe_2 relative to MoS_2 is sufficient to shrink the phase space for electron–phonon scattering. In the AB configuration, spin–orbit coupling pushes the valence band edge of WSe_2 to the energetically-isolated K point, whereas the band edge in MoS_2 is still the Γ point. The spin–orbit-induced shift of the K point valence band removes possible K–Γ electron–phonon scattering and therefore enhances hole lifetimes, which are clearly very sensitive to band-edge position.

In general, the spin-related effects on dynamics are opposite for the MoS_2 and WSe_2 systems, which is distinctly different from the behavior observed for monolayer TMDCs.[32] In addition, the magnitude of the lifetimes are at least 50% lower than the corresponding monolayer case, regardless of spin considerations. The maximum valley coherence time calculated at room temperature is only 1 picosecond in WSe_2, which is 4 times smaller than the time determined for the same monolayer system.[32] However, hole lifetimes are still observed to be larger than electron lifetimes in general, consistent with the trends in the monolayer case. The bilayer structures can be said to introduce more scattering channels than the monolayer case, given that the band edge is not isolated at K in many of the systems studied. van der Waals effects *via* interlayer coupling have previously been shown to have an impact on predicted dynamics.[41] The nature of the band edges (especially compared to the monolayer case) are crucial to explain the smaller lifetimes predicted for the TMDCs studied. Additionally, the alignment of the layers and their corresponding symmetries impact the nature and strength of interlayer and spin–orbit coupling, which manifests in the substantially different dynamics observed, as is visible when comparing the left- and right-hand side of Fig. 2.

C. Mobility

In addition to lifetimes, we can also evaluate the predicted carrier mobilities, which are also governed by electron–phonon interactions. Using the same general formalism used in eqn (3), we compute the momentum-relaxing lifetimes $\tau^{\mathrm{p}}_{\mathbf{k}n}$ by accounting for the angle between the electron velocities before and after scattering. This is given by:

$$\left(\tau^{\mathrm{p}}_{\mathbf{k}n}\right)^{-1} = \frac{2\pi}{\hbar}\sum_{n'\alpha\pm}\int_{\mathrm{BZ}}\frac{\Omega \mathrm{d}\mathbf{k}'}{(2\pi)^2}\delta\left(\varepsilon_{\mathbf{k}'n'} - \varepsilon_{\mathbf{k}n} \mp \hbar\omega_{\mathbf{q}\alpha}\right) \times \left[n_{\mathbf{q}\alpha} + \frac{1}{2} \mp \left(\frac{1}{2} - f_{\mathbf{k}'n'}\right)\right]\left|g^{\mathbf{q}\alpha}_{\mathbf{k}'n',\mathbf{k}n}\right|^2 \times \left(1 - \frac{\mathbf{v}_{\mathbf{k}n}\cdot\mathbf{v}_{\mathbf{k}'n'}}{\left|\mathbf{v}_{\mathbf{k}n}\right|\left|\mathbf{v}_{\mathbf{k}'n'}\right|}\right), \tag{4}$$

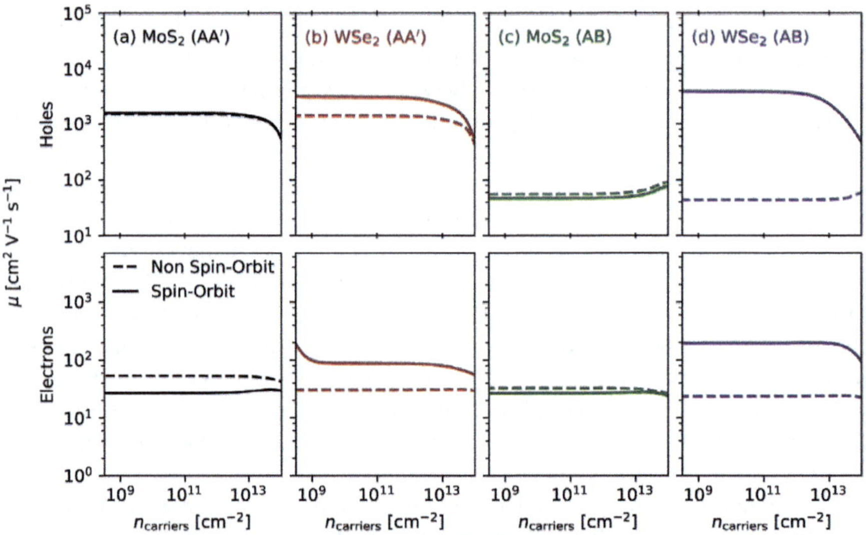

Fig. 3 (a–d) Calculated mobilities of AA′ and AB-stacked transition metal dichalcogenides. In general, the spin–orbit interactions more significantly affect the mobilities in WSe_2 than in MoS_2, which is a result of its larger predicted splitting and the relative energies of the valleys at each of the band edges. Both the AA′ and AB configurations of WSe_2 show increased mobilities with spin–orbit coupling for the hole and electron cases. Similarly to the lifetime in Fig. 2, the AB orientation introduces a larger difference between spin–orbit and non spin–orbit predictions than AA′, where there is over an order of magnitude increase for the same reasons described for Fig. 2. Compared to calculated monolayer mobilities, both configurations of the TMDCs have mobilities smaller than those of their monolayer counterparts, except in the case of the electron mobility of AB-stacked WSe_2, which is slightly larger in the bilayer compared to the monolayer when spin–orbit interactions are considered.

which is identical to eqn (3) except for the angular dependence, denoted by the band velocities $\mathbf{v}_{\mathbf{k}n}$ (defined as $\mathbf{v} \equiv \partial\varepsilon/\partial\mathbf{k}$) of the initial $\mathbf{k}n$ and final $\mathbf{k}'n'$ states. After calculating $\tau^{\mathrm{p}}_{\mathbf{k}n}$, we determine the mobility by solving the linearized Boltzmann equation using a full-band relaxation-time approximation:

$$\bar{\mu}(\varepsilon_{\mathrm{F}}) = \frac{e}{|n(\varepsilon_{\mathrm{F}})|}\sum_{n}\int_{\mathrm{BZ}}\frac{g_{\mathrm{s}}\mathrm{d}\mathbf{k}}{(2\pi)^{2}}\frac{\partial f_{\mathbf{k}n}(\varepsilon_{\mathrm{F}})}{\partial\varepsilon_{\mathbf{k}n}}(\mathbf{v}_{\mathbf{k}n}\otimes\mathbf{v}_{\mathbf{k}n})\tau^{\mathrm{p}}_{\mathbf{k}n}. \tag{5}$$

Here g_{s} is the spin degeneracy of the band (=1 with and =2 without spin–orbit coupling). The derivative of the Fermi function effectively selects carriers which contribute to transport at a specified doping level given as ε_{F}. We further define the density $n(\varepsilon_{\mathrm{F}})$ for a given Fermi energy as:

$$n(\varepsilon_{\mathrm{F}}) = \sum_{n}\int_{\mathrm{BZ}}\frac{g_{\mathrm{s}}\mathrm{d}\mathbf{k}}{(2\pi)^{2}}f_{\mathbf{k}n}(\varepsilon_{\mathrm{F}}) - n_{0}. \tag{6}$$

This quantity is positive for n-type semiconductors and negative for p-type semiconductors. The quantity n_0 represents the number density of carriers in the neutral DFT calculation.

Using this formalism, we vary the Fermi-level and evaluate eqn (4)–(6) in order to determine the mobility as plotted in Fig. 3. Similarly to the lifetime behavior observed, the role of spin–orbit coupling in the predicted mobilities is systematically different when comparing MoS_2 and WSe_2 for both the AA′ and AB stacking configurations. The mobilities are similar for electrons and holes in these regions, and spin reduces the electron mobility in both stacking configurations. On the other hand, the mobility in WSe_2 is highly dependent on spin–orbit coupling for both electrons and holes. In the hole case, spin-splitting isolates the K point from Γ at the valence band edge, while in the electron case it isolates Q from K, both cases helping to remove scattering channels and improve mobility. The enhanced electron mobility is novel from the monolayer results, as here the Q point is energetically more separated from K than it was in the monolayer. Despite this, all of these calculated mobilities are smaller than those predicted in the monolayer. The only slight exception to this is the electron mobility in the AB configuration of WSe_2, which is slightly larger than that in the monolayer when spin–orbit coupling effects are included.

D. Hot carriers

Information about electron–electron and electron–phonon coupling allows us to better understand the optical response. This is especially important for TMDCs, which are efficient sources of long-lived inter-layer excitons[42] and single photon emitters[43–46] exhibiting widely tunable properties when embedded in multilayer heterostructures.[20,47,48] While we do not explicitly account for excitonic effects in our calculations, these results hint at the importance of direct transitions, and our calculated response can provide a qualitative understanding of the nature of favorable excitation pathways. We calculate the complex dielectric function for each monolayer by including both direct and phonon-assisted transitions:

$$\bar{\varepsilon}(\omega) = \bar{\varepsilon}_{\mathrm{direct}}(\omega) + \bar{\varepsilon}_{\mathrm{phonon}}(\omega). \tag{7}$$

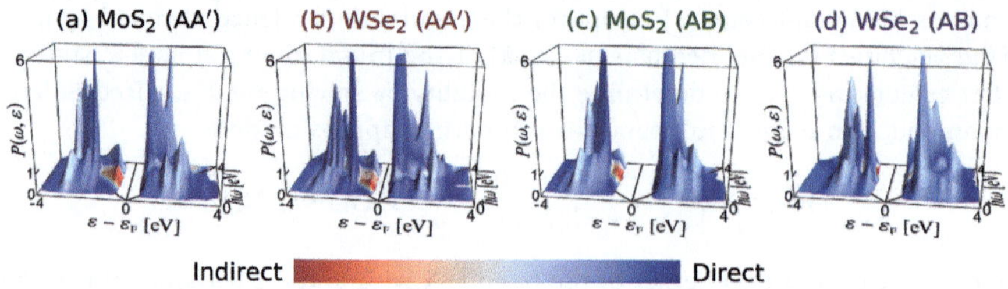

Fig. 4 (a–d) Hot carrier generation profiles of AA′ and AB-stacked transition metal dichalcogenides. The profiles are normalized and show the likelihood of hot carriers being generated from direct *vs.* indirect transitions. In all cases, despite the indirect nature of the band gap, transitions are dominated by direct transitions as opposed to indirect. This is an interesting yet surprising result from our calculations. The indirect transitions at the valence band edge suggest that Γ valleys can be populated, but their overall density of states limits the quantity of transitions to this energy. Meanwhile, direct transitions associated with the K point dominate the hot carrier generation mechanisms for the photon energy range considered.

Fermi's Golden rule is used to calculate the imaginary part of the direct contribution *via*:[49]

$$\hat{\lambda}\cdot\mathrm{Im}\,\bar{\varepsilon}_{\mathrm{direct}}(\omega)\cdot\hat{\lambda} = \frac{4\pi^2e^2}{\omega^2}\int_{\mathrm{BZ}}\frac{\mathrm{d}k}{(2\pi)^2}\sum_{n'n}\left(f_{kn}-f_{kn'}\right)\times\delta\left(\varepsilon_{kn'}-\varepsilon_{kn}-\hbar\omega\right)\left|\hat{\lambda}\cdot\mathbf{v}^{\mathbf{k}}_{n'n}\right|^2, \tag{8}$$

which describes electronic transitions from initial and final electronic states which involve a photon of energy $\hbar\omega$. The variable $\hat{\lambda}$ is a test unit vector oriented in the same direction as the field, which enables for directionally-dependent characterization of the dielectric response. We capture the response from a field oriented perpendicular to the bilayer. Indirect transitions can also contribute to optical excitation and hot carrier generation. In order to evaluate the electron–phonon scattering involved in these processes, the phonon-assisted process is captured using second-order perturbation theory:[50]

$$\begin{aligned}\hat{\lambda}\cdot\mathrm{Im}\,\bar{\varepsilon}_{\mathrm{phonon}}(\omega)\cdot\hat{\lambda} = {} & \frac{4\pi^2e^2}{\omega^2}\int_{\mathrm{BZ}}\frac{\mathrm{d}\mathbf{k}'\mathrm{d}\mathbf{k}}{(2\pi)^4}\sum_{n'n\alpha\pm}\left(f_{\mathbf{k}n}-f_{\mathbf{k}'n'}\right)\\ & \times\left(n_{\mathbf{q},\alpha}+\frac{1}{2}\mp\frac{1}{2}\right)\delta\left(\varepsilon_{\mathbf{k}'n'}-\varepsilon_{\mathbf{k}n}-\hbar\omega\mp\hbar\omega_{\mathbf{q},\alpha}\right)\\ & \times\left|\hat{\lambda}\cdot\sum_{n_1}\left(\frac{g^{\mathbf{q},\alpha}_{\mathbf{k}'n',\mathbf{k}n_1}\mathbf{v}^{\mathbf{k}}_{n_1n}}{\varepsilon_{\mathbf{k}n_1}-\varepsilon_{\mathbf{k}n}-\hbar\omega+i\eta}+\frac{\mathbf{v}^{\mathbf{k}'}_{n'n_1}g^{\mathbf{q},\alpha}_{\mathbf{k}'n_1,\mathbf{k}n}}{\varepsilon_{\mathbf{k}'n_1}-\varepsilon_{\mathbf{k}n}\mp\hbar\omega_{\mathbf{q},\alpha}+i\eta}\right)\right|^2.\end{aligned} \tag{9}$$

Here all symbols are as they have previously been defined, and once more the summation over $\pm$ ensures that both phonon emission and absorption processes are captured. The δ-function ensures energy conservation if broadened to Lorentzians whose widths are related to the initial and final carrier linewidths, and the η quantity is to ensure singular contributions are eliminated. These quantities are again evaluated on a dense BZ grid generated *via* our *ab initio* Wannier function techniques.

By calculating these quantities we capture the relative contributions to the optical response and can therefore map the carrier distribution profile as a function of both carrier and optical excitation energy. This is plotted in Fig. 4. Here, despite the indirect band gap of each of these materials, hot carriers are predicted to be generated *via* direct transitions almost exclusively. In the MoS_2 profiles, it is clear that indirect excitations of holes to Γ are predicted, although their small overall magnitude suggests they will not contribute significantly to excitation transitions. In both TMDCs, the AB configuration introduces a greater fraction of indirect transitions, however, in both cases the materials are expected to have a majority of direct transitions. Similar behavior has also been experimentally observed in optical studies of bilayer MoS_2 (ref. 51) and WSe_2.[52] In both cases, a substantial contribution to the emission spectra is found to be from direct transitions as opposed to indirect. These experimental results qualitatively agree with the theoretical results found here. Taken altogether, we confirm the preference for direct transitions even for these indirect gap bilayers.

IV. Conclusions

We investigate bilayer TMDCs MoS_2 and WSe_2 *via* an *ab initio* treatment of electron–electron and electron–phonon interactions. Specifically, we describe the complex interplay of spin, layer and valley indexing involved in two different stacking orientations of each bilayer. We find that unlike in the monolayer case, the dynamics are highly TMDC-specific. Based on detailed electronic structure calculations, both electron and hole dynamics (and mobilities) in bilayer WSe_2 are found to be significantly enhanced by spin–orbit coupling in the AB configuration. However, MoS_2 dynamics and mobilities are barely impacted by spin–orbit coupling effects. Stacking orientation was explored and found to be TMDC-specific, as longer lifetimes were found for MoS_2 in the AA′ configuration, compared to the slightly preferred AB configuration in WSe_2. These calculations and results emphasize the significance of understanding the nature of the band edges in determining dynamics, which are highly dependent on interlayer coupling. More generally, these results suggest that effective mechanisms for tuning the orientation and energy of the relevant valleys at the band edges such as layer orientation or strain would be ideal in creating an optimized material for valleytronic application with long-lived states. We also investigate hot carrier generation profiles, which indicate that each of these systems absorb mostly *via* direct transitions, despite their indirect gap nature. The implications of these results are important in understanding bilayers, their spin–valley–layer coupling and relation to symmetry *via* stacking orientation. Understanding coupled spin, valley, layer and symmetry effects in other bilayer $MoSe_2$ and WS_2 will further our understanding of the phenomena observed, and more broadly will help in understanding the underlying nature of quantum material heterostructures and larger multilayer stacks.

Conflicts of interest

There are no conflicts to declare.

Acknowledgements

All authors acknowledge support from the STC Center for Integrated Quantum Materials (CIQM) NSF grant number DMR-1231319 and from the Army Research Office MURI (*Ab Initio* Solid-State Quantum Materials) grant number W911NF-18-1-0431. This work used resources of the National Energy Research Scientific Computing Center, a DOE Office of Science User Facility, as well as resources at the Research Computing Group at Harvard University.

References

1 A. Splendiani, L. Sun, Y. Zhang, T. Li, J. Kim, C.-Y. Chim, G. Galli and F. Wang, *Nano Lett.*, 2010, **10**, 1271.
2 A. M. Jones, H. Yu, N. J. Ghimire, S. Wu, G. Aivazian, J. S. Ross, B. Zhao, J. Yan, D. G. Mandrus, D. Xiao, W. Yao and X. Xu, *Nat. Nanotechnol.*, 2013, **8**, 634.
3 T. Yan, X. Qiao, X. Liu, P. Tan and X. Zhang, *Appl. Phys. Lett.*, 2014, **105**, 101901.
4 K. F. Mak, K. He, J. Shan and T. F. Heinz, *Nat. Nanotechnol.*, 2012, **7**, 494.
5 C. Mai, A. Barrette, Y. Yu, Y. G. Semenov, K. W. Kim, L. Cao and K. Gundogdu, *Nano Lett.*, 2014, **14**, 202.
6 J. R. Schaibley, H. Yu, G. Clark, P. Rivera, J. S. Ross, K. L. Seyler, W. Yao and X. Xu, *Nat. Rev. Mater.*, 2016, **1**, 16055.
7 A. Molina-Sánchez, D. Sangalli, L. Wirtz and A. Marini, *Nano Lett.*, 2017, **17**, 4549.
8 A. K. Geim and I. V. Grigorieva, *Nature*, 2013, **499**, 419.
9 R. Frisenda, E. Navarro-Moratalla, P. Gant, D. P. D. Lara, P. Jarillo-Herrero, R. V. Gorbachev and A. Castellanos-Gomez, *Chem. Soc. Rev.*, 2018, **47**, 53.
10 D. N. Basov, R. D. Averitt and D. Hsieh, *Nat. Mater.*, 2017, **16**, 1077.
11 H. Boschker and J. Mannhart, *Annu. Rev. Condens. Matter Phys.*, 2017, **8**, 145.
12 Y. Cao, V. Fatemi, S. Fang, K. Watanabe, T. Taniguchi, E. Kaxiras and P. Jarillo-Herrero, *Nature*, 2018, **556**, 43.
13 A. Ramasubramaniam, D. Naveh and E. Towe, *Phys. Rev. B: Condens. Matter Mater. Phys.*, 2011, **84**(15), 205325.
14 A. M. van der Zande, J. Kunstmann, A. Chernikov, D. A. Chenet, Y. You, X. Zhang, P. Y. Huang, T. C. Berkelbach, L. Wang, F. Zhang, M. S. Hybertsen, D. A. Muller, D. R. Reichman, T. F. Heinz and J. C. Hone, *Nano Lett.*, 2014, **14**, 3869.
15 Y.-H. Zhao, F. Yang, J. Wang, H. Guo and W. Ji, *Sci. Rep.*, 2015, **5**, 8356.
16 M.-L. Lin, Q.-H. Tan, J.-B. Wu, X.-S. Chen, J.-H. Wang, Y.-H. Pan, X. Zhang, X. Cong, J. Zhang, W. Ji, P.-A. Hu, K.-H. Liu and P.-H. Tan, *ACS Nano*, 2018, **12**, 8770.
17 S. Fang, R. Kuate Defo, S. N. Shirodkar, S. Lieu, G. A. Tritsaris and E. Kaxiras, *Phys. Rev. B: Condens. Matter Mater. Phys.*, 2015, **92**(20), 205108.
18 R. K. Defo, S. Fang, S. N. Shirodkar, G. A. Tritsaris, A. Dimoulas and E. Kaxiras, *Phys. Rev. B*, 2016, **94**, 155310.
19 S. Carr, D. Massatt, S. Fang, P. Cazeaux, M. Luskin and E. Kaxiras, *Phys. Rev. B*, 2017, **95**, 075420.
20 H. Fang, C. Battaglia, C. Carraro, S. Nemsak, B. Ozdol, J. S. Kang, H. A. Bechtel, S. B. Desai, F. Kronast, A. A. Unal, G. Conti, C. Conlon, G. K. Palsson,

M. C. Martin, A. M. Minor, C. S. Fadley, E. Yablonovitch, R. Maboudian and A. Javey, *Proc. Natl. Acad. Sci. U. S. A.*, 2014, **111**, 6198.

21 J. Zhang, H. Hong, J. Zhang, H. Fu, P. You, J. Lischner, K. Liu, E. Kaxiras and S. Meng, *Nano Lett.*, 2018, **18**, 6057.

22 X. Fan, D. J. Singh and W. Zheng, *J. Phys. Chem. Lett.*, 2016, **7**, 2175.

23 J. Kim, C. Jin, B. Chen, H. Cai, T. Zhao, P. Lee, S. Kahn, K. Watanabe, T. Taniguchi, S. Tongay, M. F. Crommie and F. Wang, *Sci. Adv.*, 2017, **3**, e1700518.

24 W. Zhao, Z. Ghorannevis, L. Chu, M. Toh, C. Kloc, P.-H. Tan and G. Eda, *ACS Nano*, 2013, **7**, 791.

25 J. Lee, K. F. Mak and J. Shan, *Nat. Nanotechnol.*, 2016, **11**, 421.

26 S. Wu, J. S. Ross, G.-B. Liu, G. Aivazian, A. Jones, Z. Fei, W. Zhu, D. Xiao, W. Yao, D. Cobden and X. Xu, *Nat. Phys.*, 2013, **9**, 149.

27 B. Zhu, H. Zeng, J. Dai, Z. Gong and X. Cui, *Proc. Natl. Acad. Sci. U. S. A.*, 2014, **111**, 11606.

28 M. Xia, B. Li, K. Yin, G. Capellini, G. Niu, Y. Gong, W. Zhou, P. M. Ajayan and Y.-H. Xie, *ACS Nano*, 2015, **9**, 12246.

29 S. Wang, A. Robertson and J. H. Warner, *Chem. Soc. Rev.*, 2018, **47**, 6764.

30 A. M. Jones, H. Yu, J. S. Ross, P. Klement, N. J. Ghimire, J. Yan, D. G. Mandrus, W. Yao and X. Xu, *Nat. Phys.*, 2014, **10**, 130.

31 S. M. Shinde, K. P. Dhakal, X. Chen, W. S. Yun, J. Lee, H. Kim and J.-H. Ahn, *NPG Asia Mater.*, 2018, **10**, e468.

32 C. J. Ciccarino, T. Christensen, R. Sundararaman and P. Narang, *Nano Lett.*, 2018, **18**(9), 5709–5715.

33 R. Sundararaman, K. Letchworth-Weaver, K. A. Schwarz, D. Gunceler, O. Yalcin and T. Arias, *SoftwareX*, 2017, **6**, 278.

34 J. P. Perdew, A. Ruzsinszky, G. I. Csonka, O. A. Vydrov, G. E. Scuseria, L. A. Constantin, X. Zhou and K. Burke, *Phys. Rev. Lett.*, 2008, **100**, 136406.

35 S. Grimme, *J. Comput. Chem.*, 2006, **27**, 1787.

36 R. Sundararaman and T. A. Arias, *Phys. Rev. B: Condens. Matter Mater. Phys.*, 2013, **87**, 165122.

37 I. Souza, N. Marzari and D. Vanderbilt, *Phys. Rev. B: Condens. Matter Mater. Phys.*, 2001, **65**, 035109.

38 X. Fan, C.-H. Chang, W. T. Zheng, J.-L. Kuo and D. J. Singh, *J. Phys. Chem. C*, 2015, **119**, 10189.

39 Z. Y. Zhu, Y. C. Cheng and U. Schwingenschlögl, *Phys. Rev. B: Condens. Matter Mater. Phys.*, 2011, **84**(15), 153402.

40 F. Ladstädter, U. Hohenester, P. Puschnig and C. Ambrosch-Draxl, *Phys. Rev. B: Condens. Matter Mater. Phys.*, 2004, **70**, 235125.

41 P. Narang, L. Zhao, S. Claybrook and R. Sundararaman, *Adv. Opt. Mater.*, 2017, **5**, 1600914.

42 P. Rivera, J. R. Schaibley, A. M. Jones, J. S. Ross, S. Wu, G. Aivazian, P. Klement, K. Seyler, G. Clark, N. J. Ghimire, J. Yan, D. G. Mandrus, W. Yao and X. Xu, *Nat. Commun.*, 2015, **6**, 6242.

43 A. Srivastava, M. Sidler, A. V. Allain, D. S. Lembke, A. Kis and A. İmamoğlu, *Nat. Nanotechnol.*, 2015, **10**, 491.

44 C. Chakraborty, L. Kinnischtzke, K. M. Goodfellow, R. Beams and A. N. Vamivakas, *Nat. Nanotechnol.*, 2015, **10**, 507.

45 M. Koperski, K. Nogajewski, A. Arora, V. Cherkez, P. Mallet, J.-Y. Veuillen, J. Marcus, P. Kossacki and M. Potemski, *Nat. Nanotechnol.*, 2015, **10**, 503.

46 C. Chakraborty, K. M. Goodfellow and A. Nick Vamivakas, *Opt. Mater. Express*, 2016, **6**, 2081.

47 C. Chakraborty, K. M. Goodfellow, S. Dhara, A. Yoshimura, V. Meunier and N. Vamivakas, *Nano Lett.*, 2017, **17**, 2253.

48 C. Chakraborty, L. Qiu, K. Konthasinghe, A. Mukherjee, S. Dhara and N. Vamivakas, *Nano Lett.*, 2018, **18**, 2859.

49 R. Sundararaman, P. Narang, A. S. Jermyn, W. A. Goddard III and H. A. Atwater, *Nat. Commun.*, 2014, **5**, 5788.

50 A. M. Brown, R. Sundararaman, P. Narang, W. A. Goddard and H. A. Atwater, *ACS Nano*, 2016, **10**, 957.

51 A. K. M. Newaz, D. Prasai, J. I. Ziegler, D. Caudel, S. Robinson, R. F. Haglund Jr and K. I. Bolotin, *Solid State Commun.*, 2013, **155**, 49.

52 H. Zeng, G.-B. Liu, J. Dai, Y. Yan, B. Zhu, R. He, L. Xie, S. Xu, X. Chen, W. Yao and X. Cui, *Sci. Rep.*, 2013, **3**, 1608.

Faraday Discussions

PAPER

Direct hot-carrier transfer in plasmonic catalysis

Priyank V. Kumar, *[a] Tuomas P. Rossi, *[b] Mikael Kuisma, *[c] Paul Erhart *[b] and David J. Norris *[a]

Received 16th October 2018, Accepted 15th November 2018

DOI: 10.1039/c8fd00154e

Plasmonic metal nanoparticles can concentrate optical energy and enhance chemical reactions on their surfaces. Plasmons can interact with adsorbate orbitals and decay by directly exciting a carrier from the metal to the adsorbate in a process termed the direct-transfer process. Although this process could be useful for enhancing the efficiency of a chemical reaction, it remains poorly understood. Here, we report a preliminary investigation employing time-dependent density-functional theory (TDDFT) calculations to capture this process at a model metal–adsorbate interface formed by a silver nanoparticle (Ag_{147}) and a carbon monoxide molecule (CO). Direct hot-electron transfer is observed to occur from the occupied states of Ag to the unoccupied molecular orbitals of CO. We determine the probability of this process and show that it depends on the adsorption site of CO. Our results are expected to aid the design of more efficient metal–molecule interfaces for plasmonic catalysis.

1 Introduction

Plasmonic photocatalysis aims to carry out chemical reactions directly on the surface of a plasmonic metal nanoparticle efficiently and selectively.[1–5] This approach exploits the properties of localized surface plasmons (LSPs), which are collective charge density oscillations induced in a plasmonic metal nanoparticle upon interaction with light. Over time, the LSPs decay and generate energetic charge carriers, commonly referred to as hot electrons and holes. The resulting electrons can be transferred to the unoccupied molecular orbitals of the adsorbate, resulting in the formation of a transient negative ion complex. Since the absorbate is pushed onto a different potential energy surface, forces are induced

[a]Optical Materials Engineering Laboratory, ETH Zurich, 8092 Zurich, Switzerland. E-mail: pkumar@ethz.ch; dnorris@ethz.ch

[b]Department of Physics, Chalmers University of Technology, 41296 Gothenburg, Sweden. E-mail: tuomas.rossi@chalmers.se; erhart@chalmers.se

[c]Department of Chemistry, Nanoscience Center, University of Jyväskylä, 40014 Jyväskylä, Finland. E-mail: mikael.j.kuisma@jyu.fi

on the adsorbate nuclei. This causes the atoms to move, thus triggering chemical transformations on the surface of a plasmonic metal.[2]

Plasmonic catalysis is advantageous in many ways. Since LSPs can concentrate light and enhance optical energy absorption, they can increase the concentration of hot carriers, leading to improved external quantum efficiency (EQE) values.[6] This provides a platform for the utilization of low-intensity sunlight as a renewable source of energy.[7] In addition, being fundamentally different from thermal catalysis in terms of its mechanism, plasmonic catalysis could enable new reaction outcomes and help tailor product selectivity.[4]

In plasmonic catalysis, the hot-carrier transfer process represents a critical step. It can occur *via* two mechanisms:[2,3,8–11] (i) indirect transfer, where the hot carriers are first produced in the metal nanoparticle, and eventually transferred to the adsorbate (Fig. 1a), or (ii) direct transfer (or chemical interface damping),[12,13] where the plasmon decays by directly exciting an electron from the occupied states of the metal to the unoccupied orbitals of the adsorbate (direct hot-electron transfer, Fig. 1b). Plasmons can also decay by directly exciting an electron from the occupied states of the adsorbate to the unoccupied states of the metal, a transition that leads to a direct hot-hole transfer to the adsorbate. Since hot carriers generated in the metal can quickly relax back to the Fermi level *via* carrier–carrier interactions before they are transferred to the adsorbate, studies have constantly pointed out that the indirect transfer process can be inefficient.[14–17] This problem can be avoided in the direct transfer process.[14] As such, controlling and favoring the direct transfer process at the expense of the indirect transfer process could potentially lead to higher EQEs.

Plasmonic catalysis, however, continues to suffer from poor EQEs (less than 1%) and poor control over product selectivity.[2,18,19] This is because although the direct transfer process is known to be more efficient, it has been poorly understood at the atomic scale. As a result, no significant efforts have been made to discover ways to promote this process and tap into its potential for plasmonic catalysis.

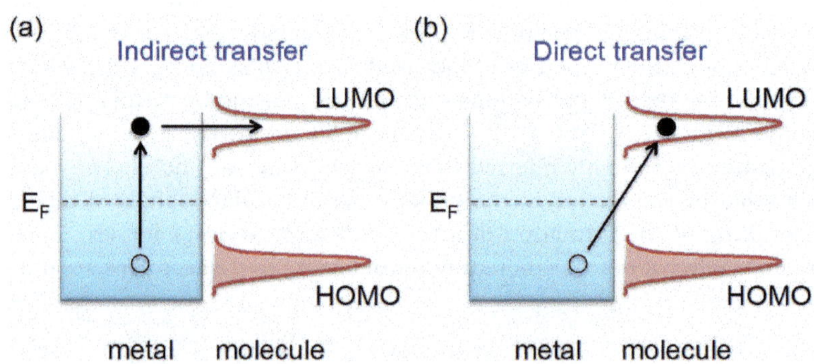

Fig. 1 Schematic representation of the hot-electron transfer process from the metal to the adsorbate *via* (a) indirect transfer and (b) direct transfer. E_F, HOMO and LUMO correspond to the Fermi level, the highest occupied molecular orbital and the lowest unoccupied molecular orbital, respectively. The filled and hollow circles denote electrons and holes, respectively. A similar scheme can be drawn for the hot-hole transfer process.

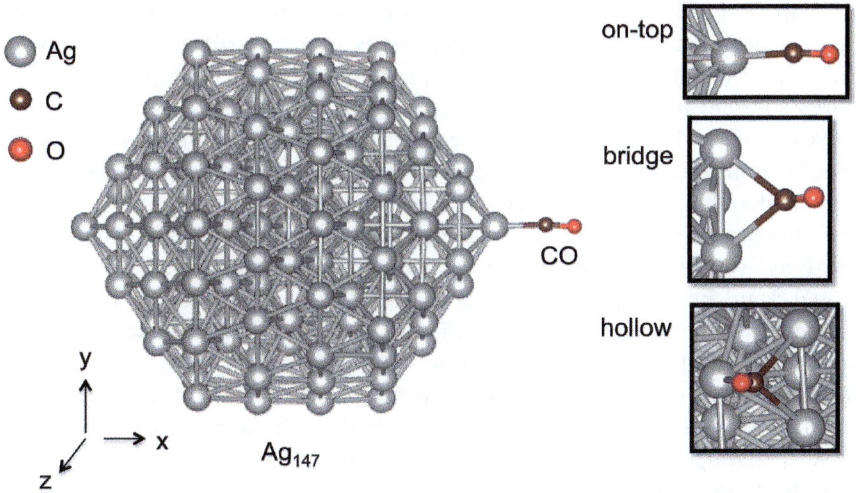

Fig. 2 Structural model of the Ag_{147}–CO on-top configuration used in our calculations. Similar structural models for the Ag_{147}–CO bridge and hollow configurations were used. Close-ups of the binding sites (on-top, bridge and hollow) are also shown.

Here, we report a preliminary *ab initio* investigation based on real-time time-dependent density-functional theory[20] (RT-TDDFT) to capture and study the direct transfer process at a model metal–adsorbate interface formed by a metallic Ag_{147} nanoparticle and a CO molecule. The rationale for choosing a computational approach comes from the fact that the two hot-carrier transfer processes occur simultaneously, and isolating individual processes in experiments can be a daunting task. Furthermore, since these processes occur on ultrafast timescales (1 fs to 100 ps),[8,10] interface-design to promote the direct transfer process *via* experiments can be complicated and time-consuming.

We demonstrate that the plasmon formed in the Ag_{147}–CO system due to an applied laser pulse induces a direct hot-electron transfer from the occupied states of Ag to the unoccupied molecular orbitals of CO. We study three adsorption configurations of CO: on-top, bridge and hollow adsorption sites (see Fig. 2). We find the probabilities of the direct hot-electron transfer process to be 0.9%, 2.0% and 1.8%, respectively, thus showing that the direct transfer process depends on the adsorption configuration of a molecule. We also reveal the direct hot-hole transfer process in these configurations, albeit with much smaller probability values. Overall, our study demonstrates an approach for the capture and quantification of the direct transfer process at a given metal–molecule interface. It also provides a framework for rational metal–molecule interface design for plasmonic catalysis.

2 Methods

The Ag_{147}–CO interface models used in our study, with CO adsorbed at three different sites (on-top, bridge and hollow), are shown in Fig. 2. The interface normal for all three configurations lies along the *x*-axis. The icosahedral Ag_{147}

nanoparticle previously studied by some of the authors[21,22] is known to support a strong plasmon. Since previous experiments have investigated the selective oxidation of CO directly on plasmonic Ag nanoparticles, we chose the CO molecule as a model adsorbate in our study.[6,23]

We described the ground-state of the system using DFT.[24,25] We used the GPAW code,[26,27] employing the projector augmented-wave method,[28] and the ASE code.[29] We included 11, 4 and 6 valence electrons to model the Ag, C and O atoms, respectively, while the lower-lying states were treated as a frozen core. Wave functions were represented using a linear combination of atomic orbitals (LCAO),[30] and a grid spacing parameter of 0.3 Å and vacuum extent of at least 6 Å around the system were used. First, the Ag_{147} and CO structures were relaxed individually until the residual forces were less than 0.03 eV $Å^{-1}$. Then, the combined Ag_{147}–CO structures were constructed and relaxed until the largest residual force was less than 0.03 eV $Å^{-1}$. During this relaxation step, the atoms of the Ag_{147} nanoparticle remained fixed, whilst the CO molecule was allowed to relax. In these relaxation calculations, we used the Perdew–Burke–Ernzerhof (PBE) exchange-correlation (XC) functional[31] and the default double-ζ polarized (DZP) basis sets.

The optical properties were computed using the LCAO-RT-TDDFT method as implemented in GPAW.[21,22] In these calculations, we used the Ag basis set including diffuse basis functions derived from the 5p orbital of the Ag atom, which is known to describe plasmons sufficiently accurately in Ag nanoparticles.[21,32] We employed the orbital-dependent GLLB-SC XC potential,[33] and used a Fermi–Dirac smearing of 0.05 eV to facilitate the ground-state convergence. More details on the methodology can be found in ref. 22.

3 Results and discussion

The plasmon resonances were determined by computing photoabsorption spectra using the δ-kick technique[34] with the external electric field applied along the *x*-axis. Fig. 3a shows the photoabsorption spectra of the individual Ag_{147} and CO systems, as well as those of the three different adsorption configurations considered. A Gaussian broadening of 0.2 eV full width at half maximum was applied to the spectra. One can observe a strong plasmon peak at 3.71 eV for Ag_{147}, which is consistent with previous calculations.[22] As expected, given a computed HOMO–LUMO gap value of 7.9 eV, CO does not exhibit any photoabsorption over the energy range plotted. For the Ag_{147}–CO systems considered here, the adsorption of CO causes only a small change in the plasmon peak position, shifting it to 3.74 eV.

To study hot-carrier generation and transfer at the Ag_{147}–CO interface, we excite the systems along the *x*-axis with a laser pulse at the plasmon frequency. The external electric field is $E(t) = E_0 \sin(\omega_p(t - t_0))\exp(-\sigma^2(t - t_0)^2/2)$, where t represents time, E_0 is the field intensity (assumed to be weak to probe the linear response regime), ω_p is the pulse frequency (set to a plasmon frequency of 3.74 eV) and σ corresponds to the pulse width (set to a value corresponding to a pulse duration of $\sim$10 fs).

The applied laser pulse and dipole-moment response of the Ag_{147}–CO (on-top configuration) system as a function of time are shown in Fig. 3b. The dipole moment presents large oscillations in response to the 10 fs laser pulse (starting at

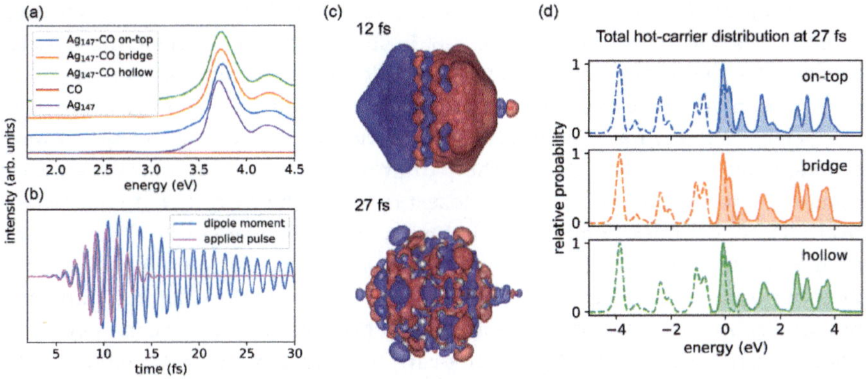

Fig. 3 (a) Photoabsorption spectra of the Ag_{147}–CO (on-top, bridge and hollow), Ag_{147}, and CO systems. (b) The applied pulse and corresponding dipole moment response in the time-domain for the Ag_{147}–CO on-top configuration. (c) Induced charge densities at 12 and 27 fs. Positive (red) and negative (blue) isosurfaces are shown using the same isovalues in both the plots. (d) Total hot-electron (solid lines and shaded) and hot-hole (dashed lines and unshaded) distributions for the three Ag_{147}–CO configurations computed at 27 fs.

~5 fs and ending at ~15 fs). We observe that the dipole moment reaches a maximum at ~12 fs and attains a steady minimum value beyond 27 fs. The induced charge densities in real space at these times are shown in Fig. 3c. While the charge density at 12 fs shows coherent oscillations, the charge-density distribution at 27 fs loses coherence. This suggests that the plasmon formed at 12 fs has decayed, and that hot carriers have been generated in the system.

Following ref. 35, we compute the resulting hot-carrier distribution at a given time, t, using the transition probability $P_{ia}(t) = \left|\delta\rho^{x}_{ia}(t)/\sqrt{(f_i - f_a)}\right|^2$, where the Kohn–Sham (KS) transition density matrix $\delta\rho^{x}_{ia}(t) = \rho^{x}_{ia}(t) - \rho^{x}_{ia}(0)$ is constructed on the basis of occupied (i) and unoccupied (a) KS states, and f_i and f_a correspond to the ground-state occupations of the KS states i and a, respectively.[22,35] Then, the probabilities for creating an electron in state a, and a hole in state i are given by $P_a(t) = \sum_i P_{ia}(t)$ and $P_i(t) = \sum_a P_{ia}(t)$, respectively.[35] These occupation probabilities are presented with respect to the eigenvalues of the states as the hot-electron distribution $P_e(E,t) = \sum_a P_a(t)G_a(E)$ with a Gaussian broadening function $G_a(E)$, and as the analogously-computed hot-hole distribution $P_h(E,t)$, respectively.

Fig. 3d shows the total hot-carrier distributions, $P_{e/h}(E)$, in the three Ag_{147}–CO systems at 27 fs, *i.e.* after the plasmon has decayed. We observe that a large concentration of holes is generated at around −4 eV, with a corresponding accumulation of electrons close to the Fermi level. This is attributed to interband (d → sp) transitions commonly observed in Ag nanoparticles. The hot-hole and hot-electron distributions at other energy levels on the other hand are the result of intraband transitions as well as transitions across the Ag–CO interface (which constitute the direct transfer process).

To separate the direct transfer contributions, we follow the approach used recently by some of the authors in ref. 36. In short, for direct hot-electron transfer we consider the partial distribution $P_{\mathrm{e/h}}^{\mathrm{Ag}\rightarrow\mathrm{CO}}$ denoting transitions from the occupied states of Ag to the unoccupied states of CO. This quantity is constructed from the weighted transition probability $P_{ia}^{\mathrm{Ag}\rightarrow\mathrm{CO}} = w_i^{\mathrm{Ag}} w_a^{\mathrm{CO}} P_{ia}$, where the Ag/CO wave function weights, $w_n^{\mathrm{Ag/CO}}$, denote the partial norm of the ground-state KS wave function $\psi_n(\mathbf{r})$ in the spatial region closest to either the Ag or CO atoms, respectively. These complementary spatial regions cover the whole space, so that the total norm $w_n^{\mathrm{Ag}} + w_n^{\mathrm{CO}} = 1$ for all wave functions, n. The corresponding probability distributions, $P_{\mathrm{e/h}}^{\mathrm{Ag}\rightarrow\mathrm{CO}}(E)$, are obtained as above by using the weighted transition probability. Analogous expressions are used for the other processes, namely direct hot-hole generation in CO, $P_{\mathrm{e/h}}^{\mathrm{Ag}\rightarrow\mathrm{CO}}$, hot-carrier generation in the Ag nanoparticle, $P_{\mathrm{e/h}}^{\mathrm{Ag}\rightarrow\mathrm{Ag}}$, and intra-adsorbate excitation,[23,37] $P_{\mathrm{e/h}}^{\mathrm{CO}\rightarrow\mathrm{CO}}$, which is found to be negligible in the present case.

The total probability of a particular partial process is calculated as the ratio of the integral over the partial hot-carrier distribution ($P_{\mathrm{e/h}}^{\mathrm{Ag}\rightarrow\mathrm{CO}}$ for example) to the integral of the total hot-carrier distribution ($P_{\mathrm{e/h}}$). This probability value is given as a percentage in the discussion below and is distinct from the previously mentioned transition probability between two particular KS states or the hot-carrier probability distributions.

Fig. 4a shows the direct hot-electron transfer distributions, $P_{\mathrm{e/h}}^{\mathrm{Ag}\rightarrow\mathrm{CO}}(E)$, for the three adsorption configurations of CO. The distributions reveal that the direct transfer process is capable of generating hot electrons with energies up to the pulse frequency (~4 eV) in the CO molecule. A larger fraction of hot electrons with energies greater than 2 eV can be seen in the case of the bridge and hollow configurations, as compared to the on-top configuration. Furthermore, the probability of this process, computed at 27 fs, shows that the direct hot-electron transfer is largest when CO is adsorbed in the bridge configuration (2.0%), followed by the hollow (1.8%) and on-top (0.9%) configurations (Fig. 4b), thus demonstrating that the direct hot-electron transfer depends on the adsorption configuration of the adsorbate.

Fig. 4c shows the direct hot-hole transfer distributions, $P_{\mathrm{e/h}}^{\mathrm{CO}\rightarrow\mathrm{Ag}}(E)$. It is interesting to note that the direct hot-hole transfer contribution is non-zero at the Ag–CO interface. The probabilities of this process at 27 fs are found to be 0.2%, 0.5% and 0.5% for the on-top, bridge and hollow configurations, respectively (Fig. 4d). These small but non-zero values are attributed to transitions arising from the hybridized states (of the HOMO and LUMO with the states of the Ag nanoparticle) emerging below the Fermi level upon bond formation.

Since the Ag nanoparticle is much larger in volume than the CO molecule, a major fraction of hot carriers (97–99%) is generated *via* the partial process, $P_{\mathrm{e/h}}^{\mathrm{Ag}\rightarrow\mathrm{Ag}}$ (part of the indirect transfer process). If the number of CO molecules adsorbed on the Ag_{147} nanoparticle is increased or if a smaller Ag nanoparticle is chosen, this fraction is expected to decrease. Furthermore, to check whether the choice of the plasmon decay time point (27 fs) has any influence on the results, we computed the probability values of the direct transfer processes over the time range 27–30 fs (Fig. 4b and d). Our results reveal only small variation in the probability values, showing that the plasmon decay has reached a steady state at 27 fs.

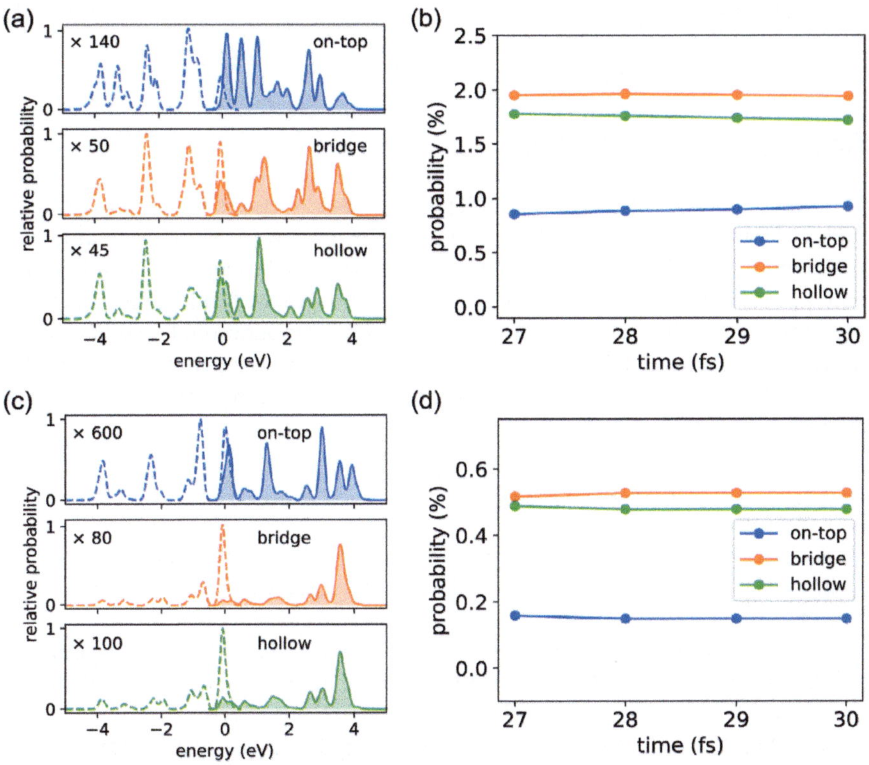

Fig. 4 (a) Hot-electron (solid lines and shaded) and hot-hole (dashed lines and unshaded) distributions generated in CO and Ag sub-systems, respectively, due to the direct hot-electron transfer process from the metal to the adsorbate. (b) Probability of the direct hot-electron transfer process over the time range 27–30 fs. (c and d) as (a and b) but for the direct hot-hole transfer process.

4 Conclusions

In summary, we have employed a RT-TDDFT approach to model and quantify the direct transfer process at a metal–adsorbate (Ag_{147}–CO) interface, a process that is difficult to isolate and assess quantitatively in experiments. For a Gaussian laser pulse tuned to the plasmon frequency with a duration of 10 fs, we observed direct hot-electron transfer from the occupied states of Ag to the unoccupied molecular orbitals of CO, with probabilities of 0.9%, 2.0% and 1.8% for CO adsorbed at the on-top, bridge and hollow sites, respectively. We also observed direct hot-hole transfer and found this to occur with much lower probability. Thus, our study not only shows that direct hot-electron transfer at Ag–CO interfaces is possible but also suggests that CO molecules adsorbed at bridge and hollow sites are more likely to undergo a hot-electron induced photochemical reaction, given all other steps are similar across the three configurations. Our approach could help in the understanding and enhancement of direct transfer transitions, thus paving the way for efficient plasmonic catalysis.

Conflicts of interest

There are no conflicts to declare.

Acknowledgements

P. V. K. thanks the Marie Curie ETH Zurich Postdoctoral Fellowship for the financial support. T. P. R. and P. E. acknowledge support from the Knut and Alice Wallenberg Foundation, the Swedish Foundation for Strategic Research (SSF), and the Swedish Research Council. M. K. acknowledges the funding received from the Academy of Finland under grant No. 295602. Computational resources were provided by the ETH High-Performance Computing Cluster (Euler), and by the Swedish National Infrastructure for Computing at PDC (Stockholm), NSC (Linköping) and C3SE (Gothenburg).

Notes and references

1 S. Linic, P. Christopher and D. B. Ingram, *Nat. Mater.*, 2011, **10**, 911–921.
2 S. Linic, U. Aslam, C. Boerigter and M. Morabito, *Nat. Mater.*, 2015, **14**, 567–576.
3 M. J. Kale, T. Avanesian and P. Christopher, *ACS Catal.*, 2014, **4**, 116–128.
4 M. L. Brongersma, N. J. Halas and P. Nordlander, *Nat. Nanotechnol.*, 2015, **10**, 25–34.
5 X.-C. Ma, Y. Dai, L. Yu and B.-B. Huang, *Light: Sci. Appl.*, 2016, **5**, e16017.
6 U. Aslam, S. Chavez and S. Linic, *Nat. Nanotechnol.*, 2017, **12**, 1000–1005.
7 P. Christopher, H. Xin and S. Linic, *Nat. Chem.*, 2011, **3**, 467–472.
8 K. Wu, J. Chen, J. R. McBride and T. Lian, *Science*, 2015, **349**, 632–635.
9 C. Boerigter, R. Campana, M. Morabito and S. Linic, *Nat. Commun.*, 2016, **7**, 10545.
10 S. Tan, A. Argondizzo, J. Ren, L. Liu, J. Zhao and H. Petek, *Nat. Photonics*, 2017, **11**, 806–812.
11 L. Yan, F. Wang and S. Meng, *ACS Nano*, 2016, **10**, 5452–5458.
12 M. J. Kale and P. Christopher, *Science*, 2015, **349**, 587–588.
13 O. A. Douglas-Gallardo, M. Berdakin and C. G. Sánchez, *J. Phys. Chem. C*, 2016, **120**, 24389–24399.
14 B. Foerster, A. Joplin, K. Kaefer, S. Celiksoy, S. Link and C. Sönnichsen, *ACS Nano*, 2017, **11**, 2886–2893.
15 R. Sundararaman, P. Narang, A. S. Jermyn, W. A. Goddard III and H. A. Atwater, *Nat. Commun.*, 2014, **5**, 5788.
16 A. M. Brown, R. Sundararaman, P. Narang, W. A. Goddard and H. A. Atwater, *ACS Nano*, 2016, **10**, 957–966.
17 J. Ma, Z. Wang and L.-W. Wang, *Nat. Commun.*, 2015, **6**, 10107.
18 G. Tagliabue, A. S. Jermyn, R. Sundararaman, A. J. Welch, J. S. DuChene, R. Pala, A. R. Davoyan, P. Narang and H. A. Atwater, *Nat. Commun.*, 2018, **9**, 3394.
19 U. Aslam, V. G. Rao, S. Chavez and S. Linic, *Nat. Catal.*, 2018, **1**, 656–665.
20 E. Runge and E. K. U. Gross, *Phys. Rev. Lett.*, 1984, **52**, 997–1000.
21 M. Kuisma, A. Sakko, T. P. Rossi, A. H. Larsen, J. Enkovaara, L. Lehtovaara and T. T. Rantala, *Phys. Rev. B*, 2015, **91**, 115431.

22 T. P. Rossi, M. Kuisma, M. J. Puska, R. M. Nieminen and P. Erhart, *J. Chem. Theory Comput.*, 2017, **13**, 4779–4790.
23 M. J. Kale, T. Avanesian, H. Xin, J. Yan and P. Christopher, *Nano Lett.*, 2014, **14**, 5405–5412.
24 P. Hohenberg and W. Kohn, *Phys. Rev.*, 1964, **136**, B864–B871.
25 W. Kohn and L. J. Sham, *Phys. Rev.*, 1965, **140**, A1133–A1138.
26 J. J. Mortensen, L. B. Hansen and K. W. Jacobsen, *Phys. Rev. B*, 2005, **71**, 035109.
27 J. Enkovaara, C. Rostgaard, J. J. Mortensen, J. Chen, M. Dułak, L. Ferrighi, J. Gavnholt, C. Glinsvad, V. Haikola, H. A. Hansen, H. H. Kristoffersen, M. Kuisma, A. H. Larsen, L. Lehtovaara, M. Ljungberg, O. Lopez-Acevedo, P. G. Moses, J. Ojanen, T. Olsen, V. Petzold, N. A. Romero, J. Stausholm-Møller, M. Strange, G. A. Tritsaris, M. Vanin, M. Walter, B. Hammer, H. Häkkinen, G. K. H. Madsen, R. M. Nieminen, J. K. Nørskov, M. Puska, T. T. Rantala, J. Schiøtz, K. S. Thygesen and K. W. Jacobsen, *J. Phys.: Condens. Matter*, 2010, **22**, 253202.
28 P. E. Blöchl, *Phys. Rev. B: Condens. Matter Mater. Phys.*, 1994, **50**, 17953–17979.
29 A. Larsen, J. Mortensen, J. Blomqvist, I. Castelli, R. Christensen, M. Dulak, J. Friis, M. Groves, B. Hammer, C. Hargus, E. Hermes, P. Jennings, P. Jensen, J. Kermode, J. Kitchin, E. Kolsbjerg, J. Kubal, K. Kaasbjerg, S. Lysgaard, J. Maronsson, T. Maxson, T. Olsen, L. Pastewka, A. Peterson, C. Rostgaard, J. Schiøtz, O. Schütt, M. Strange, K. Thygesen, T. Vegge, L. Vilhelmsen, M. Walter, Z. Zeng and K. W. Jacobsen, *J. Phys.: Condens. Matter*, 2017, **29**, 273002.
30 A. H. Larsen, M. Vanin, J. J. Mortensen, K. S. Thygesen and K. W. Jacobsen, *Phys. Rev. B*, 2009, **80**, 195112.
31 J. P. Perdew, K. Burke and M. Ernzerhof, *Phys. Rev. Lett.*, 1996, **77**, 3865–3868.
32 T. P. Rossi, S. Lehtola, A. Sakko, M. J. Puska and R. M. Nieminen, *J. Chem. Phys.*, 2015, **142**, 094114.
33 M. Kuisma, J. Ojanen, J. Enkovaara and T. T. Rantala, *Phys. Rev. B: Condens. Matter Mater. Phys.*, 2010, **82**, 115106.
34 K. Yabana and G. F. Bertsch, *Phys. Rev. B: Condens. Matter Mater. Phys.*, 1996, **54**, 4484–4487.
35 T. P. Rossi, P. Erhart and M. Kuisma, unpublished.
36 P. V. Kumar, T. P. Rossi, D. M. Dafcik, D. Reichmuth, M. Kuisma, P. Erhart, M. J. Puska and D. J. Norris, *ACS Nano*, 2019, DOI: 10.1021/acsnano.8b08703.
37 E. Kazuma, J. Jung, H. Ueba, M. Trenary and Y. Kim, *Science*, 2018, **360**, 521–526.

Faraday Discussions

PAPER

Generation of hot electrons in nanostructures incorporating conventional and unconventional plasmonic materials†

Tianji Liu,[ab] Lucas V. Besteiro,[ac] Zhiming Wang[a] and Alexander O. Govorov*[ab]

Received 2nd October 2018, Accepted 30th October 2018
DOI: 10.1039/c8fd00145f

The generation of energetic electrons is an effect occurring in any plasmonic nanostructure. However, the number of electrons with high energies generated optically in a plasmonic nanostructure can be relatively small. This is an intrinsic property of the collective plasmon excitations in a Fermi gas of electrons. But the choices of material and geometry have a great impact on the generation rate, and are therefore crucial for designing a nanostructure with a large rate of generation of energetic (hot) electrons. Here we test different plasmonic materials from the point of view of the generation of hot electrons (HEs). Our choice of materials includes both strongly-plasmonic materials (Au, Ag, Cu and Al) and crystals with strongly broadened plasmonic resonances (Pt, TiN and ZrN). Regarding the choice of geometry, we consider two types of nanostructures, single nanocrystals deposited over a dielectric substrate and metastructure absorbers, observing interesting opto-electronic properties. For single nanocrystals, the rate of HE generation is strongly material-dependent since the HE generation rate strongly depends on several physical parameters such as plasmonic enhancement, plasmonic resonance wavelength, Fermi energy, *etc.* Interestingly, the plasmonic meta-absorbers exhibit a different behaviour. The strongly-plasmonic metals, such as Au, Ag, Cu or Al, show very similar performances, while the materials with damped plasmon resonances demonstrate diverse and reduced rates of HE generation. The physical reason for these different behaviours lies in the dielectric functions of these materials. In the metastructures, plasmonic resonances are in the infrared region and the strongly-plasmonic materials behave as an almost ideal metal, whereas the second group of materials exhibits strong dissipation. This makes the responses from the metastructures made of crystals with damped plasmons strongly dependent on the choice of material. The physical principles described in our study can be useful for designing metastructures and nanodevices based on HEs, which can be used in photo-chemistry and opto-electronics.

[a]*Institute of Fundamental and Frontier Sciences, University of Electronic Science and Technology of China, Chengdu, China*

[b]*Department of Physics and Astronomy, Ohio University, Athens, USA. E-mail: govorov@ohio.edu*

[c]*Centre Énergie Matériaux et Télécommunications, Institut National de la Recherche Scientifique, 1650 Boul. Lionel Boulet, Varennes, QC J3X 1S2, Canada*

† Electronic supplementary information (ESI) available. See DOI: 10.1039/c8fd00145f

1. Introduction

Hot electron (HE) generation is an active field of research in modern nanotechnology. This field is motivated by two main types of applications: plasmon-enhanced photochemistry[1–9] and photodetectors involving plasmon-induced photocurrents.[10–18] Recent theories have shown that the choice of geometry for a nanostructure can be crucial to induce large numbers of photo-generated HEs.[19–23] Another design parameter to explore is the choice of plasmonic material of the nanostructure. Overall, plasmonic materials with a longer mean free path of electrons can generate HEs more efficiently. This has been shown, for example, for the cases of Ag and Au,[22,24] where Ag nanostructures typically generate HEs more efficiently than Au ones. Here we explore further these related properties by considering two broad categories of available plasmonic materials: (1) strongly-plasmonic materials (SPMs), such as Au, Ag, Cu and Al and (2) materials with damped plasmonic resonances (MDPR) such as Pt, TiN and ZrN. Furthermore, we investigate here two important types of available nanostructures: single nanocrystals and metastructures.

We carry out this research with the tools provided using our previously developed theory, as described in ref. 19–22 and already applied to explain or motivate experimental work in the field.[25,26] The formalism used in this paper describes the generation of HEs at the surfaces and hot spots, where the linear momentum of an electron is not conserved and the excitation of high energy electrons (or HEs) becomes possible. This formalism is very convenient since it allows us to compute the classic electromagnetic response of nanostructures of arbitrary shape using fast and reliable commercial software, such as COMSOL Multiphysics (version 5.1). The formalism obtains the rate of optical generation of excited carriers through intra-band transitions in Drude-like metals. Such excitations dominate the optical responses in the yellow, red and IR spectral intervals of typical plasmonic materials. This HE generation mechanism involves practically all electronic properties of the plasmonic material used to build a nanostructure. In this study, we focus on two important types of nanostructures: single nanocrystals and metamaterial absorbers. We show that the rates of HE generation are strongly dependent on the geometry and material of the nanostructure. Metamaterial absorbers typically produce more HEs due to the strong infrared hot spots occurring at the metal gap and the correspondingly enhanced gap plasmons. Furthermore, the SPMs exhibit much higher peak rates in the HE spectra, whereas the MDPR show broader lines with smaller peak values. However, the MDPR can still provide significant spectrally integrated rates of HE production, which can be useful for solar light harvesting. This paper discusses the aforementioned properties based on theoretical calculations for a range of nanosystems.

2. Formalism for nonthermal energetic (hot) electrons

In Fig. 1, we show the models of the systems investigated in this study. We focus on the case of single nanocrystals on a glass substrate and on the metastructures. Both cases assume a periodic arrangement of nanocrystals with a period of

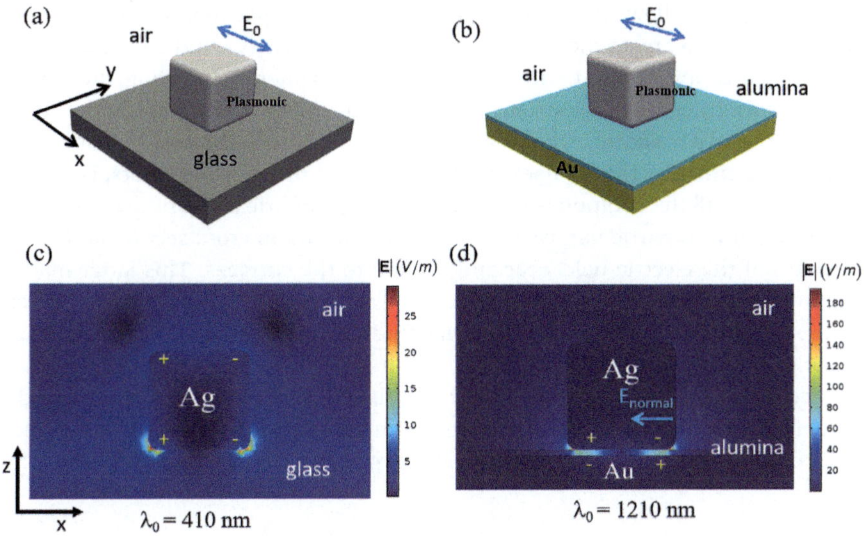

Fig. 1 (a and b) Geometrical models of the nanostructures considered in this paper. (c and d) Normalized electromagnetic field (magnitude) maps of the nanostructures. The incident plane wave is polarized along the *x*-direction and is taken at the plasmon resonance wavelengths. For the Ag–glass structure, one can see the field pattern from the dipolar plasmon excitation. For the NC-mirror metastructures, we observe the typical pattern associated with a gap plasmon excitation. In panel (d), we also show the normal electric field (E_{normal}) at the surface, responsible for the surface mechanism of generation of HEs.

500 nm in both *x* and *y* directions. The main geometry used here is a nanocube (NC), either deposited on a glass substrate or forming a metastructure together with a continuous Au layer and an alumina spacer, as shown in Fig. 1a and b, respectively. The use of this type of metastructures as HE generators has been recently investigated in ref. 25 and 27. The defining property of these metastructures (Fig. 1b) is the presence of strong plasmonic hot spots associated with a gap plasmon. Fig. 1b depicts such gap-like plasmonic excitation. It has been shown[25,27] that such metastructures with infrared hot spots exhibit very unusual ultra-fast dynamics. In contrast with a number of previous experiments[28] the transient optical signals in such metastructures exhibit a very strong component with an ultra-fast decay time of the order of 50 fs. Using a theoretical model, we explained this interesting observation in terms of the HE generation effect.[25,27] We have shown that such nanostructures with IR gap plasmons should generate HEs very efficiently and much more strongly than NCs on a glass substrate. Overall, it is currently well established that many optical properties are strongly enhanced in the geometry of a plasmonic meta-absorber.[18,29]

In our computational study, we first deal with the classical optical properties of NCs using the empirical dielectric constants of metals in COMSOL. The optical dielectric functions for the plasmonic materials were taken from the following literature: Au,[30] Ag,[30] Cu,[31] Al,[32] Pt,[33] TiN[34] and ZrN.[35] For the dielectric constant of the substrates in our structures, we took the refractive indices of glass ($n = 1.52$) and sapphire (Al_2O_3, $n = 1.77$). In all our models, we maintained the geometrical model previously used by us in ref. 25. Our goal was to maintain a geometrical consistency that allows the comparison of the different plasmonic materials used in our models to create the plasmonic resonators. The NCs and substrate in the metastructure are covered with thin layers of polymer, mimicking the experimental structures.[25] An important parameter for the metastructure is the thickness of the Al_2O_3 spacer, which was taken to be 8 nm. The total gap between a nano-antenna and the Au back-mirror, including the polymer layers, is 12 nm. The ESI† has all the computational details. The geometrical size of the plasmonic NC is 150 nm. In particular, we looked at the absorption cross section and at the strength of the electric field near and normal to the surfaces. This latter magnitude will allow us to obtain the generation rates of HEs, using our theoretical framework to connect the classical electrodynamical results with a quantum description of the system.

The fundamental magnitude describing the electrodynamical response of the system is the electric field, and in using a general representation of the magnitude as a complex vector field, its physical value is given by the real part of the complex field:

$$\mathbf{E} = \mathbf{E}_\omega \cdot e^{-i\omega t} + \mathbf{E}^*_\omega \cdot e^{+i\omega t},$$

where $\mathbf{E}_\omega$ is the complex vector of the electric field in the system.

For the absorbed power Q_{abs} and the absorption cross section $\sigma_{abs}(\omega)$, we write

$$Q_{abs} = \left\langle \int_{NPs} dV \mathbf{j}(t) \cdot \mathbf{E}(t) \right\rangle_{time} = \mathrm{Im}(\varepsilon_{metal}) \frac{\omega}{2\pi} \int_{NPs} dV \ \mathbf{E}_\omega \cdot \mathbf{E}^*_\omega, \quad \sigma_{abs}(\omega) = \frac{Q_{abs}}{I_0}, \tag{1}$$

where $\mathbf{j}(t)$ is the AC current in the NCs and ε_{metal} is the dielectric constant of the plasmonic NC. The incident flux I_0 is given by the following standard formula: $I_0 = c_0\sqrt{\varepsilon_0} \cdot E_0{}^2/(2\pi)$. In the following calculations, we will use the flux $I_0 = 2.5 \times 10^8$ W cm^{-2}; this value is typical for time-resolved experiments.[25] Here E_0 denotes the amplitude of the electric field in the incident electromagnetic wave, c_0 is the speed of light in vacuum and ε_0 is the vacuum electric permittivity.

The physical magnitude leading to the generation of hot plasmonic electrons is the electric field component normal to the surface, taken near the surface inside a nanocrystal (Fig. 1d). Typical distributions of hot electrons in nanocrystals are shown in Fig. 2c and d. These distributions were calculated in ref. 22 using the quantum formalism previously developed by us. The distribution has two types of excited carriers: the Drude electrons and the nonthermal HEs. The physical mechanism for the generation of HEs in this picture is the following: the linear momentum of an electron near the surface is not conserved and, therefore, the electromagnetic field is able to create excited electrons and holes with large energies $\sim\hbar\omega$. Generation of such energetic (hot) electrons is therefore a surface effect. Simultaneously, the Drude electrons in our picture (Fig. 2c and d) represent the bulk effect of the excitation of electrons in the Fermi sea near the Fermi energy, E_F. These electrons and the related dissipation are well described by the classical Drude theory. The HEs generated with energies above the Fermi sea are excited to energies in the interval $E_F < \varepsilon < E_F + \hbar\omega$, according to the law of conservation of energy. For the excited states below the Fermi surface, we observe negative values of the rates in Fig. 2c and d. It means that the external optical

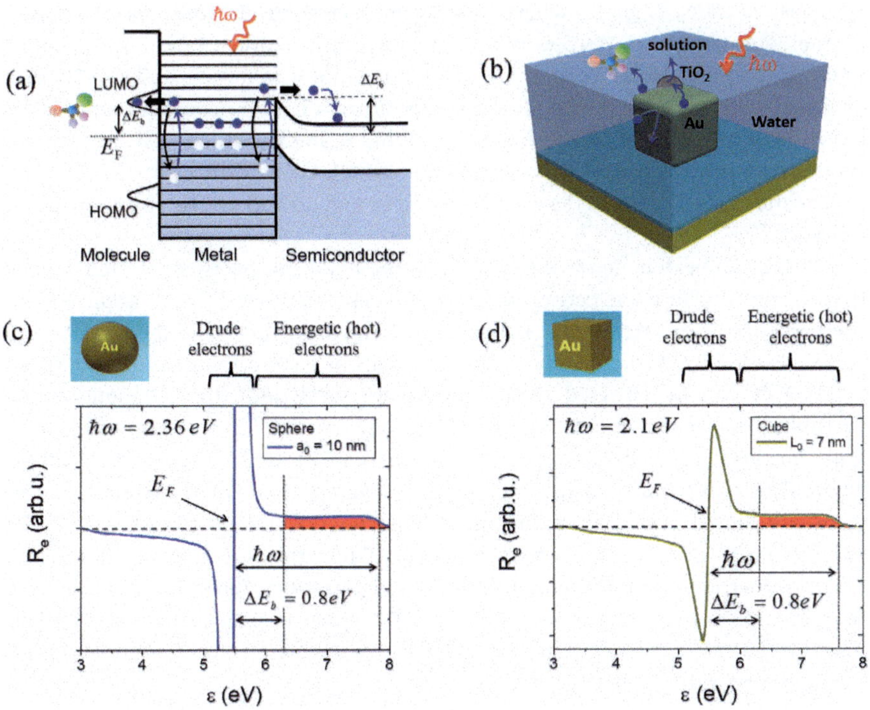

Fig. 2 (a) Energy diagram with the Fermi sea of electrons and the related processes of excitation, extraction and relaxation in a plasmonic nanocrystal. (b) Illustration of the mechanisms of extraction of HEs from a metal nanocrystal. HEs can be transferred to a semiconductor nanoparticle, or to a reacting molecule. (c) Energy-resolved spectrum of the HE generation rate in the case of a Au-nanosphere with a diameter of 10 nm. (d) Energy-resolved spectrum of the HE generation rate in the case of a Au-nanocube with a length of 7 nm.

excitation creates empty states (holes) in the interval $E_F - \hbar\omega < \varepsilon < E_F$. Drude electrons lead to the so-called frictional heating of a nanocrystal and have lower energies, $\sim k_BT$. Under CW illumination, the nonequilibrium steady state of a nanocrystal will include both Drude and nonthermal hot electrons. The HEs extend over the energy interval $E_F < \varepsilon < E_F + \hbar\omega$. However, in applications such as photochemistry or Schottky photodetectors, only high-energy HEs with energies $\varepsilon > \Delta E_b$ participate in the device operation, where ΔE_b is the barrier potential shown in Fig. 2a. In Fig. 2c and d, we show the high-energy HEs as red areas; these electrons can propagate to a neighbouring semiconductor or tunnel to a molecule adsorbed at the surface. Such areas in Fig. 2c and d correspond to the interval $E_F + \Delta E_b < \varepsilon < E_F + \hbar\omega$.

Fig. 2c and d show the excitation rates per electron energy, having units of 1/(eV s). Then, the spectral rate for HE generation should be integrated over the energy in the interval $E_F < \varepsilon < E_F + \hbar\omega$. In this way, we can obtain the total number of HEs generated per second. We will denote this rate as $\text{Rate}_{\text{hot electrons}}$. The rate of generation of high-energy electrons (with $\varepsilon > \Delta E_b$) will be given by the parameter $\text{Rate}_{\text{high energy}}$. The number of HEs in the steady state for a nanocrystal will depend on their generation rate and relaxation time. Approximately, δN_{hot}

electrons $\sim \tau_e \cdot \text{Rate}_{\text{hot electrons}}$, where τ_e is the characteristic electron relaxation time of energetic electrons. Here we will focus on the functions $\text{Rate}_{\text{hot electrons}}$ and $\text{Rate}_{\text{high energy}}$ and compute them for several plasmonic materials and two distinct geometries. These results will allow us to perform a comparison between the HE generation capabilities of different plasmonic systems, which will be of great interest to inform the development of efficient HE-based devices.

The methods to detect HEs are illustrated in Fig. 2b and can be summarized as follows:

(1) HEs can be detected using ultra-fast time-resolved spectroscopy that probes the non-equilibrium dielectric function of a nanostructure[25,27,36–40] arising from the presence of excited photocarriers in the metal. In this case, the rate $\text{Rate}_{\text{hot electrons}}$ can be used to characterize the strength of generation of ultra-fast HEs.

(2) HEs can be injected into a nanoscale semiconductor component (for example, TiO_2 nanoparticles) attached to the surface of a plasmonic nano-antenna (Fig. 2b).[1,26,41] Then, these electrons can be transferred to a solution, where they will create radicals or other reactive species. For these electrons to transit to the semiconductor, they must have energies larger than the Schottky barrier at the metal–semiconductor barrier. Then, the injection of HEs to the semiconductor should be characterized by the parameter $\text{Rate}_{\text{high energy}}$. We note here that plasmonic nanoantennas and HEs injection can also be involved in photodetectors in which plasmonic HEs are injected into a semiconductor through a Schottky barrier.[8,14,42,43]

(3) HEs can be transferred directly to a reacting molecule near the surface.[41,44–46] This can occur in a gas reactor or on the surface of a plasmonic antenna in solution.

Our calculations in this paper are relevant to all cases described above. However, the structures that we model are made to directly resemble the experiments with time-resolved transient spectroscopy (case 1). In this case, our structures are exposed to air (Fig. 1a and b). In the photochemical experiments, like the case 2, the metastructure should be in solution (Fig. 2b). This case can also be computed by simply adding the dielectric function of the solution into the model.

For the HE rates, we have the following quantum equation:[22,24]

$$\text{Rate}_{\text{hot electrons}} = \sum_{i=1,2} \frac{2}{\pi^2} \times \frac{e^2 E_{\text{F},i}{}^2}{\hbar} \frac{1}{(\hbar\omega)^3} \int_{S_i} |E_{\text{normal}}(\theta,\varphi)|^2 \text{d}s \tag{2}$$

where the summation is taken over the nanocrystal ($i = 1$) and the substrate ($i = 2$). Here $E_{\text{normal}}(\theta,\varphi)$ is the electric field normal to the surface and inside the metal and S_i denotes the surface of the element. In fact, the surface integral should be taken over all surfaces, including the 2D metal mirror in our metastructures (Fig. 1b). The rate of photo-generation of electrons with high energies, $\varepsilon > \Delta E_b$, is smaller than the total rate: $\text{Rate}_{\text{high energy}} < \text{Rate}_{\text{hot electrons}}$. The equation of $\text{Rate}_{\text{high energy}}$ is given by:[22,24]

$$\text{Rate}_{\text{high-energy}} = \frac{2}{\pi^2} \times \frac{e^2 E_{\text{F},1}{}^2}{\hbar} \frac{(\hbar\omega - \Delta E_b)}{(\hbar\omega)^4} \int_{S_1} |E_{\text{normal}}(\theta,\varphi)|^2 \text{d}s \tag{3}$$

Here the integration is taken only over the nanocrystal antenna (surface S_1 in the integral), since the injection process occurs only at the nanocrystal's surface in our system containing the metastructure (Fig. 2b). Another difference between eqn (2) and (3) is the multiplier $(\hbar\omega - \Delta E_b)/\hbar\omega$. In the infrared region, the number of injected HEs will be strongly reduced due to this coefficient. In the numerical calculations we used a typical value for the barrier potential, $\Delta E_b = 0.8$ eV.

$\text{Rate}_{\text{hot electrons}}$ also characterizes the strength of quantum effects in our structures. To analyse the role of quantum surface scattering in plasmonic nanostructures, we need to compare $\text{Rate}_{\text{hot electrons}}$ with the rate of absorption of photons, $\text{Rate}_{\text{absorption, photons}} = Q_{\text{abs,tot}}/\hbar\omega$. The ratio of the above rates gives the quantum parameter of the plasmon, as introduced by us in ref. 22 and 24.

3. Numerical data for the hot electron generation

We now look at the computational data generated using our formalism based on eqn (2). This formalism quantifies the efficiency of generation of HEs in a particular nano-architecture. We now consider two distinct cases, the antenna–glass system (Fig. 3) and the metastructure system composed of a NC and a Au mirror (Fig. 4).

3.1 Antenna–glass system

In the antenna–glass system, the main resonances involve mostly the dipolar and multipolar excitations (Fig. 3). The dipolar and multipolar plasmon resonances become strongly affected by the asymmetric dielectric environment (surrounding air and glass substrate), which makes the plasmonic spectrum of the system more complex. In Fig. 3a and b, we show two characteristic cases of the relevant surface-field maps for both Ag- and Au-NCs. These surface-field maps also indicate the spatial distributions for the HE generation rates on the surface, as $R_{\text{hot electrons}} \propto |E_{\text{normal}}(\theta,\varphi)|^2$, and here $R_{\text{hot electrons}}$ is the surface density of the generation of HEs. In these maps, we observe multipolar plasmons with charges localized in different surface regions. In contrast with the metastructure geometry discussed below, the NC–glass system does not have well-defined gap modes. One can see Fig. 1c and d for the comparison between the NC–glass mode and the typical gap-like mode in the metastructure. In the NC–glass system, the mode is localized near the lower vertices of the NC because of the effect of the dielectric substrate (see also the insets in Fig. 3a and b). In the metastructure, the field is really localized in the gap between the NC and the metal mirror. A more detailed picture of the gap mode, computed for different plasmonic modes, can be found in our recent paper.[25] For the Ag-NCs, we see also strong characteristic resonances in the UV region, which come from a mixture of plasmon resonances and inter-band excitations in Ag. In Fig. 3c, we show the remaining cases for Cu, Al, Pt, TiN and ZrN. Here the main observation is that the peak rate of generation depends strongly on a material system. Later in Fig. 6a, we will summarize the rates at the plasmonic peaks. Overall, the dependence of the peak rates on a material system is complex since the rate involves at least three factors: the plasmonic enhancement, the plasmonic frequency and the Fermi energy (see eqn (2)). Below we will discuss it further.

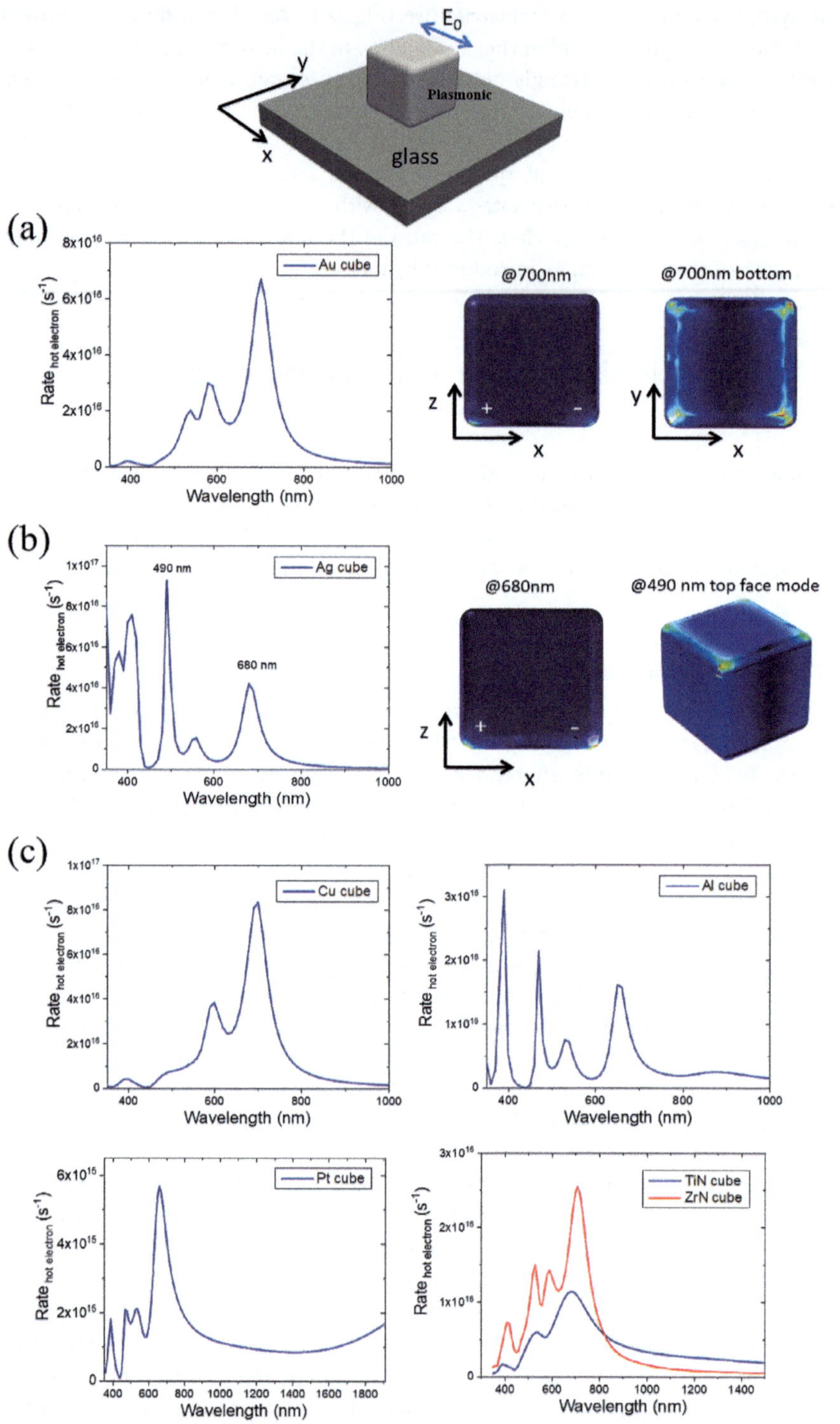

Fig. 3 Rates of HE generation for the NC-glass structure for Au (a), Ag (b) and the other materials (c). The insets on the right show the surface maps for the generation of HEs for some selected resonances. The surface density of the HE generation rates is proportional to $|E_{normal}|^2$ at the surface inside the NC.

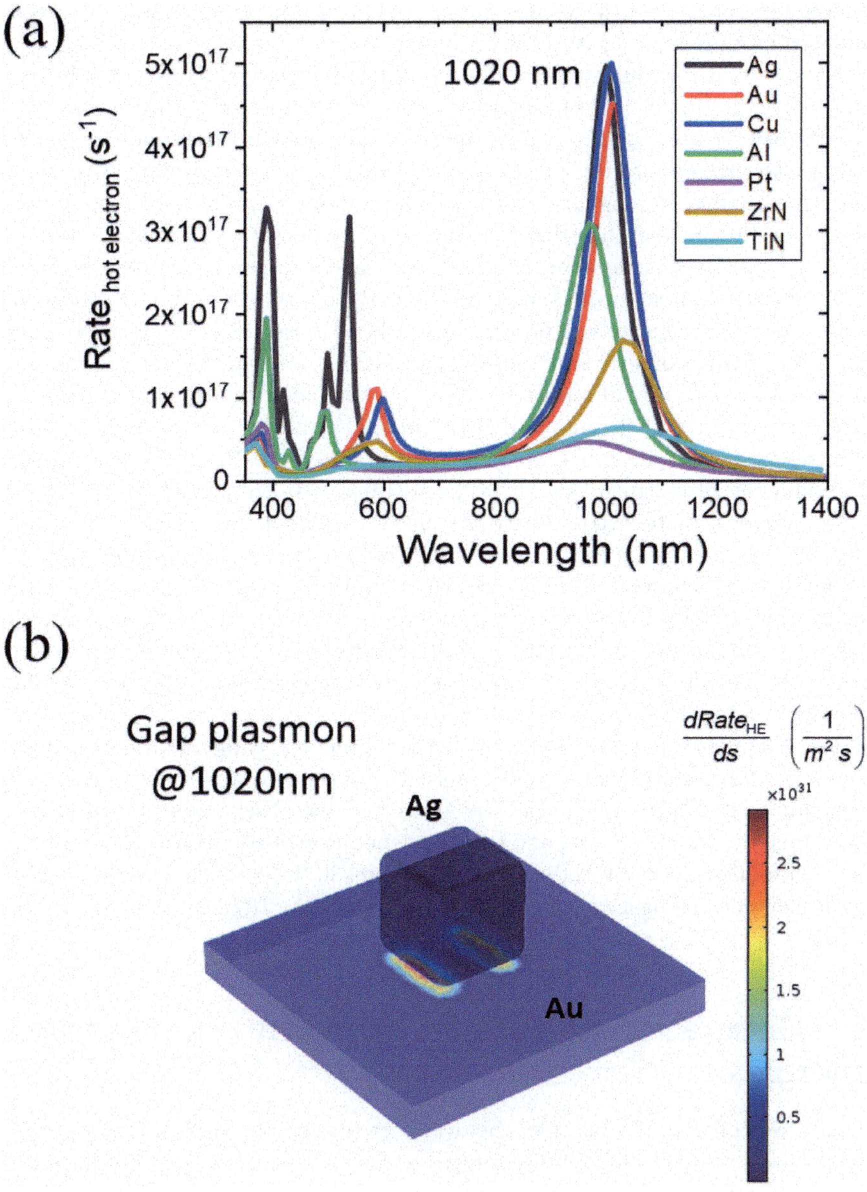

Fig. 4 (a) Rates of HE generation for the metastructure NC-spacer-mirror. (b) Surface density map of the HE generation rate for the gap plasmon mode in the Ag–Au metastructure.

3.2 Metastructure

The case of the metastructure geometry looks somewhat simpler (Fig. 4) since it has only one dominant mode, located in the infrared (IR) region. This main plasmonic resonance near 1000 nm is the gap plasmon. The short-wavelength peaks are due to multipolar plasmons and the inter-band excitations in Au and in the other metals. Ref. 25 has the detailed assignment of the multiple plasmonic peaks that appear in the NC-mirror metastructure for the Ag–Au case.

Now, if we direct our attention to the peak values in Fig. 4a we will observe one interesting property. The metals with strong and narrow plasmon resonances (SPMs) form a group with high rates of generation. Those are Au, Ag, Cu and Al. The HE generation peak rates for those metals are similar. In particular, the metals Au, Ag and Cu show remarkably similar main peaks in Figure 4a. In contrast, the other plasmonic materials (Pt, TiN and ZrN), which have typically weaker and broader resonances, have smaller HE generation rates and their peak values depend strongly on the material. A physical explanation for the observed properties for the systems with SPMs (Au, Ag and Cu) lies in their geometry and materials properties. Strongly-plasmonic metals in the metamaterial configuration produce resonances that depend mostly on geometry (the size of the gap and NC). These resonances are in the IR region and the metals can be considered as ideal; in other words, these metals have the following properties: $|\mathrm{Re}[\varepsilon]| \gg 1$ and $|\mathrm{Re}[\varepsilon]| \gg |\mathrm{Im}[\varepsilon]|$. In the ESI,† we summarized the dielectric functions used in this paper. Thus, the gap-plasmon frequency depends mostly on the geometrical properties of the metastructure geometry. Simultaneously, the other plasmonic materials, like Pt, TiN and ZrN, have large imaginary components of the dielectric function ($|\mathrm{Re}[\varepsilon]| \sim |\mathrm{Im}[\varepsilon]|$) at the relevant wavelengths and, therefore, cannot be considered as ideal metals. In this case, the optoelectronic properties should be material-dependent.

Fig. 5 depicts the generation rates for the structural components of the two characteristic metastructures, Ag–Au and TiN–Au. Namely, we plot separately the generation rates in the NC and in the Au mirror. We observe that, for the good plasmonic metals (like Au, Ag and Cu), the generation of HEs is equally strong in both components. However, the generations rates in the components can be very dissimilar when using a weakly-plasmonic material, like TiN (see Fig. 5, the lower panel).

4. Generation of hot electrons with over-barrier energies

Finally, we consider the rates for the over-barrier electrons. The difference with the previous data is only in the factor $(\hbar\omega - \Delta E_\mathrm{b})/\hbar\omega$. Because of this multiplier, the structures with plasmon resonances in the blue region become preferable. However, despite this factor, the metamaterial geometry still provides high generation rates for the over-barrier carriers (Fig. 6c,d). We observe that the parameter $\mathrm{Rate}_{\text{high-energy}}$ is typically higher for the metastructures. The physical reasons for this are the following: (1) the gap plasmon has very strong fields that generate HEs very efficiently and (2) the quantum generation rate always has the factor $(\hbar\omega)^{-4}$ that gives advantage to nanostructures with plasmon resonances in the IR interval. The origin of this quantum factor and the related derivation can be found in our recent paper.[22]

In Figure 6c we can also see that Ag generates over-barrier HEs in the NC-glass system very efficiently. This is because this metal has very strong and narrow resonances and its plasmon resonances are in the high energy side

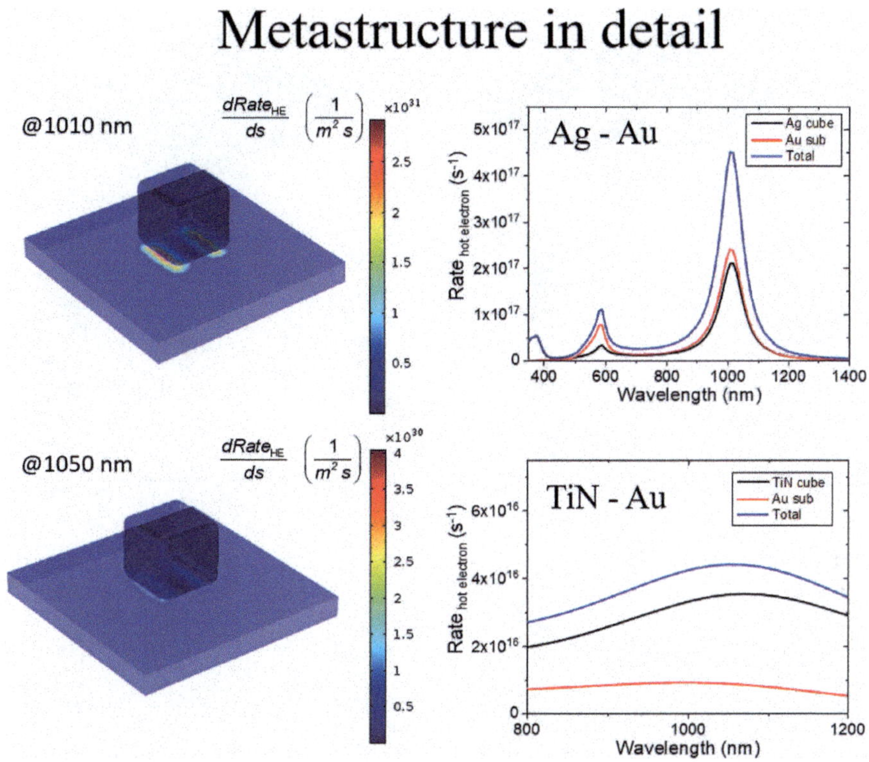

Fig. 5 Rates of HE generation for the components of the metastructures. The curves show the rates for the NC and for the Au–substrate. For different NC material systems, we observe different ratios for the rates of the components. This behaviour comes from the relationships between the dielectric functions of the pairs of materials involved in these metastructures.

of the visible spectrum. The latter comes from the over-barrier energy factor $(\hbar\omega - \Delta E_b)$ in eqn (3).

5. Physical principles for the generation of large numbers of energetic electrons

Here we will briefly comment on the results for different materials summarized in the final Figure 6. The rates of generation of nonthermal HEs and highly-excited (over-barrier) electrons are given by eqn (2) and (3). The following physical factors contribute to these rates:

(1) The rates depend on the presence of plasmonic hot spots. In the metastructures, such hot spots appear in the gap region and are especially strong. Therefore, the rates for the HE generation are the largest for the plasmon-gap structures (Figures 6a and b).

(2) The rates for Ag are typically the largest (Figures 6a and b). Ag has a long electron mean free path and strong plasmonic resonances leading to large enhancement factors. Therefore, Ag-based nanostructures are very efficient in the generation of HEs.[24,25]

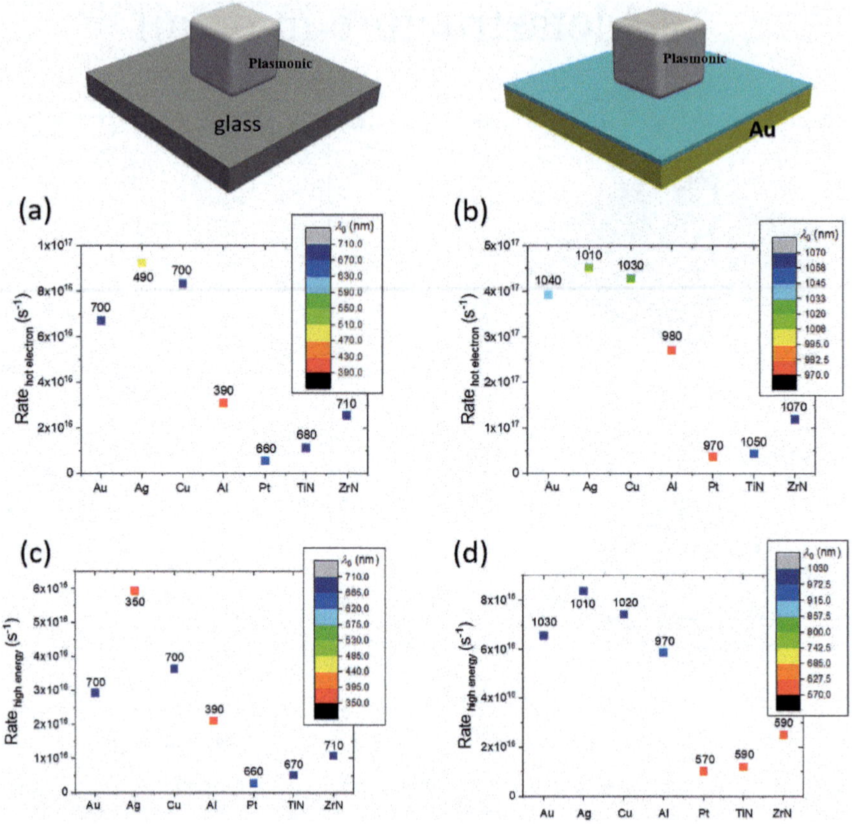

Fig. 6 (a and b) The rates of HE generation at the peaks for different structures and materials. (c and d) The same for the rates of generation of high-energy (over-barrier) carriers for the structures considered in this study. $\Delta E_b = 1$ eV was taken as the barrier potential.

(3) Cheap plasmonic materials, like TiN and ZrN, look very promising. The spectral HE generation rates in such materials with broad plasmonic resonances are relatively small, but by being broad and potentially in the IR, their integrated generation rates under solar illumination can be significant.[8]

(4) Both the HE and high-energy electron generation rates include the quantum factor $(\hbar\omega)^{-3}$. This factor favours plasmon resonances in the IR. This is why the metastructures with strong gap plasmons, which are typically in the IR, are able to generate HEs so efficiently. This was observed in the time-resolved experiments that showed an anomalous ultra-fast component.[25,27]

(5) Regarding the rate $\text{Rate}_{\text{high-energy}}$, this magnitude involves the energy multiplier $(\hbar\omega - \Delta E_b)/\hbar\omega$. This multiplier resembles the Fowler law, which describes the electron injection through a metal–semiconductor boundary in a semi-classical approach. Of course, this factor tends to reduce the high-energy rates for all the studied systems, but the Ag-based structures are the ones that respond most efficiently to this limitation (Fig. 6c and d). Simultaneously, even

with this factor, the metastructures with gap plasmons in the IR are still very good generators of HEs (Fig. 6d).

6. Conclusions

We have examined various plasmonic materials for the generation of hot and high-energy electrons, finding a variety of generally applicable physical insights of relevance for designing efficient systems for hot electron generation. Metastructures with a plasmonic mirror generate HEs much more efficiently than single nanocrystals. Materials with broadened plasmonic peaks are less efficient for the HE generation devices, yet these materials (TiN and ZrN) are very interesting because some of them are much cheaper than traditional plasmonic metals and can still provide a significant number of HE when considering the aggregate production under wide-spectrum illumination. In a metastructure made of strongly-plasmonic metals, the peak generation rates may not depend on the material system since the gap plasmon is mostly determined by the geometry. Finally, the proposed material designs can be interesting for experimental realizations. Possible experimental detection methods for high-energy and hot electrons include nonlinear time-resolved spectroscopy, photo-chemistry and photocurrent measurements.

Conflicts of interest

There are no conflicts to declare.

Acknowledgements

T. L. was financially supported by the Institute of Fundamental and Frontier Sciences, the University of Electronic Science and Technology of China. Z. W. was supported by the National Basic Research Program of China (Project 2013CB933301) and the National Natural Science Foundation of China (Project 51272038). This work was also supported *via* the 1000-talent Award of Sichuan, China (A. O. G.). A. O. G. holds Chang Jiang (Yangtze River) Chair Professorship in China.

References

1 Y. Tian and T. Tatsuma, *J. Am. Chem. Soc.*, 2005, **127**, 7632–7637.
2 S. K. Cushing, J. Li, F. Meng, T. R. Senty, S. Suri, M. Zhi, M. Li, A. D. Bristow and N. Wu, *J. Am. Chem. Soc.*, 2012, **134**, 15033–15041.
3 K. Wu, J. Chen, J. R. McBride and T. Lian, *Science*, 2015, **349**, 632–635.
4 A. Naldoni, F. Riboni, U. Guler, A. Boltasseva, V. M. Shalaev and A. V. Kildishev, *Nanophotonics*, 2016, **5**, 112–133.
5 H. Robatjazi, S. M. Bahauddin, C. Doiron and I. Thomann, *Nano Lett.*, 2015, **15**, 6155–6161.
6 P. Christopher and M. Moskovits, *Annu. Rev. Phys. Chem.*, 2017, **68**, 379–398.
7 L. Weng, H. Zhang, A. O. Govorov and M. Ouyang, *Nat. Commun.*, 2014, **5**, 4792.

8 A. Naldoni, U. Guler, Z. Wang, M. Marelli, F. Malara, X. Meng, L. V. Besteiro, A. O. Govorov, A. V. Kildishev, A. Boltasseva and V. M. Shalaev, *Adv. Opt. Mater.*, 2017, **5**, 1601031.

9 Z. Liu, W. Hou, P. Pavaskar, M. Aykol and S. B. Cronin, *Nano Lett.*, 2011, **11**, 1111–1116.

10 A. Akbari, R. N. Tait and P. Berini, *Opt. Express*, 2010, **18**, 8505–8514.

11 S. V. Boriskina, H. Ghasemi and G. Chen, *Mater. Today*, 2013, **16**, 375–386.

12 P. Narang, R. Sundararaman and H. A. Atwater, *Nanophotonics*, 2016, **5**, 96–111.

13 S. V. Zhukovsky, V. E. Babicheva, A. V. Uskov, I. E. Protsenko and A. V. Lavrinenko, *Plasmonics*, 2014, **9**, 283–289.

14 M. W. Knight, H. Sobhani, P. Nordlander and N. J. Halas, *Science*, 2011, **332**, 702–704.

15 M. A. Mangold, C. Weiss, M. Calame and A. W. Holleitner, *Appl. Phys. Lett.*, 2009, **94**, 161104.

16 H. Chalabi, D. Schoen and M. L. Brongersma, *Nano Lett.*, 2014, **14**, 1374–1380.

17 F. Pelayo García de Arquer, A. Mihi and G. Konstantatos, *Nanoscale*, 2015, **7**, 2281–2288.

18 W. Li, Z. J. Coppens, L. V. Besteiro, W. Wang, A. O. Govorov and J. Valentine, *Nat. Commun.*, 2015, **6**, 8379.

19 A. O. Govorov, H. Zhang and Y. K. Gun'Ko, *J. Phys. Chem. C*, 2013, **117**, 16616–16631.

20 A. O. Govorov, H. Zhang, H. V. Demir and Y. K. Gun'Ko, *Nano Today*, 2014, **9**, 85–101.

21 L. V. Besteiro and A. O. Govorov, *J. Phys. Chem. C*, 2016, **120**, 19329–19339.

22 L. V. Besteiro, X.-T. Kong, Z. Wang, G. Hartland and A. O. Govorov, *ACS Photonics*, 2017, **4**, 2759–2781.

23 G. V. Hartland, L. V. Besteiro, P. Johns and A. O. Govorov, *ACS Energy Lett*, 2017, **2**, 1641–1653.

24 X. T. Kong, Z. Wang and A. O. Govorov, *Adv. Opt. Mater.*, 2017, **5**, 1600594.

25 M. E. Sykes, J. W. Stewart, G. M. Akselrod, X. T. Kong, Z. Wang, D. J. Gosztola, A. B. F. Martinson, D. Rosenmann, M. H. Mikkelsen, A. O. Govorov and G. P. Wiederrecht, *Nat. Commun.*, 2017, **8**, 986.

26 A. Sousa-Castillo, M. Comesaña-Hermo, B. Rodríguez-González, M. Pérez-Lorenzo, Z. Wang, X.-T. Kong, A. O. Govorov and M. A. Correa-Duarte, *J. Phys. Chem. C*, 2016, **120**, 11690–11699.

27 H. Harutyunyan, A. B. F. Martinson, D. Rosenmann, L. K. Khorashad, L. V. Besteiro, A. O. Govorov and G. P. Wiederrecht, *Nat. Nanotechnol.*, 2015, **10**, 770–774.

28 A. Crut, P. Maioli, N. Del Fatti and F. Vallée, *Chem. Soc. Rev.*, 2014, **43**, 3921–3956.

29 P. Yu, L. V. Besteiro, Y. Huang, J. Wu, L. Fu, H. H. Tan, C. Jagadish, G. P. Wiederrecht, W. Zhiming and A. O. Govorov, *Adv. Opt. Mater.*, 2018, 1800995.

30 P. B. Johnson and R. W. Christy, *Phys. Rev. B*, 1972, **6**, 4370.

31 H.-J. Hagemann, W. Gudat and C. Kunz, *J. Opt. Soc. Am.*, 1975, **65**, 742–744.

32 K. M. McPeak, S. V. Jayanti, S. J. P. Kress, S. Meyer, S. Iotti, A. Rossinelli and D. J. Norris, *ACS Photonics*, 2015, **2**, 326–333.

33 A. D. Rakić, A. B. Djurišić, J. M. Elazar and M. L. Majewski, *Appl. Opt.*, 1998, **37**, 5271–5283.

34 J. Pflüger, J. Fink, W. Weber, K. P. Bohnen and G. Crecelius, *Phys. Rev. B*, 1984, **30**, 1155–1163.

35 M. Veszelei, K. Andersson, C.-G. Ribbing, K. Järrendahl and H. Arwin, *Appl. Opt.*, 1994, **33**, 1993–2001.

36 S. K. Cushing, C.-J. Chen, C. L. Dong, X.-T. Kong, A. O. Govorov, R.-S. Liu and N. Wu, *ACS Nano*, 2018, **12**, 7117–7126.

37 J. Cheng, Y. Li, M. Plissonneau, J. Li, J. Li, R. Chen, Z. Tang, L. Pautrot-d'Alençon, T. He, M. Tréguer-Delapierre and M.-H. Delville, *Sci. Rep.*, 2018, **8**, 14136.

38 S. Tan, A. Argondizzo, J. Ren, L. Liu, J. Zhao and H. Petek, *Nat. Photonics*, 2017, **11**, 806–812.

39 M. Taghinejad, H. Taghinejad, Z. Xu, Y. Liu, S. P. Rodrigues, K.-T. Lee, T. Lian, A. Adibi and W. Cai, *Adv. Mater.*, 2018, **30**, 1704915.

40 T. Heilpern, M. Manjare, A. O. Govorov, G. P. Wiederrecht, S. K. Gray and H. Harutyunyan, *Nat. Commun.*, 2018, **9**, 1853.

41 Z. Xu, M. Quintanilla, F. Vetrone, A. O. Govorov, M. Chaker and D. Ma, *Adv. Funct. Mater.*, 2015, **25**, 2950–2960.

42 Y. Zhang, C. Yam and G. C. Schatz, *J. Phys. Chem. Lett.*, 2016, **7**, 1852–1858.

43 A. Naldoni, T. Montini, F. Malara, M. M. Mróz, A. Beltram, T. Virgili, C. L. Boldrini, M. Marelli, I. Romero-Ocaña, J. J. Delgado, V. Dal Santo and P. Fornasiero, *ACS Catal.*, 2017, **7**, 1270–1278.

44 E. Cortés, W. Xie, J. Cambiasso, A. S. Jermyn, R. Sundararaman, P. Narang, S. Schlücker and S. A. Maier, *Nat. Commun.*, 2017, **8**, 14880.

45 P. Wang, A. V. Krasavin, M. E. Nasir, W. Dickson and A. V. Zayats, *Nat. Nanotechnol.*, 2018, **13**, 159–164.

46 Q. Zhang, X. Jin, Z. Xu, J. Zhang, U. F. Rendón, L. Razzari, M. Chaker and D. Ma, *J. Phys. Chem. Lett.*, 2018, **9**, 5317–5326.

Faraday Discussions

PAPER

Assistance of metal nanoparticles in photocatalysis – nothing more than a classical heat source

Yonatan Sivan,[*ad] Ieng Wai Un[abd] and Yonatan Dubi[cd]

Received 4th October 2018, Accepted 2nd November 2018
DOI: 10.1039/c8fd00147b

In a recent paper, we derived a self-consistent theory of the steady-state electron distribution of a metal under continuous wave illumination which treats thermal and non-thermal effects on the same footing. Here, we re-derive the main analytical results of that study from very simple arguments, and draw a series of conclusions which contradict claims made in previous studies of the steady-state distribution. In particular, we show that the faster chemical reactions reported in many previous papers are extremely unlikely to originate from high energy non-thermal electrons. Instead, the faster reactions very likely originate from a purely thermal effect.

I. Introduction

In recent years, a central focus of research into nanoplasmonics has been on finding ways to exploit the enhanced absorption exhibited by nanoplasmonic systems. Two main routes are discussed in the literature. First, it was shown that plasmonic nanostructures can be efficient sources of heat on the nanoscale,[1,2] a research topic that is usually referred to as thermo-plasmonics. This has enabled a wide range of emerging applications at different temperature ranges, such as photothermal imaging,[3,4] cancer treatment,[5] plasmonic photovoltaics,[6] water boiling, sanitation and super-heating,[7–9] solvothermal (hydro-thermal) chemistry,[10] thermo-photovoltaics,[11,12] diffusive switching,[13] thermoelectrics,[14] plasmon-assisted chemical vapor deposition[15] and heat-assisted magnetic recording,[16] which may involve temperatures even higher than 2000 K.

Clearly, the generation of heat is a result of internal thermalization processes that convert the high energy non-thermal electron distribution generated initially by the absorption of photons to a thermal distribution. In that context, a second way to exploit heat has emerged – it was hypothesized that the initially generated

[a]Unit of Electro-Optics Engineering, Ben-Gurion University, Israel. E-mail: sivanyon@bgu.ac.il
[b]Joan and Irwin Jacobs TIX Institute, National Tsing Hua University, Taiwan
[c]Department of Chemistry, Ben-Gurion University, Israel
[d]Ilse Katz Center for Nanoscale Science and Technology, Ben-Gurion University, Israel

high energy non-thermal (frequently (ill-)referred to as "hot") electrons could be used as a means to catalyze photochemical reactions,[17–23] as well as in artificial photosynthesis,[24,25] photo-detection,[26–30] frequency upconversion[31,32] *etc.* Specifically, a large number of experimental studies have reported faster chemical reactions in the presence of illuminated metal nanoparticles.[19,20,33] However, it was not clear whether this acceleration of chemical reactions occurred due to the increased population of high energy carriers, or due to the mere (consequent) increase in temperature; in other words – is it thermal or non-thermal effects that are responsible for the faster chemistry? This is important to know because although heating is frequently used in catalysis, it is accompanied by various undesired side effects (such as the enhancement of additional parasitic reactions)[18] and incurs significant practical and financial complications for real-life applications. As a result, heating is generally avoided in practice.

Very few attempts have been made to distinguish experimentally between thermal and non-thermal effects. This should be done by reproducing the same temperature rise caused by laser illumination through the use of purely external heating. Clearly, it is very difficult to measure temperature on the nanoscale with sufficient spatial resolution; indeed, thermometry methods for metal nanostructures have emerged only in the last few years (*e.g.* ref. 34–36). An alternative determination of the temperature *via* numerical simulation is simple for a single nanoparticle (NP),[37,38] but non-trivial in the presence of many nanoparticles due to the long range inter-particle thermal interactions[39,40] which give rise to (highly) non-uniform temperatures on multiple length-scales. For example, in ref. 41 the control experiment showed that the chemistry is sensitive to the external temperature, *i.e.* the metal does more than just enhance photon absorption.[42] However, the control experiment was carried out at a temperature measured far away from the illuminated nanoparticles, and this was most likely lower than the actual nanoparticle temperature, as noted later on by the same group of authors[43] and by others.[44] Even if the temperature is known, it is not easy to perform meaningful control experiments due to the difficulty in reproducing the non-uniform temperature distribution across the typical suspension volume with an external heat source. In pulsed experiments, this is even more difficult due to the transient nature of the temperatures and the differences between the electron and lattice temperatures.

In addition to all these practical complications, there has been also a fundamental controversy regarding the definition of temperature for metal nanostructures under illumination. Indeed, one could naively argue that upon illumination, the system is out of equilibrium, meaning that temperature – inherently an equilibrium property – cannot be well defined.[45] Accordingly, while temperatures were defined after the initial non-equilibrium stage that follows ultrafast illumination (see for example ref. 46 and 47), all previous theoretical studies of continuous wave (CW) illumination treated the non-equilibrium in great detail, but ignored the possibility of an increase in the electron and phonon temperatures to above the environment temperature.

In order to reconcile this paradox, in ref. 48 we suggested a unique solution for the determination of the steady-state temperature developing in a metal nanostructure, based only on energy conservation and basic thermodynamics. Specifically, we developed a self-contained theory for the photo-generation of non-thermal energetic carriers in metal nanostructures based on a quantum-like version of the Boltzmann equation (BE).[46,47,49–57] To the BE we added, in a self-consistent way, electron–phonon scattering and energy leakage to the environment *via* phonon–phonon coupling, such that the electron and phonon (lattice) temperatures in the plasmonic nanostructure can rise above the ambient temperature. The steady-state solution of this set of equations ensures energy conservation – the power flowing into the metal due to photon absorption is exactly balanced by heat leakage to the environment. Since there is obviously only one pair of values for the electron and phonon temperatures for which this happens, the approach of ref. 48 enables a determination of the temperature(s) of a system which is out of equilibrium in a unique and unambiguous way. This approach also enables a comparison, for the first time to our knowledge, of the relative importance of thermal and non-thermal effects.

The simulations showed that the population of non-thermal energetic electrons and holes[88] can increase dramatically under illumination, yet this process is extremely inefficient, as the majority of the absorbed energy leads to heating; the electron and phonon temperatures are found to be similar, thus justifying the use of the (classical) single temperate heat model.[2] Somewhat surprisingly, we found that just above (below) the Fermi energy, the non-equilibrium consists of holes (electrons), rather than the other way around; we showed that this behaviour is due to the dominance of e–ph collisions.

The results of ref. 48 were based on a rather complicated and slowly converging numerical procedure involving standard but cumbersome scattering integrals. This approach also lacks deep physical insights. In the current manuscript, we provide several simple estimates of the numerical results obtained in ref. 48 that partially obviate the need for the heavy/lengthy numerical procedure performed so far. We also apply the results of ref. 48 to show that, unlike what is frequently claimed in the literature, the density of high energy non-thermal carriers is, to leading order at least, independent of the NP size and shape, and that in fact, small spherical NPs are the least efficient source of high energy non-thermal carriers. Furthermore, we develop a simple model for photocatalysis which shows that it is highly unlikely that the faster chemical reactions observed experimentally are due to the generation of non-thermal, high energy carriers in the metal. The faster reactions much more likely originate from a purely thermal effect. Finally, we describe several experiments that have to be carried out to confirm our predictions.

II. A simple calculation of the electron distribution and temperatures

In this section, we briefly reintroduce the theoretical model reported in ref. 48 and show how to obtain some of its numerical results without going though a full-scale numerical calculation, based on standard, physically-sensible assumptions.

The Boltzmann equation (BE) is

$$\frac{\partial f(\mathcal{E}); T_e, T_{ph}}{\partial t} = \underbrace{\left(\frac{\partial f}{\partial t}\right)}_{\text{photon absorption}} + \underbrace{\left(\frac{\partial f}{\partial t}\right)}_{\text{e-e collisions}} + \underbrace{\left(\frac{\partial f}{\partial t}\right)}_{\text{e-ph collisions}}. \tag{1}$$

Here, f is the electron distribution function at an energy, $\mathcal{E}$, electron temperature, T_e and phonon temperature, T_{ph}, representing the population probability of electrons in a system characterized by a continuum of states within the conduction band. The first term on the right-hand-side (RHS) of eqn (1) describes the excitation of conduction electrons due to photon absorption (see Appendix: A for its explicit form). The second term on the RHS of eqn (1) describes the energy relaxation due to collisions between electrons and phonons (see ref. 48 and 54 for its explicit form). The third term on the RHS of eqn (1) represents the thermalization induced by e–e collisions, *i.e.* the convergence of the non-thermal population into the thermalized Fermi–Dirac distribution, given by

$$f^{\mathrm{T}}(\mathcal{E};T_{\mathrm{e}})=\left(1+\mathrm{e}^{(\mathcal{E}-\mathcal{E}_{\mathrm{F}})/k_{\mathrm{B}}T_{\mathrm{e}}}\right)^{-1}. \tag{2}$$

Here, k_{B} is the Boltzmann constant and $\mathcal{E}_{\mathrm{F}}$ is the Fermi energy.[89] The explicit form of the e–e relaxation term used in our formulation can be found in ref. 48. Note that eqn (1) does not account for interband transitions. In that sense, it is suitable for describing Ag NPs, but is relatively less accurate for Au NPs. See ref. 47 and 48 for a detailed justification of this model.

We now split the electron distribution into a thermal and non-thermal part, namely, we define $\Delta f^{\mathrm{NT}}\equiv f(\mathcal{E},T_{\mathrm{e}},T_{\mathrm{ph}})-f^{\mathrm{T}}(\mathcal{E},T_{\mathrm{e}})$ in ref. 48. We then multiply the resulting equations by the product of the electron energy, $\mathcal{E}$, and the density of electron states, $\rho_{\mathrm{e}}(\mathcal{E})$, and integrate over the electron energy. The resulting equations describe the dynamics of the 3 energy channels, namely, non-thermal electron energy, thermal energy and phonon energy:

$$\frac{\mathrm{d}\mathcal{U}_{\mathrm{e}}^{\mathrm{NT}}}{\mathrm{d}t}=-W_{\mathrm{e-e}}-W_{\mathrm{e-ph}}^{\mathrm{NT}}+W_{\mathrm{ex}}^{\mathrm{NT}},$$

$$\frac{\mathrm{d}\mathcal{U}_{\mathrm{e}}^{\mathrm{T}}}{\mathrm{d}t}=W_{\mathrm{e-e}}-W_{\mathrm{e-ph}}^{\mathrm{T}}+W_{\mathrm{ex}}^{\mathrm{T}},$$

$$\frac{\mathrm{d}\mathcal{U}_{\mathrm{ph}}^{\mathrm{T}}}{\mathrm{d}t}=W_{\mathrm{e-ph}}^{\mathrm{T}}+W_{\mathrm{ph-env}}+W_{\mathrm{e-ph}}^{\mathrm{NT}}.$$

Here, $\mathcal{U}_{\mathrm{e}}^{\mathrm{NT}}=\int\mathcal{E}\rho_{\mathrm{e}}(\mathcal{E})[f-f^{\mathrm{T}}]\mathrm{d}\mathcal{E}$ is the total energy of the non-thermal electron sub-system, and we define the energy change rates due to photo-excitation $\left(W_{\mathrm{ex}}^{\mathrm{NT}}\equiv\int\mathcal{E}\rho_{\mathrm{e}}(\mathcal{E})\left[\left(\frac{\mathrm{d}f}{\mathrm{d}t}\right)_{\mathrm{ex}}-\left(\frac{\mathrm{d}f^{\mathrm{T}}}{\mathrm{d}t}\right)_{\mathrm{ex}}\right]\mathrm{d}\mathcal{E}\text{ and }W_{\mathrm{ex}}^{\mathrm{T}}\equiv\int\mathcal{E}\rho_{\mathrm{e}}(\mathcal{E})\left(\frac{\mathrm{d}f^{\mathrm{T}}}{\mathrm{d}t}\right)_{\mathrm{ex}}\mathrm{d}\mathcal{E}\right)$, e–e collisions $\left(W_{\mathrm{e-e}}\equiv\int\mathcal{E}\rho_{\mathrm{e}}(\mathcal{E})\left(\frac{\partial f}{\partial t}\right)_{\mathrm{e-e}}\mathrm{d}\mathcal{E}\right)$, e–ph collisions $\left(W_{\mathrm{e-ph}}^{\mathrm{NT}}\equiv-\int\mathcal{E}\rho_{\mathrm{e}}(\mathcal{E})\left[\left(\frac{\partial f}{\partial t}\right)_{\mathrm{e-ph}}-\left(\frac{\partial f^{\mathrm{T}}}{\partial t}\right)_{\mathrm{e-ph}}\right]\mathrm{d}\mathcal{E}\text{ and }W_{\mathrm{e-ph}}^{\mathrm{T}}\equiv-\int\mathcal{E}\rho_{\mathrm{e}}(\mathcal{E})\left(\frac{\partial f^{\mathrm{T}}}{\partial t}\right)_{\mathrm{e-ph}}\mathrm{d}\mathcal{E}\right)$, and the coupling of energy to the environment ($W_{\mathrm{ph\text{-}env}}$).

A. Determining the steady-state temperatures

We now show how the steady-state electron and phonon temperatures can be determined without the need for the lengthy exact numerical calculations of the Boltzmann equation.

In order to avoid the heavy numerical computation of the BE and the rate equations above, we adopt several standard approximations. First, we approximate $W_{e\text{-}e}$ and $W^{NT}_{e\text{-}ph}$ as $\Gamma_e\mathcal{U}_e^{NT}$ and $\Gamma_{ph}\mathcal{U}_e^{NT}$, where Γ_e and Γ_{ph} are the effective time scales that represent the decay of the non-thermal energy due to e–e and e–ph collisions, respectively. Such an estimate is customary within the approach known as the (extended) Two Temperature Model ((e)TTM), initially introduced phenomenologically[58] and later derived from a classical point of view in ref. 59. As an initial guess, we can set $\Gamma_e^{-1} \sim 0.5$ ps and $\Gamma_{ph}^{-1} \sim 1$ ps.[53,58–60] We also replace $W^T_{e\text{-}ph}$ with $G_{e\text{-}ph}(T_e - T_{ph})$, an approximation which has been shown to be an excellent approximation for $T_e < 3000$ K.[46,54,61–63] Following this, the above rate equations become

$$\frac{d\mathcal{U}_e^{NT}}{dt} = -\Gamma_e\mathcal{U}_e^{NT} - \Gamma_{ph}\mathcal{U}_e^{NT} + W_{ex}^{NT},$$

$$\frac{d\mathcal{U}_e{}^T}{dt} = \Gamma_e\mathcal{U}_e^{NT} - G_{e-ph}\left(T_e - T_{ph}\right) + W_{ex}{}^T,$$

$$\frac{d\mathcal{U}_{ph}}{dt} = G_{e-ph}\left(T_e - T_{ph}\right) - G_{ph-env}\left(T_{ph} - T_{env}\right) + \Gamma_{ph}\mathcal{U}_e^{NT},$$

where we also set $W_{ph\text{-}env} = G_{ph\text{-}env}(T_{ph} - T_{env})$. This set of equations in fact provides a self-consistent derivation of the (e)TTM from the BE; similar derivations were previously suggested in ref. 64–66, 90 and 91.

For CW illumination, the system is in a steady-state. Thus, $\mathcal{U}_e^{NT} = W_{ex}^{NT}/(\Gamma_e + \Gamma_{ph})$,

$$\frac{\Gamma_e}{\Gamma_e + \Gamma_{ph}}W_{ex}^{NT} + W_{ex}^T = G_{e-ph}\left(T_e - T_{ph}\right)$$

and

$$G_{e-ph}\left(T_e - T_{ph}\right) = G_{ph-env}\left(T_{ph} - T_{env}\right) - \frac{\Gamma_{ph}}{\Gamma_e + \Gamma_{ph}}W_{ex}^{NT}.$$

Therefore, *via* simple algebra, we can derive the following equations:

$$T_e = T_{ph} + \frac{\frac{\Gamma_e}{\Gamma_e + \Gamma_{ph}}W_{ex}^{NT} + W_{ex}^T}{G_{e-ph}}, \tag{3}$$

$$T_{ph} = T_{env} + \frac{W_{ex}^{NT} + W_{ex}^T}{G_{ph-env}}. \tag{4}$$

Eqn (3) and (4) show that T_e and T_{ph} grow linearly with the incident intensity (*via* the absorbed power density, eqn (A1)–(A6)). The difference between the temperatures is also proportional to the absorbed power density. Most importantly, the difference between the two temperatures is independent of the particle size and coupling to the environment; it instead depends only on the internal details of the electron subsystem and on the e–ph coupling coefficient, $G_{\text{e-ph}}$.

B. High energy non-thermal carrier density

In order to determine the density of high energy non-thermal carriers, we exploit the fact that e–ph interactions are important only near the Fermi energy (see specifically ref. 48, Fig. S2(b)). The reason for that, by the FLT, is that the energy lost by a high energy non-thermal electron in a collision is a significant part of its excess energy with respect to the Fermi energy. Since that energy is much higher than the phonon energies, they are not generated during collisions of high energy non-thermal electrons. Thus, the electron population far away from the Fermi energy can be determined simply by balancing the electron excitation rate with the e–e collision rate. Indeed, we show in Appendix: A that the density of high energy non-thermal carriers is given by

$$\Delta f^{\text{NT}}(\mathcal{E}) \sim \tau_{\text{e-e}}(\mathcal{E}) R |E|^2, \tag{5}$$

where $R \equiv \frac{4\varepsilon_0 \varepsilon''_{\text{m}}(\omega_{\text{p}})}{\hbar} \frac{\mathcal{E}_{\text{F}}}{3\mathcal{E}_{\max} n_{\text{e}}}$ (definitions of the various parameters can be found in Appendix: A) and E is the local electric field. This is found to be in excellent agreement with the numerical results (see Fig. 1).

Eqn (5) not only provides a simple way to compute the high energy, non-thermal carrier population, but also provides fundamental insight into the physics of non-thermal carrier generation. Specifically, it shows that the non-thermal carrier population is independent of the ambient temperature, phonon temperature and the temperature of the electrons, as well as of the thermal properties of the system. This means that previous studies of the steady-state electron distribution (*e.g.* ref. 67–70) could have been used to calculate quantitatively the high energy non-thermal carrier density correctly.[92] However, they could not have been used to obtain the distribution near the Fermi energy because of the neglect of the temperature rise; accordingly, the relative amount of power going to the thermal and non-thermal channels has not been evaluated thus far. This also means that previous approaches incorrectly account for the role of interband transitions which, due to the need to bridge the bandgap energy, cause an increase in the non-thermal carrier distribution relatively close to the

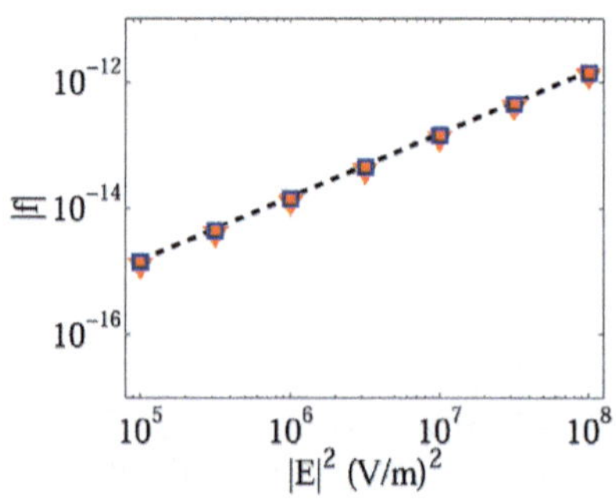

Fig. 1 The population $f(\mathcal{E})$ of holes at $\mathcal{E} = 1.8$ eV ($<\hbar\omega_{\text{p}}$) below the Fermi level (blue rectangles) and electrons at $\mathcal{E} = 1.8$ eV above the Fermi level (orange triangles), compared with the estimate (5) (black dashed line).

Fermi energy.[68] In fact, assuming that the number of excess electrons generated near the Fermi energy is comparable to that in the high energy shoulders, it is most likely that their contribution would be negligible with respect to the thermal changes. Accordingly, one should re-examine all claims about the relative importance of interband transitions for the generation of non-thermal carriers.[71]

A further implication of the accuracy of the estimate (5) is the independence of the density of high energy non-thermal carriers on the shape and size of the metal nanoparticle. Indeed, since it has been shown *via* fully quantum mechanical calculations that the e–e collision rates are independent of the particle size (*e.g.* ref. 47 and 72), it follows from eqn (5) that the density depends on the NP size and shape only *via* the value of the local field. This claim is corroborated by the numerical results of Section III below & Fig. 2. This result is in contradiction to the claims that have frequently appeared in previous studies. It should be noted that the metal interface may support long-lived surface states which, in turn, might allow for a higher density of high energy non-thermal carriers. However, the importance of such states should be weighted by their relative number, which is surely small, especially for spherical nanoparticles of more than a few nm in size.

Finally, we note that the estimate (5) is essentially identical to the result for the population of the excited state in a system with discrete energy states (*e.g.* a two level system) in the weak excitation limit. This means that $(\tau_{\text{e-e}}R)^{-1/2}$ is essentially the effective saturation field for the metal, $|E_{\text{sat}}| = \sqrt{\text{ZI}_{\text{sat}}/(2n)}$.[93] Then, assuming that the absorptivity of a single metal atom is comparable to that of a single gas atom or a semiconductor atom, it follows that since the relaxation rate of the metal electron is about 6 orders of magnitude faster compared with the relaxation of a gas or semiconductor atom, then the square of the saturation field, $|E_{\text{sat}}|^2$, of the metal will be 6 orders of magnitude higher than in an atomic gas or a semiconductor. For $I_{\text{sat}} \sim 10^7$ W cm^{-2} (typical for a gas or semiconductor atom), it is easy to see that $|E_{\text{sat}}|^2 \sim 100\ \Omega\cdot10^7$ W cm^{-2} $= 10^{13}$ V^2 m^{-2}. In Appendix: A, we obtained $|E_{\text{sat}}|^2 \sim 2.5 \times 10^{19}$ V^2 m^{-2}, which is in excellent agreement with the above estimate.

This shows that accounting for the discreteness of the energy states (as for the atomic physics case) or neglecting it (as in our approach) has, at most, a minor effect on the resulting density of high energy electrons. This conclusion is the opposite of the conclusions drawn in previous studies.

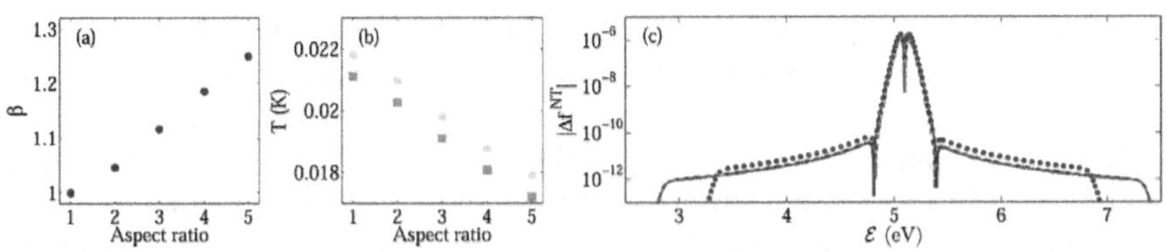

Fig. 2 (a) Geometric rescaling factor of the thermal conductivity of the host for metal nanorods with different aspect ratios.[37] (b) Electron (yellow dots) and phonon (green squares) temperature rises of nanorods with different aspect ratios but identical wavelength (2.5 eV), volume and averaged field ($|\bar{E}|^2 = 10^8$ (V m^{-1})^2). (c) Deviation from the equilibrium distribution for the aspect ratios 1 : 1 (blue solid line) and 1 : 5 (red dashed line), respectively. The green dotted line shows the deviation from the equilibrium distribution for a different wavelength (1.75 eV) with an aspect ratio of 1 : 1.

III. Dependence on nanoparticle size and shape

Previous studies have claimed that particles with sharp edges, for which regions of strong electric fields (also confusingly referred to as "hot spots"[94]) of electric fields occur, are superior for the purpose of generating high energy, non-thermal carriers. Unfortunately, however, these claims were based on what we believe to be a somewhat improper comparison.

Specifically, while previous comparisons were carried out for particles of the same volume (and probably the same incident intensity), the changes in size and shape incur changes in the resonance position and quality, therefore resulting in a change in the local electric field within the NP. Thus, the differences observed in such comparisons ensue partially (if not completely) from these purely classical electromagnetic effects, meaning that it is not clear what conclusion can be made from such a comparison.

In that sense, a better suited comparison should be made for particles with the same volume and the same average internal (squared) fields, $\int_V |E|^2 \mathrm{d}V$. Such a comparison would also allow one to isolate the role of additional quantum mechanical effects from the purely classical electrodynamic considerations associated with the quality and position of the optical response. Since we observed in ref. 48 that away from the Fermi energy the density of high energy carriers is proportional to the square of the local field, $f \sim |E|^2$, it is immediately obvious that, to leading order, the total number of carriers at a given energy (away from the Fermi energy) is the same for NPs with the same (average) internal field, regardless of their shape and of the electric field distribution.

Some differences may still exist due to field and temperature gradients. These, however, are expected to be weak – the charge currents created by the electric field non-uniformity are assumed to be weak (due to the fast e–e collisions and the high conductance of the metal); heat currents are definitely weak – indeed, the temperature uniformity is excellent due to the high thermal conductivity within the metal. Either way, the neglect of these currents is, to the best of our knowledge, common to all previous studies, and will accordingly be adopted here as well.

Having said that, we should emphasize that the electron distribution near the Fermi energy (hence, the electron temperature) might be very sensitive to the shape and size of the NP. Indeed, the particle size and shape affect the coupling efficiency of heat to the environment, manifested *via* the phonon-environment coupling coefficient $G_{\text{ph-env}}$ (introduced in Section A). In ref. 37, Baffou *et al.* used exact numerical simulations to provide accurate empirical expressions for the effective thermal coupling coefficient to the environment for various generic particle shapes. In particular, they showed that since a spherical nanoparticle has a minimal surface area-to-volume ratio, it is optimal in terms of holding on to the heat generated within it. Accordingly, one can expect that for a given average (squared) local field (which is directly proportional to the absorbed power, see eqn (A4)), a sphere will be the hottest among all other particles. Thus, overall, while the number of non-thermal carriers generated within a sphere is the same as in any other particle (for a given average local field and particle volume), it exhibits the poorest relative efficiency of thermal *vs.* non-thermal energy partition.

These expectations are confirmed by numerical solutions of the Boltzmann eqn (1) for Ag rods with various aspect ratios but with the same volume. As the input electromagnetic field, we used a fixed value representing the averaged square of the local electric field rather than the field itself (see our justification above). One can see that the lowest temperature is attained for a rod with a 1 : 5 aspect ratio due to its 30% stronger coupling coefficient of heat to the environment (Fig. 2). The population of the high energy, non-thermal carriers is, however, the same for the different rods.

Our approach also shows the differences in the non-thermal distributions near the Fermi-energy for two different wavelengths. Specifically, the change in wavelength affects the shoulder width (due to the change in $\hbar\omega_p$) and height (due to the change in the density of states at $\mathcal{E}$ and p_{abs} with the illumination wavelength[95]). For example, $\phi_{A,1.75\ eV}/\phi_{A,2.25\ eV} \approx 1.3$ and $N_A(1.75\ eV)/N_A(2.25\ eV) = \varepsilon''_m(1.75\ eV)/\varepsilon''_m(2.25\ eV) = 1.35$ (ref. 73) for the same average field. As a result, $|\Delta f^{NT}_{1.75\ eV}|/|\Delta f^{NT}_{2.25\ eV}| = (N_A(1.75\ eV)\phi_{A,1.75\ eV})/(N_A(2.25\ eV)\phi_{A,2.25\ eV}) \approx 1.76$, which is in excellent agreement with the numerical results.

IV. Non-thermal carriers and photocatalysis

We now wish to make the connection between our theory, in which the (non-thermal) carrier distribution is evaluated, and photocatalysis, namely the enhancement of a reaction rate due to the presence of an illuminated nanoparticle (NP). The first stage is to understand the nature of the photocatalytic process, and its relation to the electron distribution in the NP. To this end, we follow the reasoning (based on the specific example of hydrogen molecule dissociation) presented in ref. 41, as depicted in Fig. 3. Consider a reaction pathway defined by a free energy curve in some reaction coordinate (blue solid line) leading from the initial state to the product state (in H_2 dissociation the initial and product states are $H_2\sigma_g$ and the dissociated molecule 2H, respectively,

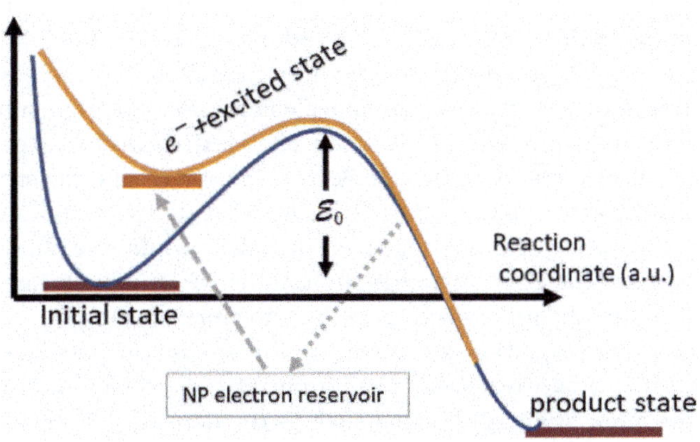

Fig. 3 Schematic illustration of the photocatalytic process. The molecule absorbs an electron from the NP, thus going from the initial state to the excited state. The barrier that has to be crossed to reach the product state is much smaller now, resulting in a much higher reaction rate. During the relaxation process the excess electron returns to the NP.

and the reaction coordinate is the H–H bond length). This reaction is limited by the energy barrier $\mathcal{E}_0$ ($\sim$2.3 eV for H_2 dissociation), which is practically unreachable at room temperature.

Now, the molecular system is placed in the vicinity of a NP, from which electrons with energy $\mathcal{E}_{\rm in}$ can hop onto the molecule (dashed gray arrow), thus generating a charged excited state (the state $H_2^{\delta-}\sigma_{\rm u}^*$ in the H_2 dissociation example). In this sense, the NP is nothing but an electron reservoir available to the molecular system. The excited state molecule has a different reaction path (solid orange line), with a smaller energy barrier for the reaction. This reduction of the energy barrier is the essence of the catalytic process. At some point along the reaction path, the excess electron returns to the NP with energy $\mathcal{E}_{\rm out}$ (dotted gray arrow), and the system returns to the neutral-state reaction path. The concept of "hot electron-assisted photocatalysis" is based on the concept that illumination of the NP will dramatically increase the probability of electrons (in the NP) having energy $\mathcal{E}_{\rm in}$, thus making their excess energy available for the chemical reaction to occur.

We can now try to relate this description to the evaluation of the electronic distribution function in the NP. For the electrons in the NP, the nearby molecule that undergoes the reaction can be considered as a source of irreversible energy loss, whereby electrons come in with energy $\mathcal{E}_{\rm in}$ and leave, after some reaction time, $\tau_{\rm r}$, with energy $\mathcal{E}_{\rm out}$. Therefore, within a Boltzmann equation description of the electron distribution, this term can be written as

$$\left(\frac{\mathrm{d}f}{\mathrm{d}t}\right)_{\rm react} \sim \gamma_{\rm r} f(\mathcal{E}_{\rm in})(1 - f(\mathcal{E}_{\rm out})), \tag{6}$$

where $\gamma_{\rm r} = \tau_{\rm r}^{-1}$ is the bare reaction rate, *i.e.* the rate at which the chemical system undergoes the reaction after being excited by an additional electron (which is the rate-limiting step). More realistically, one can assume that there is some broadening to the energies $\mathcal{E}_{\rm in,out}$, resulting in a term that looks like

$$\left(\frac{\mathrm{d}f}{\mathrm{d}t}\right)_{\rm react} \sim \gamma_{\rm r} \int \mathrm{d}\mathcal{E} \int \mathrm{d}\mathcal{E}' g(\mathcal{E}; \mathcal{E}_{\rm in}, \sigma) g(\mathcal{E}; \mathcal{E}_{\rm out}, \sigma) f(\mathcal{E}) \left(1 - f(\mathcal{E}')\right), \tag{7}$$

where $g(\mathcal{E}; \mathcal{E}_0, \sigma)$ is a Gaussian centered around $\mathcal{E}_0$ with width σ. Eqn (6) is reproduced if the limit $\sigma \to 0$ is taken, *i.e.*, taking g to be a δ-function.

In principle, eqn (7) should be inserted into the full Boltzmann equation for the distribution function (1). However, since chemical kinetics are substantially slower than electronic processes, $\gamma_{\rm r}$ is much smaller than any of the other terms (photo-excitation, electron–electron and electron–phonon scattering terms) in the Boltzmann equation. It follows that in order to determine the steady-state distribution, the $\gamma_{\rm r}$-term can be neglected. In other words, the chemical reaction does not affect the electron distribution in the NP. The photocatalytic reaction rate is therefore defined by eqn (6) or (7), and since $1 - f(\mathcal{E}_{\rm out}) \sim 1$, it can be approximated by $\gamma_{\rm r} f(\mathcal{E}_{\rm in})$. This is a rather intuitive result; the photocatalytic rate is proportional to the bare reaction rate multiplied by the probability of finding an electron at the relevant energy (factors such as the density of states go into the proportionality constant). The conclusion to be drawn from this simple set of arguments is straightforward and unavoidable; if this mechanism indeed describes the experiments, then the reaction rate should increase in proportionality to the increase in the high-energy part of the electron distribution. In ref. 48 as well as above (see eqn (5)), we have shown that the high-energy part of the electron distribution increases by $\sim$10 orders of magnitude (and maybe more). This is nowhere near the increase in reaction rate observed for any experiment reporting metal NP-assisted photocatalysis; specifically, it is far higher than what was observed in ref. 74, which found a mild 6-fold increase in the reaction rate. We therefore conclude that "hot carriers" are extremely unlikely to be responsible for the observed increase in the reaction rate. The strong dependence of the photocatalytic rate on the substrate implies that the reason for the faster chemistry is merely heating. Indeed, faster reactions were observed with a glass substrate, which has a much lower thermal conductivity (in comparison with the TiO_2 substrate used in ref. 41), thus giving rise to higher temperatures in the nanoparticles.

V. Summary & outlook

The simplified analysis of Section II enables an effortless determination of the electron distribution function, the electron temperature and the phonon temperature for metals under steady-state illumination. This should motivate a re-examination of all previous recent theoretical studies of the non-equilibrium in metallic nanostructures that did not account for thermal effects. It should also be the basis for improved modelling, *e.g.* by accounting for more realistic band structures (thus enabling a comparison of different metals), considering more complex nanostructures (involving different materials, especially semiconductors that provide an independent source of "hot" carriers[42]), adding (secondary) physical mechanisms (surface states, carrier tunneling, field and temperature gradients *etc.*), and studying the competing contributions of thermal and non-thermal effects to the nonlinear response of metals *etc.*

Our results also enable a re-evaluation of the interpretations of previous experiments. In particular, since the model reported in Section IV predicts dramatically faster chemistry (many orders of magnitude faster reaction rates), the absence of such an effect in the experimental data reported thus far implies that it is extremely unlikely that the faster chemistry reported was indeed due to the generation of high energy, non-thermal carriers. This conclusion should induce a more careful examination of the origin and energy scale of the faster reactions in those studies, which are far less sensitive to the presence of the metal and illumination than predicted in Section IV. Most likely, the mechanism responsible for the faster reactions will turn out to be mere regular heating. Conversely, our theory can provide insight into the design of experiments in which high energy, non-thermal carriers are dominant.

In this vein, we hope that our results will motivate a quantitative comparison of the theoretical predictions with experimental data. The latter should include measurements of the metal temperature using one of the emerging techniques (*e.g.*, ref. 34–36, 75–80) together with electron photo-emission spectroscopy measurements.[81–83] Ideally, these measurements should be carried out for a single particle (such as in ref. 42, 81, 84 and 85) in order to exclude multi-particle heating interactions, which are relatively hard to quantify.[40,41]

Appendix: A. An estimate of the quantum mechanical excitation term for CW illumination of a metal nanoparticle

We employ the elegant expression proposed in ref. 57; namely, we define $A(\mathcal{E};\omega)$ such that $A(\mathcal{E};\omega)\mathrm{d}\omega\mathrm{d}\mathcal{E}$ is the (joint) probability of photon absorption of frequency between ω and $\omega + \mathrm{d}\omega$ for the final energy $\mathcal{E}$ measured with respect to the bottom of the band at $\mathcal{E} = 0$. We define this probability as

$$A(\mathcal{E}_{\text{final}} = \mathcal{E};\omega) = \frac{n_{\text{A}}(\omega)}{N_{\text{A}}}\frac{D_{\text{J}}(\mathcal{E},\mathcal{E}-\hbar\omega)\rho_{\text{J}}(\mathcal{E},\mathcal{E}-\hbar\omega)}{\int D_{\text{E}}(\mathcal{E},\mathcal{E}-\hbar\omega)\rho_{\text{J}}(\mathcal{E},\mathcal{E}-\hbar\omega)\mathrm{d}\mathcal{E}}, \tag{A1}$$

where $D_{\text{J}}(\mathcal{E}_{\text{final}},\mathcal{E}_{\text{initial}})$ is the squared magnitude of a transition matrix element for the electronic process $\mathcal{E}_{\text{initial}}\rightarrow\mathcal{E}_{\text{final}}$. Furthermore, ρ_{J} corresponds to the population-weighted density of pair states,

$$\rho_{\text{J}}(\mathcal{E}_{\text{final}},\mathcal{E}_{\text{initial}}) = [f(\mathcal{E}_{\text{initial}})\rho_{\text{e}}(\mathcal{E}_{\text{initial}})][1 - f(\mathcal{E}_{\text{final}})\rho_{\text{e}}(\mathcal{E}_{\text{final}})], \tag{A2}$$

and $\rho_{\text{e}} = \dfrac{3n_{\text{e}}}{2\mathcal{E}_{\text{F}}}\sqrt{\dfrac{\mathcal{E}}{\mathcal{E}_{\text{F}}}}$ is the density of states of a free electron gas;[50] n_{e} is the electron density. Finally, $n_{\text{A}}(\omega)$ is the number density of absorbed $\hbar\omega$ photons per unit time between ω and $\omega + \mathrm{d}\omega$, and $N_{\text{A}} = \int \mathrm{d}\omega n_{\text{A}}(\omega)$ is the total number density of absorbed photons per unit time. For CW illumination, this is given by

$$N_{\text{A}} = \frac{\langle p_{\text{abs}}\rangle}{\hbar\omega}, \tag{A3}$$

where the averaged absorbed optical power density (in units of W m^{-3}) is given by the Poynting vector,[86] namely

$$\langle p_{\text{abs}}\rangle = \omega\varepsilon''(\omega, T_{\text{e}}, T_{\text{ph}})\langle\vec{E}(\vec{r},t)\cdot\vec{E}(\vec{r},t)\rangle_{t,\vec{r}}, \tag{A4}$$

where ε'' is the imaginary part of the metal permittivity, and the spatio-temporal averaging, $\langle\rangle t,\vec{r}$, is performed over a single optical cycle such that only the time-independent component remains and is over the NP volume.

The net change of electronic population at energy $\mathcal{E}$ per unit time and energy at time t due to absorption is $N_{\text{A}}\phi_{\text{A}}$, where

$$\phi_{\text{A}}(\mathcal{E};\omega) = \int_0^{\infty}\mathrm{d}\omega[A(\mathcal{E};\omega) - A(\mathcal{E}+\hbar\omega;\omega)] \tag{A5}$$

describes the total (probability of a) population change at energy $\mathcal{E}$ per unit time and energy at time t. Altogether,

$$\left(\frac{\partial f}{\partial t}\right)_{\text{ex}} = \frac{N_{\text{A}}\phi_{\text{A}}(E)}{\rho_{\text{e}}(E)}, \tag{A6}$$

meaning that electron number conservation is ensured, $\int \mathrm{d}\mathcal{E}\rho_{\text{e}}(\mathcal{E})(\partial f/\partial\mathcal{E})_{\text{ex}} \sim \int \mathrm{d}\mathcal{E}\phi_{\text{A}}(\mathcal{E}) = 0$.

For CW illumination, $n_{\text{A}} = N_{\text{A}}\delta(\omega - \omega_{\text{p}})$ and $N_{\text{A}} = 2\varepsilon''_{\text{m}}(\omega_{\text{p}}, T_{\text{e}}, T_{\text{ph}})|E(\omega_{\text{p}})|^2/\hbar$. For the low intensities considered here, one can ignore the temperature dependence of the metal permittivity.[38,87]

Table 1 Parameters used in the simulations and the values chosen for (low quality)[73] Ag

Parameter	Parameter symbol	Value
Photon wavelength	λ	2.25 eV
Metal permittivity	$\varepsilon_{Ag}(\lambda)$	$-8.5 + 1.8i$[73]
Fermi energy	$\mathcal{E}_F$	5.1 eV
Conduction band width	$\mathcal{E}_{max}$	9 eV
Chemical potential	μ	5.1 eV
Electron density	n_e	5.86×10^{28} m^{-3}
Ambient temperature	T_{amb}	297 K
Electron mass	m_e	9.1×10^{31} kg

For simplicity, we now approximate ρ_e by its value at $\mathcal{E}_F$. In the same spirit, we define the normalized transition matrix element $\overline{D} \equiv \int D_J(E, E - \hbar\omega_p)\rho_J(E, E - \hbar\omega_p)dE/D_J(E, E - \hbar\omega_p)$ and ignore its energy-dependence. Thus, we also find that $\overline{D} \cong \int D_J(E + \hbar\omega_p, E)\rho_J(E + \hbar\omega_p, E)dE/D_J(E + \hbar\omega_p, E)$, meaning that $\overline{D} \approx \int \rho_J dE \approx \rho_e^2(E_F)/E_{max}$. In this case, the excitation rate (A6) can be approximated by

$$\left(\frac{\partial f}{\partial t}\right)_{ex} = \frac{N_A}{\overline{D}}\rho_e(E_F)B(E) = \frac{2\varepsilon''_m}{\hbar}\frac{B(E)|E|^2}{E_{max}\rho_e(E_F)} = \mathrm{R}B(E)|E|^2, \tag{A7}$$

where $B(\mathcal{E}) = f(\mathcal{E} - \hbar\omega_p)(1 - f(\mathcal{E})) - f(\mathcal{E})(1 - f(\mathcal{E} + \hbar\omega_p))$ is a dimensionless O(1) function representing the 2 $\hbar\omega_p$ step-like energy dependence of the quantum excitation term and $R \equiv \frac{4\varepsilon_0\varepsilon''_m(\omega_p)}{3\hbar n_e}\frac{E_F}{E_{max}} \approx 2 \times 10^{-6}$ m^2 V^{-2} s^{-1} (see the parameter definitions and values in Table 1). Thus, we can estimate the non-thermal carrier density as

$$\Delta f \equiv f - f^T(E, T_{env}) \approx \tau_{e-e}R|E|^2 = 4 \times 10^{-20}|E|^2\left[\mathrm{V}^2\ \mathrm{m}^{-2}\right]. \tag{A8}$$

Conflicts of interest

There are no conflicts to declare.

References

1 A. O. Govorov and H. H. Richardson, Generating heat with metal nanoparticles, *Nano Today*, 2007, **2**, 30–38.

2 G. Baffou and R. Quidant, Thermo-plasmonics: using metallic nanostructures as nano-sources of heat, *Laser Photonics Rev.*, 2013, **7**, 171–187.

3 D. Boyer, P. Tamarat, A. Maali, B. Lounis and M. Orrit, Photothermal imaging of nanometer-sized metal particles among scatterers, *Science*, 2002, **297**, 1160–1163.

4 V. P. Zharov and D. O. Lapotko, Photothermal imaging of nanoparticles and cells, *IEEE J. Sel. Top. Quantum Electron.*, 2005, **11**, 733–751.

5 M. W. Dewhirst, B. L. Viglianti, M. Lora-Michiels, M. Hanson and P. J. Hoopes, Basic principles of thermal dosimetry and thermal thresholds for tissue damage from hyperthermia, *Int. J. Hyperthermia*, 2003, **19**, 267–294.

6 H. A. Atwater and A. Polman, Plasmonics for improved photovoltaic devices, *Nat. Mater.*, 2010, **9**, 205–213.

7 G. Baffou, J. Polleux, H. Rigneault and S. Monneret, Super-heating and micro-bubble generation around plasmonic nanoparticles under CW illumination, *J. Phys. Chem. C*, 2014, **118**, 4890–4898.

8 O. Neumann, A. S. Urban, J. Day, S. Lal, P. Nordlander and N. J. Halas, Solar vapor generation enabled by nanoparticles, *ACS Nano*, 2013, 7, 42–49.

9 Z. Fang, Y. R. Zhen, O. Neumann, A. Polman, F. J. Garcia de Abajo, P. Nordlander and N. J. Halas, Evolution of light-induced vapor generation at a liquid-immersed metallic nanoparticle, *Nano Lett.*, 2013, **13**, 1736–1742.

10 H. M. L. Robert, F. Kundrat, E. Bermúdez-Ure na, H. Rigneault, S. Monneret, R. Quidant, J. Polleux and G. Baffou, Light-assisted solvothermal chemistry using plasmonic nanoparticles, *ACS Omega*, 2016, **1**, 2–8.

11 S. Molesky, C. J. Dewalt and Z. Jacob, High temperature epsilon-near-zero and epsilon-near-pole metamaterial emitters for thermophotovoltaics, *Opt. Express*, 20113, **21**, A96–A110.

12 J. Liu, U. Guler, A. Lagutchev, A. V. Kildishev, O. Malis, A. Boltasseva and V. M. Shalaev, Quasi-coherent thermal emitter based on refractory plasmonic materials, *Opt. Mater. Express*, 2014, **5**, 2721–2728.

13 J. B. Khurgin, G. Sun, W. T. Chen, W.-Y. Tsai and D. P. Tsai, Ultrafast thermal nonlinearity, *Sci. Rep.*, 2015, **5**, 17899.

14 E. Rousseau, A. Siria, G. Jourdan, S. Volz, F. Comin, J. Chevrier and J.-J. Greffet, Radiative heat transfer at the nanoscale, *Nat. Photonics*, 2009, **3**, 514–517.

15 D. A. Boyd, L. Greengard, M. Brongersma, M. Y. El-Naggar and D. G. Goodwin, Plasmon-assisted chemical vapor deposition, *Nano Lett.*, 2006, **6**, 2592–2597.

16 U. Guler, A. Boltasseva and V. M. Shalaev, Refractory plasmonics, *Science*, 2014, **334**, 263.

17 K. Watanabe, D. Menzel, N. Nilius and H.-J. Freund, Photochemistry on metal nanoparticles, *Chem. Rev.*, 2006, **106**, 4301–4320.

18 A. Naldoni, F. Riboni, U. Guler, A. Boltasseva, V. M. Shalaev and A. V. Kildishev, Solar-powered plasmon-enhanced heterogeneous catalysis, *Nanophotonics*, 2016, **5**, 112–133.

19 S. Linic, P. Christopher and D. B. Ingram, Plasmonic-metal nanostructures for efficient conversion of solar to chemical energy, *Nat. Mater.*, 2011, **10**, 911–921.

20 C. Clavero, Plasmon-induced hot-electron generation at nanoparticle/metal-oxide interfaces for photovoltaic and photocatalytic devices, *Nat. Photonics*, 2014, **8**, 95–103.

21 G. Baffou and R. Quidant, Nanoplasmonics for chemistry, *Chem. Soc. Rev.*, 2014, **43**, 3898.

22 M. L. Brongersma, N. J. Halas and P. Nordlander, Plasmon-induced hot carrier science and technology, *Nat. Nanotechnol.*, 2015, **10**, 25.

23 M. Moskovits, The case for plasmon-derived hot carrier devices, *Nat. Nanotechnol.*, 2015, **10**, 6.

24 S. Mubeen, J. Lee, N. Singh, S. Kraemer, G. D. Stucky and M. Moskovits, An autonomous photosynthetic device in which all charge carriers derive from surface plasmons, *Nat. Nanotechnol.*, 2013, **8**, 247–251.

25 K. Ueno, T. Oshikiri, X. Shi, Y. Zhong and H. Misawa, Plasmon-induced artificial photosynthesis, *Interface Focus*, 2014, **5**, 20140082.

26 I. Goykhman, B. Desiatov, J. Khurgin, J. Shappir and U. Levy, Locally oxidized silicon surface-plasmon Schottky detector for telecom regime, *Nano Lett.*, 2011, **11**, 2219–2224.

27 I. Goykhman, B. Desiatov, J. Khurgin, J. Shappir and U. Levy, Waveguide based compact silicon Schottky photodetector with enhanced responsivity in the telecom spectral band, *Opt. Express*, 2012, **20**, 28594.

28 M. W. Knight, Y. Wang, A. S. Urban, A. Sobhani, B. Y. Zheng, P. Nordlander and N. J. Halas, Embedding plasmonic nanostructure diodes enhances hot electron emission, *Nano Lett.*, 2013, **13**, 1687–1692.

29 W. Li and J. Valentine, Harvesting the loss: Surface plasmon-based hot electron photodetection, *Nanophotonics*, 2016, **6**, 177–191.

30 A. Giugni, B. Torre, A. Toma, M. Francardi, M. Malerba, A. Alabastri, R. Proietti Zaccaria, M. I. Stockman and E. Di Fabrizio, Hot-electron nanoscopy using adiabatic compression of surface plasmons, *Nat. Nanotechnol.*, 2013, **8**, 845–852.

31 G. V. Naik and J. A. Dionne, Photon upconversion with hot carriers in plasmonic systems, *Appl. Phys. Lett.*, 2015, **107**, 133902.

32 G. V. Naik, A. J. Welch, J. A. Briggs, M. L. Solomon and J. A. Dionne, Hot-carrier-mediated photon upconversion in metal-decorated quantum wells, *Nano Lett.*, 2017, **17**, 4583–4587.

33 X. Zhou, G. Liu, J. Yu and W. Fan, Surface plasmon resonance-mediated photocatalysis by noble metal-based composites under visible light, *J. Mater. Chem.*, 2012, **22**, 21337–21354.

34 J. T. Hugall and J. J. Baumberg, Demonstrating photoluminescence from Au is electronic inelastic light scattering of a plasmonic metal: The origin of SERS backgrounds, *Nano Lett.*, 2015, **15**, 2600–2604.

35 X. Xie and D. G. Cahill, Gold nanoparticles as absolute nano-thermometers, *Appl. Phys. Lett.*, 2016, **109**, 183104.

36 A. Carattino, M. Caldarola and M. Orrit, Gold nanoparticles as absolute nano-thermometers, *Nano Lett.*, 2017, **18**, 874–880.

37 G. Baffou, R. Quidant and F. J. Garcia de Abajo, Nanoscale control of optical heating in complex plasmonic systems, *ACS Nano*, 2010, **4**, 709–716.

38 Y. Sivan and S.-W. Chu, Nonlinear plasmonics at high temperatures, *Nanophotonics*, 2017, **6**, 317–328.

39 H. H. Richardson, M. T. Carlson, P. J. Tandler, P. Hernandez and A. O. Govorov, Experimental and theoretical studiesof light-to-heat conversion and collective heating effects in metal nanoparticle solutions, *Nano Lett.*, 2009, **9**, 1139–1146.

40 G. Baffou, P. Berto, E. Bermudez Urena, R. Quidant, S. Monneret, J. Polleux and H. Rigneault, Photoinduced heating of nanoparticle arrays, *ACS Nano*, 2013, 7, 6478–6488.

41 S. Mukherjee, L. Zhou, A. Goodman, N. Large, C. Ayala-Orozco, Y. Zhang, P. Nordlander and N. J. Halas, Hot-electron-induced dissociation of H2 on gold nanoparticles supported on SiO2, *J. Am. Chem. Soc.*, 2014, **136**, 64–67.

42 S. Tan, A. Argondizzo, J. Ren, L. Liu, J. Zhao and H. Petek, Plasmonic coupling at a metal/semiconductor interface, *Nat. Photonics*, 2017, **11**, 806–812.

43 L. Zhou, D. F. Swearer, C. Zhang, H. Robatjazi, H. Zhao, L. Henderson, L. Dong, P. Christopher, E. A. Carter, P. Nordlander and N. J. Halas, Quantifying hot carrier and thermal contributions in plasmonic photocatalysis, *Science*, 2018, **362**, 69.

44 X. Zhang, X. Li, M. E. Reish, D. Zhang, N. Q. Su, Y. Gutiërrez, F. Moreno, W. Yang, H. O. Everitt and J. Liu, Plasmon-enhanced catalysis: Distinguishing thermal and nonthermal effects, *Nano Lett.*, 2018, **18**, 1714–1723.

45 A. Puglisi, A. Sarracino and A. Vulpiani, Temperature in and out of equilibrium: a review of concepts, tools and attempts, *Phys. Rep.*, 2017, **709–710**, 1–60.

46 L. D. Pietanza, G. Colonna, S. Longo and M. Capitelli, Non-equilibrium electron and phonon dynamics in metals under femtosecond laser pulses, *Eur. Phys. J. D*, 2007, **45**, 369–389.

47 J. R. M. Saavedra, A. Asenjo-Garcia and F. Javier Garcia de Abajo, Hot-electron dynamics and thermalization in small metallic nanoparticles, *ACS Photonics*, 2016, **3**, 1637–1646.

48 Y. Dubi and Y. Sivan, "Hot electrons" in metallic nanostructures - non-thermal carriers or heating?, *Nat. Photonics*, 2018, under review.

49 J. M. Ziman, *Principles of the theory of solids*. Cambridge University Press, 1972.

50 N. W. Ashcroft and N. D. Mermin, *Solid state physics*, Brooks/Cole, 1976.

51 D. Pines and P. Nozieres, *The theory of quantum liquids*, Benjamin, New York, 1966.

52 M. Dressel and G. Grüner, *Electrodynamics of solids - optical properties of electrons in matter*, Cambridge University Press, 2002.

53 R. H. M. Groeneveld, R. Sprik and A. Lagendijk, Femtosecond spectroscopy of electron-electron and electron-phonon energy relaxation in Ag and Au, *Phys. Rev. B: Condens. Matter Mater. Phys.*, 1995, **51**, 11433–11445.

54 N. Del Fatti, C. Voisin, M. Achermann, S. Tzortzakis, D. Christofilos and F. Valleé, Nonequilibrium electron dynamics in noble metals, *Phys. Rev. B: Condens. Matter Mater. Phys.*, 2000, **61**, 16956–16966.

55 B. Rethfeld, A. Kaiser, M. Vicanek and G. Simon, Ultrafast dynamics of nonequilibrium electrons in metals under femtosecond laser irradiation, *Phys. Rev. B: Condens. Matter Mater. Phys.*, 2002, **65**, 214303.

56 P. Grua, J. P. Morreeuw, H. Bercegol, G. Jonusauskas and F. Valleé, Electron kinetics and emission for metal nanoparticles exposed to intense laser pulses, *Phys. Rev. B: Condens. Matter Mater. Phys.*, 2003, **68**, 035424.

57 M. Kornbluth, A. Nitzan and T. Seidman, Light-induced electronic non-equilibrium in plasmonic particles, *J. Chem. Phys.*, 2013, **138**, 174707.

58 R. W. Schoenlein, W. Z. Lin, J. G. Fujimoto and G. L. Eesley, Femtosecond studies of nonequilibrium electronic processes in metals, *Phys. Rev. Lett.*, 1987, **58**, 1680–1683.

59 E. Carpene, Ultrafast laser irradiation of metals: Beyond the two-temperature model, *Phys. Rev. B: Condens. Matter Mater. Phys.*, 2006, **74**, 024301.

60 G. Della Valle, M. Conforti, S. Longhi, G. Cerullo and D. Brida, Real-time optical mapping of the dynamics of nonthermal electrons in thin gold films, *Phys. Rev. B: Condens. Matter Mater. Phys.*, 2012, **86**, 155139.

61 P. E. Hopkins, Contributions of inter and intra-band excitations to electron heat capacity and electron-phonon coupling in noble metals, *J. Heat Transfer*, 2010, **132**, 014504.

62 A. Giri and P. E. Hopkins, Transient thermal and nonthermal electron and phonon relaxation after short-pulsed laser heating of metals, *J. Appl. Physiol.*, 2015, **118**, 215101.

63 A. M. Brown, R. Sundararaman, P. Narang, W. A. Goddard and H. A. Atwater, Ab initio phonon coupling and optical response of hot electrons in plasmonic metals, *Phys. Rev. B*, 2016, **94**, 075120.

64 J. K. Chen, D. Y. Tzou and J. E. Beraun, A semiclassical two-temperature model for ultrafast laser heating, *Int. J. Heat Mass Transfer*, 2006, **49**, 307–316.

65 V. V. Baranov and V. V. Kabanov, Theory of the electron relaxation in metals excited by an ultrashort optical pump, *Phys. Rev. B: Condens. Matter Mater. Phys.*, 2014, **84**, 125102.

66 J. Zhou, N. Li and R. Yang, An electrohydrodynamics model for non-equilibrium electron and phonon transport in metal films after ultra-short pulse laser heating, *Eur. Phys. J. B*, 2015, **88**, 156.

67 A. Manjavacas, J. G. Liu, V. Kulkarni and P. Nordlander, Plasmon-induced hot carriers in metallic nanoparticles, *ACS Nano*, 2014, **8**, 7630–7638.

68 L. V. Besteiro, X.-T. Kong, Z. Wang, G. Hartland and A. O. Govorov, Understanding hot-electron generation and plasmon relaxation in metal nanocrystals: Quantum and classical mechanisms, *ACS Photonics*, 2017, **4**, 2759–2781.

69 T. Gong and J. N. Munday, Materials for hot carrier plasmonics, *Opt. Mater. Express*, 2015, **5**, 2501.

70 S. Dal Forno, L. Ranno and J. Lischner, Material, size and environment dependence of plasmon-induced hot carriers in metallic nanoparticles, *J. Phys. Chem. C*, 2018, **122**, 8517–8527.

71 J. Zhao, S. C. Nguyen, R. Ye, B. Ye, H. Weller, G. A. Somorjai, A. P. Alivisatos and F. D. Toste, A comparison of photocatalytic activities of gold nanoparticles following plasmonic and interband excitation and a strategy for harnessing interband hot carriers for solution phase photocatalysis, *ACS Cent. Sci.*, 2017, **3**, 482.

72 A. M. Brown, R. Sundararaman, P. Narang, W. A. Goddard and H. A. Atwater, Nonradiative plasmon decay and hot carrier dynamics: Effects of phonons, surfaces, and geometry, *ACS Nano*, 2016, **10**, 957–966.

73 S. T. Sundari, S. Chandra and A. K. Tyagi, Temperature dependent optical properties of silver from spectroscopic ellipsometry and density functional theory calculations, *J. Appl. Phys.*, 033515, **114**, 2013.

74 S. Mukherjee, F. Libisch, N. Large, O. Neumann, L. V. Brown, J. Cheng, J. Britt Lassiter, E. A. Carter, P. Nordlander and N. J. Halas, Hot electrons do the impossible: Plasmon-induced dissociation of H2 on Au, *Nano Lett.*, 2013, **13**, 240–247.

75 M. Honda, Y. Saito, N. I. Smith, K. Fujita and S. Kawata, Nanoscale heating of laser irradiated single gold nanoparticles in liquid, *Opt. Express*, 2011, **19**, 12375–12383.

76 M. Mecklenburg, W. A. Hubbard, E. R. White, R. Dhall, S. B. Cronin, S. Aloni and B. C. Regan, Nanoscale temperature mapping in operating microelectronic devices, *Science*, 2015, **347**, 629.

77 G. Kucsko, P. C. Maurer, N. Y. Yao, M. Kubo, H. J. Noh, P. K. Lo, H. Park and M. D. Lukin, Nanometre-scale thermometry in a living cell, *Nature*, 2013, **500**, 54–58.
78 S. Sadat, A. Tan, Y. J. Chua and P. Reddy, Nanoscale thermometry using point contact thermocouples, *Nano Lett.*, 2010, **10**, 2613–2617.
79 K. Kim, B. Song, V. Fernández-Hurtado, W. Lee, W. Jeong, L. Cui, D. Thompson, J. Feist, M. T. Homer Reid, F. J. García-Vidal, J. Carlos Cuevas, E. Meyhofer and P. Reddy, Radiative heat transfer in the extreme near field, *Nature*, 2015, **528**, 387–391.
80 Y.-K. Tzeng, P.-C. Tsai, H.-Y. Liu, O. Y. Chen, H. Hsu, F.-G. Yee, M.-S. Chang and H.-C. Chang, Time-resolved luminescence nanothermometry with nitrogen-vacancy centers in nanodiamonds, *Nano Lett.*, 2015, **15**, 3945–3952.
81 M. Lisowski, A. P. Loukakos, U. Bovensiepen, J. Stähler, C. Gahl and M. Wolf, Ultra-fast dynamics of electron thermalization, cooling and transport effects in Ru(001), *Appl. Phys. A*, 2004, **78**, 165–176.
82 M. Bauer, A. Marienfeld and M. Aeschlimann, Hot electron lifetimes in metals probed by time-resolved two-photon photoemission, *Prog. Surf. Sci.*, 2015, **90**, 319–376.
83 J. Vogelsang, J. Robin, B. J. Nagy, P. Dombi, D. Rosenkranz, M. Schiek, P. Gross and C. Lienau, Ultrafast electron emission from a sharp metal nanotaper driven by adiabatic nanofocusing of surface plasmons, *Nano Lett.*, 2015, **15**, 4685–4691.
84 Y. Levartovsky and E. Gross, High spatial resolution mapping of chemically-active self-assembled n-heterocyclic carbenes on Pt nanoparticles, *Faraday Discuss.*, 2016, **5188**, 345.
85 C.-Y. Wu, W. J. Wolf, Y. Levartovsky, H. A. Bechtel, M. C. Martin, F. D. Toste and E. Gross, High-spatial-resolution mapping of catalytic reactions on single particles, *Nature*, 2017, **541**, 511–515.
86 J. D. Jackson, *Classical electrodynamics*, Wiley & Sons, 3rd edn, 1998.
87 I. Gurwich and Y. Sivan, A metal nanosphere under intense continuous wave illumination - a unique case of non-perturbative nonlinear nanophotonics, *Phys. Rev. E*, 2017, **96**, 012212.
88 Negative values of $f - f^T$ are referred to as holes, regardless of their position with respect to the Fermi energy. This nomenclature is conventional within the literature[51]
89 Note that we ignore here the difference between the Fermi energy and the chemical potential; we verified *via* simulations that the difference between them is truly negligible in all of the cases we studied.
90 Note that the eTTM does not capture the increase in the rate of energy transfer to the lattice during the thermalization time, as discussed in ref. 53. However, this effect should have, at most, a modest quantitative effect on the issues discussed in the current work.
91 Note that the differences between eqn (A) and the more conventional TTM extend only to the initial non-thermal regime, for which the temperature is anyhow not well defined.
92 Yet, in these papers, the authors did not attempt to claim more than a qualitative prediction of the "hot electron" density.
93 Z being the material impedance of the gas/semiconductor and n its refractive index.

94 Indeed, due to the high thermal conductivity of the metal, the temperature itself is uniform across the NPs, meaning that the "hot" spot is not hotter than other regions.

95 In principle, the matrix element may also change with wavelength. However, this effect is not taken into account in the current calculation.

Faraday Discussions

PAPER

Out-of-equilibrium electron dynamics of silver driven by ultrafast electromagnetic fields – a novel hydrodynamical approach

Andrea Marini,*[a] Alessandro Ciattoni[b] and Claudio Conti[cd]

Received 15th October 2018, Accepted 31st October 2018
DOI: 10.1039/c8fd00153g

We investigate the ultrafast nonlinear response of silver upon excitation by infrared electromagnetic radiation pulses with a duration of a few femtoseconds. By adopting the Landau weak coupling approach to account for electron–electron and electron–phonon collisions, we solve the Boltzmann equation through the method of moments obtaining a novel set of hydrodynamical equations describing the ultrafast nonlinear dynamics of electrons in silver. While the novel hydrodynamical model that was obtained reduces to the Drude model for small intensities of the driving field, it predicts that absorption saturates for large but experimentally attainable peak intensities of the order of GW cm^{-2}. Our results are important for absorption mitigation in plasmonic devices, with potential impact for low-loss plasmonic waveguides and interconnects.

1 Introduction

The optical absorption of metals stems mainly from two distinct processes: resonant absorption due to interband transitions and nonresonant absorption ensuing from intraband electron collisions.[1] Among the noble metals, gold and copper undergo interband dynamics at optical frequencies, thus implying an inherently high absorption. On the contrary, silver undergoes interband dynamics only at ultraviolet frequencies, and thus its optical and infrared responses are affected mainly by intraband electron oscillations and collisions. As of today, silver still remains the best plasmonic material for plasmonic applications in optical and infrared frequency domains. However, absorption in current plasmonic devices such as plasmonic waveguides,[2,3] interconnects,[4] switches and modulators[5] is far too high for practical applications. Thanks to the reduced electron density, graphene,[6,7] oxides and nitrides,[8–11] and polar dielectrics[12] are

[a]*Department of Physical and Chemical Sciences, University of L'Aquila, Via Vetoio, 67100 L'Aquila, Italy. E-mail: andrea.marini@aquila.infn.it; Fax: +39 0862433033; Tel: +39 0862433093*
[b]*Consiglio Nazionale delle Ricerche, CNR-SPIN, Via Vetoio 10, 67100 L'Aquila, Italy*
[c]*Institute for Complex Systems (ISC-CNR), Via dei Taurini 19, 00185, Rome, Italy*
[d]*Department of Physics, University Sapienza, Piazzale Aldo Moro 5, 00185, Rome, Italy*

promising plasmonic materials that exhibit smaller loss, although they operate mainly in the infrared. Amplification schemes embedding externally pumped active inclusions have been proposed for the suppression of absorption, *e.g.* surface plasmon amplification through stimulated emission of radiation (SPASER),[13,14] amplification of surface plasmon polaritons (SPPs),[15–21] and plasmonic nanolasers.[22–25] Further attempts to mitigate absorption in plasmonic materials exploit the reduction of surface roughness[26] and self-induced-transparency plasmon solitons for the suppression of interband absorption.[27] Current theoretical and experimental investigations on plasmonics are mainly focused on optical pulses with time durations of hundreds of femtoseconds or higher, *e.g.* second- and third-harmonic generation,[28–31] optical thermo-modulation,[32–36] all-optical switching,[37,38] self-phase modulation[39] and ultraviolet (UV) generation in tapered plamonic capillaries filled with highly nonlinear gases.[40] The ultrafast dynamics of electrons in plasmonic materials generally involve complex dynamics encompassing electron heating and the following exchange of energy from hot electrons to the lattice *via* electron-phonon scattering.[32–36] In this respect, the exploitation of ultrafast electron dynamics at the femtosecond regime is a promising strategy for quenching absorption and heating since it enables the achievement of high peak intensity with the reduced transfer of energy to hot electrons.

Herein, we derive a novel model based on the Fokker–Planck–Landau (FPL) formalism[41] – a well-established approach in out-of-equilibrium (OOE) statistical mechanics – for the modelling of the ultrafast dynamics of electrons in silver upon excitation by ultrashort pulses of electromagnetic (EM) radiation with a time duration of a few femtoseconds. By calculating analytically the Landau collisional integral in the weak coupling limit, we obtain a novel set of hydrodynamical equations accounting for both the local and non-local electron dynamics in silver, as well as for nonlinear damping. In particular, owing to nonlinearity, we find that damping saturates for OOE electron average velocities higher than the thermal velocity, and that optical absorption in silver can be highly reduced by utilising ultrashort pulses with a high peak intensity and a time duration that is smaller than the electron collision time. In practice, for such ultrashort pulse durations, the probability that an electron undergoes a collision is much smaller, and the transfer of optical energy to silver heating is highly reduced. In terms of the electromagnetic field peak intensity, we find that for infrared frequencies the absorption saturation intensity is of the order of GW cm^{-2}. We further study the electron ballistics and the dependence of the complex χ_3^{coll} on the angular frequency, electron collision time and relaxation time, finding that χ_3^{coll} scales with the square of the photon angular frequency. Our results open up new possibilities for the suppression of ohmic losses in metals and thus for the development of novel plasmonic devices with reduced absorption and enhanced operational functionality.

2 Out-of-equilibrium electron dynamics in silver

Owing to the highly detuned interband transitions, at optical and infrared carrier frequencies of the driving EM radiation pulses we model the electron dynamics in silver using classical kinematic theory, where the electron plasma is regarded as a dilute gas of N electrons enclosed in a box of volume V and immersed in a homogeneous background of ions with constant density $n_0 = N/V$ (equivalent to the lattice density), mass M, and equilibrium temperature T_0. The time dependent electron distribution function $f(\boldsymbol{r},\boldsymbol{w},t)$ is defined in such a way that $f(\boldsymbol{r},\boldsymbol{w},t)\mathrm{d}\boldsymbol{r}\mathrm{d}\boldsymbol{w}$ corresponds to the number of electrons with mass m that, at time t, have positions lying within a volume element $\mathrm{d}\boldsymbol{r}$ about $\boldsymbol{r}$ and velocities lying within a velocity-space element $\mathrm{d}\boldsymbol{w}$ about $\boldsymbol{w}$, so that $\int f(\boldsymbol{r},\boldsymbol{w},t)\mathrm{d}\boldsymbol{r}\mathrm{d}\boldsymbol{w} = N$. Since ions are much heavier than electrons ($M \gg m$), we neglect their motion upon EM excitation and assume that their time-independent distribution function is $f_0(\boldsymbol{r},\boldsymbol{w}) = n_0(M/2\pi k_\mathrm{B}T_0)^{3/2}\exp[-Mw^2/2k_\mathrm{B}T_0]$, where k_B is the Boltzmann constant, so that the total charge of the system vanishes ($e\int \mathrm{d}\boldsymbol{r}\mathrm{d}\boldsymbol{w}(f_0 - f) = 0$). The evolution of $f(\boldsymbol{r},\boldsymbol{w},t)$ is ruled by the Boltzmann equation

$$\partial_t f + \boldsymbol{w}\cdot\nabla_r f + \frac{1}{m}\boldsymbol{F}_w(\boldsymbol{r},t)\cdot\nabla_w f = (\partial_t f)_\mathrm{coll}, \tag{1}$$

where $\boldsymbol{F}_\mathbf{w}(\boldsymbol{r},t) = -e\boldsymbol{E}(\boldsymbol{r},t) - e\boldsymbol{w}\times\boldsymbol{B}(\boldsymbol{r},t)$, $-e$ is the electron charge, $\boldsymbol{E}(\boldsymbol{r},t)$ and $\boldsymbol{B}(\boldsymbol{r},t)$ are the external electric and magnetic fields, and $(\partial_t f)_\mathrm{coll} = (\partial_t f)_\mathrm{coll}^\mathrm{el-el} + (\partial_t f)_\mathrm{coll}^\mathrm{el-ph}$ is the total collision rate resulting from electron–electron $[(\partial_t f)_\mathrm{coll}^\mathrm{el-el}]$ and electron–phonon $[(\partial_t f)_\mathrm{coll}^\mathrm{el-ph}]$ collisions. Such collision rates in the Boltzmann Stosszahlansatz of molecular chaos are complex integral expressions involving the Coulomb differential scattering cross-sections of electron–electron and electron–ion collisions. In principle, the numerical solution of the Boltzmann equation provides the optical response of the system. However, this approach is computationally expensive and does not provide much physical insight into the nonlinear dynamics of the system. In turn, in order to account for the effect of electron collisions, in the so-called relaxation time approximation (RTA) the collision integral is usually approximated by $(\partial_t f)_\mathrm{coll} \simeq -(f - f_0)/\tau_\mathrm{eff}$, where τ_eff is an effective relaxation time fitted from the experimental results. However, the RTA neglects the nonlinear dependency of $(\partial_t f)_\mathrm{coll}$ over $f(\boldsymbol{r},\boldsymbol{w},t)$, which we evaluate here through the Landau weak coupling assumption,[41] where electron collisions with a large deflection angle are neglected whilst retaining all grazing-angle contributions leading to the so-called Fokker–Planck–Landau (FPL) nonlinear damping operators (see Appendix). Thus, we solve eqn (1) following the method of moments, which leads to a hierarchy of hydrodynamical equations (HDEs) that are fully equivalent to the Boltzmann equation.[41] The relevant physical moments for describing the OOE dynamics of hot electrons are the number density $n(\boldsymbol{r},t) = \int \mathrm{d}\boldsymbol{w}f(\boldsymbol{r},\boldsymbol{w},t)$, the current density $n(\boldsymbol{r},t)\boldsymbol{v}(\boldsymbol{r},t) = \int \mathrm{d}\boldsymbol{w}\boldsymbol{w}f(\boldsymbol{r},\boldsymbol{w},t)$, and the energy density $(3/2)n(\boldsymbol{r},t)k_\mathrm{B}T_\mathrm{e}(\boldsymbol{r},t) = (m/2)\int \mathrm{d}\boldsymbol{w}|\boldsymbol{w} - \boldsymbol{v}|^2 f(\boldsymbol{r},\boldsymbol{w},t)$. Thus, we truncate the hierarchy by neglecting higher-order moments obtaining the HDEs

$$\partial_t n + \nabla\cdot(n\boldsymbol{v}) = 0, \tag{2a}$$

$$\partial_t \boldsymbol{v} + (\boldsymbol{v}\cdot\nabla)\boldsymbol{v} + \frac{3k_\mathrm{B}}{mn}\nabla(nT_\mathrm{e}) = \frac{1}{m}\boldsymbol{F}_\mathrm{eff} - \gamma\mathcal{F}_\gamma(v, T_\mathrm{e})\boldsymbol{v}, \tag{2b}$$

$$\partial_t T_\mathrm{e} + \frac{2}{3}T_\mathrm{e}\nabla\cdot\boldsymbol{v} + \boldsymbol{v}\cdot\nabla T_\mathrm{e} = Q(v, T_\mathrm{e}), \tag{2c}$$

where $\boldsymbol{F}_\mathrm{eff} = -e\boldsymbol{E} - e\boldsymbol{v}\times\boldsymbol{B}$ is the external effective force, γ is the linear damping rate, while $\mathcal{F}_\gamma(v,T_\mathrm{e})$ and $Q(v,T_\mathrm{e})$ are involved functions accounting for nonlinear current damping and hot electron heating dynamics (see Appendix). The HDEs reported above in eqn (2) constitute a substantial extension of the traditional

HDEs used to model the EM response of plasmonic materials,[42–44] where the current damping is a constant ($\mathcal{F}_\gamma(v,T_e) \simeq 1$) obtained within the RTA. The nonlinear dependencies of $\mathcal{F}_\gamma$ and Q are depicted in Fig. 1a and b at ambient lattice temperature ($T_0 = 300$ K) and by adjusting the constants n, M, γ to fit the silver constants $\omega_P = 14$ fs^{-1}, $\gamma = 32$ ps^{-1}, and $\gamma_{th} = 1.4$ ps^{-1},[45,46] obtained by setting $n_0 = 6.16 \times 10^{22}$ cm^{-3} and $M = 4.07 \times 10^{-26}$ g. We emphasize that in the limit of small electron heating, $T_e - T_0 \ll T_0$, and velocity, $|v/v_{T_0}| \ll 1$ ($v_{T_0} = \sqrt{2k_BT_0(1/m + 1/M)}$), $\mathcal{F}_\gamma \simeq 1$ (see Fig. 1) and $Q \simeq -\gamma_{th}(T_e - T_0)$, where γ_{th} is the hot electron relaxation rate describing relaxation towards equilibrium *via* electron–phonon scattering, thus recovering the RTA limit.

2.1 Electron ballistics

In order to understand the collision dynamics predicted by our model, we focus here on the local response, for which an initial understanding can be grasped by initially ignoring the external driving field and focusing only on plasma relaxation. In Fig. 2 we plot the temporal evolution (obtained by a numerical integration of eqn (2) by a fourth-order Runge–Kutta algorithm) of (a) $v(t)$ and (b) $\Delta T(t) = T_e(t) - T_0$ at a fixed initial temperature $T_e(t = 0) = T_0$ and at several initial velocities $v(t = 0) = v_0$. Note that when v_0 overcomes v_{T_0}, the damping rate towards equilibrium $v = 0$ is heavily quenched (see Fig. 2a). Intuitively, damping quenching for large macroscopic velocities can be understood as a reduced electron deflection upon collision (originating damping) owing to the increased kinetic energy. It should also be noted that, as a consequence of damping quenching, $\Delta T(t)$ saturates at a large v_0/v_{T_0} due to the reduced rate of energy transfer (see Fig. 2b).

2.2 Saturation of absorption

Our model indicates that the damping, $\gamma\mathcal{F}_\gamma$, is nonlinear and quenches when $v/v_{T_0} \gg 1$ (see Fig. 1a). We wish to emphasize that the nonlinear damping rate, $\gamma\mathcal{F}_\gamma$, in the HDEs obtained in eqn (2) ensues only from $(\partial_t f)^{\text{el-ph}}_{\text{coll}}$ since electron–electron collisions conserve total momentum and energy,[41] thus implying that $(\partial_t f)^{\text{el-el}}_{\text{coll}} = 0$, which we obtain analytically. From a quantum perspective, electron–electron collisions in metals can contribute to damping through an Umklapp process involving a reciprocal lattice vector, but such a process is quite inefficient

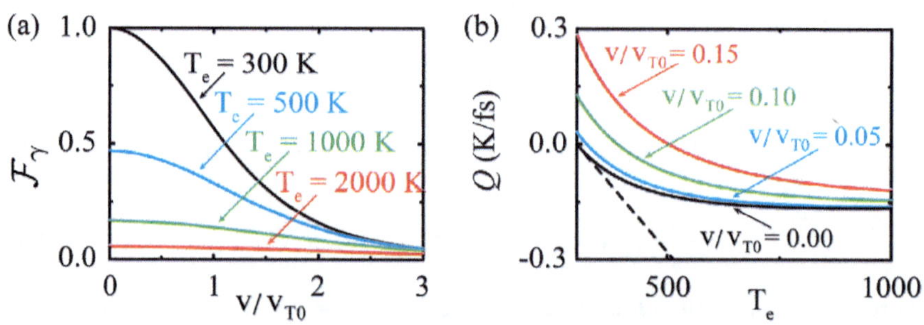

Fig. 1 Nonlinear dependency of (a) dimensionless damping $\mathcal{F}_\gamma$ and (b) heating rate Q on v/v_{T_0} and T_e, where $v_{T_0} = 96.4$ km s^{-1} at $T_0 = 300$ K. The dashed black line in (b) indicates the linear limit ($v \to 0$) of the plasma temperature relaxation $Q \simeq -\gamma_{th}(T_e - T_0)$.

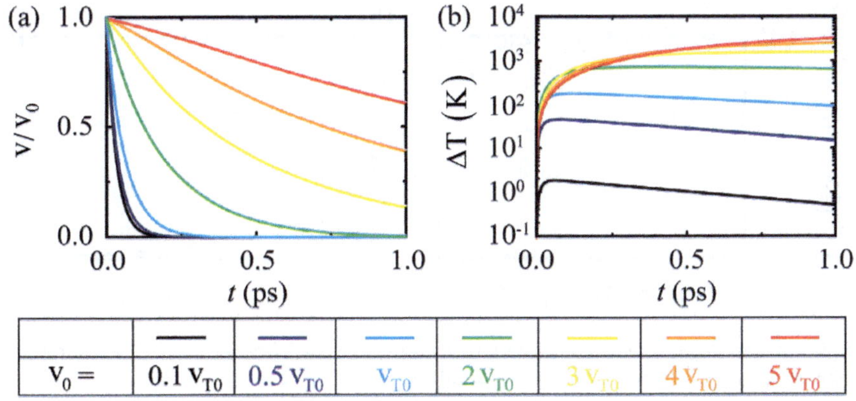

Fig. 2 Temporal evolution of (a) normalized average velocity, v/v_0, and (b) temperature increase, $\Delta T = T_e - T_0$, for a plasma with an initial temperature of $T_0 = 300$ K and with the velocities, v_0, indicated in the legend.

for most EM frequency ranges and becomes comparable with electron–phonon scattering only at optical frequencies.[1] Damping quenching implies that EM absorption quenches accordingly at intensities of the driving field larger than the saturation intensity I_S, which can be derived by assuming a monochromatic field, $\boldsymbol{E}(t) = \mathrm{Re}[\boldsymbol{E}_0\,\mathrm{e}^{-\mathrm{i}\omega t}]$, and solving eqn (2) perturbatively by setting $\boldsymbol{v} \simeq \mathrm{Re}[(\boldsymbol{v}_1 + \boldsymbol{v}_3|\boldsymbol{E}_0|^2)\mathrm{e}^{-\mathrm{i}\omega t}] + \mathcal{O}(|\boldsymbol{E}_0|^4)$ and $T_e \simeq T_0 + T_1|\boldsymbol{E}_0|^2 + \mathcal{O}(|\boldsymbol{E}_0|^4)$, and by neglecting higher order harmonics and powers of $|\boldsymbol{E}_0|^2$. By substituting the electron mean velocity in the current density, $\boldsymbol{J}(t) = -n_0 e\boldsymbol{v}(t)$, we find that the plasmonic material polarization is $\boldsymbol{P} = \int \boldsymbol{J}(t)\mathrm{d}t \simeq \varepsilon_0 \mathrm{Re}[(\chi_{\mathrm{Drude}} + \chi_3^{\mathrm{coll}}|\boldsymbol{E}_0|^2)$ $\boldsymbol{E}_0\,\mathrm{e}^{-\mathrm{i}\omega t}]$, where

$$\chi_{\mathrm{Drude}} = -\omega_P{}^2/\omega(\omega + \mathrm{i}\gamma), \tag{3a}$$

$$\chi_3^{\mathrm{coll}} = -\mathrm{i}a\frac{\gamma\omega_P{}^2(\omega - \mathrm{i}\gamma)^2}{\omega(\omega^2 + \gamma^2)^3}, \tag{3b}$$

and

$$a = e^2(20\gamma - \gamma_{\mathrm{th}})(2\gamma_{\mathrm{th}}{}^{-1} - \gamma^{-1})/(80mk_{\mathrm{B}}T_0). \tag{4}$$

In Fig. 3a and b we plot the real and imaginary parts of the complex, χ^{coll}_3, and the absorption saturation intensity, $I_{\mathrm{sat}} = -\varepsilon_0 c\,\mathrm{Im}\,\chi_{\mathrm{Drude}}/4\,\mathrm{Im}\,\chi_3$, respectively, as a function of the wavelength, λ. It should be noted that both $\mathrm{Re}\,\chi_3^{\mathrm{coll}}$ and $\mathrm{Im}\,\chi_3^{\mathrm{coll}}$ are negative in the infrared region of the spectrum, thus implying that for the limit of low driving intensity, electron collisions in silver induce a defocusing nonlinearity and a reduction in absorption. We find that I_{sat} depends quadratically over the angular frequency $\omega = 2\pi c/\lambda$ for the limit $\omega \gg \gamma$, and that when the EM intensity is $I_0 = I_{\mathrm{sat}}$, the effective susceptibility, $\chi_{\mathrm{eff}} = \chi_{\mathrm{Drude}} + \chi_3^{\mathrm{coll}}|\boldsymbol{E}_0|^2$, is such that $\mathrm{Im}\,\chi_{\mathrm{eff}} = \mathrm{Im}\,\chi_{\mathrm{Drude}}/2$, meaning that the effective EM absorption coefficient is two times smaller than the linear one.

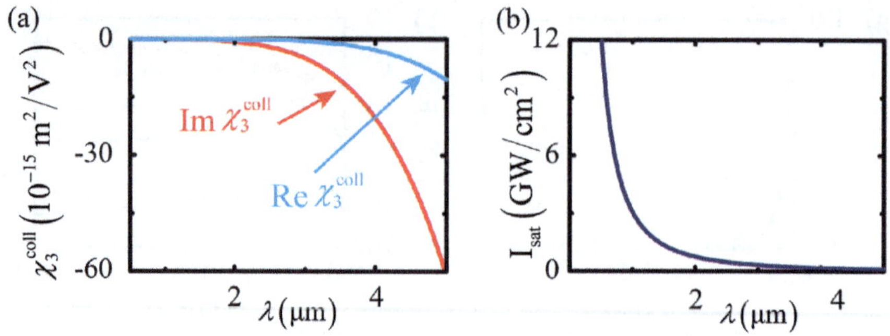

Fig. 3 (a) Real (cyan curve) and imaginary (red curve) parts of χ_3^{coll}, and (b) saturation intensity, I_{sat}, as a function of the wavelength, λ.

3 Outlook and perspectives

In conclusion, starting from the Boltzmann equation in the weak coupling assumption, we have developed a novel set of hydrodynamical equations that are able to describe the ultrafast nonlinear dynamics of electrons in silver. We find that the collision-induced damping quenches for ultrashort pulses at infrared carrier wavelengths and peak intensities of the order of GW cm^{-2}, thus leading to absorption saturation. Although ultrafast nonlinear dynamics in metals have been experimentally investigated by several groups, typically these studies have focused on EM pulses with a time duration of > 100 fs. Furthermore, the inherently high dispersion of silver leads to efficient temporal broadening of the EM pulse over propagation, whilst intraband absorption saturation has not yet been observed. We conclude that dispersion management techniques and temporal plasmon solitons constitute promising strategies for counterbalancing temporal dispersion and enabling absorption mitigation in silver-based plasmonic devices, thus opening up novel avenues for the development of low-loss plasmonic circuits.

4 Appendix – derivation of hydrodynamical equations

In the Landau weak coupling assumption[41] we neglect electron–electron and electron–phonon collisions with large deflection angles while retaining all grazing-angle contributions, and we recast the RHS of eqn (1) into the so-called Fokker–Planck–Landau (FPL) diffusion and drift terms

$$(\partial_t f)_{\text{coll}}^{\text{el–el}} = \frac{C_{\text{ee}}}{m^2}\nabla_w\cdot\{[\mathcal{D}\cdot\nabla_w - 2\nabla_w\alpha]f\}, \tag{5a}$$

$$(\partial_t f)_{\text{coll}}^{\text{el–ph}} = \frac{C_{\text{ei}}}{m^2}\nabla_w\cdot\left\{\left[\mathcal{D}_0\cdot\nabla_w - 2\frac{m}{M}\nabla_w\alpha_0\right]f\right\}, \tag{5b}$$

where C_{ee} and C_{ei} are two constants that account for the electron–electron and electron–ion Coulomb interactions, $\mathcal{D}(\beta) = \nabla_w\nabla_w\beta$ and $\mathcal{D}_0 = \mathcal{D}(\beta_0)$ are diffusion tensors, and $\alpha(\boldsymbol{r},\boldsymbol{w},t) = \int d\boldsymbol{w}_1 f(\boldsymbol{r},\boldsymbol{w}_1,t)/|\boldsymbol{w}-\boldsymbol{w}_1|$ and $\beta(\boldsymbol{r},\boldsymbol{w},t) = \int d\boldsymbol{w}_1 f(\boldsymbol{r},\boldsymbol{w}_1,t)|\boldsymbol{w}-\boldsymbol{w}_1|$

are the so-called Rosenbluth potentials,[41] $\alpha_0 = \alpha(f_0)$, and $\beta_0 = \beta(f_0)$. We then solve eqn (1) and (5) by following the method of moments, which enables us to obtain a hierarchy of hydrodynamical equations that are fully equivalent to the FPLE.[41] Since we are only interested in the zero [electron density $\int d\boldsymbol{w} f(\boldsymbol{r},\boldsymbol{w},t) = n(\boldsymbol{r},t)$, such that $\int d\boldsymbol{r} n(\boldsymbol{r},t) = N$], first [current density $\int d\boldsymbol{w}\boldsymbol{w} f(\boldsymbol{r},\boldsymbol{w},t) = n(\boldsymbol{r},t)\boldsymbol{v}(\boldsymbol{r},t)$], and second [energy density $(m/2)\int d\boldsymbol{w}|\boldsymbol{w} - \boldsymbol{v}|^2 f(\boldsymbol{r},\boldsymbol{w},t) = (3/2)n(\boldsymbol{r},t)k_B T_e(\boldsymbol{r},t)$] moments, we truncate the hierarchy by neglecting higher-order moments, thus obtaining the solution

$$f(\boldsymbol{r},\boldsymbol{w},t) \simeq \frac{n(\boldsymbol{r},t)m^{3/2}}{[2\pi k_B T_e(\boldsymbol{r},t)]^{3/2}} e^{-\frac{m|\boldsymbol{w}-\boldsymbol{v}(\boldsymbol{r},t)|^2}{2k_B T_e(\boldsymbol{r},t)}}. \tag{6}$$

Thus, the moments $n(\boldsymbol{r},t)$, $\boldsymbol{v}(\boldsymbol{r},t)$, and $T_e(\boldsymbol{r},t)$ satisfy eqn (2), whereby the nonlinear expressions

$$\mathcal{F}_\gamma(\boldsymbol{r},t) = \frac{3{v_{T_0}}^3}{2v^2 v_T}\left[\mathcal{G}(v/v_T) - e^{-\frac{v^2}{{v_T}^2}}\right], \tag{7a}$$

$$Q(\boldsymbol{r},t) = \gamma_{th}\frac{M{v_{T_0}}^3}{2k_B v_T}\left[\mathcal{G}(v/v_T) - \frac{T_e {v_{T_0}}^2}{T_0 {v_T}^2} e^{-\frac{v^2}{{v_T}^2}}\right], \tag{7b}$$

are calculated analytically by the Rosenbluth potentials and the integration of every moment equation, where

$$\mathcal{G}(v/v_T) = \sqrt{\pi} v_T \operatorname{erf}(v/v_T)/2v, \tag{8a}$$

$$v_T(T_e) = \sqrt{2k_B(T_0/M + T_e/m)}, \quad v_{T_0} = v_T(T_0), \tag{8b}$$

$$\gamma_{th} = 2m\gamma/(m+M). \tag{8c}$$

Conflicts of interest

There are no conflicts to declare.

Acknowledgements

A.M. acknowledges the support received from the Rita Levi Montalcini Fellowship (grant number PGR15PCCQ5), funded by the Italian Ministry of Education, Universities and Research (MIUR), and from the Marie Curie Individual Fellowship OUTNANO (grant number 746774). A. C. acknowledges the U.S. Army International Technology Center Atlantic for their financial support (grant number W911NF-14-1-0315). C.C. acknowledges the H2020 QuantERA Quomplex project (grant number 731743). A.M. acknowledges useful discussions with A. V. Zayats, C. Rizza and A. Puglisi.

Notes and references

1 J. B. Khurgin, *Nat. Nanotechnol.*, 2015, **10**, 2–6.

2 R. H. Ritchie, E. T. Arakawa, J. J. Cowan and R. N. Hamm, *Phys. Rev. Lett.*, 1968, **21**, 1530–1533.

3 A. V. Zayats, I. I. Smolyaninov and A. A. Maradudin, *Phys. Rep.*, 2005, **408**, 131–314.

4 E. Ozbay, *Science*, 2006, **311**, 189–193.

5 M. Kauranen and A. V. Zayats, *Nat. Photonics*, 2012, **6**, 737–748.

6 F. Bonaccorso, Z. Sun, T. Hasan and A. C. Ferrari, *Nat. Photonics*, 2010, **4**, 611–622.

7 F. J. G. de Abajo, *ACS Photonics*, 2014, **1**, 135–152.

8 G. V. Naik, J. Kim and A. Boltasseva, *Opt. Mater. Express*, 2011, **1**, 1090–1099.

9 J. Kim, G. V. Naik, A. V. Gavrilenko, K. Dondapati, V. I. Gavrilenko, S. Prokes, O. J. Glembocki, V. M. Shalaev and A. Boltasseva, *Phys. Rev. X*, 2013, **3**, 041037.

10 G. V. Naik, V. M. Shalaev and A. Boltasseva, *Adv. Mater.*, 2013, **25**, 3264–3294.

11 U. Guler, V. M. Shalaev and A. Boltasseva, *Mater. Today*, 2015, **18**, 227–237.

12 J. D. Caldwell, L. Lindsay, V. Giannini, I. Vurgaftman, T. L. Reinecke, S. A. Maier and O. J. Glembocki, *Nanophotonics*, 2015, **4**, 44–68.

13 D. J. Bergman and M. I. Stockman, *Phys. Rev. Lett.*, 2003, **90**, 027402.

14 M. I. Stockman, *J. Opt.*, 2010, **12**, 024004.

15 M. A. Noginov, G. Zhu, M. Mayy, B. A. Ritzo, N. Noginova and V. A. Podolskiy, *Phys. Rev. Lett.*, 2008, **101**, 226806.

16 M. A. Noginov, G. Zhu, A. M. Belgrave, R. Bakker, V. M. Shalaev, E. E. Narimanov, S. Stout, E. Herz, T. Suteewong and U. Wiesner, *Nature*, 2009, **460**, 1110–1113.

17 A. Marini, A. V. Gorbach, D. V. Skryabin and A. V. Zayats, *Opt. Lett.*, 2009, **34**, 2864–2866.

18 P. M. Bolger, W. Dickson, A. V. Krasavin, L. Liebscher, S. G. Hickey, D. V. Skryabin and A. V. Zayats, *Opt. Lett.*, 2010, **35**, 1197–1199.

19 A. Marini and D. V. Skryabin, *Phys. Rev. A*, 2010, **81**, 033850.

20 I. D. Leon and P. Berini, *Nat. Photonics*, 2010, **4**, 382–387.

21 A. Marini, D. V. Skryabin and B. Malomed, *Opt. Express*, 2011, **19**, 6616–6622.

22 Y.-J. Lu, J. Kim, H.-Y. Chen, C. Wu, N. Dabidian, C. E. Sanders, C.-Y. Wang, M.-Y. Lu, B.-H. Li, X. Qiu, W.-H. Chang, L.-J. Chen, G. Shvets, C.-K. Shih and S. Gwo, *Science*, 2012, **337**, 450–453.

23 J. Y. Suh, C. H. Kim, W. Zhou, M. D. Huntington, D. T. Co, M. R. Wasielewski and T. W. Odom, *Nano Lett.*, 2012, **12**, 5769–5774.

24 A. Yang, T. B. Hoang, M. Dridi, C. Deeb, M. H. Mikkelsen, G. C. Schatz and T. W. Odom, *Nat. Commun.*, 2015, **6**, 6939.

25 P. Melentiev, A. Kalmykov, A. Gritchenko, A. Afanasiev, V. Balykin, A. S. Baburin, E. Ryzhova, I. Filippov, I. A. Rodionov, I. A. Nechepurenko, A. V. Dorofeenko, I. Ryzhikov, A. P. Vinogradov, A. A. Zyablovsky, E. S. Andrianov and A. A. Lisyansky, *Appl. Phys. Lett.*, 2017, **111**, 213104.

26 Y. Wu, C. Zhang, N. M. Estakhri, Y. Zhao, J. Kim, M. Zhang, X. Liu, G. K. Pribil, A. Alú, C. Shih and X. Li, *Adv. Mater.*, 2014, **26**, 6106–6110.

27 A. Marini and F. Biancalana, *Phys. Rev. Lett.*, 2013, **110**, 243901.

28 I. I. Smolyaninov, D. L. Mazzoni, J. Mait and C. C. Davis, *Phys. Rev. B: Condens. Matter Mater. Phys.*, 1997, **56**, 1601.

29 B. Lamprecht, J. R. Krenn, A. Leitner and F. R. Aussenegg, *Phys. Rev. Lett.*, 1999, **83**, 4421–4424.

30 A. Bouhelier, M. Beversluis, A. Hartschuh and L. Novotny, *Phys. Rev. Lett.*, 2003, **90**, 013903.

31 P. Ginzburg, A. V. Krasavin and A. V. Zayats, *New J. Phys.*, 2013, **15**, 013031.

32 E. Carpene, *Phys. Rev. B: Condens. Matter Mater. Phys.*, 2006, **74**, 024301.

33 N. Rotenberg, A. D. Bristow, M. Pfeiffer, M. Betz and H. M. Van Driel, *Phys. Rev. B: Condens. Matter Mater. Phys.*, 2007, **75**, 155426.

34 H. Baida, D. Mongin, D. Christofilos, G. Bachelier, A. Crut, P. Maioli, N. Del Fatti and F. Vallée, *Phys. Rev. Lett.*, 2011, **107**, 057402.

35 M. Conforti and G. Della Valle, *Phys. Rev. B: Condens. Matter Mater. Phys.*, 2012, **85**, 245423.

36 A. Marini, M. Conforti, G. Della Valle, H. W. Lee, T. X. Tran, W. Chang, M. A. Schmidt, S. Longhi, P. S. Russell and F. Biancalana, *New J. Phys.*, 2013, **15**, 013033.

37 A. V. Krasavin, T. P. Vo, W. Dickson, P. M. Bolger and A. V. Zayats, *Nano Lett.*, 2011, **11**, 2231–2235.

38 W. Li, B. Chen, C. Meng, W. Fang, Y. Xiao, X. Li, Z. Hu, Y. Xu, L. Tong, H. Wang, W. Liu, J. Bao and Y. R. Shen, *Nano Lett.*, 2014, **14**, 955–959.

39 I. D. Leon, J. E. Sipe and R. W. Boyd, *Phys. Rev. A*, 2014, **89**, 013855.

40 I.-Y. Park, S. Kim, J. Choi, D.-H. Lee, Y.-J. Kim, M. F. Kling, M. I. Stockman and S.-W. Kim, *Nat. Photonics*, 2011, **5**, 677–681.

41 R. Balescu, *Equilibrium and non-equilibrium statistical mechanics*, John Wiley & Sons, New York, 1975.

42 M. Scalora, M. A. Vincenti, D. de Ceglia, V. Roppo, M. Centini, N. Akozbek and M. J. Bloemer, *Phys. Rev. A*, 2010, **82**, 043828.

43 C. Cirací, E. Poutrina, M. Scalora and D. R. Smith, *Phys. Rev. B: Condens. Matter Mater. Phys.*, 2012, **86**, 115451.

44 M. Moeferdt, T. Kiel, T. Sproll, F. Intravaia and K. Busch, *Phys. Rev. B*, 2018, **97**, 075431.

45 W. Cai and V. Shalaev, *Optical Metamaterials*, Springer-Verlag, New York, 2010.

46 S. Link and M. A. El-Sayed, *J. Phys. Chem. B*, 1999, **103**, 8410–8426.

Faraday Discussions

DISCUSSIONS

Theory of hot electrons: general discussion

Javier Aizpurua, Francesca Baletto, Jeremy Baumberg, Phillip Christopher, Bart de Nijs, Preeti Deshpande, Yuri Diaz Fernandez, Laura Fabris, Simon Freakley, Sylwester Gawinkowski, Alexander Govorov, Naomi Halas, Romain Hernandez, Bartlomiej Jankiewicz, Jacob Khurgin, Mikael Kuisma, Priyank Vijaya Kumar, Johannes Lischner, Jie Liu, Andrea Marini, Reinhard J. Maurer, Niclas Sven Mueller, Matteo Parente, Jeong Y. Park, Stephanie Reich, Yonatan Sivan, Giulia Tagliabue, Laura Torrente-Murciano, Madasamy Thangamuthu, Xiaofei Xiao and Anatoly Zayats

DOI: 10.1039/C9FD90012H

Sylwester Gawinkowski opened discussion of the paper by Javier Aizpurua: How many electrons are transferred by one ultrashort pulse? If you are thinking about applications in electronics, then it seems that this number may be too small. Is simply increasing the pulse energy and therefore significantly increasing the number of such emitted electrons possible?

Javier Aizpurua responded: From our calculations, we find that the average number of electrons transmitted by each pulse, for these particular conditions, is around 0.1 e for ∼100 nm long nanowires. This is small in terms of conduction, but it is measurable, and indeed our experimental colleagues at Konstanz and Luxembourg seem to obtain precisely this amount for such a situation. One should be careful when increasing the energy of the pulse. If you refer to the frequency of the pulse, then resonant effects can play a role and the current established is more complex. If you refer to an increase in power (amplitude increase), plasmonic gaps can handle a limited amount of power before melting, therefore there are limits to this too. We believe that the current rate is a good start for producing a more ambitious roadmap for ultrafast pulse-driven coherent optoelectronics.

Niclas Sven Mueller remarked: I am excited about the large gap sizes that you consider in your paper. These are much larger than 1 nm, where one would usually expect the onset of a tunnelling current for CW illumination. What is the scaling of the electron emission current with the gap size? What pump fluences are needed for a detectable net current?

Javier Aizpurua replied: You are right. Our range of distances is large enough so as to avoid the tunneling regime, and all of the current considered in our discussion is associated with ballistic transport of the electrons emitted by the strong optical field (note that this is the Fowler-Nordheim regime, which is completely different from photoemission by multiphoton absorption). The dependence of the current on the gap size can be derived from the quiver amplitude that we outline in our discussion: $X_{\mathrm{q}} = E_0/\omega^2$. This expression basically states that the quiver amplitude depends linearly on the amplitude of the field, which can be connected with the square root of the power applied in the pulse multiplied by an enhancement factor of the gap, G. As the gap distance, d_{gap}, is increased, the local field enhancement is reduced as $1/d_{\mathrm{gap}}$. Therefore, there will be a point at which the field induced in the gap will produce a very small quiver amplitude compared to the gap size ($X_{\mathrm{q}} \ll d_{\mathrm{gap}}$), and thus the electrons will have very little chance to reach the electrode on the other side of the gap. The pump fluences needed to start detecting the current in the gap require a few picojoules per pulse, which generate fields of a few V nm^{-1} in gaps of about 6 nm.

Mikael Kuisma asked: I have a question about your 2D equations of motion for a 3D system, where the kinetic energy in the *z*-direction is omitted and the *k*-point integration is replaced with varying occupation numbers. Is this an approximation of an exact formulation? I am asking this because we have observed that plasmons are sensitive to the electron-hole topology close to the Fermi level, *i.e.* long wavelength density fluctuations from holes just below the Fermi surface to electrons just above.

Looking at Equation 2 in your paper (DOI: 10.1039/c8fd00158h), I can see that one has omitted the kinetic energy towards the z-direction, which naively corresponds to infinite effective mass for electrons in the z-direction since $\lim_{m_z \to \infty} \frac{1}{2m_z} p_z^2 = 0$. With the *z*-kinetic energy, the quadratic band dispersion due to the *z* translation symmetry of each band is given as $E_{\mathbf{k}_z n} = E_{\Gamma n} + \frac{1}{2m_e} k_z^2$. Now, in the irreducible response function, the poles are OK since the eigenvalue differences are conserved:

$$\chi_0(\mathbf{r}, \mathbf{r}', \omega) = \sum_{nn'} \int_0^\infty \mathrm{d}k_z \left(f_{k_z n} - f_{k_z n'}\right) \frac{\psi_n^*(\mathbf{r})\psi_n(\mathbf{r})\psi_n(\mathbf{r}')\psi_{n'}^*(\mathbf{r}')}{\omega + E_{k_z n} - E_{k_z n'} + i\eta}$$

$$= \sum_{nn'} \int \mathrm{d}k_z \left(f_{k_z n} - f_{k_z n'}\right) \frac{\psi_n^*(\mathbf{r})\psi_n(\mathbf{r})\psi_n(\mathbf{r}')\psi_{n'}^*(\mathbf{r}')}{\omega + E_{\Gamma n} - E_{\Gamma n'} + \underbrace{\left(\frac{1}{2m_e}k_z^2 - \frac{1}{2m_e}k_z^2\right)}_{0} + i\eta}$$

However, what about the occupation numbers when each band crosses the Fermi level at a different k_z? Upon such a crossing, the nature of excitations changes, as excitations from this band are no longer available, but this band can now accept excitations.

After taking a closer look at the equations of motion, I am now convinced that your approach is not only exact, but actually very intriguing. The exactness is actually easy to see; if we plug in the zero Kelvin Fermi distribution,

$f_{k_z n} = \Theta\left(E_f - \left(E_{\Gamma n} + \frac{1}{2m_e}k_z^2\right)\right)$, we find that the Heaviside function changes integration boundaries and then results in your occupation numbers:

$$\chi_0(\mathbf{r}, \mathbf{r}', \omega) = \sum_{nn'} \int_{\sqrt{2(E_f - E_{\Gamma n})m_e}}^{\sqrt{2(E_f - E_{\Gamma n'})m_e}} \mathrm{d}k_z \frac{\psi_n^*(\mathbf{r})\psi_n(\mathbf{r})\psi_n(\mathbf{r}')\psi_{n'}^*(\mathbf{r}')}{\omega + E_{\Gamma n} - E_{\Gamma n'} + i\eta}$$

$$= \sum_{nn'} \left(\sqrt{2(E_f - E_{\Gamma n})m_e} - \sqrt{2(E_f - E_{\Gamma n'})m_e}\right) \frac{\psi_n^*(\mathbf{r})\psi_n(\mathbf{r})\psi_n(\mathbf{r}')\psi_{n'}^*(\mathbf{r}')}{\omega + E_{\Gamma n} - E_{\Gamma n'} + i\eta}$$

My mistake is that I have been too used to thinking in terms of Grassman manifolds, *i.e.* the manifold of density matrices with eigenvalues of 1 and 0 only. Naturally, since your occupation values each have a distinct value, this correctly accounts for excitations from a partly occupied state to a less occupied state. In these 2D effective equations, instead of one Grassman manifold per k-point to represent the possible states of the system, one needs only one generalized flag manifold (*i.e.* the space of all density matrices with the given non integer occupations). Maybe this idea relating to the topology of electronic excitations could be used to analyze the collectivity of plasmons under continuous symmetries.

Javier Aizpurua answered: To the best of our knowledge, the equations of motion are correct. Although we numerically solve the 2D time-dependent problem, its formulation results from the reduction of the full 3D TDDFT equations in the Kohn–Sham form. Indeed, the invariance of the problem along the z-direction allows for a trivial factorization of the effect of the z-motion as $e^{ikz}e^{-i\left(\frac{k_z^2}{2}\right)t}$. This leads to the corresponding weight factors, as given in the paper.

Johannes Lischner enquired: The method you present combines a potential from Maxwell's equation with a solution of the Kohn–Sham equation. It appears as if this procedure double counts electron–electron interactions; they are first included through the dielectric function in the Maxwell equation solver and then again as the Hartree and exchange–correlation terms in the Kohn–Sham solver. Would it not be more accurate and simpler to solve the Kohn–Sham equations without electron–electron interactions? This would possibly allow for the modelling of larger systems and thus avoid the need to artificially localize the electron in an infinite wire (instead of modelling the full bow-tie structure).

Javier Aizpurua responded: In the current system discussed in our paper, the system is small enough to perform directly and exclusively the full quantum many-body calculation where an incident Gaussian pulse has been considered as the external perturbation that induces the response of the system. However, I mentioned in my presentation that, in practice, one often needs to treat larger systems (for example bow-tie antennas of several hundreds of nanometers), and in that case, one could implement an alternative strategy by decomposing the problem into two separate calculations: one involving the classical solution of the induced fields at the gap for the whole large system, and a second quantum calculation, more reduced to the area where current is produced (a few square nanometers), to properly account for all of the quantum aspects of the dynamics of the transport along the optical cycle.

We are pretty sure that in this procedure there is no double counting of the field enhancement. If one implemented TDDFT calculations with the aim of reproducing experimental data, the incident pulse could be set such that the self-consistent field in the junction would be equal to the field obtained by solving classical Maxwell's equations for the actual experimental field profile. However, it should be noted that in our current discussion the incident pulse considered in the TDDFT calculations is a perfectly Gaussian one, and not an experimental one.

We have tested the use of this splitting of the problem into classical and quantum calculations, and made sure that we were not double counting the electron interactions and their corresponding fields. In particular, we compared the results of the full calculation and the split calculations for a small system, and found excellent agreement. I presume that the option you suggest (removing the electron–electron interactions) would rather jeopardize the results, and would not provide such an improvement in the size of the doable systems. Therefore, we prefer to adopt this split strategy which has thus far proven to be valid. We are implementing such an approach and expect to publish the results soon.

Naomi Halas questioned: In the experiments you are describing in your theoretical study of nanogaps, is a small bias voltage being applied, and is that not responsible for the currents being measured?

Javier Aizpurua replied: In this purely theoretical paper we consider a plasmonic gap illuminated with a Gaussian sub-cycle pulse. No constant fields were applied. Thus, the effects reported in this work result from the sensitivity of the optical field emission to the CEP of the incident pulse, and the photocurrents measured in the experiments are produced by the incoming photons. It would however be interesting to bias the cavity, and observe to what extent the current can be controlled by the application of such an external bias.

Jeremy Baumberg queried: I am looking to gain an intuitive understanding as to why, when the maximum field is in one direction, the current is in the other direction. A positive voltage should result in a positive current. Is there a sign convention issue, and are you perhaps looking at electron flux, not current?

Javier Aizpurua answered: Yes, the sign is an issue of sign convention. We are looking at the electron current density. Thus, the electric field oriented in the positive direction of the x-axis produces electron currents flowing in the negative direction of the x-axis. This positive electric field amplitude will extract electrons from the right-hand side electrode with negative velocity towards the left-hand side electrode, as the charge of the electrons is negative. This is what we show in Fig. 4 of our paper (DOI: 10.1039/c8fd00158h), where every polarization of the field moves electrons in the opposite direction within the gap.

Andrea Marini remarked: Your paper reports experimental evidence of optical rectification by a plasmonic gap supported by a nice and convincing theoretical explanation of the process, which arises from the highly nonlinear potential inside the gap. Can you please comment on the opportunities offered by plasmonic gaps concerning high-harmonic generation (HHG)? Did you observe HHG in your theoretical calculations or in the experiments carried out?

Javier Aizpurua responded: We would like to stress that no experimental results are reported in our theoretical work in the current paper. I only mentioned that this theoretical work can explain the results of experiments that are still to come. In the current work, we focused on the effect of the plasmonic gap on the generation of DC current as a response to an ultrafast single-cycle pulse. However, in the past we have also studied the non-linear response of plasmonic gaps as a response to continuous optical illumination, and yes, we have identified a strong non-linear response including HHG. Furthermore, we have identified that the non-linear response can be associated either with the intrinsic non-linearity of the metal for large gap separations, or with the non-linear tunneling current generated at very narrow gaps. You can find information on this in the papers by G. Aguirregabiria *et al.*[1] and Marinica *et al.*[2]

1 G. Aguirregabiria, D. C. Marinica, R. Esteban, A. K. Kazansky, J. Aizpurua and A. G. Borisov, *Phys. Rev. B*, 2018, **97**, 115430.
2 D. C. Marinica, A. K. Kazansky, P. Nordlander, J. Aizpurua and A. G. Borisov, *Nano Lett.*, 2012, **12**, 1333–1339.

Jacob Khurgin said: Thank you for your great talk. I have a few short questions:

(1) Have you considered the effect of polarization of light?

(2) Is it possible to trace the electrons in the lateral (yz) plane after they exit the metal?

(3) What if you use two metals with different work functions ('rectenna')?

(4) Can you relate your work to that carried out early on by Baranova *et al.*[1] on directional photoemission for the fields where the average cube is not zero?

1 N. B. Baranova and B. Ya. Zel'dovich, *J. Opt. Soc. Am. B*, 1991, **8**, 27–32.

Javier Aizpurua replied: (1) In our description we consider the polarization of light to be perpendicular to the interfaces forming the gap, as this is the component that maximizes the field emission. If one was to polarize the pulse along the walls of the gap, the ejected electrons would not end up at the opposite antenna arm. Moreover, field enhancement would be non-operative, meaning that a higher amplitude of the incident pulse would be required to reach the optical field ionization regime.

(2) Yes, in our calculations the axis of symmetry is the z-axis, where the dynamics are assumed to be translationally invariant. However, in the y direction the electrons follow a motion dictated by the boundary conditions across the gap and the local fields, thus this aspect of the motion is well considered in our calculations. In practice, a gap antenna is usually cut in the z direction (usually a few tens of nanometers), therefore the relevant motion when it comes to establishing the net current in the device can be assumed to be exclusively due to the transverse motion, with a relatively small contribution of lateral components.

(3) The use of metals with different work functions would be interesting. We have not modelled this explicitly, but one could anticipate that the effect will be either to enhance or mitigate the effect of the main peak of the incident pulse, depending on whether the polarization of the pulse peak extracts electrons from one or the other metal.

(4) The process that you refer to is particularly focused on the interference of two photons at frequency ω and one at frequency 2ω, which can produce directional emission originating from the non-linear response of the material. In our case, the source of directionality is clearly the polarization of the incident field, as pointed out above.

Anatoly Zayats enquired: There may be 2 different kinds of rectification in the illuminated tunnel junction. If the tunnelling junction consists of 2 different metals, the current–voltage curve will be asymmetric, which will automatically result in the rectification of the electric field of the incident light. On the other hand, in the considered geometry under ultrashort pulse illumination, nonlinear optical effects may be important, leading to nonlinear rectification, *viz.* second-order susceptibility. Which rectification did you consider?

Javier Aizpurua answered: The sources of optical rectification, *i.e.* a DC component of current arising from an AC excitation, that you mention are the standard ones, originating either from asymmetry in the material, which unbalances the flow of current at both sides of a heterogeneous metallic junction, or from an association with the second-order non-linear tensor response of the material. Additionally, for ultrasmall gaps there would be a source of nonlinearity associated with the tunneling current itself.[1] In our discussion, however, the source of the DC component of the current is the optical field emission (also quoted as strong AC field ionization) associated with the strong asymmetry of the single-cycle pulse itself. Different to multiphoton ionization, the optical field emission and thus the electron transport across the gap is sensitive to the carrier envelope phase (CEP). This CEP sensitivity is well known in atomic physics and in strong field emission from metallic tips. Thus, our 'source' of rectification is different and relies on the single-cycle temporal profile of the pulse.

1 G. Aguirregabiria, D. C. Marinica, R. Esteban, A. K. Kazansky, J. Aizpurua and A. G. Borisov, *Phys. Rev. B*, 2018, **97**, 115430.

Francesca Baletto asked: Does the geometry (*e.g.* curvature) of the cylinders affect your results? How about in the case of a square nano-antenna in front of a cylindrical one? Do you think there is the opportunity to design plasmonic gaps?

Javier Aizpurua responded: As far as the junction is symmetric, the results reported here will not change qualitatively since the underlying physics of the strong optical field emission are very robust, provided that the field enhancement in the junction is sufficient to drive the system in this regime of strong field emission. For weak fields, one deals with photoemission by multiphoton absorption which is not sensitive to the CEP of the pulse. For asymmetric junctions like the one you have pointed out, *i.e.* a square nano-antenna in front of a cylindrical one, the net electron transfer will ultimately be influenced by the different geometries and, *a priori*, both regimes, *i.e.* optical field emission and multiphoton photoemission, will occur. Another aspect is the influence of the macroscopic shape (larger scale structures). The actual shape of the nanoantenna forming the gap sets the plasmonic resonance, and thus it will establish the spectral region in which the incident pulse might behave differently. The response of the gap near the plasmonic resonance introduces a phase shift in the induced field at those frequencies, and thus its spectral proximity to the energy of the incident pulse is a very relevant aspect. In this respect, the design of the shape and separation distance of the gap can be engineered to produce different spectral responses. In the current work we have focused on a small cylindrical system in order to focus on the physics of strong-field emission, rather than on the macroscopic plasmonic resonance.

Bart de Nijs opened discussion of Stephanie Reich's paper: In your system, you rely on *s*-polarized light to excite dark modes in your self-assembled gold structures. However, if you send polarized light through a high (0.9) NA objective there will also be a nearly equal amount of *p*-polarized light after the objective. How does this affect your system, especially when multiple layers are probed where *p*-polarized light can potentially excite a vertical dipole in the stacked gold structures?

Stephanie Reich replied: We worried about this point as well and therefore performed experiments with 20× objectives (NA of 0.25). There is no difference between the spectra measured with 100× and 20× objectives despite the much larger spot size and smaller NA, as shown in Fig. 6 of our paper (DOI: 10.1039/c8fd00149a). We thereby ruled out a significant contribution of *p*-polarized radiation. We also note that the *z*-polarized plasmonic excitations are at much higher frequency in the bilayer system.

Preeti Deshpande remarked: In this work you have shown that the generation of dark plasmons is possible with a particle size of 40–50 nm. Will the system work if we increase the particle size to 140–150 nm in diameter? Does the field retardation in large size particles assist or inhibit dark plasmon generation?

What is the role of spacing between the particle layers? How will the system behave if we reduce the separation to sub-nm gaps?

Will the system work for other plasmonic materials, *e.g.* silver?

Stephanie Reich answered: The dark mode can also be excited in particle bilayers made up of nanoparticles with diameters of 100–200 nm. The activation mechanism has no upper limit. If the size of the nanoparticles themselves becomes comparable to the light wavelength, we expect the excitation of layer modes that combine the quadrupole excitation of the individual particles. We expect a very rich excitation spectrum in this situation. For the excitation and extraction of hot electrons however, smaller nanoparticles are preferred. Energetic electrons inside large nanoparticles will not reach the surface before losing their energy. It is thus of limited use for energy harvesting, catalysis and other applications that we have in mind for hot electrons.

The spacing between the particles creates plasmonic hotspots. We expect them to contribute to dephasing and thus increase the number of highly excited hot electrons. Reducing the nanoparticle separation below 0.5 nm will likely lift the localization of the electrons on the nanoparticles. The system will then behave like a piece of bulk metal.

The excitation of dark modes will occur for any plasmonic material (silver, aluminum, highly doped semiconductors *etc.*). There is nothing material specific in our reasoning. Of course, the frequency of the dark mode will strongly depend on the choice of material.

Yuri Diaz Fernandez said: I have two questions. (1) Have you tried different shapes of particles?

(2) What is the role of defects and how reproducible and scalable is your sample preparation method?

Stephanie Reich responded: (1) No, we have not. The plasmonic excitations that we consider are eigenmodes of the two-dimensional nanoparticle lattice. Other shapes will make it much harder to produce films with long range order.

2) Defects and stacking faults did not affect our experiments. We scanned large areas of the sample and reproducibly obtained identical spectra (see Fig. 6 of our paper). Most recent films extended over hundreds of microns. Increasing the size even further is the subject of ongoing work.

Bartlomiej Jankiewicz asked: Isn't the stabilizing layer you used for the Au NPs influencing the efficiency of hot-electron generation? If you use, for example, Cu NPs instead of AuNPs, would there be a big drop in hot electron generation?

Stephanie Reich replied: Our key point is that the use of dark instead of bright modes for hot electron generation eliminates an undesired relaxation channel. Radiative relaxation, as occurs for bright modes, will carry away large quanta of energy (1–3 eV depending on the wavelength) without producing any hot electrons. Dark modes exhibit less radiative decay, *i.e.* more excited electrons remain available for hot electron generation. Any other parameter that affects hot electron generation, such as the choice of material, stabilizing layers, crystalline quality and so forth, will equally affect the dark and bright plasmon modes.

Jeremy Baumberg remarked: These are beautiful structures! I am trying to clarify your nomenclature, and am wondering about the use of the word 'dark'. This system is far from a dimer, where the dark mode has the field transverse to each nanoparticle and not in the gaps. Can you show an additional plot of the field distribution at the low energy mode resonance in your bilayer system, in order to see where the light is concentrated? This would help in the understanding of whether it is concentrated in horizontal and vertical gaps (and thus rather similar to existing gap modes, *e.g.* in dimer-on-mirror work).

Essentially, since dark modes can arise from symmetry or poor coupling, is this quasistatically dark or some other sort of 'dark'? How does the symmetry of light absorption to light emission work here? If you can excite the mode strongly then surely it must give about half absorption and half emission?

Stephanie Reich answered: The near-fields are concentrated in the vertical gaps. Electromagnetic hotspots form between the nanoparticles within the layer. For the plasmons considered in our paper, there is no sizable near-field between two layers.

Your second question points towards the fascinating aspects of ordered nanoparticle crystals. In our paper we defined 'bright' as plasmon modes with a finite dipole moment and 'dark' as plasmon modes with a vanishing dipole moment. This corresponds to the standard optical selection rules that you will find in textbooks on the symmetry analysis of crystals and which are widely applied to plasmonic systems as well. Standard symmetry analysis assumes that (i) the electromagnetic field has only a dipolar component and (ii) the light wavelength is infinite compared to the extension of the unit cell. While the second assumption is well justified for normal solids and molecules, it breaks down for the nanoparticle system we considered here, due to the large nanoparticle size. In our paper we performed a symmetry analysis for optical absorption by nanoparticle crystals under the assumption that the wavelength of light is identical to the translational periodicity along the crystal *c*-axis (hexagonal close-packed structure). Light at this wavelength excites gerade representations, among them the E_{2g} mode discussed in our paper. In the framework of your question, the mode is 'quasistatically' dark. In fact, we should expect rich phenomena in the optical spectra of ordered nanoparticle arrays due to field retardation, because plasmon excitation will depend on the propagation direction in addition to polarization direction.

A dipole-active bright mode (quasistatically bright mode) will radiate light in any direction, thus making radiative decay one of the prime relaxation channels for such a plasmon. The dark E_{2g} mode discussed in our paper can only radiate normal to the gold surface, as explained in the previous paragraph. The strongly limited solid angle of radiation suppresses radiative decay.

Naomi Halas commented: How does the dark mode you observe in your bilayer nanoparticle films relate to hot electrons, and what are you measuring that relates to hot electron generation in these materials?

Stephanie Reich responded: The dark plasmon is a source of excited electrons. Essentially, all processes for dark plasmons that lead from plasmon excitation to hot electron generation are identical to optically active bright modes, except for the missing radiative decay. This is an important advantage of dark modes, because radiative decay carries away large quanta of energy (1–3 eV depending on the wavelength) without producing any excited electrons. Tailoring hot electron extraction from dark modes follows the same principles as for bright modes, *i.e.* increasing surface scattering, producing hot spots as places for dephasing and so forth.

We measured the generation of hot electrons using transient absorption spectroscopy. This technique excites plasmon modes with a strong pump pulse and follows the decay of the temperature rise on a ps time scale using a probe pulse. The number of hot electrons created with an infrared pump pulse increased strongly in the bilayer compared to that in a monolayer, which is due to the excitation of the dark plasmon in the bilayer system. We have recent data that shows an increase of 50–60% of hot electron generation by the dark mode in comparison to that of the bright plasmon mode.

Romain Hernandez commented: How much more efficient is it to generate hot electrons by the dark mode rather than the bright mode?

In your experiment, a gold mirror is placed below the bilayer in order to enhance absorption. How can you extract hot carriers from the bilayer efficiently?

Stephanie Reich replied: We do not have a final answer to the first question. Our most recent (but still preliminary) transient absorption measurements showed an increase of 50–60% in hot electron population between spots with several layers (dark mode) and a monolayer (bright mode). This compares the generation rate at the same pump energy. On the other hand, we estimated that for 50 nm particles, the decay rate that contributes to hot electrons may be on the order of 130 meV. When we relate this value to the measured FWHM of the dark and bright modes, a factor of two appears to be an upper limit for the generation of hot electrons. This question is particularly difficult to answer for gold, because the bright mode overlaps too strongly with the interband transitions. Simulations and experiments on silver or aluminum layers would be highly beneficial.

With regards to the 'bilayer above a gold mirror' configuration, we introduced this configuration in order to further increase absorption by dark plasmons. It is not a required configuration for our experiments. Even in this configuration, the gold bilayer absorbs light through the generation of dark plasmons. The electrons may be extracted from the upper surface of the bilayer system.

Reinhard J. Maurer opened discussion of the paper by Priyank V. Kumar: You have observed direct hot-electron transfer from the metal to the adsorbate with probabilities of between 1–2%. One would imagine that the contribution of local density-of-states around the Fermi level that arises due to the CO molecular states would also be in the range of a few percentage points of the total DOS. From your current understanding, do you think that the electronic coupling magnitude or the local DOS will be more relevant in defining the effectiveness of particle-to-molecule electron transfer? If the local DOS dominates, one would assume that this ratio decreases with particle size. What dependence with respect to particle size have you observed or would you expect?

Priyank Vijaya Kumar responded: This is a good question. The probability values depend on two quantities: matrix elements and the joint density of states (DOS). From our current understanding, both these quantities play a role. However, we do not currently know what quantity dominates in which system. Having said that, in the few systems we have investigated, we have observed that the ratio decreases with particle size, as you mentioned. This suggests that the local DOS could have an upper hand in these systems. Therefore, one could simply compute the joint density of states as you mentioned and obtain a preliminary estimate for direct transfer.

Jeong Y. Park queried: In your theoretical model, you are focusing on direct hot electron transfer. However, from your results, 98% of the electrons are dissipated; perhaps some of them would go through indirect hot electron transfer. Can you comment on the possible probability of indirect hot electron transfer?

Priyank Vijaya Kumar answered: It is true that some of the 98% of the total electrons excited through plasmon decay could transfer to the molecule constituting indirect hot-electron transfer. It is difficult at this stage to estimate a probability since we have to account for electron–electron and electron–phonon scattering in our calculations, which has not yet been done.

Anatoly Zayats asked: Regarding plasmon lifetime, do you know how the lifetime depends on the size of nanoparticle and type of material? Will it be shorter for gold nanoparticles?

Priyank Vijaya Kumar replied: As the metal nanoparticle size increases, the plasmonic character of the excitation becomes stronger due to an increase in the number of electrons in the system. As such, we observe that the lifetime of the plasmon increases.

As for gold nanoparticles, the d-band is closer to the Fermi level than in the case of silver nanoparticles. This leads to stronger screening of the plasmon and a faster decay. Therefore, the lifetime is observed to be shorter.

Simon Freakley enquired: With one CO molecule you show that the distribution of transitions results in 1% direct injection to the CO. Will this scale linearly with increasing CO coverage on the particle and do you expect any cooperative effects with surface coverage that could influence the mechanism?

Do you see any difference in the behaviour of the CO depending on where it is sitting? Is one position more favourable for dissociation, activation or desorption upon illumination?

Priyank Vijaya Kumar responded: We certainly expect the probability of direct transfer to increase with increasing CO coverage. It is difficult to say whether this increase would be linear since there are many effects that influence the mechanism, such as the orientation of the molecule with respect to the applied electric field, hybridization of different CO molecules, interaction between CO molecules *etc.* As we have shown in the paper, we do observe different direct-transfer probabilities for different CO binding sites. From these probabilities alone, it would be far-fetched to state that a particular position would be more favorable for chemical transformation upon illumination since there are multiple other processes that occur which eventually result in a transformation.

Yuri Diaz Fernandez asked: Have you tried any polarization dependence calculations for this system? Can you also comment on the adsorption energy of CO on each site and how these energies relate to the experimental scenario?

Priyank Vijaya Kumar answered: Since our paper presents a preliminary investigation, we placed the CO molecule along the direction of the applied electric field (x-axis). The probability of the direct-transfer process certainly depends on the orientation of the CO molecule with respect to the applied electric field. If we change the direction of the applied electric field, *i.e.* along the y-axis, the probability is close to zero since the plasmon is perpendicular to the CO molecule and the relevant dipole transition cannot be excited.

Phillip Christopher queried: With the real time approach you used here is it possible to characterize which adsorbate states were involved in direct interactions with the plasmon oscillation and decay process?

Priyank Vijaya Kumar replied: Yes, the developed method is able to point out which adsorbate states are involved in the direct transfer transition.

Matteo Parente remarked: In your simulations you depict your system as a cluster of Ag on the surface of which one molecule of CO is adsorbed. In reality, however, Ag oxidizes. Therefore, from a more realistic point of view, your system would be a Ag cluster with a thin layer of Ag_2O and a molecule of CO adsorbed on top of the Ag_2O. Are you planning to take the silver oxide layer into account in your simulations, or do you have any evidence for explaining why the presence of this layer would not influence the mechanisms of charge transfer?

Priyank Vijaya Kumar responded: The oxide layer of Ag would certainly affect the mechanisms of charge transfer. Our simulations represented a proof-of-concept to show that we can now indeed capture direct-transfer transitions in plasmonic catalysis, and to shed some light on the physics of this mechanism. In order to start comparing with experimental investigations, thorough knowledge of the system under consideration is necessary, as you have mentioned. We hope that our computational tool can help in this regard in the future.

Laura Torrente-Murciano said: My question follows previous discussion about the effect of the degree of CO coverage and the effect of the contact area between the plasmonic and semiconductor particles. Based on that, would you maximise the direct transfer by covering the plasmonic particle with a semiconductor in a core–shell manner? As you mentioned that there is an optimum semiconductor size in the example presented in your paper, what do you think would be the optimum size of the shell?

Priyank Vijaya Kumar answered: Based on our current understanding, I would expect that we would maximize the direct transfer by forming a core–shell structure like you describe if the plasmonic properties of the core metallic nanoparticle remain unaffected. I'm not sure that we mentioned that there is an optimum semiconductor size in our paper. The optimum shell thickness depends on a few things. On the one hand, increasing shell thickness improves hybridization with the core and thus the direct transfer, but this effect should flatten out after a particular thickness is reached. On the other hand, increasing thickness could deteriorate the plasmonic properties of the core particle, which would result in a lower number of hot carriers and thus less charge transfer to the semiconductor. One has to weigh up these two factors and estimate an optimal shell thickness.

Johannes Lischner remarked: In your calculations, you describe the direct transfer of hot electrons into adsorbate states. This transfer is significant because of the strong chemical interactions between the nanoparticle and the adsorbate resulting in hybridized orbitals. Your calculations do not include indirect transitions resulting from the thermalization of excited electrons which are initially not resonant with the molecular orbitals. However, it appears likely that these indirect transitions are also significant because of the strong chemical interactions between adsorbate and nanoparticle.

Priyank Vijaya Kumar replied: I agree with your statement. It is likely that the indirect transfer rate increases due to strong chemical interactions between the adsorbate and nanoparticle. In order to verify this, one could further propagate our system and simulate the direct transfer by taking into account electron–electron and electron–phonon scattering.

Madasamy Thangamuthu communicated: (1) What happens to the HOMO–LUMO gap (7.9 eV) of CO after adsorption on Ag? Will it decrease to a smaller value?

(2) As you can see that silver can generate electrons with a maximum energy of 4 eV, if the HOMO–LUMO energy of CO does not change significantly then it is difficult to understand the energetics of the charge transfer. Please can you comment on this?

(3) Would this system work under a CW laser instead of a pulsed laser? Did you try any simulations using CW?

(4) What are the dimensions of the Ag nanoparticles? Does the number 147 represent the size?

Priyank Vijaya Kumar communicated in reply: (1) The HOMO and LUMO states of CO become hybridized with the states of the Ag nanoparticle. This is observed to slightly decrease the HOMO–LUMO gap of CO upon adsorption.

(2) What matters is the LUMO-Fermi level gap rather than the HOMO–LUMO gap when it comes to hot-electron-transfer based reactions. Here the LUMO-Fermi level gap is less than 4 eV, therefore there is potential for electrons to get transferred.

(3) We did not attempt any simulations using a CW laser, but we expect the physics to remain the same.

(4) The number 147 corresponds to the number of Ag atoms in the nanoparticle, and so is related to the size. The 147-atom Ag particle is about 1.5 nm in diameter.

Xiaofei Xiao returned to the discussion of the papers by Javier Aizpurua and Stephanie Reich: I found that in your projects, there exist small gaps. Indeed, more and more publications are focusing on very small gaps. For example, some of them are around 2 nm to 0.5 nm. However, the commonly used methods in simulations such as FDTD and FEM are those for classical electromagnetism. I have two questions. First, when should we consider quantum effects in our system? Second, when both quantum effects and classical electromagnetic phenomena exist, how do we carry out the simulation?

Javier Aizpurua answered: This question is very relevant for small gaps, as you say. We introduced in the past[1] a way to introduce quantum effects in ultrasmall gaps, making those quantum effects compatible with larger scale classical calculations. We referred to that effective method as a quantum-corrected model (QCM), which is nowadays extensively used to address quantum effects of tunneling and so on, in large plasmonic antennas. In the plasmonic gaps that we consider in this paper, the separation distance is maintained between 3 and 6 nm, therefore we certainly minimize quantum effects in this situation. One might still have non-local effects that can be properly addressed with effective model calculations, or by re-normalizing the local classical response with the Feibelman parameter. Nevertheless, for a gap of 3 nm, such as that considered here, classical calculations describe the optical response very well.

1 R. Esteban, A. G. Borisov, P. Nordlander and J. Aizpurua, *Nat. Commun.*, 2012, **3**, 825.

Stephanie Reich responded: Quantum effects become noticeable for gaps below 0.5 nm. With the 2 nm gaps used in our system, we remain in the regime of classical electromagnetism. There are hybrid approaches for *ab initio* calculations of plasmonic systems that consider quantum effects, but they remain restricted to smaller particles.

Jacob Khurgin opened discussion of the paper by Alexander O. Govorov: I have some inconvenient questions and remarks:

(1) Where does the energy of the absorbed plasmon go if not in hot carriers?

(2) I am very familiar with the Lindhard formula which describes the real part of epsilon adequately but uses phenomenological parameters to describe the absorption.

(3) Consider a simple picture of the standard photo-effect in a vacuum. The incident light is normal to the metal surface. Therefore no Landau damping can take place as it requires a longitudinal field. Nevertheless, hot carriers are clearly generated and the Fowler formula works well including the cut-off, without ever mentioning Landau damping. How does it work if the only carriers are generated near the Fermi surface?

Alexander Govorov replied: I think that your questions are very standard, and I do not see them as inconvenient. My answers are as follows:

(1) The total energy is conserved. Light energy turns into electronic energy. Electrons create phonons and, therefore, the electronic energy turns into heat. This is the standard picture. This mechanism gives the Drude formula, which is widely used and reported in almost all textbooks on solid state physics.

(2) The Lindhard formula is one of the key results in the physics of metals. It gives a good quantitative description without dissipation. With dissipation, it gives a correct physical picture.

(3) The first line of our paper (DOI: 10.1039/c8fd00145f) is "The generation of energetic electrons is an effect occurring in any plasmonic nanostructure." My picture, and the picture given in textbooks (Drude model), is the following; the plasmon is made up mostly of many low-energy electrons (I call these carriers Drude electrons), since our system is weakly excited. A number of high-energy electrons are always present due to non-conservation of linear momentum near the surfaces (surface-scattering mechanism of plasmon decay) and due to defects/phonons *via* second-order processes. In large nanocrystals, the number of Drude electrons is much larger than the number of non-thermalized (photo-generated) high-energy electrons. In my later responses to Mikael Kuisma and Johannes Lischner, I provide a detailed description of the electronic structure of a plasmon, together with some textbook examples, and so I will not repeat it here. It is important to ask questions that are directly related to the discussed effects and not to mix different physical regimes. The Fowler law is for UV light, $\lambda < 250$ nm, whilst I discuss the visible, red and IR regimes (>500 nm) where the Drude formula is well applied. For UV photons (>5 eV), it is known that the inter-band transition effects are very strong. The vertical transitions in the k-space with the excitation of hot electrons are possible for the UV photons because of the band structure effects. Other reasons for the generation of hot electrons in the case

described in your question include the presence of a non-planar surface and defects/impurities and phonons *via* 2nd-order processes. As an additional note, for the effect of the generation of hot electrons near the surface, we use the term 'surface-induced decay of plasmons', as is also used in many related papers. This effect was calculated by Kreibig in 1975.[1]

1 L. Genzel, T. P. Martin and U. Kreibig, *Z. Phys. B: Condens. Matter*, 1975, **21**, 339–346.

Mikael Kuisma remarked: I think I know where the missing energy is (I am comparing Fig. 1e of ref. 1 with lots of low energy electrons to Fig. 3 of ref. 2, where the electron and hole distributions are equal, but are shifted exactly with the photon energy $\hbar\omega$). It is in the energy of the electric field of the plasmon. We have carried out real-time TDDFT simulations of plasmons in nanoparticles using coherent light pulses.[3] When the plasmon is formed we see these Drude electron–hole transitions near the Fermi level. They are in collective motion and all of them contribute to the induced surface density, hence causing a strong electric field. The energy stored in this field is the missing energy. To further scrutinize this, since the random phase approximation is a large part of the TDDFT response, after a quarter cycle there is a different number of these Drude excitations due to non-particle-conserving operators in RPA (aa and $a^\dagger a^\dagger$), and the energy is still conserved even when the electric field energy goes to zero since it is in the current.

On the other hand, we have found that the hot-carrier transitions are generated by this plasmon-induced electric field, and also that these transitions form such a strong background density of states that the hot-carrier generation may be treated using Fermi's golden rule. As long as the collective motion of the Drude electrons is present, the plasmon will get absorbed into the hot-carrier transitions. Again, the energy is conserved, since these hot-carrier transitions have the energy of the photon. The fact that these Drude electrons appear in Professor Govorov's hot-carrier distributions and in the non-energy conservation brought up by Professor Khurgin seems to reflect the presence of plasmons.

1 L. V. Besteiro, X.-T. Kong, Z. Wang, G. Hartland and A. O. Govorov, *ACS Photonics*, 2017, **4**, 2759–2781.
2 R. Sundararaman, P. Narang, A. S. Jermyn, W. A. Goddard III and H. A. Atwater, *Nat. Commun.*, 2014, **5**, 5788.
3 D. B. Tanner, *Optical Effects in Solids*, Cambridge University Press, London, 2019.

Alexander Govorov responded: Thank you for your excellent comments. I think that I agree with some of your central comments, and those comments are fully consistent with the interpretation I gave after my talk.

I would like to make some remarks about plasmons, the Drude model and Drude electrons. The response of electron gas in real metals (like Au and Ag) is well described by the Drude dielectric function in the red and IR intervals. In the blue interval, we need to add the inter-band transitions. Overall, the Drude dielectric function describes very well the intra-band transitions. The Drude dielectric function comes from the response of mobile excited electrons and, in my papers, I call such excited electrons 'Drude carriers'. Another suitable name for such electrons could be 'intraband excitations'. As I will explain and illustrate below, such electrons and holes (or electron–hole pairs in the Fermi sea) have, of course, low-excitation energies and are generated near the Fermi level.

The above picture of excited electrons and holes with low-excitation energies (e–h pairs in the vicinity of the Fermi level) is the standard and well-established picture given in solid-state textbooks as well as in many research papers. Below, I will provide some examples from 'classical' textbooks.

Electrons in metals in the first approximation are free carriers, and it has been proven that a free electron cannot absorb a photon quantum $\hbar\omega$. The reason for this is that the linear momentum cannot be not conserved for such an absorption process and, therefore, the transition matrix element is equal to zero. Therefore, the application of Fermi's golden rule for a free electron without dissipation or scattering gives zero. However, the electric field still acts strongly on the electron; an electron becomes accelerated in the classical electric field, and this acceleration effect is very strong. In the *k*-space (momentum space), the Fermi sea of electrons becomes shifted.[1] Therefore, the Drude model was proposed, which describes the acceleration and motion of free electrons in metals. The acceleration of the center of mass of a free electron in an electric field can be described using the classical Newton equation or with the quantum equations of motion, momentum and coordinate (the results will of course be similar). This is the spirit of the Drude conductivity that is the central equation in the transport theory of an electron gas in metals and doped semiconductors.[1,2]

Fig. 8.4 in a recent book by David B. Tanner[1] illustrates the effect of acceleration of the Fermi sea in the presence of an external field. When the electrons get accelerated and then occupy excited states of momenta above the Fermi level, they of course scatter by phonons and defects. This scattering limits the effect of acceleration. *Via* scattering by phonons, the weakly-excited electrons (we assume weak fields and a linear regime) lose energy; this scattering process is described well by the transport rate given by Fermi's golden rule, when an electron makes the following transition with the creation of one phonon: $E1 \rightarrow E2 + \hbar\omega_{\text{phonon}}$. For this, one can refer to a famous textbook by Gerald. D. Mahan[3] or many other textbooks. Fig. 8.4 in David Tanner's book[1] shows the scattering process in the excited (accelerated) Fermi sea.

The above physical picture produces excited electrons near the Fermi level, and I refer to such carriers as Drude electrons in our paper. In our paper, we used the quantum master equation for the density matrix to derive the related distributions of Drude electrons. However, the Boltzmann equation gives the same results for the Drude electrons and holes, since the process is quasi-classical in its nature. David Tanner's book[1] offers a nice description for the derivation within the Boltzmann formalism and show the solution. The excited electrons are located near the Fermi sea.

In the quantum approach used in our paper, we also obtained a number of non-thermalized high-energy electrons due to scattering by the surface and in hot spots. Near the surface and in the plasmonic hot spots, the linear momentum of the electron is not conserved, and we observe a number of high-energy electrons with energies of $E_F + \hbar\omega$. In this case, the electron can absorb the photon quantum directly. This effect is well known and is given by the theory presented by Kreibig et al;[4] this theory describes the decay of plasmons due to surface scattering (generation of high-energy electrons near the surface).

In 3D boundless systems, a small number of high-energy electrons appear due to the weak second-order processes involving phonons and defects (non-conservation of momentum due to phonons and defects). However, in 3D, the majority of plasmonic electrons are Drude-like and have small excitation energies. Textbooks typically show only the low-energy (Drude) electrons that give the Drude model, because these electrons and currents govern the overall picture.[1,2,5]

One important example of a solution of the Boltzmann equation for the plasmonic wave can be found in a book written by Charles Kittel.[5] This example is based on the Boltzmann formalism and it again shows the distribution of excited electrons in the vicinity of the Fermi level. The reason why the excited electrons should be near the Fermi level is simple; the plasmonic system is weakly excited, *i.e.* in the linear regime.

With regards to your comment about the missing energy, the energy of the electromagnetic field turns into kinetic and potential energy of the electrons. In the first step, an electron gets accelerated and receives extra kinetic energy. The electron then scatters and gives energy to the lattice. The total energy is, of course, conserved. You also mentioned the "energy of the electric field of the plasmon" which looks like potential energy, and which is one form of energy in the system. Your observation is good and is fully consistent with the physics of the plasmon and with the 3D picture of a plasmon. In large nanocrystals, the majority of excited electrons will be Drude electrons with small energies near the Fermi level. I am looking forward to seeing your results as a published paper or as a posted pre-print. I also agree with your comment "They are in collective motion..."

With regards to your comment that "The energy stored in this field is the missing energy", here we need here 2 energies: the kinetic energy (KE) and potential energy (PE). In a harmonic oscillation, these energies are equal. Regarding your comment beginning "To further scrutinize this...", here one should be careful. Any many-body Hamiltonian has only the terms with the conservation of the number of particles, like $a^{\dagger}a$ and $aa^{\dagger}$, or $a^{\dagger}a^{\dagger}aa$ (see the textbooks written by Mahan[3] or Platzman and Wolff[6]). Finally, with regards to your comment beginning "The fact that these Drude electrons...", yes the Drude electrons form the plasmon. The total energy is conserved at all times perfectly, of course. The 3D case of the plasmon is well studied and documented in textbooks.

1 D. B. Tanner, *Optical Effects in Solids*, Cambridge University Press, London, 2019.
2 F. Wooten, *Optical Properties of Solids*, Academic Press, New York and London, 1972, ch. 4.
3 G. D. Mahan, *Many-Particle Physics*. Kluwer Academic/Plenum Publishers, New York, 3rd edn, 2000, sections 8.1–8.3, Scattering by impurities and phonons.
4 U. Kreibig and M. Vollmer, *Optical Properties of Metal Clusters*; Springer, Berlin, 1995, vol. 25.
5 C. Kittel, *Quantum Theory of Solids*, John Wiley and Sons, New York, 2nd edn, 1987, ch 16: Electrodynamics of metals, section: Mathematical Theory of the Anomalous Skin Effect, eqns. 28, 33 and 35. These equations tell us that the non-equilibrium energy-distribution function under illumination, which is obtained from the dynamic Boltzmann transport equation, is $\delta(E - E_F)$, which is a good approximation for low temperature plasma ($E_F \gg kT$).
6 P. M. Platzman and P. A. Wolff, *Waves and Interactions in Solid State Plasmas*, Academic Press, New York, 1973.

Yonatan Sivan said: I would like to comment on energy conservation in the problem of the non-equilibrium electron distribution in a metal nanostructure under steady-state illumination, due to the various contradicting claims that have been made during this discussion.

Most theoretical studies of this problem consider only the electron system. Its energy increases due to photon absorption and decreases due to energy transfer to the phonons. However, the usual approach whereby the latter effect is described using a fixed e–ph collision rate discards a lot of the relevant physics. Indeed, the rate of energy transfer from the electrons to the phonons depends on the temperature difference between the systems (as we demonstrated in our earlier theory paper,[1] the non-equilibrium component does not affect this too much). This difference depends on the rate of energy transfer from the phonons to the environment, which itself depends on the particle size and shape. Thus, using a fixed e–ph collision rate for different particles and/or different illumination levels introduces severe quantitative errors, which invalidate the claims made regarding the dependence of hot electron numbers on these parameters. We believe that the only paper that has taken this issue into account is our own. Attempts to manually adjust the electron temperature can partially amend those quantitative errors, but are unlikely to yield self-consistent temperatures, unless they adopt the proper relations between the electron and phonon temperature. This can be done using the 2 temperature model that arises from our formulation (DOI: 10.1039/c8fd00147b).

Thus, overall, the energy goes to three different channels; part of it goes into raising the electron temperature, another part gets transferred to the phonons (and heats them up as well) and part establishes the non-equilibrium distribution of the electrons. As we demonstrated in our theory paper,[1] the amount of power going to the last channel (hot electrons) is incredibly small in comparison with the amount of power that results in the heating of the electrons and phonons. For pulsed illumination, which has been studied in many more theory papers, these considerations are not important; the energy of the electron system is not conserved, because the time scale for energy transfer to the environment is much slower than the pulse duration or energy transfer to the phonons.

1 Y. Dubi and Y. Sivan, 2018, arXiv:1810.00565 [physics.optics].

Alexander Govorov responded: During the meeting there were active discussions on the conservation of energy when light interacts with an electronic system. As I described after my presentation, the total energy should of course be conserved. Optical energy turns into electronic energy and the electronic energy turns into thermal heat *via* phonon emission. Our theoretical calculations conserve the total energy. The interesting question at the meeting was regarding the appearance of large numbers of low-energy electrons (Drude electrons) in the CW regime of excitation, which is the intrinsic/fundamental property of the plasmon wave. Textbooks and research papers on 3D plasmons describe this very well. These Drude electrons give the Drude dielectric constant that we use every day in our calculations of the photonic properties of metal nanostructures. In my responses to other questions, I have described the mechanism of the Drude response in detail; this is due to the acceleration of an electron in the field. The low-energy electron–hole pairs (Drude electrons) create the plasmonic currents and they are not connected by the energy $\hbar\omega$ (Fermi's golden rule with the delta function cannot be directly applied here), since the excited electrons (Drude electrons) are created *via* a different mechanism, *i.e. via* acceleration (the Drude model and picture). The acceleration can be calculated quasi-classically (using the Boltzmann equation or Newton equation) or *via* quantum mechanics (quantum density matrix); the results will be the same (please see my previous comments and discussion of some textbook examples). Regarding your comment "As we demonstrated in our theory paper, the amount of power going to the last channel (hot electrons) is incredibly small in comparison…"; my results are different. In our calculations, we do see essential numbers of nonthermal hot electrons and also essential power absorbed to create such electrons. Importantly, the above comment does not agree with the well-recognized theory reported by Uwe Kreibig.[1] Kreibig's theory has been applied to many experiments, and there are hundreds of related theoretical papers that have used and developed further the concept of Kreibig's decay term for the plasmon (the decay is due to surface scattering). One can refer to citations of the original Kreibig papers and his book.[1]

The number of non-thermalized hot electrons with high energies ($\sim\hbar\omega$ above the Fermi level) can be essential for small nanoparticles and for nanocrystals with hot spots; the energy rate for the creation of non-thermalized hot electrons with high energies can be comparable to the Drude heat (creation of nearly-thermal electrons) for small nanoparticles and for nanocrystals with hot spots. For small nanoparticles, this was shown by Kreibig. For nanocrystals with hot spots, this has been shown in our papers. In our paper, we also gave (perhaps for the first time) the distributions of non-equilibrium electrons including the hot (non-thermalized) high-energy electrons and the Drude (nearly-thermalized) low-energy electrons near the Fermi level. The Drude (nearly-thermalized) low-energy electrons near the Fermi level give the plasmonic currents and charges. The hot (non-thermalized) high-energy electrons give the plasmon decay due to surface scattering, owing to non-conservation of linear momentum at the surface and in hot spots. Our paper (DOI: 10.1039/c8fd00145f) reports results for the linear regime when the temperature increase of the lattice is assumed to be small. Correspondingly, the electron gas is slightly out of equilibrium (the CW regime of continuous driving) and the electron system is being excited, having Drude electrons (plasmonic currents) with thermal excitation energies.

1 U. Kreibig and M. Vollmer, *Optical Properties of Metal Clusters*; Springer, Berlin, 1995, vol. 25.

Johannes Lischner queried: Textbook derivations of Fermi's golden rule point out that energy is not conserved at short times. When light is absorbed by an interacting many-electron system, the relevant time scale is set by the quasiparticle lifetimes. If those are short, it appears as if energy is not conserved. Where does it go?

Alexander Govorov answered: Thank you for your question. This comment is in relation to time-dependent perturbation theory or Fermi's golden rule applied to a quantum system. Indeed, Fermi's golden rule assumes conservation of energy in the form of the delta function at long times. The probability of a transition has a multiplier of $\delta(E1 - E2 + \hbar\omega_{\text{photon}})$ or similar. The total energy of the whole system (electrons + electromagnetic field) is conserved at all times, of course. We see this always in quasi-classical calculations (Boltzmann theory), classical calculations (Drude model) and quantum calculations (a master equation for the density matrix). The light energy is converted into the energy of electrons. Electrons then get accelerated and emit phonons, causing heating of the lattice. Therefore, we have 3 forms of energy (light energy, energy of electrons and heat) and the total energy is conserved. Physically, the plasmon oscillation comes from the free electrons moving in an oscillating external field, and this is a classical process. For example, we immediately observe that the plasmon frequency is a classical parameter. In other words, the acceleration of free electrons in an electric field is a classical process and, therefore, the quantum Fermi's golden rule is not needed to describe such a process of acceleration of a free electron. However, Fermi's golden rule is indeed involved in the kinetic coefficient for the e–phonon scattering rate; when we consider the emission of a single phonon by an accelerated electron in a Fermi sea, we indeed use conservation of energy during the collision event: $\delta(E1 - E2 - \hbar\omega_{\text{phonon}})$. This picture can be seen for example in Fig. 8.4 in David Tanner's book on the physics of metals.[1]

1 D. B. Tanner, *Optical Effects in Solids*, Cambridge University Press, London, 2019.

Jacob Khurgin enquired: According to your Lindhard-based theory the wavevector q is small. However, in order to conserve energy the phonons must be involved. Phonons have a large wavevector. Therefore, how is momentum conserved?

Alexander Govorov replied: I previously replied to related questions. Again, the total energy of the crystal + the field is conserved. This mechanism can be found in textbooks. It is called free electron absorption and is where the related Drude model is derived from. Regarding your query about the small wavevector q, unfortunately the Lindhard formula is not mine, and is instead a famous result reported by Lindhard. However, yes, the momentum of light q is small. With regards to your comment beginning "However, in order to conserve energy...", photons are indeed involved and have small momenta. In terms of how momentum is conserved, electrons give their momenta to phonons and to the crystal *via* defects.

Priyank Vijaya Kumar said: From your talk and ongoing discussion, I understand that there are two types of electrons: Drude electrons when plasmons are present, and hot electrons that are generated when a plasmon decays. We know that hot electrons are useful, however, can we use the Drude electrons to carry out any useful work *e.g.* chemical reactions? If so, is there any evidence for this?

Alexander Govorov responded: Thank you for this excellent question, which is a fundamental one. Plasmonics exhibits large losses when we try to apply it to devices. Yes, plasmons in nanocrystals are mostly made from low-energy electron–hole pairs in the Fermi sea. This is a picture reported in textbooks. However, we can generate a number of highly-excited electrons with energies $\sim E_F + \hbar\omega$ above the Fermi level when we strongly break the conservation of linear momentum of electrons near surfaces and hot spots. The use of thermalized electrons (with low energies after a pulse) and Drude electrons (under CW illumination) is not easy, but is possible. We need to extract these electrons somehow. One possibility is to construct a Schottky junction with a low barrier energy under applied voltage and try to extract such low-energy electrons. A previous paper has reported in this direction for a TiN–TiO_2junction (please see

Figure 4d in this paper).[1] Alternatively, one could utilize these electrons in some kind of thermoelectric device.

1 A Naldoni, U. Guler, Z. Wang, M. Marelli, F. Malara, X. Meng, L. V. Besteiro, A. O. Govorov, A. V. Kildishev, A. Boltasseva and V. M. Shalaev, *Adv. Opt. Mater.*, 2017, **5**, 1601031.

Mikael Kuisma communicated in response to Priyank Vijaya Kumar: As long as the Drude electrons are collective, the answer is no with single electron–hole-pair generation, but perhaps with electron–electron scattering this could be possible. The situation is similar but opposite in BCS superconductors. Instead of collective repulsive Coulomb interactions, in superconductors there is a collective attractive interaction and one cannot break the coherence of the ground state with thermal excitations, because one needs to also supply the remaining energy cost in order to break the attractive coherence equal to the superconducting gap.

For coherent plasmon excitation, one cannot have a plasmon scattered into low-energy electron–hole transitions without gaining even more energy from breaking the coherence. In any case, the outcome of any process scattering the plasmon should conserve energy. Such processes exists with Drude electrons as one of their elements, *i.e.* multi electron–hole processes exist in electron–electron scattering.

Alexander Govorov further commented: These are some interesting comments, and we can discuss them further. However, regarding your comment "As long as the Drude electrons are collective, the answer is no...", I think that we can use the energy of low-energy Drude electrons if we can extract them or if we can transfer the energy from them. Of course, it is much more challenging to use low-energy electrons then high-energy electrons. Please refer to my direct response to the question posed by Dr Kumar.

Jeremy Baumberg remarked: In the pulsed regime I am worried about the hot electron yield without damaging the samples. In realistic systems we find that even at 1–2 orders of magnitude less than 10^8 W cm^{-2}, the atoms move around in the plasmonic gaps and this can be directly seen. There is evidence for this in other experiments also, and it is much worse for short pulses than long pulses or CW. Furthermore, would your model recommend always using many pulses or CW if you are doing photochemistry?

Alexander Govorov replied: Thank you for your comment/question. As far as I know from the comments of my experimental collaborators, these levels of optical flux (10^8–10^9; $\sim$80 fs pulse) do not noticeably change the crystal structure. The positions of the plasmonic peaks and the absorption spectra themselves do not change after these intense fs pulses. I believe that they mentioned this to me, but one is welcome to ask them personally. Overall, the systems under study[1,2] look robust. The systems have no real gaps, but they have spacers. One system[1] has spacers made of TiO_2 and sapphire. The other system[2] has a spacer of sapphire and a thin polymer layer on the surface of the Ag cube. The spacers are relatively thick ($\sim$8 nm), but the length of the plasmonic gap is long ($\sim$100 nm). It could be that the motion of single atoms does not occur in reference 1 since there is not an open metal surface in the gap. In reference 2, perhaps, the atomic motions cannot change the spectra.

1 H. Harutyunyan, A. B. F. Martinson, D. Rosenmann, L. K. Khorashad, L. V. Besteiro, A. O. Govorov and G. P. Wiederrecht, *Nat. Nanotechnol.*, 2015, **10**, 770–774.
2 M. E. Sykes, J. W. Stewart, G. M. Akselrod, X.-T. Kong, Z. Wang, D. J. Gosztola, A. B. F. Martinson, D. Rosenmann, M. H. Mikkelsen, A. O. Govorov and G. P. Wiederrecht, *Nat. Commun.*., 2017, **8**, 986.

Jacob Khurgin opened discussion of Yonatan Sivan's paper: When one considers the excitation of hot carriers under normal conditions, it is very simple to see that at any given time we have on average less than one plasmon per nanoparticle and, therefore, at any given time we have just one hot carrier pair generated which loses its energy *via* electron–electron scattering. My question is how can we then talk about 'electron temperature'? In my view, the way to proceed is just to trace what happens with the energy after a few collisions, and the dependence will clearly be different from the Fermi- or Boltzmann-like exponential.

Yonatan Sivan responded: (1) Our formulation (nor any of the other theoretical descriptions of the problem we are familiar with) does not employ second quantization, so we do not talk about 'plasmons' and their decay, but rather about a direct excitation of e–h pairs by a photon.

(2) From the standpoint of statistical mechanics and within the semi-quantum approach we adopt (where we do not discretize the electron levels and numbers), the number of high energy non-thermal electrons is indeed extremely small.

Mikael Kuisma further commented: I think that Professor Khurgin raises a good point. Even though there is only one hot electron excitation on average and not a statistical distribution, one still needs to treat it as a quantum mechanical superposition. Say, after a coherent laser pulse, a nanoparticle system is in quantum superposition with the ground state and lots of many-body wavefunctions with one hot electron excitation, the ground state is still the most probable. Therefore, the hot electron distribution is not statistical, but rather quantum mechanical with typical interference of amplitude effects.

Now, at some point for the practical purposes of simulating photoreactions, one would like to collapse this superposition of hot electrons and perform, for example, surface hopping with single hot electron excitation of the system. However, the quantum superposition argument goes even further to a Schrödinger's cat-like scenario. Consider a nanoparticle with N reactants, which after a coherent light pulse is either in state A or B. The final state of the full electron and nuclear wavefunction is a superposition of 2^N states (neglecting degrees of freedom other than the state of the reactants). In practice, this is of course computationally impossible, so the challenge for the surface science and quantum chemistry community is to find an answer to the question "when do you collapse the wave-function so that you can simulate these systems in practice?"

Yonatan Sivan answered: This question is about aspects of the problem that we did not address. I will nevertheless try my best to answer it. The formulation we adopted is semi-quantum, in the sense that the photon absorption term in our

Boltzmann equation is derived from Fermi's golden rule and is incoherent, *i.e.* it is proportional to $|E|^2$ (this is different to the work of Andrea Marini (DOI: 10.1039/c8fd00153g), where this term is purely classical: coherent and linear in E). This ensures that integer quanta of electromagnetic energy, indeed $\hbar\omega$, are being absorbed, in-line with the photoelectric effect. This causes the creation of the non-thermal, high energy (so-called 'hot') electron shoulders that are seen in our distribution plots (See Fig. 2(c) in DOI: 10.1039/c8fd00147b). We also employed a similar formulation for the e–ph interactions. We believe that for particles bigger than a few nm this approach captures the essential aspects of the problem. Indeed, a more complete quantum mechanical formulation of this problem was adopted by the Govorov group, where they employ a density matrix formulation and discretize all electron energy states. They showed that the semi-quantum form we adopted reproduces nicely the 'hot' electron shoulders, obtained in the more complete quantum mechanical formulation for particle sizes as small as 2 nm (at least when looking at more than one particle at a time, as is done in almost all experiments).

Jacob Khurgin commented in response to Mikael Kuisma: I agree with this. However, where I see the problem is that after a couple of scatterings the hot carriers will be redistributed and this distribution will not be Fermi-like. I agree with the superposition description. What I find problematic is introducing some kind of effective electron temperature.

Naomi Halas said: There is no difference experimentally in most cases, photocatalysis being one of those cases, between a hot non-thermal electron distribution and a hot thermal electron distribution. However, it appears that theorists are making a distinction between the two and it seems as though you are only calling 'hot electrons' the electrons in a hot non-thermal electron distribution. Is this the case?

Yonatan Sivan replied: We find this question rather surprising. After all, the latest paper from the Halas group is titled "Quantifying hot carrier and thermal contributions in plasmonic photocatalysis".[1] As we understand it, the title (and contents of the paper) implies a very clear goal: to distinguish between reaction rates under external heating (responsible just for the 'thermal contribution', *i.e.* thermocatalysis) and under laser illumination (in which case, according to the authors, there is an additional contribution of the high energy, non-thermal 'hot carriers/electrons', *i.e.* photocatalysis). The nomenclature is explained in more detail in a later response to Jeremy Baumberg, and the difference in distributions is shown specifically in Figs. 1 and 2 of our paper (DOI: 10.1039/c8fd00147b). The methodology of these works was to try to reproduce the temperature distribution in a reaction created by laser illumination as accurately as possible, such that by subtracting the thermocatalytic contribution one can isolate the effects associated with the non-thermal ('hot') electrons, as the title of the 2018 Science paper[1] clearly implies. This goal and methodology is the same as that of some previous papers by the same group[2,3] as well as by other groups.

We strongly believe that it is important to distinguish between thermal and non-thermal effects, as there is also a mechanistic difference between them (as described in the papers by the Halas group).[1,2] If the effect is purely thermal, then the energy for the reaction comes from the thermal energy of the substrate (vibrations), although electrons can still play a role in the catalytic process (*e.g.* by reducing the reaction barrier). However, even if the reaction takes place through an electron-charging reaction path,[2] the electrons required for this path need to have high energy. In this case, one can ask: do these high-energy electrons that participate in the reaction come from the tails of the Fermi distribution, or do they come from a non-thermal distribution? Put differently, is the number of high-energy electrons available for the reaction higher because the distribution is at a sufficiently high temperature, or because the distribution is non-thermal (and hence there is the addition of high-energy electrons to the thermal distribution)?

This is not a semantic question, but a question of orders of magnitude, as there is a very small number of high-energy electrons in thermal distributions, but this number can increase by many orders of magnitude when there are non-thermal 'hot electrons'. Therefore, if there were high-energy non-thermal ('hot') electrons participating in the reaction, one would expect a huge orders-of-magnitude increase in the reaction rate; unfortunately this has never been observed experimentally. If it were high-energy thermal electrons, then the reaction rate would be indistinguishable from the regular thermal reaction path (because it is determined by the same Arrhenius function). This last point, that the experimental data can be described by an Arrhenius plot, was exactly what we showed in the manuscript referred to by Professor Halas.[4] This demonstrates that her claims of quantifying the different contributions of thermal effects and non-thermal electrons are not justified.

In any case, all of the above implies that the distinction Professor Halas mentions in her question has in fact been adopted by her team as well, and not just by theorists. We believe, unlike Professor Halas, that there are experimental ways to distinguish between them, *e.g.* by looking at photoemission from the metals, single particle spectroscopy and temperature measurements (such as those carried out by the Gross group, Maier group, Dionne group, *etc.*), or by carrying out clever control experiments (such as those carried out by Jie Liu's group). It is now up to experimentalists to demonstrate unambiguously that a given experiment shows one and not the other. As we demonstrated in our poster and arXiv paper,[4,5] this seems to not be the case for many past experiments, including Professor Halas' own attempts.

1 L. Zhou, D. F. Swearer, C. Zhang, H. Robatjazi, H. Zhao, L. Henderson, L. Dong, P. Christopher, E. A. Carter, P. Nordlander and N. J. Halas, *Science*, 2018, **362**, 69–72.
2 S. Mukherjee, F. Libisch, N. Large, O. Neumann, L. V. Brown, J. Cheng, J. Britt Lassiter, E. A. Carter, P. Nordlander and N. J. Halas, *Nano Lett.*, 2013, **13**, 240–247.
3 S. Mukherjee, L. Zhou, A. M. Goodman, N. Large, C. Ayala-Orozco, Y. Zhang, P. Nordlander and N. J. Halas, *J. Am. Chem. Soc.*, 2014, **136**, 64–67
4 Y. Sivan, I. W. Un and Y. Dubi, 2019, arXiv:1902.03169 [physics.chem-ph].
5 Y. Sivan, J. Baraban, I. W. Un and Y. Dubi, Comment on "Quantifying hot carrier and thermal contributions in plasmonic photocatalysis", *Science*, 2019, **364**, eaaw9367.

Alexander Govorov commented further: Thank you for these interesting comments, which should be useful for the current discussion. Regarding your comment beginning "There is no difference experimentally in most cases...", this is a very good point to discuss. I think that this strongly depends on the situation and conditions. If the excitation is broad-band (solar-like spectrum or similar), the difference between the hot non-thermal electrons and the hot thermal electrons can indeed be small. However, I think that for monochromatic optical excitation, the difference can be very strong. Hot non-thermal electrons under weak excitation (such as in photodetectors or photochemistry) have a threshold energy above the Fermi level, $E = E_{\mathrm{Fermi}} + \hbar\omega$. In contrast, hot thermal electrons have a smooth Boltzmann energy distribution. If we have an injection process into a barrier (TiO_2 for photochemistry) or to a surface molecule level with a well-defined energy, the above two distributions will give very different spectral responses. One example is the Fowler law that has the threshold photon energy, $\hbar\omega > E_{\mathrm{barriers}}$. With regards to your comment "it appears that theorists are making a distinction...", yes, our theoretical papers distinguish between non-thermalized and thermalized hot carriers. Indeed, we (in the papers where I am a co-author) often refer to non-thermalized hot carriers as simply 'hot carriers'. This is indeed misleading, and I have commented on this at many conferences. In an attempt to somehow correct this situation, we have explained the term 'hot carriers' in our recent work.[1,2,3] I guess that the terms 'hot non-thermalized carriers' and 'hot thermalized carriers' could be considered, or simply 'non-thermalized carriers'.

1 In ref. 2 is the following paragraph: "However, what are usually termed "hot" electrons fall into two disparate populations of excited carriers: "nonthermal" electrons with an initial stepwise distribution extending up to the photon energy from the Fermi level and possessing an indefinable temperature, and "thermal" electrons with a quasi-equilibrated Fermi–Dirac distribution and an electronic temperature elevated above that of the surrounding lattice."

2 M. E. Sykes, J. W. Stewart, G. M. Akselrod, X.-T. Kong, Z. Wang, D. J. Gosztola, A. B. F. Martinson, D. Rosenmann, M. H. Mikkelsen, A. O. Govorov and G. P. Wiederrecht, *Nat. Commun.*., 2017, **8**, 986.

3 DOI: 10.1039/c8fd00145f

Andrea Marini further responded to Naomi Halas: Regarding my paper (DOI: 10.1039/c8fd00153g), yes, in this case the electron distribution is out-of-equilibrium and non-thermal since, in addition to the increased electron temperature, the mean velocity of the electron plasma distribution is finite and oscillates following the external driving electromagnetic field. In our paper we show that the mean velocity can reach high values that are sufficient to generate collision quenching, which is the main reason for the predicted saturation of absorption.

Jeremy Baumberg remarked: I would first like to comment on Professor Halas' question regarding the question of what to call 'hot'; can we not get a hot Fermi thermalized distribution just by heating? The argument is that the material system falls apart for bulk heating. However, having hot-spots at this temperature keeps it viable, but allows spot catalysis. Is this scalable? Do you refer to this as hot electron effects?

I have a further question. I like your use of Occam's razor, aiming to explain large numbers of phenomena just by heating if that is possible. Your work sets a challenge to experimental groups, in that it suggests that there are many highly localised hot spots in a complex nanoparticle/nanostructured sample, where high temperature thermal catalysis operates. Since catalysis exponentially depends on barrier heights, gross measurements pick out these hot spots only (rather like the original problem of single molecule SERS). Can you suggest any experimental approaches to quantify the number and characteristics of these hot spots?

Yonatan Sivan replied: Firstly, with regards to 'hot' *vs.* heat; as I understand things, 'hot' means non-thermal, namely, a distribution (rather than a single carrier) that differs from a Fermi–Dirac (FD) distribution (which by definition is thermal). The distribution in the latter case is defined by the temperature of the electron system. In that sense, we have inherited/adopted a highly confusing nomenclature; I suggest (and tend) to stick to thermal/non-thermal.

Now, for pulsed illumination, a non-thermal distribution is created upon absorption, but is then converted into a thermal distribution within 500fs to a few ps (depending on who you ask). There might then still be aspects of the problem that people confusingly refer to as 'hot'—be it the difference between the electron and the lattice temperatures (bound to disappear within the first few/tens of ps; I like calling this situation quasi-equilibrium; electron and lattice systems are equilibrated within themselves but not with each other), the differences between the lattice and environmental temperatures (a situation I like to call internal equilibrium), or the differences in temperature between different parts of a (macroscopic, multi-particle) sample. All of these purely thermal situations are different from what I would call complete/external equilibrium, where all systems and all regions of space have the same temperature. I believe that the literature is coming into agreement on the differences between the first, non-thermal scenario (which requires a microscopic, even quantum mechanical, description) and the other thermal scenarios, which can all be modelled using 'lumped', averaged, and even classical heat equations. Certainly, the attempts reported in recent papers for distinguishing between photocatalysis and thermocatalysis (see recent work by Halas, Maier, Cortes and Schlücker, as well as perhaps that by Dionne and Giessen) are aimed exactly at distinguishing between what I call thermal and non-thermal effects. In that sense, I think that calling the latter 'hot' electron effects is anywhere between unintended to an intended way to keep things vague. I hope that the outcome of this meeting will be to try to clarify and standardize the nomenclature.

Finally, with regards to your question about suggestions of experimental approaches, I have already advocated several times for single particle experiments and wavelength dependent studies (searching for a potential drop of the non-thermal electron driven effect when going beyond the $\hbar\omega$ wide non-thermal electron shoulders). Other suggestions were raised by Guillaume Baffou at NFO in August 2018. Among other things, he mentioned the different scaling of non-thermal and thermal effects with beam size and/or particle densities. His work is not yet published, however I believe it is being reviewed.

Naomi Halas addressed Yonatan Sivan and Jeremy Baumberg: Why do you think that a purely thermal model that predicts temperatures well above the melting temperature of the metal is a valid one?

Yonatan Sivan answered: This question refers to our temperature-shifted Arrhenius model which is, it should be noted, not a part of our *Faraday Discussions* paper, but rather was derived after the submission of that paper. It is described in a recent arXiv paper[1] that has been submitted for publication. The model is nothing but a typical Arrhenius model for the rate of chemical reactions, namely

$$R = R_0 \exp\left[\frac{-E_a}{k_B T(I_{inc})}\right]$$

where the reaction temperature $T(I_{\text{inc}})$ depends linearly on the incoming intensity I_{inc} *via* heating induced by photon absorption. Importantly, due to the exponential sensitivity of the reaction to the temperature, the temperature that results from the fitting of the data corresponds to the maximum temperature inside the reactor. Indeed, the calculations carried out in that paper (as well as the measurements reported in several other papers, *e.g.* a recent paper from Jie Liu's group)[2] show a rather significant temperature inhomogeneity inside the reactor.

In our paper we initially applied this model to 2 papers by the Halas group[3,4] and one by the Linic group.[5] First, we showed that the temperature measured in those experiments was lower than the actual temperature of the reaction. Second, we used the experimental data to deduce the maximal temperature rise of the reactor. In those three papers, the deduced temperature rise was limited to a few tens of degrees, well below the melting temperature of the metal nanoparticles. The model fits the experimental measurements perfectly with the bare minimum number of fit parameters.

In the last part of our arXiv paper[1] to which Professor Halas refers, we applied the temperature-shifted Arrhenius model to the 2018 *Science* paper by the Halas group.[6] Here, we show again why the authors underestimated the reaction temperature, compute the reactor temperature using the sample description provided by the authors, and extract what we believe is the correct temperature from their own data for low intensities. As we have explained in a recently accepted 'Technical Comment' of ours in *Science*,[9] this prediction of higher temperatures is in accordance with the underestimation of the temperature due to the incorrect high emissivity employed in this work (namely, the use of the default 0.95 value rather than the values of 0.02–0.2 associated with the materials in the pellet).

As for the previous papers, the agreement of our model with the experimental data is essentially perfect for average intensities up to about 4 W cm^{-2} for which the temperature rise is moderate. However, unlike in the previous papers, higher illumination intensities were employed, such that the temperature rise was now much higher (at least several hundreds of degrees). For these higher temperatures, our prediction slightly overestimates the reaction rate (hence, the temperature). These two observations imply that the linear dependence of the temperature on the incoming intensity assumed so far should be replaced by a nonlinear model. Extensive experimental data and modelling obtained and carried out by us previously,[7,8] as well as various high temperature ellipsometry studies of metals,[10,11] show that for such high temperatures, the temperature rise is strongly nonlinear and, in particular, it is much slower than the linear prediction we have employed so far. A quadratic fit to the experimental data given in Fig. 1d of the 2018 *Science* paper by the Halas group[6] shows that the maximal temperature reached is ~1700 K for the highest intensity used ($\langle I \rangle = 9.6$ W cm^{-2}). This is significantly lower than what one would deduce from the linear model used for the low intensity data.

As for the question of melting, for small nanoparticles this is not a trivial issue. Indeed, it may not be precise to refer to the nanoparticles as being solid even at temperatures modestly above room temperature (*e.g.* for gold). Instead they are unstable, in the sense that the atoms continuously migrate and the nanoparticle internal morphology fluctuates with time between various nearly degenerate states; these effects might depend, among other aspects, on the environment and to the best of our knowledge have not been characterized for Cu–Ru nanoparticles.

Nevertheless, let us assume that the nanoparticles do undergo a well-defined phase transition, and furthermore that the melting temperature is significantly lower than the bulk melting temperature of Cu (but not of Ru). First and foremost, we must bear in mind that the temperature we extract from the experimental data is the maximum within the reactor. As mentioned earlier and shown in a paper of ours,[1] the temperatures of other parts of the reactor (most of its volume, in practice) are significantly lower, potentially by more than 50%. Thus, the question we should ask ourselves is how melting of the nanoparticles in a small part of the reactor affects the observed reaction rate. Overall, we tend to say that the effect will be, at most, rather small. This conjecture relies on several arguments:

1. The antenna–reactor nanoparticles in the pellet were unstructured and perhaps fluxional even at room temperature. Even if melting causes shape modifications to each particle, it will have a very small effect on average, and will not affect their properties in any deleterious way.

2. Due to the presence of the surrounding oxide support and an inert atmosphere well in excess of the Cu vapor pressure, any such melting is not expected to lead to any significant or irreversible effects on the nanoparticles. The particles then solidify once the light is turned off.

3. Despite the above, even if we do assume that the melted layer drips off/evaporates, moves, merges with other particles *etc.*, this will only cause a slight change in the overall temperature distribution. Indeed, the absorption/heat source might become slightly distorted (its center shifted to a lower position within the layer). However, since clearly the final temperature distribution is determined primarily by the heat diffusion, the overall temperature distribution (and therefore the reaction rate) would change only slightly. This has been confirmed by extensive numerical simulations that we have been performing recently (to be published in the near future).

Either way, the possibility of morphological changes due to laser illumination should be accounted for in future experiments at high temperatures. In particular, one would want to measure the emission spectrum from the heated nanoparticle composite in a spectrally and even spatially resolved manner, in order to isolate the emission from the parts of the sample that supposedly melted. One could also characterize the nanoparticles before and after illumination to explore any such effects—in the absence of any such data, we are left only with speculation about what changes, if any, the nanoparticles undergo. This is in sharp contrast to the catalytic activity of the nanoparticles, where the experimental data and our model make it abundantly clear that the observations can be explained by purely thermal effects.

1 Y. Sivan, I. W. Un and Y. Dubi, 2019, arXiv:1902.03169 [physics.chem-ph].

2 X. Zhang, X. Li, M. E. Reish, D. Zhang, N. Q. Su, Y. Gutiérrez, F. Moreno, W. Yang, H. O. Everitt and J. Liu, *Nano Lett.*, 2018, **18**, 1714–1723.
3 S. Mukherjee, F. Libisch, N. Large, O. Neumann, L. V. Brown, J. Cheng, J. Britt Lassiter, E. A. Carter, P. Nordlander and N. J. Halas, *Nano Lett.*, 2013, **13**, 240–247.
4 S. Mukherjee, L. Zhou, A. M. Goodman, N. Large, C. Ayala-Orozco, Y. Zhang, P. Nordlander and N. J. Halas, *J. Am. Chem. Soc.*, 2014, **136**, 64–67.
5 P. Christopher, H. Xin, A. Marimuthu and S. Linic, *Nat. Mater.*, 2012, **11**, 1044–1050.
6 L. Zhou, D. F. Swearer, C. Zhang, H. Robatjazi, H. Zhao, L. Henderson, L. Dong, P. Christopher, E. A. Carter, P. Nordlander and N. J. Halas, *Science*, 2018, **362**, 69–72.
7 Y. Sivan and S.-W. Chu, *Nanophotonics*, 2017, **6**, 317–328.
8 I. Gurwich and Y. Sivan, *Phys. Rev. E*, 2017, **96**, 012212.
9 Y. Sivan, J. Baraban, I. W. Un and Y. Dubi, Comment on "Quantifying hot carrier and thermal contributions in plasmonic photocatalysis", *Science*, 2019, **364**, eaaw9367.
10 H. Reddy, U. Guler, A. V. Kildishev, A. Boltasseva and V. M. Shalaev, *Opt. Mater. Express*, 2016, **6**, 2776–2802.
11 P.-T. Shen, Y. Sivan, C.-W. Lin, H.-L. Liu, C.-W. Chang and S.-W. Chu, *Opt. Express*, 2016, **24**, 19254–19263.

Jeremy Baumberg further replied to Naomi Halas' question: This is an extremely interesting question. One response is that we already have clear data from anti-Stokes to Stokes emission intensity ratios that molecular temperatures in a thermal model are 1000 K or larger, far above the melting point of the attached Au, but it is not trivially molten.[1,2] One explanation for this is that vibrational pumping produces such situations, but this is not a thermally equilibrated situation. Another important consideration is that the idea of a melting temperature for these nanoparticles is not what you are alluding to here, since even at room temperature it is well known that the surface mobility of the surface atoms is far higher than that of the bulk, and in the materials community these are considered to be a monolayer or so of molten atoms. As the temperature rises, these surface atoms increase even more dramatically in mobility, and bulk melting begins gradually from the surface. Hence the picture of a 'melting temperature' is not very robust in this system. Effectively the temperature varies through the nanoparticles.

It is thus clear that in these nanostructures, it is impossible to define a single uniform temperature, and efforts to calibrate chemistry to a single temperature are not going to capture the full complexity. Since catalysis is happening at specific sites on the irregular nanoparticle surfaces and junctions, it is perfectly possible for these to be hotter than the 'melting temperature', but also this is not needed in any model as far as I can see, since the inhomogeneity of temperature means that local thermal-enhanced chemistry is strongly possible. I believe that we have to improve our creation of structures and devices that allow us to probe this more carefully in order to make strong claims.

1 F. Benz, M. K. Schmidt, A. Dreismann, R. Chikkaraddy, Y. Zhang, A. Demetriadou, C. Carnegie, H. Ohadi, B. de Nijs, R. Esteban, J. Aizpurua and J. J. Baumberg, *Science*, 2016, **354**, 726–729.
2 D. R. Ward, D. A. Corley, J. M. Tour and D. Natelson, *Nat. Nanotechnol.*, 2011, **6**, 33–38.

Giulia Tagliabue queried: In your equation you have no term accounting for the removal of carriers from the system. It is thus not surprising that all the energy is dissipated as heat and the final distribution has a very low fraction of hot-carriers. However, for plasmonic photocatalysis applications we expect to be continuously removing a number of hot-carriers through either indirect or direct transfer to an adsorbate or semiconductor, and this term is not present in your model. Do you have any insight as to how the steady-state distribution would change if you were to account for the energy removed through hot-carrier extraction?

Yonatan Sivan responded: This is an excellent question! The theoretical approach on which our *Faraday Discussions* paper relies aims to understand what happens inside the metal only. Accordingly, it accounts only for the main mechanism (e–h excitation by photons, e–e and e–h collisions); all other effects, including in particular electron tunnelling out, are far slower, as has been shown by many authors, including by Jacob Khurgin in his latest *ACS Photonics* paper.[1] The implications for photocatalysis were realized after completion of the paper. At the moment we can predict, based on the existing results, that the tunneling will be a small perturbation that will not change the electron distribution significantly. Thus, we predict that the spectrum of the electrons that tunnel will be the same as the distribution. We do plan to verify this prediction using numerical simulations in the very near future.

1 M. Grajower, U. Levy and J. B Khurgin, *ACS Photonics*, 2018, **5**, 4030–4036.

Alexander Govorov provided a general comment: One advanced model, which describes the dynamics of excited electrons in nanocrystals, is called a 3-temperature model. It is a set of kinetic differential equations for three parameters: the number of non-thermalized high-energy (hot) electrons (these electrons come from direct absorption of one light quantum), electronic temperature and lattice temperature. During and after a pump pulse, the lattice temperature does not increase much, but the electronic temperature can be high. In the current literature, one can see numbers like $dT \sim 100$ K–4000 K, depending on the intensity and the system.

Phillip Christopher responded: The work of E. Carpene[1] is relevant here.

1 E. Carpene, *Phys. Rev. B*, 2006, **74**, 024301.

Anatoly Zayats commented further: If electron temperature is the same as lattice temperature, I would not consider such electrons as hot electrons, even if their temperature is higher than room temperature. My view is that hot electrons may be non-thermalized or thermalized, but should have a temperature higher than that of the lattice. For these reasons, in my view, the heating of metal on a hot-plate will have different effects to optical excitation.

Phillip Christopher addressed Yonatan Sivan: I have 2 comments on your paper and conclusions. My first comment is related to the use of a steady state distribution of hot charge carriers and phonons to understand the mechanisms of photocatalysis by continuous wave excitation of surface plasmons. It has been shown many times in quite detailed experiments that adsorbate-derived electronic states can interact with plasmons during their coherent oscillation and directly upon dephasing; these processes are often described as chemical interface damping. It has been shown that a significant fraction of absorbed photons derived from plasmon dephasing occur due to direct interactions with adsorbates when metal particles are in 'chemical environments', *i.e.* dephasing populates normally unpopulated adsorbate-derived electronic states. As a result, it is important to consider the temporal evolution of plasmon excitation and dephasing, and the potential for interactions between charge carriers and adsorbates through these processes in order to have a complete picture of how photoexcitation of plasmonic particles could facilitate catalytic processes. This is not captured in the steady state picture presented here.

My second comment is related to the broad-reaching conclusion of your paper that it is extremely unlikely that any reports of 'faster' chemistry are associated with non-thermal mechanisms. There have been many reports on the photodetection of hot carrier injection from plasmonic structures to surrounding materials with quite high quantum yields—much higher than would be possible based on the predictions made using the steady state model presented in your paper. Similarly, there have been reports of redox chemical conversions that require the transfer of charge. In non-redox catalytic processes, there have been reports of plasmon excitation leading to non-thermal reaction selectivity and anomalously large kinetic isotope effects (an increased temperature would decrease kinetic isotope effects within the formalism of transition state theory) in reactions where the rate was enhanced by plasmon excitation. While there certainly may be cases where enhanced catalytic reactivity is related to plasmon-induced heating, there is also overwhelming evidence that non-thermal mechanisms of energy exchange to species at the metal interface do occur under continuous wave illumination. The discrepancy between implications derived from the predictions of the model presented in your paper and the countless reports of plasmon-mediated non-thermal interactions with interfacial species is likely a result of the chemical interface damping interactions described above.

Yonatan Sivan replied: Thank you for your question. Let me start by re-iterating our main claims. What we find, using what we believe is the first ever complete quantitative theory for the (indeed, semi-quantum; see my previous response to Dr Kuisma) steady-state electron distribution, is that the generation of high energy non-thermal electrons is an incredibly inefficient process; only one out of every 1–100 billion absorbed photons generates a high energy electron and hole pair. Furthermore, a Fermi's golden rule calculation shows that if a certain chemical reaction is enhanced by these high energy non-thermal electrons, then one should expect the reaction to become faster by the ratio of the population of high energy electrons with light to that without light. Since this ratio can be 10–15 orders of magnitude, and since no one has ever observed such a massive improvement of reaction rates in a photocatalysis experiment, an unavoidable conclusion is that in the latter experiments the faster reaction does not originate from the high energy electrons. Instead, the reaction was likely enhanced due to a purely thermal origin, *i.e.* due to the exponential sensitivity of the reaction to the number of electrons near the Fermi energy, those which obey Fermi–Dirac statistics. Indeed, as we showed in our previous work,[1] essentially all of the energy absorbed (save for the tiny amount of energy that establishes the non-equilibrium population) ends up just heating the metal. The thermal explanation is also found to be in line with our re-interpretation of the results of some of the more famous experimental papers, where we showed that the temperature measurements underestimated the actual temperature of the reaction. In particular, our thermal model can also capture the kinetic isotope effect as described previously.[2]

Nevertheless, what we show is not unequivocal proof that thermal effects are responsible for the catalytic reactions reported in those papers. Instead, we claimed that by virtue of Ockham's razor, our far simpler, well-established and fully-quantitative classical thermal explanation is much more likely to explain the experimental findings compared with the sophisticated, qualitative-only quantum mechanical explanation provided in your own papers, as well as the purely empirical (thus, theory-free) reports by the Halas group. We believe this argument is particularly convincing in light of the incorrect temperature readings in those papers. In other words, this implies that the "evidence that non-thermal mechanisms of energy exchange to species at the metal interface do occur under continuous wave illumination" should undergo a re-examination rather than being considered "overwhelming", as you have said.

In that sense, our results do not mean that effects associated with these high energy non-thermal electrons were never observed. They are surely responsible for photo-detection experiments, as we heard during the opening lecture, for example (DOI: 10.1039/c9fd00001a), and might provide an explanation for some of the selectivity in reaction catalysis reported by Jie Liu's group and surely others (although selectivity may also be, in some cases, explained by thermal effects). The effects you describe in your question are also potential signatures of hot electron action. However, our results prove that 'hot' electron effects are far more subtle (weaker) than previously thought.

As we obviously did not examine all of the papers that have been published on this topic, we believe this is now a challenge that the community has to confront—to re-examine all previous "countless reports of plasmon-mediated non-thermal interactions with interfacial species" (to quote from your question) with an awareness of previous mistakes in temperature evaluation, using our simple theory for non-thermal electron population, with the extensive temperature calculations we provide for complex particle clusters, and more importantly, adopting more careful temperature measurements (such as those carried out by Jie Liu's group[4] and by Cortes *et al.*[5]).

We do plan to contribute to this effort. As I previously stated in my response to Dr Tagliabue's question, our first step would be to add electron tunneling to our equations, and see how much it affects the final temperature and what the overall efficiency of tunneling is. We could then also model the effects you describe in your question. It is already clear to us that we would need the assistance of people whose expertise is in those chemical effects, and of experimentalists, who can provide actual evidence and the necessary parameters for modelling. We would be delighted to collaborate with anyone who is interested, and to lend our modelling services. We would be delighted to be convinced that non-thermal effects are playing a role, whether *via* charge transfer, chemical interface damping, Schottky barrier crossing or any other mechanism, as these mechanisms might have a true impact on energy related challenges to humanity. However, it should be emphasized that in light of the above, the level of proof that shall be required will be higher than that provided in the many papers you refer to. In particular, obtaining unequivocal proof of hot electron action might be difficult using some of the approaches employed so far, namely, looking at macroscopic suspensions of metal nanoparticles and performing external heating control experiments. Indeed, since the reaction rate is exponentially sensitive to the local temperature, any tiny variation of the temperature (spatial or temporal) will be massively amplified and will dominate the overall reaction; thus, any small difference in the temperature distribution in the photocatalysis and thermocatalysis experiments will be incorrectly interpreted as due to 'hot electrons'. This makes control experiments with macroscopic particles nearly useless in their ability to identify 'hot' electron effects. Alternative suggestions for control experiments are discussed in our recent arXiv paper.[3]

1 Y. Dubi and Y. Sivan, 2018, arXiv:1810.00565 [physics.optics].
2 P. Christopher, H. Xin, A. Marimuthu and S. Linic, *Nat. Mater.*, 2012, **11**, 1044–1050.
3 Y. Sivan, I. W. Un and Y. Dubi, 2019, arXiv:1902.03169 [physics.chem-ph].
4 X. Zhang, X. Li, M. E. Reish, D. Zhang, N. Q. Su, Y. Gutiérrez, F. Moreno, W. Yang, H. O. Everitt and J. Liu, *Nano Lett.*, 2018, **18**, 1714–1723.
5 E. Pensa, J. Gargiulo, A. Lauri, S. Schlücker, E. Cortés and S. A. Maier, *Nano Lett.*, 2019, **19**, 1867–1874.

Laura Fabris remarked: In your manuscript you seem to discount the role of shape, whilst many of us have shown experimentally that the shape does indeed make a difference in the efficiencies of photocatalytic systems. When in your manuscript you discuss the role of shape, you seem to comment on the results of several papers, but do not address them directly in the citations, therefore it is hard for one to determine which works you are referring to. I believe that removing the role of shape leads to a rather simplistic approach of the problem, and the way you suggest that spherical nanoparticles are less efficient than anisotropic ones just because of heat partition seems limiting. Can you comment in more detail regarding your view of the role of shape by also citing the literature that you use as a benchmark in your manuscript?

Yonatan Sivan responded: This is an excellent question, which addresses a rather complicated issue. First, I would like to note that we never (*a priori*) dismissed the role of particle shape altogether in (photo)catalysis experiments, but rather say something far more limited in scope—namely, that nanoparticle shape has no effect on determining the number of non-thermal high energy electrons. This is a result of our solution to the optical (*i.e.* Maxwell) and electronic (*i.e.* Boltzmann) equations, which in fact makes perfect sense. Indeed, e–e collisions are so fast that any gradients of charge created by an electric field (and their gradients) necessarily disappear almost completely in the steady-state. For the same reason, there are essentially no temperature gradients in metal nanoparticles, simply because the thermal conductivity in the metal is so high (see, for example, the discussion by Baffou, Quidant and G. de Abajo quoted in our manuscript (DOI: 10.1039/c8fd00147b). Admittedly, our formulation does not account for the possibility of field or temperature gradients, but rather we assume they are negligibly small by looking only at averaged quantities. The same assumption underlies all theory papers on this problem that we are familiar with (*e.g.* those by Govorov, Nordlander, Louie, Lischner *etc.*). We do plan to improve our formulation to prove this point, hopefully in the near future.

What the above implies is that the immense body of experimental evidence on the faster photocatalysis associated with 'pointy' particles has its origin in a chemical (rather than optical or electronic) consideration. For example, this can be associated with the different lattice facet exposed at a corner or weaker atom binding which makes that region of space more reactive. As Javier Aizpurua pointed out, we should at some point complement our optical+electronic models to account for such effects. Needless to say, this is a difficult task that currently lies beyond the capabilities of even the best theory groups.

Regarding the comparison of heat generation in a sphere and a rod, if you believe our claims that the photocatalysis ensues due to regular heating (at least in the papers we criticized from the opening lecture (DOI: 10.1039/c9fd00001a), our poster and the slides I showed during my own talk), then our results imply that metal spheres will be better (thermo)catalysts in comparison with rods. This conclusion again excludes any additional 'chemical' considerations. The efficiency we discussed in our paper is however associated with a different aspect—namely, the relative amount of power that ends up generating non-thermal high energy electrons in comparison with the absorbed power that ends up as regular heating. In that respect, the spheres are the least efficient, as they lose the smallest amount of heat to the environment. This is, however, a purely classical effect. In that respect, our conclusions differ from claims you can find in the literature (see, for example, work by the Govorov group[1]).

1 L. V. Besteiro, X.-T. Kong, Z. Wang, G. Hartland and A. O. Govorov, *ACS Photonics*, 2017, **4**, 2759–2781.

Reinhard J. Maurer commented: Your paper (DOI: 10.1039/c8fd00147b) finds that the data of a number of experimental works can be interpreted by assuming local surface heating (of both electrons and the lattice) rather than non-thermal electron-driven effects. Not considering many other works, where significant deviations of reaction paths from thermal conditions have been seen, this hypothesis could, in principle, be valid. However, it is important to stress that this does not mean that the corresponding adsorbate reaction dynamics will behave the same as they would in the case of a thermally equilibrated surface. Even if the experimentally observed effects would only arise from local surface heating and an elevated, yet thermal, distribution of hot electrons, the exchange of energy between the surface and the adsorbate that induces the corresponding enhancement of surface reactions is highly non-thermal. Our own work[1,2] and the works of many others in the field have clearly shown that the nonadiabatic energy transfer between adsorbate motion and hot electrons is highly directional and specific to particular vibrational modes. As such, just because the increased energy at the surface is distributed according to equilibrium statistics, this would not mean that the same is true for the adsorbate motion it induces. This is particularly true in realistic gas-flow experiments where reacting molecules do not have a sufficient surface residence time to fully equilibrate. Simple fits on reaction rates, either by varying adsorption energies or apparent temperatures, will not provide sufficient detail to capture the complex picture behind hot-electron-enhancement and, therefore, will not be able to inform how such effects can be controlled and optimised. Recent molecular beam scattering and ultrafast dynamics studies[3] provide clear evidence of hot electron effects in reaction dynamics and provide the necessary data to develop more advanced theoretical models.

1 M. Askerka, R. J. Maurer, V. S. Batista and J. C. Tully, *Phys. Rev. Lett.*, 2016, **116**, 217601.
2 R. Maurer, B. Jiang, H. Guo and J. C. Tully, *Phys. Rev. Lett.*, 2017, **118**, 256001.

3 K.-I. Inoue, K. Watanabe, T. Sugimoto, Y. Matsumoto and T. Yasuike, *Phys. Rev. Lett.*, 2016, **117**, 186101.

Yonatan Sivan replied: We appreciate your comment, and would like to clarify some points raised in it:

(1) Our paper(s) and poster demonstrated unequivocally that one can use thermal Arrhenius theory to explain some of the results of main articles in the field of plasmonic photocatalysis, by reproducing the data with unprecedented accuracy. Notably, some of these papers did not offer any quantitative theory. This shows that the claims made in these papers for "hot electron-generated photocatalysis" are not supported by the data. This has to be recognized as our main message.

(2) As we discussed in our paper and poster, some of these papers (but not all) also had serious flaws in their measurement procedures, which we now understand, leading them to the wrong conclusions, without even testing the standard theory at all.

(3) Our thermal interpretation of the data of those papers is in accordance with a simple Boltzmann equation theory for the steady-state non-equilibrium electron distribution in a metal. As was explained in our slides, our theory is, to date, the only one for which energy is conserved in the electron–phonon-environment system.

Having addressed these three points, I will now go on to address the rest of your remark. Never, in our papers nor our lecture, did we argue that all experiments in photocatalysis can be explained by thermal effects, nor do we think that. However, now that it is clear that thermal effects can lead to apparent photocatalysis, we argue that it should always be considered at least as an option. Take for example experiments which show catalytic selectivity. Selectivity can be explained as a thermal effect if the energy barriers of two reaction paths are close, but not too close. Is this the case, or are there really hot-electron dynamics in such experiments? One has to provide clear evidence one way or another.

We state that in order to have thermal effects, the molecules on the surface have to be in (quasi)thermal equilibrium with the surface, which happens at a time-scale defined by the thermal conductivities between molecules and surfaces (which is determined by a complicated combination of vibration–electron and vibration–vibration couplings, the two channels with which energy can go from the surface to the reactants). This requires that the gas flow in the devices is slow enough. This is implicitly assumed, as was clearly stated in our discussion on reaction rates in our paper (DOI: 10.1039/c8fd00147b), where the derivation of reaction rates from the Boltzmann equation assumes that the reaction times are much slower than the thermalization time-scale. Our own estimations clearly show (and will soon be published) that indeed, in the experiments we consider, the gas flow rate is more than slow enough to allow thermalization. Clearly, this is the reason why experiments that aim to explore hot-electron 'dynamics' use either molecular beams or ultrafast light pulses, since in these cases there is indeed no time for thermalization. In optics terms, this is the difference between pulsed illumination (or situations with rapidly moving molecules, like those you have studied) and continuous illumination (studied by us and in the papers we have criticized). Regarding your statement beginning "Even if the experimentally observed effects...", this is only correct if the adsorbates do not thermalize with the surface. If that were true, it would have an experimental signature. Since we unambiguously proved that there is no apparent signature, one can conclude that in the experiments we discuss, there indeed seems to be thermalization of the adsorbents with the nanoparticle surface.

Jie Liu remarked: (1) I feel that there is a small disconnect between physicists and chemists at this meeting. We need to think about the processes in NPs together with their chemical environment. They form an integrated system rather than act separately.

(2) If plasmonic nanoparticles are heated to a super high temperature by light *via* a plasmonic effect, and they have an effect on the chemical reaction, is this plasmonic catalysis or is it just thermal catalysis?

Yonatan Sivan responded: (1) I agree, yet from the theory point of view even accounting correctly just for the nanoparticles is rather non-trivial and controversial. We should indeed also account for additional chemical aspects at a later stage, when there is an agreed formulation for treating the solid state and quantum mechanical aspects of the problem. We believe that our formulation provides the simplest comprehension of such a formulation.

(2) It is 'plasmon-assisted' catalysis, but the right question, in my opinion, is whether the mechanism responsible for faster reactions relies on thermal or non-thermal electrons, in other words, whether there is a difference between laser heating which pushes the system out of equilibrium (even though by an extremely small amount) and heating by a uniform external bath, in which case, the electron distribution is purely thermal. This also seems to be the exact motivation behind the recent thermocatalysis experiments conducted by many groups, including those of Halas, Maier, Martin, Baumberg and surely others. As we showed in our poster and slides, and in two papers[1,2] the results of some of the most famous papers in the field were incorrectly interpreted (in our opinion) as originating from non-thermal electrons, and can instead be shown to indicate the second scenario—that the reaction is accelerated by purely thermal electrons, such that the laser heating approach will suffer from all of the difficulties that make heating an unfavourable solution for catalysis.

1 Y. Sivan, I. W. Un and Y. Dubi, 2019, arXiv:1902.03169 [physics.chem-ph].
2 Y. Sivan, J. Baraban, I. W. Un and Y. Dubi, Comment on "Quantifying hot carrier and thermal contributions in plasmonic photocatalysis", *Science*, 2019, **364**, eaaw9367.

Alexander Govorov opened discussion of the paper by Andrea Marini: This is excellent work on plasmonic hydrodynamics. In your picture, the Fermi sea moves as a whole with an average time-dependent velocity. This average velocity of the Fermi sea is described by the Drude model in the linear regime. We then observe the excited electrons and holes with low energies near the Fermi sea, which I referred to in my talk as Drude electrons. Such electrons form the plasmon wave. I guess that your approach, being intrinsically non-linear, can be applied to many current time-resolved experiments where an electron gas in a nano-crystal becomes highly excited. Could you please comment more on how one can apply your theory to experimental situations?

Andrea Marini replied: Under experimental conditions, such as for a nanocrystal, the theory provided enables the calculation of the nonlinear and non-local infrared response of the electron plasma through a full numerical solution of Maxwell's equations coupled with a hydrodynamical model. Our paper illustrates analytically that Ohmic losses can be reduced by exciting the nanocrystal with ultrashort pulses of peak intensities of the order of GW cm^{-2} owing to collision quenching. Such analytical predictions, obtained for a homogeneous plasma, are expected to also hold for a nanocrystal. However, in such nanostructures Landau damping will play a relevant role, and the simultaneous dynamics resulting from Landau damping and electron collisions need to be clarified in future investigations (it is however expected that Landau damping will pose a lower limit to absorption).

Javier Aizpurua queried: You show in your paper a nice description of electron–electron and electron–phonon coupling in the description of the response of the electron gas. How could this be extended to include surfaces, where the Landau damping is substantial? In other words, how can you properly include dynamical screening in a semi-infinite metallic interface? For instance, I am thinking of the application of this to plasmonic waveguides. Furthermore, how can we treat the responses of metals that are not so free-electron, such as Al or Na? How can we extend your formalism to treat materials with d-electron bands, such as Au or Ag?

Andrea Marini answered: In the hydrodynamical model the Landau damping is accounted for by the evolution equation of the mean velocity $\boldsymbol{v}$ in the term $(\boldsymbol{v}\cdot\nabla)\boldsymbol{v}$. In turn, the numerical solution of the hydrodynamical model in a semi-infinite interface, a plasmonic waveguide, or in an arbitrary nanostructure will incorporate such an effect. The extension of this model to metals with resonant d-bands would require non-perturbative recalculation of the collisional integral (which in our reported results was developed for a classical plasma through the Fokker-Planck-Landau theory of electron collisions) to the quantum regime and its embedding in semiclassical Maxwell-Bloch equations. Usually in the literature this is done perturbatively through an effective linear damping in the Lindblad formulation. Thus, such a formulation needs to be extended to the nonlinear regime in order to account for the nonlinear damping arising from electron collisions.

Conflicts of interest

There are no conflicts to declare.

Faraday Discussions

PAPER

Gap-plasmon enhanced water splitting with ultrathin hematite films: the role of plasmonic-based light trapping and hot electrons†

Aveek Dutta,[ab] Alberto Naldoni,*[c] Francesco Malara,[d] Alexander O. Govorov,[ef] Vladimir M. Shalaev[ab] and Alexandra Boltasseva*[ab]

Received 6th October 2018, Accepted 7th November 2018
DOI: 10.1039/c8fd00148k

Hydrogen is a promising alternative renewable fuel for meeting the growing energy demands of the world. Over the past few decades, photoelectrochemical water splitting has been widely studied as a viable technology for the production of hydrogen utilizing solar energy. A solar-to-hydrogen (STH) efficiency of 10% is considered to be sufficient for practical applications. Amongst the wide class of semiconductors that have been studied for their application in solar water splitting, iron oxide (α-Fe_2O_3), or hematite, is one of the more promising candidate materials, with a theoretical STH efficiency of 15%. In this work, we show experimentally that by utilizing gold nanostructures that support gap-plasmon resonances together with a hematite layer, we can increase the water oxidation photocurrent by two times over that demonstrated by a bare hematite film at wavelengths above the hematite bandgap. Moreover, we achieve a six-fold increase in the oxidation photocurrent at near-infrared wavelengths, which is attributed to hot electron generation and decay in the gap-plasmon nanostructures. Theoretical simulations confirmed that the metamaterial geometry with gap plasmons that was used allows us to confine electromagnetic fields inside the hematite semiconductor and to enhance the surface photochemistry.

[a]*School of Electrical and Computer Engineering, Purdue University, IN-47907, USA. E-mail: aeb@purdue.edu*
[b]*Birck Nanotechnology Center, Purdue University, West Lafayette, IN-47907, USA*
[c]*Regional Center for Advanced Technologies and Materials, Olomouc-78371, Czech Republic. E-mail: alberto.naldoni@upol.cz*
[d]*CNR-Istituto di Scienze e Tecnologie Molecolari, Via Golgi 19, 20133 Milan, Italy*
[e]*Institute of Fundamental and Frontier Sciences, University of Electronic Science and Technology of China, Chengdu 610054, China*
[f]*Department of Physics and Astronomy, Ohio University, Athens OH-45701, USA*

† Electronic supplementary information (ESI) available. See DOI: 10.1039/c8fd00148k

Introduction

The ever-increasing population of the world has resulted in a growing demand for energy sources. Due to declining fossil fuel reserves, the need for alternate renewable sources of energy has never been greater. In recent years, hydrogen has been proposed as an alternate fuel resource for the production of clean and efficient electric power. It is also a well-established fuel resource in industrial processes such as petroleum recovery and refining, ammonia synthesis, methanol production and metal fabrication, and is used in the electronics and food industries.[1] Photoelectrochemical (PEC) water splitting utilizing solar energy and earth-abundant semiconductors is one of the most widely researched areas for hydrogen production.[2] PEC water splitting dates back to the early 1970s and the pioneering work of Fujishima and Honda, who demonstrated overall water splitting using a PEC cell with a single crystalline TiO_2 (rutile) anode and a Pt cathode under UV irradiation.[3] Since then, there have been several studies carried out on water splitting that have reached acceptable limits of the solar-to-hydrogen (STH) efficiency (such as an STH of 12.4% reported in 1998)[4]. Several recent articles have reviewed the advances in PEC solar water splitting reported to date.[5,6]

In order to achieve efficient water splitting, a semiconductor electrode has to satisfy some critical requirements. The semiconductor band gap should feature appropriate valence and conduction band positions relative to water redox potentials, and should demonstrate high light absorption, chemical stability and commercial viability. Electrochemical theory states that a material with a bandgap of 1.23 eV should be sufficient for PEC water splitting. However, due to potential losses and the kinetic overpotentials required to drive hydrogen and oxygen evolution reactions in a PEC cell, the energy range for sustaining overall water splitting is generally reported to be 1.6–2.4 eV.[5,7] A variety of semiconductor materials, such as TiO_2, $BiVO_4$, Cu_2O, TaON and WO_3 amongst others, have been investigated as candidate electrodes for PEC water splitting.[8] Among these semiconductors, α-Fe_2O_3 (hematite) is one of the more extensively studied semiconductor materials for PEC water splitting. It is an earth-abundant, non-toxic and photochemically stable semiconductor. The progress made in PEC water splitting using hematite has been outlined in several review articles.[9–11]

Hematite has a band gap of 2.0–2.2 eV, which is ideally suited for PEC water splitting.[12] The maximum solar energy conversion efficiency for hematite is 14–16%.[13] However, the reported STH efficiency of hematite is very low, which is mainly due to the short excited state lifetime (<10 ps)[14,15] and the short hole diffusion length of ~2–4 nm.[16,17] The majority of photoexcited electrons are lost in hematite through electron–hole recombination processes within the first few picoseconds, and only holes close to the semiconductor electrolyte interface contribute to the oxygen evolution reaction. Furthermore, hematite is an indirect bandgap material, and requires a relatively thick layer of active material for total solar absorption. This competes with the small hole diffusion lengths, which call for a thin active layer. The poor electrical conductivity of hematite also limits charge transport in the material.[18] This can however be overcome by elemental doping. Reports of Si[19] and Ti[20] doping in hematite in order to increase the photocurrent of water oxidation can be found in the literature. These dopants can significantly increase the photocurrent and reduce the photocurrent onset potential. Si doping can result in photocurrents of around 2.2 mA cm^{-2} at 1.23 V *vs.* Reference Hydrogen Electrode (RHE).[21] Ti doping can cause a negative offset of 0.2 V of the onset photocurrent potential.[22] The slow kinetics of water oxidation at the hematite surface is also a cause of the low solar conversion efficiency of hematite. Surface catalysts such as IrO_2 can significantly enhance the photocurrent.[23]

One possible way to improve the oxidation photocurrent and address the trade-off between light absorption and hole diffusion length is to use resonant nanostructures to enhance the light collection efficiency of the active hematite layer. It was previously shown that optical Mie resonances and guided modes in a nanostructured hematite layer can increase the photocurrent by three times.[24] Plasmonics – optics with metal nanostructures – can also be used to efficiently trap light inside the active layer for PEC water splitting. Several reviews have reported the progress made in plasmon-enhanced PEC water splitting.[25–29] The field of plasmonics deals with the collective oscillation of free electrons in metals under light illumination.[30] These collective oscillations, or plasmons, have wavevectors much larger than free space light and can confine light or electromagnetic energy in sub-wavelength dimensions. Consequently, plasmons lead to high electromagnetic field density in nanoscale dimensions.[31,32] The field of plasmonics thus allows us to tailor light collection and propagation at the nanoscale, and has been used to increase the photocurrent of hematite water oxidation. An increase of over an order of magnitude in the photocurrent demonstrated by hematite nanorods due to plasmon-induced light trapping from a gold nanohole array has recently been reported.[33] Several other studies have also reported plasmon-enhanced photocatalytic water splitting with hematite.[34–36]

In this work, we report on plasmon-assisted oxidation photocurrent enhancement in ultrathin (10–20 nm) hematite films. Since hematite has a very low hole diffusion length, it is preferable to use thin films with enhanced absorption in order to harvest the maximum possible solar energy for water splitting. The enhancement reported in this work is due to plasmon resonances and optical scattering in a gold–hematite–gold sandwich structure. Here we report the design of a gold-based plasmonic hematite photoelectrode, which demonstrates an approximately two-fold increase and six-fold increase in the water oxidation photocurrent, above and below the hematite bandgap, respectively. The below-bandgap enhancement is believed to be due to the generation of hot electrons from plasmon decay. This technique has been widely reported to be useful in broadening the absorption spectrum of semiconductors for use in solar energy harvesting.

Material growth, characterization and fabrication

The hematite films were grown using Pulsed Laser Deposition (PLD). Fig. 1a shows a detailed flowchart of the fabrication process. First, a 200 nm-thick gold layer was deposited on a Si(100) substrate with a 6 nm-thick Al_2O_3 adhesion layer. A 15 nm-thick layer of hematite was then deposited using a commercially available 99.9% pure Fe_2O_3 target. The hematite growth temperature was around 500 °C in a 30 m Torr O_2 ambient, and the PLD had a 1.5–2 J cm^{-2} laser fluence. Gold nanodisks were fabricated on top of the hematite with e-beam lithography and lift-off using PMMA. The details of the fabrication process are provided in the

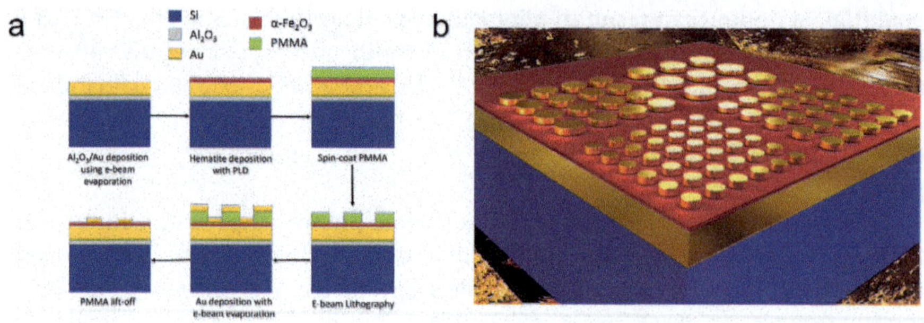

Fig. 1 (a) Flowchart showing the fabrication steps for realizing the gold-hematite gap-plasmon electrode. (b) Schematic representation of a gap-plasmon array of different nanodisk diameters and periods on top of hematite-gold (blue-silicon, yellow-gold and red-hematite).

ESI.† Fig. 1b shows a schematic of the fabricated sample that contained four arrays of gold nanostructures with various diameters and array periods. The structures support so-called gap-plasmon resonances, which are characterized by high field confinement in a dielectric/semiconductor sandwiched between the two metals (gold in this case). This gap geometry has been widely studied in the literature,[37–40] and has also been used previously for photocatalytic enhancement.[41] Here we use a gold gap-plasmon structure to enhance the light trapping and oxidation photocurrent in thin hematite films.

Raman analysis was used to identify the phase of hematite grown *via* PLD. Fig. 2a shows the Raman spectrum, which was collected using a DXR Raman spectroscope (Thermo Scientific, USA) with a laser at 633 nm. A total of eight vibration modes were detected in the hematite samples grown *via* PLD. The peaks, at 224, 242, 297, 408, 491 and 610 cm^{-1},[42] correspond respectively to $A_{1g}(1)$, $E_g(1)$, unresolved $E_g(2)$–$E_g(3)$, $E_g(4)$, $A_{1g}(2)$ and $E_g(5)$ normal modes, which were predicted by factor group analysis for the crystal space group $D^3{}_{d6}$. We associate the peaks at 224, 242 and 297 cm^{-1} with the motion of iron cations, while those at 408, 491 and 610 cm^{-1} are associated with the motion of oxygen anions.[43] In nanocrystalline hematite, a peak at ~660 cm^{-1} is further detected, owing to the lack of long-range order. Due to structural disorder generated by the presence of lattice defects,[44] small grain size[45] or the incorporation of dopants,[45–51] the Raman selection rules were relaxed. This allowed us to detect the IR-active E_u mode.

Fig. 2b shows a High Resolution Transmission Electron Microscopy (HRTEM) image of the sample. The inset of Fig. 2b shows the Selected Area Electron Diffraction (SAED) pattern for the hematite flake. Two crystal planes, namely (110) and (006), can be easily seen. The area chosen for SAED image collection is shown in red along with the lattice constants for the (110) and (006) planes. From analysing the Raman and TEM data, we can conclude that the PLD-grown hematite films were polycrystalline in nature.

Fig. 3a shows the SEM images of the fabricated gold nanodisks on hematite. Four different nanodisk diameters (*D*) and nanodisk pitch sizes (*P*) were utilized. These diameter–pitch (*D*–*P*) combinations were 100–200, 125–200, 150–300 and 200–300 in nanometers. All of the fabricated nanodisks showed a small amount of gold flakes, which is a common feature of e-beam lift-off with single layer PMMA.

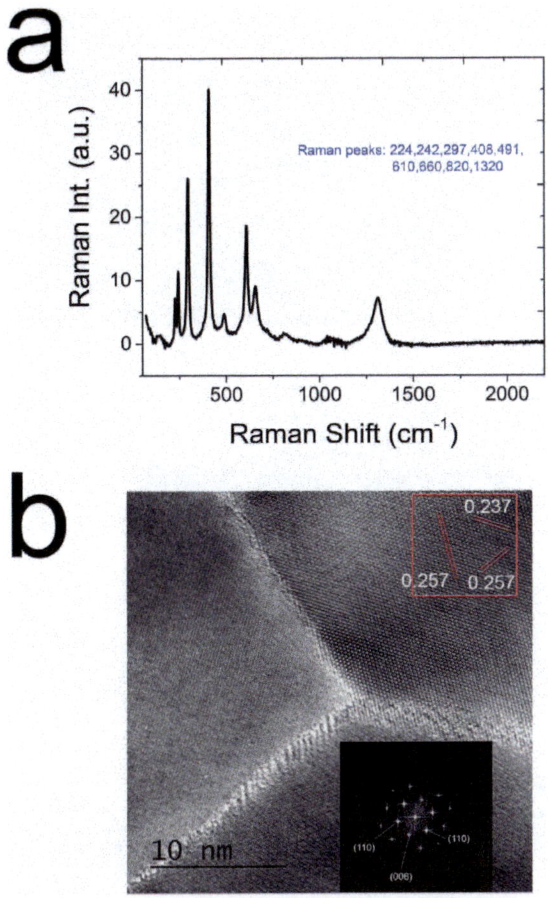

Fig. 2 (a) Raman spectroscopy data for the hematite films grown on gold *via* pulsed laser deposition. The inset lists the positions of the Raman peaks in cm^{-1}. (b) High resolution TEM image of the hematite thin film on gold. The SAED (Selected Area Electron Diffraction) image is shown in the inset. The red box highlights the area chosen for SAED. The inter-plane distances between the (110) and (006) planes are also stated in the red box.

Removing the flakes requires the use of more complex bi-layer PMMA e-beam lithography, which was not used in the current fabrication. Fig. 3b shows the absorption spectra of all of the nanodisk arrays. Fig. 3b also shows the absorption measured for a bare hematite film on gold for comparison purposes. The two arrays with 100 nm and 125 nm nanodisk diameters and a 200 nm period show absorption peaks at wavelengths of 820 nm and 970 nm, respectively. These peaks correspond to the excitation of gap-plasmon resonances, as is evident from Fig. S3a in the ESI† obtained *via* simulation. The two arrays with disk diameters of 150 nm and 200 nm and a 300 nm period show absorption peaks at 1120 nm, for the former, and 730 nm and 1550 nm, for the latter. The peak at 1120 nm for the nanodisk with a 150 nm diameter and at 1550 nm for the 200 nm diameter nanodisk are gap-plasmon resonances. The absorption peak at 730 nm for the 200 nm diameter nanodisk is a higher order gap-plasmon resonance. This is easily verified in the electric field map plots obtained *via* simulation, as shown in

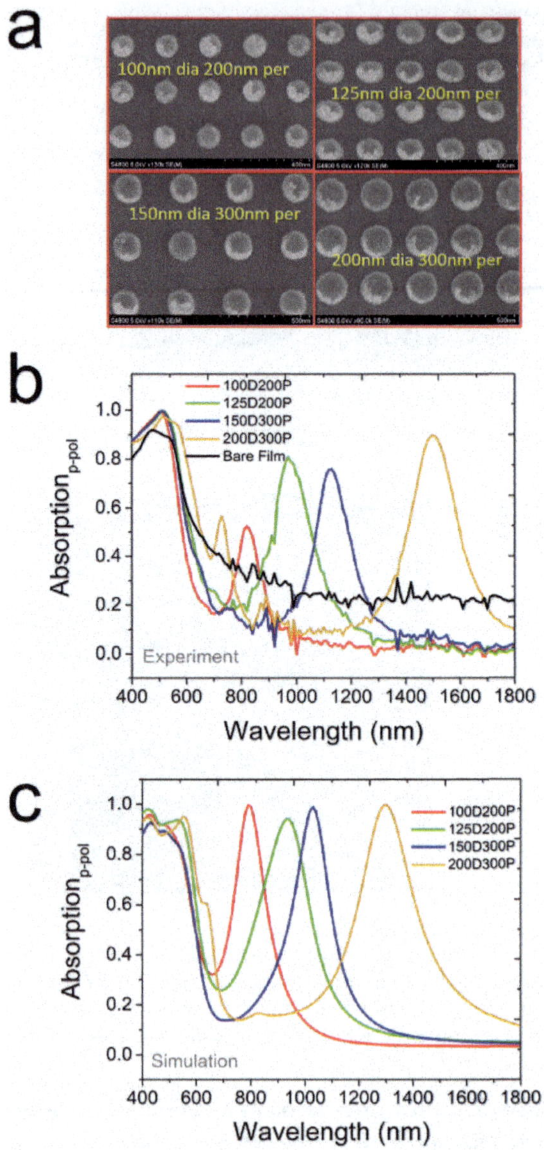

Fig. 3 (a) SEM images of the fabricated Au nanodisks for the gap-plasmon arrays with four different disk diameters (dia: diameter, per: period). (b) Experimentally measured absorption spectra, using *p*-polarized light at 20° incidence for all of the arrays. The measured absorption of the bare hematite film on gold is also shown. (c) Simulated absorption spectra, using *p*-polarized light at 20° incidence for all of the arrays. *D*: diameter of the nanodisk in nanometers and *P*: period of the array in nanometers.

Fig. S3b.† Fig. 3c shows the simulated absorption spectra for all of the nanodisk arrays. In all cases (experiment and simulation), the illumination was with *p*-polarized light at an angle of 20°. Simulations were carried out *via* a finite element method using the commercially available software, COMSOL. All of the gap-plasmon resonances demonstrate field confinement inside the hematite semiconductor. The observed and simulated gap-plasmon peaks follow the same trend; the plasmonic resonance wavelength increases with increasing disk diameter. This is typical for gap-plasmons.[37] The simulations were carried out with the optical constants of hematite, as shown in Fig. S1b in the ESI.† The optical constants for the 15 nm-thin hematite film were obtained *via* Variable Angle Spectroscopic Ellipsometry. Some degree of numerical discrepancy between the measured and calculated wavelengths arises, we believe, due to possible uncertainties in the optical dielectric constants and thicknesses of the hematite and gold layers, since the hematite layer was deposited at a high temperature at which the underlying gold may not be stable. Therefore, the properties of gold could be modified, and, in addition, the gold-hematite top boundary may not be perfectly flat (Fig. S2†).

Results and discussion

The gap-plasmon electrode was tested in a three electrode PEC setup. The gap-plasmon electrode had all of the four different arrays of nanodisks described above, with each array being a 500 μm × 500 μm square. Furthermore, every array of a particular nanodisk diameter and pitch was replicated four times so as to give as large an area coverage as is possible to fabricate within a reasonable timeframe. A bare hematite film electrode was also measured for comparison purposes. The three-electrode PEC cell used a silver/silver chloride electrode (Ag/AgCl/Sat. KCl) as the reference electrode, and a high surface area Pt mesh as the counter electrode. The potential (E) referred to the RHE scaled through the Nernst equation:

$$E_{\mathrm{RHE}} = E_{\mathrm{AgCl}} + 0.197\ \mathrm{V} + 0.059\ \mathrm{pH} \tag{1}$$

where E_{AgCl} is the measured electrode potential *vs.* the reference electrode used and 0.197 is the reference electrode standard potential *vs.* RHE. In this manuscript, all measurements are reported with respect to the RHE. The measurements were carried out in 1 M NaOH aqueous solution at pH 13.6. Linear sweep voltammetry curves were measured at a scan rate of 10 mV s^{-1}. PEC measurements were carried out using a PGSTAT204 Autolab potentiostat. At least three electrodes of each type were fabricated and tested. All electrodes demonstrated similar characteristics. Fig. 4a shows the results of the measured photocurrent for the PEC cell. The red curve corresponds to the photocurrent measured for the gap-plasmon electrode whereas the black curve corresponds to the photocurrent measured for a bare hematite film on gold. The red dashed curve corresponds to the dark current, which is negligible. It can be seen that the gap-plasmon electrode exhibits about twice the photocurrent of that of the bare hematite film at 1.23 V *vs.* RHE. There is also a cathodic shift of approximately 0.4 V in the onset potential for the gap-plasmon electrode. This may be due to the catalytic effect of the gold nanodisks. We will discuss more on this later.

Fig. 4b shows the chopped photocurrent measurements performed at 1.5 V *vs.* RHE on the gap-plasmon (red curve) and bare hematite (black curve) electrodes. Once again, the gap-plasmon electrode exhibits twice the photocurrent in comparison with that of the bare film. The increase and decrease in photocurrent with the switching on and off of the incident light clearly demonstrates that solar-driven photocatalysis is the dominant response of both of the electrodes. That

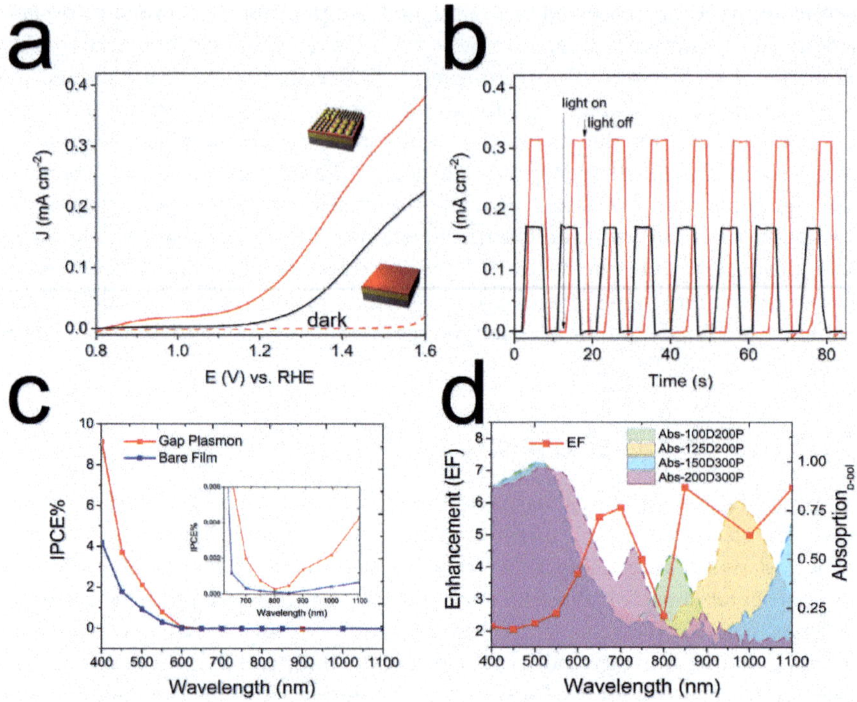

Fig. 4 (a) Linear sweep voltammetry measurements with a 3 electrode PEC cell for gold-hematite gap-plasmon and bare hematite films (red curve: gap-plasmon, black: bare film, red dashed: dark current). (b) Chopped photocurrent measured at 1.5 V *vs.* RHE for the two samples. The photocurrent from the gap-plasmon resonators is almost 2 times that of the bare flat hematite electrode. (c) Incident Photon Conversion Efficiency (IPCE) for the gap-plasmon electrode and bare hematite electrode at 1.5 V of applied bias *vs.* RHE. Inset: IPCE at near-IR wavelengths (d) Enhancement Factor (EF) of the IPCE for the gap-plasmon electrode with all four nanostructure arrays and a bare film. Above the hematite bandgap, the gap-plasmon electrode exhibits approximately twice the efficiency of the bare electrode. Below the bandgap, the increase in efficiency is up to six/seven-fold. The experimentally measured absorption plots for the four different arrays are also shown. We observe a complex, oscillating behavior for the EF in the IR spectral region where the plasmonic resonances are located.

gold nanodisks have a catalytic influence on the photocurrent due to better charge separation in the space charge layer has already been reported in literature. This effect has been shown to contribute to the photocurrent on top of the plasmon-induced enhancement.[52,53] We also attribute the cathodic shift of the onset potential for the gap-plasmon electrode to the catalytic effects of the gold nanodisk. However, to better understand the influence of this catalytic effect, we conducted Electrochemical Impedance Spectroscopy (EIS) analysis on the gap-plasmon and bare film electrodes. The gap-plasmon electrode exhibited a 20% reduction in the charge transfer resistance in comparison with that of the bare film electrode. This means that the gap-plasmon electrode exhibits better separation of the generated charge carriers, which would lead to some increase in the measured photocurrent over that of the bare film electrode. Nevertheless, this increased charge separation capability does not sufficiently explain the doubling of the measured photocurrent for the gap-plasmon electrode, which we will discuss now.

To better understand the photocurrent enhancement for the gap-plasmon electrode, chronoamperometry measurements at 1.5 V of applied bias were carried out on both of the electrodes. Fig. 4c plots the Incident Photon Conversion Efficiency (IPCE), as obtained from these measurements. IPCE is calculated as

$$\text{IPCE} = \frac{J_{\text{meas}} \times 1240 \times 100}{\lambda \times I_0} \tag{2}$$

where J_{meas} is the measured photocurrent at wavelength, λ, of incident intensity, I_0. Looking at Fig. 4c, it can be clearly seen that the gap-plasmon electrode exhibits approximately double the efficiency of that of the bare hematite electrode, especially until 600 nm, *i.e.* above the hematite band gap. This may be due to off-resonant light scattering by the gold nanodisks and light reflection from the gold back-mirror. ESI Fig. S4† shows the maximum electric field amplitude enhancement inside the hematite film for each of the four different arrays of gold nanodisks compared to a bare hematite film as a function of the incident wavelength. It can be seen that the presence of the nanodisk arrays results in a higher electric field intensity inside the hematite at above-bandgap wavelengths, which may result in an increase in the photocurrent. The inset of Fig. 4c shows a zoomed-in version of the IPCE plot at visible (red)-to-infrared wavelengths. Fig. 4d shows the Enhancement Factor (EF) in the IPCE as a function of the wavelength. The Enhancement factor is calculated as

$$\text{EF} = \frac{\text{IPCE}_{\text{GPE}}}{\text{IPCE}_{\text{BFE}}} \tag{3}$$

where IPCE_{GPE} corresponds to the IPCE of the gap-plasmon electrode and IPCE_{BFE} corresponds to the IPCE of the bare film electrode. At around a 1 μm wavelength, the gap-plasmon electrode exhibits a six-fold increase in the water oxidation photocurrent in comparison to that of the bare hematite electrode. This increase is believed to be due to plasmon decay and "hot" electron generation. As shown in Fig. 4d, the gap-plasmon resonator electrode supports plasmon modes at infrared wavelengths. These plasmons decay into "hot" electrons, which are non-equilibrium intraband excitations above the Fermi level in the conduction band of the metal. This also leaves "hot" holes in the corresponding valence band. These "hot" carriers thermalize due to electron–electron scattering (over fs timescales) followed by electron–phonon scattering (over ps timescales). These "hot" carriers can be harvested before thermalization when they cross the Schottky barrier at the semiconductor–metal interface and contribute to the electrochemical process, which in this case is water splitting. "Hot" electron generation and their dynamics have been extensively studied by the plasmonics community. Several review articles and research papers have reported theoretical and experimental studies on hot electron dynamics in plasmonic metals.[54–61] An increase of over an order of magnitude in the photocurrent of TiO_2 due to hot electrons has been reported in the literature.[62] Other reports on hot electron induced photocurrent enhancement can also be found.[63–66] It should be noted that hematite only supports oxygen evolution in a PEC cell that requires holes as the charge carriers. Recently it has been reported that Au nanoparticles are very suitable for hot hole generation.[67] This is mainly due to the fact that transition metals (like Au or Cu) demonstrate hot carrier generation from the d-bands which sit approximately 2 eV below the Fermi level. Hence these metals produce more energetic hot holes than hot electrons and hence are more suited to drive OER.[68] As Fig. 4d shows qualitative agreement between the plasmon absorption peaks of the different nanodisk arrays and the measured EF at near-IR wavelengths, we believe that in the sub-bandgap region of hematite we are witnessing plasmon generated hot hole induced enhanced photoactivity.

Conclusion

We demonstrated plasmon-assisted photocurrent enhancement in ultrathin hematite films. The hematite films were grown *via* pulsed laser deposition on a gold film on Si. A gap-plasmon based gold–hematite–gold structure was used in order to achieve the photocurrent enhancement. Our measurements show a two-fold increase in the photocurrent above the hematite bandgap, which is mainly attributed to the enhanced scattering by the gold nanodisks and back reflection from the gold mirror. Below the bandgap, we also observe a wavelength-dependent increase in the hematite photocurrent. This is attributed to plasmon decay and subsequent hot hole generation, which contributed to the photocurrent measured at near IR wavelengths. Our results provide a possible pathway to enhance the light trapping and photocatalytic response of ultrathin hematite films, thereby addressing the challenging trade-off of small carrier diffusion length and low light absorption. To the best of our knowledge, this is the first time that a gap-plasmon based design has been used to achieve this effect experimentally in ultrathin hematite films. Improvement in hematite film growth by using different back metals such as Pt is believed to lead to a further increase in the oxidation photocurrent. The role of geometry in plasmon-induced hot carrier generation has already been elucidated and can be used to enhance the photocurrent response.[68] Through the use of different plasmonic metals such as TiN,[63] or using plasmonic nanostructures in the strong-coupling regime,[69] or using different mechanisms such as direct hot electron injection,[70,71] one should also be able to tailor the photocurrent response of these ultrathin hematite films. Finally, by utilizing the design of different plasmonic materials and/or nanostructure geometry and shape, it should also be possible to shift the plasmon resonance into the visible region above the hematite bandgap. This should lead to an even greater enhancement in the photocurrent due to increased interband excitation of the charge carriers.

Conflicts of interes

There are no conflicts to declare.

Acknowledgements

The authors would like to acknowledge the funding received from AFOSR (FA9550-17-1-0243). The authors would like to thank Prof. Peilin Liao for their useful suggestions and discussions and Bradlee Beauchamp for their help with the AFM measurements. A. N. gratefully acknowledges the support by the Operational Programme Research, Development and Education – European Regional Development Fund, project no. CZ.02.1.01/0.0/0.0/15_003/0000416 of the Ministry of Education, Youth and Sports of the Czech Republic.

References

1 V. Dal Santo, A. Gallo, A. Naldoni, M. Guidotti and R. Psaro, *Catal. Today*, 2012, **197**, 190–205.

2 A. J. Bard and M. A. Fox, *Acc. Chem. Res.*, 1995, **28**, 141–145.

3 A. Fujishima and K. Honda, *Nature*, 1972, **238**, 37–38.

4 O. Khaselev and J. A. Turner, *Science*, 1998, **280**, 425–427.

5 M. G. Walter, E. L. Warren, J. R. McKone, S. W. Boettcher, Q. Mi, E. A. Santori and N. S. Lewis, *Chem. Rev.*, 2010, **110**, 6446–6473.

6 L. Yuan, C. Han, M.-Q. Yang and Y.-J. Xu, *Int. Rev. Phys. Chem.*, 2016, **35**, 1–36.

7 J. R. McKone, N. S. Lewis and H. B. Gray, *Chem. Mater.*, 2014, **26**, 407–414.

8 K. Sivula and R. van de Krol, *Nat. Rev. Mater.*, 2016, **1**, 15010.

9 D. A. Wheeler, G. Wang, Y. Ling, Y. Li and J. Z. Zhang, *Energy Environ. Sci.*, 2012, **5**, 6682.

10 A. G. Tamirat, J. Rick, A. A. Dubale, W.-N. Su and B.-J. Hwang, *Nanoscale Horiz.*, 2016, **1**, 243–267.

11 K. Sivula, F. Le Formal and M. Grätzel, *ChemSusChem*, 2011, **4**, 432–449.

12 F. Le Formal, M. Grätzel and K. Sivula, *Adv. Funct. Mater.*, 2010, **20**, 1099–1107.

13 A. B. Murphy, P. R. F. Barnes, L. K. Randeniya, I. C. Plumb, I. E. Grey, M. D. Horne and J. A. Glasscock, *Int. J. Hydrogen Energy*, 2006, **31**, 1999–2017.

14 A. G. Joly, J. R. Williams, S. A. Chambers, G. Xiong, W. P. Hess and D. M. Laman, *J. Appl. Phys.*, 2006, **99**, 053521.

15 N. J. Cherepy, D. B. Liston, J. A. Lovejoy, H. Deng and J. Z. Zhang, *J. Phys. Chem. B*, 1998, **102**, 770–776.

16 D. L. Stojić, M. P. Marčeta, S. P. Sovilj and Š. S. Miljanić, *J. Power Sources*, 2003, **118**, 315–319.

17 J. H. Kennedy and K. W. Frese, *J. Electrochem. Soc.*, 1978, **125**, 709.

18 G. Goodlet, S. Faty, S. Cardoso, P. Freitas, A. M. Simões, M. G. Ferreira and M. Da Cunha Belo, *Corros. Sci.*, 2004, **46**, 1479–1499.

19 I. Cesar, A. Kay, J. A. Gonzalez Martinez and M. Grätzel, *J. Am. Chem. Soc.*, 2006, **128**, 4582–4583.

20 P. Zhang, A. Kleiman-Shwarsctein, Y.-S. Hu, J. Lefton, S. Sharma, A. J. Forman and E. McFarland, *Energy Environ. Sci.*, 2011, **4**, 1020.

21 A. Kay, I. Cesar and M. Grätzel, *J. Am. Chem. Soc.*, 2006, **128**, 15714–15721.

22 G. Wang, Y. Ling, D. A. Wheeler, K. E. N. George, K. Horsley, C. Heske, J. Z. Zhang and Y. Li, *Nano Lett.*, 2011, **11**, 3503–3509.

23 S. D. Tilley, M. Cornuz, K. Sivula and M. Grätzel, *Angew. Chem., Int. Ed.*, 2010, **49**, 6405–6408.

24 S. J. Kim, I. Thomann, J. Park, J.-H. Kang, A. P. Vasudev and M. L. Brongersma, *Nano Lett.*, 2014, **14**, 1446–1452.

25 T. G. U. Ghobadi, A. Ghobadi, E. Ozbay and F. Karadas, *ChemPhotoChem*, 2018, 161–182.

26 M. Valenti, M. P. Jonsson, G. Biskos, A. Schmidt-Ott and W. A. Smith, *J. Mater. Chem. A*, 2016, **4**, 17891–17912.

27 N. Wu, *Nanoscale*, 2018, **10**, 2679–2696.

28 J. B. Lee, S. Choi, J. Kim and Y. S. Nam, *Nano Today*, 2017, **16**, 61–81.

29 A. Naldoni, F. Riboni, U. Guler, A. Boltasseva, V. M. Shalaev and A. V. Kildishev, *Nanophotonics*, 2016, **5**, 112–133.

30 S. A. Maier, *Plasmonics: fundamentals and applications*, Springer, 2007.
31 S. A. Maier and H. A. Atwater, *J. Appl. Phys.*, 2005, **98**, 011101.
32 J. A. Schuller, E. S. Barnard, W. Cai, Y. C. Jun, J. S. White and M. L. Brongersma, *Nat. Mater.*, 2010, **9**, 193–204.
33 J. Li, S. K. Cushing, P. Zheng, F. Meng, D. Chu and N. Wu, *Nat. Commun.*, 2013, **4**, 2651.
34 I. Thomann, B. A. Pinaud, Z. Chen, B. M. Clemens, T. F. Jaramillo and M. L. Brongersma, *Nano Lett.*, 2011, **11**, 3440–3446.
35 E. Thimsen, F. Le Formal, M. Grätzel and S. C. Warren, *Nano Lett.*, 2011, **11**, 35–43.
36 H. Gao, C. Liu, H. E. Jeong and P. Yang, *ACS Nano*, 2012, **6**, 234–240.
37 M. G. Nielsen, D. K. Gramotnev, A. Pors, O. Albrektsen and S. I. Bozhevolnyi, *Opt. Express*, 2011, **19**, 19310.
38 P. Ginzburg, D. Arbel and M. Orenstein, *Opt. Lett.*, 2006, **31**, 3288–3290.
39 S. I. Bozhevolnyi and J. Jung, *Opt. Express*, 2008, **16**, 2676.
40 P. Neutens, P. Van Dorpe, I. De Vlaminck, L. Lagae and G. Borghs, *Nat. Photonics*, 2009, **3**, 283–286.
41 D. O. Sigle, L. Zhang, S. Ithurria, B. Dubertret and J. J. Baumberg, *J. Phys. Chem. Lett.*, 2015, **6**, 1099–1103.
42 D. L. A. de Faria and F. N. Lopes, *Vib. Spectrosc.*, 2007, **45**, 117–121.
43 I. V. Chernyshova, M. F. Hochella Jr and A. S. Madden, *Phys. Chem. Chem. Phys.*, 2007, **9**, 1736.
44 S.-H. Shim and T. S. Duffy, *Am. Mineral.*, 2002, **87**, 318–326.
45 I. Cesar, K. Sivula, A. Kay, R. Zboril and M. Grätzel, *J. Phys. Chem. C*, 2009, **113**, 772–782.
46 R. N. Bhowmik, G. Vijayasri and R. Ranganathan, *J. Appl. Phys.*, 2014, **116**, 123905.
47 Y.-S. Hu, A. Kleiman-Shwarsctein, A. J. Forman, D. Hazen, J.-N. Park and E. W. McFarland, *Chem. Mater.*, 2008, **20**, 3803–3805.
48 J. Liu, C. Liang, H. Zhang, S. Zhang and Z. Tian, *Chem. Commun.*, 2011, **47**, 8040.
49 P. Kumar, P. Sharma, R. Shrivastav, S. Dass and V. R. Satsangi, *Int. J. Hydrogen Energy*, 2011, **36**, 2777–2784.
50 M. Rahman, N. Wadnerkar, N. J. English and J. M. D. MacElroy, *Chem. Phys. Lett.*, 2014, **592**, 242–246.
51 D. Varshney and A. Yogi, *J. Mol. Struct.*, 2011, **995**, 157–162.
52 Z. Liu, W. Hou, P. Pavaskar, M. Aykol and S. B. Cronin, *Nano Lett.*, 2011, **11**, 1111–1116.
53 Y. Fang, Y. Jiao, K. Xiong, R. Ogier, Z.-J. Yang, S. Gao, A. B. Dahlin and M. Käll, *Nano Lett.*, 2015, **15**, 4059–4065.
54 P. Narang, R. Sundararaman and H. A. Atwater, *Nanophotonics*, 2016, **5**, 96–111.
55 M. L. Brongersma, N. J. Halas and P. Nordlander, *Nat. Nanotechnol.*, 2015, **10**, 25–34.
56 A. O. Govorov, H. Zhang and Y. K. Gun'ko, *J. Phys. Chem. C*, 2013, **117**, 16616–16631.
57 J. G. Liu, H. Zhang, S. Link and P. Nordlander, *ACS Photonics*, 2017, **5**, 2584–2595.

58 H. Harutyunyan, A. B. F. Martinson, D. Rosenmann, L. K. Khorashad, L. V. Besteiro, A. O. Govorov and G. P. Wiederrecht, *Nat. Nanotechnol.*, 2015, **10**, 770–774.

59 M. E. Sykes, J. W. Stewart, G. M. Akselrod, X.-T. Kong, Z. Wang, D. J. Gosztola, A. B. F. Martinson, D. Rosenmann, M. H. Mikkelsen, A. O. Govorov and G. P. Wiederrecht, *Nat. Commun.*, 2017, **8**, 986.

60 T. Heilpern, M. Manjare, A. O. Govorov, G. P. Wiederrecht, S. K. Gray and H. Harutyunyan, *Nat. Commun.*, 2018, **9**, 1853.

61 S. Ishii, S. L. Shinde, W. Jevasuwan, N. Fukata and T. Nagao, *ACS Photonics*, 2016, **3**, 1552–1557.

62 C. Ng, J. J. Cadusch, S. Dligatch, A. Roberts, T. J. Davis, P. Mulvaney and D. E. Gómez, *ACS Nano*, 2016, **10**, 4704–4711.

63 A. Naldoni, U. Guler, Z. Wang, M. Marelli, F. Malara, X. Meng, L. V. Besteiro, A. O. Govorov, A. V. Kildishev, A. Boltasseva and V. M. Shalaev, *Adv. Opt. Mater.*, 2017, **5**, 1601031.

64 F. Su, T. Wang, R. Lv, J. Zhang, P. Zhang, J. Lu and J. Gong, *Nanoscale*, 2013, **5**, 9001.

65 X. Zhang, Y. Liu, S.-T. Lee, S. Yang and Z. Kang, *Energy Environ. Sci.*, 2014, **7**, 1409.

66 Y. Nishijima, K. Ueno, Y. Yokota, K. Murakoshi and H. Misawa, *J. Phys. Chem. Lett.*, 2010, **1**, 2031–2036.

67 S. Dal Forno, L. Ranno and J. Lischner, *J. Phys. Chem. C*, 2018, **122**, 8517–8527.

68 R. Sundararaman, P. Narang, A. S. Jermyn, W. A. Goddard III and H. A. Atwater, *Nat. Commun.*, 2014, **5**, 5788.

69 P. Zeng, J. Cadusch, D. Chakraborty, T. A. Smith, A. Roberts, J. E. Sader, T. J. Davis and D. E. Gómez, *Nano Lett.*, 2016, **16**, 2651–2656.

70 R. Long and O. V. Prezhdo, *J. Am. Chem. Soc.*, 2014, **136**, 4343–4354.

71 K. Wu, J. Chen, J. R. McBride and T. Lian, *Science*, 2015, **349**, 632–635.

Faraday Discussions

PAPER

Tuning the SERS activity and plasmon-driven reduction of *p*-nitrothiophenol on a Ag@MoS$_2$ film

Peng Miao,[a] Yan Ma,[a] Mengtao Sun,[b] Jing Li*[c] and Ping Xu*[a]

Received 30th September 2018, Accepted 28th November 2018
DOI: 10.1039/c8fd00139a

The combination of plasmonic metal nanostructures with semiconductors has been widely applied in plasmon-driven photocatalysis. Here, a uniform Ag@MoS$_2$ hybrid film is fabricated by depositing MoS$_2$ onto a thin Ag film *via* the pulsed laser deposition (PLD) technique. The thickness and crystallinity of MoS$_2$ can be adjusted by controlling the PLD deposition time and temperature, respectively. With the assistance of surface-enhanced Raman scattering (SERS) analysis, the Raman enhancement uniformity of the substrate and plasmon-driven reaction of *p*-nitrothiophenol (PNTP) dimerizing into *p,p′*-dimercaptobenzene (DMAB) are carefully studied on Ag@MoS$_2$ film substrates with different MoS$_2$ crystallinities. The Raman enhancement decreases with increased MoS$_2$ thickness, due to the weakened electromagnetic field enhancement as suggested by finite-difference time-domain (FDTD) simulations. The increased crystallinity of MoS$_2$ can efficiently accelerate the hot electron transfer process, resulting in the enhancement of SERS activity and the improved efficiency of the plasmon-driven reaction. This study may pave the way for the design of other uniform metal–semiconductor hybrids for use as SERS substrates and photocatalysts.

Introduction

Surface plasmon resonance (SPR) on the surfaces of noble metals at the nanoscale has attracted wide attention owing to its potential applications in surface enhanced Raman scattering (SERS), thermotherapy and photocatalysis.[1–4] In particular, the hot electrons generated on the surfaces of plasmonic metal (*e.g.* Ag and Au) nanostructures from SPR can drive some chemical reactions,[5–8] and the SERS technique can be used for *in situ* monitoring of the dynamic reaction

[a]*School of Chemistry and Chemical Engineering, Harbin Institute of Technology, Harbin 150001, China. E-mail: pxu@hit.edu.cn*

[b]*Beijing Key Laboratory for Magneto-Photoelectrical Composite and Interface Science, Center for Green Innovation, School of Mathematics and Physics, University of Science and Technology Beijing, Beijing 100083, China*

[c]*Key Laboratory of Photochemical Conversion and Optoelectronic Materials, Technical Institute of Physics and Chemistry, Chinese Academy of Sciences, Beijing 100190, China. E-mail: lijingsp@mail.ipc.ac.cn*

process.[9] Thus far, plasmon-driven catalytic reactions of the photodimerization of *p*-nitrothiophenol (PNTP) or *p*-aminothiophenol (PATP) into *p*,*p*′-dimercaptobenzene (DMAB) have been mostly studied on plasmonic metal nanostructures.[7,10,11] It has been reported that the laser power, reaction atmosphere and SERS substrate can strongly affect the rate and efficiency of such reactions.[12–15] For the reduction reaction of PNTP into DMAB, hot electrons originating from the decay of SPR are considered to play an essential role in this plasmon-driven reaction.[13,16] However, the lifetime of the hot electrons is very short due to fast decay, which strongly restricts the efficiency of the hot electron induced reduction reactions.[17] One promising way to effectively transfer and prolong the lifetime of hot electrons is to combine metals with semiconductor materials,[18–21] thus benefiting from the coupling between plasmons and excitons in semiconductors.

Plasmonic metal nanostructures combined with semiconductors, such as TiO_2,[22–24] h-BN,[25] graphene[26–28] and MoS_2,[17,29] have been utilized to investigate plasmon-driven reactions. When semiconductors with a large bandgap (h-BN) cover the Ag nanoparticles, this can prevent oxidation of the metal, whilst the probability of reduction reactions occurring will be compromised because h-BN can totally block the transfer of hot electrons.[25] In contrast, graphene can greatly facilitate the transfer of hot electrons on a metal surface, thus significantly prolonging the lifetime of the hot electrons and increasing the reaction efficiency for PNTP dimerizing into DMAB.[26] MoS_2 is a promising hot electron acceptor owing to its considerable bandgap (between 1–2 eV) and various traps at interfaces.[20,30] Compared to the bare Ag substrate, monolayer Ag@MoS_2 nanoparticle hybrids have been reported to strongly improve the efficiency of conversion of PNTP into DMAB based on a plasmon-exciton co-driven pathway.[17] Our previous work has revealed that both the thickness and crystallinity of MoS_2 can influence the hot electron transfer process.[29] However, most of these 2D materials and metals are first fabricated individually and then combined *via* a transfer method,[17,25] which presents the difficulty of guaranteeing a well-defined interface structure and ensuring the uniformity and repeatability of substrates. Moreover, the detailed mechanisms of plasmon-exciton interaction also remain to be deeply understood.

In this work, we demonstrate the fabrication of a uniform Ag@MoS_2 film by depositing MoS_2 on a Ag film *via* the pulsed laser deposition (PLD) technique.[31] A Ag film (thickness of about 12 nm) with a uniform particle size distribution grown by the electron beam evaporation method is used as the deposition substrate.[32] The uniformity of the SERS signals has been confirmed through the use of Raman mapping technology. The SERS enhancement ability and the efficiency of the plasmon-driven reactions are strongly affected by the thickness and crystallinity of the MoS_2 layer (Scheme 1). This work might provide a new way to design uniform substrates through the combination of semiconductors and metal nanostructures for SERS and plasmon-driven reaction applications.

Experimental

Preparation of the Ag film and Ag@MoS_2 hybrid nanostructures

Ag nanoparticles with a thickness of 12 nm were deposited onto 300 nm SiO_2/Si substrates *via* electron beam evaporation with a deposition rate of 0.1–0.2 $Å\ s^{-1}$.

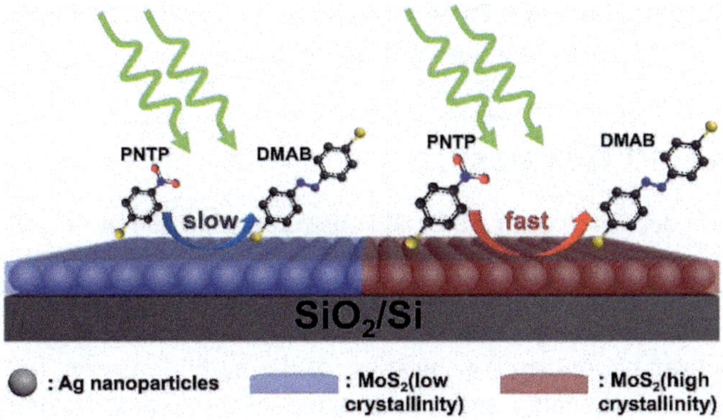

Scheme 1 Schematic diagram of the surface plasmon-driven conversion of PNTP into DMAB on a Ag@MoS_2 film with different MoS_2 crystallinity.

MoS_2 films with different thicknesses were coated on the Ag nanoparticles *via* the pulsed laser deposition (PLD) technique. MoS_2 (99.9%) purchased from Alfa Aesar was used as the target material. The MoS_2 films were deposited at 10^{-7} Torr and the deposition temperature was controlled from 20 °C to 500 °C. A pulsed KrF gas excimer laser ($\lambda = 248$ nm, 3 Hz, 1–2 J cm^{-2}) was used.

Raman and SERS measurements

A Rhodamine 6G (R6G) ethanol solution at different concentrations (10^{-5} to 10^{-8} M) and PNTP (10^{-3} M) were used for SERS detection and the SP-induced reduction reaction, respectively. The Ag@MoS_2 film on SiO_2/Si substrates were immersed in the solution for more than 1 h, and the substrates were then washed several times with ethanol and dried in air. SERS spectra were collected using a Renishaw inVia Raman microscope under 532 nm laser excitation. The laser power was about 0.65 mW and a 50× objective lens was used. The laser spot size was about 1 μm and the acquisition time was 10 s.

Characterization

Scanning electron microscope (SEM) images of Ag nanoparticles were obtained using an FEI HELIOS NanoLab 600i instrument at an accelerating voltage of 20 kV. Raman spectra were collected using a Renishaw inVia Raman microscope under 532 nm laser excitation. Atomic force microscopy (AFM) images were obtained using a Bruker Dimension ICON system in the ScanAsyst mode.

Finite-difference time-domain (FDTD) simulations

Electrodynamics simulations were carried out using the FDTD method in order to study the optical characteristics of the Ag@MoS_2 structure. In the simulations, the Ag NPs array was placed between the glass substrate and the MoS_2 film, with periodic boundary conditions applied for the horizontal axis and perfectly matched layer (PML) boundary conditions for the vertical axis. The refractive indexes of the MoS_2 film, Ag NPs and glass substrate are obtained from the data reported by C. Yim *et al.*,[33] Johnson and Christy,[34] and Palik,[35] respectively. The diameter and gap of the Ag NPs were set according to the experimental result of the Ag NPs array obtained *via* the AFM method.

Results and discussion

A Ag film was deposited onto a SiO_2/Si substrate with a 300 nm thick oxide layer *via* electron beam evaporation.[32] Fig. 1 shows an SEM image of the as-prepared Ag film consisting of Ag nanoparticles that are highly uniform in size and well dispersed on the surface of the SiO_2/Si substrate. Atomic force microscopy (AFM) was also utilized to confirm the thickness of the Ag film. As seen in Fig. 2a, the AFM image shows the fairly uniform thickness of the film. From analysis of the height plots of three different areas in Fig. 2b, we can determine that the average heights of these lines are 12.8 nm, 12.3 nm and 11.3 nm, respectively. The uniform distribution of Ag nanoparticles and nanogaps between the particles can ensure the uniformity of "hot spot" and SERS signals.[1,2]

MoS_2 films were deposited onto the SiO_2/Si substrate with or without Ag nanoparticles *via* the pulsed laser deposition (PLD) method.[29,31] The thickness and crystallinity were adjusted by controlling the deposition time and the temperature of the substrates. According to our previous work,[29] the thicknesses of the MoS_2 films were about 4, 8, 11, 15, 27 and 50 nm for the different deposition times of 5 min, 10 min, 20 min, 30 min, 40 min and 1 h, respectively. Fig. 3 shows the Raman spectra of the MoS_2 films with different deposition times on SiO_2/Si substrates at a deposition temperature of 500 °C. Obvious Raman peaks at 372 and 402 cm^{-1}, corresponding to 2H–MoS_2, can be seen in all of the spectra,[19] thus indicating good crystallinity at this deposition temperature. Moreover, the peak at 520 cm^{-1}, corresponding to the Si substrate, becomes weaker as the deposition time increases from 5 min to 2 h, thus confirming the increase in the thickness of the MoS_2 films.

In order to evaluate the Raman enhancement of the Ag@MoS_2 films with different MoS_2 layer thicknesses, rhodamine 6G (R6G) dye (10^{-6} M) was used as a probe molecule for SERS detection. A large-area Raman mapping image at 610 cm^{-1} of R6G (Fig. 4b) was obtained for the Ag@MoS_2 film substrates (Fig. 4a),

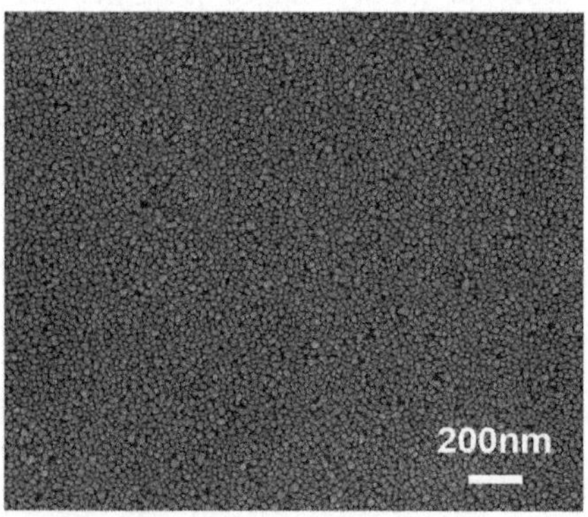

Fig. 1 SEM image of a Ag film consisting of uniform nanoparticles on a SiO_2/Si substrate prepared *via* the electron beam evaporation technique.

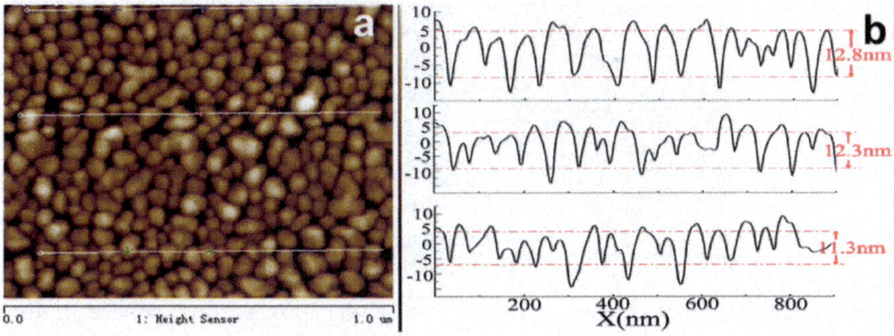

Fig. 2 (a) AFM image of Ag nanoparticles on a SiO_2/Si substrate. (b) Height plots of the three lines shown in (a). The thickness of the Ag film layer was estimated to be 12 nm.

with uniform Raman signals demonstrating the uniformity of the Ag@MoS_2 substrates. It can be seen in Fig. 5a that the remarkable Raman signals of R6G at 610, 773, 1185, 1311, 1362, 1506, 1570 and 1647 cm^{-1} are observed on the Ag@MoS_2 film substrate with a PLD deposition time of 5 min.[27] The intensities of these peaks gradually decrease with the increasing thickness of the MoS_2 layer, with the peaks even disappearing at a PLD deposition time over 40 min, as the thicker MoS_2 layer weakens the surface plasmon resonance (SPR) induced electromagnetic enhancement effect of the Ag nanoparticles.[17] The effect of the crystallinity of the MoS_2 film was also investigated in order to compare the enhancement effect of the Ag@MoS_2 film substrates (Fig. 5b) with the same MoS_2 layer thickness (PLD time = 10 min, 8 nm). Significantly increased Raman signals of R6G can be observed with increasing PLD temperatures from 20 to 500 °C. In particular, when the crystallization temperature is higher than 300 °C, the enhancement ability of the Ag@MoS_2 film substrate remains at almost the same level. We suggest that the SERS effect of the Ag@MoS_2 film originates from

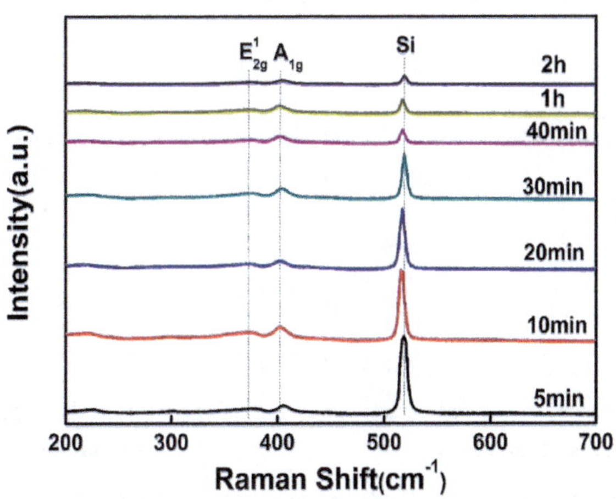

Fig. 3 Raman spectra of MoS_2 films on SiO_2/Si substrates with different deposition times at a deposition temperature of 500 °C.

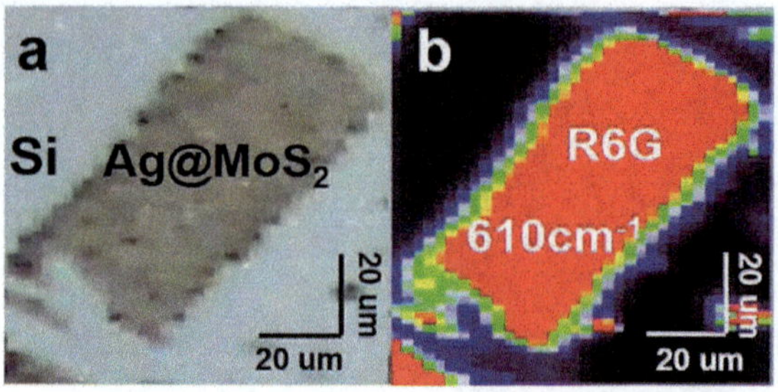

Fig. 4 (a) Optical image of $Ag@MoS_2$ films on a SiO_2/Si substrate (PLD time of 10 min, 500 °C). (b) Raman mapping image of the area shown in (a) for the vibration mode of R6G at 610 cm^{-1}.

a combination of the surface plasmon resonance of the Ag nanoparticles and the charge transfer between MoS_2 and adsorbed molecules, and increasing the crystallinity of MoS_2 may largely increase the charge transfer between adsorbed molecules and substrates.[29]

To further confirm the influence of MoS_2 on surface plasmon driven reactions on $Ag@MoS_2$ films, a typical reduction reaction of PNTP dimerizing into DMAB was carried out on $Ag@MoS_2$ substrates with different MoS_2 crystallinities (the thickness of MoS_2 was controlled by using the same deposition time of 10 min, 8 nm) (Fig. 6) and on bare Ag nanoparticles (Fig. 7). The time-dependent SERS spectra for the $Ag@MoS_2$ substrates clearly indicate the conversion of PNTP (10^{-3} M) into DMAB, as evidenced by the decreasing intensity of the Raman peak at 1330 cm^{-1}, attributed to the $\nu(NO_2)$ of PNTP, and the emerging bands at 1391 and 1430 cm^{-1}, attributed to the ν(N=N) of the newly formed DMAB.[36,37] As the reaction continues, the intensities of the ν(N=N) peaks at 1391 and 1430 cm^{-1} become stronger and the $\nu(NO_2)$ peak at 1330 cm^{-1} gradually decreases, thus indicating that the amount of the product DMAB is increasing during the reaction. When the deposition temperature of PLD is at 20 °C, 100 °C and 200 °C, the intensity of the nitro group is still higher than that of the ν(N=N) peaks of DMAB even after a reaction time of 120 s, thus indicating a very low conversion efficiency.

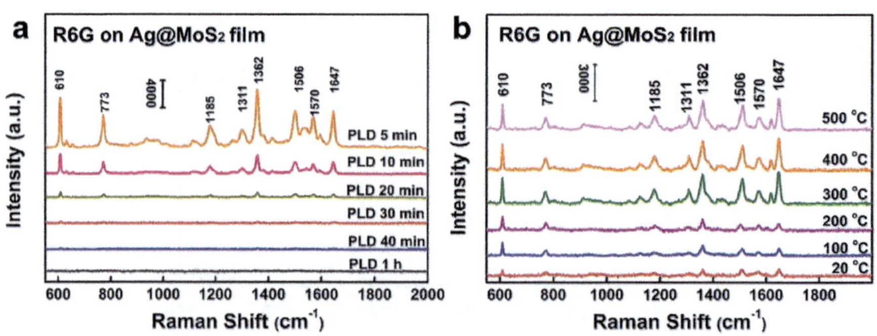

Fig. 5 Raman spectra of R6G (10^{-6} M) on (a) $Ag@MoS_2$ films with different deposition times at the same deposition temperature of 500 °C, and (b) on $Ag@MoS_2$ films with different deposition temperatures at the same deposition time of 10 min.

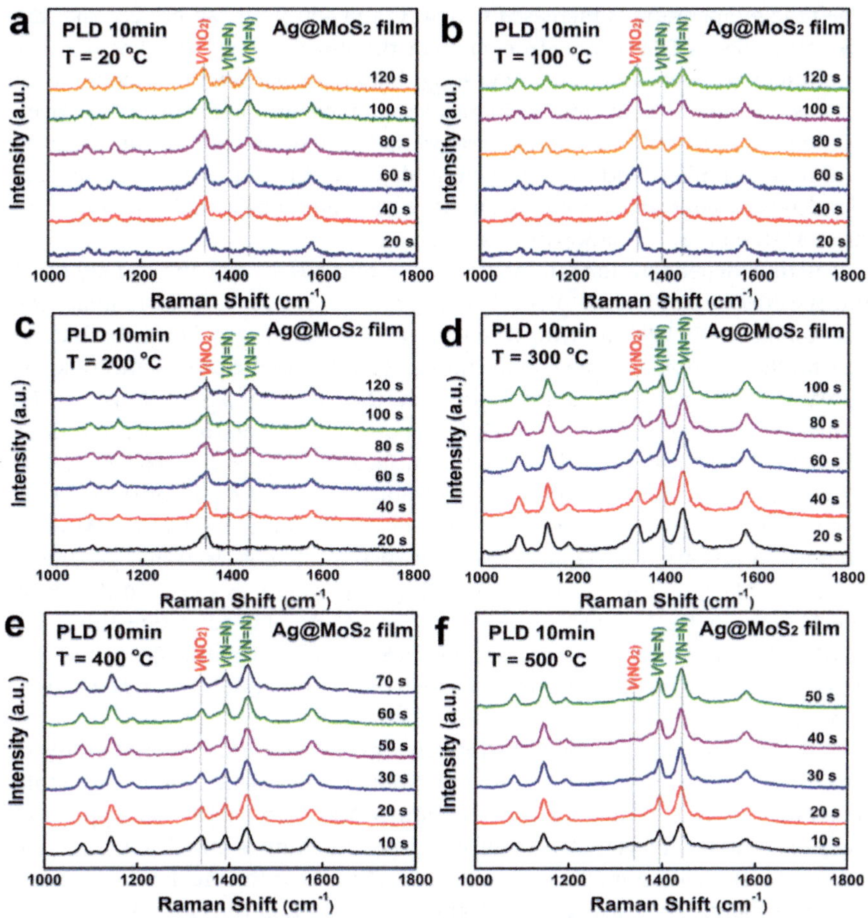

Fig. 6 Time-dependent Raman spectra of PNTP dimerizing into DMAB on Ag@MoS_2 substrates at different deposition temperatures (a–f, $T = 20$, 100, 200, 300, 400 and 500 °C) with the same deposition time (PLD time = 10 min) under 532 nm laser excitation (0.65 mW).

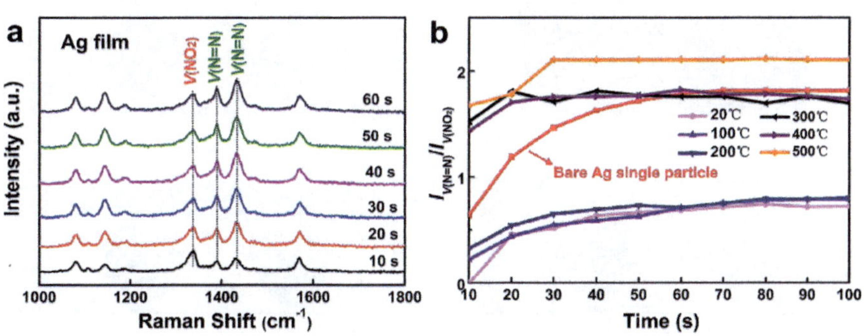

Fig. 7 (a) Time-dependent Raman spectra of PNTP dimerizing into DMAB on a bare Ag film. (b) Time-dependent Raman intensity ratios of the ν(N=N) and $\nu(NO_2)$ bands for the Ag@MoS_2 substrates prepared at different PLD deposition temperatures at the same deposition time of 10 min (thickness of 8 nm) under 532 nm laser excitation (0.65 mW).

When the temperature is increased to 300 °C and 400 °C, it can be seen that the ν(N=N) peaks of DMAB are stronger than the nitro peak of PNTP at the very beginning of the reaction, thus suggesting that an increase in the crystallinity of MoS_2 is conducive to an improvement in the efficiency of this plasmon-driven reaction. Moreover, when the temperature is increased to 500 °C, the reaction is almost complete within a few seconds, and the nitro peak disappears completely. In contrast, it can be seen that PNTP cannot be completely converted into DMAB on bare Ag nanoparticles (Fig. 7a).

The time-dependent Raman intensity ratio of the ν(N=N) band at 1430 cm^{-1} and the $\nu(NO_2)$ band at 1330 cm^{-1}, $I_{\nu(N=N)}/I_{\nu(NO_2)}$, can be used to estimate the reaction rate of PNTP dimerizing into DMAB.[13,16] It can be seen in Fig. 7b that the time for the reaction to reach equilibrium gets shorter as the crystallinity of MoS_2 increases. Moreover, in comparison with the Ag nanoparticles, when the temperature is higher than 300 °C the reaction efficiency on the Ag@MoS_2 films can surpass the reaction efficiency on the surface of the silver nanoparticles, thus indicating that a higher crystallinity is favorable for promoting the progress of the reaction. Considering that the reaction of PNTP dimerizing into DMAB is a reduction process that requires the participation of "hot electrons" generated *via* surface plasmon decay, the MoS_2 with high crystallinity is shown to efficiently transfer the "hot electrons" from the surface of Ag and accelerate this reaction.[29,38,39] In contrast, with decreasing MoS_2 crystallinity, the generated hot electrons may be restrained, meaning that the plasmon-induced reaction efficiency is reduced.

In order to better understand the influence of the thickness and crystallinity of MoS_2 on the Raman enhancement and plasmon-driven reaction on the Ag@MoS_2 substrate, 3D finite-difference time-domain (FDTD) simulation was applied to study the optical characteristics of the Ag@MoS_2 structure. A middle cross-section view of the simulation model is shown in Fig. 8a, where the incident plane wave propagates along the *z* axis direction and its polarization direction is parallel to the *x* axis. In order to investigate the SPR characteristics of the Ag@MoS_2

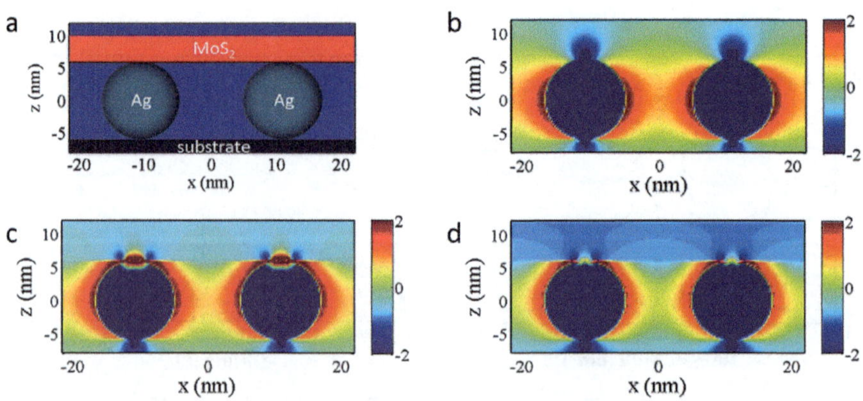

Fig. 8 (a) Structural model of the Ag@MoS_2 structure used in the FDTD simulations. In the simulations, we set the diameter of the Ag NPs and the gap between Ag NPs as constants of 12 nm and 10 nm, respectively. Electric field distributions (logarithmic scale) of the Ag NP array (b) without a MoS_2 film, (c) with a 4 nm MoS_2 film, and (d) with an 8 nm MoS_2 film.

structure, we calculated the electric field distributions by varying the thickness of the MoS_2 film (0 nm, 4 nm and 8 nm), with the results shown in Fig. 8b–d. The electric field distributions clearly show that there are strong electromagnetic enhancements above the Ag NPs as the thickness of the MoS_2 film is increased from 0 nm to 4 nm. However, a further increase to 8 nm weakens the electromagnetic enhancement. Therefore, an increased thickness of MoS_2 will have a negative effect on the Raman enhancement on the Ag@MoS_2 substrates. Increasing the crystallinity of the MoS_2 film can reduce the amount of the grain boundary, thus resulting in decreased electron scattering. Thus, there will be a smaller heat effect for the electron transfer process, and the efficiency of the plasmon-driven reaction can be improved.

Conclusions

In conclusion, we have studied the SERS activity and plasmon-driven reaction of PNTP dimerizing into DMAB on uniform Ag@MoS_2 film substrates prepared by depositing MoS_2 on a Ag film *via* a pulsed laser deposition (PLD) method. Large-area Raman mapping technology confirmed the uniformity of the hybrid SERS substrates. The Raman enhancement on the Ag@MoS_2 film substrates, as indicated by the FDTD simulations, decreases upon increasing the thickness of the MoS_2 film due to the weakened electromagnetic field. The reaction rate and efficiency of the conversion of PNTP to DMAB are greatly accelerated when the deposition temperature is increased from room temperature to 500 °C (with an increase in the crystallinity of MoS_2). At a deposition temperature higher than 300 °C, the reaction efficiency on the Ag@MoS_2 films can surpass the reaction efficiency on the surface of the Ag nanoparticles. Considering the high Raman signal uniformity and high catalytic activity on the Ag@MoS_2 film substrates, our results suggest that the hybrid metal–semiconductor SERS substrates hold great promise for photocatalytic reactions and sensors.

Conflicts of interest

There are no conflicts to declare.

Acknowledgements

We are thankful for the financial support received from the National Natural Science Foundation of China (21471039, 21571043, 21671047, 21871065), and the Natural Science Foundation of Heilongjiang Province (B2015001).

References

1 S.-Y. Ding, E.-M. You, Z.-Q. Tian and M. Moskovits, *Chem. Soc. Rev.*, 2017, **46**, 4042–4076.
2 S. Lal, N. K. Grady, J. Kundu, C. S. Levin, J. B. Lassiter and N. J. Halas, *Chem. Soc. Rev.*, 2008, **37**, 898–911.
3 S.-Y. Ding, J. Yi, J.-F. Li, B. Ren, D.-Y. Wu, R. Panneerselvam and Z.-Q. Tian, *Nat. Rev. Mater.*, 2016, **1**, 16021.
4 W. Xie and S. Schlücker, *Nat. Commun.*, 2015, **6**, 7570.

5 Z. Zheng, T. Tachikawa and T. Majima, *J. Am. Chem. Soc.*, 2015, **137**, 948–957.
6 Z. Zhang, P. Xu, X. Yang, W. Liang and M. Sun, *J. Photochem. Photobiol., C*, 2016, **27**, 100–112.
7 Y. Fang, Y. Li, H. Xu and M. Sun, *Langmuir*, 2010, **26**, 7737–7746.
8 W. Xie, B. Walkenfort and S. Schlucker, *J. Am. Chem. Soc.*, 2013, **135**, 1657–1660.
9 X. Tang, W. Cai, L. Yang and J. Liu, *Nanoscale*, 2014, **6**, 8612–8616.
10 Y.-F. Huang, M. Zhang, L.-B. Zhao, J.-M. Feng, D.-Y. Wu, B. Ren and Z.-Q. Tian, *Angew. Chem., Int. Ed.*, 2014, **53**, 2353–2357.
11 W. Lin, E. Cao, L. Zhang, X. Xu, Y. Song, W. Liang and M. Sun, *Nanoscale*, 2018, **10**, 5482–5488.
12 L. Kang, P. Xu, B. Zhang, H. Tsai, X. Han and H.-L. Wang, *Chem. Commun.*, 2013, **49**, 3389–3391.
13 P. Xu, L. Kang, N. H. Mack, K. S. Schanze, X. Han and H.-L. Wang, *Sci. Rep.*, 2013, **3**, 2997.
14 C. H. Choi, L. Lin, S. Gim, S. Lee, H. Kim, X. Wang and W. Choi, *ACS Catal.*, 2018, **8**, 4241–4256.
15 Q. Ding, M. Chen, Y. Fang, Z. Zhang and M. Sun, *J. Phys. Chem. C*, 2017, **121**, 5225–5231.
16 N. C. Brandt, E. L. Keller and R. R. Frontiera, *J. Phys. Chem. Lett.*, 2016, **7**, 3179–3185.
17 X. Yang, H. Yu, X. Guo, Q. Ding, T. Pullerits, R. Wang, G. Zhang, W. Liang and M. Sun, *Materials Today Energy*, 2017, **5**, 72–78.
18 M. Achermann, *J. Phys. Chem. Lett.*, 2010, **1**, 2837–2843.
19 Y. Kang, S. Najmaei, Z. Liu, Y. Bao, Y. Wang, X. Zhu, N. J. Halas, P. Nordlander, P. M. Ajayan, J. Lou and Z. Fang, *Adv. Mater.*, 2014, **26**, 6467–6471.
20 Y. Yu, Z. Ji, S. Zu, B. Du, Y. Kang, Z. Li, Z. Zhou, K. Shi and Z. Fang, *Adv. Funct. Mater.*, 2016, **26**, 6394–6401.
21 L. Cui, X. Ren, X. Yang, P. Wang, Y. Qu, W. Liang and M. Sun, *J. Raman Spectrosc.*, 2016, **47**, 877–883.
22 J. Chu, P. Miao, X. Han, Y. Du, X. Wang, B. Song and P. Xu, *ChemCatChem*, 2016, **8**, 1819–1824.
23 J. Wang, R. A. Ando and P. H. C. Camargo, *Angew. Chem., Int. Ed.*, 2015, **127**, 7013–7016.
24 Q. Ding, R. Li, M. Chen and M. Sun, *Appl. Mater. Today*, 2017, **9**, 251–258.
25 G. Kim, M. Kim, C. Hyun, S. Hong, K. Y. Ma, H. S. Shin and H. Lim, *ACS Nano*, 2016, **10**, 11156–11162.
26 Z.-G. Dai, X.-H. Xiao, W. Wu, Y.-P. Zhang, L. Liao, S.-S. Guo, J.-J. Ying, C.-X. Shan, M.-T. Sun and C.-Z. Jiang, *Light: Sci. Appl.*, 2015, **4**, e342.
27 N. Zhang, L. Tong and J. Zhang, *Chem. Mater.*, 2016, **28**, 6426–6435.
28 E. Cao, X. Guo, L. Zhang, Y. Shi, W. Lin, X. Liu, Y. Fang, L. Zhou, Y. Sun, Y. Song, W. Liang and M. Sun, *Adv. Mater. Interfaces*, 2017, **4**, 1700869.
29 Y. Shen, P. Miao, C. Hu, J. Wu, M. Gao and P. Xu, *ChemCatChem*, 2018, **10**, 3520–3525.
30 W. Zhao, S. Wang, B. Liu, I. Verzhbitskiy, S. Li, F. Giustiniano, D. Kozawa, K. P. Loh, K. Matsuda, K. Okamoto, R. F. Oulton and G. Eda, *Adv. Mater.*, 2016, **28**, 2709–2715.
31 C. Hu, X. Wang, P. Miao, L. Zhang, B. Song, W. Liu, Z. Lv, Y. Zhang, Y. Sui, J. Tang, Y. Yang, B. Song and P. Xu, *ACS Appl. Mater. Interfaces*, 2017, **9**, 18362–18368.

32 H. Sun, H. Liu and Y. Wu, *Appl. Surf. Sci.*, 2017, **416**, 704–709.
33 C. Y. Yim, M. O'Brien, N. McEvoy, S. Winters, I. Mirza, J. G. Lunney and G. S. Duesberg, *Appl. Phys. Lett.*, 2014, **104**, 103114.
34 P. B. Johnson and R. W. Christy, *Phys. Rev. B: Solid State*, 1972, **6**, 4370–4379.
35 E. D. Palik, *Handbook of Optical Constants of Solids*, Elsevier Inc., Academic Press, 1997.
36 E. M. van Schrojenstein Lantman, O. L. Gijzeman, A. J. Mank and B. M. Weckhuysen, *ChemCatChem*, 2014, **6**, 3342–3346.
37 L. B. Zhao, M. Zhang, Y. F. Huang, C. T. Williams, D. Y. Wu, B. Ren and Z. Q. Tian, *J. Phys. Chem. Lett.*, 2014, **5**, 1259–1266.
38 S. Zu, B. Li, Y. Gong, Z. Li, P. M. Ajayan and Z. Fang, *Adv. Opt. Mater.*, 2016, **4**, 1463–1469.
39 E. Cao, W. Lin, M. Sun, W. Liang and Y. Song, *Nanophotonics*, 2018, **7**, 145–167.

Faraday Discussions

PAPER

The role of a plasmonic substrate on the enhancement and spatial resolution of tip-enhanced Raman scattering†

Mahfujur Rahaman, *[a] Alexander G. Milekhin, [bc] Ashutosh Mukherjee, [a] Ekaterina E. Rodyakina, [bc] Alexander V. Latyshev,[bc] Volodymyr M. Dzhagan [ad] and Dietrich R. T. Zahn [a]

Received 1st October 2018, Accepted 7th November 2018

DOI: 10.1039/c8fd00142a

Since the first report in the early 2000s, there have been several experimental configurations that have demonstrated enhancement and spatial resolution of tip-enhanced Raman spectroscopy (TERS). The combination of a plasmonic substrate and a metallic tip is one suitable approach to achieve even higher enhancement and lateral resolution. In this contribution, we demonstrate TERS on a monolayer of MoS_2 on an array of Au nanodisks. The Au nanodisks were prepared by electron beam writing. Thereafter, MoS_2 was transferred onto the plasmonic substrate *via* the exfoliation technique. We witness an unprecedented enhancement and spatial resolution in the experiments. In the TERS image a ring-like shape is observed that matches the edges of the nanodisks. TERS enhancement at the edges is about 170 times stronger than at the center of the nanodisks. For a better understanding of the experimental results, finite element method (FEM) simulations were employed to simulate the TERS image of the MoS_2/plasmonic heterostructure. Our calculations show a higher electric field concentration at the edges that exponentially decays to the center. Therefore, it reproduces the ring-like shape of the experimental image. Moreover, the calculations suggest a TERS enhancement of 135 at the edges compared to the center, which is in very good agreement with the experimental data. According to our calculations, the spatial resolution is also increased at the edges. For comparison, FEM simulations of a tip–flat metal substrate system (conventional gap-mode TERS) were carried out. The calculations confirmed a 110 times stronger enhancement at the edges of the nanodisks than that of conventional gap-mode TERS and explained the experimental

[a] *Semiconductor Physics, Chemnitz University of Technology, D-09107 Chemnitz, Germany. E-mail: mahfujur.rahaman@physik.tu-chemnitz.de*

[b] *Rzhanov Institute of Semiconductor Physics RAS, Lavrentiev Ave. 13, 630090 Novosibirsk, Russia*

[c] *Novosibirsk State University, Pirogov 2, 630090 Novosibirsk, Russia*

[d] *Lashkaryov Institute of Semiconductors Physics, National Academy of Sciences of Ukraine, 03028 Kyiv, Ukraine*

† Electronic supplementary information (ESI) available. See DOI: 10.1039/c8fd00142a

maps. Our results provide not only a deeper understanding of the TERS mechanism of this heterostructure, but can also help in realizing highly efficient TERS experiments using similar systems.

Introduction

Non-destructive Raman spectroscopy is widely used to obtain information about the molecular composition, chemical bonds, crystal structure, phase composition, and chemical environment of the sample under investigation.[1–6] This technique relies on the inelastic scattering of light predominantly by the vibrational modes (phonons) of the material. However, the spatial resolution of this technique is usually diffraction limited and amounts to several hundreds of nanometers in the visible spectral range. However, with the advent of nanotechnology, there has been a growing demand to probe the physical properties of materials with a spatial resolution down to a few nanometers, and even down to the sub-nanometer range.[7] This major limitation of Raman spectroscopy has been overcome through the introduction of tip-enhanced Raman scattering (TERS), as proposed by several groups[8–11] in 2001. Since then, this method has been rapidly developed into an important experimental tool for materials analysis on the nanoscale.[12–15] TERS combines the advantages of Raman spectroscopy with the imaging capabilities of scanning probe microscopy (SPM) using the localized surface plasmon resonance (LSPR) of a sharp metallic tip. The spatial resolution of a TERS experiment depends on the tip radius and the coupling between the tip and the sample.[16] When using a suitable excitation laser energy, the electric field in the vicinity of the tip is strongly enhanced due to LSPR and the lightning rod effect.[14] The confined local electric field at the tip apex enhances the Raman signal from the specimen underneath and thus surpasses two major drawbacks of conventional Raman spectroscopy by both increasing the Raman sensitivity and breaking the diffraction limit.

Over the last decade, TERS has been used for organic molecules not only to study the physical properties of the materials, but also to aid in the understanding of the underlying physics of the technique involving enhancement and spatial resolution.[17–25] TERS has also been used to study the physical properties of inorganic materials such as CdS,[26] CdSe[27] and GaAs films,[28] local strain in Si,[29] GaN,[30] Ge[31] and GeSi,[32] spin-wave modes in FeO_3,[33] and, very recently, 2D materials.[34–40] Since TERS is a fairly new technique, reports are being published intermittently addressing different technical issues to improve the sensitivity (recent papers and comprehensive reviews can be found in ref. 16, 34 and 41–46). The most significant advantages of TERS are the improvement of Raman enhancement and spatial resolution, and one suitable way of improving these parameters is to use a metallic substrate.[47–50] In such a system, an image dipole is created on the metal surface and thus the local electric field becomes stronger and confined in a narrow lateral space (this system is also known as a gap-mode configuration). However, in our previous reports we showed that a plasmonic substrate can enhance the TERS sensitivity dramatically due to the coupling between the tip and the plasmonic structures.[39,40] The local electric field is particularly pronounced on the sharp metallic edges of the plasmonic substrate, where the tip–substrate effective dimer is the strongest (also known as hot spots).

Usually, the size of the hot spots is very small in comparison to the volume of the metal nanostructures. The formation of hot spots in metal nanostructures is a widely used concept in surface-enhanced Raman scattering (SERS).[46,51] By creating such hot spots between a sharp metallic tip and plasmonic nanostructures, we can improve the sensitivity in a TERS experiment. This model system can be used for a wide range of samples from inorganic/organic semiconductors to biomolecules. However, for the use of such a system in a more routine way, it is essential to understand the hot spot formation leading to the TERS enhancement, especially due to the fact that the far field coupling can be different to the near field coupling.[52] In far field coupling, the plasmonic structures are only sensitive to the excitation wavelength. However, in near field coupling, the interaction is dominated by the tip (in a dimer configuration) and controls the confinement parameters. In this contribution, we investigated this effect using Au nanodisks as plasmonic structures. Since two dimensional (2D) materials exhibit strong coupling to plasmonic fields,[53–55] they can be used as probing materials for the TERS mechanisms in such configurations. Besides their unique physical properties, these 2D materials are stable and straightforward to be prepared and handled. Therefore, monolayer MoS_2 is used as the probing material in the current study. The experimental findings are compared with the results obtained from FEM simulations.

Experimental details

TERS experiments

Details of the preparation of the sample and tip (a typical SEM image of the tip is shown in Fig. S1;† the tip radius is approximately 40 nm), as well as of the TERS experiments, are given in ref. 27 and 39. In brief, TERS imaging was performed on a NanoRaman Platform in side-illumination geometry from HORIBA Scientific (formerly AIST-NT). The incident and scattered light was collected by a long working distance objective (100×, 0.7 N.A.) inclined at an angle of 65° with respect to the normal of the sample surface. The atomic force microscope (AFM) was operated in the intermittent contact mode during the measurements of the topography (optical and AFM images of a MoS_2 flake are presented in Fig. S2 and S3†), while TERS spectra were acquired with the tip touching the substrate and with the contact time defined by the spectral acquisition time of 0.4 s. An electron multiplying charge coupled detector was used for the detection of the scattered light dispersed by a 600 l per mm grating. The monolayer thickness of the flake was confirmed by analysis of the Raman and PL measurements presented in Fig. S4.†

FEM simulations

All FEM calculations were performed using the COMSOL Multiphysics 5.2a simulation software in both 2D and 3D platforms. The calculations were performed solving Maxwell's equations in the frequency domain as in eqn (1):

$$\nabla \times {\mu_\mathrm{r}}^{-1}(\nabla \times E) - {k_0}^2\left(\varepsilon_\mathrm{r} - \frac{\mathrm{j}\sigma}{\omega\varepsilon_0}\right)E = 0 \quad (1)$$

where ω, μ_r, ε_r and σ correspond to the excitation frequency, relative permeability, relative permittivity and electrical conductivity, respectively. Assuming the tip in the simplest form of a sphere, an image dipole is induced in the substrate, as shown in Fig. 1. The local electric field, E_{loc}, can be expressed as a function of the excitation field, E_{far}, as follows:[56,57]

$$E_{\text{loc}} = \left(\frac{2R + d}{d}\right) E_{\text{far}} \tag{2}$$

With $d \approx 0$, eqn (2) reads as

$$E_{\text{loc}} \approx \left(\frac{2R}{d}\right) E_{\text{far}} \tag{3}$$

Therefore, one can expect a dramatic enhancement of the local electric field when the tip moves close to the sample. The full width at half maximum (FWHM) of the confined electric field is then given by:[16,56,57]

$$w_{\text{loc}} = 2\sqrt{Rd} \tag{4}$$

The minimum elemental mesh size was chosen to be 0.1 nm for accurate electric field distribution calculation with a maximum size limited by $\lambda/5$; here λ is the wavelength. The optical constants of gold were taken from ref. 58. Maxwell's equations were solved for the scattered field with the incident electric field polarization matching the tip long axis and the experimental configuration as discussed in the next section. Linearly polarized light at 1 V m^{-1} was used for excitation. A perfectly matched layer (PML) was used as the boundary domain to absorb all outgoing photons.

Results and discussion

TERS imaging of the MoS_2/plasmonic heterostructure

A schematic of the TERS measurement is presented in Fig. 2a. The *p*-polarized light illuminates the tip at an angle of 65° such that the incoming electric field is parallel to the tip longitudinal axis. Therefore, it enhances the field underneath the tip apex to obtain a TERS signal from the monolayer (1L)-MoS_2. A representative TERS spectrum of 1L-MoS_2 was acquired from the heterostructure and is

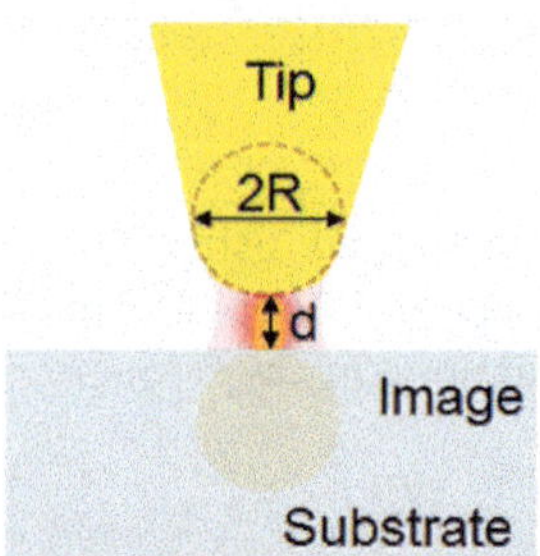

Fig. 1 Schematic of image dipole formation between the tip and substrate. The gradient of the image dipole enhances the local electric field in the gap mode TERS configuration.

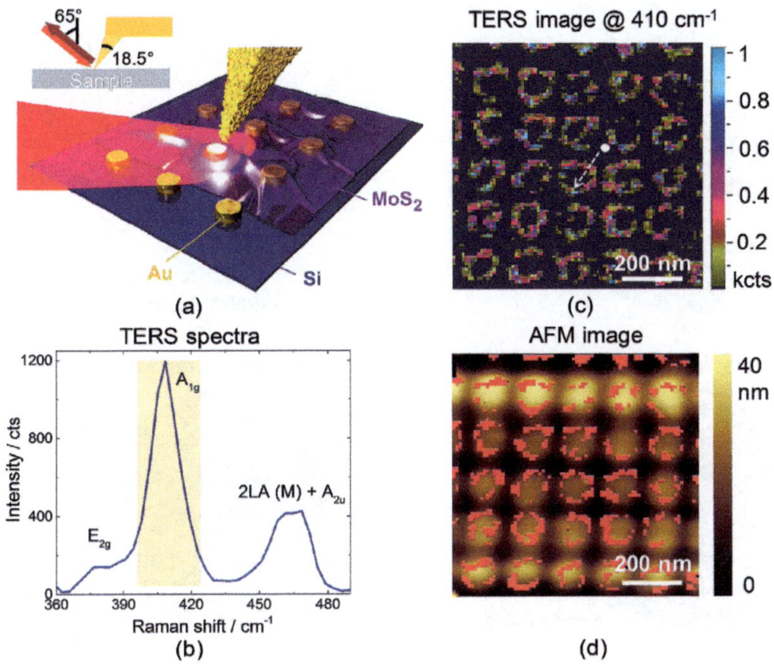

Fig. 2 (a) Schematic of the TERS experiment. The experimental geometry is shown in the inset. (b) A representative TERS spectrum of 1L-MoS_2 on a Au nanocylinder array using 785 nm excitation and a homemade Au coated silicon tip. (c) A spatial map of the Raman intensity of the A_{1g} mode. The white dot with the arrowed line indicates the pixel from which the TERS spectrum was acquired in (b). The TERS map was acquired within the area marked by the shaded rectangular box around the A_{1g} mode in (b). (d) Corresponding AFM topography image of the TERS map. The topography image is superimposed by the maximal intensity map shown in (c). This figure has been reproduced from ref. 39 with permission from the Royal Society of Chemistry.

presented in Fig. 2b. The most pronounced feature observed around 405 cm^{-1} (Fig. 2b) refers to the out-of-plane A_{1g} mode of MoS_2. Details of the peak description are provided in ref. 39. The shaded rectangle indicates the wave-number range from which the TERS map shown in Fig. 2c was built. The spectral acquisition time was 0.4 s and the map area is $0.7 \times 0.7\ \mu m^2$ with a step size of 10 nm. The TERS image in Fig. 2c reveals that the enhanced Raman signal from 1L-MoS_2 forms a pattern of ordered rings. The corresponding topography image with superimposed TERS intensity is presented in Fig. 2d. The superimposed image reveals that the rings perfectly coincide with the Au nanodisks. This indicates that the TERS signal predominantly originates from the edges of the Au nanodisks. In order to understand the origin of such a strong enhancement from the edges of the cylinders, we calculated the electric field distribution for a Au nanocylinder of the same size and shape ($R = 55$ nm, $h = 50$ nm) using FEM. The simulation results are presented in Fig. 3. A schematic of the simulation configuration is shown in Fig. 3a. Linearly polarized light at a wavelength of 785 nm was used for excitation. The incoming light beam has an angle of incidence of 65° with respect to the normal of the sample surface as in the experimental configuration.

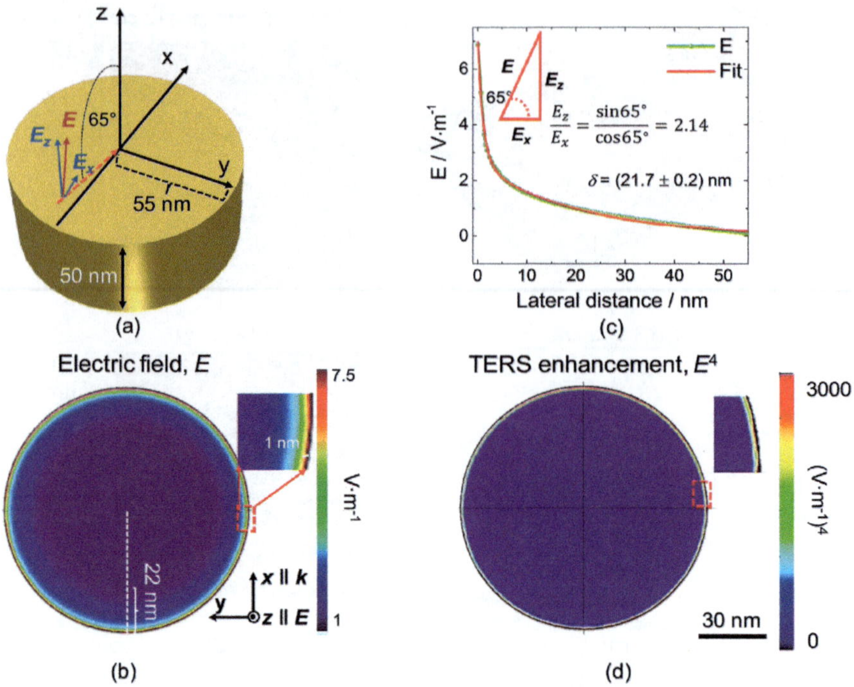

Fig. 3 (a) Schematic of the simulation model used for the Au nanocylinder. (b) Calculated local electric field distribution in the Au cylinder. The strongest field is observed at the sharp edge of the cylinder. The electric field gradually decays towards the center due to the skin effect. The inset image shows a magnified section of the edges taken from the rectangular area marked by a red box. The white dashed line indicates the direction of the cross section of the electric field taken and presented in (c). The cross section decays exponentially towards the center. The inset image shown in (c) corresponds to the electric field component of the excitation. The ratio of the components is 2.14. The cross section of the electric field is fitted using eqn (6). (d) TERS enhancement image taken using the 4th power of the local electric field. The inset image shows the magnified edges taken from the rectangular area marked by a red box.

The incidence geometry allows the electric field to predominantly act parallel to the cylinder axis (z axis); the ratio of the out-of-plane to the in-plane electric field is 2.14. Fig. 3b displays the electric field distribution in the vicinity of the Au nanodisk. The top view of the nanodisk in Fig. 3b reveals that the electric field is strongly localised at the edges and gradually decreases towards the center. However, the strongest enhancement follows the sharp edge of the disk with a drastic decrease within 1 nm from the edge, as shown in the inset of Fig. 3b. This phenomenon can be explained by the combination of edge and skin effects due to electron conduction induced by the optical frequency.[59–61] Since the electric field is proportional to the density of electrons, the field strength is strongest at the sharp edge, *i.e.* at the edges of the nanodisk. This effect is also known as the lightning rod effect and leads to local field enhancement (non-resonance effect).[62,63] The other effect is localized surface plasmon resonance (LSPR), which depends on the excitation wavelength. The conduction electrons predominantly oscillate along the direction of the cylinder axis. However, due to the frequency of the electric field, they move to the surface of the cylinder. Hence, the electric field distribution inside the cylinder forms a ring due to both edge and skin effects, being strongest at the edges and weakest at the center. The skin depth of Au at optical frequency can be calculated as[60]

$$\delta = \frac{c}{\omega_p} \tag{5}$$

where ω_p is the plasma frequency of Au. Using the plasma frequency of Au, $\omega_p = 2\pi c/\lambda_p$, with $\lambda_p = 138$ nm,[64] the skin depth of Au is determined to be $\delta = 22$ nm. This means that the electric field decays to 1/e at a distance of 22 nm towards the center of the cylinder. In order to verify this statement, a cross section was taken from the edge to the center of the cylinder, as shown in Fig. 3b (white dashed line), and the electric field cross section is shown in Fig. 3c. As can be seen from the cross section, the electric field decays exponentially towards the center. To determine the skin depth, the field was fitted using the following equation:[65]

$$E_{loc} = E_{edge}e^{-ax} + \text{sqrt}((E_{lx}e^{-bx})^2 + (E_{lz}e^{-x/\delta})^2) \tag{6}$$

Three local electric field components were used in the fitting, with $E_{edge}\,e^{-ax}$ and $E_{lx}\,e^{-bx}$ being the field gradients due to electric field components at the edge and along the x direction. The terms a and b correspond to the decay constants. The ratio between E_z and E_x is kept at 2.14. The skin depth of Au was determined to be (21.7 $\pm$ 0.2) nm from the fitting, which is in very good agreement with the value determined from eqn (5).

Since the TERS enhancement is proportional to the 4th power of the electric field, we also acquired the 4th powered image of the electric field, as shown in Fig. 3b. The TERS enhancement image is presented in Fig. 3d. The image reveals a sharp enhancement along the edges, making it a ring like structure. Therefore, one would also expect a significant enhancement of the TERS signal along the edges of the cylinder forming the ring. Similar results were obtained by Etxarri *et al.*[66] and Habteyes *et al.*[67] using an inclined incident field in which the electric field components (E_x and E_z) are comparable to each other. Therefore, the scattered field from the Au cylinders created an asymmetric circle with the strongest contribution aligned to the direction of the in-plane component of the electric field. In our simulations, we matched the experimental configuration with an electric field inclined at an angle of 65° from the normal of the sample surface. In such a configuration, the ratio between the out-of-plane and in-plane components of the field is 2.14. Such a strong contribution of the out-of-plane electric field results in an almost symmetric ring-like shape of the electric field distribution on the Au cylinder. This is in very good agreement with the TERS image experimentally observed for 1L-MoS_2 on Au nanodisks, shown in Fig. 2c.

Influence of the Au nanodisks on TERS sensitivity

According to Fig. 2 and 3, it is clear at this stage why one should observe a ring-like TERS image for such nanodisks. However, what is not known is (1) how the TERS sensitivity is distributed over the Au cylinder and how it compares to the so-called gap-mode TERS, and (2) the role of the Au cylinder in defining the spatial resolution. In order to address these issues, we took a high resolution TERS image of two Au cylinders with a step size of 5 × 5 nm^2 as shown in Fig. 4. Fig. 4a and b display the TERS image and corresponding AFM topography image taken simultaneously during the TERS measurements. Interestingly, the TERS image does not show an ideal ring pattern along the edges of the Au cylinder, as was expected from a consideration of Fig. 2 and 3. This can be understood by looking at the AFM image shown in Fig. 4b. A topographical deviation is clearly visible in the top half of the cylinder where the TERS image also deviates from the ring-like pattern along the edges. Therefore, one can conclude that the deformation of the flake on top of the Au cylinder impedes the ideal circular pattern formation. To determine the spatial resolution we took two cross sections along the dashed lines shown in both images. The corresponding cross sections are presented in Fig. 4c along with 1st order derivatives and Gaussian fits of them. As can be seen in Fig. 4c, the spatial resolution of the TERS image (5 nm) is much better than that of the AFM topography image (18 nm). However, in the experiments, the edges of the Au cylinders were smoothened by a MoS_2 "carpet" (see Fig. 4b). Moreover, the experimental spatial resolution was also limited by the step size of 5 nm used in the TERS imaging. Therefore, the best possible spatial resolution should be even better than the experimentally determined value of 5 nm. For a similar system we recently reported a spatial resolution of 2.3 nm.[39] In order to understand the strength of coupling (related to TERS sensitivity) on different positions of the Au cylinder we compare two TERS spectra (one at the edge and the other at the center, as indicated by the two circles in Fig. 4a). Two representative TERS spectra in the spectral range around the A_{1g} mode are presented in Fig. 4c. The difference in peak position for the two spectra is related to local heterogeneities and is addressed in ref. 39. As expected, the TERS intensity is much higher at the edges

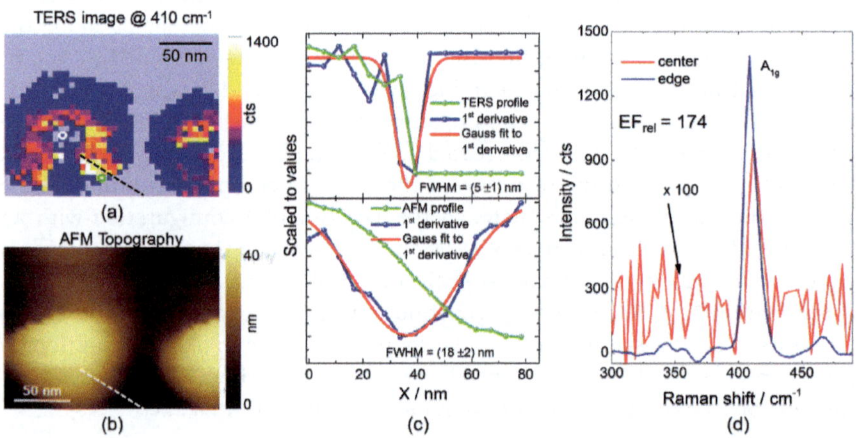

Fig. 4 (a) High resolution TERS map of the A_{1g} mode, and (b) corresponding AFM topography image of 1L-MoS_2 on Au nanodisks as shown in Fig. 2. The step size of the mapping is 5 × 5 nm^2. (c) Two cross sections were taken along the dashed lines shown in (a) and (b), representing the TERS and AFM profiles, respectively. The 1st order derivatives of the profiles were fitted with Gaussian profiles to determine the respective spatial resolutions. (d) Two TERS spectra taken from the TERS image shown in (a) marked by white and green circles (one at the center and the other at the corner) in order to determine the enhancement ratio.

than at the center of the cylinder. The intensity ratio (I_{edge}/I_{center}) is determined to be approximately 170. To verify the relative enhancement ratio, we built a similar tip–sample model and calculated the local electric enhancement at different positions of the Au nanodisks. The simulation results are presented in Fig. 5. The tip is tilted by 25° from the normal of the sample surface in order to align with the electric field of the exciting light, as shown in the figure. The electric field is enhanced by a factor of 10 at the center of the Au cylinder, whereas it is enhanced by a factor of 34 at the edges (Fig. 5a and b). Therefore, the relative enhancement ratio ($ER_{rel} = (E_{edge}/E_{center})^4$) is determined to be 134. This is in very good agreement with the experimental observation considering the fact that the experiments were performed with the tip in contact with the sample. In the case of a sub-nanometer tip–sample gap, an additional quantum mechanical effect is added to the interaction.[44,68,69] Classical simulation tools like COMSOL are not equipped with mathematical functions to handle such an effect. Therefore all simulations were performed at a tip–sample gap of 1 nm. It is obvious from eqn (3) that the enhancement would be even higher for a smaller tip–sample gap. Hence, the calculated ER_{rel} could be expected to be even higher under the experimental conditions used. Another interesting aspect is the difference between TERS sensitivity on Au cylinders and on a flat Au film, especially as it is well known that on a metal surface the TERS sensitivity is enhanced due to the so-called gap-mode configuration (see Fig. 1). To address this question, we also calculated the local electric field distribution on a flat Au surface, as shown in Fig. 5c. The electric field is enhanced by a factor of 10.5 right under the tip apex on the metal substrate. Thus, the enhancement ratio ($ER = (E_{edge}/E_{flat\ Au})^4$) was determined to be 110. Hence, one can obtain significantly higher TERS sensitivity on a plasmonic substrate in comparison to the so-called gap-mode TERS configuration.

Since the TERS enhancement depends on the strength of the tip–sample coupling, the spatial resolution should also be influenced by this interaction. We thus took cross sections of the 4th power of the calculated electric field, as shown in Fig. 5d. As can be seen, the lateral distribution of the 4th power of E_{loc} is narrower at the edges compared to at the center. The lateral distribution of $E_{loc}{}^4$ at the center maintains a symmetrical peak shape. However, this is not the case at the edge. This deformation originates from the asymmetric lateral distribution of

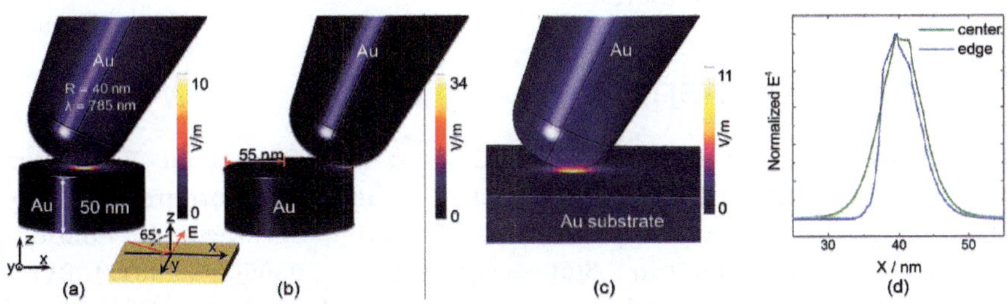

Fig. 5 FEM simulations of the Au–Au tip–sample system. (a) Tip at the center, (b) at the corner or edges of the Au cylinder, and (c) on a flat Au substrate. (d) Lateral distribution of the 4th power of E_{loc} of the geometries shown in (a) and (b). The FWHM of E_{loc} when the tip is at the corner is smaller than that when the tip is at the center, thus confirming better spatial resolution at the edges.

the field at the edges, as shown in Fig. 5b. Gaussian fitting of E_{loc}^4 at the center gives a FWHM value of (7.2 ± 0.1) nm, whereas asymmetric Gaussian fitting at the corner gives a FWHM value of (5.1 ± 0.1) nm. The fitted graphs are presented in Fig. S6.† The higher lateral resolution of the edges can be explained by a consideration of the structural shape of the cylinder and the local field coupling. Due to the sharp edges, the lightning rod effect enhances the field confinement and thus provides increased sensitivity. Additionally, by perfectly matching the LSPR conditions of the tip–sample dimer system, one can also markedly tune the TERS sensitivity. Therefore, we can say that the TERS resolution not only depends on the product of *Rd* (according to eqn (4)), but also on the size and shape of the substrate. Upon comparing the calculated result with the experimental value shown in Fig. 4c, one can see that the spatial resolutions are comparable. In practical situations, the lateral resolution is limited by the step size and may even be better than 5 nm, while in the calculations we are limited by a tip–sample gap of 1 nm.

Conclusions

In summary, we combined finite element method (FEM) simulations and experiments in order to understand the enhancement and spatial resolution of a tip-enhanced Raman scattering (TERS) system. We observed the strongest TERS enhancement at the edges and the weakest enhancement at the center of the Au nanodisks, thus leading to the formation of ring-like patterns aligned with the nanodisk rims in the TERS images. FEM simulations were used to calculate the electric field distribution and produce a TERS image of the combined tip–nanodisk system. Both the enhancement and spatial resolution were found to be better at the edges than at the center of the Au cylinders. Very good agreement is observed between the experimental and theoretical observations. Our present understanding of the TERS method can be used as a benchmark for future experiments using similar configurations. Moreover, the knowledge gained can also be extended to routine material characterization using such a system.

Conflicts of interest

There are no conflicts to declare.

Acknowledgements

We acknowledge the funding received from the Cluster of Excellence "Center for Advancing Electronics Dresden" (cfaed). The authors gratefully acknowledge the financial support received from the Volkswagen Foundation, MERGE project (TU Chemnitz), State assignment (no. 0306-2016-0017), RFBR (projects 18-02-00615_a, and 19-52-12041_NNIO_a), and the Ministry of Education and Science of the Russian Federation. The authors are thankful to Mr A. S. Medvedev for the Au deposition. The SEM studies and fabrication of the metal nanostructures were carried out using the equipment of CCU "Nanostructures" at the Rzhanov Institute of Semiconductor Physics SB RAS.

References

1 S. Sil and S. Umapathy, Raman spectroscopy explores molecular structural signatures of hidden materials in depth: Universal Multiple Angle Raman Spectroscopy, *Sci. Rep.*, 2014, **4**, 5308.

2 J. Zhang, M. Li, Z. Feng, J. Chen and C. Li, UV Raman Spectroscopic Study on TiO_2. I. Phase Transformation at the Surface and in the Bulk, *J. Phys. Chem. B*, 2006, **110**, 927–935.

3 K. Kneipp, H. Kneipp, I. Itzkan, R. R. Dasari and M. S. Feld, Ultrasensitive Chemical Analysis by Raman Spectroscopy, *Chem. Rev.*, 1999, **99**, 2957–2976.

4 A. Chehaidar, R. Carles, A. Zwick, C. Meunier, B. Cros and J. Durand, Chemical bonding analysis of a-SiC:H films by Raman spectroscopy, *J. Non-Cryst. Solids*, 1994, **169**, 37–46.

5 K. Larsson and R. P. Rand, Detection of changes in the environment of hydrocarbon chains by Raman spectroscopy and its application to lipid-protein systems, *Biochim. Biophys. Acta, Lipids Lipid Metab.*, 1973, **326**, 245–255.

6 M. Ocaña, J. V. Garcia-Ramos and C. J. Serna, Low-Temperature Nucleation of Rutile Observed by Raman Spectroscopy during Crystallization of TiO_2, *J. Am. Ceram. Soc.*, 1992, **75**, 2010–2012.

7 A. Tarun, N. Hayazawa and S. Kawata, Tip-enhanced Raman spectroscopy for nanoscale strain characterization, *Anal. Bioanal. Chem.*, 2009, **394**, 1775–1785.

8 R. M. Stöckle, Y. D. Suh, V. Deckert and R. Zenobi, Nanoscale chemical analysis by tip-enhanced Raman spectroscopy, *Chem. Phys. Lett.*, 2000, **318**, 131–136.

9 N. Hayazawa, Y. Inouye, Z. Sekkat and S. Kawata, Metallized tip amplification of near-field Raman scattering, *Opt. Commun.*, 2000, **183**, 333–336.

10 M. S. Anderson, Locally enhanced Raman spectroscopy with an atomic force microscope, *Appl. Phys. Lett.*, 2000, **76**, 3130–3132.

11 H. Xu, J. Aizpurua, M. Käll and P. Apell, Electromagnetic contributions to single-molecule sensitivity in surface-enhanced Raman scattering, *Phys. Rev. E: Stat. Phys., Plasmas, Fluids, Relat. Interdiscip. Top.*, 2000, **62**, 4318–4324.

12 P. Verma, Tip-Enhanced Raman Spectroscopy: Technique and Recent Advances, *Chem. Rev.*, 2017, **117**, 6447–6466.

13 N. Jiang, D. Kurouski, E. A. Pozzi, N. Chiang, M. C. Hersam and R. P. Van Duyne, Tip-enhanced Raman spectroscopy: From concepts to practical applications, *Chem. Phys. Lett.*, 2016, **659**, 16–24.

14 N. Kumar, S. Mignuzzi, W. Su and D. Roy, Tip-enhanced Raman spectroscopy: principles and applications, *EPJ Techniques and Instrumentation*, 2015, **2**, 9.

15 X. Wang, D. Zhang, K. Braun, H.-J. Egelhaaf, C. J. Brabec and A. J. Meixner, High-Resolution Spectroscopic Mapping of the Chemical Contrast from Nanometer Domains in P3HT:PCBM Organic Blend Films for Solar-Cell Applications, *Adv. Funct. Mater.*, 2010, **20**, 492–499.

16 M. Richard-Lacroix, Y. Zhang, Z. Dong and V. Deckert, Mastering high resolution tip-enhanced Raman spectroscopy: towards a shift of perception, *Chem. Soc. Rev.*, 2017, **46**, 3922–3944.

17 A. B. Zrimsek, N. Chiang, M. Mattei, S. Zaleski, M. O. McAnally, C. T. Chapman, A.-I. Henry, G. C. Schatz and R. P. Van Duyne, Single-Molecule Chemistry with Surface- and Tip-Enhanced Raman Spectroscopy, *Chem. Rev.*, 2017, **117**, 7583–7613.

18 T. Deckert-Gaudig, D. Kurouski, M. A. B. Hedegaard, P. Singh, I. K. Lednev and V. Deckert, Spatially resolved spectroscopic differentiation of hydrophilic and hydrophobic domains on individual insulin amyloid fibrils, *Sci. Rep.*, 2016, **6**, 33575.

19 S. Jiang, Y. Zhang, R. Zhang, C. Hu, M. Liao, Y. Luo, J. Yang, Z. Dong and J. G. Hou, Distinguishing adjacent molecules on a surface using plasmon-enhanced Raman scattering, *Nat. Nanotechnol.*, 2015, **10**, 865.

20 R. Zhang, Y. Zhang, Z. C. Dong, S. Jiang, C. Zhang, L. G. Chen, L. Zhang, Y. Liao, J. Aizpurua, Y. Luo, J. L. Yang and J. G. Hou, Chemical mapping of a single molecule by plasmon-enhanced Raman scattering, *Nature*, 2013, **498**, 82.

21 M. D. Sonntag, J. M. Klingsporn, L. K. Garibay, J. M. Roberts, J. A. Dieringer, T. Seideman, K. A. Scheidt, L. Jensen, G. C. Schatz and R. P. Van Duyne, Single-Molecule Tip-Enhanced Raman Spectroscopy, *J. Phys. Chem. C*, 2012, **116**, 478–483.

22 J. Steidtner and B. Pettinger, Tip-Enhanced Raman Spectroscopy and Microscopy on Single Dye Molecules with 15 nm Resolution, *Phys. Rev. Lett.*, 2008, **100**, 236101.

23 E. Bailo and V. Deckert, Tip-Enhanced Raman Spectroscopy of Single RNA Strands: Towards a Novel Direct-Sequencing Method, *Angew. Chem., Int. Ed.*, 2008, **47**, 1658–1661.

24 W. Zhang, B. S. Yeo, T. Schmid and R. Zenobi, Single Molecule Tip-Enhanced Raman Spectroscopy with Silver Tips, *J. Phys. Chem. C*, 2007, **111**, 1733–1738.

25 B. Pettinger, B. Ren, G. Picardi, R. Schuster and G. Ertl, Nanoscale Probing of Adsorbed Species by Tip-Enhanced Raman Spectroscopy, *Phys. Rev. Lett.*, 2004, **92**, 096101.

26 D. Mehtani, N. Lee, R. D. Hartschuh, A. Kisliuk, M. D. Foster, A. P. Sokolov and J. F. Maguire, Nano-Raman spectroscopy with side-illumination optics, *J. Raman Spectrosc.*, 2005, **36**, 1068–1075.

27 E. Sheremet, A. G. Milekhin, R. D. Rodriguez, T. Weiss, M. Nesterov, E. E. Rodyakina, O. D. Gordan, L. L. Sveshnikova, T. A. Duda, V. A. Gridchin, V. M. Dzhagan, M. Hietschold and D. R. T. Zahn, Surface- and tip-enhanced resonant Raman scattering from CdSe nanocrystals, *Phys. Chem. Chem. Phys.*, 2015, **17**, 21198–21203.

28 P. G. Gucciardi and J.-C. Valmalette, Different longitudinal optical—transverse optical mode amplification in tip enhanced Raman spectroscopy of GaAs(001), *Appl. Phys. Lett.*, 2010, **97**, 263104.

29 Y. Saito, M. Motohashi, N. Hayazawa, M. Iyoki and S. Kawata, Nanoscale characterization of strained silicon by tip-enhanced Raman spectroscope in reflection mode, *Appl. Phys. Lett.*, 2006, **88**, 143109.

30 R. Matsui, P. Verma, T. Ichimura, Y. Inouye and S. Kawata, Nanoanalysis of crystalline properties of GaN thin film using tip-enhanced Raman spectroscopy, *Appl. Phys. Lett.*, 2007, **90**, 061906.

31 J. S. Reparaz, N. Peica, R. Kirste, A. R. Goñi, M. R. Wagner, G. Callsen, M. I. Alonso, M. Garriga, I. C. Marcus, A. Ronda, I. Berbezier, J. Maultzsch, C. Thomsen and A. Hoffmann, Probing local strain and composition in Ge nanowires by means of tip-enhanced Raman scattering, *Nanotechnology*, 2013, **24**, 185704.

32 N. Hayazawa, M. Motohashi, Y. Saito, H. Ishitobi, A. Ono, T. Ichimura, P. Verma and S. Kawata, Visualization of localized strain of a crystalline thin layer at the nanoscale by tip-enhanced Raman spectroscopy and microscopy, *J. Raman Spectrosc.*, 2007, **38**, 684–696.
33 R. D. Rodriguez, E. Sheremet, T. Deckert-Gaudig, C. Chaneac, M. Hietschold, V. Deckert and D. R. T. Zahn, Surface- and tip-enhanced Raman spectroscopy reveals spin-waves in iron oxide nanoparticles, *Nanoscale*, 2015, **7**, 9545–9551.
34 Y. Zhang, D. V. Voronine, S. Qiu, A. M. Sinyukov, M. Hamilton, Z. Liege, A. V. Sokolov, Z. Zhang and M. O. Scully, Improving resolution in quantum subnanometre-gap tip-enhanced Raman nanoimaging, *Sci. Rep.*, 2016, **6**, 25788.
35 K. K. H. Smithe, A. V. Krayev, C. S. Bailey, H. R. Lee, E. Yalon, Ö. B. Aslan, M. Muñoz Rojo, S. Krylyuk, P. Taheri, A. V. Davydov, T. F. Heinz and E. Pop, Nanoscale Heterogeneities in Monolayer $MoSe_2$ Revealed by Correlated Scanning Probe Microscopy and Tip-Enhanced Raman Spectroscopy, *ACS Appl. Nano Mater.*, 2018, **1**, 572–579.
36 D. V. Voronine, G. Lu, D. Zhu and A. Krayev, Tip-Enhanced Raman Scattering of MoS_2, *IEEE J. Sel. Top. Quantum Electron.*, 2017, **23**, 138–143.
37 K.-D. Park, O. Khatib, V. Kravtsov, G. Clark, X. Xu and M. B. Raschke, Hybrid Tip-Enhanced Nanospectroscopy and Nanoimaging of Monolayer WSe_2 with Local Strain Control, *Nano Lett.*, 2016, **16**, 2621–2627.
38 W. Bao, N. J. Borys, C. Ko, J. Suh, W. Fan, A. Thron, Y. Zhang, A. Buyanin, J. Zhang, S. Cabrini, P. D. Ashby, A. Weber-Bargioni, S. Tongay, S. Aloni, D. F. Ogletree, J. Wu, M. B. Salmeron and P. J. Schuck, Visualizing nanoscale excitonic relaxation properties of disordered edges and grain boundaries in monolayer molybdenum disulfide, *Nat. Commun.*, 2015, **6**, 7993.
39 A. G. Milekhin, M. Rahaman, E. E. Rodyakina, A. V. Latyshev, V. M. Dzhagan and D. R. T. Zahn, Giant gap-plasmon tip-enhanced Raman scattering of MoS_2 monolayers on Au nanocluster arrays, *Nanoscale*, 2018, **10**, 2755–2763.
40 M. Rahaman, R. D. Rodriguez, G. Plechinger, S. Moras, C. Schüller, T. Korn and D. R. T. Zahn, Highly Localized Strain in a MoS_2/Au Heterostructure Revealed by Tip-Enhanced Raman Spectroscopy, *Nano Lett.*, 2017, **17**, 6027–6033.
41 S. Kawata, T. Ichimura, A. Taguchi and Y. Kumamoto, Nano-Raman Scattering Microscopy: Resolution and Enhancement, *Chem. Rev.*, 2017, **117**, 4983–5001.
42 T. Deckert-Gaudig, A. Taguchi, S. Kawata and V. Deckert, Tip-enhanced Raman spectroscopy – from early developments to recent advances, *Chem. Soc. Rev.*, 2017, **46**, 4077–4110.
43 Z. Zhang, S. Sheng, R. Wang and M. Sun, Tip-Enhanced Raman Spectroscopy, *Anal. Chem.*, 2016, **88**, 9328–9346.
44 W. Zhu, R. Esteban, A. G. Borisov, J. J. Baumberg, P. Nordlander, H. J. Lezec, J. Aizpurua and K. B. Crozier, Quantum mechanical effects in plasmonic structures with subnanometre gaps, *Nat. Commun.*, 2016, **7**, 11495.
45 J. Stadler, B. Oswald, T. Schmid and R. Zenobi, Characterizing unusual metal substrates for gap-mode tip-enhanced Raman spectroscopy, *J. Raman Spectrosc.*, 2013, **44**, 227–233.
46 H. Wei and H. Xu, Hot spots in different metal nanostructures for plasmon-enhanced Raman spectroscopy, *Nanoscale*, 2013, **5**, 10794–10805.

47 Z. Yang, J. Aizpurua and H. Xu, Electromagnetic field enhancement in TERS configurations, *J. Raman Spectrosc.*, 2009, **40**, 1343–1348.

48 N. Behr and M. B. Raschke, Optical Antenna Properties of Scanning Probe Tips: Plasmonic Light Scattering, Tip–Sample Coupling, and Near-Field Enhancement, *J. Phys. Chem. C*, 2008, **112**, 3766–3773.

49 N. Kumar, W. Su, M. Vesely, B. M. Weckhuysen, A. J. Pollard and A. J. Wain, Nanoscale chemical imaging of solid-liquid interfaces using tip-enhanced Raman spectroscopy, *Nanoscale*, 2018, **10**, 1815–1824.

50 N. Chiang, X. Chen, G. Goubert, D. V. Chulhai, X. Chen, E. A. Pozzi, N. Jiang, M. C. Hersam, T. Seideman, L. Jensen and R. P. Van Duyne, Conformational Contrast of Surface-Mediated Molecular Switches Yields Ångstrom-Scale Spatial Resolution in Ultrahigh Vacuum Tip-Enhanced Raman Spectroscopy, *Nano Lett.*, 2016, **16**, 7774–7778.

51 A. Shiohara, Y. Wang and L. M. Liz-Marzán, Recent approaches toward creation of hot spots for SERS detection, *J. Photochem. Photobiol., C*, 2014, **21**, 2–25.

52 Y. Liu, L. Zhao, X. Li, Z. Zeng, P. Wang, L. Zhang and Y. Fang, Investigation of plasmon properties of silver microsphere array demonstrated experimentally by tip-enhanced Raman spectroscopy, *Appl. Surf. Sci.*, 2018, **428**, 900–905.

53 Y. Li, Z. Li, C. Chi, H. Shan, L. Zheng and Z. Fang, Plasmonics of 2D Nanomaterials: Properties and Applications, *Adv. Sci.*, 2017, **4**, 1600430.

54 Y. Yu, Z. Ji, S. Zu, B. Du, Y. Kang, Z. Li, Z. Zhou, K. Shi and Z. Fang, Ultrafast Plasmonic Hot Electron Transfer in Au Nanoantenna/MoS_2 Heterostructures, *Adv. Funct. Mater.*, 2016, **26**, 6394–6401.

55 S. Butun, S. Tongay and K. Aydin, Enhanced Light Emission from Large-Area Monolayer MoS_2 Using Plasmonic Nanodisc Arrays, *Nano Lett.*, 2015, **15**, 2700–2704.

56 L. Gunnarsson, E. J. Bjerneld, H. Xu, S. Petronis, B. Kasemo and M. Käll, Interparticle coupling effects in nanofabricated substrates for surface-enhanced Raman scattering, *Appl. Phys. Lett.*, 2001, **78**, 802–804.

57 K. D. Sattler, *Handbook of Nanophysics: Nanoparticles and Quantum Dots*, CRC press, vol. 3, 2016.

58 P. B. Johnson and R. W. Christy, Optical Constants of the Noble Metals, *Phys. Rev. B: Solid State*, 1972, **6**, 4370–4379.

59 V. V. Temnov, G. Armelles, U. Woggon, D. Guzatov, A. Cebollada, A. Garcia-Martin, J.-M. Garcia-Martin, T. Thomay, A. Leitenstorfer and R. Bratschitsch, Active magneto-plasmonics in hybrid metal–ferromagnet structures, *Nat. Photonics*, 2010, **4**, 107.

60 P. W. Gilberd, The anomalous skin effect and the optical properties of metals, *J. Phys. F: Met. Phys.*, 1982, **12**, 1845.

61 G. E. H. Reuter and E. H. Sondheimer, The theory of the anomalous skin effect in metals, *Proc. R. Soc. London, Ser. A*, 1948, **195**, 336–364.

62 A. V. Ermushev, V. M. Boris, V. A. Oleĭnikov and A. V. Petukhov, Surface enhancement of local optical fields and the lightning-rod effect, *Quantum Electron.*, 1993, **23**, 435.

63 P. F. Liao and A. Wokaun, Lightning rod effect in surface enhanced Raman scattering, *J. Chem. Phys.*, 1982, **76**, 751–752.

64 S. Berciaud, L. Cognet, P. Tamarat and B. Lounis, Observation of Intrinsic Size Effects in the Optical Response of Individual Gold Nanoparticles, *Nano Lett.*, 2005, **5**, 515–518.

65 S. H. Brewer and S. Franzen, Optical properties of indium tin oxide and fluorine-doped tin oxide surfaces: correlation of reflectivity, skin depth, and plasmon frequency with conductivity, *J. Alloys Compd.*, 2002, **338**, 73–79.

66 A. García-Etxarri, I. Romero, F. J. García de Abajo, R. Hillenbrand and J. Aizpurua, Influence of the tip in near-field imaging of nanoparticle plasmonic modes: Weak and strong coupling regimes, *Phys. Rev. B: Condens. Matter Mater. Phys.*, 2009, **79**, 125439.

67 T. G. Habteyes, S. Dhuey, K. I. Kiesow and A. Vold, Probe-sample optical interaction: size and wavelength dependence in localized plasmon near-field imaging, *Opt. Express*, 2013, **21**, 21607–21617.

68 K. J. Savage, M. M. Hawkeye, R. Esteban, A. G. Borisov, J. Aizpurua and J. J. Baumberg, Revealing the quantum regime in tunnelling plasmonics, *Nature*, 2012, **491**, 574.

69 R. Esteban, A. G. Borisov, P. Nordlander and J. Aizpurua, Bridging quantum and classical plasmonics with a quantum-corrected model, *Nat. Commun.*, 2012, **3**, 825.

Faraday Discussions

PAPER

Hot electron-driven photocatalysis and transient absorption spectroscopy in plasmon resonant grating structures

Yi Wang,†[a] Lang Shen,†[b] Yu Wang,[b] Bingya Hou,[c] George N. Gibson,[ef] Nirakar Poudel,[c] Jihan Chen,[c] Haotian Shi,[a] Ernest Guignon,[f] Nathaniel C. Cady,[g] William D. Page,[f] Arturo Pilar,[f] Jahan Dawlaty[a] and Stephen B. Cronin *[acd]

Received 1st October 2018, Accepted 12th December 2018

DOI: 10.1039/c8fd00141c

Plasmon resonant grating structures provide an effective platform for distinguishing between the effects of plasmon resonant excitation and bulk metal absorption *via* interband transitions. By simply rotating the polarization of the incident light, we can switch between resonant excitation and non-resonant excitation, while keeping all other parameters of the measurement constant. With light polarized perpendicular to the lines in the grating (*i.e.*, TE-polarization), the photocatalytic reaction rate (*i.e.*, photocurrent) is measured as the angle of the incident laser light is tuned through the resonance with the grating. Here, hot holes photoexcited in the metal are used to drive the oxygen evolution reaction (OER), producing a measurable photocurrent. Using TE-polarized light, we observe sharp peaks in the photocurrent and sharp dips in the photoreflectance at approximately 9° from normal incidence, which corresponds to the conditions under which there is good wavevector matching between the incident light and the lines in the grating. With light polarized parallel to the grating (*i.e.*, TM), we excite the grating structure non-resonantly and there is no angular dependence in the photocurrent or photoreflectance. In order to quantify the lifetime of these hot carriers, we performed transient absorption spectroscopy of these plasmon resonant grating structures. Here, we observe one feature in the spectra corresponding to interband transitions and another feature associated with the plasmon resonant mode in the

[a]*Department of Chemistry, University of Southern California, Los Angeles, CA 90089, USA. E-mail: scronin@usc.edu*

[b]*Mork Family Department of Chemical Engineering and Materials Science, University of Southern California, Los Angeles, CA 90089, USA*

[c]*Ming Hsieh Department of Electrical Engineering, University of Southern California, Los Angeles, CA 90089, USA*

[d]*Department of Physics and Astronomy, University of Southern California, Los Angeles, CA 90089, USA*

[e]*Department of Physics, University of Connecticut, Storrs, CT 06269, USA*

[f]*Ciencia Inc., East Hartford, CT 06108, USA*

[g]*SUNY Polytechnic Institute, Colleges of Nanoscale Science & Engineering, Albany, NY 12203, USA*

† Authors contributed equally.

grating. Both features decay over a time scale of 1–2 ps. The spectral responses of grating structures fabricated with Ag, Al, and Cu are also presented.

Introduction

In the early days of plasmonic catalysis, around 2006, several papers reported photothermally-driven processes, including plasmon resonant chemical vapor deposition (PRCVD).[1,2] Here, Boyd *et al.* demonstrated plasmon assisted chemical vapor deposition (PACVD) of PbO nanowires and TiO_2 microstructures by irradiating 23 nm Au nanoparticles using a metalorganic $Pb(C_{11}H_{19}O_2)_2$ Pb precursor and a $Ti[OCH(CH_3)_2]_2[C_{11}H_{19}O_2]_2$ Ti precursor, respectively.[1] In a similar study, Hung *et al.* demonstrated plasmon resonant chemical vapor deposition of Fe_2O_3 and carbon nanotubes in a CO environment containing trace amounts of iron pentacarbonyl.[2] Here, *in situ* Raman spectroscopy measurements indicated that temperatures in excess of 1180 °C were generated using this approach.

A few years later, around 2010, plasmon resonant enhancement of redox electrochemical processes were demonstrated (including water splitting) by coupling strongly plasmonic metal nanoparticles (*e.g.*, Ag and Au) to strongly catalytic metal oxide semiconductors (*e.g.*, TiO_2).[3–6] Several of these papers have been cited over 500 times, indicating the strong interest amongst the scientific community in plasmon-driven photoelectrochemistry. The initial interpretation of this plasmon resonant enhancement of photoelectrochemical processes was based on local field enhancement of sub-bandgap defect-mediated absorption within the TiO_2 semiconducting material. At this point, large scale electromagnetic simulations using the finite difference time domain (FDTD) method had become commercially available and provided a high level of rigor for these nanoscale metal/semiconductor systems. Despite the high degree of accuracy of these electromagnetic simulations, however, several experimental studies reported enhancement factors that were substantially higher than theoretical predictions. For example, Liu *et al.* reported a 66-fold increase in the photocurrent of a Au/TiO_2 system at a wavelength of 633 nm, which corresponds to a photon energy that is roughly half the bandgap of this material, exceeding that which can be explained by simple local field enhancement.[5]

Around 2013, plasmon-induced dissociation of molecular hydrogen was demonstrated in oxide-supported Au nanoparticles in a process that could only be explained by the generation of hot electrons within the metal and subsequent Feshbach resonance.[7,8] Several theoretical studies ensued calculating the properties of hot electrons in plasmonic nanostructures.[9–11] However, a vast majority of the experimental papers on the subject of hot electron-driven photochemistry entail metal nanostructures in contact with a semiconducting material, which presents some additional complexity in the interpretation of charge transfer and energetics, as described above.[12] Park *et al.* reported hot electron-driven CO oxidation using hybrid metal–semiconductor nanostructures, and reported time-resolved transmission spectroscopy showing hot carrier relaxation on ps time scales.[13] DuChene *et al.* reported hot hole-driven photoelectrochemical CO_2 reduction using a plasmonic Au/p-GaN photocathode.[14]

In the work presented here, we explore photoelectrochemistry on all-metal nanostructures, which provide a direct means of exploring hot electrons in photoelectrochemical processes. By excluding semiconductor materials from this study, a more straightforward interpretation of data can be achieved regarding behaviour of photoexcited charge in metals. Here, we utilize plasmon resonant gratings to distinguish plasmon resonant excitation from bulk metal absorption *via* interband transitions. In addition to photoelectrochemical and photoreflectance measurements, we have performed transient absorption spectroscopy in order to establish the time scale over which these hot carriers relax back to equilibrium at the Fermi energy. Electromagnetic simulations of these grating structures using the finite difference time domain method provide a detailed picture of the local field distribution attained under resonant excitation conditions and provide a theoretical estimate for the plasmon resonant enhancement factors in these structures.

Experimental

In this work, we fabricate two types of grating structures. In the first structure, a corrugated surface is first patterned on a silicon substrate with a period of 500 nm using photolithography and reactive ion etching. A 50 nm thick film of Au is then deposited onto this structure, as shown in the SEM image of Fig. 1a. Fig. 1c shows the reflectance spectrum of one of these grating structures together with that of a bulk Au film. A clear resonance can be seen at a photon energy of 2.0 eV, corresponding to the plasmon resonance of the grating. In a second type of grating structure, we pattern 250 nm thick lines in a Au film with a 500 nm period using photolithography and reactive ion etching, as shown in Fig. 6a. The lithography was performed using an ASML deep UV stepper lithography tool. These grating structures were patterned on a transparent BK-7 substrate, which enables transient absorption spectroscopy to be readily performed.

Photoelectrochemical measurements were performed using a standard 3-terminal potentiostat. In order to measure the small photocurrents associated with the relatively short-lived hot electrons and holes, which are on the order of μA, we use an AC lock-in technique. Here, the incident laser light is modulated using an optical chopper wheel. The current monitor output of the potentiostat is measured by a lock-in amplifier, which is synchronized with the optical chopper, as illustrated in Fig. 2a. The grating sample (*i.e.*, working electrode) is mounted on a rotational stage, as illustrated in Fig. 2b, which enables us to collect angular dependent photocurrent and photoreflectance spectra. A 1/2-wave plate is used to rotate the polarization of the incident laser beam. Fig. 2c illustrates the hot hole-driven photochemical process in which a hot hole is injected into the solution driving the OER reaction.

Results and discussion

Before attempting to achieve hot electron-driven photochemistry on plasmon resonant grating structures, we applied the AC lock-in technique depicted in Fig. 2a on a 100 nm thick Au film. Here, we are simply exciting the bulk interband transitions in Au without any plasmon resonance.[15] Fig. 3 shows the AC photocurrent observed by irradiating this bare Au film using the optically modulated AC lock-in technique in a pH = 7, 0.5 M Na_2SO_4 solution. Here, we see a peak in the photocurrent around 1.5 V *vs.* NHE. This peak corresponds to the tuning of the

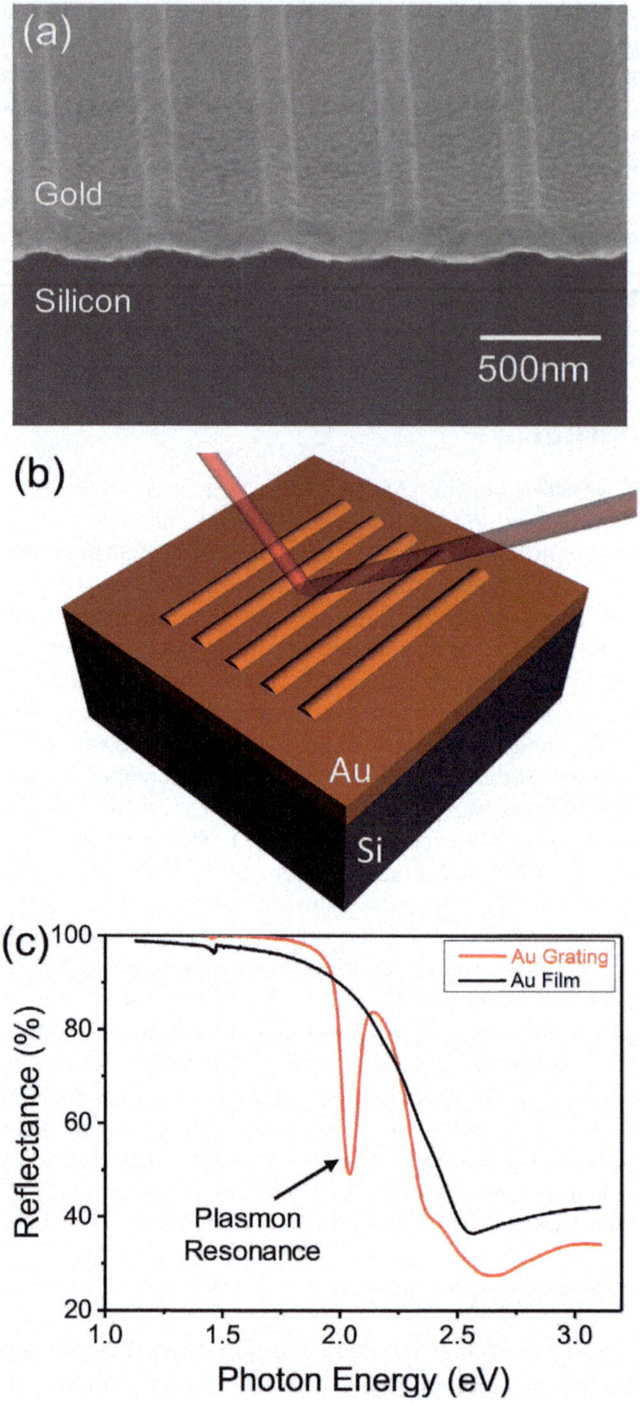

Fig. 1 (a) Scanning electron microscope (SEM) image of the corrugated Au grating deposited on an underlying silicon substrate. (b) Schematic diagram of the variable angle setup for measuring the photoreflectance and photoelectrochemical current. (c) Photoreflectance spectra for a flat Au film and the corrugated grating depicted in (a).

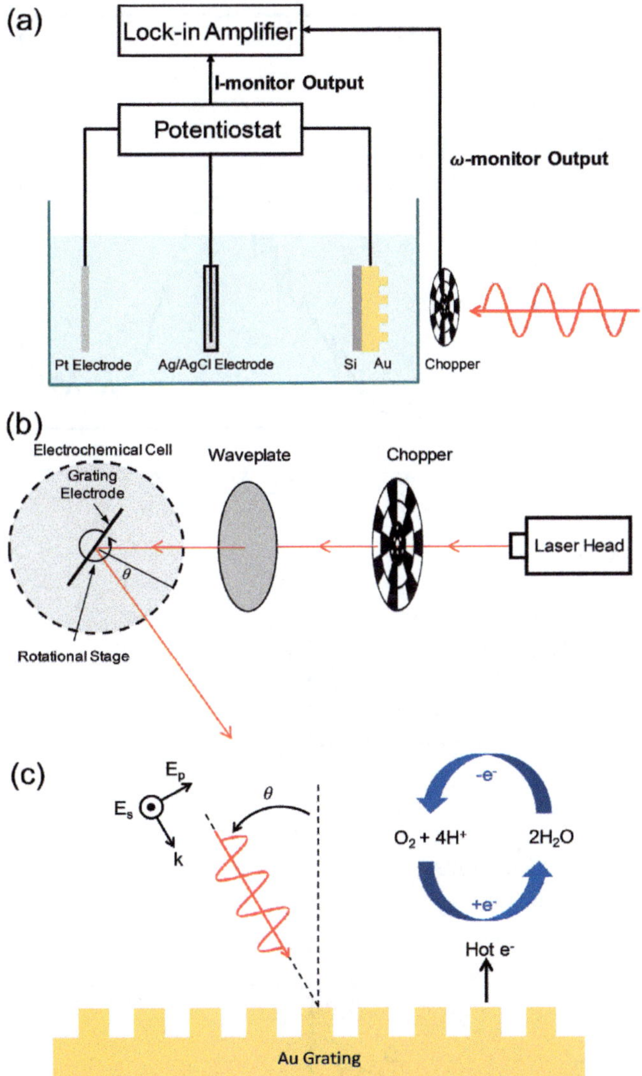

Fig. 2 Schematic diagrams for (a) the photoelectrochemical setup, (b) angle-dependent photocatalytic measurements, and (c) the grating structure for the photoelectrochemical setup.

relatively narrow distribution of hot holes (illustrated in Fig. 3b) through a resonance with the redox potential of the oxygen evolution reaction. These results demonstrate the effectiveness of the AC lock-in technique for measuring photocurrent produced on a bare metal surface for basic water splitting reactions.

Having demonstrated these capabilities on a bare metal film, we then explored these reactions on plasmon resonant nanostructures in order to amplify this effect. For the plasmon resonant grating structures shown in Fig. 1, we have measured the AC photocurrent and photoreflectance as a function of the incident angle. This data, shown in Fig. 4, exhibits a sharp peak in the photocurrent as well

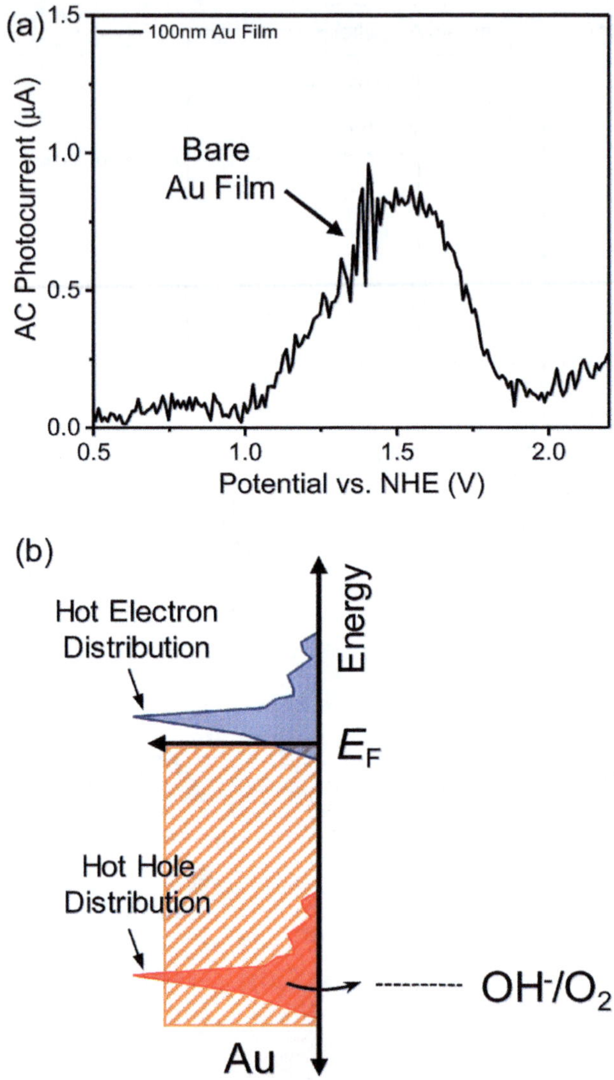

Fig. 3 (a) AC photocurrent measured by irradiating a 100 nm thick gold film in a pH = 7, 0.5 M Na_2SO_4 solution with optically modulated 532 nm wavelength light. (b) Energy band diagram indicating the relatively narrow hot electron (blue) and hot hole (red) distributions.

as a dip in the photoreflectance around an incident angle of 9° from normal incidence. This corresponds to the conditions under which we are resonantly exciting the grating and there is good wavevector matching between the incident light and the lines on the grating. It should be noted that this strong angular dependence of the photocurrent and photoreflectance is only seen when illuminating with TE-polarized light, that is, light polarized perpendicular to the lines on the grating. For TM-polarized light, there is no angle dependence in either the photocurrent or photoreflectance. Here, the non-resonant TM excitation of the metal grating generates a photocurrent of approximately 0.17 μA, due to bulk

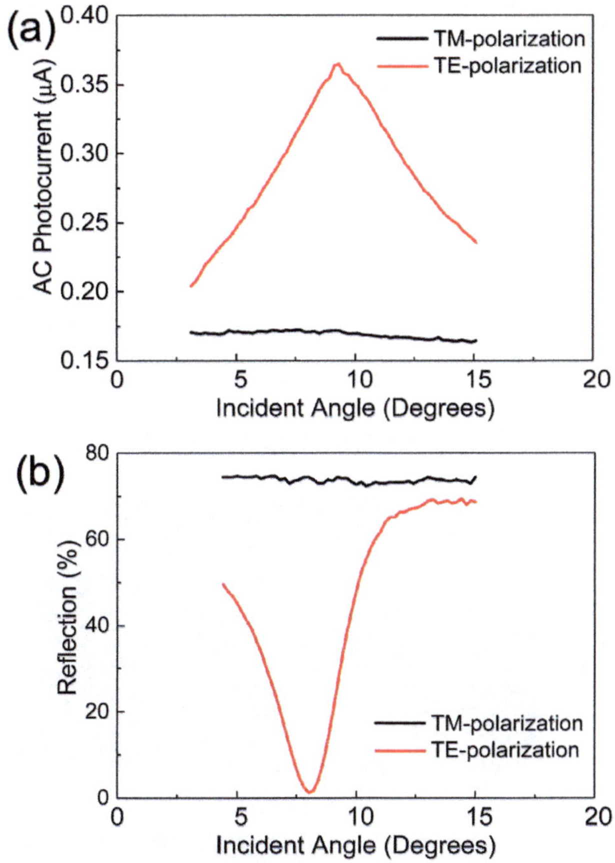

Fig. 4 (a) AC photocurrent and (b) photoreflectance spectra for the 500 nm period Au grating structure shown in Fig. 1a plotted as a function of incident angle for both TM- and TE-polarized light.

interband transitions in the Au. These interband transitions have been calculated by Sundararaman *et al.* and occur near the L- and X-points in the Brillouin zone.[9] Under resonant excitation with TE-polarized light at an incident angle of 9° from normal incidence, we observe a roughly 2.1-fold increase in the photocurrent due to hot electrons generated in the plasmon resonant mode of the grating. This gain factor is consistent with electromagnetic simulations of these grating structures, as described below in Fig. 5.

In order to provide a microscopic picture of the electromagnetic response of these plasmon resonant grating structures, we performed finite difference time domain (FDTD) simulations, which provide the complete electric field and charge distributions of our grating structures under various conditions of excitation both as a function of polarization and incident angle of illumination. Fig. 5a shows the calculated photoreflectance plotted as a function of incident angle for both TM- and TE-polarized light. These spectra agree remarkably well with our experimentally observed spectra (plotted in Fig. 4b), showing a sharp dip in the photoreflectance at 9° from normal incidence under TE-polarized illumination.

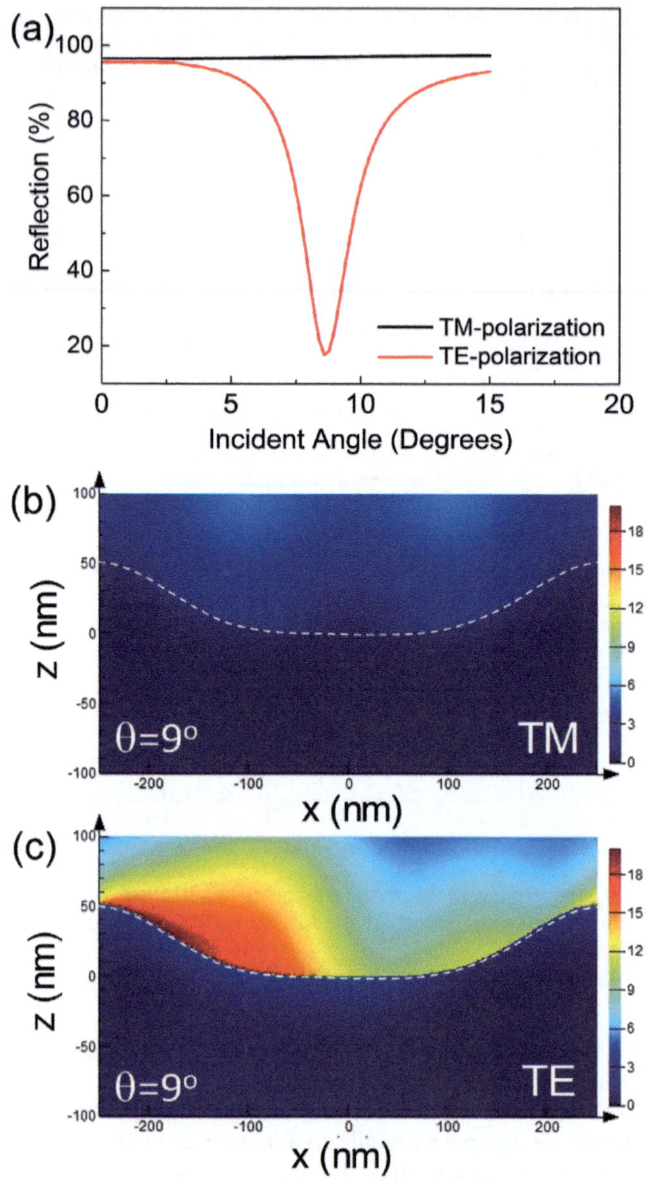

Fig. 5 (a) Calculated photoreflectance spectra for the 500 nm period grating shown in Fig. 1a for both TM- and TE-polarized light. Electric field intensity distributions calculated at an incident angle of 9° from surface normal for (b) TM-polarized and (c) TE-polarized light. Here, the Au/water interface is indicated by the white dashed lines.

Under TM-polarized illumination, the grating simply behaves like a mirror and exhibits no angular dependence. The electric field intensity distributions calculated at 9° incidence are plotted in Fig. 5b and c for the TM- and TE-polarizations, respectively. Here, the electric field intensity is plotted on the colour scale indicated on the side of the figure. This is a linear scale corresponding to the multiplicative factor of enhancement with respect to the incident electric field intensity of the incident light. For the case of resonant excitation with TE-polarized light, the peak field intensity (corresponding to red on this scale) is approximately 18 times higher than the incident electric field intensity. Under TM-polarized excitation, shown in Fig. 4b, no field enhancement is achieved, and a vast majority of the incident photons are simply reflected off the metal surface along with a small amount of bulk Au interband absorption, as described above. Based on these FDTD simulations, we can estimate a plasmonic enhancement factor based on the following equation:

$$\mathrm{EF} = \frac{\displaystyle\int_{f(x,y)-10\ \mathrm{nm}}^{f(x,y)} |E_{\mathrm{TE}}|^2 \mathrm{d}x\mathrm{d}y\mathrm{d}z}{\displaystyle\int_{f(x,y)-10\ \mathrm{nm}}^{f(x,y)} |E_{\mathrm{TM}}|^2 \mathrm{d}x\mathrm{d}y\mathrm{d}z}$$

where E_{TE} and E_{TM} are the electric field strengths under TE-polarized and TM-polarized illumination, respectively. Here, we integrate over the corrugated surface topography $f(x,y)$ from the surface of the Au grating, indicated by the dashed lines in the figure, down 10 nm below the surface, corresponding roughly to the hot electron diffusion length. Upon performing this integral, we obtain a theoretical plasmon resonant enhancement factor of 15.9×. This is considerably higher than the enhancement observed experimentally, indicating that significant losses occur as these hot electrons/holes are transferred to the ions in solution.[16] These losses are not accounted for in these simulations.

One of the open questions in the field of hot electrons is how long these optically-excited hot carriers live in metals, and to what extent their energy and charge can be harvested before they decay back to equilibrium at the Fermi energy. In order to address this open question, we performed transient absorption spectroscopy (TAS). This is a pump–probe technique, in which a 50 fs pump pulse is first used to optically excite hot electrons and hot holes in the material, and then broadband absorption spectra are collected as a function of time. Since this technique requires transparent samples, we used optically transparent gratings shown in Fig. 6 consisting of 300 nm wide lines with a 500 nm period patterned into a 50 nm thick Au film on a BK-7 substrate using reactive ion etching. The UV-vis transmission spectrum is shown in Fig. 6b, which exhibits a resonant feature around 550 nm.

The transient absorption spectra of these plasmon resonant grating structures are plotted for both the TM-polarized, non-resonant excitation (Fig. 7a) and TE-polarized, resonant excitation (Fig. 7b). Here, the pump wavelength is approximately 560 nm and the broadband absorption spectra are collected over a wavelength range from 400 nm to 600 nm. For non-resonant, TM-polarized excitation, we see one dominant feature in these spectra around 500 nm corresponding to interband transitions within the bulk Au material. These interband hot carrier excitations decay over a time scale of 1–2 ps. It is, in fact, these interband transitions that give rise to gold's yellow colour. Fig. 7c and d show the electric field distributions of similar grating structures illuminated with TM- and TE-polarized light, respectively. When illuminated on resonance with TE-polarized light, the local electric fields are enhanced by roughly a factor of 10. However, it should be noted that these local fields, and hence the hot electron–hole pair generation, are confined within the top few nanometers of the gold surface, which is

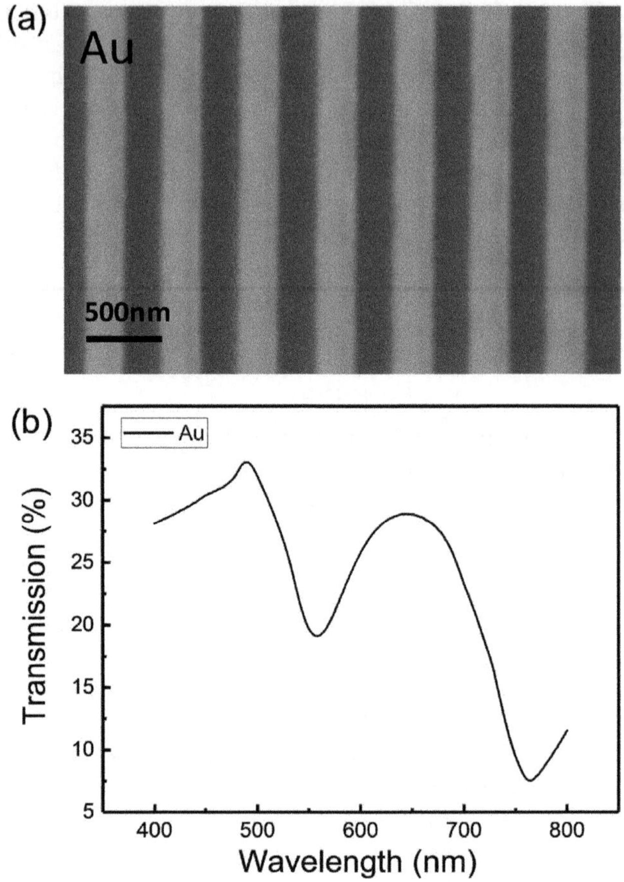

Fig. 6 (a) SEM image and (b) UV-vis transmission spectrum of a Au grating with a 250 nm line width and 500 nm period.

advantageous for photocatalysis, since it is difficult for these hot electrons and holes to diffuse over long distances in metals without relaxing back to equilibrium. Fig. 7b shows the transient absorption spectra of the resonantly excited grating structure under TE-polarized illumination. These spectra were obtained under the same conditions as those in Fig. 7a, the only change being the rotated polarization of the incident light. Here, we see the same feature at 500 nm corresponding to the interband transitions in bulk Au, which decay over a time scale of 1–2 ps. In addition to this interband transition feature, however, we also see the emergence of a new feature at 540 nm, which corresponds to the plasmon resonant mode of the grating structure. The photo-excited hot carriers associated with this plasmon resonance also decay over a time scale of 1–2 ps. Here, the plasmon resonance gives rise to ground state bleaching, which makes this feature appear as a negative change in transmission ($\Delta T/T$). After the Au is excited by the pump laser, there are fewer plasmon electrons in the pumped sample and more electrons in the excited states than the non-pumped sample. Since the transient absorption spectrum is a result of pumped probe minus non-pumped probe,

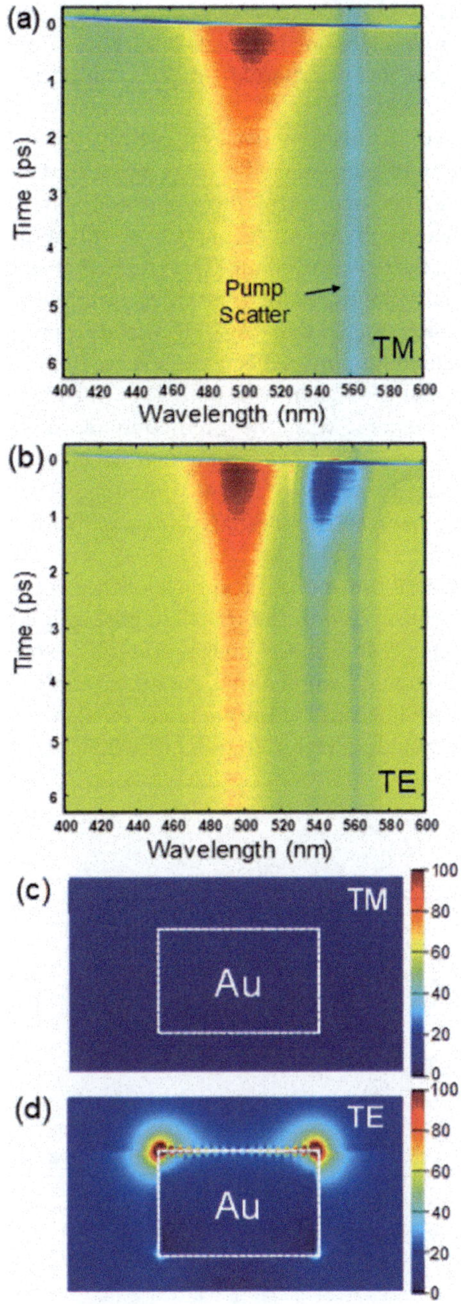

Fig. 7 Transient absorption spectra of the Au grating structure shown in Fig. 6 taken with (a) TM-polarized and (b) TE-polarized light. (c and d) Electromagnetic simulations plotting the cross-sectional electric field intensity distributions of grating structures illuminated with TE-polarized and TM-polarized light.

therefore the plasmon resonance peak appears to be negative. Upon closer inspection, one can notice a slight time-dependent blueshift of this plasmon resonant mode. This aspect of the data enables us to further distinguish these two absorption mechanisms. This blueshift may provide additional insight into the nonlinear response of hot electrons in metals and charge transfer in this system. Similar blueshifts have been reported in TiO_2-supported Au nanoparticle systems.[17]

Based on our FDTD simulations, we can predict the behaviour of grating structures consisting of different metals such as Ag, Al, and Cu. Fig. 8a shows the simulated photoreflectance spectra of the same corrugated grating structures shown in Fig. 1 calculated with Au, Ag, Al, and Cu. Based on these calculations, we expect Ag to produce a narrower resonance than Au and at slightly lower incident angles than Au. Similarly, we expect Al to produce a narrower resonance than Ag and at a lower incident angle than Ag. Fig. 8b–d show the experimentally measured angular dependent photoreflectance spectra of the corrugated gratings fabricated with Au, Ag, and Al taken with both TM- and TE-polarized light. Here we see qualitative agreement with the prediction of the FDTD calculations. Namely, the Ag resonance is narrower than that of Au and occurs at a lower incident angle than that of Au. Also, the Al resonance occurs a lower incident angle than that of Ag, however, its lineshape is not in fact narrower than that of Ag.

We have also evaluated the spectral response of the semi-transparent grating structures (similar to those used in the transient absorption measurements) for Al, Ag, and Cu metals. Fig. 9 shows SEM images and UV-vis spectra of these Au, Ag, Al, and Cu gratings. It should be noted that the lithographic processing of the Ag gratings did not work well because of the chemical sensitivity of Ag, which readily forms an oxide that is soluble in water. In order to achieve the 200–300 nm feature

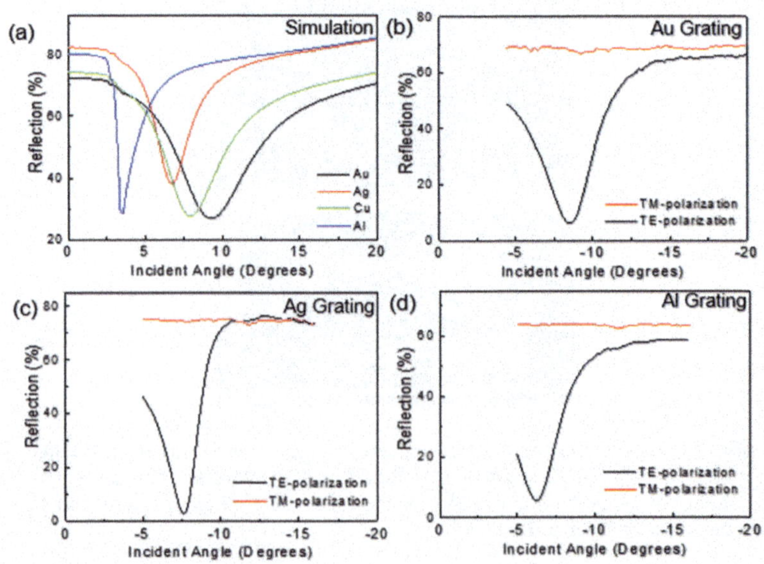

Fig. 8 (a) Calculated photoreflectance spectra of Au, Ag, Cu, and Al gratings with geometry depicted in Fig. 1a. Experimentally measured photoreflectance spectra of (b) Au, (c) Ag, and (d) Al gratings taken with both TE- and TM-polarized light.

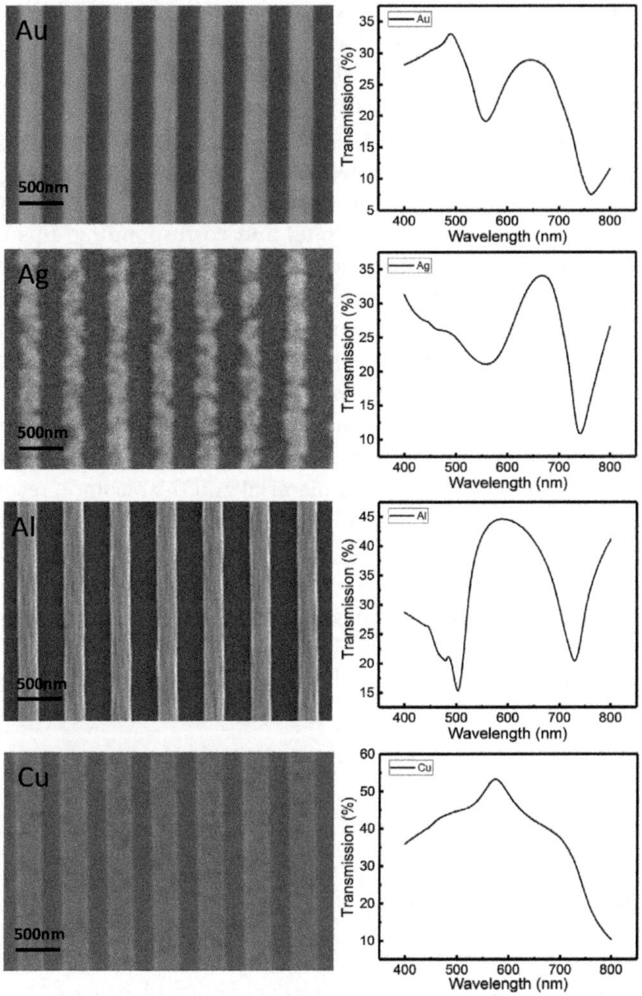

Fig. 9 SEM images and UV-vis transmission spectra of semi-transparent Au, Ag, Al, and Cu gratings.

sizes using photolithography, an anti-reflective coating is deposited on top of the Ag surface. As one of the last steps in the processing, a reactive ion etch is used to remove this anti-reflective coating. Unfortunately, however, this etch step oxidized the Ag and then partially dissolved the material, as is readily apparent in the SEM image of Fig. 9. Despite this poor quality, the spectral response of the Ag is still somewhat sharp, as evident in the UV-vis spectrum shown in Fig. 9. The grating peaks for the Cu are not as apparent because of the strong interband transitions in this metal, which give rise to its reddish colour.

Conclusions

In conclusion, we have developed a method for measuring the relatively small photocurrents produced by hot electrons and holes that are photoexcited in bulk metal films. By patterning these metals into plasmon resonant grating structures, we demonstrate enhanced photoelectrochemical performance in these hot-electron/hole mediated processes. The plasmon resonant grating structures enable us to compare interband absorption and plasmon resonant excitation by simply rotating the polarization of the incident light, while leaving all other parameters of the measurement constant. At a fixed laser wavelength, we can tune through the plasmon resonance of the grating. Sharp peaks in the photoelectrochemical current (*i.e.*, reaction rate) and corresponding dips in the photoreflectance of the grating are observed around 9° from normal incidence. Under non-resonant photoexcitation, we explore photocatalysis driven by bulk metal interband absorption. Under resonant excitation, we explore photocatalysis driven by surface plasmon modes, which are largely confined to the first few nanometers of the Au surface. Transient absorption spectroscopy performed under resonant and non-resonant conditions reveals that there are two mechanisms of photoexcitation of hot electrons/holes. Both the non-resonant excitation of interband transitions in the bulk Au material and the plasmon resonant mode relax back to equilibrium over a timescale of 1–2 ps. In addition to being spectrally distinct, the plasmon resonant mode produces a negative change in the optical transmission ($\Delta T/T < 0$) and exhibits a time-dependent blueshift associated with the local enhanced fields of the plasmon mode. In addition to Au, gratings fabricated with Ag and Al exhibit similar spectral features presenting additional promising platforms for hot electron and hot hole driven photocatalysis.

Conflicts of interest

There are no conflicts to declare.

Acknowledgements

This research was supported by ARO Award No. W911NF-17-1-0325 (Yi W.), the U.S. Department of Energy, Office of Science, Office of Basic Energy Sciences, under Award DE-SC0019322 (Yu W.), NSF Award No. CBET-1512505 (L. S.), AFOSR Grant No. FA9550-15-1-0184 (B. H.), DOE Award No. DE-FG02-07ER46376 (N. P.), ACS-PRF Grand No. 55993-ND5 (J. C.), and ARO Award No. W911NF-14-1-0228 (H. S.).

Notes and references

1 D. A. Boyd, L. Greengard, M. Brongersma, M. Y. El-Naggar and D. G. Goodwin, Plasmon-assisted chemical vapor deposition, *Nano Lett.*, 2006, **6**, 2592–2597.

2 W. H. Hung, I.-K. Hsu, A. Bushmaker, R. Kumar, J. Theiss and S. B. Cronin, Laser directed growth of carbon-based nanostructures by plasmon resonant chemical vapor deposition, *Nano Lett.*, 2008, **8**, 3278–3282.

3 P. Christopher, H. Xin and S. Linic, Visible-light-enhanced catalytic oxidation reactions on plasmonic silver nanostructures, *Nat. Chem.*, 2011, **3**, 467–472.

4 D. B. Ingram, P. Christopher, J. L. Bauer and S. Linic, Predictive Model for the Design of Plasmonic Metal/Semiconductor Composite Photocatalysts, *ACS Catal.*, 2011, **1**, 1441–1447.

5 Z. Liu, W. Hou, P. Pavaskar, M. Aykol and S. B. Cronin, Plasmon resonant enhancement of photocatalytic water splitting under visible illumination, *Nano Lett.*, 2011, **11**, 1111–1116.

6 D. B. Ingram and S. Linic, Water splitting on composite plasmonic-metal/semiconductor photoelectrodes: evidence for selective plasmon-induced formation of charge carriers near the semiconductor surface, *J. Am. Chem. Soc.*, 2011, **133**, 5202–5205.

7 S. Mukherjee, F. Libisch, N. Large, O. Neumann, L. V. Brown, J. Cheng, J. B. Lassiter, E. A. Carter, P. Nordlander and N. J. Halas, Hot electrons do the impossible: plasmon-induced dissociation of H_2 on Au, *Nano Lett.*, 2013, **13**, 240–247.

8 S. Mukherjee, L. Zhou, A. M. Goodman, N. Large, C. Ayala-Orozco, Y. Zhang, P. Nordlander and N. J. Halas, Hot-electron-induced dissociation of H_2 on gold nanoparticles supported on SiO_2, *J. Am. Chem. Soc.*, 2014, **136**, 64–67.

9 R. Sundararaman, P. Narang, A. S. Jermyn, W. A. Goddard III and H. A. Atwater, Theoretical predictions for hot-carrier generation from surface plasmon decay, *Nat. Commun.*, 2014, **5**, 5788.

10 A. M. Brown, R. Sundararaman, P. Narang, W. A. Goddard III and H. A. Atwater, Nonradiative Plasmon Decay and Hot Carrier Dynamics: Effects of Phonons, Surfaces, and Geometry, *ACS Nano*, 2016, **10**, 957–966.

11 P. Narang, R. Sundararaman and H. A. Atwater, Plasmonic hot carrier dynamics in solid-state and chemical systems for energy conversion, *Nanophotonics*, 2016, **5**, 96.

12 S. M. Kim, S. W. Lee, S. Y. Moon and J. Y. Park, The effect of hot electrons and surface plasmons on heterogeneous catalysis, *J. Phys.: Condens. Matter*, 2016, **28**, 254002.

13 J. Y. Park, S. M. Kim, H. Lee and Nedrygailov II, Hot-electron-mediated surface chemistry: toward electronic control of catalytic activity, *Acc. Chem. Res.*, 2015, **48**, 2475–2483.

14 J. S. DuChene, G. Tagliabue, A. J. Welch, W. H. Cheng and H. A. Atwater, Hot Hole Collection and Photoelectrochemical CO_2 Reduction with Plasmonic Au/p-GaN Photocathodes, *Nano Lett.*, 2018, **18**, 2545–2550.

15 B. Hou, L. Shen, H. Shi, R. Kapadia and S. B. Cronin, Hot electron-driven photocatalytic water splitting, *Phys. Chem. Chem. Phys.*, 2017, **19**, 2877–2881.

16 L. Shen, G. N. Gibson, N. Poudel, B. Hou, J. Chen, H. Shi, E. Guignon, N. C. Cady, W. D. Page and A. Pilar, Plasmon resonant amplification of hot electron-driven photocatalysis, *Appl. Phys. Lett.*, 2018, **113**, 113104.

17 A. Aiboushev, F. Gostev, I. Shelaev, A. Kostrov, A. Kanaev, L. Museur, M. Traore, O. Sarkisov and V. Nadtochenko, Spectral properties of the surface plasmon resonance and electron injection from gold nanoparticles to TiO_2 mesoporous film: femtosecond study, *Photochem. Photobiol. Sci.*, 2013, **12**, 631–637.

Faraday Discussions

PAPER

Enhancing hot electron generation and injection in the near infrared *via* rational design and controlled synthesis of TiO_2–gold nanostructures†

Supriya Atta,[a] Fuat E. Celik[b] and Laura Fabris*[c]

Received 11th October 2018, Accepted 21st November 2018

DOI: 10.1039/c8fd00152a

Plasmonic nanostructure/semiconductor composites are receiving great interest as powerful photocatalytic platforms able to increase solar energy conversion efficiency compared to more traditional approaches. The possibility to grow a thin titania shell onto the gold nanoparticle, thus substantially increasing the metal–semiconductor area of contact, is expected to be ideal for photocatalytic water reduction, especially if the titania (TiO_2) coating displays limited thickness and high crystallinity. We argue however that the morphology of the underlying gold nanoparticle and the quality of the interface are the main drivers of photocatalytic performance. Herein, we show how we can synthesize TiO_2-coated gold nanostar- and gold nanorod-based photocatalysts and identify the most important design parameters that one should be focusing on for the optimization of hot electron-based photocatalysts. In addition to nanoparticle morphology and interface quality, we determine that the integrated absorptivity of the plasmon band and the uniformity and crystallinity of the semiconductor shell are important, even though to a lesser extent. These results may prove interesting not only to increase production rates in hydrogen evolution reactions or other chemical conversions, but also to decouple and understand additional mechanisms driving photocatalysis, other than the sequential, hot electron mediated one, as we reported before.

Introduction

Nanostructured photocatalytic systems based on noble metal/semiconductor pairs have been attracting increasing attention for their potential to improve

[a]*Rutgers University, Department of Chemistry and Chemical Biology, 123 Bevier Road, Piscataway, NJ 08854, USA*

[b]*Rutgers University, Department of Chemical and Biochemical Engineering, 98 Brett Road, Piscataway, NJ 08854, USA*

[c]*Rutgers University, Department of Materials Science and Engineering, 607 Taylor Road, Piscataway, NJ 08854, USA. E-mail: lfabris@soe.rutgers.edu*

† Electronic supplementary information (ESI) available: Additional TEM and UV-Vis characterization. See DOI: 10.1039/c8fd00152a

our ability to leverage solar energy for chemical conversion or clean fuel generation.[1–6] Au/TiO_2 pairs, in particular, have shown promise for the enhancement in photocatalytic production of hydrogen from water, also known as the hydrogen evolution reaction (HER).[7] While the majority of the reported approaches have involved the use of Au nanorods/TiO_2 as the photocatalyst and visible light as the source of illumination for HER,[8,9] we have recently reported that gold nanostars/TiO_2 systems, illuminated under broad- or near infrared (NIR) illumination conditions, provide substantially enhanced yields of hydrogen, under specific design conditions.[10] Our results appear to validate the hypothesis that, under visible/NIR illumination, the reduction of water in gold nanostars/TiO_2 systems occurs *via* a sequential mechanism mediated by hot electron generation in the plasmonic nanoparticle and subsequent transfer into the semiconductor catalyst.[11] While other concurring or alternative mechanisms might be taking place, such as the direct transfer of electrons from the valence band to the conduction band of TiO_2 under UV illumination (with the plasmonic nanoparticle acting as the cocatalyst), or the direct excitation of charge transfer transitions,[12] for photocatalysts such as the one we proposed, in which the plasmonic nanoparticle is entirely coated by the semiconductor, and the highest conversion enhancement occurs under NIR illumination, the latter mechanisms are likely to be less plausible. However, for the sequential mechanism to occur with high efficiency, the generated hot electrons need to be rapidly transferred from the plasmonic nanoparticle into the semiconductor. This should take place before they thermalize (within 200 fs)[13] to a high temperature thermal distribution *via* electron–electron scattering processes,[14–16] and subsequently further relax (within a few ps)[17] to lower electron temperatures *via* electron–phonon coupling. These timelines can be met only if we ensure that an intimate contact between the noble metal nanoparticle and the semiconductor exists, and is possibly free of defects, such as occurs *via* epitaxial growth.[10] Furthermore, the hot electrons should be generated at locations in which they can easily escape the metal, which are most likely positions of high curvature, such as the tips of gold nanostar spikes.[18] Interestingly, these are also locations at which the generation of hot electrons is maximized, due to the local enhancement of the electric field as well as the breaking of momentum conservation, which occurs where the field is highly nonuniform.[11] However, because in our assumption the semiconductor should be deposited as a shell around the nanoparticle, a limited TiO_2 shell thickness should also be ensured, to avoid electron quenching. In our case, we were able to epitaxially grow the TiO_2 shell to be only 4 nm in thickness.[10] Finally, to ensure that TiO_2 is catalytically active, the shell needs to be crystalline, and ideally constituted by a mix of rutile and anatase for optimized alignment with the energy levels involved in the HER.[10] So far, from a design standpoint, Au/TiO_2 nanoparticle systems have been the most promising ones among the various metal/semiconductor pairs, due in particular to the morphology dependence of the localized surface plasmon resonance (LSPR) band in gold and the ease of shape tunability that has been enabled through years of discovery and optimization of synthetic protocols.[19,20] Most of the previously reported Au/TiO_2 photocatalytic systems have been limited, with few exceptions, to spherical nanoparticles under UV illumination conditions. However, these plasmonic materials suffer from limited LSPR tunability, which still leaves the majority of the solar spectrum untapped.[4,5] To fully utilize the solar spectrum, it is important to tune the morphology of the Au nanoparticles, while keeping all the previous design constraints met. Unfortunately, one of the main limitations for the colloidal synthesis of TiO_2-coated Au nanoparticles is the poor stability of the starting plasmonic nanoparticles under the reaction conditions necessary to synthesize the TiO_2 shell; in particular, their sensitivity to the high temperatures at which TiO_2 crystallization occurs has made it prohibitively difficult to realize all the conditions described before at the same time, as high temperatures lead to Au atom migration and nanoparticle reshaping, as we observed.[10]

Building from our previous observations, herein we assess the effect of each design parameter on the photocatalytic efficiency of Au nanorod/TiO_2 systems in HER reactions. We focus in particular on (1) nanoparticle morphology, (2) LSPR tunability, (3) interface properties, and (4) TiO_2 shell quality, including uniformity, thickness, and crystallinity. We are interested in particular in dissecting how meeting (or not) one of the design constraints reported above improves (or not) the performance of the photocatalyst. We report on a new method to synthesize crystalline TiO_2 without initial addition of a polymer, whose role is to stabilize the nanorods in organic solvent prior to exposure to the titanium precursor, but that drastically affects electron transfer at the interface as it interferes with the epitaxial growth thus hindering electron pathways. Importantly, our low temperature protocol generates crystalline TiO_2 without modifying the morphology of the nanoparticle. We investigated the photocatalytic activity of TiO_2 coated gold nanorods toward HER under broad illumination conditions and compared it to that of nanostars with the same LSPR position, observing an initial higher performance of the nanostars. While this trend inverts after 20 minutes, possibly due to heating and reshaping of the nanostar tips, these results underscore once again the importance of the nanostar morphology as the main parameter affecting performance. Furthermore, because our nanorods, synthesized in the absence of any polymer overcoating as stabilizers, show identical performance to similar systems reported in the literature, characterized by a complete TiO_2 shell grown in the presence of polymer,[8] we confirm our hypothesis that capping agents left at the interface hinder hot electron mobility and are deleterious to photocatalytic performance.

Results and discussion

The protocol for the synthesis of TiO_2 shells on gold nanorods has been reported in the literature,[4,21] and is characterized by three steps. In the first step, gold nanorods are protected by a polymer (poly(sodium 4-styrenesulfonate), PSS) to prevent agglomeration in organic solvent, where the growth of the shell occurs. Then, the TiO_2 shell is grown on the gold nanorods, in the presence of the TiO_2 precursor $TiCl_3$. In the third step, amorphous TiO_2 is crystallized at 450 °C. The high temperature treatment, while substantially improving the crystallinity, leads to the generation of Ti^{3+} species and oxygen vacancies.[22] However, the presence of the polymer layer creates a barrier for electron transfer from gold into the semiconductor, likely reducing the performance of the photocatalyst. We therefore propose here a new low temperature protocol for the synthesis of a crystalline TiO_2 shell on gold nanorods in the absence of a polymer overcoating.

Traditional protocols cannot be carried out in these conditions as the nanorods precipitate and coalesce in organic solvents, even when mildly polar, such as isopropanol. Our water-based method relies on the use of titanium isopropoxide (TTIP) as the Ti precursor and is run at 70 °C for two days. It is well known that the gold nanorod LSPR maximum should redshift upon TiO_2 shell formation due to the increase in the dielectric function of the surrounding medium in going from water to TiO_2. As we suspected, in the absence of a polymer overcoating the LSPR of the gold nanorods disappeared in isopropanol, consistent with nanorod aggregation and precipitation (Fig. S1†). Transmission electron microscopy (TEM) micrographs show substantial aggregation of the gold nanorods (Fig. S1a†), inconsistent with mere aggregation induced by water drying or the electron beam. We believe instead that the aggregation is entirely due to the increase in critical micelle concentration (CMC) for cetyltrimethylammonium bromide (CTAB), the surfactant employed for the synthesis and stabilization of the nanorods, in water/alcohol mixtures. The increase in CMC leads to the unzipping of the CTAB micelle and consequent nanorod aggregation.[23] On the contrary, by running the reaction in water, we avoided nanorod aggregation and were able to grow a, albeit not complete, TiO_2 shell on the nanorods. TEM characterization shows that while the TiO_2 shell that results is not uniform, the nanorod morphology is retained (Fig. 1a–e). The presence of the TiO_2 shell is further confirmed by the redshift and reduction in intensity of the LSPR band (Fig. 1c, f), as we observed before.[10] Importantly, while the growth of the shell was not possible in pure water, the addition of acetic acid, aimed at reducing the rate of hydrolysis of TTIP, led to its formation. Following our previously reported protocol for crystallization of the TiO_2 shell on gold nanostars, we heated the reaction mixture to 70 °C and maintained it at this temperature for 2 days. High resolution TEM characterization (HRTEM) shows crystallization of TiO_2 with formation of anatase, although amorphous TiO_2 is still present (Fig. 2b).

Gold nanorods of increasing aspect ratio, and therefore with redshifted LSPR maxima (870 nm, 910 nm, and 980 nm), were synthesized to understand the effect

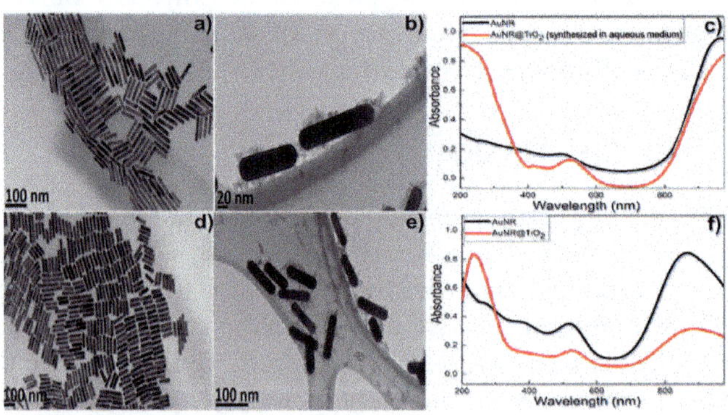

Fig. 1 (a) TEM micrograph of CTAB-capped gold nanorods characterized by a LSPR at 940 nm. (b) TEM micrograph of the same nanorods after TiO_2 coating realized in aqueous medium, showing that the morphology of the nanorods is retained. (c) UV-Vis extinction spectra show that a 50 nm red shift occurs when the TiO_2 coating is realized *via* TTIP in aqueous medium. (d) TEM micrograph of CTAB-capped gold nanorods characterized by a LSPR at 840 nm. (e) TEM micrograph of the same nanorods after TiO_2 coating. As in the previous example, the coating appears not to be forming uniformly around the nanorods. (f) UV-Vis extinction spectra show that a 30 nm red shift occurs when the TiO_2 coating is realized *via* TTIP in aqueous medium.

of resonant position on photocatalytic efficiency. An LSPR maximum at a shorter wavelength should generate a larger population of highly energetic electrons (proportional to integrated plasmon absorbance scaled for solar flux at each wavelength[24]) available for transfer into the semiconductor and subsequent photocatalytic reduction. As we have observed before,[20] redshifted LSPR bands are more intense and are correlated to increased electric near fields. In our study, however, nanorods with a 910 nm LSPR show substantially broadened and less intense resonances, which will be interesting to correlate integrated plasmon absorptivity to hot electron generation and photocatalytic efficiency.

TiO_2-coated gold nanostars were also synthesized to compare our nanorod-based platform to a system that we have previously proven to be an excellent photocatalyst for HER.[10] Gold nanostars were synthesized to possess an LSPR maximum at 980 nm after TiO_2 coating, to study how the morphology of the nanoparticles and the quality of the coating affect the photocatalytic performance at equal absorption maxima (Fig. 3). In Fig. 3c we can observe a TEM micrograph of the Au nanostars upon coating with TiO_2, in which the presence of a rather uniform shell can be observed. On the other hand, Fig. 3c also shows the presence of isolated TiO_2 particles that were difficult to remove and may impact the activity of the photocatalyst. Interestingly, when comparing the UV-Vis extinction spectra of the TiO_2-coated nanorods and nanostars with coinciding LSPR maxima at 980 nm, one can observe how the extinction spectrum for the nanostars shows a more intense TiO_2 band compared to that of the nanorods, consistent with the presence of the isolated TiO_2 nanoparticles. More importantly, the LSPR band for the gold nanostars is substantially broadened compared to that of the nanorod counterpart. This broadening, and reduced intensity, may be detrimental with respect to the photocatalyst performance, considering the correlation between the electric field intensity and the number of hot electrons generated, as explained above.

By running hydrogen evolution reactions *via* broad spectrum illumination, we observed that the hydrogen productivity of crystalline TiO_2-coated gold nanorods increased with an increasing plasmon resonance wavelength from 870 to 980 nm (Fig. 4). By illuminating gold nanorods with 870 nm LSPR, 0.37 mmol H_2 per gram of catalyst was produced in 1 hour, whereas by illuminating the nanorods with

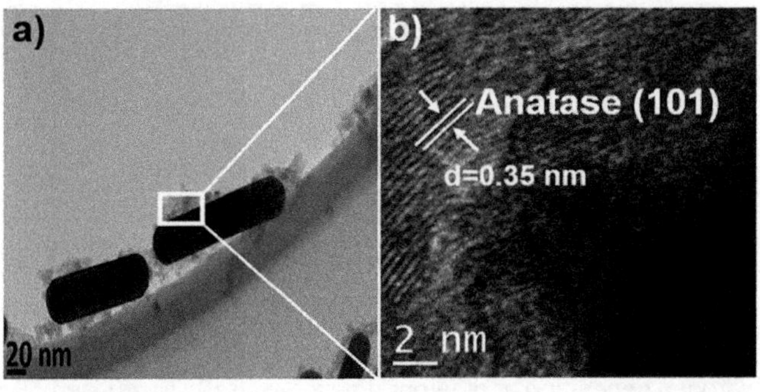

Fig. 2 HRTEM micrographs of TiO_2-coated gold nanorods. In (a) the morphology of the rod is retained upon coating with TiO_2, even though the shell is not complete. In (b) the high magnification micrograph provides evidence of crystallization into anatase TiO_2 (*d* spacing 0.35 nm) even though the presence of an amorphous region is still evident.

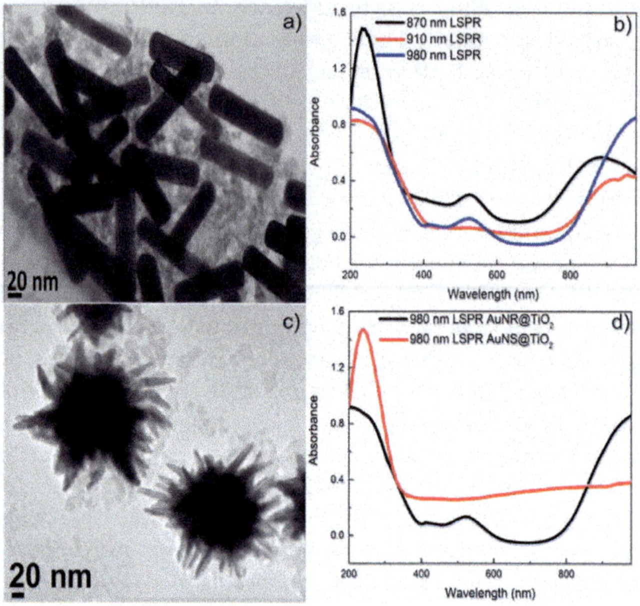

Fig. 3 (a) TEM image of CTAB capped gold nanorods where the plasmon resonance comes at around 910 nm which is a mixture of 870 and 980 nm LSPR gold nanorods. (b) UV-Vis absorption spectra of three different gold nanorod samples. (c) TEM image of TiO_2 coated gold nanostars. (d) UV-Vis absorption spectra of TiO_2 coated gold nanostars and nanorods where the plasmon resonance comes at 980 nm.

980 nm LSPR, 0.7 mmol H_2 g^{-1} was instead produced, with an almost doubled yield. This result appears to correlate nicely with the integrated plasmon absorptivity for the two nanorod batches, even though they are counterintuitive if one simply looks at the position of the LSPR band. We then monitored the activity of the gold nanorod batch with 910 nm LSPR, composed of a polydispersed mixture of particles. Interestingly, the hydrogen productivity of mixed TiO_2 coated

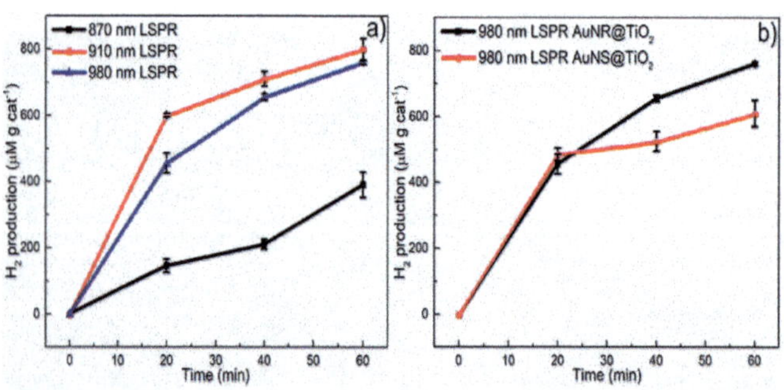

Fig. 4 (a) Hydrogen generation at different time intervals (0, 20, 40, and 60 minutes) for water reduction at broad spectrum illumination over three different TiO_2 coated gold nanorods having LSPR maxima at 870 nm (black), 980 nm (blue), and 910 nm (red). (b) Hydrogen evolution comparison for TiO_2 coated gold nanorods and nanostars having LSPR maxima at 980 nm.

gold nanorods was higher than that of the 870 nm and 980 nm LSPR samples, possibly due to a panchromatic effect, as shown by Mubeen *et al.*, who demonstrated that the photocatalytic efficiency of a mixture of AuNR–TiO_2–Pt Janus nanorods increased compared to the efficiency of the individual rods, possibly due to the ability of the mixture to absorb a broader range of solar radiation.[24] Our results compare fairly well with those reported by Kou *et al.*,[8] who showed *ca.* 320 μmol g^{-1} h^{-1} of H_2 produced by their Au nanorod–TiO_2 system, in which a full titania shell was grown onto the nanorods but in the presence of a polymer overcoating. This result suggests that the presence of a complete semiconductor layer around the nanorod is drastically affected by the lack of an epitaxial interface, which will likely not grow because of the interference of the polymer. One therefore should not only pay attention to the completeness of the shell but also that no interfering molecules are trapped between metal and semiconductor, thus affecting electron mobility.

Photocatalytic water reduction studies were carried out under broad-spectrum illumination conditions also to compare the performance of gold nanostars and gold nanorods, using 1.5 mg of TiO_2 coated gold nanoparticles in each study, where methanol was again used as the sacrificial agent for hole capture and oxygen generation (Fig. 4b). The photocatalytic activity of the TiO_2 coated gold nanostars and nanorods was roughly comparable at 20 minutes (with the nanostars behaving only slightly better), with the nanorods instead performing better at longer times. This shift in performance could be due to localized heating occurring at the tips of the nanostars under illumination, which may cause increased Au atom mobility at the tips and local reshaping, therefore rendering the electric field intensity at these positions lower over time, and decreasing the overall generation of hot electrons and injection into TiO_2. This evidence suggests that, under broad spectrum illumination, and in the absence of heat-induced reshaping, light absorption *via* the gold nanostar generates more hot electrons than in nanorods, primarily because each gold nanostar possesses an average of 100 spikes, as shown by us and others, where the enhanced electric field concentrates.[18,20] What emerges from the results at 20 minutes correlates the extinction spectra of the gold nanostars (Fig. 3d) to their performance. In fact, if one were only to advocate for the plasmon absorptivity to be the main determining factor in the photocatalytic performance, the conversion rates observed for the (non-optimized) nanostars reported here would not support this hypothesis. On the contrary it appears to be indeed the presence of the spikes that is the important morphology feature affecting the performance of TiO_2-coated Au nanostars as photocatalysts in HER. While it could be argued that the nanorods do not possess a complete semiconductor shell compared to the nanostars, the results reported above, comparing different nanorod systems, should further strengthen the role of the spiked morphology as the relevant parameter with respect to hot electron generation and photocatalytic performance.

Taken all together, these results confirm our hypothesis that various materials design parameters need to be taken into account to synthesize effective photocatalysts for hot electron-induced chemical conversions. While we have herein focused on hydrogen evolution reactions, these considerations should be generalizable to a large extent. The main features that we have analyzed in this work include (1) the morphology of the nanoparticles, (2) the position and integrated absorptivity of the LSPR, (3) the quality of the metal–semiconductor interface, (4) the uniformity of the TiO_2 shell and (5) its crystallinity. While nanorods and nanostars can be synthesized to possess the same LSPR, it is the presence of the spikes on the nanostars that drastically affects the performance of the photocatalyst, similarly to what we observed before with respect to the magnitude of the extinction coefficient.[26] The integrated absorptivity appears to be the second most important parameter, with higher absorptivity values correlating to higher yields of hot electrons and therefore to improved performance. The quality of the interface should also be kept high, ideally free of intermediate coating layers (*e.g.* polymers) that can hinder electron mobility. Ideally, as we showed previously,[10] the interface should be synthesized epitaxially. While we have not studied the role of the semiconductor shell thickness and its crystallinity in this work (all our samples showed a mix of anatase and amorphous titania), we have determined that the presence of a full semiconductor shell on the plasmonic nanomaterial is not a fundamental driver of performance.

Conclusions

In this study, we have synthesized TiO_2-coated gold nanorods and nanostars and investigated their photocatalytic properties with respect to several materials design parameters. Interestingly, we have observed that the photocatalytic reaction under broad spectrum illumination leads to higher H_2 production even when the plasmon resonance of the gold nanorods is shifted closer to 1000 nm, as long as the integrated absorptivity of the LSPR band is high. We have also confirmed the results of Moskovits and coworkers,[24] who showed that a panchromatic device, in which gold nanorods of various aspect ratios are mixed together, yields a higher H_2 production if the nanorods are tuned to absorb a broader spectrum of solar radiation. The catalytic activity of TiO_2-coated gold nanostars was found to be higher than that of TiO_2-coated gold nanorods within the first 20 minutes of reaction, with the trend reversing at longer times, probably due to local heating at the spike tip and atomic restructuring. Overall however, it is the spiked morphology of the gold nanostars, which leads to high generation of hot electrons in proximity of the tips, that is the main driver of photocatalytic performance, closely followed by the presence of an intimate contact between the metal and the semiconductor, not hindered by residual by-products of the synthesis. Therefore, synthetic efforts should focus on achieving these two important results. Because, as we reported before, additional mechanisms, other than the sequential one, might be at play with respect to the generation and use of hot electrons for chemical conversion, we believe that the design rules proposed here might be ideally suited to producing nanostructured photocatalysts and aid in decoupling, understanding, and leveraging each of them. Finally, the application of gold nanostar systems to drive other chemical reactions with infrared illumination should be explored, importantly to understand their suitability for not only increasing conversion rates but also selectivity.

Experimental

Materials

Silver nitrate (99%), sodium borohydride (≥96%), titanium isopropoxide (TTIP), and hydroquinone (99%) were purchased from Sigma Aldrich. Gold(III) chloride trihydrate ($HAuCl_4 \cdot 3H_2O$) and L-(+)-ascorbic acid were purchased from Acros Organics. Cetyltrimethylammonium bromide (CTAB) was purchased from GFS Chemicals. All these chemicals were used without purification. Ultrapure water (18.2 MΩ cm) was used in all preparations. All glassware was aqua regia cleaned before each synthesis.

Instrumentation

Extinction spectra were collected using an S.I. Photonics Fiber Optic CCD Array UV-Vis Spectrophotometer using a quartz cuvette with a 1 cm path length. The nanoparticle morphology was determined using a Topcon 002B transmission electron microscope depositing the nanoparticle suspension on Ted Pella Inc. PELCO TEM grids. HRTEM analysis was performed on a JEOL 2010 F high-resolution transmission electron microscope. The spike length and thickness of the TiO_2 shell of the nanorods were measured using ImageJ software. Particle morphology was analysed using Gatan Digital Micrograph (TM) 3.11.1 for GMS 1.6.1. The values of average *d* spacing were obtained from Fourier transform analysis of high-magnification images.

Synthesis of seeds

Gold seeds were prepared following a reported procedure.[25] Briefly, 5 mL of a freshly prepared 0.01 M sodium borohydride solution was dissolved in 5 mL of a freshly prepared 0.01 M solution of sodium hydroxide. 460 μL of the sodium borohydride aqueous solution was added to 10 mL of a 0.1 M CTAB solution containing 0.5 mM $HAuCl_4$ under rapid stirring. The color of the solution changed from yellowish to light brown.

Synthesis of nanorods

10 mL of a 0.5 mM solution of $HAuCl_4$ was added to the 0.1 M CTAB solution. 70 μL of 0.1 M $AgNO_3$ solution was added to the mixture, followed by addition of 500 μL of a 0.1 M hydroquinone aqueous solution. The resulting mixture was stirred for a minute. Then, 160 μL of seed solution was added to the growth solution. The growth solution was aged overnight. Gold nanorods were purified by centrifugation at 10 000 rpm and redispersed in 5 mL of Milli-Q water.

Synthesis of TiO_2 coated gold nanorods

TiO_2 coated gold nanorods were synthesized using TTIP as the precursor in aqueous environment. Briefly, 300 mL of gold nanorods were synthesized and purified by centrifugation at 10 000*g* for 1 hour. They were then redispersed in 100 mL Milli-Q water. 27 mL acetic acid and 250 μL of TTIP were added to the solution and stirred for 12 hours at room temperature. The resulting solution was then heated at 70 °C to crystallize the TiO_2 shell. The solution was washed with Milli-Q water twice before use.

Synthesis of TiO_2 coated gold nanostars

TiO_2 coated gold nanostars were synthesized by following our reported method.[10] Briefly, 300 mL of surfactant free gold nanostars were synthesized and centrifuged at 5000*g* for 15 minutes. They were then redispersed in 60 mL of isopropanol.

6 mL acetic acid and 250 μL of TTIP were added to the mixture and stirred for 30 minutes. The resulting solution was stirred for 12 hours at room temperature and then heated at 70 °C to crystallize the TiO_2 shell. The solution was washed with a 1 : 1 MilliQ water and ethanol mixture twice and then dried to produce a powder.

Photocatalytic water reduction

The photocatalytic activity of the samples was assessed by following our previously reported procedure.[10] Briefly, a 4 mL solution of TiO_2 coated gold nanorods (which is equivalent to 1.5 mg) was dispersed in 1 mL of Milli-Q water and 0.5 mL methanol. The reactor was illuminated by a 150 W xenon ozone-free Arc lamp (Newport) with an approximate flux of 3000 and brightness of 150 cd mm^{-2} using 20 V and 7.5 A of power. We have investigated the catalytic activity under broad spectrum illumination (200–1500+ nm) conditions. The hydrogen production rate was monitored at 20 minute intervals for an hour with a vessel temperature of 45 °C. A 300 μL aliquot of the reactor headspace gas was injected into the TCD of a gas chromatograph (Agilent 7890B). Hydrogen production was reported per gram of catalyst for all samples.

Conflicts of interest

There are no conflicts to declare.

References

1 A. Naldoni, V. M. Shalaev and M. L. Brongersma, *Science*, 2017, **356**, 908.
2 L. Lu, Z. Luo, T. Xu and L. Yu, *Nano Lett.*, 2013, **13**, 59.
3 S. Yu, A. J. Wilson, G. Kumari, X. Zhang and P. K. Jain, *ACS Energy Lett.*, 2017, **2**, 2058.
4 A. Tanaka, S. Sakaguchi, K. Hashimoto and H. Kominami, *ACS Catal.*, 2013, **3**, 79.
5 T. Hirakawa and P. V. Kamat, *J. Am. Chem. Soc.*, 2005, **127**, 3928.
6 H. Zhu, X. Chen, Z. Zheng, X. Ke, E. Jaatinen, J. Zhao, C. Guo, T. Xie and D. Wang, *Chem. Commun.*, 2009, 7524.
7 D. A. Panayotovb and J. R. Morrisa, *Surf. Sci. Rep.*, 2016, **71**, 77.
8 S. F. Kou, W. Ye, X. Guo, X. F. Xu, H. Y. Sun and J. Yang, *RSC Adv.*, 2016, **6**, 39144.
9 B. Wu, D. Liu, S. Mubeen, T. T. Chuong, M. Moskovits and G. D. Stucky, *J. Am. Chem. Soc.*, 2016, **138**, 1114.
10 S. Atta, A. M. Pennington, F. E. Celik and L. Fabris, *Chem*, 2018, **4**, 2140.
11 G. V. Hartland, L. V. Besteiro, P. Johns and A. O. Govorov, *ACS Energy Lett.*, 2017, **2**, 1641.
12 M. J. Kale, T. Avanesian, H. Xin, J. Yan and P. Christopher, *Nano Lett.*, 2014, **14**, 5405.
13 C. Voisin, N. D. Fatti, D. Christofilos and F. Valleé, *J. Phys. Chem. B*, 2001, **105**, 2264.
14 T. Hertel, E. Knoesel, M. Wolf and G. Ertl, *Phys. Rev. Lett.*, 1996, **76**, 535.
15 S. Ogawa and H. Petek, *Surf. Sci.*, 1996, **357**, 585.

16 M. Wolf, *Surf. Sci.*, 1997, **377**, 343.
17 G. V. Hartland, *Chem. Rev.*, 2011, **111**, 3858.
18 A. Sousa-Castillo, M. C. Hermo, B. R. Gonzalez, M. P. Lorenzo, Z. Wang, X. T. Kong, A. O. Govorov and M. A. C. Duarte, *J. Phys. Chem. C*, 2016, **120**, 11690.
19 B. Nikoobakht and M. A. El-Sayed, *Chem. Mater.*, 2003, **15**, 1957.
20 S. Atta, T. V. Tsoulos and L. Fabris, *J. Phys. Chem. C*, 2016, **120**, 20749.
21 N. Zhou, L. Polavarapu, N. Gao, Y. Pan, P. Yuan, Q. Wang and Q.-H. Xu, *Nanoscale*, 2013, **5**, 4236–4241.
22 K. Yanagisawa and J. Ovenstone, *J. Phys. Chem. B*, 1999, **103**, 7781.
23 A. S. D. S. Indrasekara, R. C. Wadams and L. Fabris, *Part. Part. Syst. Charact.*, 2014, **31**, 819.
24 S. Mubeen, J. Lee, D. Liu, G. D. Stucky and M. Moskovits, *Nano Lett.*, 2015, **15**, 2132.
25 L. Vigderman and E. R. Zubarev, *Chem. Mater.*, 2013, **25**, 1450.
26 T. V. Tsoulos, L. Han, J. Weir, H. L. Xin and L. Fabris, *Nanoscale*, 2017, **9**, 3766.

Faraday Discussions

PAPER

Enhanced hot electron generation by inverse metal–oxide interfaces on catalytic nanodiode†

Hyosun Lee,[a] Sinmyung Yoon,[b] Jinwoung Jo,[cd] Beomjoon Jeon,[ae] Taeghwan Hyeon,[cd] Kwangjin An*[b] and Jeong Young Park*[ae]

Received 27th September 2018, Accepted 7th November 2018
DOI: 10.1039/c8fd00136g

Identifying the electronic behavior of metal–oxide interfaces is essential for understanding the origin of catalytic properties and for engineering catalyst structures with the desired reactivity. For a mechanistic understanding of hot electron dynamics at inverse oxide/metal interfaces, we employed a new catalytic nanodiode by combining Co_3O_4 nanocubes (NCs) with a Pt/TiO_2 nanodiode that exhibits nanoscale metal–oxide interfaces. We show that the chemicurrent, which is well correlated with the catalytic activity, is enhanced at the inverse oxide/metal (CoO/Pt) interfaces during H_2 oxidation. Based on quantitative visualization of the electronic transfer efficiency with chemicurrent yield, we show that electronic perturbation of oxide/metal interfacial sites not only promotes the generation of hot electrons, but improves catalytic activity.

1. Introduction

Most heterogeneous catalysts have been designed in the form of combined nanostructures of metals and oxides to achieve acceptable catalytic performance.[1–3] However, there are still questions about how the catalysts modulate the reaction kinetics and which portions of the complex nanostructure function as the real active sites.[3] Among the various configurations, the metal–oxide interface has received intensive attention as a special active site with unique electrical and chemical properties that influence catalytic performance, such as activity,

[a]*Center for Nanomaterials and Chemical Reactions, Institute for Basic Science (IBS), Daejeon 34141, Republic of Korea. E-mail: jeongypark@kaist.ac.kr*

[b]*School of Energy and Chemical Engineering, Ulsan National Institute of Science and Technology (UNIST), Ulsan 44919, Republic of Korea. E-mail: kjan@unist.ac.kr*

[c]*Center for Nanoparticle Research, Institute for Basic Science (IBS), Seoul 08826, Republic of Korea*

[d]*School of Chemical and Biological Engineering, Institute of Chemical Processes, Seoul National University, Seoul 08826, Republic of Korea*

[e]*Department of Chemistry and Graduate School of EEWS, Korea Advanced Institute of Science and Technology (KAIST), Daejeon 34141, Republic of Korea*

† Electronic supplementary information (ESI) available. See DOI: 10.1039/c8fd00136g

selectivity, and stability.[4–10] In particular, several research groups have focused on fundamentally understanding the electronic effects at the metal–oxide interface because it is postulated to be the main feature that controls the catalytic properties.[11–15] However, it remains challenging to identify the electronic structure of metal–oxide interfaces in complex nanocatalysts because there is a lack of definite evidence.

Since the enhancement of catalytic activity from the presence of oxide–metal interfaces was first suggested by Schwab, the strong metal–support interaction (SMSI) effect has been important for achieving enhanced catalytic activity for heterogeneous catalysts.[16,17] Recent studies of CO oxidation on Au/TiO_2 suggest that the perimeter between the Au nanoparticles and the TiO_2 is catalytically reactive.[18] The reaction rate in the oxide–metal model system has been shown to depend on the oxidation state of the supporting oxide, the free metal surface area, and the number of sites at the interface between the metal and the support.[19] In the work by Boffa *et al.*[20] using rhodium deposited on a large assortment of oxides, a remarkable 14-fold increase in the turnover rate for CO_2 hydrogenation was observed in the presence of three oxides: TiO_x, NbO_x, and TaO_x. The activity was highest when the oxide–metal interface area was at a maximum, which occurred at the oxide coverage of a half monolayer.

One of the most effective strategies for verifying the specific electronic features of the metal–oxide interface is quantitative detection of charge transfer using metal–semiconductor catalytic nanodiodes.[21–24] In these devices, both the thin metal layer (<15 nm, *i.e.* the mean free path of electrons) and the low Schottky barrier height at the metal–semiconductor interface make it possible to collect hot electrons with high energy (1–3 eV) generated on the catalytic nanodiodes during surface reactions.[14,25] Recently, the dynamics of hot electrons in various nanocatalysts were investigated through real-time detection of hot electrons using catalytic nanoparticle (NP)/Au/TiO_2 nanodiodes; a clear correlation between catalytic activity and the electronic properties of different sizes and compositions of nanocatalysts was clearly demonstrated.[26,27] Earlier works combining nanocatalysts and hot electron generation exhibit photon-induced hot electrons that cause a dramatic change in the catalytic activity.[28–31] Recently, oxide/metal (CoO/Pt) interfacial sites formed by the surface segregation of PtCo bimetallic NPs were proven to be a major factor in improving the overall catalytic performance for H_2 oxidation.[27]

As a follow-up study, we demonstrate here a new design for a catalytic nanodiode composed of an inverse oxide/metal structure, a Co_3O_4 nanocube (NC)/Pt/TiO_2 nanodiode, to gain in-depth insight into the effect of the oxide/metal interface on catalytic activity. In the current system, well-defined interfacial areas of cobalt oxide and metallic platinum are systematically controlled, unlike for the previously studied PtCo bimetallic NPs. Using the Co_3O_4 NCs/Pt/TiO_2 nanodiode, hot electron transport at the inverse oxide/metal interface is investigated during the exothermic H_2 oxidation reaction. Surprisingly, when the inverse oxide/metal (CoO/Pt) interfaces are created on the catalytic nanodiodes, both the catalytic activity and hot electron transfer improved significantly, demonstrating that the electronic properties of the oxide/metal interfacial sites are indeed beneficial to the catalytic properties.

2. Experimental details

2.1. Synthesis of cobalt oxide NCs

In a typical synthesis of Co_3O_4 NCs, 0.37 g (1 mmol) of cobalt(II) perchlorate (Sigma-Aldrich) was dissolved in a mixture of 2.7 g (10 mmol) of oleylamine (Sigma-Aldrich) and 15 mL of 1-octanol (Sigma-Aldrich). The solution was heated to 120 °C in air. Just before the temperature reached 120 °C, 0.7 mL of deionized water was rapidly injected. The solution was maintained at 120 °C for 2 h. The reaction mixture was then cooled to room temperature and the precipitate was collected by centrifugation. The resulting Co_3O_4 NCs were redissolved in hexane and chloroform. The acetone wash, centrifugation and redispersion process was repeated three times.

2.2. Fabrication of the Co_3O_4 NCs/Pt/TiO_2 nanodiode

To fabricate the Pt/TiO_2 Schottky nanodiodes, 200 nm thick Ti films were directly deposited onto a 4-inch SiO_2 (500 nm)/p-type Si wafer in a 4×6 mm^2 pattern using electron beam evaporation. The Ti films were then thermally oxidized in air at 450 °C for 2.5 h, generating a TiO_2 layer with the appropriate Fermi level. On the electrode, a 50 nm thick Ti film and a 150 nm thick Au film were deposited sequentially with an area of 5×5 mm^2. The Ti layer forms an ohmic contact with the TiO_2 film. Next, a 10 nm thick Pt thin film was deposited with an area of 2×3 mm^2, which creates a Schottky contact with an active area of 4 mm^2 between the Pt and TiO_2 films. Finally, to assemble the inverse oxide/metal interface structure, we used spin coating to deposit a 2D monolayer of Co_3O_4 NCs on the active surface of the Pt/TiO_2 nanodiode.

2.3. Measurement of catalytic activity and chemicurrent

To measure the reaction rates, we performed the atmospheric H_2 oxidation reaction in a high-vacuum bath reactor (1 L) with a base pressure of 2.0×10^{-8} Torr. After loading the sample on a 25 mm standard pyrolytic boron nitride heater, the evacuated reactor chamber was charged with 15 Torr of H_2 and 745 Torr of O_2 at room temperature. To activate the thermal catalytic reactions, the catalysts were heated from 30 to 90 °C while monitoring the temperature with a K-type thermocouple. The reaction gas mixture was recirculated using a Metal Bellows circulation pump through the reaction line at a rate of 2 L min^{-1}. To analyze the conversion rates of the chemical reactions, the gas mixture inside the reactor was continuously monitored using a gas chromatography system (DS iGC 7200) equipped with a thermal conductivity detector. To separate the mixture of reaction gases, we used a 6 ft long and 1/8″ outer diameter stainless steel column with a mesh size of 80/100. The measured reaction rates were estimated as the turnover frequency (TOF) of H_2O molecules produced per platinum surface site per reaction time (second). Measurement of the current was carried out simultaneously using a Keithley 2400 sourcemeter.

3. Results and discussion

The inverse oxide/metal interface was fabricated on the catalytic nanodiodes by depositing a monolayer of Co_3O_4 NCs on the surface of the Pt/TiO_2 nanodiode

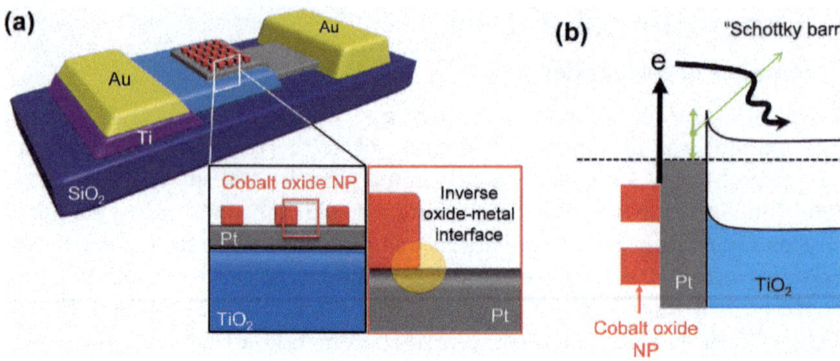

Fig. 1 (a) Schematic of the Co_3O_4 NCs/Pt/TiO_2 catalytic nanodiode. The rectifying contact is formed at the interface between the Pt and TiO_2 layers. (b) Energy band diagram for the Co_3O_4 NCs/Pt/TiO_2 nanodiode. Hot electrons energetic enough to overcome the Schottky barrier can be detected as a steady-state current.

(Fig. 1(a)). Cubic Co_3O_4 was chosen to produce a well-defined interface between the cobalt oxide and the platinum metal. In this architecture, the oxide/metal interfaces were expected to act as additional active sites for the catalytic reaction along with the pure metal sites, which helps to clarify the effect of the oxide/metal interface on the catalytic properties. Because the inverse oxide/metal structures were formed on the surface of a thin Pt film (*i.e.* 10 nm, which is less than the mean free path of an electron) contacting a TiO_2 layer, hot electrons generated by the chemical reaction on the catalytic active sites, *i.e.* metal surfaces and oxide/metal interfaces, could be extracted across the Schottky barrier at the Pt/TiO_2 interface without significant decay caused by electron–electron or electron–phonon interactions. The energy diagram for detecting hot electrons using the inverse oxide/metal catalytic nanodiode is illustrated in Fig. 1(b).

The Co_3O_4 NCs were prepared following the previously reported method;[32] their size, shape, and uniformity were confirmed by transmission electron microscopy (TEM). As shown in Fig. 2(a), the as-synthesized Co_3O_4 NCs have an average size of 9.8 ± 1.1 nm and have sharp edges. A high-resolution TEM (HR-TEM) image also reveals a lattice spacing of 2.86 Å, which corresponds to the (220) plane in a spinel Co_3O_4 crystal. Interestingly, CoO (200) facet is also identified on the surface of the NCs (Fig. 2(b)). The powder X-ray diffraction (XRD) pattern indicates that the major crystal structure of the NCs is Co_3O_4 (Fig. 2(c)).

To confirm the monolayer of Co_3O_4 NCs deposited on the Pt/TiO_2 nanodiodes, we characterized the surface morphology of the Pt/TiO_2 nanodiodes using scanning electron microscopy (SEM). As shown in Fig. 3(a) and (b), it is obvious that two-dimensional arrays of Co_3O_4 NCs are uniformly deposited on the thin-film Pt surface. Cross-sectional high-angle annular dark-field scanning transmission electron microscopy (HAADF-STEM) and STEM-energy dispersive X-ray spectroscopy (STEM-EDS) were used to identify the formation of interfaces between the Co_3O_4 NCs and the Pt film (Fig. 3(c) and (d)). In particular, the detailed chemical composition of the Co_3O_4 NCs/Pt/TiO_2 nanodiodes was observed in the STEM-EDS mapping images, thus demonstrating that well-defined inverse oxide/metal interfaces were formed on the device.

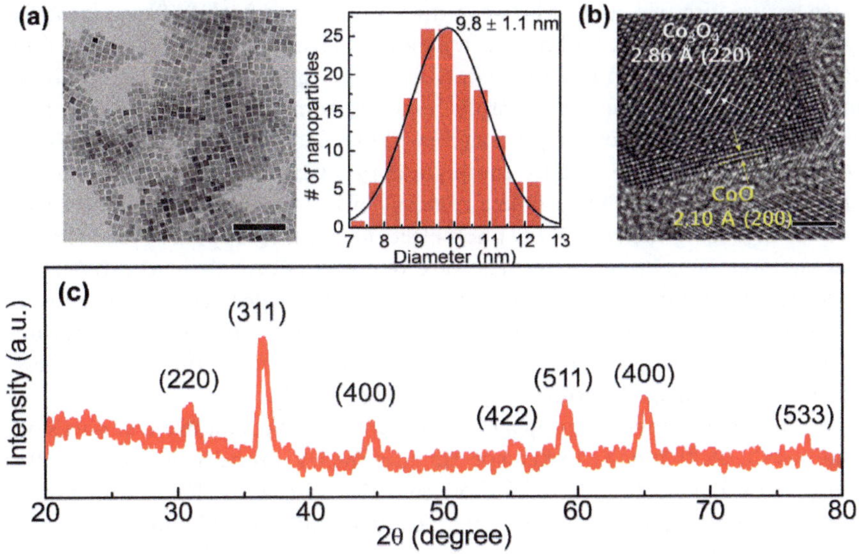

Fig. 2 (a) TEM image (scale bar is 100 nm) and size distribution histogram of the as-synthesized Co_3O_4 NCs. (b) HR-TEM image of the Co_3O_4 NCs (scale bar is 2 nm) showing thin layers of CoO, and (c) XRD pattern of the as-synthesized Co_3O_4 NCs.

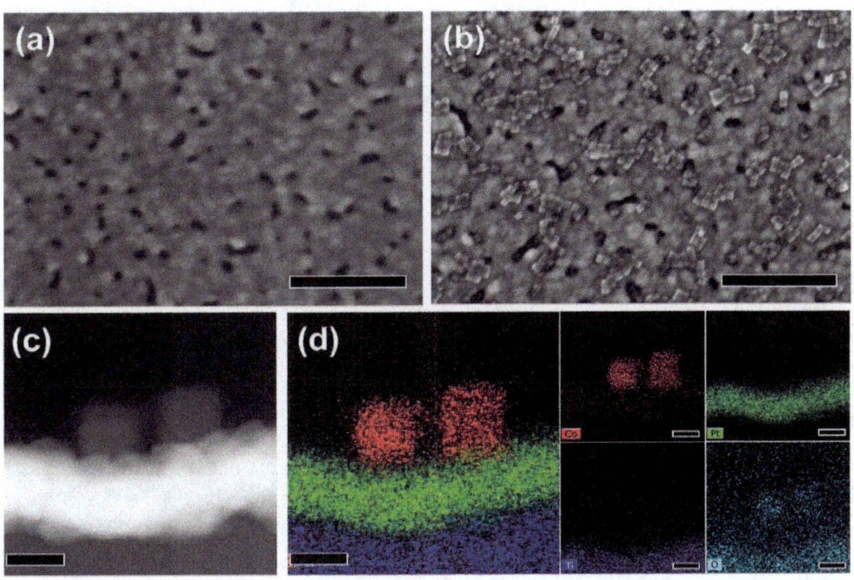

Fig. 3 Scanning electron microscopy (SEM) images of the 10 nm Pt thin film on the TiO_2 layer (a) before and (b) after deposition of the Co_3O_4 NCs. Scale bars are 200 nm. (c) Cross-sectional high-angle annular dark-field scanning transmission electron microscopy (HAADF-STEM), and (d) STEM-energy dispersive X-ray spectroscopy (STEM-EDS) mapping images of the Co_3O_4 NCs/Pt/TiO_2 interface. The elemental mapping includes Co K (red), Pt L (green), Ti K (blue), and O K (cyan). Scale bars are 10 nm.

To ensure intimate contact between the Co_3O_4 NCs and the Pt film, we carried out additional UV–ozone (O_3) treatment in air for 3 h after deposition of the Co_3O_4 NCs (Fig. 4(a)). Through this process, the organic capping molecules, *e.g.* oleylamine and 1-octanol, were successfully removed from around the Co_3O_4 NCs. As reported elsewhere,[33,34] ozone produced by ultraviolet light (wavelength from 185 to 257 nm) oxidizes the organic compounds into CO_2 or H_2O on the surface of NPs, which results in the complete removal of the organic capping molecules without damaging the NPs. X-ray photoelectron spectroscopy (XPS) results show that the organic capping molecules were removed effectively by the UV–ozone treatment; the C : Pt and the N : Pt atomic ratios clearly decreased after exposure to the UV–ozone treatment for 3 h (Fig. S1(a)†), which is consistent with previous results.[35] A SEM image taken after the treatment also revealed that the individual NCs were not damaged during the process (Fig. S1(b)†).

To identify the chemical states of the platinum and cobalt oxide surfaces, we obtained the Pt 4f and Co 2p XPS spectra after the UV–ozone treatment (Fig. 4(b) and (c)). Deconvoluted XPS data show that the surface of the Pt thin film is predominantly metallic, while the cobalt oxide is composed of a mixture of Co^{2+} and Co^{3+} oxidation states, which is comparable to the original chemical states of the surface before the UV–ozone treatment (Fig. S2†) and confirms that there were no significant changes in the chemical states of either the platinum or cobalt oxide surfaces while removing the organic capping molecules using UV–ozone

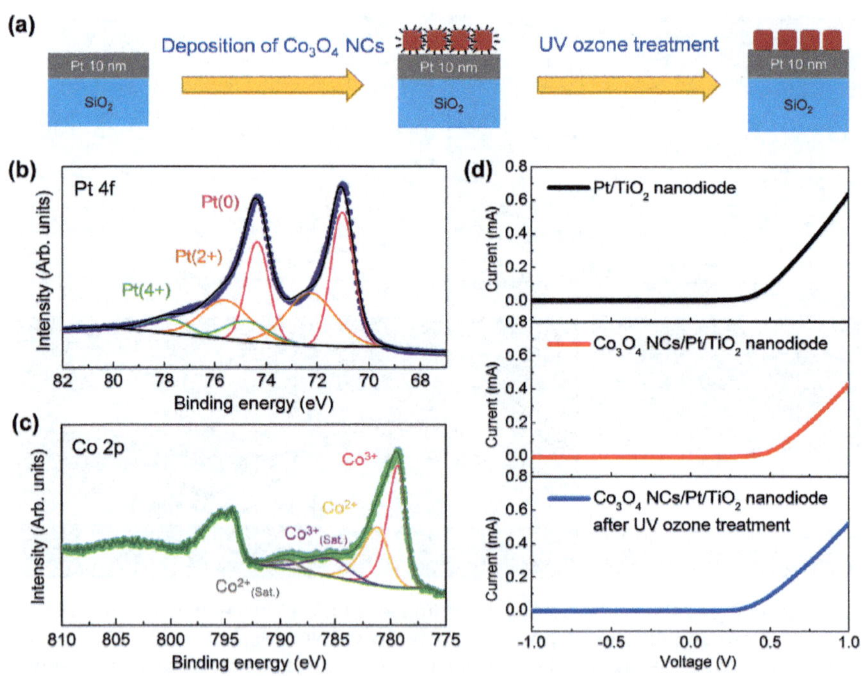

Fig. 4 (a) Schematic showing the sample preparation procedure. (b) Pt 4f and (c) Co 2p XPS spectra at the surface of the Co_3O_4 NCs/Pt/TiO_2 nanodiode after 3 h of UV–ozone treatment. (d) Current–voltage (I–V) curves measured on the Pt/TiO_2 (black), Co_3O_4 NCs/Pt/TiO_2 (red), and UV-treated Co_3O_4 NCs/Pt/TiO_2 (blue) nanodiodes.

treatment. Intimate contact at the Co_3O_4 NC/Pt interfaces was also achieved on the catalyst surface. Because the surface of the Co_3O_4 NCs is terminated by 3–4 CoO layers (Fig. 2(b)), it is worth noting that CoO/Pt interfaces were formed on the UV-treated Co_3O_4 NCs/Pt/TiO_2 nanodiodes. The electrical properties of the different catalytic nanodiodes, *i.e.* the Pt/TiO_2 nanodiode, the Co_3O_4 NCs/Pt/TiO_2 nanodiode, and the UV-treated Co_3O_4 NCs/Pt/TiO_2 nanodiode, were investigated using current–voltage (*I*–*V*) curves. As shown in Fig. 4(d), all the devices have identical rectifying properties at constant reaction conditions, indicating that neither the deposition of the Co_3O_4 NCs nor the UV–ozone treatment had an effect on the 0.7 eV Schottky barrier height formed at the Pt/TiO_2 interface.

To investigate the flow of hot electrons produced during nonadiabatic dissipation of chemical energy on the catalytic surfaces, we monitored the current signals using the catalytic nanodiodes during H_2 oxidation at temperatures of 30–90 °C. Considering the possibility for additional current created by the thermoelectric effect on the device, we compared the current signals measured under a mixture of H_2 (15 Torr) + O_2 (745 Torr) with the signals obtained in pure O_2 (760 Torr) (Fig. S3†). The difference between the currents with and without the catalytic reaction is clearly attributed to hot electron generation on the catalytic surface during H_2 oxidation, *i.e.* the chemicurrent (Fig. S3†).

Fig. 5(a) shows the chemicurrent densities obtained from the Pt/TiO_2 nanodiode, the Co_3O_4 NCs/Pt/TiO_2 nanodiode, and the UV-treated Co_3O_4 NCs/Pt/TiO_2 nanodiode. Here, the chemicurrent signals were normalized by the Pt surface area because Co_3O_4 NCs do not show any activity for H_2 oxidation at temperatures below 100 °C. As shown in Fig. 5(a), the chemicurrent density measured from the Co_3O_4 NCs/Pt/TiO_2 nanodiode was smaller than that from the Pt/TiO_2 nanodiode across the entire temperature range, indicating that the organic compounds surrounding the Co_3O_4 NCs hindered the catalytically active Pt surface as well as the CoO/Pt interface. However, after removing the organic molecules from the surface of the Co_3O_4 NCs/Pt/TiO_2 nanodiode through the UV–ozone treatment, *i.e.* the UV-treated Co_3O_4 NCs/Pt/TiO_2 nanodiode, the chemicurrent signal was significantly enhanced even compared to the values obtained from the Pt/TiO_2

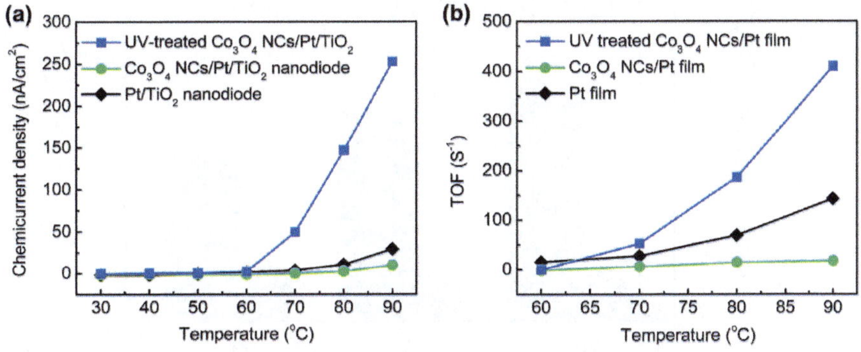

Fig. 5 (a) Chemicurrent density associated with the H_2 oxidation reaction obtained from the Pt/TiO_2, Co_3O_4 NCs/Pt/TiO_2, and UV-treated Co_3O_4 NCs/Pt/TiO_2 nanodiodes. (b) Catalytic activity, *i.e.* TOF, for the H_2 oxidation reaction on the Pt, Co_3O_4 NCs/Pt, and UV-treated Co_3O_4 NCs/Pt films.

nanodiode. Since the only difference between the UV-treated Co_3O_4 NCs/Pt/TiO_2 nanodiode and the Pt/TiO_2 nanodiode was the presence of the oxide/metal interfaces between the Co_3O_4 NCs and the Pt on the catalysts, it was concluded that the increased hot electron flows were generated at the CoO/Pt interface.

To demonstrate whether the oxide/metal interfacial sites play a critical role in determining a catalytic process or not, we also investigated the TOF of the catalyst as a function of temperature at identical reaction conditions. To measure the reaction rates for H_2 oxidation, all the samples, *i.e.* the 10 nm Pt film, the Co_3O_4 NCs/Pt film and the UV-treated Co_3O_4 NCs/Pt film, were prepared on a SiO_2 substrate. As shown in Fig. 5(b), when the oxide/metal (CoO/Pt) interfaces existed on the Pt surface (the UV-treated Co_3O_4 NCs/Pt film), the catalytic activity improved substantially, compared with the pure Pt surface (Pt film). The lowest catalytic performance was observed in the Co_3O_4 NCs/Pt film, suggesting that the organic capping molecules blocked the active Pt sites and prevented direct contact between the Co_3O_4 NCs and the Pt film. The trend correlating the TOFs of the different catalysts with the measured chemicurrents demonstrated that the oxide/metal interfacial sites were highly active for the H_2 oxidation reaction, particularly when compared with the other catalysts.

To address the unique electronic features of the oxide/metal interface, we estimated the chemicurrent yield, which is the probability of hot electron detection per surface reaction on the catalytic nanodiodes, thus enabling quantitative visualization of the electronic transfer efficiency on the catalysts. The chemicurrent yield is calculated based on the equation $\alpha = q_0\frac{AN_{pt}TOF}{I_{ch}}$, where α is the chemicurrent yield, q_0 is the elementary charge, A is the active area of the catalyst, and I_{ch} is chemicurrent. As shown in Fig. 6, there are significant increases in the chemicurrent yield for the UV-treated Co_3O_4 NCs/Pt/TiO_2 nanodiode ($3.1 \pm 0.6 \times 10^{-6}$) compared with that of the Pt/TiO_2 nanodiode ($6.3 \pm 0.2 \times 10^{-5}$). It is clear that efficient charge transport occurs at the oxide/metal (CoO/Pt) interfaces of the Co_3O_4 NCs and the Pt.

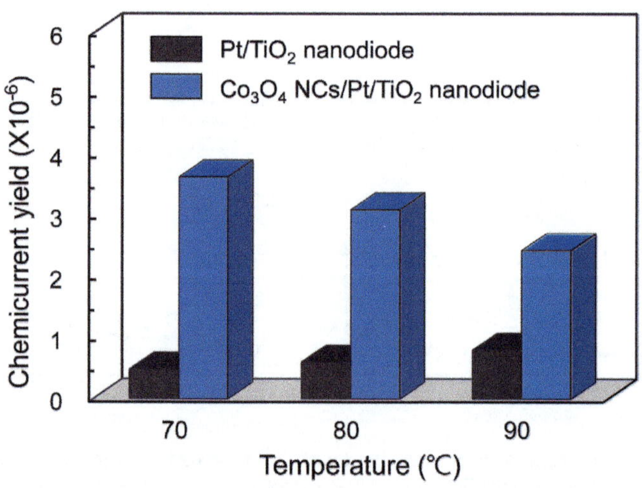

Fig. 6 Chemicurrent yield obtained from the Pt/TiO_2 and UV-treated Co_3O_4 NCs/Pt/TiO_2 nanodiodes during the H_2 oxidation reaction.

These results are in line with previous observations, confirming that the CoO/Pt interface formed on the PtCo bimetallic NPs plays a decisive role in enhancing catalytic activity by facilitating hot electron transfer.[27] In addition, based on density functional theory calculations, a lower activation energy is required for the H_2 oxidation reaction at the CoO/Pt interfaces formed on the PtCo bimetallic NPs, resulting in higher catalytic performance.[27] Furthermore, higher exothermic energy was observed at the interface between the CoO and the Pt, which is consistent with the higher chemicurrent yield on PtCo nanoparticles compared with pure Pt nanoparticles.[27] Therefore, we conclude that the high catalytic activity observed on the UV-treated Co_3O_4 NCs/Pt film catalyst can indeed be attributed to the existence of active CoO/Pt interfacial sites that allow for efficient charge transfer because of their unique electronic structure. Overall, direct measurement of hot electrons using catalytic nanodiodes sheds light on the fundamental mechanism of heterogeneous catalysis at metal–oxide interfaces and provides potential applications for designing new catalysts and devices.

4. Conclusion

In summary, we investigated hot electron dynamics at the inverse oxide/metal interface using a newly designed catalytic nanodiode composed of a 2D array of Co_3O_4 NCs on a Pt/TiO_2 nanodiode to clarify the effect of the oxide/metal interface on catalytic properties. The organic capping molecules on the surface of the Co_3O_4 NCs/Pt film were removed through UV–ozone treatment, forming an intimate CoO/Pt interface on the catalyst. We found a dramatic increase in catalytic activity when using the UV-treated Co_3O_4 NCs/Pt film catalysts, as confirmed by higher chemicurrent and TOF values during H_2 oxidation compared with those of the Pt film. The chemicurrent yield calculated based on both chemicurrent and TOF was significantly improved in the UV-treated Co_3O_4 NCs/Pt film. We thus conclude that the catalytic activity was improved by the presence of intimate CoO/Pt interfacial sites promoting electronic transfer.

Conflicts of interest

There are no conflicts to declare.

Acknowledgements

This work was supported by the Institute for Basic Science (IBS-R004 and IBS-R006-D1) and the Ministry of Science and ICT (2018R1A1A1A05079555).

References

1 K. An and G. A. Somorjai, Nanocatalysis I: Synthesis of Metal and Bimetallic Nanoparticles and Porous Oxides and Their Catalytic Reaction Studies, *Catal. Lett.*, 2014, **145**, 233–248.

2 E. Gross and G. A. Somorjai, The Impact of Electronic Charge on Catalytic Reactivity and Selectivity of Metal-Oxide Supported Metallic Nanoparticles, *Top. Catal.*, 2013, **56**, 1049–1058.

3 G. A. Somorjai and J. Y. Park, Molecular factors of catalytic selectivity, *Angew. Chem., Int. Ed.*, 2008, **47**, 9212–9228.

4 J. A. Rodriguez, P. Liu, J. Graciani, S. D. Senanayake, D. C. Grinter, D. Stacchiola, J. Hrbek and J. Fernandez-Sanz, Inverse Oxide/Metal Catalysts in Fundamental Studies and Practical Applications: A Perspective of Recent Developments, *J. Phys. Chem. Lett.*, 2016, **7**, 2627–2639.

5 S. Kattel, P. Liu and J. G. Chen, Tuning Selectivity of CO_2 Hydrogenation Reactions at the Metal/Oxide Interface, *J. Am. Chem. Soc.*, 2017, **139**, 9739–9754.

6 J. C. Matsubu, S. Zhang, L. DeRita, N. S. Marinkovic, J. G. Chen, G. W. Graham, X. Pan and P. Christopher, Adsorbate-mediated strong metal-support interactions in oxide-supported Rh catalysts, *Nat. Chem.*, 2017, **9**, 120–127.

7 G. Zhao, F. Yang, Z. Chen, Q. Liu, Y. Ji, Y. Zhang, Z. Niu, J. Mao, X. Bao, P. Hu and Y. Li, Metal/oxide interfacial effects on the selective oxidation of primary alcohols, *Nat. Commun.*, 2017, **8**, 14039.

8 M. Cargnello, V. V. Doan-Nguyen, T. R. Gordon, R. E. Diaz, E. A. Stach, R. J. Gorte, P. Fornasiero and C. B. Murray, Control of metal nanocrystal size reveals metal-support interface role for ceria catalysts, *Science*, 2013, **341**, 771–773.

9 K. Qadir, B. T. Quynh, H. Lee, S. Y. Moon, S. H. Kim and J. Y. Park, Tailoring metal-oxide interfaces of inverse catalysts of TiO_2/nanoporous-Au under hydrogen oxidation, *Chem. Commun.*, 2015, **51**, 9620–9623.

10 P. Mehta, J. Greeley, W. N. Delgass and W. F. Schneider, Adsorption Energy Correlations at the Metal–Support Boundary, *ACS Catal.*, 2017, **7**, 4707–4715.

11 F. Calaza, C. Stiehler, Y. Fujimori, M. Sterrer, S. Beeg, M. Ruiz-Oses, N. Nilius, M. Heyde, T. Parviainen, K. Honkala, H. Hakkinen and H. J. Freund, Carbon dioxide activation and reaction induced by electron transfer at an oxide-metal interface, *Angew. Chem., Int. Ed.*, 2015, **54**, 12484–12487.

12 M. Kitano, S. Kanbara, Y. Inoue, N. Kuganathan, P. V. Sushko, T. Yokoyama, M. Hara and H. Hosono, Electride support boosts nitrogen dissociation over ruthenium catalyst and shifts the bottleneck in ammonia synthesis, *Nat. Commun.*, 2015, **6**, 6731.

13 Y. Bai, W. Zhang, Z. Zhang, J. Zhou, X. Wang, C. Wang, W. Huang, J. Jiang and Y. Xiong, Controllably interfacing with metal: a strategy for enhancing CO oxidation on oxide catalysts by surface polarization, *J. Am. Chem. Soc.*, 2014, **136**, 14650–14653.

14 J. Y. Park, S. M. Kim, H. Lee and I. I. Nedrygailov, Hot-electron-mediated surface chemistry: toward electronic control of catalytic activity, *Acc. Chem. Res.*, 2015, **48**, 2475–2483.

15 K. An, S. Alayoglu, N. Musselwhite, S. Plamthottam, G. Melaet, A. E. Lindeman and G. A. Somorjai, Enhanced COoxidation rates at the interface of mesoporous oxides and Pt nanoparticles, *J. Am. Chem. Soc.*, 2013, **135**, 16689–16696.

16 J. Y. Park, L. R. Baker and G. A. Somorjai, Role of hot electrons and metal-oxide interfaces in surface chemistry and catalytic reactions, *Chem. Rev.*, 2015, **115**, 2781–2817.

17 G. M. Schwab, Boundary-layer catalysis, *Angew. Chem., Int. Ed.*, 1967, **6**, 375.

18 T. Fujitani and I. Nakamura, Mechanism and active sites of the oxidation of CO over Au/TiO_2, *Angew. Chem., Int. Ed.*, 2011, **123**, 10326–10329.

19 K. Hayek, M. Fuchs, B. Klötzer, W. Reichl and G. Rupprechter, Studies of metal–support interactions with "real" and "inverted" model systems: reactions of CO and small hydrocarbons with hydrogen on noble metals in contact with oxides, *Top. Catal.*, 2000, **13**, 55–66.

20 A. Boffa, C. Lin, A. Bell and G. A. Somorjai, Promotion of CO and CO_2 hydrogenation over Rh by metal oxides: the influence of oxide Lewis acidity and reducibility, *J. Catal.*, 1994, **149**, 149–158.

21 H. Lee, I. I. Nedrygailov, C. Lee, G. A. Somorjai and J. Y. Park, Chemical-reaction-induced hot electron flows on platinum colloid nanoparticles under hydrogen oxidation: impact of nanoparticle size, *Angew. Chem., Int. Ed.*, 2015, **54**, 2340–2344.

22 I. I. Nedrygailov, C. Lee, S. Y. Moon, H. Lee and J. Y. Park, Hot Electrons at Solid-Liquid Interfaces: A Large Chemoelectric Effect during the Catalytic Decomposition of Hydrogen Peroxide, *Angew. Chem., Int. Ed.*, 2016, **55**, 10859–10862.

23 H. Lee, I. I. Nedrygailov, S. W. Lee and J. Y. Park, Isotope Effect of Hot Electrons Generated on Pt Nanoparticle Surfaces Under H_2 and D_2 Oxidation, *Top. Catal.*, 2018, **61**, 915–922.

24 K. C. Goddeti, H. Lee, B. Jeon and J. Y. Park, Enhancing hot electron collection with nanotube-based three-dimensional catalytic nanodiode under hydrogen oxidation, *Chem. Commun.*, 2018, 8968–8971.

25 I. I. Nedrygailov and J. Y. Park, The nature of hot electrons generated by exothermic catalytic reactions, *Chem. Phys. Lett.*, 2016, **645**, 5–14.

26 H. Lee, I. I. Nedrygailov, Y. K. Lee, C. Lee, H. Choi, J. S. Choi, C. G. Choi and J. Y. Park, Graphene-Semiconductor Catalytic Nanodiodes for Quantitative Detection of Hot Electrons Induced by a Chemical Reaction, *Nano Lett.*, 2016, **16**, 1650–1656.

27 H. Lee, J. Lim, C. Lee, S. Back, K. An, J. W. Shin, R. Ryoo, Y. Jung and J. Y. Park, Boosting hot electron flux and catalytic activity at metal-oxide interfaces of PtCo bimetallic nanoparticles, *Nat. Commun.*, 2018, **9**, 2235.

28 S. M. Kim, D. Park, Y. Yuk, S. H. Kim and J. Y. Park, Influence of hot carriers on catalytic reaction; Pt nanoparticles on GaN substrates under light irradiation, *Faraday Discuss.*, 2013, **162**, 355.

29 S. Mukherjee, L. Zhou, A. M. Goodman, N. Large, C. Ayala-Orozco, Y. Zhang, P. Nordlander and N. J. Halas, Hot-electron-induced dissociation of H_2 on gold nanoparticles supported on SiO_2, *J. Am. Chem. Soc.*, 2014, **136**, 64–67.

30 U. Aslam, S. Chavez and S. Linic, Controlling energy flow in multimetallic nanostructures for plasmonic catalysis, *Nat. Nanotechnol.*, 2017, **12**, 1000.

31 Y. Zhang, S. He, W. Guo, Y. Hu, J. Huang, J. R. Mulcahy and W. D. Wei, Surface-plasmon-driven hot electron photochemistry, *Chem. Rev.*, 2017, **118**, 2927–2954.

32 M. H. Oh, T. Yu, S.-H. Yu, B. Lim, K.-T. Ko, M.-G. Willinger, D.-H. Seo, B. H. Kim, M. G. Cho and J.-H. Park, Galvanic replacement reactions in metal oxide nanocrystals, *Science*, 2013, **340**, 964–968.

33 S. Yoon, K. Oh, F. Liu, J. H. Seo, G. A. Somorjai, J. H. Lee and K. An, Specific Metal–Support Interactions between Nanoparticle Layers for Catalysts with Enhanced Methanol Oxidation Activity, *ACS Catal.*, 2018, **8**, 5391–5398.

34 C. Aliaga, J. Y. Park, Y. Yamada, H. S. Lee, C.-K. Tsung, P. Yang and G. A. Somorjai, Sum frequency generation and catalytic reaction studies of

the removal of organic capping agents from Pt nanoparticles by UV–ozone treatment, *J. Phys. Chem. C*, 2009, **113**, 6150–6155.

35 J. Y. Park, C. Aliaga, J. R. Renzas, H. Lee and G. A. Somorjai, The Role of Organic Capping Layers of Platinum Nanoparticles in Catalytic Activity of CO Oxidation, *Catal. Lett.*, 2009, **129**, 1–6.

Faraday Discussions

DISCUSSIONS

New materials for hot electron generation: general discussion

Javier Aizpurua, Jeremy Baumberg, Alexandra Boltasseva, Phillip Christopher, Emiliano Cortes, Stephen B. Cronin, Bhavesh Kumar Dadhich, Bart de Nijs, Preeti Deshpande, Yuri Diaz Fernandez, Laura Fabris, Sylwester Gawinkowski, Alexander Govorov, Naomi Halas, Junyang Huang, Bartlomiej Jankiewicz, Rifat Kamarudheen, Jacob Khurgin, Tae Kyung Lee, Julien Mahin, Andrea Marini, Reinhard J. Maurer, Niclas Sven Mueller, Jeong Y. Park, Mahfujur Rahaman, Sebastian Schlücker, Zachary Schultz, Yonatan Sivan, Giulia Tagliabue, Madasamy Thangamuthu, Hongxing Xu and Anatoly Zayats

DOI: 10.1039/C9FD90013F

Naomi Halas opened a general discussion of the paper by Alexandra Boltasseva: I am very happy that Alexandra has led the way in addressing the need for sustainable plasmonic materials. This is particularly relevant in the photocatalysis field since we have a limited supply of Pt and Pd on this planet.

Alexandra Boltasseva answered: The comment is highly appreciated. Our efforts have been simply to expand the existing material database for plasmon-driven hot electron generation so as to provide the community with more choices for specialized applications.

Zachary Schultz noted: Incorporating new materials into your devices should enable new chemical reactions to be investigated. Can you comment on any new reactions you have studied or insights your devices are providing into other possible reactions?

Alexandra Boltasseva responded: These devices can also be used for different photoelectrochemical reactions such as degradation of organic pollutants, alcohol oxidation, and some heterogeneous chemical reactions involving organic molecules. Also, the use of emerging plasmonic materials such as titanium nitride, zirconium nitride and MXenes such as titanium carbide offer the possibility to explore photocatalysis over a wide wavelength spectrum for various applications.

Bart de Nijs enquired: In your experiments you use nanodiscs on a metal film with a thin insulator spacer; these provide a very high confinement perpendicular to the surface but very little lateral confinement. For example, using spheres instead of discs will provide strong field confinements in all directions resulting in a much higher field but a smaller hotspot. Would your system benefit from lateral confinement as well or is the perpendicular confinement sufficient?

Alexandra Boltasseva replied: Any form of confinement of the electromagnetic field would in general benefit photocatalysis. For lateral confinement, the semiconductor needs to be along the sidewalls of the nanodisk as well, so as to facilitate photocatalysis. Such a design, although difficult to fabricate, would also increase water splitting efficiency. See, for example, work from Gao *et al.*[1]

1 H. Gao, C. Liu, H. E. Jeong and P. Yang, *ACS Nano*, 2012, **6**, 234-240.

Bartlomiej Jankiewicz asked: Are there any specific reasons for using hematite instead of TiO_2? Are there any advantages of hematite compared to titanium dioxide? The reasons for using hematite that you have mentioned in your article and presentation apply also to TiO_2.

Alexandra Boltasseva responded: Yes, there is indeed a lot of work on hot electron generation for water splitting in TiO_2. We chose hematite as it has a bandgap in the visible and therefore can provide a larger photocurrent.

Niclas Sven Mueller remarked: My question concerns the choice of plasmonic material in your study. For the gold nanostructures you presented pronounced plasmonic resonances that are spectrally well below the band gap of hematite. To my understanding this will lead to no considerable increase in the photon conversion efficiency when you integrate over the entire spectrum. Did you consider using silver instead of gold? I would not expect that oxidation at the Ag/hematite interface will be an issue as you can directly deposit silver on top of the hematite film by e-beam lithography.

Alexandra Boltasseva answered: Yes, the fact that we used gold nanodisks that gave resonances below the hematite bandgap leads to a lower photocurrent. We did investigate the possibility of using plasmonic structures that would enable us to push the resonance above the bandgap and reap the benefits of high field confinement. However, we did not use silver as it would oxidise when used in a photo-electrochemical cell.

Jeremy Baumberg asked: I am trying to understand how your water splitting works at the nanoscale details. Hole transport is a few nm, but your hematite is trapped inside the narrow gap. Where is the hydrogen being produced and how do reactants and products get in or out? Is the hematite porous or completely solid as in your TEMs? Can water get in? More generally the optical field is concentrated in inaccessible parts of the hematite so is this only a fringing field effect? If so, can we do better with a different geometry and what would this be?

Alexandra Boltasseva clarified: In our photoelectrochemical system, we used a three-electrode cell where O_2 is evolved at the hematite/Au photoanode, while H_2 is produced at the Pt counterlectrode. The third electrode is the reference electrode. The reactant is water, which is split to give H_2 and O_2 (the products), that could be separated by gas membrane technology in a real system.

Hematite is completely solid as shown by TEM images reported in the manuscript. Water cannot get inside the solid. The optical fields are concentrated in the inaccessible parts of the hematite. So the measured photocurrent enhancement is only due to those hot carriers, generated from plasmon decay, that contribute to the surface chemical reaction (oxidation of water).

We do believe that the fact that hematite is sandwiched between the gold mirror and gold nanodisks hinders the hot electron injection. We are currently working on a different geometry that would enhance hot electron injection into the semiconductor and hope that we can report our results in the near future.

Madasamy Thangamuthu enquired: A constant voltage of 1.5 V is applied for photocurrent measurement; is there any reason behind this value? The literature uses different numbers – I don't understand why and how I can choose the right potential. Is it to offset the electrical double layer potential formed at the electrode–electrolyte interface?

Alexandra Boltasseva replied: Chronoamperometry measurements can be done at any potential and the choice is somewhat arbitrary. However, the potential is usually chosen such that it corresponds to the generation of a plateau photocurrent in the J–V curve. Depending on the shape of the J–V curve one can choose different potentials which would provide photocurrents representing the photoanode performance working in different regimes.

Hongxing Xu queried: Which structure is ideal for photoelectron generation? Is a rough surface or a sharp edge good?

Alexandra Boltasseva replied: I am not sure what exactly is the difference between rough surfaces and sharp edges. In general, strong electromagnetic fields at the surface can lead to an enhanced photoelectric effect from the metal surface. There are many articles in the literature which discuss the role of geometry and surfaces on photoelectron generation. For example, ref. 1.

1 A. M. Brown, R. Sundararaman, P. Narang, W. A. Goddard III and H. A. Atwater, *ACS Nano*, 2016, **10**, 957–966.

Jacob Khurgin commented: It is important that the efficiency of any process involving photo (plasmon) generated hot carriers depends on two factors:

(A) Absorption efficiency – *i.e.* what percentage of the incoming line is absorbed;

(B) Emission efficiency – *i.e.* what percentage of carriers generated by absorbing photons (directly or *via* plasmons) actually come out and do useful work.

It seems to me that often we are concerned with the first and in my view it is not all that important as in the end, once you have enough catalyst such as TiO_2 a significant portion of light will be absorbed anyway and there is no shortage of titanium. Whether it is 80% or 100% makes a very small difference in the overall efficiency given that efficiency (B) is measured in a fraction of a percent. Therefore, we should use plasmons judiciously, for example if we could extend the absorption edge into longer wavelength regions. Making fancy structures just because we can only redistributes absorption but it also introduces additional loss in the metal.

Giulia Tagliabue asked: Do you have any estimate of the Schottky barrier between gold and hematite? I would expect that depending on this parameter and the photon energy, a fraction of the hot-holes generated in gold could be transferred to the p-type hematite. Do you have any estimate for this process and what would be its contribution in the visible regime *versus* the infrared regime?

Alexandra Boltasseva responded: By using the Mott-Schottky rule (work function of Au = 5.1 eV; electron affinity of hematite = 4.85 eV; see work by Hankin *et al.*[1]) we estimate a Schottky barrier height of 0.25 eV. This value tells us that by using this diode configuration we might also collect NIR hot carriers. In this system, hematite in an n-type semiconductor and under photoelectrochemical conditions holes migrate toward the electrolyte while electrons move into the external circuit. Therefore, hot holes generated in plasmonic Au nanodisks travel toward the electrolyte and may directly react with water (*i.e.* the catalytic effect of the plasmonic metal) to oxidize it.

1 A. Hankin, J. C. Alexander and G. H. Kelsall, *Phys. Chem. Chem. Phys.*, 2014, **16**, 16176–16186.

Sylwester Gawinkowski remarked: You have shown that the electromagnetic properties of ceramic materials can be changed by changing their composition and they can mimic the plasmonic properties of metals. The compositions of such ceramic materials are fundamentally limited by the limited number of elements in the periodic table. Is there any other fundamental limit and how similar can such ceramic materials be to perfect metals?

Alexandra Boltasseva answered: The optical properties of ceramic materials will of course depend on the composition chosen. For example while titanium nitride and zirconium nitride are highly metallic (carrier concentration $\sim 10^{22}$ cm^{-3}), other nitrides like tantalum and hafnium nitride do not have such high carrier concentration. Eventually it boils down to how the atomic arrangement in the lattice provides a high enough density of states at the Fermi level with relatively high mobility to get good plasmonic metal-like behavior.

Jeremy Baumberg opened a general discussion of the paper by Mahfujur Rahaman: There is another explanation for your TERS rings, which I believe is more likely. You have strong excitons in the monolayer TMD, which are oriented in-plane (you can see many papers reporting this). In your plasmonic tip-disk geometry, the field orientation is normally perpendicular to the disk so does not couple to these excitons. However, at the disk edges there is an increasing in-plane component which couples to the excitons. Although your SERS laser is at a longer wavelength, there is still a very strong SERS enhancement from the resonant electronic states of the exciton (again extensive literature exists, for instance on spin-flip Raman in semiconductors in the 1980s-90s). Hence you are seeing TERRS (tip-enhanced resonant Raman scattering) which in this geometry appears as rings where there is an in-plane field. One way to check would be to look at hBN which has no exciton. Have you tried this?

Mahfujur Rahaman replied: Many thanks for your suggestion. However, we disagree that in plane excitons are a more likely cause of our TERS enhancement. We used an excitation laser of 785 nm (1.58 eV) which is much below the optical band gap (1.9 eV) of monolayer MoS_2. Therefore, if there is an excitonic contribution due to the reason you mentioned that would not be that significant. Typical as measured TERS spectra taken at the edge of the Au disk and on Si are shown here in Fig. 1. As one can see, at the edge the TERS signal of monolayer MoS_2 is much stronger resulting in the TERS images we have presented in our work. Additionally, if there was any excitonic effect then one would also expect an increased background which is clearly missing in the spectra.

Javier Aizpurua commented: When interpreting the TERS signal in terms of local fields, your paper shows a description based on a modelling of sharp edges of the structures. How realistic are these sharp edges that you modeled? Figure 2(c) in your paper shows a very symmetric image profile around the disks. Wouldn't one expect more directional enhancement, following the polarization of

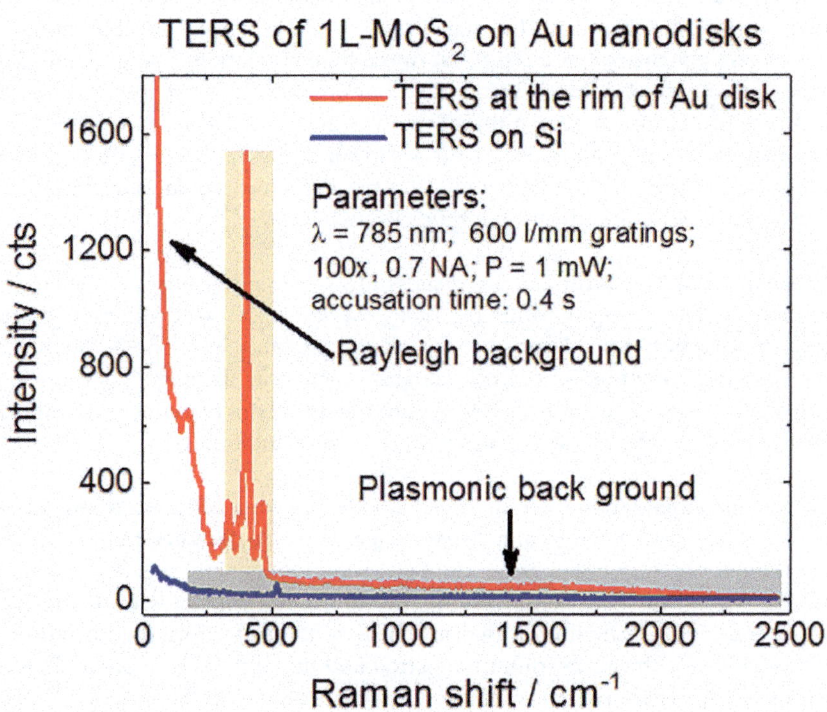

Fig. 1 Typical as measured TERS spectra taken at the edge of the Au disk and on Si.

the incident light? Why do you not have a more asymmetric field enhancement? In that TERS image, the intense signal spreads over almost 50 nm – but that is not what you get from the ultrasharp edges. Is there a certain roughness that might spread the signal and explain the contrast?

Mahfujur Rahaman responded: In our TERS image we have observed a ring-like shape following the rim of the Au disks as is shown in Fig. 2 in our paper. In the experiments the shape of the TERS images were systematic and reproducible. To understand the TERS images we introduced an FEM simulation in which we modelled a Au disk of similar dimensions to those used in the experiment. We also excited the disk in the same experimental configuration. The simulated TERS image shows a ring-like shape along the edges of the disk. However, you are right that in the simulation the edge of the Au disk was sharp resulting in a very narrow width of the TERS distribution. In reality, the edge of the Au disks may not be that sharp and therefore the electric field distribution may not be identical. Hence there is a deviation from the experimental situation to the theoretical model. However, the purpose of the simulation is to understand the TERS behaviour (which causes the ring-like shape), which the simulation successfully explained. The angle of incidence of the excitation and collection was 65°. This angle was chosen in order to maximize the direction of the excitation electric field parallel to the tip long axis. In this configuration the ratio between the out-of-plane and in-plane components of the electric field is 2.14. Also, one shall expect TERS enhancement due to the out-of-plane component if we only consider the tip. However, the in-plane component will introduce a slight asymmetry in the ring shape on the Au disk which we also observe in the simulation, though in the experimental images the asymmetric pattern was hardly observed. This may be because of the weak contribution of the in-plane component. You are right that in the high resolution experimental TERS images we observed a more broad distribution. This may be because of the roughness of the Au surface on top of the disk as you mentioned. Moreover, the experiment was performed when the tip was in contact with the MoS_2 film meaning a slight deformation of the film was also expected which could also change the interaction between the tip and the Au disk.

Zachary Schultz continued the discussion: In your response to Javier, you commented that the tip deforms the MoS_2 overlayer. Can you comment on the force provided by the tip? From your schematic, the MoS_2 layer should be further from the gold structures in the middle and touch near the edges. Is the deformation force enough to push the MoS_2 onto the Au pillars to maintain a uniform distance between the tip, MoS_2, and gold structures underneath?

Mahfujur Rahaman answered: The TERS spectra were acquired when the tip was in contact with the MoS_2 film. This is a special mode designed in the HORIBA NanoRaman platform to maximize the TERS signal. In general the AFM system is operating in intermittent contact mode meaning a soft contact with the sample. Now in the TERS configuration the system permits you to hold the tip onto the probe surface for the accusation time period and this repeats in every oscillation cycle of the tip. Therefore, the amount of the force acting on the film by the tip is minimal and you can control that by changing the tip–sample distance. However, in the default condition optimized by the system, the exerting force is in the range of a few nN. However, for this current experiment we did not measure the exact value.

Thanks to the very good flexibility of the MoS_2 film and also the very good affinity to Au, the MoS_2 film nicely follows the morphology of the Au nanostructures. This is also confirmed by our AFM and SEM images of the sample (please see our previous work[1,2]). Unless there is any twisting or wrinkle, MoS_2 has a homogeneous coverage. However, you are right that the slight force exerted by the tip can also help to keep the tip–sample distance uniform.

1 M. Rahaman, R. D. Rodriguez, G. Plechinger, S. Moras, C. Schüller, T. Korn, and D. R. T. Zahn, *Nano Lett.*, 2017, **17**, 6027–6033.
2 A. G. Milekhin, M. Rahaman, E. E. Rodyakina, A. V. Latyshev, V. M. Dzhaganc and D. R. T. Zahn, *Nanoscale*, 2018, **10**, 2755–2763.

Sebastian Schlücker remarked: In UHV-TERS on molecules, colleagues from Hefei in China have reported sub-nm spatial resolution which equals intramolecular resolution. What is the best spatial resolution known for TERS on semiconductor samples?

Mahfujur Rahaman replied: For organic molecules the reported spatial resolution is about 0.5 nm.[1] For semiconducting CNTs the reported spatial resolution is about 0.7 nm.[2] For 2D semiconductors we have achieved a spatial resolution of 2.3 nm.[3]

1 R. Zhang, Y. Zhang, Z. C. Dong, S. Jiang, C. Zhang, L. G. Chen, L. Zhang, Y. Liao, J. Aizpurua, Y. Luo, J. L. Yang and J. G. Hou, *Nature*, 2013, **498**, 82–86.
2 M. Liao, S. Jiang, C. Hu, R. Zhang, Y. Kuang, J. Zhu, Y. Zhang and Z. Dong, *Nano Lett.*, 2016, **16**, 4040–4046.
3 A. G. Milekhin, M. Rahaman, E. E. Rodyakina, A. V. Latyshev, V. M. Dzhaganc and D. R. T. Zahn, *Nanoscale*, 2018, **10**, 2755–2763.

Preeti Deshpande asked: How can you decouple the effects due to strain and doping from the Raman spectra of MoS_2 since both affect the peak position? Also could you give some reference works for doing this?

Mahfujur Rahaman answered: MoS_2 has two well-known first order Raman modes in the form of in-plane E2g and out-of-plane A1g. Due to vibrational symmetry these two modes are sensitive to strain and doping differently. For more details you can check any strain and doping related study of MoS_2. Perhaps the two articles by Lloyd *et al.*[1] and Chakraborty *et al.*[2] would be helpful.

In our work we have investigated the peaks to decouple strain, local temperature rise, and hot electron doping. Since the in-plane E2g mode is more sensitive to strain and non-sensitive to electron–phonon coupling *via* doping we focused on this particular mode for strain measurement. However, high frequency modes around 450 cm^{-1} are also sensitive to strain and we monitored these peaks' evolution to quantify the strain in the MoS_2 system with local temperature rise. The out-of-plane A1g mode was then monitored to quantify the hot electron doping effect. For more details please check our recent papers.[3,4]

1 D. Lloyd, X. Liu, J. W. Christopher, L. Cantley, A. Wadehra, B. L. Kim, B. B. Goldberg, A. K. Swan, and J. S. Bunch, *Nano Lett.*, 2016, **16**, 5836–5841.
2 B. Chakraborty, A. Bera, D. V. S. Muthu, S. Bhowmick, U. V. Waghmare and A. K. Sood, *Phys. Rev. B*, 2012, **85**, 161403(R).

3 A. G. Milekhin, M. Rahaman, E. E. Rodyakina, A. V. Latyshev, V. M. Dzhaganc and D. R. T. Zahn, *Nanoscale*, 2018, **10**, 2755–2763.
4 M. Rahaman, R. D. Rodriguez, G. Plechinger, S. Moras, C. Schüller, T. Korn, and D. R. T. Zahn, *Nano Lett.*, 2017, **17**, 6027–6033.

Tae Kyung Lee enquired: (1) Is there an effect of the curvature of MoS_2 on the TERS intensity or the generation of hot electrons?

(2) Is there a MoS_2 phase transition caused by the hot electrons of the Au nanocylinder? And how does this phase transition affect the TERS intensity?

Mahfujur Rahaman responded: Since MoS_2 was exfoliated on a Au nanodisk array, it created a carpet-like structure on top, which one can also see in the AFM and SEM images.[1] From the topography it is clear that MoS_2 follows the Au disk morphology resulting in curvature. We have quantified the strain distribution in the MoS_2 curvatures using TERS. Regarding the TER intensity, what we observed is a more or less homogeneous distribution of the TERS intensity along the disk rims or edges. Using FEM simulation we explained that this enhancement is related to the plasmonic field created in the hot spots between the tip and the disk. Moreover, we also observed a strong hot electron doping in MoS_2 of the order of 10^{13} cm^{-2}. Due to this strong doping we observed that MoS_2 underwent a phase transition from 1H to 1T. And interestingly, this phase transition was highly localized in the hot spots where we observed the strongest TERS intensity as well. The work we have presented in this article is the continuation of the work we published previously.[1] All the physical phenomenon related to the MoS_2 properties were discussed in that contribution. In the present contribution we have investigated the physical process of TERS enhancement and therefore no discussion regarding the properties of MoS_2 is included. We would kindly request that you read our previous paper[1] in which we quantified the strain, local temperature rise and plasmonic hot electron doping effects on MoS_2.

1 A. G. Milekhin, M. Rahaman, E. E. Rodyakina, A. V. Latyshev, V. M. Dzhaganc and D. R. T. Zahn, *Nanoscale*, 2018, **10**, 2755–2763.

Naomi Halas questioned: What are some of the real challenges where we could use more theoretical focus? What would it take to really advance the field?

Mahfujur Rahaman replied: Many thanks for this question. As an experimental physicist working on 2D material systems, I think one of the key areas in this field is the control of the local optical properties of 2D semiconductors, which is one of the black boxes to understand for nanoelectronic devices. Tip enhanced optical (TEO) measurements not only help us to understand local properties of the materials but can also help us to tune or modify the local properties as well. In particular, if we operate the TEO measurements in the quantum mechanical regime by reducing the tip–sample distance to a few angstrom, then for example hot electrons surely have a significant say in the local physical properties. This is very important for 2D semiconductors since they are ultra sensitive to any change in local environment. I think it is an interesting field for both experimentalists and theoreticians to work together.

Alexandra Boltasseva contributed: To really advance the field of plasmon-induced hot electrons, I feel the field should branch out to other areas of science and technology such as photostimulation in biology, spintronics and magnetic memory *etc.* For example a couple of reports of plasmon decay and subsequent thermal heating on neuronal action potentials can be found in the literature.[1] Study of hot electron mediated localized neuronal firings can be of immediate significance to the bio-community. Similarly hot electron mediated ultrafast magnetization dynamics has also been reported in the literature.[2] Studying hot electron mediated processes in magnetic materials could open up an avenue for plasmonic applications in spintronics and magnetic memory beyond HAMR. Needless to say that these explorations would demand a lot from theoreticians to take their existing knowledge of hot electron dynamics and apply it to scientific fields different from semiconductor/metal mediated photochemistry.

1 J. L. Carvalho-de-Souza, J. S. Treger, B. Dang, S. B. H. Kent, D. R. Pepperberg and F. Bezanilla, *Neuron*, 2015, **86**, 207–217.
2 Y. Yang, R. B. Wilson, J. Gorchon, C.-H. Lambert, S. Salahuddin and J. Bokor, *Sci. Adv.*, 2017, **3**, e1603117.

Mahfujur Rahaman addressed the room: In our TERS experiments we use Au nanodisks which have an LSPR around 610 nm. The LSPR position is also confirmed by the SERS experiments. However, when we brought down the Au tip for the TERS experiments we observed the strongest enhancement for 785 nm. We know that in a dimer situation, due to the restoring force, the combined LSPR should red shift. However, in our case the red shift is quite strong. We do not have any theoretical study of the observed strong red shift of the system. If any one can help us in the understanding of this phenomenon it would be really great.

Javier Aizpurua noted: Our contribution in this Faraday Discussion is focused on the optical field emission in plasmonic gaps subjected to intense single-cycle optical pulses, however, we have studied the plasmonic response of nanoparticle dimers and gaps since many years ago. I can refer to our seminal paper on single molecule spectroscopy in gaps, where we already reported the redshift of the gap plasmon,[1] or to the theoretical paper that described this effect for very narrow gaps.[2] The understanding of the plasmonic redshift in a gap is clear at this point as originating from the interaction of the surface charges at both sides of the gap (either dipole–dipole interaction for a large separation, or multipole-multipole interaction for narrower gaps). If one considers the SERS and TERS signal as approximately proportional to the 4th power of the local field enhancement, the spectral shifts that you find might be consistent with the use of the very narrow gaps that you achieve in your TERS configuration (the tip–sample gap). If this separation does not fully explain the enhancement, one might need to account for other more subtle effects connected with non-locality, charge transfer mechanisms, or optomechanical effects which can produce extra enhancement of the TERS signal.

1 H. Xu, J. Aizpurua, M. Käll and P. Apell, *Phys. Rev. E*, 2000, **62**, 4318–4324.
2 I. Romero, J. Aizpurua, G. W. Bryant and F. J. García de Abajo, *Opt. Express*, 2006, **14**, 9988–9999.

Javier Aizpurua continued: How plasmonic spectral shifts happen due to tip-substrate interactions is clear. However, the reduction or enhancement of the Raman signal could be due to many effects which can go beyond the local field enhancement of the plasmonic antenna. One can find resonant effects, shifts between the near field and the far field. Sometimes a local field description is very good, but for others one needs to include quantum effects. Also, often one needs to include the presence of the molecule or the excitonic material and consider its electronic structure as part of the whole response and enhancement. In other words, I would like to note that accurate modelization of Raman signals from molecular species often requires going beyond electromagnetics.

Mahfujur Rahaman answered: Many thanks for the comments. When we have performed SERS on these Au disks, we observed the strongest signals for the 638 nm laser and no signal for the 785 nm laser. Therefore, our initial assumption was that we could perform TERS better using the 638 nm laser since it is in resonance with the MoS_2 and also the disks. However, after conducting TERS it turned out that the 638 nm laser had little impact in TERS and the 785 nm laser had the strongest. Since we do not have a laser of higher wavelength than 785 nm we could not try this. This phenomenon surprised us a lot since the shift is quite big if we only think of LSPR of the combined system only. However, you are absolutely right in your comments that it may not only be about plasmonic enhancement. We are now investigating this issue and hopefully will be able to get an answer.

Alexander Govorov remarked: In our studies of hot electron (HE) generation, we found that size, shape and material matter. Since hot-electron generation is a quantum surface effect, small sizes of nanocrystals are preferable. The efficiency of HE generation increases strongly for nanocrystals with complex shapes and hot spots, such as nanostars, nanocubes and metamaterial absorbers with gap plasmons. Nonthermal HEs with high energies are generated due to non-conservation of linear momentum, and such processes are very efficient near surfaces and in plasmonic hot spots. Since the HE generation is a quantum process, metals with a longer mean-free path will work better. For example, silver is more efficient than gold. The quantum efficiency of HE generation also increases in the infrared spectral interval because the quantum amplitudes of HE generation grow for small photon quanta. Detailed description for the above properties can be found in our papers.

To be more specific, the energy efficiency of creation of high-energy hot electrons in nanocrystals (NCs) with simple shapes (nanospheres and nanorods) is given by the ratio[1] l_{mfp}/a, where l_{mfp} is the mean fee path and a is the size of a nanocrystal. This simple equation for the efficiency can also be derived from the equation for the surface-induced decay rate of a plasmon in a NC.[2,3] For this, we need to take a ratio between the surface-decay rate of a plasmon and the decay rate of a plasmon in the bulk. Overall, the effect of decay of plasmons due to the generation of high-energy electrons near the surfaces has a long history.[2,3]

1 L. V. Besteiro, X.-T. Kong, Z. Wang, G. Hartland, and A. O. Govorov, *ACS Photonics*, 2017, **4**, 2759–2781.
2 L. Genzel, T. P. Martin and U. Kreibig, *Z. Phys. B: Condens. Matter*, 1975, **21**, 339–346.

3 U. Kreibig and M. Vollmer, *Optical Properties of Metal Clusters*, Springer-Verlag Berlin Heidelberg, 1995, vol. 25.

Jacob Khurgin responded: Agreed. In my paper it is clearly stated that in the end it is the surface to volume ratio that must be maximized for maximum HE emission efficiency.

Jacob Khurgin commented: In my view the most important work for theorists is to study hybridization at the surface of the metal. There lies the key to efficiency.

Emiliano Cortes noted: I totally agree that this is one of the most interesting directions for the field, not only because of efficiency but more important to me for selectivity of the generated products. Our current knowledge of the electronic structure of adsorbed molecules is very limited, more so if we want to access the excited states.

Anatoly Zayats asked: In the case of the film on top of the disk structure probed by the tip, the excitation of the localised plasmon will lead to a strong field enhancement and may result in photoelectron injection from the gold disks to the film. This may change the observed spectra. Have you considered these effects?

Mahfujur Rahaman replied: Yes we have considered that and you are absolutely right about hot electron injection. Due to strong hot spot formation between the tip and the Au disks (which sandwiched the monolayer MoS_2 in between), we observed selective doping. The doping sites were concentrated on the hot spots only. In our previous publication[1] we have reported this phenomenon. Due to strong hot electron doping, we observed a structural phase change from the semiconducting 1H to the metallic 1T phase. Thanks to our very good spatial resolution observed (2.3 nm), we were able to locate those doping sites.

1 A. G. Milekhin, M. Rahaman, E. E. Rodyakina, A. V. Latyshev, V. M. Dzhaganc and D. R. T. Zahn, *Nanoscale*, 2018, **10**, 2755–2763.

Naomi Halas enquired: Do theorists have any suggestions about what experimentalists should be measuring?

Javier Aizpurua responded: The experimental effort to identify the role of plasmon-induced hot electron chemistry is tremendous and extremely useful, so far. In the same way that theory lacks an entire compact methodology to address the whole dynamical process including hot electron generation in a large nanoparticle, and later injection in a particular adsorbate (mainly due to the multiscale nature of the problem both in time and space), I also have the feeling that it might be good to experimentally try to access the dynamics of chemical reactions more *in situ* and in real time. Couldn't we try to combine ultrafast techniques which could provide some information on the dynamics of both aspects of the problem? We could use linear optical absorption and scattering techniques, as we do nowadays, together with other condensed-matter techniques such as photoemission spectroscopy (PES), or scanning-tunneling microscopy (STM)-induced fluorescence and Raman scattering. And all this could be merged with pump and probe techniques. Of course these techniques address more academic situations, and might be far from the real world of interesting chemical reactions in more realistic situations, but it might be interesting to identify good testing situations, and to merge efforts from the chemistry community, the ultrafast optics community and the condensed matter community. I realize that this is not trivial at all, and at least as challenging as the theoretical efforts needed, or even more.

Alexander Govorov commented: For the rates and efficiencies of generation of nonthermal high-energy (hot) electrons, both size and shape are important. In nanocrystals of small sizes, one should expect higher efficiencies of hot electrons generated *via* the surface-scattering mechanism, since the surface-to-volume ratio is larger. Also, the shape of a nanocrystal matters since it determines the position and strength of the plasmonic resonance and also an intensity of the electric fields inside a nanocrystal; this internal electric field generates hot electrons near the surfaces where the electron momentum is not conserved. Plasmonic nanocrystals with complex shapes and hot spots are good generators of hot electrons. Some examples are nanostars and nanostructures with narrow plasmonic gaps (a nanocrystal near a mirror and a nanoparticle dimer).

Jeremy Baumberg addressed Alexandra Boltasseva: In your plasmonic enhanced systems you need to use very thin layers of photocatalytic materials – whose properties are different to bulk. Do you see more photodegradation when you make them this thin (we have seen this in many systems)? How are the properties changed by the additional quantum confinement effects, or increased surfaces which change the film reconstructions and strain? Are there ways to characterise these without TEM (which only looks at a tiny fraction of the film)?

Alexandra Boltasseva answered: No, we did not see any degradation in our hematite films. Indeed, one of the strengths of hematite is its high stability under PEC conditions. We tried only $\sim$15nm hematite thickness where we have not studied any quantum confinement effects. Going to 2D hematite might provide unique properties that could be enhanced/modified by interaction with surface plasmons. See for example ref. 1.

1 A. Puthirath Balan, S. Radhakrishnan, C. F. Woellner, S. K. Sinha, L. Deng, C. L. Reyes, B. M. Rao, M. Paulose, R. Neupane, A. Apte, V. Kochat, R. Vajtai, A. R. Harutyunyan, C. W. Chu, G. Costin, D. S. Galvao, A. A. Martí, P. A. van Aken, O. K. Varghese, C. S. Tiwary, A. Malie Madom Ramaswamy Iyer, P. M. Ajayan, *Nat. Nanotechnol.*, 2018, **13**, 602–609.

Junyang Huang commented: The efficiency of the solid-state configuration used here for water splitting is, to some degree, limited by the fact that water molecules are not able to freely move into the plasmonic hotspot. However, if the equivalent system could be constructed in the colloidal suspension (*i.e.* making many small divisions of reactors which are stable in the aqueous phase), do you think the accessibility of the hotspot and surface to volume ratio could be significantly improved and therefore enhance the efficiency of the system? What are the major challenges associated with transforming catalysis into the colloidal system?

Alexandra Boltasseva replied: Photoelectrochemical systems based on thin films and colloidal suspensions of nanostructures are two different technological approaches in photocatalysis. In general, the former enables a higher solar-to-hydrogen (STH) efficiency (*i.e.* the current benchmark is around 20%), while the latter provides a much lower STH efficiency, with the current benchmark being around 1% for a solar water splitting colloidal catalyst. Besides, the major challenges associated with using a colloidal system are related to oxygen/hydrogen separation, catalyst recyclability and recovery. Most of all, in colloidal systems it is very challenging to achieve a high charge carrier lifetime and spatial separation, which strongly limits the overall efficiency. Otherwise, PEC systems allow us to separate energetics and kinetics constraints by using two electrodes to drive oxygen and hydrogen evolution, respectively, with the additional benefit of applying a bias to control the water splitting process. We think that for some chemical reactions the colloidal approach is more convenient in terms of hot spot configuration, and the use of hot carriers can be more convenient than in thin films.

Emiliano Cortes asked: The idea of semi-metals as alternative materials for plasmonic applications sounds interesting. However, our control on patterning/colloidal synthesis for traditional plasmonic metals is highly advanced compared to that for semi-metals. What is the state-of-the-art and the challenges for patterning or nanostructuring semi-metals?

Alexandra Boltasseva responded: Significant progress has been reported in the literature regarding the nano-patterning of semi-metals like titanium nitride and zirconium nitride. Titanium nitride is also compatible with standard CMOS processes. See for example, Li *et al.*[1] and Chaudhuri *et al.*[2]

1 W. Li, U. Guler, N. Kinsey, G. V. Naik, A. Boltasseva, J. Guan, V. M. Shalaev and A. V. Kildishev, *Adv. Mater.*, 2014, **26**, 7959–7965.
2 K. Chaudhuri, A. Shaltout, D. Shah, U. Guler, A. Dutta, V. M. Shalaev and A. Boltasseva, *ACS Photonics*, 2019, **6**, 99–106.

Jeremy Baumberg opened a general discussion of the paper by Stephen B. Cronin: I am puzzled by your data on the grating reflectivity and photocurrent *vs.* angle in Fig. 4 in your paper. They are not the same – the peaks are shifted, and of different widths. Is it obvious why this is? I suggest this tells us something about where the photocatalytic activity is going on – where on the profile the photocatalysis is happening. You can calculate back from the photocurrent to the near-field generation position, and thus localise where is most active. I wonder if on the sloping edges of the deposition that there are more step-edges and higher roughness creating improved photocatalysis? Can you measure this with STM/AFM?

Stephen B. Cronin answered: This is indeed a puzzling phenomenon, however, one that we observe consistently over repeated measurements with different samples. I think it is unlikely that roughness varies much along the period of the grating. And, while it is likely that the photocurrent is produced predominately in the high field region of the grating (see Fig. 5c in our paper), and thus not produced uniformly over the grating surface, I don't think this will result in a shift in or broadening of the resonant angle, since the periodicity is still the same. It is worth pointing out that photoreflection is an instantaneous process, whereas the hot electron injection is finite. As such, there can be other mechanisms of damping in the system that lead to these shifts and increased broadening of the peaks.

Yonatan Sivan asked: Among other things, you showed the change in transmissivity of a metal grating under optical illumination by a strong pump. Can you give us a historical perspective please by comparing to the numerous earlier studies of the same configuration (*e.g.*, by the van Driel group, the Vallee group, and many others)?

Stephen B. Cronin replied: Work from the van Driel group[1] is probably most closely related. Here, ultrafast control of grating-assisted light coupling to surface plasmons was explored, in which 810 nm pulses produced a shift of the 570 nm plasmon resonance by ~0.75 nm with a reflectivity change up to 6% and decay time of ~1 ps. These DR/R spectra show similar profiles which have a positive and negative component, centered around the plasmon resonance. Vallee's group[2] has measured the ultrafast nonlinear response of metal nanoparticles and nanorods to be of the order of 1-2 ps. These data are understood on the basis of a model describing the wavelength and time-dependent modifications of the object material dielectric function. Perhaps most recently, Planken's group[3] looked at the effect of metallic bilayers on the ultra-fast dynamics of grating structures, and found that the presence of a buried dissimilar metal layer strongly affects these relaxation times. While the general relaxation time of these transient absorption features is consistent with our results (~1-2 ps), what hasn't been reported before in these time-resolved studies are distinctive features associated with resonant and non-resonant excitation presented here.

1 N. Rotenberg, M. Betz and H. M. van Driel, *Opt. Lett.*, 2008, **33**, 2137-2139.
2 T. Stoll, P. Maioli, A. Crut, N. Del Fatti and F Vallée, *Eur. Phys. J. B*, 2014, **87**, 260–278.
3 S. Edward, A. Antoncecchi, H. Zhang, H. Sielcken, S Witte and P. C. M. Planken, *Opt. Express*, 2018, **26**, 23380–23396.

Niclas Sven Mueller enquired: Do you have an explanation why interband transitions cause a positive signal in the transient absorption spectra? This would imply that the absorption probability by interband transitions increases upon illumination by a pump pulse.

Stephen B. Cronin responded: Upon pump excitation (560 nm), 6s band electrons are excited to empty states above the Fermi level, creating an electron-hole population in the 6s band around the Fermi level. The absorption peak around 520 nm corresponds to the transition from the filled 5d band to holes in the 6s band, therefore the transient signal is positive on the spectrum. A slight decrease in the signal should be expected at higher energy probes due to blocking of absorption by electrons above the Fermi level. In our experiments with other pump wavelengths (680 nm, for example) that reduction in absorption is observed at higher probe energies. However, for the 560 nm pump, the reduction in absorption due to electrons above the Fermi level is small. Further studies are needed in order to obtain a complete understanding of these subtleties.

Andrea Marini noted: Your paper nicely reports a hot electron-driven transient absorption spectroscopy in plasmon resonant grating structures. Such a nonlinear process has already been reported before in the reflection by thin films and in propagating surface plasmon polaritons in gold nanowires. However, in such previous investigations, the surface plasmon resonance was playing a minor role and the reflectance/transmission modulation arises from the Fermi smearing of conduction electrons that modulate metallic absorption and dispersion. Can you please comment on the role of the surface plasmon resonance in your experimental findings? Does the reflectance modulation arise from a nonlinear shift of the plasmon resonance itself?

Stephen B. Cronin answered: The transient absorption signal of the surface plasmon resonance is the result of the "Fermi smearing" phenomenon. In detail, upon pump excitation, electrons in the 6sp band are excited to the empty state above Fermi level. The instantaneous excitation by the pump will result in a non-Fermi–Dirac distribution around the Fermi level. After a short relaxation time, about 50 fs, a new "hotter" Fermi–Dirac distribution is formed through electron-electron scattering. Then the energy will dissipate to the metal lattice through electron–phonon scattering and phonon–phonon scattering, of which the lifetime is greater than 100 ps. The surface plasmon oscillation frequency is a function of the density of electrons in the metal and the mean free path of electrons in the Drude free electron model. "Fermi smearing" upon pump excitation will result in a hot Fermi–Dirac electron distribution. This will lead to higher average kinetic energy. According to the Bohm Gross electron dispersion equation, increasing thermal energy of electrons results in a higher plasmon resonance frequency and the absorption is broadened. In the spectrum, the transient absorption signal appears to be a large depletion and two wings on the sides, which is a result from subtracting the broadened transient signal from steady state plasmon resonance absorption.

Madasamy Thangamuthu remarked: The photocurrent shown is difficult to understand. Is the photocurrent generation due to the electron transfer or hole transfer? If your answer is due to the electron transfer, then is it photoanodic (electrons injected into the back metal contact) or photocathodic (electrons injected into the electrolyte solution)?

Stephen B. Cronin replied: We observe AC photocurrent under both oxidation (positive electrode potentials) and reduction (negative electrode potentials) conditions. For oxidation, the charge transfer process can be thought of as an electron transferred out of the adsorbed molecule into the hole state in the metal electrode. For reduction, the charge transfer can occur as an electron being injected into a reactant adsorbed on the surface of the electrode or directly into the electrolyte solution. However, the barrier for producing a solvated electron is most likely too high to contribute significantly to the observed photocurrent.

Naomi Halas questioned: The spectral structure is really quite different for one polarity compared to the other – do you have any insight on how this relates to the DOS?

Stephen B. Cronin responded: The spectral structure difference comes from whether electronic surface plasmons are excited. Electronic surface plasmons are excited when the probe is in p-polarization with respect to the grating, while they cannot be excited when the probe is in s-polarization to the grating. Other theoretical studies have reported on the density of states of Au.[1] The electronic density of states of gold can be described as being composed of a broad and diffuse 6s-p band spanning the Fermi level, and a 5d band with a larger density of states several eV below the Fermi level. We think that the pump laser excites electrons from the 6s-p band below the Fermi level to empty states above the Fermi level, generating holes and electrons in the 6s-p bands. Two excited state absorption peaks were observed in the spectra. The peak around 510 nm is the excess absorption corresponding to excitation of 5d electrons to 6s-p holes. The peak around 470nm corresponds to electrons excited from the 5d band to empty states above the Fermi level by the probe. Since the broadness of the depletion and absorption peaks are similar, one can make a rough calculation of how deep the 5d-band is with respect to the Fermi level. In the TA data, the spectra cross zero at around 480 nm, which implies that the 5d band in Au is 2.6 eV below the Fermi level. This finding is consistent with other theoretical DOS studies.[2]

1 L-M. Yang, M. Dornfeld, T. Frauenheima and E. Ganzb, *Phys. Chem. Chem. Phys.*, 2015,**17**, 26036–26042.
2 U. Kreibig and M. Vollmer, *Optical Properties of Metal Clusters*, Springer-Verlag Berlin Heidelberg, 1995.

Zachary Schultz said: Do you see changes in the overpotential at which the reaction, either H^+ reduction or O_2 evolution, proceeds with the plasmonic catalyst? This should also be affected by the solution pH.

Stephen B. Cronin answered: The electrolyte condition was pH = 7, 0.5M Na_2SO_4 solution. We haven't yet explored these processes in solutions with different pH values for HER and OER, however, that would be an interesting set of experiments to perform in order to gain a better understanding of any changes in the overpotential of these reactions.

Yuri Diaz Fernandez opened a general discussion of the paper by Laura Fabris: I hope we will all agree that in the field of anisotropic nanoparticles, we commonly address materials as "gold" nanorods or "gold" nanostars, despite the fact that silver may be incorporated to different extents into these materials, depending on the synthetic conditions. Do you have any comment on the effect of silver on the nucleation of TiO_2 at the surface of these materials and on the role of silver/gold alloys on the efficiency of hot electron generation in anisotropic nanoparticles?

Laura Fabris replied: This is an excellent point. Currently, the community is starting to become more clear on the actual presence of non-negligible amounts of silver on anisotropic nanoparticles, which originate from the need to employ

$AgNO_3$ to induce anisotropy during growth. We have recently shown[1] that monolayers of Ag (in addition to the presence of Ag^+) can be clearly measured using STEM-EDX, and that the position and characteristics of these monolayers depend on the initial concentrations of $AgNO_3$ used. Clearly, it is possible that Ag will influence the growth of TiO_2. While we have not yet studied this effect, we will be looking into it in the near future. As for how the presence of Ag may influence hot electron generation, this will likely not come from differences in LSPR due to the presence of Ag. It will likely be due to Ag atoms inducing strain at the interface or representing defects that are obstacles to electron transfer across the interface.

1 S. Atta, M. Beetz and L. Fabris, *Nanoscale*, 2019, **11**, 2946–2958.

Bhavesh Kumar Dadhich asked: Is it possible to produce hot electrons and photo-thermal effects using a non-coherent light source?

Laura Fabris answered: Yes, indeed these experiments were carried out under illumination of a Xe arc lamp, which is a non-coherent light source.

Bart de Nijs commented: The presented star shape is an out of equilibrium structure for gold nanoparticles. When you heat these structures they will attempt to revert back to a spherical shape. Have you compared absorption spectra before and after the experiments to see if there have been any significant changes to the optical properties of the structures?

Laura Fabris replied: Indeed, nanostars will likely revert into spheres if treated at high temperatures *in operando*. In fact, an excellent paper from the groups of Bals and Liz-Marzan[1] shows, by using TEM tomography with *in situ* heating, that the shape of gold nanostars is rapidly modified due to atom migration as the temperature is increased. This is why we need to run the TiO_2 reaction at 70 degrees. However, TiO_2 is likely offering good thermal protection to the particles once they are completely coated. Monitoring the LSPR position after photocatalytic conversion does not show substantial differences leading us to hypothesize that TiO_2 represents an excellent barrier against reshaping.

1 H. Vanrompay, E. Bladt, W. Albrecht, A. Béché, M. Zakhozheva, A. Sánchez-Iglesias, L. M. Liz-Marzán and S. Bals, *Nanoscale*, 2018, **10**, 22792–22801.

Bartlomiej Jankiewicz commented: You have mentioned in your article that plasmonic materials with spherical geometry suffer from limited LSPR tunability, which still leaves the majority of the solar spectrum untapped. However, as has been shown by the work of Prof. Halas on spherical core-shell nanostructures you may tune their optical properties in a broad spectral range simply by changing their geometry. You have also mentioned that you are not able to obtain a full shell for Au nanorods, while you are able to do that for Au nanostars. This is likely related to the coating procedures in which you are using an alcohol solvent for stars and water for nanorods. Organic titania precursors undergo fast hydrolysis in the presence of even small amounts of water, which makes the reaction with them quite difficult to control. However, you may try to use less reactive titania precursors such as TBT. Why do you heat your samples for two days at 70 degrees instead of doing that *e.g.* for a few hours at 90 degrees? Based on the extinction spectra modification of Au nanorods with TiO_2 doesn't significantly change their optical properties. Don't you then think that the higher reactivity of the AuNR-TiO_2 structures can be related to formation of the interface between semiconductor and metal?

Laura Fabris responded: When I mentioned the limited LSPR tunability in spherical gold nanoparticles I meant solid spherical particles, which differ from Prof. Halas' nanoshells. Nanoshells have a silica core and a gold shell of variable thickness. The relative size of the core *vs.* the whole nanoparticle and the thickness of the shell are what contributes to the LSPR tunability obtained for these materials.

It is true that we were not able to obtain a full TiO_2 shell on nanorods and this is due to the need to synthesize it in a water environment, where the CTAB shell is not compromised as is observed in alcohols, where the critical micelle concentration of CTAB is modified, leading to the "unzipping" of the micelle. As you correctly point out, employing a less reactive precursor will likely improve the properties of the shell. It is not possible to synthesize the shell on nanostars at temperatures higher than 80 degrees, due to the sensitivity of the nanostar spikes to adatom migration, which is high at locations with a high radius of curvature and is further accelerated at higher temperatures. If nanostars are treated above 80 degrees, the sharpness at the tips is severely compromised, leading to shifts in LSPR and reduced field localization and, therefore, hot electron generation.

The optical properties of TiO_2-coated nanorods are not significantly altered (in LSPR position) because of the limited coverage by the semiconductor.

What we observed with respect to AuNR reactivity is that there is a compromise between the uniformity of the TiO_2 shell and the lack of polymer stabilizer. In order to synthesize the TiO_2 shell in the absence of a polymer stabilizer, one needs to run the reaction in water (as described above) leading to a compromised semiconductor shell quality. However, these results confirm the importance of leaving the interface unhindered, to improve electron transfer across it.

Rifat Kamarudheen said: In Fig. 4a of the manuscript where the hydrogen production of three different gold nanorods (with different LSPRs) are compared with respect to time, the hydrogen production for nanorods with 910 and 980 nm LSPRs starts to saturate after 20 min while the hydrogen production of the 870 nm LSPR nanorod is still increasing. How would you explain these different trends?

Laura Fabris answered: This is a good point and to be quite honest we are not sure why this is happening. We have carried out these reactions multiple times obtaining similar results. This will be an interesting point to start exploring in our upcoming experiments.

Madasamy Thangamuthu asked: From the TEM image one can clearly see the phase differences but how do you confirm this is anatase? Can it be a mixture of anatase and rutile together? Could you comment on whether it would be advantageous to have a mixture of anatase and rutile?

Laura Fabris replied: We ran XRD characterization studies where we observed the presence of peaks for both rutile and anatase. Because this synthesis is run at a low temperature, it is likely that both rutile and anatase are present in, most likely, a sea of amorphous phase. However, based on the literature, it appears that the presence of both rutile and anatase leads to improved band alignment with the hydrogen reduction half reaction, thus improving yields of hydrogen. In our previously published study[1] we also verified this hypothesis, indeed observing improved yields when both rutile and anatase were present.

1 S. Atta, A. M. Pennington, F. E. Celik and L. Fabris, *Chem*, 2018, **4**, 2140–2153.

Julien Mahin enquired: Did you notice any secondary nucleation of the titanium dioxide or does it all form on the surface of the gold nanostars?

Laura Fabris answered: Yes, we observed, in some instances, the presence of secondary nucleation sites for titania nanoparticles. However, this can be avoided by slightly modifying reaction conditions or by post-synthetic purification.

Javier Aizpurua remarked: It is fascinating to compare nanorods and stars for chemical reactivity, as they are nanoantennas which show different bulk to surface ratios, and thus, they could show very different behavior, in principle. The structures in your paper seem like canonical structures to show the role of hot electrons. I do not know if production of hydrogen is the best experiment to look at – as it shows differences of 10% in performance, according to your paper. Are there better or more suitable reactions in which we could look at the performance of more controlled and different ratios of hot electrons and holes, even if they are more academic? Do you have any plan to study this in future?

Laura Fabris responded: There can indeed be better suited reactions to look into in the future. Since I am not a catalysis expert, I welcome suggestions from the audience on interesting chemical processes to look at. Once these reactions are identified, we can define design rules to address them, in particular as it concerns choosing the ideal semiconductor catalyst and energy level alignments.

Sebastian Schlücker noted: Creating and extracting hot carriers is equally important. One of the key aspects in your presentation was to nicely demonstrate how to obtain surfactant-free anisotropic nanostars (tips). I guess this aspect has been overlooked in colloid-based plasmonic chemistry with respect to efficient hot carrier extraction. Could you please share your opinion on the general role of surface chemistry in this area with us?

Laura Fabris replied: As you point out, surface chemistry has an important role with respect to achieving efficient hot carrier extraction. Our results seem to indicate that a very intimate contact between the metallic surface of the plasmonic nanoparticle and the semiconductor is necessary to obtain hot electron injection in the latter. Our results on nanorods, although not definitive, appear to further confirm this hypothesis. In fact, in our report, we show that the yield of hydrogen in nanorods with imperfect TiO_2 shells is equal to that of nanorods reported in the literature in which a uniform titania shell was grown in the presence of a polymer overcoat between the metal and the semiconductor. Therefore, there is evidence that polymer layers or other reaction intermediates left at the interface are going to dramatically affect the performance of the photocatalytic system.

Jeong Y. Park enquired: In your interpretation, you are assuming that there is a Schottky barrier height between gold and TiO_2. Do you have some estimation of the Schottky barrier height formed at the Au–TiO_2 interface? How does it change with the shape of the Au nanostructures? How does this affect the catalytic activity? In my opinion, trying local probe measurements such as conductive atomic force microscopy on Au nanostructures on a TiO_2 substrate would help to understand the local charge transport properties.

Laura Fabris answered: We have based our experiments on a generally estimated value of 1 eV between the Au Fermi level and titania conduction band. Clearly, this may not be the case at the nanoscale, especially when the morphology of the nanoparticle is highly anisotropic. Indeed, measurements that probe the local properties will be highly beneficial and could shine new light on the process.

Yuri Diaz Fernandez opened a general discussion of the paper by Jeong Y. Park: I am curious about the CoO layers observed in your as-prepared nanoparticles. Did you still observe this lower oxidation state after ozone cleaning? Do you have any evidence of lattice strain due to mismatch between the core lattice and the surface CoO layers?

Jeong Y. Park responded: We measured the XPS of Co before and after UV ozone treatment, as shown in Fig. 4c and Fig. S2 in our paper. We found that there were no significant changes in the chemical states of either the platinum or cobalt oxide surfaces while removing the organic capping molecules using UV–ozone. Therefore, we can conclude that the oxidation states of the Co remain almost the same after UV ozone cleaning. We assume that the lattice strain from mismatch between the core lattice and the surface CoO layers is present. The effect of the lattice strain on the catalytic activity of Co oxide nanoparticles is not clear to us and it would be an interesting subject for a follow-up study.

Phillip Christopher asked: Can you use the catalytic diode approach to estimate adsorption energies of species? Because the current should be proportional to exothermicity, would this be an easy way to measure the adsorption energy of molecules?

Jeong Y. Park replied: The catalytic reaction involves multiple processes including adsorption, desorption of molecules, and chemical rearrangement. Hot electrons measured with catalytic nanodiodes can give clues about the exothermicity of a reaction, but it does not give exclusive information about the adsorption process. Therefore, it is challenging to obtain the adsorption energies of a chemical species with a catalytic diode. However, I would like to note that an earlier approach using a Schottky diode focused on the detection of hot electrons generated during the adsorption of molecules. Using this approach, Gergen *et al.* obtained a correlation between the hot electron flux and the adsorption energy.[1]

1 B. Gergen, H. Nienhaus, W. H. Weinberg and E. W. McFarland, *Science*, 2001, **294**, 2521–2523.

Reinhard J. Maurer remarked: In Fig. 5 in your paper, the chemicurrent and TOF for the three systems follow the trend: Co_3O_4 NCs/Pt < Pt < UV-treated Co_3O_4 NCs/Pt. What is your current working hypothesis to explain this trend?

Jeong Y. Park answered: The key claim in this study is that the interface between the Pt and cobalt oxide gives rise to enhanced hot electron generation. We explain this trend with a theoretical DFT calculation that shows a lower activation barrier forming for *OH formation at the Pt–cobalt oxide interface[1] and a higher exothermicity, which are responsible for the higher catalytic activity and the higher chemicurrent yield, respectively. The effect of UV treatment is to remove the capping layers that were present on the oxide nanoparticles.[2,3] Before UV treatment, this capping layer blocks the reactive sites and stops hot electron transport.

1 H. Lee, J. Lim, C. Lee, S. Back, K. An, J. W. Shin, R. Ryoo, Y. Jung and J. Y. Park, *Nat. Commun.*, 2018, **9**, 2235.
2 J. Y. Park, C. Aliaga, J. R. Renzas, H. Lee and G. A. Somorjai, *Catal. Lett.*, 2009, **129**, 1-6.
3 J. K. Nørskov, T. Bligaard, J. Rossmeisl and C. H. Christensen, *Nat. Chem.*, 2009, **1**, 37–46.

Naomi Halas commented: HER and water splitting have been around for a long time. If we took all the money that was ever spent on research into that and split it into all of us in the room we could fund our research for our career. Can we ever solve HER – what are the breakthroughs we need? Should we be looking at other avenues?

Stephen B. Cronin replied: Water splitting is a nice test reaction for developing better understanding of photocatalytic processes in general and developing appropriate materials for these processes. There are still many aspects of these photocatalytic processes that are poorly understood, and it is likely that, ultimately, the "payoff" of this large body of work will be in another reaction or liquid–solid interfacial system.

Phillip Christopher contributed: A lot of technoeconomic analyses have been done that suggest that at the largest scales it will be tough for water splitting to compete with steam methane reforming (SMR), even with close to theoretical efficiencies. Probably the use of water splitting for H_2 generation will be used mostly on small scales, where some advantages exist over SMR, unless significant carbon taxes are levied.

Zachary Schultz questioned: Should we be looking at the materials for other reactions? There have been a lot of comments and reports showing that the chemistry changes what the catalysts are useful for. Can we develop guiding principles based on fundamentals of various reactions to help guide the development of future catalysts?

Jeong Y. Park responded: In our study of hydrogen oxidation, we have used the most well-known (*i.e.*, Pt-based) catalysts. Because noble metals like Pt are quite expensive, there are numerous approaches to develop novel catalysts that are cost-effective and highly reactive. I think an important area is bimetal catalysts because of the potential for improving catalytic performance and reducing cost. I think a theoretical approach can provide additional insight to guide the development of future catalysts.[1]

1 J. K. Nørskov, T. Bligaard, J. Rossmeisl and C. H. Christensen, *Nat. Chem.*, 2009, **1**, 37–46.

Sebastian Schlücker remarked: Guiding principles for this are of course very important yet probably very challenging to obtain. In my opinion we need support from theory which is capable of describing the involved potential energy surfaces (the ground state but also the relevant excited states) as this changes from system to system (metal plus molecule). If possible at all, I expect this type of theory input therefore more likely to come from the quantum chemistry/surface science community than from the plasmonics field.

Naomi Halas noted: Absolutely, that is the future of this field. Looking at what reactors have binding affinities for reactants of interest is a start, and also, with the help of accurate calculations, determining the binding energy for bond dissociation is typically one way of utilizing the antenna particle, that is, tuning its plasmon to or near the energy of bond dissociation of an adsorbate molecule. The reactions should dictate all these specifics.

Conflicts of interest

There are no conflicts to declare.

Faraday Discussions

PAPER

Optimizing hot carrier effects in Pt-decorated plasmonic heterostructures

Jorge U. Salmón-Gamboa,* Mayela Romero-Gómez, Diane J. Roth, Matthew J. Barber, Pan Wang, Simon M. Fairclough, Mazhar E. Nasir, Alexey V. Krasavin, Wayne Dickson and Anatoly V. Zayats

Received 9th October 2018, Accepted 14th November 2018

DOI: 10.1039/c8fd00150b

Hot carrier generation by light in various semiconductors and metallic nanostructures is important for many photocatalytic and photochemical processes, including water and hydrogen splitting. Here, we report on investigations of hot electron generation and extraction from Pt decorated SiO_2–Au nanoparticles using the degradation of methylene blue dye as a test-bed. Enhanced catalytic activity was found with an increase of Pt loading on the surface of the heterostructures. The small size of the Au nanoparticles (~12 nm) decorating the silica nanoparticles reduces hot electron collisions and related thermalization processes, since charge carriers have short paths to the surface where reactions take place and where Pt is situated. The heterostructures exhibit a broad plasmonic resonance in the visible wavelength range from 500 to 700 nm and hot carrier generation predominately takes place under resonant excitation. Electron-microscopy characterization and numerical modelling have allowed the optimization of Pt coverage for hot-electron transfer, consisting of a thin Pt shell covering the Au nanoparticle with Pt nanoparticles additionally placed on top. This geometry provides an increased number of active sites for methylene blue degradation and promotes separation of charge carriers generated by plasmonic excitations in Au. Such SiO_2–Au–Pt nanoparticles are attractive for hot-electron production due to the tunability of their plasmonic resonance and enhanced catalytic activity.

1 Introduction

Heterogeneous photocatalysis by plasmonic nanoparticles is one of the most important processes using hot charge carriers.[1–4] Hydrogen generation by plasmon-assisted water splitting is amongst the most investigated processes.[5–11] The ability to dissociate diatomic hydrogen molecules at room temperature and atmospheric pressure by plasmonic nanoparticles[12–14] proved that well-designed plasmonic nanostructures are able to perform difficult chemical reactions in

Department of Physics and London Centre for Nanotechnology, King's College London, Strand, London WC2R 2LS, UK. E-mail: jorge.salmon@kcl.ac.uk

ambient conditions, which is of paramount interest for applications. Water purification from organic pollutants, using semiconductor-noble metal hybrids, has also been extensively studied, driven by the increasing demand for clean water sources needed for population growth and industrial development.[15–17]

While spherical Au nanoparticles have strong and distinct resonances in the short wavelength range, the spectral tunability and opportunity to engineer electron transfer at the interfaces are important requirements for catalytic applications. Spherical dielectric-Au core–shell nanoparticles, often referred to as Au nanoshells, are one of the simplest nano-sized plasmonic heterostructures. These can be easily synthesized in volume *via* chemical methods and offer remarkable optical properties,[18] which can be tuned by varying the size of the core dielectric or the shell thickness.[19] This gives the nanostructures the ability to absorb and scatter light throughout the visible and near-infrared spectral range. For this, Au nanoshells have been under a continuous study and their applications include biosensing, bioimaging and photothermal ablation of tissues and cells.[20] Since plasmonic resonances are sensitive to the immediate surroundings of the nanoparticle where the plasmonic field extends, modification of plasmonic resonances can be achieved by loading nanoparticles with even smaller nanoparticles of different materials.

In this paper, silica nanoparticles decorated with Au nanoparticles and Pt (from now on referred to as SiO_2–Au–Pt heterostructures, schematic shown in Fig. 1a) were designed and optimized for the enhancement of hot-electron generation and extraction. Pt–Au heterostructures are often used to achieve an enhanced catalytic performance.[9,10,21] SiO_2–Au–Pt nanoparticles are attractive for hot-electron production thanks to the tunability of their plasmonic resonance in the visible spectrum, and the use of SiO_2 cores has added the advantage of having an inert dielectric support for Au. In this case no Schottky barrier is formed, therefore more hot carriers are available for reduction processes, instead of migrating through a Schottky junction[13] if a semiconductor was used. The small size of the Au nanoparticles reduces hot electron collisions, due to short pathways to the surface, and related thermalisation processes. Furthermore, the addition of Pt on the surface of

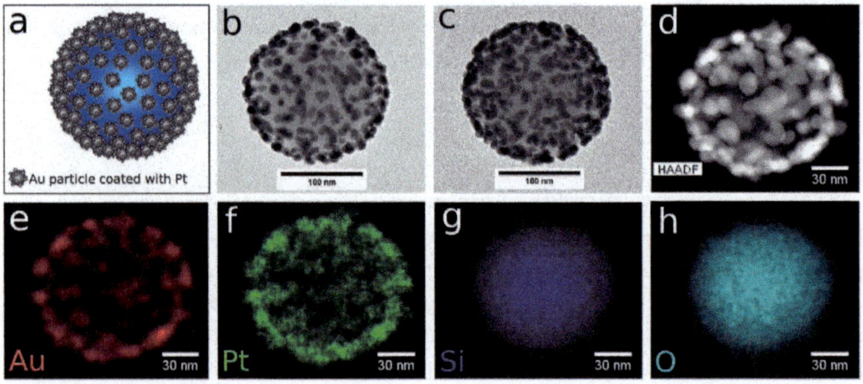

Fig. 1 (a) Schematic of the heterostructured nanoparticle. (b and c) TEM images of a single nanoparticle of SiO_2–Au and SiO_2–Au–Pt (sample Pt), respectively. (d) HAADF-STEM and (e–h) elemental mapping images of a SiO_2–Au–Pt heterostructure (sample Pt+++).

Au delays electron–hole recombination *via* charge separation, which improves the functionality of plasmonic heterostructures as exhibited by photocatalytic activity assessments of the prepared catalysts for methylene blue photodegradation.

2 Experimental methods

2.1 Chemicals

Tetraethyl orthosilicate (TEOS, 98%), 200 proof ethanol, ammonium hydroxide solution (NH_3 in H_2O, 28%), 3-aminopropyltriethoxysilane (APTES, 99%), tetrakis(hydroxymethyl)phosphonium chloride (THPC, 80% in water), Au(III) chloride trihydrate ($HAuCl_4\cdot3H_2O$, 99.9%), hexachloroplatinic(IV) acid hexahydrate ($H_2PtCl_6\cdot6H_2O$, 99%), potassium carbonate (K_2CO_3, 99.995%), formaldehyde solution (CH_2O, 37 wt% in H_2O), sodium hydroxide (NaOH, 98%), sodium chloride (NaCl, 99%), L-ascorbic acid ($C_6H_8O_6$, 99%), methylene blue solution (methylene blue, 1.5%), potassium iodide (KI, 99.5%), *p*-benzoquinone (BQ, 98%) and 2-propanol (2-Pr, 99.5%) were used as received from Sigma-Aldrich. Deionized (DI) water with a resistivity of 18.2 MΩ was used throughout the procedures, and all reactions were performed in air.

2.2 Synthesis of precursors and heterostructures

Precursor solutions

Au seeds suspension. The protocol followed for the synthesis of this solution is fully described and studied by Duff *et al.*[22] Briefly, 500 μL of 1 M NaOH was added into 45 mL of DI water while stirring at 900 rpm, followed by the addition of 1 mL of THPC (68 mM). After 5 minutes of constant stirring, 2 mL of 1 wt% $HAuCl_4$ solution was added. Immediately the solution turned to a dark brown colour and was left under stirring in the dark for a further 15 minutes. The mixture was then stored at 4 °C for a period of at least 2 weeks. The resultant solution is a colloidal suspension of 1–2 nm Au spherical seeds.

Au plating solution (K–Au). 12.42 mg of K_2CO_3 was added into 50 mL of DI water, followed by the addition of 750 μL of 1 wt% $HAuCl_4$ solution. The mixture was then stirred at 900 rpm for 30 minutes and stored at 4 °C overnight before use.

SiO_2 nanoparticles and APTES functionalization. SiO_2 nanoparticles dispersed in ethanol were synthesized following the Stöber method.[23] First, 167.5 μL of TEOS was mixed with 1.33 mL of ethanol, while 38.1 μL of 28% ammonium hydroxide was mixed with 407.8 μL of DI water and 1.05 mL of ethanol. Subsequently the two solutions were mixed under constant stirring at 300 rpm for 210 min. This process resulted in monodisperse silica nanoparticles of approximately 155 nm in diameter. The mixture was centrifuged 3 times, at 9000 rpm for 15 min, and redispersed in 5 mL of ethanol. In order to obtain amine-terminated nanoparticles, organosilane molecules (APTES) were adsorbed on the surface of silica. For this purpose, 500 μL of the prepared solution of SiO_2 nanoparticles was mixed with 5 mL of ethanol and, under stirring for 10 min at 650 rpm, 10 μL of APTES was added. The mixture was then allowed to stir at 100 rpm overnight. Subsequently, the solution was heated to 90 °C whilst stirring for one hour, along with addition of ethanol during the synthesis which prevented total evaporation of the solvent. Finally, the nanoparticles were washed three times in ethanol by centrifugation at 8500 rpm for 15 min and redispersed in 2 mL of ethanol.

SiO_2–Au nanoparticles. The procedure followed for the synthesis of SiO_2–Au nanoshells was based on the framework outlined by Wang *et al.*[24] Herein the process of obtaining incomplete nanoshells (*i.e.* SiO_2-cores decorated with discrete Au nanoparticles) is described as follows. The pH of a 3 mL solution of THPC–Au was adjusted to a pH of 3 by the addition of HCl to ensure interaction between the Au seeds and the amino-terminated silica cores. 60 μL of 1 M NaCl and 200 μL of APTES functionalized silica nanoparticles were added in the mixture, sonicated for 5 minutes and left undisturbed overnight. The mixture was centrifuged 3 times at 7500 rpm for 15 minutes and redispersed in 1 mL of DI water. The outcome of this procedure is a solution of colloidal silica cores decorated with 1–2 nm Au seeds. The resultant solution was refrigerated at 4 °C when not in use. In order to increase the size of the Au seeds attached on the surface of the cores, 3 mL of K–Au was added to 4 mL of the previously prepared seeded silica nanoparticles while stirring, followed by the addition of 50 μL of formaldehyde. The mixture was then left stirring for 15 more minutes and a change in colour, from pink to deep violet, was observed. The nanoparticles were immediately washed by centrifugation at 3000 rpm for 15 min and redispersed in 8 mL of DI water. As the result of this procedure, SiO_2–Au nanoparticles with extinction around 560 nm were obtained. The colloidal solution was then stored at room temperature.

SiO_2–Au–Pt nanoparticles. The coating of the nanoparticles with Pt requires ions provided by H_2PtCl_6, reduced by ascorbic acid, to produce the elemental form on the surface of the heterostructure. First, 200 μL of 0.1 M L-ascorbic acid and 6.5 μL of 10 mM H_2PtCl_6 were mixed with 4 mL of SiO_2–Au in a vial. The mixture was kept in a water bath at 70 °C for 3 hours and let to rest overnight. Finally, the mixture was centrifuged at 2700 rpm for 15 min and redispersed in 4 mL of DI water. Additional samples with different amounts of H_2PtCl_6 were synthesized in order to study the effect of Pt loading on the efficiency of hot electron extraction. In addition to this sample designated as "sample Pt", 13 μL (sample Pt+), 26 μL (sample Pt++) and 39 μL (sample Pt+++) of the Pt precursor were used to prepare the additional samples, under the same conditions described above. The final concentration of heterostructures of the samples was estimated to be around 0.15 nM.

2.3 Photocatalytic activity experiments

The hot carrier extraction of the synthesized heterostructures was investigated by photodegradation of methylene blue. During the degradation experiments, the absorbance maximum of methylene blue, located at 664 nm, was monitored over time, allowing the reaction kinetics to be studied. All photocatalysis studies were performed in a standard 1 cm × 1 cm fluorescence cell (Optical Glass 340–2500 nm) for 4 hours of irradiation at room temperature. A collimated supercontinuum laser, of a beam diameter of 1 cm, was used as a light source to excite the plasmonic samples, and the optical absorption of methylene blue was recorded by a spectrometer. Optical filters were used to selectively control the excitation wavelength range. For all optical measurements, mixtures composed of 467 μL of 20 μM methylene blue, 333 μL of catalyst solution and 200 μL of DI water were prepared. The control experiment was carried out with a solution of 467 μL of 20 μM methylene blue mixed with 533 μL of DI water.

2.4 Materials characterization

Transmission electron microscopy (TEM) images were taken on JEOL F200 running at 200 kV. TEM electron dispersive X-ray spectroscopy (EDS) was performed using an Oxford Instruments EDS detector. The samples were prepared by drop casting the catalysts in methanol on to Agar Scientific carbon grids. Elemental maps were taken by scanning transmission electron microscopy (STEM) with EDS on a Titan G2 ChemiSTEM operating at 200 kV.

2.5 Numerical simulations

The extinction cross-section responses of individual SiO_2–Au and SiO_2–Au–Pt structures dispersed in water were obtained from far field calculations using the transient solver utilizing finite integration technique in time domain (CST Microwave studio). Dispersionless refractive index values were considered for SiO_2 and water to be 1.4585 and 1.3325, respectively. The optical constant of Au was taken from experimentally measured tabulated values,[25] with a 5 nm mean free path correction taking into account the nanoscale size of the object, while the optical constant of Pt used was taken from experimental values reported by Werner *et al.*[26] The nanostructure was excited by a linearly polarized plane wavefront electromagnetic pulse with a broad spectrum, from 400 to 800 nm, and its response was investigated. To be able to retrieve the far field scattered light after the excitation, open space boundary conditions were set, on which minimal reflections are also ensured. Finally, from the obtained field distribution an extinction cross-section of the nanostructure was calculated.

3 Results and discussion

3.1 Characterization of heterostructures

Au nanostructures are of great interest due to the tunability of their optical extinction *via* the control of particle size, surrounding medium and morphology.[27] The type of heterostructure studied in this work features plasmon resonance in the visible spectrum and can be tuned by varying the size of the attached Au particles onto the silica matrix. Fig. 1b and c show the TEM images of single SiO_2–Au and SiO_2–Au–Pt heterostructures, respectively. The average size of the silica cores is around 155 nm, while the discrete decorating Au particles, randomly distributed over the surface of the silica core, range from 9 to 15 nm in size. The elemental mapping images of a SiO_2–Au–Pt nanoparticle synthesized with a higher amount of Pt precursor (39 μL of H_2PtCl_6) are presented in Fig. 1d–h. It can be seen that Pt preferentially deposits onto the Au surface, forming a thin layer of Pt covering the surface of each decorating Au particle. EDS measurements of the SiO_2–Au–Pt heterostructure revealed an elemental composition of Pt and Au of 2.34% and 11.20%, respectively.

The optical extinction spectra of the nanoparticles dispersed in water and measured after each fabrication step are shown in Fig. 2a. The extinction of the SiO_2 nanoparticle solution monotonically decreases with wavelength, as expected. After the deposition of Au seeds, a plasmonic resonant peak starts to appear in the visible wavelength range. The extinction spectrum of the synthesized SiO_2–Au nanoparticles exhibits a broad localized surface plasmon resonance, typically between 500–700 nm, with the peak maximum around 560 nm, as shown in

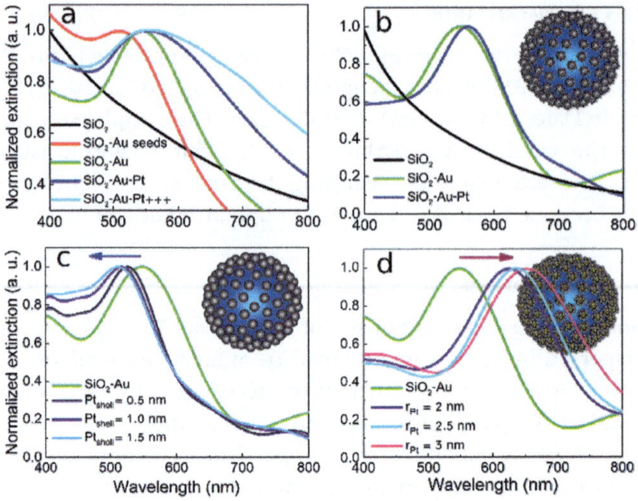

Fig. 2 (a) Experimental extinction spectra of the heterostructures at various synthesis stages. (b) Calculated extinction spectra of the heterostructure for a SiO_2 particle diameter of 150 nm, Au particle diameter $d = 15$ nm and elevation distance of the center of the Au nanoparticle from the silica surface $3d/8$, Pt shell thickness of 0.5 nm and Pt particle diameter of 2 nm. (c and d) Calculated extinction spectra for the limiting cases of Pt deposition on Au nanoparticles: Pt shells of various thicknesses (c) and Pt hemispheres of various radii (d); all the other parameters are as in (b). Insets (b–d): schematics of the modelled heterostructures used for the numerical simulations.

Fig. 2a (green curve). The resonance red shifts as the Au nanoparticle size increases which is in agreement with previous observations.[18] A further ~10 nm red shift of the resonance is observed after deposition of Pt, together with a broadening of the peak. This effect has been reported previously in unsupported Au nanoparticles covered with Pt.[21] At higher Pt loading the extinction further broadens and red shifts (Fig. 2a).

To gain insight into the experimental findings numerical simulations of the scattering of light on the studied nanostructure at different fabrication stages were performed (Fig. 2b–d). As the average distance between the individual SiO_2 nanoparticles estimated from the concentration of the solution used in the experiments was found to be ~2 μm, the particles were considered to be optically isolated and, therefore, the scattering on an individual nanostructure was numerically studied. The geometrical parameters of the nanostructure were derived from the sample TEM images and material analysis (Fig. 1). The structures were modelled considering a uniform distribution of 130 Au nanoparticles with a 15 nm diameter attached onto the surface of a SiO_2 particle having a diameter of 150 nm. The extinction cross-sections derived from the simulations are in very good agreement with the experimental observations. For simplicity, the shape of the Au particles deposited on the SiO_2 sphere were considered to be spherical, truncated by the SiO_2 particle boundary.

A significant dependence of the resonance wavelength of the system on the distance between the centre of the Au sphere and SiO_2 particle surface was observed, which is the result of the change in the position of plasmonic resonance of the Au nanoparticles with their shape. The optimal separation distance resulting in the correct position of the plasmonic peak of the overall nanostructure was found to be $3d/8$, where d is the Au sphere diameter (Fig. 2b, green curve). As derived from the elemental mapping, Pt was present only on the surface of the Au nanoparticles. To model the resulting Pt distribution several approaches were implemented: (1) a uniform Pt shell on Au nanoparticles with a thickness from 0.5 to 1.5 nm, (2) Pt hemispheres with a diameter from 2 to 3 nm on the surface of the Au nanoparticles and (3) a combination of the Pt shell and hemispheres with the same dimensions as described above. In the limit of a uniform Pt shell deposition, a strong blue shift of the resonance is observed with increasing shell thickness (Fig. 2c). With increasing Pt content, the plasmonic resonance peak moves towards the one of a purely Pt particle, in the UV spectral range. This trend is in agreement with previous calculations by Zhang *et al.*[28] On the other side, in the limit of Pt hemispheres formed on the surface of Au nanoparticles, the plasmonic resonance of the overall nanostructure is red shifted, as can be seen in Fig. 2d. This suggests a strong influence on the optical properties by the shape effects. Combining these two cases together, a very good agreement with the position of the experimentally observed extinction peak was obtained, suggesting that, in addition to the thin layer of Pt formed after deposition, small Pt clusters are formed on the surface of the Au particles. The resultant extinction spectrum of the model is presented in Fig. 2b (blue curve), while the derived nanostructure geometry replicating the experimental results is presented in the inset.

3.2 Photocatalytic activity

Photocatalytic experiments were first carried out under visible-light illumination (450–1200 nm) for a period of 4 hours, with an average power of 40 mW. First, a control experiment was performed to determine the extent of degradation of methylene blue by photolysis only. Under the same experimental conditions, photodegradation was studied in the presence of the plasmonic heterostructures SiO_2–Au and SiO_2–Au–Pt, in order to compare the effects of the different catalysts on the degradation of methylene blue. Fig. 3a–c show the time evolution of the absorbance spectra of methylene blue. It can be seen that both in absence of nanoparticles (photolysis) and with the presence of SiO_2–Au the photodegradation was similar over the time period of irradiation, indicating that the catalytic activity of the plasmonic heterostructure, under these conditions, is negligible due to fast electron–hole recombination. In contrast, the degradation rate in the presence of Pt on the catalyst is increased. These results are summarized in Fig. 3d, where the degradation percentages of the processes are presented. Pt plays an important role in determining the catalyst efficiency by providing active catalytic sites, where photogenerated hot carriers interact with adsorbates, and by preventing plasmonically-derived electron–hole recombination in Au, therefore improving catalytic performance of plasmonic nanostructures.

The effect of the Pt concentration in the SiO_2–Au–Pt heterostructures was investigated under 556–566 nm illumination, which overlaps with the localized plasmon resonance of the nanoparticles of four different catalysts: samples Pt, Pt+, Pt++ and Pt+++. As a control experiment, a methylene blue mixture without catalysts was used and resulted in a 3.7% degradation by photolysis only. An

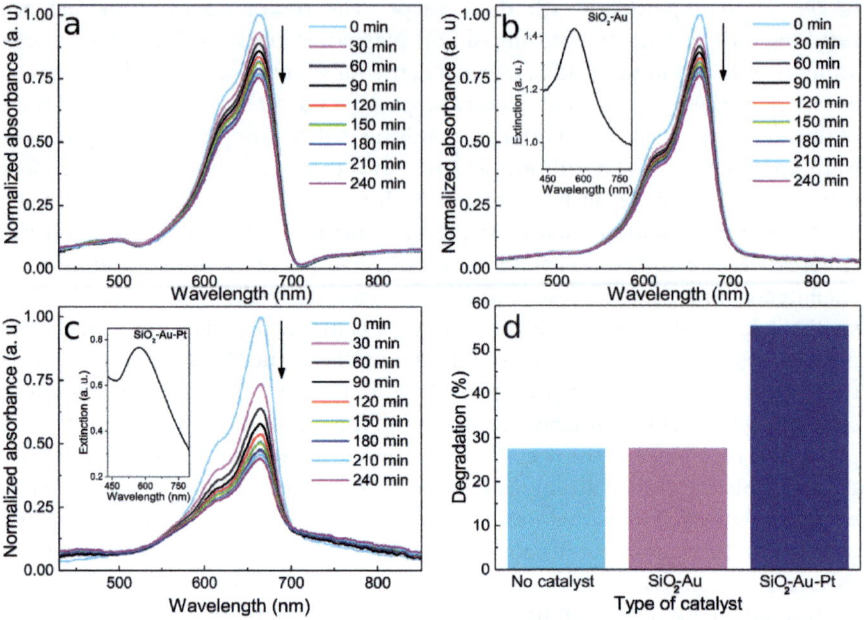

Fig. 3 Time evolution of absorption spectra of methylene blue during photodegradation, under white light irradiation: (a) photolysis, (b) SiO_2–Au and (c) SiO_2–Au–Pt heterostructures. (d) Degradation percentage of methylene blue, after 4 hours, for different types of heterostructures. Insets: extinction spectra of the heterostructures dispersed in water used in (b) and in (c).

increase of the degradation reaction rate was observed with the increase in Pt loading on the nanoparticles (Fig. 4). This results from an increased number of hot carriers extracted through the small Pt clusters, efficiently generated in the Au nanoparticles under resonant excitation. The highest efficiency was reached in the case of the SiO_2–Au–Pt+++ heterostructures, with a 23-fold increase in the percentage of degraded methylene blue compared to the control experiment (Fig. 4). Even though the addition of Pt on the heterostructure is beneficial for the

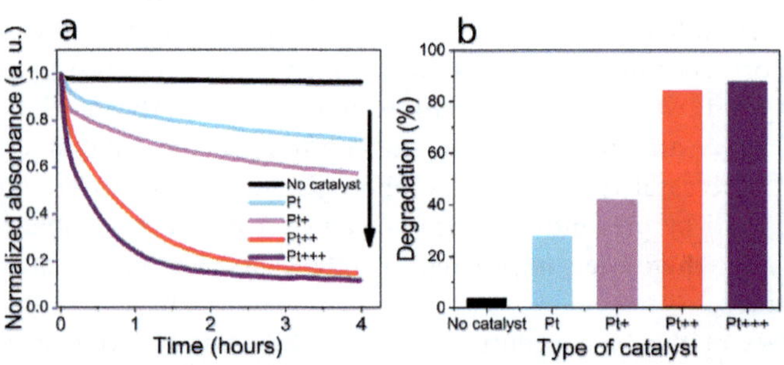

Fig. 4 (a) Time evolution of normalized absorbance and (b) absolute percentage of methylene blue degradation by the SiO_2–Au–Pt heterostructures with different Pt loadings, under 556–566 nm illumination (the amount of Pt loading increases from Pt to Pt+++).

catalytic activity, a limit of performance is expected, as a thicker layer of Pt would hinder hot carriers from reaching the surface and prevent any chemical reaction from taking place with an external adsorbate.

During this kind of photocatalytic process, highly reactive species that chemically interact with organic molecules are produced, such as superoxide anion radicals ($^{\bullet}O_2^-$) and hydroxyl radicals ($^{\bullet}OH$), due to the interaction of hot carriers with the surrounding aqueous medium. In general, the steps involved in the mechanism of photodegradation of methylene blue molecules in aqueous heterogeneous suspensions are related to hot carriers created after photon absorption by the catalyst, which are able to reach the surface and interact with externally placed molecules.[21,29] Once on the surface, excited electrons can ionize oxygen diatomic molecules (O_2) on the catalyst surface, producing superoxide anion radicals. Also, excited holes can neutralize hydroxide ions (OH^-) from water and produce hydroxyl radicals. The efficiency of these processes depends on the lifetime of the hot carriers, which can be controlled *via* the Pt loading of the Au nanoparticles. Lastly, the degradation of methylene blue can occur by chemical interaction with the generated superoxide and hydroxyl radicals.

4 Conclusions

Plasmonic SiO_2–Au–Pt hetero-nanoparticles were synthesized and their hot-electron extraction capabilities for catalytic performance assessed. It was found that the presence of Pt enhances hot-electron extraction efficiency, as compared to bare SiO_2–Au nanoparticles. Even though plasmonic semiconductor-noble metal hybrid nanostructures are under constant study, SiO_2-core based nanoparticles were chosen instead due to the dielectric properties of SiO_2, so that a Schottky barrier between metal and semiconductor is avoided, and thus more hot-carriers are available for reduction processes. The Pt loading has been shown to affect the hot-electron extraction rate, with higher quantities of Pt allowing increased catalytic activity, due to the increased carrier lifetime until highest performance was reached. This saturation of performance is expected, as a thicker layer of Pt on the Au hinders hot-carriers from reaching the surface, thus lowering their interaction rate with an external adsorbate. The comparison between TEM measurements and numerical modelling of the extinction spectra indicates that small Pt clusters are formed in addition to the Pt shells on the Au nanoparticles. This might increase the availability of charge carriers for catalysis at the active sites on the surface of Pt, improving the catalytic activity by facilitating hot-carrier extraction through the interface. Plasmonic hybrid bimetallic nanoparticles show interesting attributes for hot carrier technologies, and heterostructures significantly increase the catalytic activity of hot carriers by controlling their lifetime. The properly designed plasmonic heterostructures allow optimization and exploitation of the full potential of plasmonically derived hot-carriers.

Conflicts of interest

There are no conflicts to declare.

Acknowledgements

This work was supported by EPSRC (UK) under the Reactive Plasmonics Programme grant. J. U. S. G acknowledges a studentship from CONACYT-Mexico. M. R. G acknowledges a studentship from SENER-CONACYT-Mexico. M. R. G. is grateful to M. F. Picardi for helpful discussions of the numerical simulations. All the data supporting this research are provided in full in the results section.

References

1 S. Linic, P. Christopher and D. B. Ingram, *Nat. Mater.*, 2011, **10**, 911–921.
2 W. Hou and S. B. Cronin, *Adv. Funct. Mater.*, 2012, **23**, 1612–1619.
3 Z. Liu, W. Hou, P. Pavaskar, M. Aykol and S. B. Cronin, *Nano Lett.*, 2011, **11**, 1111–1116.
4 P. Wang, A. V. Krasavin, M. E. Nasir, W. Dickson and A. V. Zayats, *Nat. Nanotechnol.*, 2018, **13**, 159–164.
5 J. Lee, S. Mubeen, X. Ji, G. D. Stucky and M. Moskovits, *Nano Lett.*, 2012, **12**, 5014–5019.
6 S. Mubeen, J. Lee, N. Singh and M. Moskovits, *Nat. Nanotechnol.*, 2013, **8**, 247–251.
7 S. Mubeen, J. Lee, D. Liu, G. D. Stucky and M. Moskovits, *Nano Lett.*, 2015, **15**, 2132–2136.
8 B. Wu, D. Liu, S. Mubeen, T. T. Chuong, M. Moskovits and G. D. Stucky, *J. Am. Chem. Soc.*, 2016, **138**, 1114–1117.
9 Z. Zheng, T. Tachikawa and T. Majima, *J. Am. Chem. Soc.*, 2014, **136**, 6870–6873.
10 Z. Lou, M. Fujitsuka and T. Majima, *ACS Nano*, 2016, **10**, 6299–6305.
11 H. Robatjazi, S. M. Bahauddin, C. Doiron and I. Thomann, *Nano Lett.*, 2015, **15**, 6155–6161.
12 S. Mukherjee, F. Libisch, N. Large, O. Neumann, L. V. Brown, J. Cheng, J. B. Lassiter, E. A. Carter, P. Nordlander and N. J. Halas, *Nano Lett.*, 2012, **13**, 240.
13 S. Mukherjee, L. Zhou, A. M. Goodman, N. Large, C. Ayala-Orozco, Y. Zhang, P. Nordlander and N. J. Halas, *J. Am. Chem. Soc.*, 2014, **136**, 64–67.
14 L. Zhou, C. Zhang, M. J. McClain, A. Manjavacas, C. M. Krauter, S. Tian, F. Berg, H. O. Everitt, E. A. Carter, P. Nordlander and N. J. Halas, *Nano Lett.*, 2016, **16**, 1478–1484.
15 M. Pirhashemi and A. Habibi-Yangjeh, *J. Colloid Interface Sci.*, 2017, **491**, 216–229.
16 S. Dong, J. Feng, M. Fan, Y. Pi, L. Hu, X. Han, M. Liu, J. Sun and J. Sun, *RSC Adv.*, 2015, **5**, 14610–14630.
17 R. A. Rather, S. Singh and B. Pal, *J. Nanosci. Nanotechnol.*, 2017, **17**, 1210–1216.
18 B. E. Brinson, J. B. Lassiter, C. S. Levin, R. Bardhan, N. Mirin and N. J. Halas, *Langmuir*, 2008, **24**, 14166–14171.
19 M. J. García-Soto and O. González-Ortega, *Gold Bull.*, 2016, **49**, 111–131.
20 R. Bardhan, S. Lal, A. Joshi and N. J. Halas, *Acc. Chem. Res.*, 2011, **44**, 936–946.
21 D. Kumar, S. B. Lee, C. H. Park and C. S. Kim, *ACS Appl. Mater. Interfaces*, 2018, **10**, 389–399.
22 D. G. Duff, A. Baiker and P. P. Edwards, *Langmuir*, 1993, **9**, 2301–2309.
23 W. Stöber, A. Fink and E. Bohn, *J. Colloid Interface Sci.*, 1968, **26**, 62–69.

24 P. Wang, A. V. Krasavin, F. N. Viscomi, A. M. Adawi, J.-S. G. Bouillard, L. Zhang, D. J. Roth, L. M. Tong and A. V. Zayats, *Laser Photonics Rev.*, 2018, **12**, 201800179.

25 P. B. Johnson and R. W. Christy, *Phys. Rev. B*, 1972, **6**, 4370–4379.

26 W. S. M. Werner, K. Glantschnig and C. Ambroxsch-Draxl, *J. Phys. Chem. Ref. Data*, 2009, **38**, 1013–1092.

27 C. M. Cobley, J. Chen, E. C. Cho, L. V. Wang and Y. Xia, *Chem. Soc. Rev.*, 2011, **40**, 44–56.

28 C. Zhang, B. Q. Chen and Z. Y. Li, *J. Phys. Chem.*, 2015, **119**, 16836–16845.

29 A. Houas, H. Lachheb, M. Ksibi, E. Elaloui, C. Guillard and J. M. Herrmann, *Appl. Catal., B*, 2001, **31**, 145–157.

Faraday Discussions

PAPER

Photocatalytic ammonia production enhanced by a plasmonic near-field and hot electrons originating from aluminium nanostructures†

Madasamy Thangamuthu, * Christian Santschi and Olivier J. F. Martin

Received 4th October 2018, Accepted 20th November 2018
DOI: 10.1039/c8fd00146d

Ammonia production at room temperature and atmospheric pressure is in high demand to assist in energy saving and the protection of the environment worldwide, as well as to help reduce CO_2 emissions. Recently, plasmonic nanomaterials have been frequently used for solar to chemical energy conversion, which has the potential to replace existing energy-intensive industrial processes. In our approach, plasmonic aluminium nanotriangles (AlNTs) were used to investigate the impact of plasmonic effects on photocatalytic ammonia production. Plasmonic near-field coupling to a semiconductor and hot electron generation from AlNTs were studied in detail through the use of electrochemical photocurrent measurements. A narrowband LED beam with a central wavelength at 365 nm was used to illuminate the AlNTs and their hot electron generation efficiency was estimated to be $2 \times 10^{-4}\%$, resulting in an ammonia production rate of 4×10^{-5} μM h^{-1} mW^{-1} cm^{-2}, which corresponds to a quantum efficiency of $2.5 \times 10^{-5}\%$. In the case of plasmonic near-field coupling, AlNTs-embedded TiO_2 demonstrates a charge-carrier generation efficiency of 2.7%, which is ~2.3 times higher than that of bare TiO_2. The ammonia production rate of AlNTs–TiO_2 is 0.1 μM h^{-1} mW^{-1} cm^{-2} with a quantum efficiency of ~0.06%, which corresponds to ~2.4 times that of the rate demonstrated by bare TiO_2 (0.04 μM h^{-1} mW^{-1} cm^{-2}, quantum efficiency ~ 0.025%). The obtained results confirm successful ammonia production through nitrogen splitting at room temperature and under atmospheric pressure. Moreover, according to the presented results, the use of plasmonic aluminium structures remarkably improves the ammonia production rate.

1. Introduction

After photosynthesis, nitrogen fixation is the second most important chemical process in the biosphere, as nitrogen is an essential component of the building

Nanophotonics and Metrology Laboratory (NAM), Swiss Federal Institute of Technology Lausanne (EPFL), CH-1015 Lausanne, Switzerland. E-mail: madasamy.thangamuthu@epfl.ch

† Electronic supplementary information (ESI) available. See DOI: 10.1039/c8fd00146d

blocks of life, *viz.* DNA, RNA and proteins. Unfortunately, direct appropriation from its abundance in the atmosphere (78%) is impossible since the robust triple bond between two nitrogen atoms makes molecular nitrogen relatively inert. The rhizobia bacteria of the legume family and which live in nodules on plant roots possess a nitrogenase enzyme which fixes nitrogen naturally and supplies it to plants and animals in the form of ammonium, urea and nitrate.[1] This naturally occurring process is, however, not sufficient to feed the rapidly expanding population of Earth. For decades the agro-food industry has relied on nitrate fertilizers, produced via the Haber–Bosch industrial synthesis of ammonia (NH_3), to boost food production. In principle, NH_3 can be produced in the presence of catalysts at high temperature (400–600 °C) and pressure (150–200 atm), contributing to approximately 2% of the world's annual energy usage.[2] Furthermore, industrial NH_3 production consumes approximately 5% of the world's natural gas resources to produce a hydrogen feedstock, and releases large quantities of carbon dioxide (CO_2) which strongly pollutes the atmosphere, thus leading to serious environmental issues, *viz.* global warming and climate change *etc.*[3] Therefore, there is strong demand for the development of alternative, fossil-fuel-saving methods that can be utilized to produce ammonia under ambient conditions in order to save energy and the environment.

Numerous researchers have attempted to identify an alternative method to the Haber–Bosch process *via* the use of an electrochemical approach.[4] Recently, Licht *et al.* introduced an electrolysis method in which a sodium and potassium hydroxide solution mixture is bubbled in wet air through a mixture of tiny particles of iron oxide (Fe_2O_3).[5] This method consumes two-thirds of the energy consumed by the Haber–Bosch process and, even more appealingly, CO_2 emissions are significantly reduced in comparison. Although electrochemical routes reduce energy input, they are still under development due to numerous technical challenges and material issues. In another approach, inorganic metal complexes were studied to mimic nitrogenase activity for nitrogen adsorption and reduction.[6,7] However, this route is limited because of the complicated synthesis of the materials and their poor stability. Semiconductor photocatalysis for ammonia production[8–10] has also been highly studied for a few decades after the Fujishima and Honda water splitting experiments were reported.[11] To improve the quantum efficiency and reduce charge-carrier recombination, Khan *et al.* and Ranjit *et al.* studied transition-metal-loaded semiconductors.[12–15] Unfortunately, the efficiency was not significantly improved.

Recently, plasmonic nanomaterials (PNs), a new family of catalysts exhibiting fundamentally different behaviour, have been shown to be highly encouraging due to their optical and catalytic properties under visible light illumination at low intensity. The photocatalytic efficiency can be greatly enhanced by combining PNs with a semiconductor, which increases charge-carrier generation in the semiconductor through plasmonic near-field coupling.[16] Moreover, PNs can solely generate charge-carriers (electron–hole pairs) through non-radiative surface plasmon decay.[17] These so-called plasmonic hot electron/hot hole pairs (EHP) have the potential to drive chemical reactions either on the PNs or on the semiconductor surface, where plasmonic hot electrons can be extracted and transferred to the conduction band of the semiconductor. A Schottky barrier at the metal/semiconductor junction prolongates the residence time of the electrons in the conduction band and, therefore, reduces the fast recombination rate of the short living EHP.[18] So far, such a plasmonic approach has been successfully applied to improve water splitting,[19] hydrogen dissociation,[20,21] the decomposition of organic pollutants such as carbon dioxide,[22] carbon monoxide[23] and dyes,[24] and the synthesis of organic compounds.[25,26] Recently, Oshikiri *et al.* studied, using light in the visible range, hot electron generation from Au nanoparticles combined with a ruthenium co-catalyst for ammonia synthesis, and encountered a poor yield.[27] In a follow-up study they used zirconium/zirconium oxide, an efficient nitrogen adsorbing material that improved ammonia production significantly.[28] However, Au and Ag are not seminal materials since they are very expensive and unstable over long periods of time, therefore making them inappropriate for long-term use and, consequently, commercialization. Furthermore, the thermal stabilities of Au and Ag are inferior to the stability of Al, therefore rendering Au and Ag unsuitable for high temperature annealing.

In our work we used aluminium as a plasmonic material. Aluminium is a cheap material that exhibits plasmonic properties in the UV range, and is, in comparison with other plasmonic metals, more abundant on earth, highly stable and non-toxic. To the best of our knowledge, we are the first group to study the use of aluminium nanotriangles (AlNTs) for enhancing ammonia production through nitrogen splitting at room temperature and under atmospheric pressure. In order to gain more insight into the plasmonic mechanisms, two different types of photoelectrodes were fabricated to study: (i) plasmonic hot electron generation from AlNTs and (ii) plasmonic near-field coupling to semiconductor materials, such as TiO_2. The charge-carrier generation rate of the photoelectrodes under 365 nm illumination was studied *via* electrochemical photocurrent measurements. An incident photon to photocurrent generation efficiency (IPCE) of $2 \times 10^{-4}\%$ was observed for the hot electron generation by non-radiative decay of surface plasmons, resulting in an ammonia production rate of 4×10^{-5} μM h^{-1} mW^{-1} cm^{-2}, which corresponds to a quantum efficiency of $2.5 \times 10^{-5}\%$. However, after applying plasmonic near-field coupling, the IPCE increases to 2.7%, which is ~2.3 times higher than that observed when using a bare TiO_2 configuration. The measured ammonia production rate of AlNTs–TiO_2 amounted to 0.1 μM h^{-1} mW^{-1} cm^{-2} with a quantum efficiency of ~0.06%, which is ~2.4 times higher than that obtained with bare TiO_2 (0.04 μM h^{-1}, quantum efficiency ~ 0.025%).

2. Experimental

2.1. Preparation of the photoelectrodes

Plasmonic aluminium nanotriangles (AlNTs) with a side length of 85 nm were fabricated on fused silica substrates *via* colloidal lithography (see ESI† for the detailed fabrication procedure). Due to the high reactivity of Al with oxygen, a self-limiting oxide layer formed on the Al/air interface. Al was deposited *via* electron beam (e-beam) evaporation with a base pressure between 10^{-6} and 10^{-7} Torr (Leybold Optics LAB600H). The thickness and deposition rate of the Al were 40 nm and 0.5 Å s^{-1}, respectively.[29] Subsequently, the PS mask was removed by sonication in toluene for 10 min, followed by the deposition of a 100 nm thick titanium dioxide (TiO_2) thin film onto the AlNTs *via* the e-beam evaporation technique with a deposition rate of 1 Å s^{-1} prior to an annealing step at 500 °C for

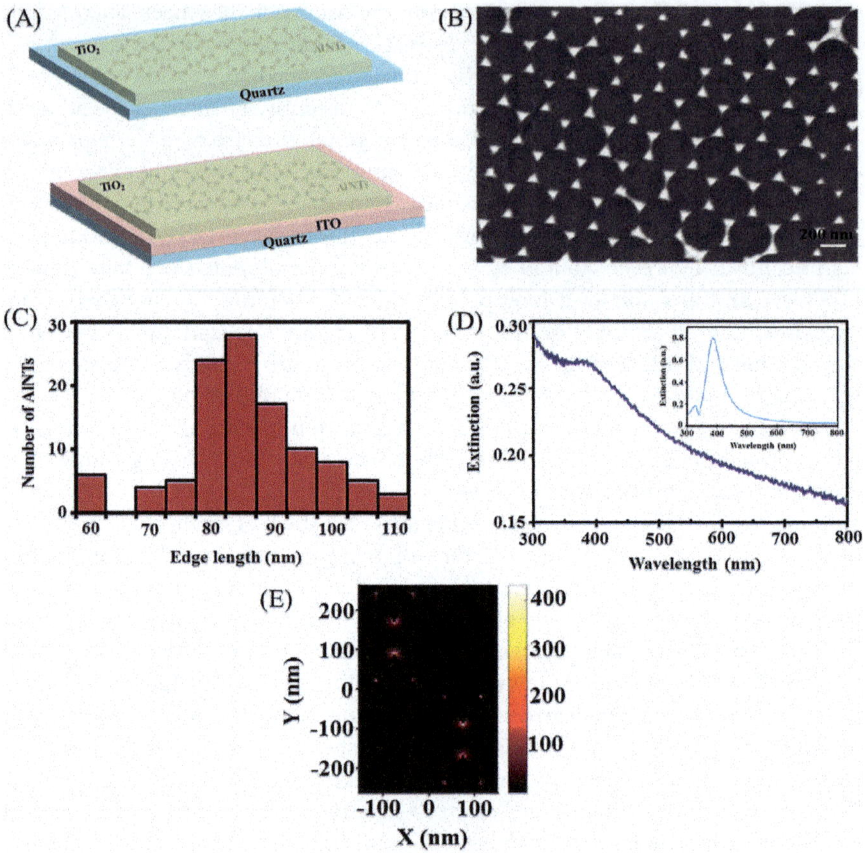

Fig. 1 Characterization of the photoelectrodes. (A) Schematics of the photoelectrodes. (B) SEM image of the AlNTs. (C) Particle size distribution of the AlNTs. (D) UV-Vis-NIR extinction spectrum and the simulated optical spectrum of the AlNTs (inset). (E) Simulated spatial distributions of the electric field enhancement for the AlNTs.

2 h under a nitrogen atmosphere. Fig. 1A shows schematics of the photoelectrodes.

2.2. Characterization

A UV-Vis-NIR spectrometer (UV-2600) with an integrating sphere was used for the extinction, absorption and scattering measurements. The electrical field distribution around the excited AlNTs was computed using the surface integral equation (SIE) method.[30] The real dimension, obtained from the scanning electron microscope (SEM) images, was introduced to calculate the optical responses of the AlNTs. All dielectric coefficients were taken from the literature.[31–33] Morphological SEM images were obtained using a Zeiss Merlin field emission scanning electron microscope (Carl Zeiss Jena GmbH) with an electron energy of 2 keV. X-ray diffraction (XRD) characterization was carried out using an Empyrean diffractometer from Panalytical equipped with a grazing incidence angle configuration and a Cu Kα radiation source ($\lambda = 0.15418$ nm) in order to study the crystalline structure of the TiO_2. The bandgap energy of the fabricated TiO_2 was calculated using a Tauc-plot. The valence band spectrum of the TiO_2 was obtained using a PHI versaprobe II scanning XPS microscope (Physical Instruments AG, Germany). The analysis was carried out using a monochromatic Al Kα X-ray source of 24.8 W with a beam size of 100 μm, and the binding energy was calibrated using a value for C 1s of 187.85 eV.

2.3. Electrochemical photocurrent study

To perform the electrochemical studies, an electrical conducting layer was added between the substrate and the AlNTs or TiO_2. A 150 nm ITO thin film was deposited *via* radio frequency magnetron sputtering (RF-MS, Pfeiffer SPIDER 600) with an RF power of 500 W and an argon flow of 15 sccm at 5×10^{-3} mbar. The AlNTs embedded in TiO_2 photoelectrodes on ITO were fabricated as described above. Electrochemical measurements were performed using a CHI 1240B potentiostat (CH Instruments, Austin, USA). A three electrode photoelectrochemical cell (Zahner instruments) consisting of a photoanode (working electrode), a Ag/AgCl reference electrode and a platinum counter electrode was used to record the photocurrent. A 0.1 M KNO_3 (pH ~ 6) solution containing 10% ethanol as an electron donor was used as the electrolyte. The working electrode potential was set to 0.15 V *vs.* Ag/AgCl for all of the experiments.

2.4. Photocatalysis experiments

A photochemical cell was used as the reaction chamber. A collimated narrowband LED beam with its central wavelength at 365 nm was used as the irradiation source. The fabricated photoelectrodes were illuminated with an intensity of 5 mW cm^{-2} on a circular area with a diameter of 1.5 cm. Unless otherwise stated, the illumination conditions were the same for all measurements. A nitrogen-saturated aqueous electrolyte solution (pH 4) containing 10% ethanol, an electron donor/hole scavenger, was used for the reaction. The solution contained ~500 μM (saturated concentration) of nitrogen, corresponding to 3×10^{16} nitrogen molecules in the reaction chamber. This concentration was held constant throughout the whole experiment by using a 100 sccm nitrogen flow. The nitrogen gas flow rates were controlled using a mass flow controller (Omega, GmbH). The rate of ammonia production was evaluated by a standard colorimetric assay using sodium salicylate, sodium hypochlorite, sodium nitroprusside and sodium hydroxide. A 500 μL sample was taken from the reaction chamber at 1 h intervals. Subsequently, 250 μL of hypochlorite solution followed by 250 μL of salicylate reagent (a mixture of sodium salicylate, nitroprusside and sodium hydroxide) were added. Sodium salicylate forms a dimer (indosalicylate) in the presence of ammonia, with an absorption peak at 650 nm. Known amounts of ammonium chloride were used to perform a calibration. The value of the ammonia concentration at $t = 0$ h was used as the baseline and subtracted from all measurements. The quantum efficiency of ammonia formation for which three photons are required to produce one ammonia molecule was determined using the following eqn (1)[34]

$$\eta_{NH_3}\ (\%) = \frac{[\text{Rate of } NH_3 \text{ formation (mol s}^{-1})] \times 3}{[\text{Incident photon flux (mol s}^{-1})]} \times 100 \qquad (1)$$

3. Results and discussion

3.1. Photoelectrodes characterization

An SEM image of the AlNTs is shown in Fig. 1B. Using ImageJ, the size distribution of the AlNTs was statistically evaluated, as shown in Fig. 1C which reveals that the average edge length of the AlNTs is approximately 85 nm, leading to a surface coverage of ∼5%. The UV-Vis-NIR extinction spectrum shows a dipolar plasmon peak at around 365 nm (Fig. 1D). The position of the plasmonic resonance wavelength matches well with the simulated spectrum (inset of Fig. 1D).

Furthermore, the SIE near-field simulations show that the highest plasmonic electric field intensities at the resonance position are generated at the edges of the AlNTs (Fig. 1E). The crystalline nature of the TiO_2 was studied before and after annealing using XRD. Fig. 2A displays the XRD pattern for TiO_2 on ITO before annealing. The crystalline peak for anatase TiO_2 is completely absent, indicating that the as-deposited TiO_2 is amorphous. The peaks observed at the 2θ angles of 30.4, 33, 35.34 and 37.65° correspond to the (222), (312), (004) and (114) polycrystalline phases, respectively, of the ITO (JCPDS: 96-101-0342) underneath. However, the XRD pattern of the annealed sample (Fig. 2B) shows an obvious peak at 25.3°, thus confirming the anatase phase (101) (JCPDS: 00-021-1272). A reorganization of the TiO_2 morphology is also evident from analysis of the SEM images (insets of Fig. 2A and B), in which a clear change in the crystal size can be observed. The optical band gaps (E_g) of TiO_2 on the fused silica and ITO substrates were calculated using Tauc plots, as shown in Fig. 2C. The quantity $(\alpha h\nu)^{1/2}$ was plotted as a function of the incident energy and, subsequently, the linear part was extrapolated (see ESI† for details). The resulting point of intersection with the abscissa at 3.18 eV corresponds to the bandgap energy of the deposited TiO_2. The measured bandgap value is in good agreement with the literature.[35] The valence band (VB) spectrum of the TiO_2 was determined *via* XPS analysis (Fig. 2D) in order to obtain the potential at the upper edge of the VB.[35] The VB spectrum shows that the valence band is located at 2.7 eV below the Fermi level, and hence the lower edge of the CB can be expected to be located at 0.48 eV above the Fermi level.

The lower conduction band edge was experimentally determined using photocurrent onset measurements,[36] as shown in Fig. 2E. Under illumination by light at 365 nm (an energy higher than the band gap), TiO_2 generates EHP, which are separated in the potential gradient space-charge layer (formed upon immersing the electrode in the electrolyte). This drives holes towards the TiO_2/electrolyte interface and electrons towards the bulk of the TiO_2 as well as to an external electrical connection (Fig. 2F). Consequently, an anodic photocurrent is observed in the TiO_2 photoelectrode which is affected by an external bias. The potential of the electrode swept in the cathodic direction (applying a more negative potential) reduces the space-charge layer thickness, thus resulting in more charge-carrier recombination and leading to a decreased photocurrent. The photocurrent becomes zero at the flatband potential. In Fig. 2E, the photocurrent reaches zero at −0.1 V with respect to Ag/AgCl, which is equal to −0.335 V *vs.* a standard hydrogen electrode (SHE). For TiO_2, recombination is practically complete 0.2 V away from the CB edge, therefore this value is added to the flatband potential in order to calculate the final CB edge value.[37] Hence, the CB edge of the TiO_2 should be located at −0.54 V above the Fermi level, which is very close

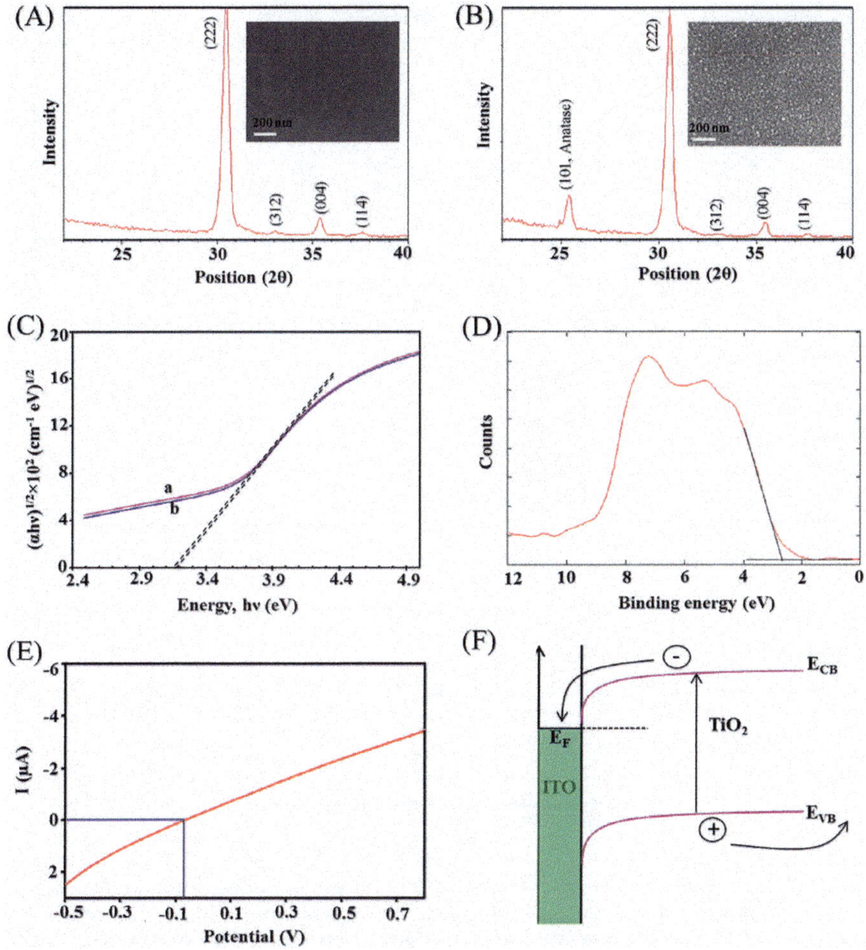

Fig. 2 XRD patterns of the TiO_2 (A) before annealing and (B) after annealing at 500 °C for 2 h under a nitrogen atmosphere (the insets show SEM images of the TiO_2 before and after annealing). (C) Tauc-plot of $(\alpha h\nu)^{1/2}$ as a function of the incident photon energy ($h\nu$) for TiO_2 on the fused silica (a) and ITO (b) substrates. (D) Valence band photoelectron spectrum of TiO_2 (anatase) after annealing. (E) Photovoltammogram of the ITO–TiO_2 electrode recorded in a KNO_3 (0.1 M; pH 6) electrolyte in the cathodic direction with a sweep rate of 2 mV s^{-1} under 365 nm irradiation. (F) Schematic showing the TiO_2 energy levels under illumination.

to the CB edge value deduced from the Tauc-plot. The band diagram of the TiO_2 is depicted in Fig. 3A.

3.2. Electrochemical study of the plasmonic effects of the AlNTs

The plasmonic effects of the AlNTs were studied by means of electrochemical photocurrent measurements. The number of charge-carriers generated in the photoelectrodes is estimated to correspond to a direct measure of their photocatalytic activity.[38] In the first step, we studied plasmonic hot electron generation from the AlNTs. Fig. 3B shows the photocurrent responses of the ITO–AlNTs and

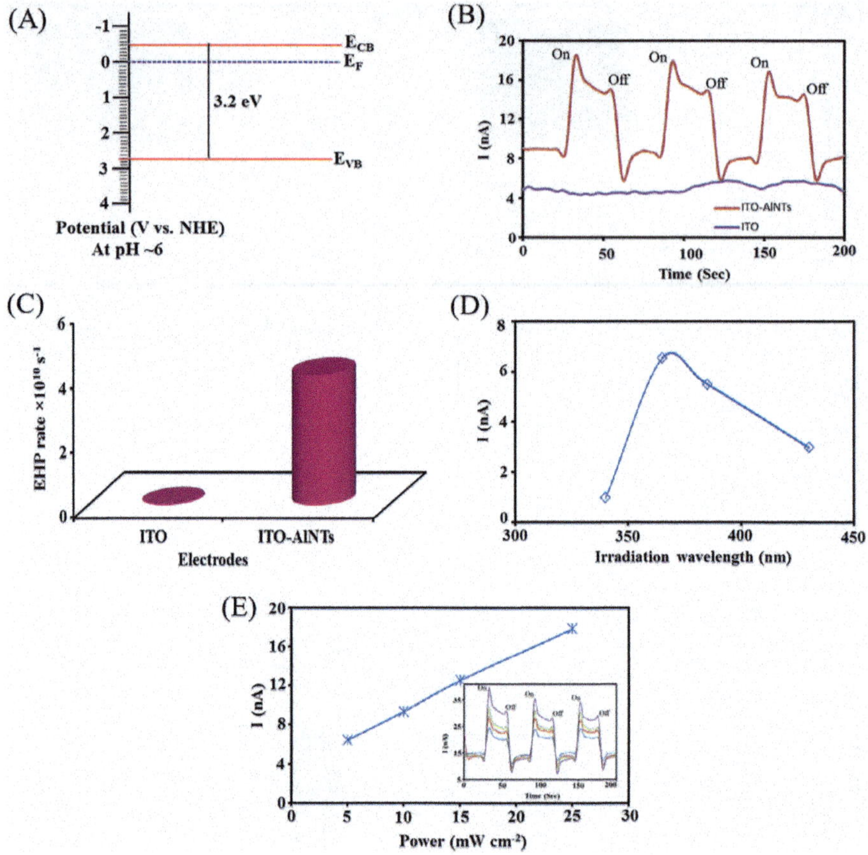

Fig. 3 (A) Energy band diagram of anatase TiO_2 on the electrochemical potential scale; NHE refers to the normal hydrogen electrode. (B) Photocurrent of bare ITO and ITO–AlNTs in a 0.1 M KNO_3 (pH ~ 6) solution containing 10% ethanol, recorded under narrowband irradiation at $\lambda = 365$ nm (5 mW cm^{-2}). (C) Charge-carrier generation rate of the photoelectrodes with and without AlNTs. (D) Wavelength dependent photocurrent of the ITO–AlNTs photoelectrode observed using 340, 365, 385 and 430 nm narrowband light sources at 5 mW cm^{-2} power. (E) Observed photocurrent as a function of light intensity. The inset shows the photocurrent responses of the ITO–AlNTs electrode at 365 nm under different light intensities (from bottom to top: 5, 10, 15 and 25 mW cm^{-2}). The electrodes were biased with 0.15 V for all of the photocurrent measurements.

bare ITO samples. The 365 nm light source was switched on and off every 30 s whilst the photocurrent response was recorded. The obtained currents were stable and reproducible. The ITO–AlNTs sample shows significant photocurrent changes of $\Delta I = 6.5$ nA under illumination, which corresponds to a generation rate of 4×10^{10} hot electrons/holes per second (Fig. 3C) and an incident photon to photocurrent conversion efficiency (IPCE), *i.e.* external quantum efficiency, of 2×10^{-4}%. This result clearly suggests that the AlNTs generate hot electrons as a result of non-radiative surface plasmon decay. The bare ITO does not demonstrate a significant change in current, thus indicating that the illumination wavelength of 365 nm does not excite the ITO. Furthermore the bandgap of ITO is large, around 4 eV, meaning that it certainly requires a higher energy (300 nm wavelength) to generate charge-carriers; this confirms that the observed 6.5 nA photocurrent is only due to the plasmonic hot electrons from the AlNTs. The wavelength dependent hot electron generation efficiency was also studied using different illumination wavelengths, as shown in Fig. 3D. The maximum current was observed at 365 nm, thus providing further evidence of hot electron generation. To gain more insight into the plasmonic response of the AlNTs, the photocurrent responses at different light intensities were studied. It is clear from looking at Fig. 3E that the AlNTs are linearly responding to the light intensity, which is a characteristic of plasmonic nanomaterials.[39]

In the second step, plasmonic near-field coupling was studied by connecting the AlNTs to TiO_2. Fig. 4A shows the photocurrent responses of the ITO–TiO_2 samples with and without AlNTs under dark and illuminated conditions. For the bare TiO_2 electrode, a photocurrent of $I = 33.2$ μA was observed, corresponding to a generation rate of 2.1×10^{14} EHP per second (Fig. 4B) and an IPCE of 1.1%. The photocurrent response increased to $I = 78$ μA for the sample with AlNTs, corresponding to a generation rate of 4.9×10^{14} EHP per second and an IPCE of 2.7%. Overall, a ~2.3 fold enhancement was observed after embedding the AlNTs into TiO_2, which is also manifested in the increase in the absorption spectrum, as shown in Fig. 4C. The wavelength dependent photocurrent enhancement of the ITO–AlNTs–TiO_2 electrode was determined and is shown in Fig. 4D, which reveals that the enhancement strongly depends on the illumination wavelength, which qualitatively follows the absorption spectrum of the AlNTs (inset of Fig. 4D). The peak position of the photocurrent enhancement and the AlNTs absorption matches very well, thus suggesting that the electromagnetic field located at the edges of the triangles, generated by the dipolar surface plasmon oscillation of the AlNTs, penetrates into the TiO_2, leading to an increased charge-carrier generation rate.

It is worth noting that the AlNTs were in direct contact with TiO_2. The Al and TiO_2 contact exhibits ohmic character since the working function of Al is inferior to that of TiO_2. Therefore, plasmonic hot electron and near-field coupling effects are indistinguishable in the studied configuration. However, the observed results suggest that the hot electron generation and extraction efficiency is very low (2×10^{-4}%) in comparison with the charge-carrier generation efficiency of the near-field coupling (2.7%), thus demonstrating that the contribution from the hot electrons is negligible towards the overall enhancement in the AlNTs–TiO_2 configuration. Furthermore, the contribution from plasmonic scattering was also studied by extracting the scattering and absorption spectra from the total extinction spectrum and carrying out measurements using an integrating sphere. Fig. 4E and F show the absorption and scattering of the AlNTs before and after combining with TiO_2. It can clearly be seen that the scattering cross section is dominant for the AlNTs prior to TiO_2 deposition; however, the absorption becomes dominant after embedding the AlNTs in TiO_2. It seems that plasmonic near-field coupling is the dominating energy transfer channel rather than scattering, which correlates well with a recent report.[40] Furthermore, we studied the thermal contribution during the photocurrent measurements at 365 nm by additionally irradiating the photoelectrodes simultaneously with a $\lambda = 650$ nm narrowband light source where aluminium absorbs light. The observed photocurrent remains stable and looks similar to Fig. 4A, thus revealing the negligible thermal contribution to the enhanced charge-carrier generation rate. The above

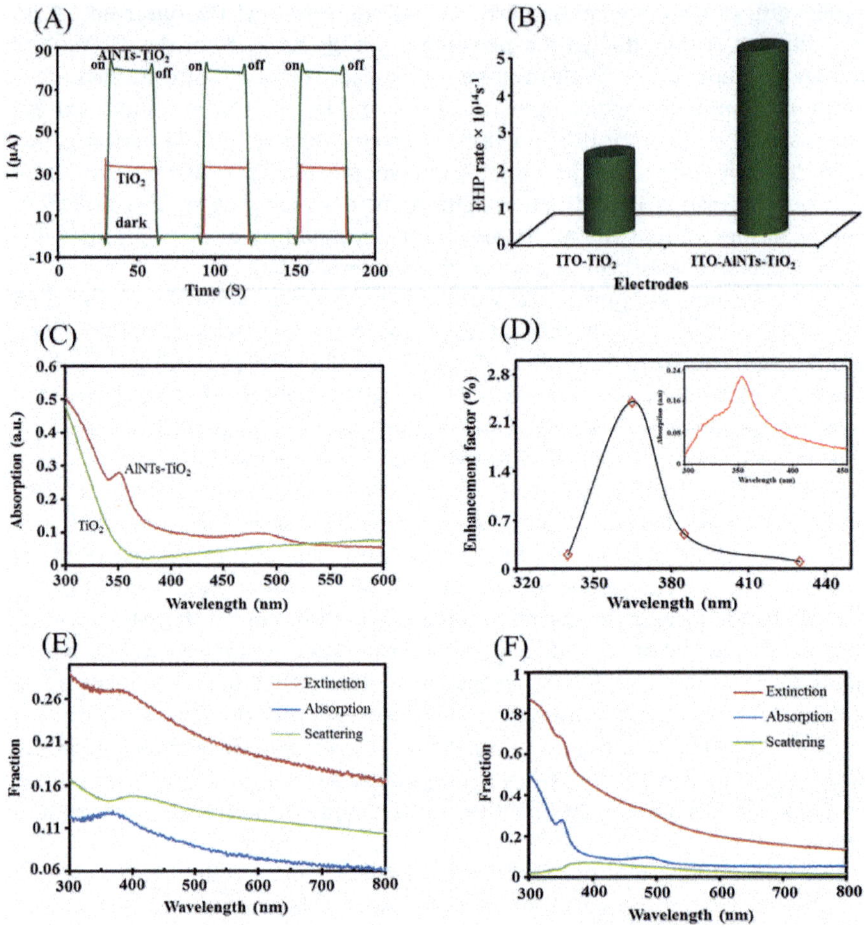

Fig. 4 (A) Photocurrent of ITO–TiO_2 and ITO–AlNTs–TiO_2 in a 0.1 M KNO_3 (pH ~ 6) solution containing 10% ethanol, recorded under narrowband irradiation at $\lambda = 365$ nm (5 mW cm^{-2}) and the dark. (B) Charge-carrier generation rate of the TiO_2 electrodes without and with AlNTs. (C) Absorption spectra of TiO_2 and AlNTs–TiO_2. (D) Wavelength dependent photocurrent enhancement of the ITO–AlNTs–TiO_2 photoelectrode observed when using 340, 365, 385 and 430 nm narrowband light sources; the inset shows the absorption spectrum of the AlNTs extracted from 4C. (E) Absorption and scattering fraction of the AlNTs. (F) Absorption and scattering fraction of AlNTs–TiO_2.

characterizations strongly indicate a predominant contribution of the plasmonic near-field coupling to the enhancement of the charge carrier generation rate.

The stabilities of the fabricated electrodes were evaluated by measuring the photocurrent responses every month over a time period of 6 months. Fig. 5A shows the photocurrent responses of ITO–AlNTs exposed to the conditions used for the hot electron measurements described above. It can clearly be seen that the current response does not change significantly over the first month and decreases subsequently, thus indicating that the hot electron generation efficiency of the AlNTs is reduced with time. It is worth noting that even after 6 months the AlNTs demonstrate 80% of their initial efficiency, thus signifying that Al is more stable

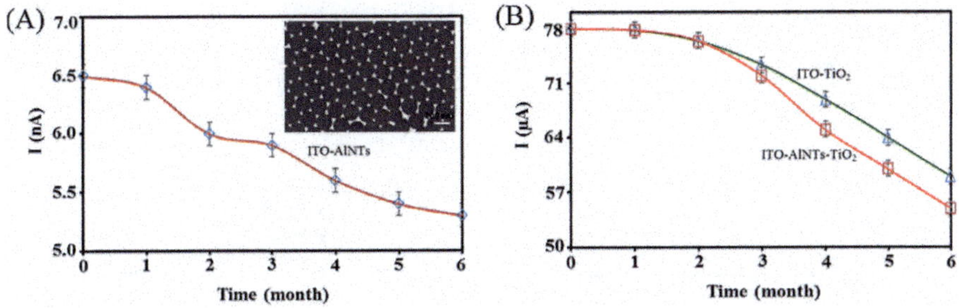

Fig. 5 Photocurrent responses of the (A) ITO–AlNTs and (B) ITO–TiO_2 and ITO–AlNTs–TiO_2 electrodes in a 0.1 M KNO_3 (pH ~ 6) solution containing 10% ethanol, recorded every month under narrowband irradiation at λ = 365 nm (5 mW cm^{-2}). The inset in Fig. 5A shows an SEM image of the AlNTs recorded 6 months after fabrication.

than Au and Ag. This is due to the fact that the self-limiting oxide (2–3 nm Al_2O_3) protects the structures from further oxidation. Furthermore, Al is insusceptible to surface diffusion, hence the shape of the structures remains, in contrast to Au and Ag,[41] conserved. The inset of Fig. 5A shows an SEM image of the AlNTs taken 6 months after fabrication. Fig. 5B shows a stability study of the ITO–TiO_2 and ITO–AlNTs–TiO_2 electrodes. The photocurrent responses of the TiO_2 and AlNTs–TiO_2 electrodes are reduced to 75% and 70%, respectively, of the initial values.

3.3. Photocatalytic ammonia production

Plasmonic hot electron-driven nitrogen splitting was investigated and the results are shown in Fig. 6A. Under dark conditions, the control experiments revealed no significant ammonia production for the fused silica (FS) without AlNTs. Under illumination, the AlNTs deposited on the fused silica substrates catalyse the production of ammonia at a rate of 4×10^{-5} µM h^{-1} mW^{-1} cm^{-2}, which corresponds to a quantum efficiency of 2.5×10^{-5}%. This clearly indicates that nitrogen splitting takes place at room temperature under atmospheric pressure. The observed ammonia production efficiency is comparable with the hot electron generation efficiency obtained from the electrochemical photocurrent measurements. Fig. 6B shows a schematic of the hot electron-driven charge injection into the nitrogen molecule. Hot electrons are generated from AlNTs as a result of non-radiative surface plasmon decay *via* Landau damping,[42] and are distributed randomly with electron energies at 0 eV to the excitation energy (365 nm) above the Fermi level. Subsequently, hot electrons possessing a suitable energy are transferred into the lowest unoccupied molecular orbital (LUMO) of the adsorbed N_2 molecule and break the N_2 triple bond. The bond breaking mechanism is still not known. The hot electron generation and ammonia production efficiencies are modest, due to the very short lifetime of the hot electrons in the femtosecond range. The hot electrons rapidly lose their energy through electron–electron scattering followed by electron–phonon interactions.[42] Furthermore, the absorption is less dominating in comparison with the scattering for 85 nm edge-length AlNTs, as inferred from Fig. 4E, thus suggesting that surface plasmon decay occurs mainly through a radiative process.

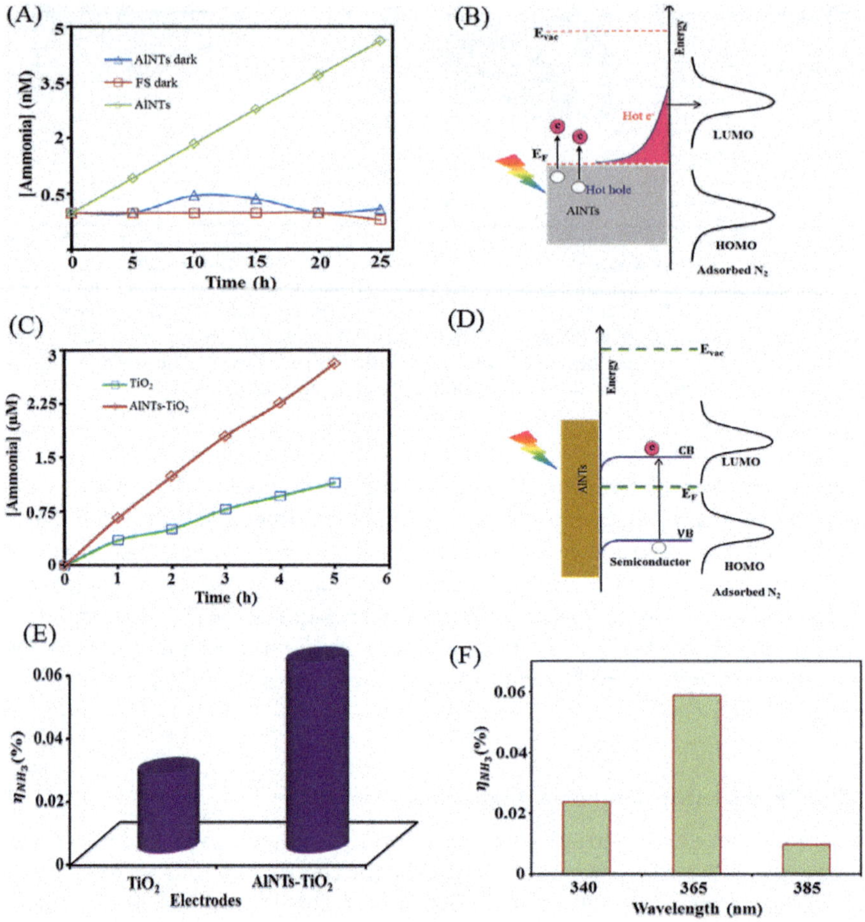

Fig. 6 (A) Plasmonic hot electron-driven ammonia production in a nitrogen-saturated solution (pH $\sim$ 4) containing 10% ethanol under dark conditions and illumination by a narrowband 365 nm beam with a 5 mW cm^{-2} power. (B) Schematic of hot electron-driven charge injection. (C) Plasmonic near-field coupling enhanced ammonia production in a nitrogen saturated solution (pH $\sim$ 4) containing 10% ethanol under illumination by a narrowband 365 nm beam with a 5 mW cm^{-2} power. (D) Schematic of plasmonic near-field coupling enhanced charge injection. (E) Apparent quantum efficiencies of the TiO_2 and AlNTs–TiO_2 photoelectrodes. (F) Wavelength-dependent quantum efficiencies of the AlNTs–TiO_2 photoelectrodes.

In a further step, the effect of near-field coupling on ammonia production was investigated. Fig. 6C shows the ammonia production on the bare TiO_2 and AlNTs–TiO_2 photoelectrodes *versus* time.

Under dark conditions, both electrodes did not exhibit significant differences, whilst under illumination significant ammonia production was observed. It can be seen that ammonia production increases linearly with time, with the production rate estimated to be 0.04 μM h^{-1} mW^{-1} cm^{-2} for bare TiO_2 and 0.1 μM h^{-1} mW^{-1} cm^{-2} for AlNTs–TiO_2. For the AlNTs–TiO_2 sample the reaction rate was observed to be ~2.5 times higher than that for TiO_2, which correlates well with the photocurrent measurements. Fig. 6D shows a schematic of the plasmonic near-field coupling from the AlNTs to TiO_2. EHP are short-living states leading to a short mean free path, on the order of 50 nm, for the electrons. Hence, the confined plasmonic near-field with its short penetration depth into TiO_2 favours the generation of EHP in the vicinity of the interface and reduces, therefore, the recombination probability. This shallowness has proved to be beneficial for nitrogen splitting. The apparent quantum efficiencies, η, of ammonia production for the TiO_2 and AlNTs–TiO_2 photoelectrodes were calculated, with the results shown in Fig. 6E. The bare TiO_2 and AlNTs–TiO_2 substrates demonstrate, η of 0.025% and 0.06%, respectively. The observed η is significantly higher than what has been previously reported.[34,43–45] Furthermore, the highest η was observed at the plasmonic resonance wavelength (Fig. 6F), revealing again the plasmonic nature of the process. In a control experiment, ammonia production without N_2 replenishment was investigated. Fig. 7A clearly shows the stagnation of ammonia production after the dissolved N_2 is exhausted.

A nitrogen splitting energy diagram is shown in Fig. 7B. The reduction of nitrogen on TiO_2 is thermodynamically possible since the conduction band potential of TiO_2 is at −0.25 V with respect to the SHE at pH 0,[46] which is higher than the potential (−0.092 V) required for N_2 splitting. In our experiments we used pH ~ 4 to ensure the stability of the plasmonic photoelectrode and to provide enough protons for ammonia formation. Under these conditions the conduction band potential is −0.486 V and the nitrogen splitting potential becomes −0.328 V (according to the Nernst equation, *i.e.* 0.059 × pH). Therefore, the energy balance allows for ammonia production under ambient conditions. The possible oxidation and reduction reactions taking place in the potential window are represented by the following equations:

$$h\nu \xrightarrow{TiO_2} e^- + h^+ \tag{2}$$

$$N_2 + 6H^+ + 6e^- \rightarrow 2NH_3\ (-0.092\ V) \tag{3}$$

$$C_2H_5OH + 2h^+ \rightarrow CH_3CHO + 2H^+\ (0.249\ V) \tag{4}$$

$$2H^+ + 2e^- \rightarrow H_2\ (0\ V) \tag{5}$$

$$4OH^- + 4h^+ \rightarrow 2H_2O + O_2\ (1.23\ V) \tag{6}$$

In addition to nitrogen reduction, the conduction band electrons can reduce protons and evolve molecular hydrogen gas (eqn (5)) since the required potential is well below the conduction band potential of TiO_2. It is well known that TiO_2 is a good photocatalyst not only for nitrogen reduction but also for hydrogen evolution,[47] because both nitrogen and hydrogen adsorb well on the TiO_2 surface. On the other hand, the holes generated in the valence band oxidize hydroxyl radicals leading to the production of water and the evolution of oxygen gas (eqn (6)). To compensate for the electrons used in nitrogen splitting, the electron donor ethanol refills the redundant holes in the VB of TiO_2 (eqn (4)).

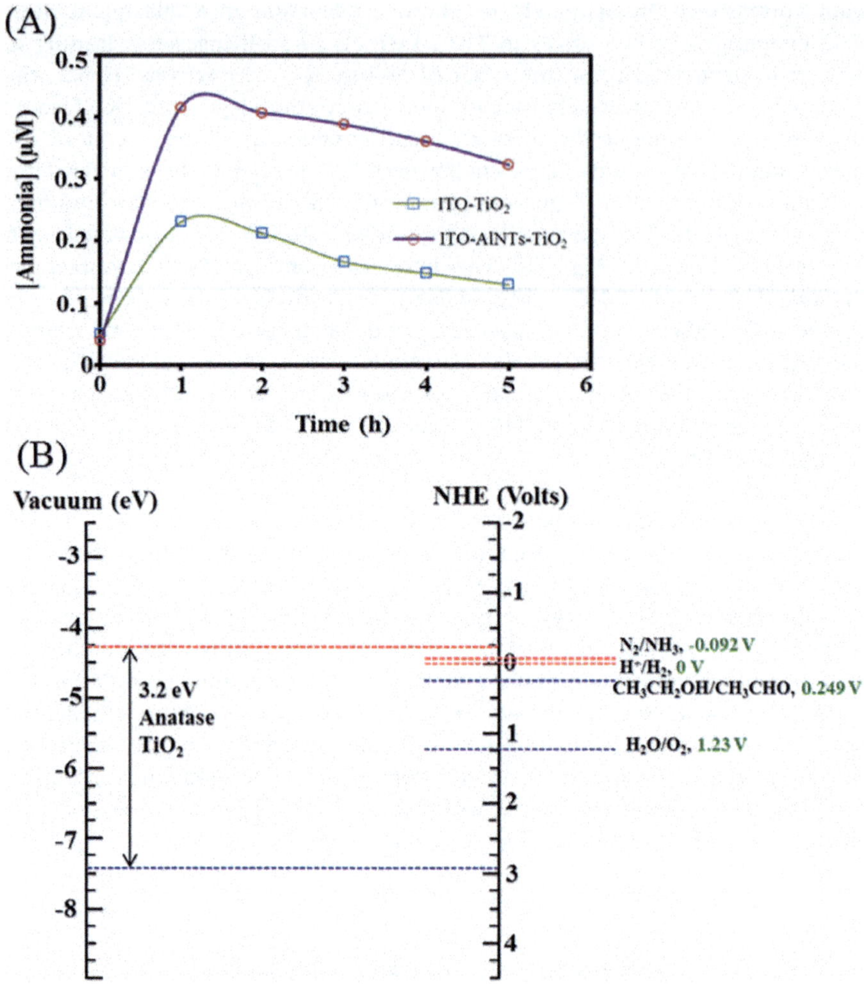

Fig. 7 (A) Ammonia production of irradiated TiO_2 and AlNTs–TiO_2 samples in the absence of a nitrogen gas flow through a nitrogen saturated solution (pH ~ 4) containing 10% ethanol under illumination by a narrowband 365 nm beam with a 5 mW cm^{-2} power. (B) Energy band alignment of anatase TiO_2 and the standard oxidation and reduction potentials of the possible reactions at the band gap energy.

4. Conclusion

We have successfully demonstrated an alternative route for ammonia synthesis through nitrogen splitting at room temperature and under atmospheric pressure using the plasmonic effects of AlNTs. The samples were first characterised *via* SEM, XPS and XRD analysis, thus providing the band energies. The plasmonic hot electron generation from the AlNTs and near-field coupling to a semiconductor were studied in detail by carrying out electrochemical photocurrent measurements. The results revealed that the overall charge-carrier generation efficiency is dominated by a plasmonic near-field coupling effect, when illuminating TiO_2 at a wavelength above the band gap energy. Hot electron-driven ammonia production was measured to be 4×10^{-5} μM h^{-1} mW^{-1} cm^{-2} with a quantum efficiency of 2.5×10^{-5}%. The AlNTs–TiO_2 photoelectrode exhibits an ammonia production rate of 0.1 μM h^{-1} mW^{-1} cm^{-2}, which is ~2.5 times higher than that of bare TiO_2 (0.04 μM h^{-1} mW^{-1} cm^{-2}), thus indicating enhanced nitrogen splitting at room temperature and under atmospheric pressure. The overall conclusion is that at wavelengths above the band gap energy, the ammonia production rate can be enhanced by a near-field stemming from a plasmonic system, whereas hot electrons play a minor role in such a configuration. The present study offers a large-area AlNTs fabrication method using colloidal lithography, and provides a cheap, efficient and up-scalable method.

Conflicts of interest

There are no conflicts of interest.

Acknowledgements

This project was supported by the Gebert Ruf Stiftung, Switzerland (Grant no. GRS-039/16).

References

1 B. M. Hoffman, D. Lukoyanov, Z.-Y. Yang, D. R. Dean and L. C. Seefeldt, *Chem. Rev.*, 2014, **114**, 4041–4062.
2 S. Ritter, *Chem. Eng. News*, 2008, **86**, 43.
3 A. R. Jawahery, *Energy Efficiency and CO_2 Emissions in Ammonia Production*, International Fertilizer Industry Association, Paris, France, 2009.
4 S. Giddey, S. P. S. Badwal and A. Kulkarni, *Int. J. Hydrogen Energy*, 2013, **38**, 14576–14594.
5 S. Licht, B. Cui, B. Wang, F.-F. Li, J. Lau and S. Liu, *Science*, 2014, **345**, 637–640.
6 J. A. Pool, E. Lobkovsky and P. J. Chirik, *Nature*, 2004, **427**, 527–530.
7 K. Arashiba, Y. Miyake and Y. Nishibayashi, *Nat. Chem.*, 2011, **3**, 120–125.
8 R. I. Bickley and V. Vishwanathan, *Nature*, 1979, **280**, 306–308.
9 O. Rusina, A. Eremenko, G. Frank, H. P. Strunk and H. Kisch, *Angew. Chem., Int. Ed.*, 2001, **40**, 3993–3995.
10 G. N. Schrauzer and T. D. Guth, *J. Am. Chem. Soc.*, 1977, **99**, 7189–7193.
11 A. Fujishima and K. Honda, *Nature*, 1972, **238**, 37–38.
12 K. T. Ranjit, T. K. Varadarajan and B. Viswanathan, *J. Photochem. Photobiol., A*, 1996, **96**, 181–185.
13 M. M. Taqui Khan and N. Nageswara Rao, *J. Photochem. Photobiol., A*, 1991, **56**, 101–111.
14 M. M. Taqui Khan, R. C. Bhardwaj, C. Bhardwaj and N. Nageswara Rao, *J. Photochem. Photobiol., A*, 1992, **68**, 137–146.
15 M. M. Taqui Khan, D. Chatterjee and M. Bala, *J. Photochem. Photobiol., A*, 1992, **67**, 349–352.
16 X.-C. Ma, Y. Dai, L. Yu and B.-B. Huang, *Light: Sci. Appl.*, 2016, **5**, e16017.
17 S. Linic, U. Aslam, C. Boerigter and M. Morabito, *Nat. Mater.*, 2015, **14**, 567–576.

18 M. W. Knight, Y. Wang, A. S. Urban, A. Sobhani, B. Y. Zheng, P. Nordlander and N. J. Halas, *Nano Lett.*, 2013, **13**, 1687–1692.
19 S. C. Warren and E. Thimsen, *Energy Environ. Sci.*, 2012, **5**, 5133–5146.
20 S. Mukherjee, L. Zhou, A. M. Goodman, N. Large, C. Ayala-Orozco, Y. Zhang, P. Nordlander and N. J. Halas, *J. Am. Chem. Soc.*, 2014, **136**, 64–67.
21 S. Mukherjee, F. Libisch, N. Large, O. Neumann, L. V. Brown, J. Cheng, J. B. Lassiter, E. A. Carter, P. Nordlander and N. J. Halas, *Nano Lett.*, 2013, **13**, 240–247.
22 S. Feng, M. Wang, Y. Zhou, P. Li, W. Tu and Z. Zou, *APL Mater.*, 2015, **3**, 104416.
23 K. Li, N. J. Hogan, M. J. Kale, N. J. Halas, P. Nordlander and P. Christopher, *Nano Lett.*, 2017, **17**, 3710–3717.
24 F. Chen, Q. Yang, C. Niu, X. Li, C. Zhang and G. Zeng, *RSC Adv.*, 2015, **5**, 63152–63164.
25 L.-B. Zhao, X.-X. Liu and D.-Y. Wu, *J. Phys. Chem. C*, 2016, **120**, 1570–1579.
26 B. Wu, J. Lee, S. Mubeen, Y.-S. Jun, G. D. Stucky and M. Moskovits, *Adv. Opt. Mater.*, 2016, **4**, 1041–1046.
27 T. Oshikiri, K. Ueno and H. Misawa, *Angew. Chem., Int. Ed.*, 2014, **53**, 9802–9805.
28 T. Oshikiri, K. Ueno and H. Misawa, *Angew. Chem., Int. Ed.*, 2016, **55**, 3942–3946.
29 K. Thyagarajan, C. Santschi, P. Langlet and O. J. F. Martin, *Adv. Opt. Mater.*, 2016, **4**, 871–876.
30 B. Gallinet, A. M. Kern and O. J. F. Martin, *J. Opt. Soc. Am. A*, 2010, **27**, 2261–2271.
31 T. A. F. König, P. A. Ledin, J. Kerszulis, M. A. Mahmoud, M. A. El-Sayed, J. R. Reynolds and V. V. Tsukruk, *ACS Nano*, 2014, **8**, 6182–6192.
32 T. Siefke, S. Kroker, K. Pfeiffer, O. Puffky, K. Dietrich, D. Franta, I. Ohlídal, A. Szeghalmi, E.-B. Kley and A. Tünnermann, *Adv. Opt. Mater.*, 2016, **4**, 1780–1786.
33 A. D. Rakić, *Appl. Opt.*, 1995, **34**, 4755–4767.
34 T. Oshikiri, K. Ueno and H. Misawa, *Angew. Chem., Int. Ed.*, 2014, **53**, 9802–9805.
35 D. O. Scanlon, C. W. Dunnill, J. Buckeridge, S. A. Shevlin, A. J. Logsdail, S. M. Woodley, C. R. A. Catlow, M. J. Powell, R. G. Palgrave, I. P. Parkin, G. W. Watson, T. W. Keal, P. Sherwood, A. Walsh and A. A. Sokol, *Nat. Mater.*, 2013, **12**, 798–801.
36 R. Beranek, *Adv. Phys. Chem.*, 2011, 1–20.
37 J. Schneider, M. Matsuoka, M. Takeuchi, J. Zhang, Y. Horiuchi, M. Anpo and D. W. Bahnemann, *Chem. Rev.*, 2014, **114**, 9919–9986.
38 X. Shi, K. Ueno, N. Takabayashi and H. Misawa, *J. Phys. Chem. C*, 2013, **117**, 2494–2499.
39 P. Christopher, H. Xin, A. Marimuthu and S. Linic, *Nat. Mater.*, 2012, **11**, 1044–1050.
40 U. Aslam, S. Chavez and S. Linic, *Nat. Nanotechnol.*, 2017, **12**, 1000–1005.
41 X. Wang, C. Santschi and O. J. F. Martin, *Small*, 2017, **13**, 1700044.
42 K. Wu, J. Chen, J. R. McBride and T. Lian, *Science*, 2015, **349**, 632–635.
43 M. Ali, F. Zhou, K. Chen, C. Kotzur, C. Xiao, L. Bourgeois, X. Zhang and D. R. MacFarlane, *Nat. Commun.*, 2016, **7**, 1–5.

44 T. Oshikiri, K. Ueno and H. Misawa, *Angew. Chem., Int. Ed.*, 2016, **55**, 3942–3946.
45 H. Zeng, S. Terazono and T. Tanuma, *Catal. Commun.*, 2015, **59**, 40–44.
46 M. Grätzel, *Nature*, 2001, **414**, 338.
47 D. B. Ingram, P. Christopher, J. L. Bauer and S. Linic, *ACS Catal.*, 2011, **1**, 1441–1447.

Faraday Discussions

PAPER

Plasmonic photocatalysis applied to solar fuels

Steven Bardey,[ab] Audrey Bonduelle-Skrzypczak,[b] Antoine Fécant,[b] Zhenpeng Cui,[c] Christophe Colbeau-Justin,[c] Valérie Caps[*a] and Valérie Keller[a]

Received 2nd October 2018, Accepted 16th November 2018

DOI: 10.1039/c8fd00144h

The induction of chemical processes by plasmonic systems is a rapidly growing field with potentially many strategic applications. One of them is the transformation of solar energy into chemical fuel by the association of plasmonic metal nanoparticles (M NPs) and a semi-conductor (SC). When the localized surface plasmon resonance (LSPR) and the SC absorption do not match, one limitation of these systems is the efficiency of hot electron transfer from M NPs to SC through the Schottky barrier formed at the M NP/SC interfaces. Here we show that high surface area 1 wt% Au/$TiO_{2\text{-UV100}}$, prepared by adsorption of a $NaBH_4$-protected 3 nm gold sol, readily catalyzes the photoreduction of carbon dioxide with water into methane under both solar and visible-only irradiation with a CH_4 *vs.* H_2 selectivity of 63%. Tuning Au NP size and titania surface area, in particular *via* thermal treatments, highlights the key role of the metal dispersion and of the accessible Au–TiO_2 perimeter interface on the direct SC-based solar process. The impact of Au NP density in turn provides evidence for the dual role of gold as co-catalyst and recombination sites for charge carriers. It is shown that the plasmon-induced process contributes up to 20% of the solar activity. The plasmon-based contribution is enhanced by a large Au NP size and a high degree of crystallinity of the SC support. By minimizing surface hydroxylation while retaining a relatively high surface area of 120 $m^2\ g^{-1}$, pre-calcining $TiO_{2\text{-UV100}}$ at 450 °C leads to an optimum monometallic system in terms of activity and selectivity under both solar and visible irradiation. A state-of-the-art methane selectivity of 100% is achieved in the hot electron process.

Introduction

The direct conversion of solar energy into valuable chemical fuels is a great challenge for clean energy production. In this context, the photo-reduction of CO_2

[a]Institut de Chimie et des Procédés pour l'Energie, l'Environnement et la Santé (ICPEES), UMR7515, 25 rue Becquerel, 67087 Strasbourg, France. E-mail: caps@unistra.fr

[b]IFP Energies Nouvelles (IFPEN), Rond-point de l'échangeur de Solaize, 69360 Solaize, France

[c]Laboratoire de Chimie Physique, CNRS UMR 8000, Univ. Paris-Sud, Université Paris-Saclay, 91405 Orsay, France

into methane and methanol has received increasing attention over the last few years.[1–3] This way of producing synthetic fuels would indeed represent a step forward towards sustainability, due to the nature of the carbon source, a major by-product of energy consumption, and its increasing availability. However, the conversion rate of photonic energy into chemical energy by semi-conductor-based photocatalysts is still too low, as current energy carriers suffer from a short lifetime (fast recombination) and use only a fraction of the solar spectrum. Hence, the development of photocatalytic materials with improved and broadened light absorption is needed in order to efficiently use sunlight as a primary energy source. Amongst the possible materials, multi-component systems containing metal nanoparticles (M NPs) interacting with a semi-conductor (SC) appear promising.[4,5] In addition to their beneficial role as co-catalyst and electron traps,[6,7] M NPs with plasmonic properties[8] may indeed improve the photocatalytic performances of the semi-conductor by plasmon-derived physical phenomena, including hot electron generation[8] and utilization.[9–13] Hot electron injection into the SC conduction band is however dependent on the Schottky barrier formed at the metal–support interface[14,15] and little is known about the nanostructuration of the optimum interaction. Energy transfer from the plasmonic NP to the SC remains low,[16] although efficient hot electron transfer by plasmon-induced interfacial charge transfer has been reported for cadmium selenide nanorods with gold tips,[17] highlighting the importance of atomic description of the interface.[18] The interface in general has been found to play a key role in many catalytic processes.[19,20] More structure–activity relationships are thus needed, which implies a precise control of the M-SC interface at the nanoscale. We have recently designed an optimized heterojunction between titania and carbon nitride with improved activity for hydrogen production from water.[21] Here we show the impact of the interface nanostructuration of gold–titania composites, in terms of accessibility of Au–TiO_2 dual perimeter sites, Au NP size and density, and titania surface functionalization and crystallinity, on their compared performances in solar and visible photocatalytic gas phase reduction of CO_2 in an optimized swept fixed-bed reactor.

Results and discussion

Under artificial solar conditions (see Experimental section), both methane and hydrogen are produced. No other products are detected, which is commonly observed in the gas phase process using water as a reductant.[22] Fig. 1 shows the typical evolution of methane and hydrogen as a function of time. As can be seen, the production rates of both methane and hydrogen increase in the first 60 to 90 minutes of reaction, at which point they reach a maximum. A decrease in methane and hydrogen production rates is then observed in the next 8.5–9 h of reaction until a plateau is reached. The following discussion will be based on average production rates, using the average of all 45 production rates monitored during the 10 h reaction (one every 14 min). In our set-up, pure anatase TiO_2 UV100 readily produces 1 μmol h^{-1} g^{-1} of methane on average under solar irradiation. This low yet significant activity obtained without metal deposition can be attributed to oxygen vacancies present on anatase crystallites[23–26] and to the low yet significant amount of UV irradiation present in the artificial solar spectrum (1%, see Experimental section), capable of efficiently activating anatase TiO_2.

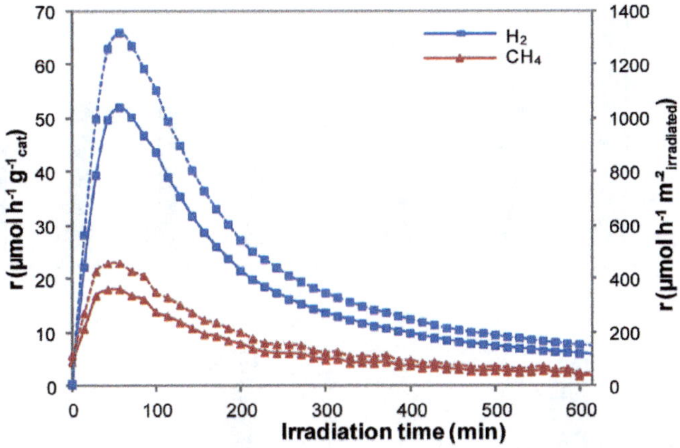

Fig. 1 H_2 (■) and CH_4 (▲) photocatalytic production rates per irradiated surface (dashed lines, right *Y* axis) and per gram (continuous lines, left axis) *versus* solar irradiation time for Au_{3nm}/TiO_2.

Oxygen vacancies are proposed to activate the particularly inert and stable CO_2 molecule by allowing CO_2 adsorption on the titania surface. Subsequent reduction of adsorbed CO_2 to CH_4 involves the *in situ* formation of the active H° reducing agent, first by water oxidation on UV-generated holes, then by reduction of the resulting protons with UV-generated electrons, following the net reaction

$$CO_2 + 8H^+ + 8e^- \rightarrow CH_4 + 2H_2O$$

Straight-forward addition of gold nanoparticles (1 wt%, 3 nm), using colloidal deposition on the high surface area semi-conductor,[27] results in a major increase in the average solar production rate to 7.1 µmol h^{-1} g^{-1}, *i.e.* 159 µL h^{-1} g^{-1}. In addition, the selectivity of the process towards CH_4 *vs.* H_2 formation significantly increases from 45% to 60% (Fig. 2b). The starting gold-based composite thus exhibits close performance to a standard established in the literature in 2009 (160 µL h^{-1} g^{-1})[28] and a high selectivity to methane, as compared with other mono-metallic Au/TiO_2 photocatalysts.[4] By acting as electron traps, Au NPs are known to enhance charge separation, *i.e.* decrease recombination occurrences, increase lifetimes of charge carriers and enhance *in situ* formation of reducing species. Au NPs are also well-known to act as co-catalysts.[6] In addition, the presence of Au NPs on the titania surface is known to create oxygen vacancies at the perimeter interface, which could facilitate CO_2 activation.

Post-calcination effect

Calcining this Au_{3nm}/TiO_2 composite in air results in a significant decrease in methane and hydrogen solar production rates with increasing calcination temperature, until no product at all is detected when calcination is performed at 900 °C (Fig. 2a). Calcination actually causes a major loss of overall surface area, accompanied with a decrease in pore volume and modification of the mean pore diameters (Fig. 3).

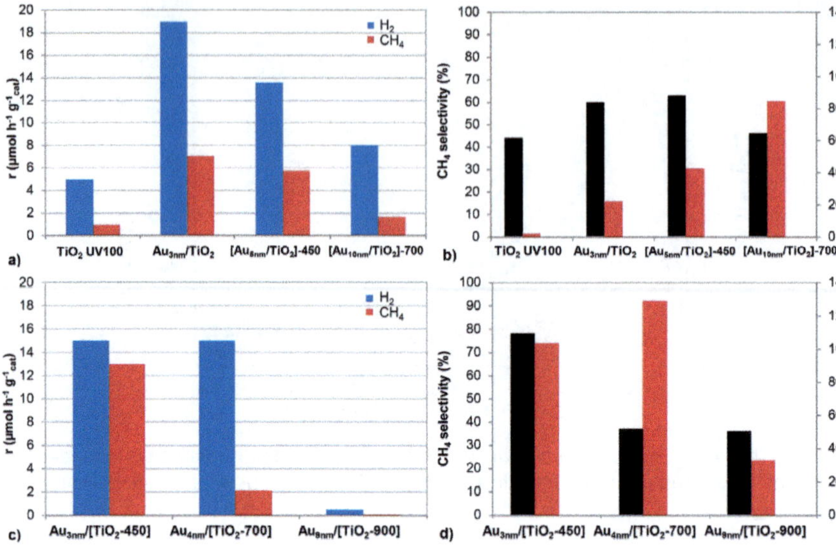

Fig. 2 (a) H_2 and CH_4 production rates for TiO_2, Au_{3nm}/TiO_2 and post-calcined Au/TiO_2 composites. The $[Au_{43nm}/TiO_2]$-900 composite showing no activity is not shown on the graph. (b) CH_4 selectivity and production rates normalized by BET specific area for post-calcined samples, (c) H_2 and CH_4 production rates for pre-calcined Au/TiO_2 composites, and (d) CH_4 selectivity and production rates normalized by BET specific area for pre-calcined samples.

Only 43%, 6% and 1.3% of the initial surface area of the Au_{3nm}/TiO_2 composite (315 $m^2\ g^{-1}$) is indeed retained after calcination at 450 °C, 700 °C and 900 °C, respectively. Hence, by normalizing the photocatalytic productions with the specific surface area of the photocatalyst, average methane production rates are found to actually increase with increasing calcination temperature, from 23 to 43 and 85 nmol $h^{-1}\ m^{-2}$, when calcining at 450 °C and 700 °C, respectively (Fig. 2b). This is attributed to the increasing degree of crystallinity of the semi-conductor particles, as evidenced by the narrowing of *e.g.* the (004) anatase reflection at $2\theta = 38°$ in the powder X-ray diffractograms of the composites (Fig. 4).

The Au–TiO_2 interface may also contribute to the increase in surface specific methane production rates, as shown by the change in the optical response of the composites upon calcination. The UV-vis spectra (Fig. 5a) indeed exhibit a major absorbance over the UV range, which is attributed to electronic transitions within the titania SC, as well as a smaller absorption in the visible range, typical of the localized surface plasmon resonance (LSPR) of gold nanoparticles. While the absorption related to bare titania ends at 400 nm, the remaining absorbance of the Au–TiO_2 composites in the 400–500 nm range is attributed to d–d interband transitions from the Au NPs. The most significant effect of calcination temperature on the optical response of the composites is the increase in the relative intensity of the SPR and the major red-shift of the SPR position. Its maximum is indeed shifted from 520 nm in Au_{3nm}/TiO_2, which is typical of free Au NPs, *i.e.* Au NPs in weak interaction with titania, to 545, 575 and up to 605 nm, when the composite is calcined at 450 °C, 700 °C and 900 °C, respectively. Although the increase in intensity is consistent with

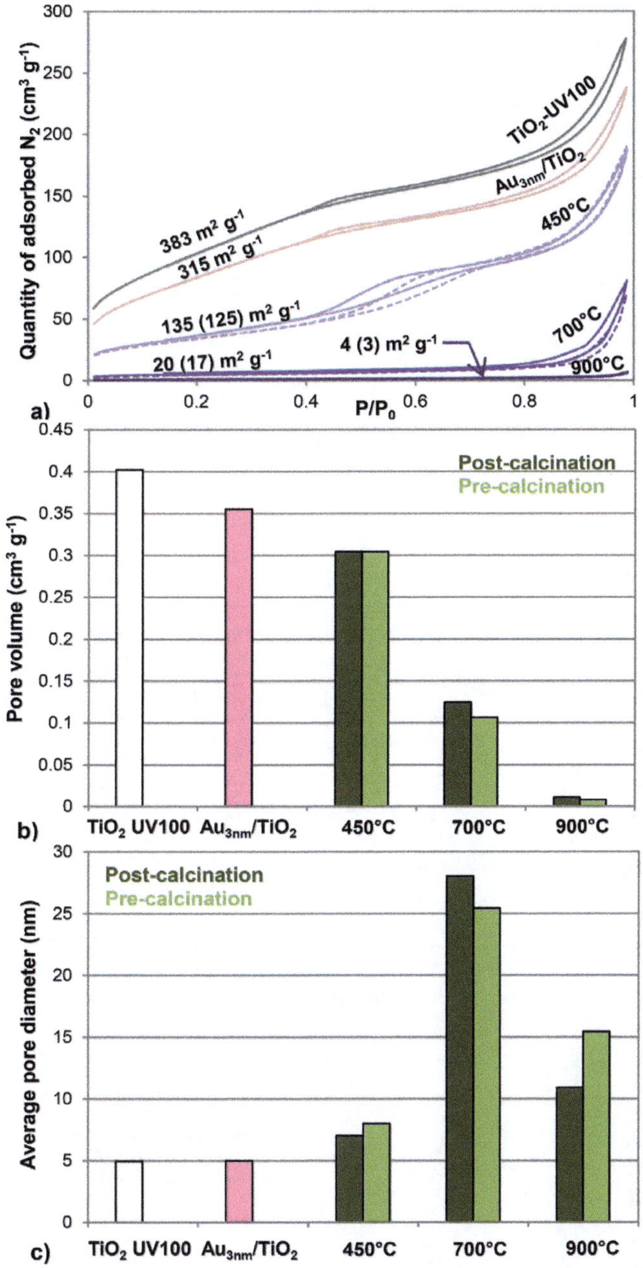

Fig. 3 (a) N_2 physisorption isotherms of TiO_2, Au_{3nm}/TiO_2 and post-calcined samples with corresponding BET surface area values (BET surface area values for pre-calcined composites are in brackets). (b) Pore volume and (c) average pore diameters for TiO_2, Au_{3nm}/TiO_2, and pre- and post-calcined composites.

the increase in Au NP size from 3 nm to 43 nm as shown by TEM studies (Fig. 6 and Table 1), the extent of the red-shift cannot be accounted for solely by a size effect.

Such a shift implies a strengthening of the metal–support interaction, which was confirmed by HRTEM studies (Fig. 7). Thermal treatment of the composite in

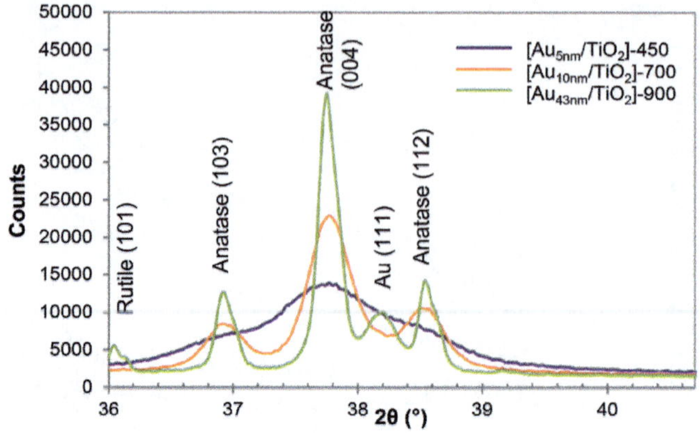

Fig. 4 Diffractograms of post-calcined samples (450 °C is purple, 700 °C orange and 900 °C green).

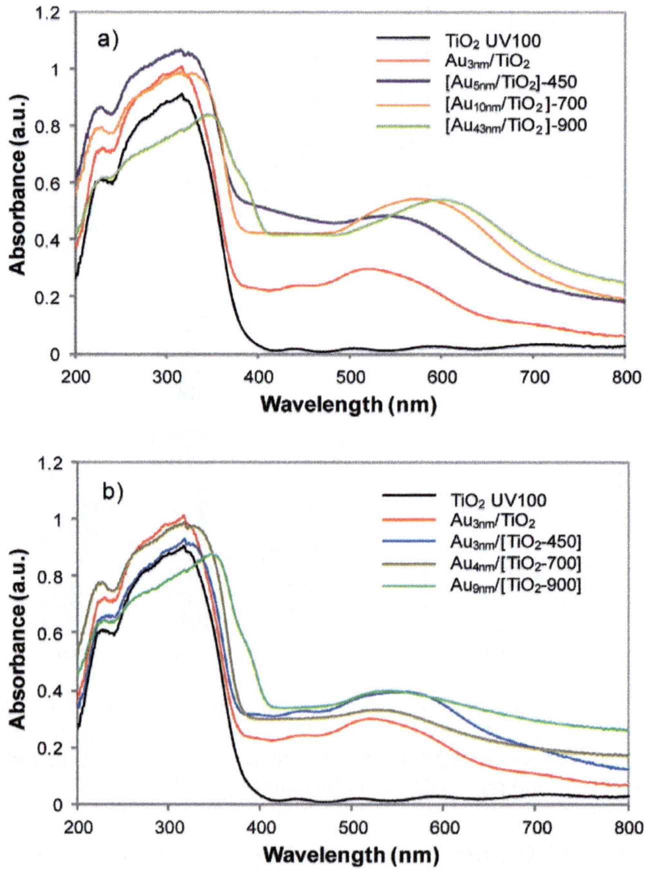

Fig. 5 UV-visible spectra of (a) TiO_2, Au_{3nm}/TiO_2 and post-calcined samples and (b) TiO_2, Au_{3nm}/TiO_2 and pre-calcined samples.

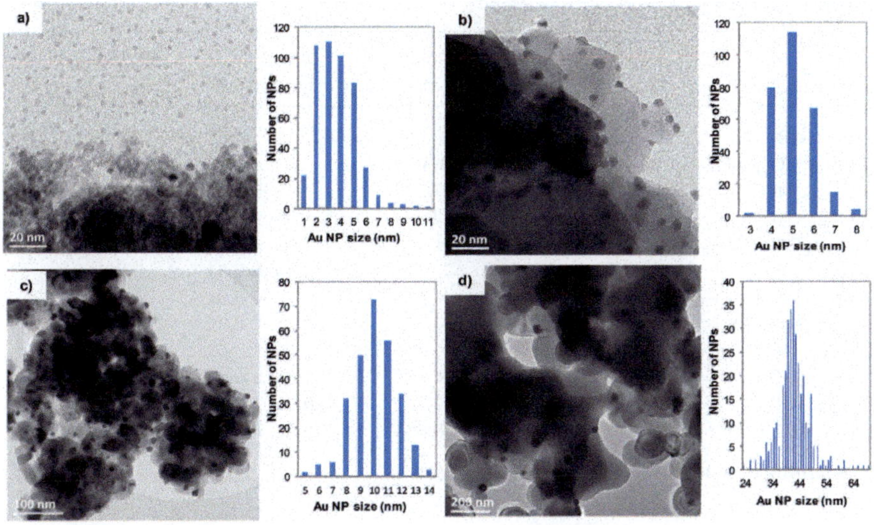

Fig. 6 Representative TEM images and Au NP size distributions of (a) Au_{3nm}/TiO_2, (b) $[Au_{5nm}/TiO_2]$-450, (c) $[Au_{10nm}/TiO_2]$-700, and (d) $[Au_{43nm}/TiO_2]$-900.

air indeed induces, from 700 °C, an interaction between gold and titania akin to the so-called strong metal–support interaction (SMSI), except it is here occurring upon oxidizing, and not reducing, heating. Inclusion of gold in the titania lattice is clearly observed in the composite treated at 700 °C, while some particles become fully encapsulated within a layer of titanium oxide upon further air treatment at 900 °C (Fig. 7). The former phenomenon has been described before for Pt/TiO_2 (ref. 29) and Au/TiO_2 (ref. 30) materials, while the latter (Au NP encapsulation) most likely contributes to the inactivity of $[Au_{43nm}/TiO_2]$-900, by lowering the metallic surface beyond the loss coming from the drastic growth of Au NPs (43 nm). It suggests that the metallic surface and the Au–TiO_2 perimeter interface sites play key roles in the photoreduction process, not from their interaction with light, but rather as direct adsorption sites for reactants or intermediates. We speculate that accessible Au-TiO_2 perimeter sites, by enhancing the number of oxygen vacancies and defects, increase the rate of CO_2 activation and that accessibility to the gold surface increases the rate of *in situ* formation of the reducing agent (H°).

TiO_2 pre-calcination effect

In order to evaluate the impact of the crystallinity of the SC (quasi-) independently from the metallic surface, titania was calcined at 450 °C, 700 °C and 900 °C before gold deposition. The $NaBH_4$-protected 3 nm gold sol was subsequently immobilized on pre-calcined SCs. Hence, variations in Au NP size with calcination temperature, as well as variations in Au–TiO_2 interactions, are minimized, as shown by the modest red shift in the LSPR position (530–560 nm, Fig. 5b). Average Au NP sizes are found to be 3.6 nm, 3.9 nm and 9.0 nm in $Au_{3nm}/[TiO_2$-450], $Au_{4nm}/[TiO_2$-700] and $Au_{9nm}/[TiO_2$-900], respectively (Experimental section, Table 1). Accordingly, the metallic dispersion, estimated on the basis of geometrical

Table 1 Summary of prepared samples

Sample name	Metal	TiO_2 pre-calcination	Post-calcination	Metal loading[a] (wt%)	M NP average size[b] (nm)
Au_{3nm}/TiO_2	Au	X	X	1.2 ± 0.03	3.6 ± 1.5
$[Au_{5nm}/TiO_2]$-450	Au	X	450 °C	1.3 ± 0.03	5.1 ± 0.9
$[Au_{10nm}/TiO_2]$-700	Au	X	700 °C	1.3 ± 0.03	10.0 ± 1.6
$[Au_{43nm}/TiO_2]$-900	Au	X	900 °C	1.3 ± 0.03	42.3 ± 6.0
$Au_{3nm}/[TiO_2$-450]	Au	450 °C	X	1.05 ± 0.03	3.6 ± 1.3
$Au_{4nm}/[TiO_2$-700]	Au	700 °C	X	0.49 ± 0.01	3.9 ± 1.0
$Au_{5nm}/[TiO_2$-700]-2	Au	700 °C	X	0.74 ± 0.01	4.7 ± 1.7
$Au_{9nm}/[TiO_2$-900]	Au	900 °C	X	0.91 ± 0.02	9.0 ± 4.2
Au_{15nm}/TiO_2	Au	X	X	1.09 ± 0.03	n.d.
Au_{50nm}/TiO_2	Au	X	X	1.30 ± 0.03	n.d.
Pt_{3nm}/TiO_2	Pt	X	X	0.77 ± 0.06	3.0 ± 0.8
$[Pt_{3nm}/TiO_2]$-700	Pt	X	700 °C	0.86 ± 0.06	n.d.

[a] Determined by ICP-AES. [b] Determined by TEM.

calculations[31] at 43% in the original sol and at 37% in the uncalcined Au_{3m}/TiO_2 composite, remains similar when the support is pre-calcined at 450 °C and 700 °C (37% and 35%, respectively). The Au–TiO_2 perimeter interface is thus expected to be unchanged in those samples. Pre-calcining TiO_2 at 900 °C, however, induces a huge drop in surface area. Longer reaction times are thus needed to load the support with a similar amount of gold (see Experimental section), which favours aggregation and sintering of the deposited NPs (TEM, not shown). In this case, the metallic dispersion decreases to 16%, but remains much higher than in the post-calcined counterpart (*ca.* 4%) and the NPs are much more accessible. Structural characterization (XRD, not shown) shows that anatase crystallites grow with increasing temperature in the same manner as in the post-calcined Au/TiO_2 composites. The specific surface area decreases accordingly: only 40%, 5.4% and 0.95% of the initial surface area of the Au_{3nm}/TiO_2 composite (315 $m^2\ g^{-1}$) is retained when TiO_2 is pre-calcined at 450 °C, 700 °C and 900 °C, respectively. Nevertheless, $Au_{3nm}/[TiO_2$-450] produces significantly more methane (and hydrogen) than the uncalcined, poorly crystallized Au_{3nm}/TiO_2 composite, setting a new standard in solar CH_4 production of 13 μmol $h^{-1}\ g^{-1}$, *i.e.* 291 μL $h^{-1}\ g^{-1}$

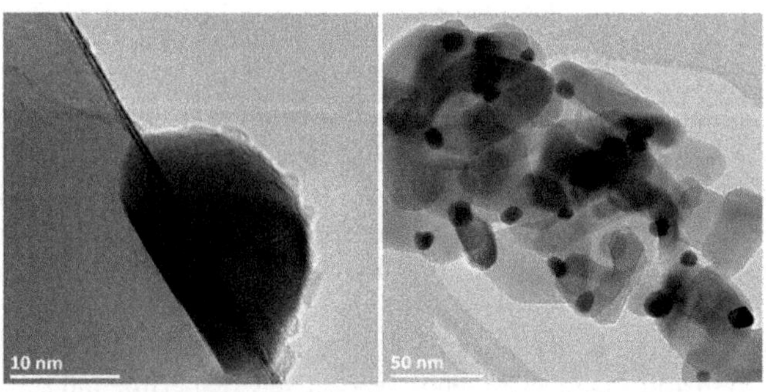

Fig. 7 HRTEM images of (left) $[Au_{43nm}/TiO_2]$-900 and (right) $[Au_{10nm}/TiO_2]$-700.

(Fig. 2c). Crystallization of the SC is thus beneficial to the photoreduction of CO_2. In fact, surface specific (*i.e.* specific surface area-normalized) CH_4 production rates increase with increasing pre-calcination temperature, *i.e.* crystallinity, of the titania support from 23 nmol h^{-1} m^{-2} in the uncalcined, poorly crystallized Au_{3nm}/TiO_2 composite to 104 and 130 nmol h^{-1} m^{-2}, when pre-calcining TiO_2 at 450 °C and 700 °C, respectively (Fig. 2d). Additionally, the surface-normalized CH_4 production of Au_{9nm}/[TiO_2-900] (33 $nmol_{CH_4}$ h^{-1} m^{-2}) remains higher than that of the uncalcined composite. Crystallization of the titania SC is thus clearly favourable to the solar photocatalytic process. An increasing degree of crystallinity may indeed increase the lifetime of the charge carriers (Time Resolved Microwave Conductivity (TRMC) measurements) by avoiding defect-induced recombination.[32] Yet, the surface-specific methane production rate observed over the most crystallized sample, Au_{9nm}/[TiO_2-900], is lower than those of the composites calcined at lower temperatures. It is attributed to the significantly decreased density of defects and oxygen vacancies in the highly crystallized structure (presence of the rutile phase, Fig. 4), which may induce less efficient adsorption of the carbon dioxide reactant per surface unit. In addition, the lower gold dispersion of Au_{9nm}/[TiO_2-900], as compared with Au_{3nm}/[TiO_2-450] and Au_{4nm}/[TiO_2-700], may also contribute to its lower BET-normalized activity.

Information about the impact of the actually accessible gold surface may be derived from comparison of the photocatalytic performances of the post-calcined and pre-calcined series. Indeed, the physico-chemical properties of the SCs are similar in both series (similar anatase crystallite size, nitrogen physisorption isotherms, surface areas, pore volumes and pore size distributions (Fig. 3)), while the accessible metallic surface is generally greater in the pre-calcined series. Clearly, the pre-calcined series is more active than the post-calcined composites under solar conditions, at each calcination temperature (Fig. 2). In particular, Au_{3nm}/[TiO_2-450] exhibits more than twice the methane production rate of [Au_{5nm}/TiO_2]-450, despite a moderate gain in accessible gold surface (0.39 *vs.* 0.35 wt%). Au_{9nm}/[TiO_2-900], despite its the low surface area and unlike its post-calcined counterpart, exhibits measurable photocatalytic activity (0.1 μmol_{CH_4} h^{-1} g^{-1}) which likely benefits from the much higher (0.15 *vs.* 0.05 wt%) and much more accessible Au surface content (no encapsulation). Finally, Au_{4nm}/[TiO_2-700] exhibits a slightly lower Au surface content (0.17 wt%) than its post-calcined counter-part (0.20 wt%), due to its peculiarly low Au loading (0.49 wt% instead of 1.1 ± 0.2 wt% in all other Au/TiO_2 composites). It nevertheless exhibits an activity gain of about 30% (2.2 µmol h^{-1} g^{-1} *vs.* 1.7 µmol h^{-1} g^{-1}), likely due to the higher accessibility of gold (no inclusion in the TiO_2 lattice). The accessible Au surface content could be marginally increased to 0.21 wt% by increasing adsorption time between the gold sol and TiO_2-700 °C (resulting in a more loaded Au_{5nm}/[TiO_2-700]-2 composite (0.74 wt%) with slightly increased NP size (4.7 nm)), with no improvement however in CH_4 production rate (1.8 µmol g^{-1} h^{-1}). Fig. 8 shows that both pre-calcined materials exhibit significantly higher Au NP densities (>400 NPs per µm^2) than [Au_{10nm}/TiO_2]-700. This suggests that gold may play a dual role in the photocatalytic process and that the catalytic role of its available surface may be partially counterbalanced by its role as a recombination center when the density of M NPs on the titania surface is high.[33]

Concerning selectivity to methane *vs.* H_2, the initial selectivity of the uncalcined Au_{3nm}/TiO_2 composite (60%) increases when applying calcination at 450 °C

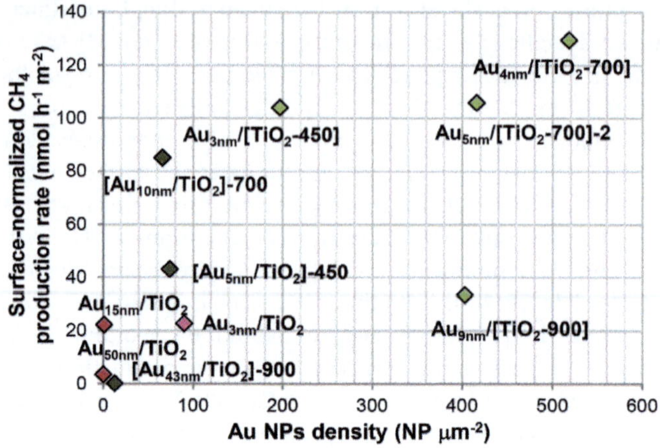

Fig. 8 BET surface normalized CH_4 production rates for uncalcined (◆, ◆), pre-calcined (◆) and post-calcined (◆) samples.

to 63% for post-calcination and to 78% for pre-calcination. It may be attributed to the loss of terminating hydroxyl functional groups of the SC surface (dehydroxylation, TGA, not shown), which contribute preferentially to H_2 production.[34] The even higher selectivity of the pre-calcined composite is attributed to the more numerous population of Au–TiO_2 interface perimeter sites due to the presence of a higher density of Au NPs (197 *vs.* 74 NP per μm^2, Fig. 8), which results from the smaller Au NP size (3.6 nm), similar loading and slightly lower surface area (125 *vs.* 135 $m^2\ g^{-1}$). The decrease in CH_4 selectivity upon further increasing the calcination temperature, *i.e.* the crystallization degree of titania, is correlated to the loss of surface defects, *i.e.* the sites that favour CO_2 over water adsorption.[24]

Au NP size effect

In order to probe the Au NP size effect, *i.e.* the effect of metallic dispersion, two commercially available PVP-protected gold colloids of 15 nm and 50 nm were loaded on the uncalcined TiO_2 UV100 support (see Experimental section) in the same amount (1.1 ± 0.2 wt%). In turn, the density of the Au NPs is also modified: it decreases significantly with increasing Au NP size. Au_{3nm}/TiO_2 and Au_{15nm}/TiO_2 lead to the same methane production rate (7.1 μmol $h^{-1}\ g^{-1}$, Fig. 9a), despite the much lower gold dispersion in Au_{15nm}/TiO_2 (10% instead of 37%). This suggests that methane production is not directly related to metal dispersion, which also implies that catalytic effects are not limiting. Such low impact of Au NP size (in the 3–12 nm range) on the photocatalytic efficiency has previously been observed in titania-based hydrogen production from water/ethanol mixtures.[35] Au_{15nm}/TiO_2 leads to the optimum internal quantum yield of this study, *i.e.* 0.012% with only 1% of UV photons emitted. Further increasing the Au NP diameter to 50 nm, *i.e.* lowering the metal dispersion to about 1%, causes a drop in the methane production rate to 1.3 μmol $h^{-1}\ g^{-1}$. It is interesting that the activity of Au_{50nm}/TiO_2 is close to that of the bare titania support. This is consistent with the fact that the density of Au NPs is particularly low in this composite (≪0.1 NP per μm^2). By increasing the Au NP size on the same support (identical surface area), the

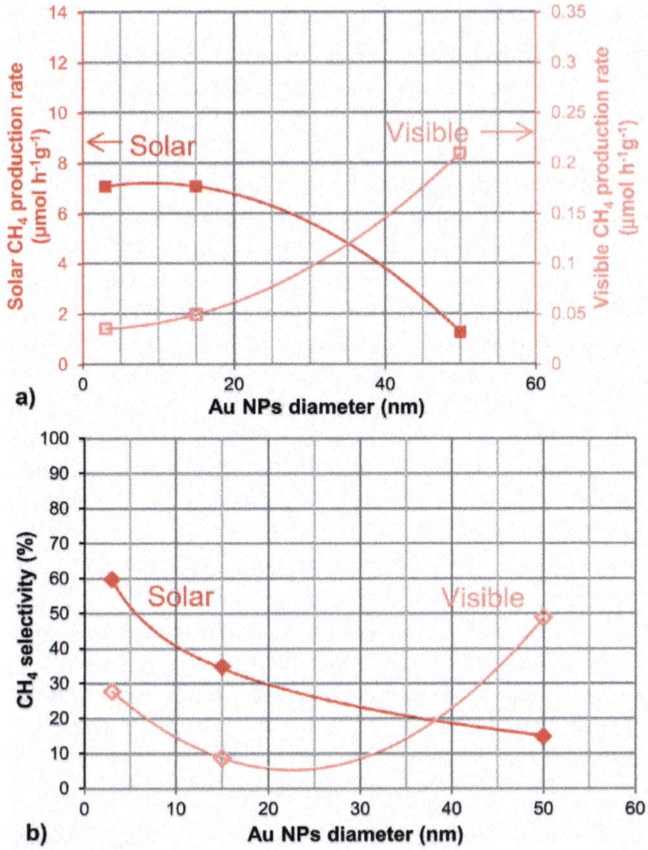

Fig. 9 (a) Comparison of visible (□) and solar (■) CH_4 production rates as a function of Au NP size. (b) Comparison of CH_4 selectivity under visible (>420 nm, ◇) and solar light (◆) as a function of Au NP size.

decrease in Au dispersion is indeed accompanied by an even more significant decrease in Au NP density, from 90 NPs per μm^2 in Au_{3nm}/TiO_2 to 1 NP per μm^2 and $\ll$0.1 NP per μm^2 in Au_{15nm}/TiO_2 and Au_{50nm}/TiO_2, respectively. At such a low Au NP density, the positive impact of gold on the titania-based photocatalytic activity is modest. Nevertheless, the selectivity to methane continuously decreases with increasing Au NP size, from 60% in Au_{3nm}/TiO_2 to 15% in Au_{50nm}/TiO_2, suggesting that larger particles favour the reaction of H° with H° rather than with carbonaceous species. Larger Au NPs have previously been found active for hydrogen production from water, suggesting that H° + H° recombination may indeed occur on the surface of large Au NPs.[36,37] More generally, larger Au NPs are reported active for radical recombination.[38,39] On the other hand, the decreasing Au NP density decreases the number of Au–TiO_2 perimeter interface sites, *i.e.* the number of CO_2 adsorption sites, which would decrease the probability for CO_2 activation and subsequent reduction to CH_4. Such a decrease in CH_4 *vs.* H_2 selectivity with increasing Au NP size is thus consistent with a reaction pathway involving H° recombination over the gold surface and H° + CO_2* reaction at the Au–TiO_2 perimeter interface.

Activity in the visible range

In order to probe the activity of the samples in the visible range, a solid filter (GG420 Farbglass SCHOTT from ITOS) that cuts-off irradiation below 420 nm was applied. Under these conditions, the overall photon flux (0.028 mol s^{-1} m^{-2}) is decreased by about 15%, to 0.024 mol s^{-1} m^{-2}. Bare UV100 is no longer active for CO_2 photoreduction and hydrocarbon or hydrogen production (if any) falls below the detection limit. The gold-containing composites, however, all exhibit measurable activity, with average methane production rates ranging from 0.02 to 0.31 μmol h^{-1} g^{-1}, *i.e.* 0.11 to 7.06 nmol h^{-1} m^{-2} (Fig. 10). This occurs in the absence of measurable photon absorption from TiO_2 anatase above 400 nm, *i.e.* in the absence of direct light-mediated activation of the titania SC (Fig. 5). This suggests that the active charge carriers are not generated directly within the SC, but rather *via* the gold nanoparticles. The optical properties of Au NPs and more specifically their LSPR in the visible range are known to induce the formation of hot electrons upon excitation at the corresponding wavelengths.[8–13] Such plasmon-derived hot electrons are reported to induce a variety of phenomena, including direct chemical reaction with adsorbates at the M NP surface[10,15,40,41] and enrichment of the conduction band of the SC by transfer of these hot carriers from the Au NP to the SC,[9–14] in addition to enhanced charge separation.[10,42] Their lifetime, which is increased by the interaction of the metal nanoparticle with the SC *via* formation of a Schottky barrier, does indeed match the chemical reaction rates.

The plasmon-related activities observed here are typically one to two orders of magnitude lower than the direct, SC-based photocatalytic process, keeping in

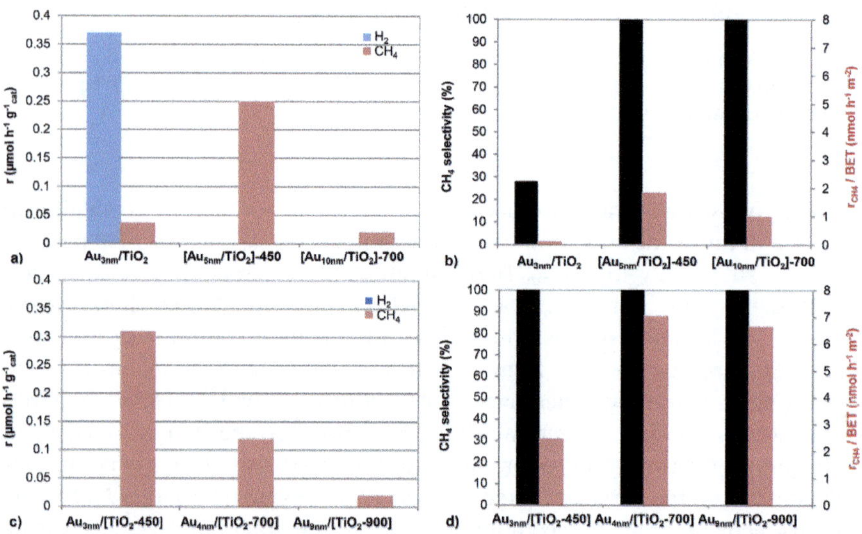

Fig. 10 (a) H_2 and CH_4 production rates for Au_{3nm}/TiO_2 and post-calcined samples under visible light (>420 nm). The $[Au_{43nm}/TiO_2]$-900 composite showing no activity is not shown on the graph. (b) CH_4 selectivity and production rates normalized by BET specific surface area for post-calcined samples under visible light (>420 nm), (c) H_2 and CH_4 production rates for pre-calcined samples under visible light (>420 nm), and (d) CH_4 selectivity and production rates normalized by BET specific area for pre-calcined samples under visible light (>420 nm).

mind that visible light irradiance with the UV-cut-off filter was 4600 W m^{-2} compared to 5100 W m^{-2} without the filter. Comparing the performance of all composites for visible CO_2 photoreduction, Au_{3nm}/[TiO_2-450] remains the most active of the series on a weight basis, producing 0.31 μmol_{CH_4} h^{-1} g^{-1} under visible light (Fig. 10c), which is about ten times more than on Au_{3nm}/TiO_2 (Fig. 10a) followed by [Au_{5nm}/TiO_2]-450 (0.25 μmol_{CH_4} h^{-1} g^{-1}) and Au_{50nm}/TiO_2 (0.21 μmol_{CH_4} h^{-1} g^{-1}) (Fig. 9a). The surface area remains an important factor, as methane production rates steadily decrease with decreasing surface area/increasing Au NP density in the pre-calcined series to 0.12 and 0.02 μmol h^{-1} g^{-1} for Au_{4nm}/[TiO_2-700] and Au_{9nm}/[TiO_2-900], respectively. However, the most crystallized samples exhibit the highest surface-normalized activities of 6.67 and 7.06 nmol h^{-1} m^{-2} for Au_{9nm}/[TiO_2-900] and Au_{4nm}/[TiO_2-700], respectively (Fig. 10d), highlighting the even higher importance of the crystallization of the SC in the plasmon-based process, as compared with the solar-induced, direct, SC-based process. Further impact of the degree of crystallization of titania can be seen by comparing "residual" activities (*i.e.* visible *vs.* solar activity ratios) in the pre-calcined series: residual activities indeed increase upon increasing calcination temperature (Fig. 11b), *i.e.* with increasing titania crystallinity (Fig. 4), with Au_{9nm}/[TiO_2-900] exhibiting the maximum residual methane production rate upon pure visible irradiation (20% of the solar activity).

The plasmon-based process also exhibits one major specificity: all calcined composites are apparently 100% selective towards methane, the amount of produced hydrogen (if any) being below the detection limit. On the other hand, uncalcined composites, except for Au_{50nm}/TiO_2, exhibit a much lower CH_4 selectivity under visible light than under solar irradiation (Fig. 9b). The beneficial effect of the SC crystallization on the CH_4 *vs.* H_2 selectivity is thus exacerbated upon visible-only irradiation. In particular, the most crystallized composites (calcined at 700 °C and 900 °C) exhibit the same complete methane selectivity as the composites calcined at 450 °C, despite the lower density of defects which is unfavourable for CH_4 adsorption. Hence the absence of unselective surface hydroxyl groups is not sufficient to account for the total selectivity to methane. Another phenomenon might contribute to the absence of hydrogen product. Molecular hydrogen is indeed known to readily dissociate over plasmon-activated Au NPs.[15,40,43] Hence, even if it was formed in the present water-mediated CO_2 photoreduction process, it would not be stable. One may also speculate that it may simply not be formed as it is in the SC-based process and that H° + H° recombination simply does not occur in the gold-driven, plasmon-based process. In addition to the instability of the hydrogen molecule under plasmonic conditions, the absence of UV irradiation, which is known to enhance formation of defects on titania,[44,45] may also lower the possibility of CO_2 adsorbing on titania. Under visible-only irradiation, CO_2 adsorption and activation would thus be restricted to the gold surface, where it would more efficiently compete with H°, especially on dehydroxylated titania surfaces which likely lower the reverse spillover rate of H°.[46] Consumption of *in situ* formed hydrogen in the direct hydrogenation of CO_2 is less likely since gold is active for the reverse water-gas shift reaction leading to CO ($CO_2 + H_2 \rightarrow CO + H_2O$) and not for CO_2 methanation leading to methane ($CO_2 + 4H_2 \rightarrow CH_4 + 2H_2O$),[47] which could not account for the CH_4 product formed here.

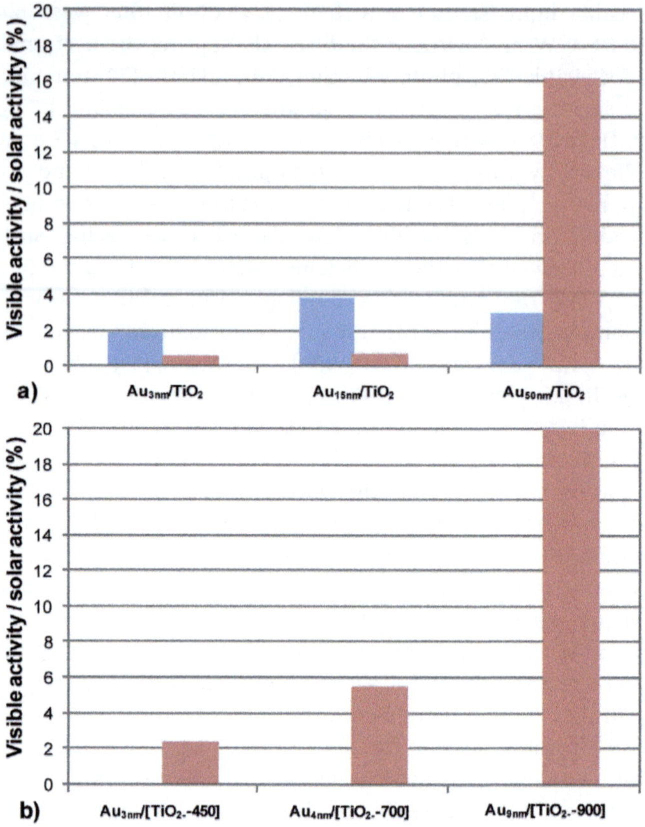

Fig. 11 Visible/solar activity ratio for uncalcined (a) and pre-calcined (b) samples.

Additionally, the pre-calcined materials (Fig. 10c and d) remain more active than their post-calcined counter-parts (Fig. 10a and b) under visible irradiation, showing that accessibility of the metal surface (and Au–TiO_2 interface) remains key in the plasmon-based process. However, the visible CH_4 production rates (Fig. 9a), as well as the visible *vs.* solar activity ratio (Fig. 11a), increase with increasing Au NP size (*i.e.* with decreasing metallic surface). Au_{50nm}/TiO_2 retains 16% of activity upon pure visible irradiation, the second highest of the series, just after the most crystallized composite and despite the poor crystallinity of the uncalcined titania support. Larger Au NPs are thus found to be more efficient for plasmon-derived CO_2 photoreduction, in agreement with previous reports.[15,48,49] This suggests that the optical properties of gold prevail over the use of the gold surface in the plasmon-induced process. The intensity of the SPR indeed increases with increasing Au NP size, enhancing hot electron generation.[49,50] In our composites, the plasmon-driven process thus appears limited by the initial charge generation, which may be due to the low metal loading used, as compared with those used in plasmonic studies.[40,48,49]

Platinum *vs.* gold

Further evidence of gold involvement in the plasmon-based process is obtained by comparison with two Pt-based photocatalysts prepared under similar conditions. Neither Pt_{3nm}/TiO_2 nor $[Pt_{3nm}/TiO_2]$-700 exhibit any detectable activity for CO_2 photoreduction under visible light–only conditions. The position of the Pt NP LSPR in the UV range (220 nm) indeed prevents any Pt-mediated CO_2 photoreduction in the visible range. However, it may account for the much higher CH_4 production rate observed over Pt_{3nm}/TiO_2 under solar irradiation of 91 μmol_{CH_4} h^{-1} g^{-1}. The one order of magnitude higher activity of Pt *vs.* Au in the uncalcined composite may be explained by Plasmon Resonance Energy Transfer (PRET) *i.e.* the ability of the LSPR to enhance the electric field intensity in the SC, *i.e.* the ability of the LSPR to enhance light absorption by the SC,[9] due to the overlap of the Pt LSPR and TiO_2 absorption bands. The higher workfunction of platinum *vs.* gold, which theoretically enhances charge separation, may also contribute to the higher solar activity of platinum. On the other hand, the activity of gold-based composites in the visible range is related to the inability of the SPR to transfer its energy to TiO_2 and to the hot carriers generated within the gold nanoparticle as a result. It is interesting that the PRET phenomenon in the SC-based photoreduction of CO_2 disappears with increasing crystallinity of titania and possible loss of accessible metallic surface *via* strong Pt–TiO_2 interactions. In the composites post-calcined at 700 °C, surface-normalized CH_4 production rates are indeed similar on Pt and Au (83 *vs.* 85 nmol h^{-1} m^{-2}). The expected higher density of Pt *vs.* Au NPs in a low surface area $[Pt_{3nm}/TiO_2]$-700 (29 m^2 g^{-1}) may also inhibit the photocatalytic activity of this composite. Additionally, the initially higher selectivity of platinum towards CH_4 *vs.* H_2 formation of 88% decreases below that of gold in the calcined composite (31% *vs.* 46%).

TRMC measurements and charge carrier dynamics

The time resolved microwave conductivity method was previously used to determine the charge carrier lifetimes in TiO_2-based composites.[21,33,51,52] The main data provided by TRMC are the maximum value of the signal (I_{max}), which indicates the number of excess charge carriers (mostly electrons) created by the pulse, including decay processes during the excitation by the laser (10 ns), and the decay ($I(t)$) due to the decrease of mobile electrons, either by recombination or trapping processes. Concerning the decay, *i.e.* the lifetime of charge carriers, a short and long range are usually analyzed. The short-range decay, arbitrarily fixed up to 40 ns after the beginning of the pulse reflects fast processes, recombination of charge carriers and/or trapping phenomena.[51,52]

At 360 nm excitation (Fig. 12), one can observe that the $[Au_{5nm}/TiO_2]$-450 and Au_{3nm}/TiO_2 samples lead to the most intense signal, followed by the TiO_2 UV100 support without Au NPs. However, the first material exhibits slightly higher charge carrier generation accompanied with a lower signal decay, characteristic of either slower charge recombination dynamics or a more pronounced electron trapping effect. Comparing those samples with the TiO_2 UV100 support, it must be underlined that Au NP deposition does not lead to an important signal intensity decrease, as very often observed after metal nanoparticle deposition and resulting from the strong electron trapping effect attributed to Au NPs. Nevertheless, this trapping effect induced by Au NP deposition may be argued for the two photocatalysts presenting the largest Au NPs, *i.e.* Au_{50nm}/TiO_2 and mainly $[Au_{10nm}/TiO_2]$-700. It is noted that absence of a signal does not mean that no

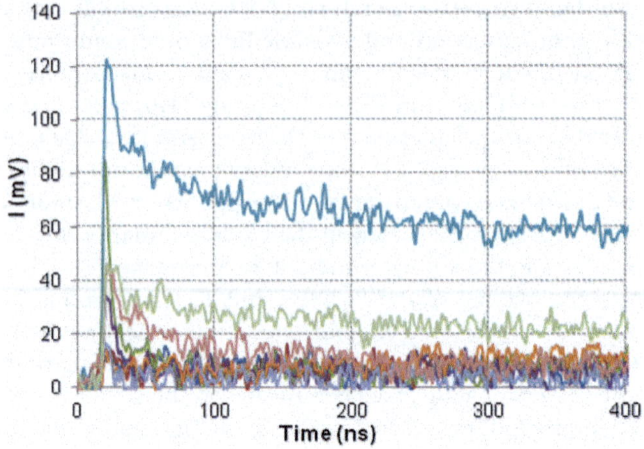

Fig. 12 TRMC signals at 360 nm on TiO_2 UV100 (pink), TiO_2 P25 (light green), Au_{3nm}/TiO_2 (dark blue), $[Au_{5nm}/TiO_2]$-450 (red), $[Au_{10nm}/TiO_2]$-700 (dark green), $Au_{3nm}/[TiO_2$-450] (dark purple), $Au_{4nm}/[TiO_2$-700] (light blue), Au_{15nm}/TiO_2 (orange) and Au_{50nm}/TiO_2 (light purple).

charges were created but it could be correlated with very rapid charge recombination after light excitation. It must also be noticed that $Au_{4nm}/[TiO_2$-700] presents specific features, with the lowest signal decay amongst all the samples. This can be correlated either with low charge carrier recombination kinetics or with higher electron trapping effects. It might be supposed that the presence of the rutile phase due to the prior thermal treatment of TiO_2 UV100 at 700 °C, in intimate contact with the anatase phase, and the resulting formation of a nano-heterojunction between both phases lead to spatial charge separation, thus limiting their recombination.

At 450 nm excitation (in the blue region of visible light, Fig. 13), $Au_{4nm}/[TiO_2$-700] also exhibits the lowest signal decay, which could be attributed to the same effect of limitation of charge recombination due to the presence of the TiO_2 rutile phase. In addition, this material yields the highest signal that may be explained by the highest generation of charges due to the tail of absorption of rutile at that wavelength of excitation, as compared with TiO_2 P25 containing *ca.* 20% of rutile. Furthermore, the signal intensity is very close to that obtained under UV-A activation. This last explanation could also be considered for $[Au_{10nm}/TiO_2]$-700. However, in this case the decay in electron mobility is higher, perhaps due to more efficient electron trapping onto larger Au NPs. It must be mentioned that TiO_2 UV100 also shows a small signal that can be associated with charge generation that might result from a defective TiO_2 structure. TiO_2 UV100 is indeed not as well crystallized as TiO_2 P25, for example. Nevertheless, those charges rapidly recombine.

At LSPR maximum excitation (or at 500 nm excitation in the absence of LSPR), and as for activation at 450 nm, $Au_{4nm}/[TiO_2$-700] reveals the highest signal associated with the highest charge carrier generation (Fig. 14). In this case, it must be underlined that the signal intensity is only slightly reduced compared to excitation at 450 nm. The other sample containing TiO_2 calcined at 700 °C, *i.e.*

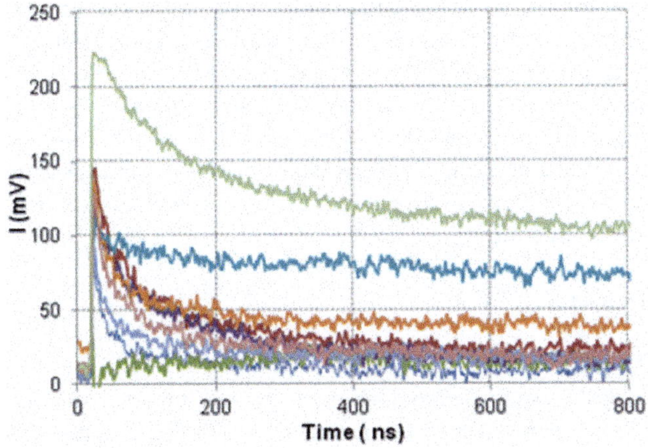

Fig. 13 TRMC signals at 450 nm on TiO_2 UV100 (pink), TiO_2 P25 (light green), Au_{3nm}/TiO_2 (dark blue), $[Au_{5nm}/TiO_2]$-450 (red), $[Au_{10nm}/TiO_2]$-700 (dark green), $Au_{3nm}/[TiO_2$-450] (dark purple), $Au_{4nm}/[TiO_2$-700] (light blue), Au_{15nm}/TiO_2 (orange) and Au_{50nm}/TiO_2 (light purple).

$[Au_{10nm}/TiO_2]$-700, also seems to generate charge carriers, however, to a lesser extent and with a rapid decrease of the signal, probably due to the higher electron trapping effect on larger Au NPs. It must be noticed here that TiO_2 P25 does not show any signal, meaning that absorption due to the presence of the rutile phase is no more effective at 500 nm. From these observations, it may be assumed that LSPR effects may take place through the injection of hot electrons from the Au NPs to the conduction band of TiO_2 nanoparticles. Additionally, the enhanced surface-normalized CH_4 production rates observed over composites calcined at 700 °C may thus be attributed to an increased charge carrier lifetime,

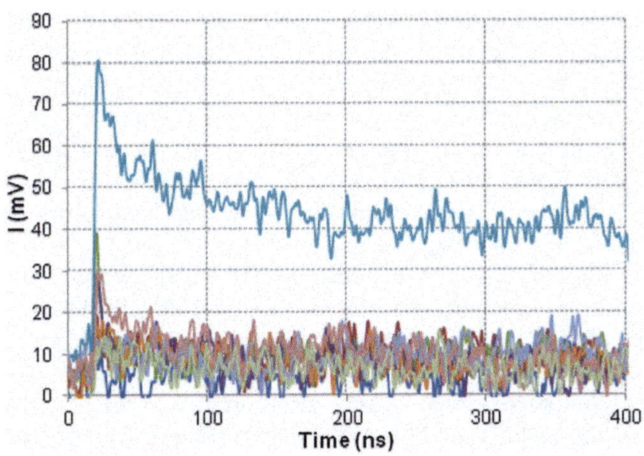

Fig. 14 TRMC signals at 500 nm on TiO_2 UV100 (pink), TiO_2 P25 (light green), Au_{3nm}/TiO_2 (dark blue), $[Au_{5nm}/TiO_2]$-450 (red), $[Au_{10nm}/TiO_2]$-700 (dark green), $Au_{3nm}/[TiO_2$-450] (dark purple), $Au_{4nm}/[TiO_2$-700] (light blue), Au_{15nm}/TiO_2 (orange) and Au_{50nm}/TiO_2 (light purple).

which may be related to the peculiar oblong shape of the titania crystallites present in these materials (Fig. 7).[53] The peculiar shape of the SC may result from restructuration related to the anatase → rutile transition and is consistent with the specificity of TiO_2-700 °C towards Au NP deposition and interaction (Fig. 5b). Finally, the absence of a signal for most of the composites further supports the limitation of the plasmon-based process by charge generation. It is expected that the resulting modest photocatalytic activities may be improved by increasing the gold content.

Experimental

Synthesis of photocatalysts

All photocatalysts were prepared through a well-known colloidal deposition method[27,54–56] but without any protecting agent, using commercially available 100% anatase titania (Hombikat UV100 from Sachtleben Chemie GmbH, 384 g m^{-2}). For all photocatalysts, a gold loading of 1 wt% was targeted. First, an aqueous solution of sodium borohydride (0.1 M, 2.54 mL) was added to an aqueous solution of $HAuCl_4$ (0.25 M, 200 μL), with $Au/NaBH_4 = 1:5$ mol mol^{-1}. The yellow gold solution immediately turned dark red upon introduction of $NaBH_4$, due to the formation of a gold sol, which is stabilized by the presence of $NaBH_4$-derived molecules in electrostatic interaction with the Au NP.[57] The support (1 g) was introduced directly after the reduction step and the mixture was stirred for 30 min for total adsorption of the gold sol onto the support. After filtration and washing (H_2O/1 L), the pink powder was dried in air at 100 °C for 24 h. The resulting Au_{3nm}/TiO_2 product was calcined for 4 h in flowing air (100 mL min^{-1}) at different temperatures: 450 °C, 700 °C and 900 °C, yielding $[Au_{5nm}/TiO_2]$-450, $[Au_{10nm}/TiO_2]$-700 and $[Au_{43nm}/TiO_2]$-900, respectively.

A “pre-calcined” sample series was also prepared using the same method, with the TiO_2 support previously calcined for 4 h in flowing air (100 mL min^{-1}) at the same temperatures (450 °C, 700 °C and 900 °C). For these samples, the contact time between the $NaBH_4$-stabilized gold sol and the support was extended in order to maximize adsorption of the gold sol on the lower surface area supports. The mixture was thus stirred for 2 h, 48 h and 48 h, when using TiO_2-450 °C, TiO_2-700 °C and TiO_2-900 °C, respectively, yielding $Au_{3nm}/[TiO_2$-450], $Au_{4nm}/[TiO_2$-700] and $Au_{9nm}/[TiO_2$-900], respectively. In the case of TiO_2-700 °C, deposition of the gold sol was subsequently tried for 96 h in order to increase the gold loading which appeared to be much lower (0.49 wt%) than in all the other composites (1.1 ± 0.2 wt%). A new composite, yet still with a low gold loading of 0.74 wt%, was obtained as $Au_{5nm}/[TiO_2$-700]-2.

Colloidal solutions of gold nanoparticles of 15 nm and 50 nm were purchased (nanoComposix, Econix, Au nanospheres, PVP) and directly deposited onto the uncalcined UV100 TiO_2 support. The colloidal solution of gold (5 mg mL^{-1}, 1 mL) and the support (0.5 g) were added to 100 mL of methanol (Carlo ERBA reagents S.A.S, 99.9%) and stirred for 2 h. After filtration and washing (MeOH/1 L), the powder was dried at 100 °C for 24 h. Two different samples were obtained: Au_{15nm}/TiO_2 and Au_{50nm}/TiO_2.

Platinum-based materials were prepared by a similar colloidal deposition method, except reduction of the platinum salt was performed in the presence of the titania support.[58] Titania (1 g) was added to an aqueous solution of H_2PtCl_6

(0.256 M, 200 μL) and stirred for 20 min. Then, an aqueous solution of sodium borohydride (0.1 M, 2.54 mL, Pt/$NaBH_4$ = 1 : 5 mol mol^{-1}) was added and the mixture was stirred for 30 min. After filtration and washing (H_2O/1 L), the grey powder was dried in air at 100 °C for 24 h. Part of the resulting Pt_{3nm}/TiO_2 composite was then calcined in flowing air (100 mL min^{-1}) for 4 h at 700 °C, yielding [Pt_{3nm}/TiO_2]-700. All samples are summarised in Table 1.

Characterization

A Micromeritics Asap 2420 porosimeter was used to obtain Nitrogen adsorption–desorption isotherms at 77 K. Materials were outgassed at 250 °C under primary vacuum for 4 h in order to desorb impurities from their surfaces and eliminate any water. Specific surface areas were calculated by the Brunauer, Emmett and Teller (BET) method in the relative pressure (p/p_0) range 0.05–0.3 and using t-plot methods.

X-ray diffraction (XRD) patterns were collected on a Bruker D8 Advance theta–theta diffractometer (Cu Kα radiation, λ = 0.154 nm), equipped with a LynxEye detector and operating at 40 kV and 40 mA. The datasets were acquired in step-scan mode over the 36–49° 2θ range, using a step interval of 0.024° and a counting time of 4 s per step. Powder samples were introduced in an alumina crucible, treated *in situ* at the desired temperature in flowing synthetic air using a heating chamber and cooled down to room temperature prior to data acquisition.

Transmission Electron Microscopy (TEM) was carried out on a JEOL 2100F microscope working at a 200 kV accelerating voltage and equipped with a probe corrector for spherical aberrations, giving a point-to-point resolution of 0.18 nm. The sample was dispersed in an ethanol solution for 2 min. A drop of the solution was subsequently deposited on a copper grid covered with a holey carbon membrane for observation. Statistical M NP size distributions were obtained on bright field images using populations of more than 200 nanoparticles and ImageJ software.[59]

UV-vis absorption spectra were recorded on a Perkin Elmer Lambda 950 Scan spectrophotometer equipped with a 100 mm diffuse reflectance integrating sphere. UV-vis spectra were done on films: powders were deposited on 2 cm diameter glass disks with the same surface concentration and using the same protocol as for the photocatalytic tests (25 g m^{-2}).

Elemental analyses of the photocatalysts were performed using Inductively Coupled Plasma Atomic Emission Spectroscopy (ICP-AES). The limit of detection of the instrument is 0.1 mg L^{-1} for Au. Analysis of the gold component of the Au/TiO_2 composites allowed us to calculate deposition yields which are defined as the ratio between the actual deposited mass of Au (deduced from ICP-AES analysis) and the mass of Au introduced in the synthesis.

The charge carrier lifetimes in the TiO_2-based composites were determined by microwave absorption experiments using the Time Resolved Microwave Conductivity method (TRMC).[51,52] The TRMC technique is based on measurement of the change of microwave power reflected by a sample, $\Delta P(t)$, induced by its laser pulsed illumination. The relative difference $\Delta P(t)/P$ can be correlated, for small perturbations of conductivity, to the difference of the conductivity $\Delta\sigma(t)$ considering the following equation:

$$\frac{\Delta P(t)}{P} = A\Delta\sigma(t) = A\,\mathrm{e}\sum_i \Delta n_i(t)\mu_i$$

where $\Delta n_i(t)$ is the number of excess charge-carriers i at time t and μ_i is their mobility. The sensitivity factor A is independent of time, but depends on different factors such as the microwave frequency or the dielectric constant. Considering that the trapped species have a small mobility, which can be neglected, Δn_i is reduced to mobile electrons in the conduction band and holes in the valence band. And in the specific case of TiO_2, the TRMC signal can be attributed to electrons because their mobility is much larger than that of the holes.[51,52]

The incident microwaves were generated by a Gunn diode of the K_α band at 30 GHz. The pulsed light source was an OPO laser (EKSPLA, NT342B) tunable from 225 to 2000 nm. It delivers 8 ns fwhm pulses with a frequency of 10 Hz. The light energy densities received by the sample were 0.8, 1.5 and 1.4–2.2 mJ cm^{-2} at 360 nm, 450 nm and at the LSPR maximum (500 nm by default), respectively.

Photocatalytic tests

The photocatalytic set-up is composed of four main parts. (1) Generation of the reaction mixture was carried out by bubbling CO_2 (Air Product, 99.5%) through a thermo-controlled water saturator maintained at 35 °C (absolute humidity = 4.3%) at 0.3 mL min^{-1}. (2) The continuously flowing reaction mixture ($CO_2 + H_2O$) flow passes through a light-transparent photoreactor (6 mL) equipped with a Hg lamp top-irradiation system, simulating artificial solar light with light concentrator (150 W Ceramic-Metal-Halide Hg Lamp). The irradiance was 46 W m^{-2} for wavelengths between 300 and 400 nm and 5100 W m^{-2} for the whole lamp spectrum. The irradiated surface into the photoreactor is 18.8 cm^2. (3) Ventilators were fixed on the photoreactor to minimize temperature increase during photocatalytic tests. The maximum temperature of the reactor was about 40 °C during test. (4) Finally, products from the reactor were analysed online using a micro-GC analytical system (Agilent 3000A SRA instrument).

To perform photocatalytic tests, 50 mg of materials were homogeneously suspended in ethanol and slowly deposited on a 50 mm diameter glass disk until dry and further thermally treated at 100 °C for 2 h. The surface concentration of the photocatalyst on the glass disc was 25 g m^{-2}. The glass disk was then inserted into the photoreactor and the set-up was purged for 5 min with a high CO_2 flow (>100 mL min^{-1}) to remove air and other gas impurities. Before starting the test, the set-up was run with a CO_2 gas flow (0.3 mL min^{-1}) without illumination to achieve dark absorption equilibrium. Finally, the lamp was switched on. The standard duration of a photocatalytic test was 10 h.

Production rates were calculated according to these equations:

$$r_X\ (\text{mol h}^{-1}\ \text{g}^{-1}) = \frac{[X] \times 10^{-6} \times (\text{flow rate}) \times 60}{V_m \times m_{\text{photocat}}}$$

$$r_X\ (\text{mol h}^{-1}\ \text{m}_{\text{irradiated}}{}^{-2}) = \frac{[X] \times 10^{-6} \times (\text{flow rate}) \times 60}{V_m \times S_{\text{irradiated}}}$$

where [X] is the concentration of CH_4 or H_2 in ppm, $m_{photocat}$ is the mass of photocatalyst, $S_{irradiated}$ is the irradiated surface (0.001885 m^2), the flow rate is 0.0003 L min^{-1} and $V_m = 24.79$ L mol^{-1} (STP conditions).

Selectivity was calculated using these equations:

$$CH_4\ \text{selectivity} = \frac{8 \times r_{CH_4}}{8 \times r_{CH_4} + 2 \times r_{H_2}}$$

$$H_2\ \text{selectivity} = 1 - [CH_4\ \text{selectivity}]$$

The production rate per specific surface area of material (r_X/BET) was calculated by dividing the production rate per gram by the BET specific surface area of each compound.

Internal quantum yields η were calculated by this formula:

$$\eta = \frac{8 \times r_{CH_4} + 2 \times r_{H_2}}{\phi_{absorbed\ \ photons}} = \frac{n_{e^-}}{\phi_{absorbed\ \ photons}}$$

where $\phi_{absorbed\ \ photons}$ is the absorbed photon flux determined by UV-vis absorbance measurements.

Conclusions

By evaluating the solar and visible photocatalytic activities of a series of gold–titania nanocomposites for the reduction of carbon dioxide with water, we have shown the contribution of a plasmon-based process, in addition to the direct UV-induced SC-based process, in the overall solar transformation. Comparison with Pt composites and bare titania SCs suggests that the visible activity, induced by gold nanoparticles only, relies on the generation of hot carriers in the M NPs. Such a hot electron-mediated process may account for up to 20% of the overall solar activity in highly crystallized composites. It is more efficient over larger Au NPs (50 nm). One major specificity of the gold-mediated visible transformation is the 100% selectivity towards methane, *i.e.* the preferential reduction of CO_2 over H_2O. Dehydroxylation of titania is crucial to achieve such total selectivity.

Conflicts of interest

There are no conflicts to declare.

Acknowledgements

Financial support from IFPEN is gratefully acknowledged. The authors also wish to thank Dris Ihiawakrim (IPCMS, CNRS UMR 7504) for performing TEM analyses. ZC and CCJ gratefully acknowledge the financial support of the Agence Nationale de la Recherche (ANR) through the UpPhotoCat project.

Notes and references

1 W.-N. Wang, J. Soulis, Y. J. Yang and P. Biswas, *Aerosol Air Qual. Res.*, 2014, **14**, 533.

2 S. N. Habisreutinger, L. Schmidt-Mende and J. K. Stolarczyk, *Angew. Chem., Int. Ed.*, 2013, **52**, 7372.

3 J. K. Stolarczyk, S. Bhattacharyya, L. Polavarapu and J. Feldmann, *ACS Catal.*, 2018, **8**, 3602.

4 S. Neaţu, J. A. Maciá-Agulló, P. Concepción and H. Garcia, *J. Am. Chem. Soc.*, 2014, **136**, 15969.

5 Q. Zhai, S. Xie, W. Fan, Q. Zhang, Y. Wang, W. Deng and Y. Wang, *Angew. Chem., Int. Ed.*, 2013, **52**, 5776.

6 J. Ran, M. Jaroniec and S.-Z. Qiao, *Adv. Mater.*, 2018, **30**, 1704649.

7 F. Zaera, *Top. Catal.*, 2018, **61**, 336.

8 A. Manjavacas, J. G. Liu, V. Kulkarni and P. Nordlander, *ACS Nano*, 2014, **8**, 7630.

9 S. C. Warren and E. Thimsen, *Energy Environ. Sci.*, 2012, **5**, 5133.

10 Y. Zhang, S. He, W. Guo, Y. Hu, J. Huang, J. R. Mulcahy and W. D. Wei, *Chem. Rev.*, 2018, **118**, 2927.

11 M. L. Brongersma, N. J. Halas and P. Nordlander, *Nat. Nanotechnol.*, 2015, **10**, 25.

12 G. V. Hartland, L. V. Besteiro, P. Johns and A. O. Govorov, *ACS Energy Lett.*, 2017, **2**, 1641.

13 P. Christopher and M. Moskovits, *Annu. Rev. Phys. Chem.*, 2017, **68**, 379.

14 Y. Shiraishi, N. Yasumoto, J. Imai, H. Sakamoto, S. Tanaka, S. Ichikawa, B. Ohtani and T. Hirai, *Nanoscale*, 2017, **9**, 8349.

15 S. Mukherjee, L. Zhou, A. M. Goodman, N. Large, C. Ayala-Orozco, Y. Zhang, P. Nordlander and N. J. Halas, *J. Am. Chem. Soc.*, 2014, **136**, 64.

16 X.-C. Ma, Y. Dai, L. Yu and B.-B. Huang, *Light: Sci. Appl.*, 2016, **5**, e16017.

17 K. Wu, J. Chen, J. R. McBride and T. Lian, *Science*, 2015, **349**, 632.

18 M. J. Kale and P. Christopher, *Science*, 2015, **349**, 587.

19 V. Caps, S. Arrii, F. Morfin, G. Bergeret and J.-L. Rousset, *Faraday Discuss.*, 2008, **138**, 241.

20 I. Ro, J. Resasco and P. Christopher, *ACS Catal.*, 2018, **8**, 7368.

21 C. Marchal, T. Cottineau, C. Colbeau-Justin, V. Caps and V. Keller, *Adv. Energy Mater.*, 2018, **8**, 1702142.

22 E. Karamian and S. Sharifnia, *J. CO2 Util.*, 2016, **16**, 194.

23 S. Bai, N. Zhang, C. Gao and Y. Xiong, *Nano Energy*, 2018, **53**, 296.

24 Z. Xiong, Z. Lei, Y. Li, L. Dong, Y. Zhao and J. Zhang, *J. Photochem. Photobiol., C*, 2018, **36**, 24.

25 Y. Ji and Y. Luo, *J. Am. Chem. Soc.*, 2016, **138**, 15896.

26 L. Liu, H. Zhao, J. M. Andino and Y. Li, *ACS Catal.*, 2012, **2**, 1817.

27 P. Lignier, M. Comotti, F. Schüth, J.-L. Rousset and V. Caps, *Catal. Today*, 2009, **141**, 355.

28 O. K. Varghese, M. Paulose, T. J. LaTempa and C. A. Grimes, *Nano Lett.*, 2009, **9**, 731.

29 M. Zhang, Z. Jin, Z. Zhang and H. Dang, *Appl. Surf. Sci.*, 2005, **250**, 29; Q. Li, K. Wang, S. Zhang, M. Zhang, J. Yang and Z. Jin, *J. Mol. Catal. A: Chem.*, 2006, **258**, 83.

30 W. Gao, S. Sivaramakrishnan, J. Wen and J.-M. Zuo, *Nano Lett.*, 2015, **15**, 2548.
31 R. Van Hardeveld and F. Hartog, *Surf. Sci.*, 1969, **15**, 189.
32 M. Pelaez, N. T. Nolan, S. C. Pillai, M. K. Seery, P. Falaras, A. G. Kontos, P. S. M. Dunlop, J. W. J. Hamilton, J. A. Byrne, K. O'shea, M. H. Entezari and D. D. Dionysiou, *Appl. Catal., B*, 2012, **125**, 331.
33 K. M. Schindler and M. Kunst, *J. Phys. Chem.*, 1990, **94**, 8222.
34 P. Akhter, M. Hussain, G. Saracco and N. Russo, *Fuel*, 2015, **149**, 55.
35 M. Murdoch, G. I. N. Waterhouse, M. A. Nadeem, J. B. Metson, M. A. Keane, R. F. Howe, J. Llorca and H. Idriss, *Nat. Chem.*, 2011, **3**, 489.
36 C. Marchal, M. Behr, F. Vigneron, V. Caps and V. Keller, *New J. Chem.*, 2016, **40**, 4428.
37 C. Marchal, A. Piquet, M. Behr, T. Cottineau, V. Papaefthimiou, V. Keller and V. Caps, *J. Catal.*, 2017, **352**, 22.
38 V. Mendez, V. Caps and S. Daniele, *Chem. Commun.*, 2009, 3116.
39 A. Fkiri, A. Mezni, C. Robert, V. Caps and L. S. Smiri, *Colloids Surf., A*, 2017, **530**, 85.
40 S. Mukherjee, F. Libisch, N. Large, O. Neumann, L. V. Brown, J. Cheng, J. B. Lassiter, E. A. Carter, P. Nordlander and N. J. Halas, *Nano Lett.*, 2013, **13**, 240.
41 U. Aslam, V. G. Rao, S. Chavez and S. Linic, *Nat. Catal.*, 2018, **1**, 656.
42 Y. Tian and T. Tatsuma, *J. Am. Chem. Soc.*, 2005, **127**, 7632.
43 G. Ganguly, M. Sultana and A. Paul, *J. Phys. Chem. C*, 2017, **121**, 15326.
44 H. Zhang, S. Kawamura, M. Tamba, T. Kojima, M. Yoshiba and Y. Izumi, *J. Catal.*, 2017, **352**, 452.
45 R. Wang, K. Hashimoto, A. Fujishima, M. Chikuni, E. Kojima, A. Kitamura, M. Shimohigoshi and T. Watanabe, *Nature*, 1997, **388**, 431.
46 H.-Y. T. Chen, S. Tosoni and G. Pacchioni, *ACS Catal.*, 2015, **5**, 5486.
47 X. Zhang, X. Li, D. Zhang, N. Q. Su, W. Yang, H. O. Everitt and J. Liu, *Nat. Commun.*, 2017, **8**, 14542.
48 J. S. DuChene, B. C. Sweeny, A. C. Johnston-Peck, D. Su, E. A. Stach and W. D. Wei, *Angew. Chem., Int. Ed.*, 2014, **126**, 8021.
49 K. Qian, B. C. Sweeny, A. C. Johnston-Peck, W. Niu, J. O. Graham, J. S. DuChene, J. Qiu, Y.-C. Wang, M. H. Engelhard, D. Su, E. A. Stach and W. D. Wei, *J. Am. Chem. Soc.*, 2014, **136**, 9842.
50 X.-T. Kong, Z. Wang and A. O. Govorov, *Adv. Opt. Mater.*, 2017, **5**, 1600594.
51 O. A. Tahiri Alaoui, C. Herissan, M. E. Le Quoc, M. Zekri, S. Sorgues, H. Remita and C. Colbeau-Justin, *J. Photochem. Photobiol., A*, 2012, **242**, 34.
52 C. Colbeau-Justin, M. Kunst and D. Huguenin, *J. Mater. Sci.*, 2003, **38**, 24.
53 Y. Ohsaki, N. Masaki, T. Kitamura, Y. Wada, T. Okamoto, T. Sekino, K. Niihara and S. Yanagida, *Phys. Chem. Chem. Phys.*, 2005, **7**, 4157.
54 S. Mishra, V. Mendez, E. Jeanneau, V. Caps and S. Daniele, *Eur. J. Inorg. Chem.*, 2013, 500.
55 V. Mendez, K. Guillois, S. Daniele, A. Tuel and V. Caps, *Dalton Trans.*, 2010, **39**, 8457.
56 S. Biella, F. Porta, L. Prati and M. Rossi, *Catal. Lett.*, 2003, **90**, 1.
57 C. Deraedt, L. Salmon, S. Gatard, R. Ciganda, R. Hernandez, J. Ruiz and D. Astruc, *Chem. Commun.*, 2014, **50**, 14194.
58 K. Guillois, L. Burel, A. Tuel and V. Caps, *Appl. Catal., A*, 2012, **415–416**, 1.
59 W. Rasband, http://rsb.info.nih.gov/ij/.

Faraday Discussions

PAPER

Unearthing the factors governing site specific rates of electronic excitations in multicomponent plasmonic systems and catalysts†

Steven Chavez, Vishal Govind Rao and Suljo Linic *

Received 2nd October 2018, Accepted 8th November 2018
DOI: 10.1039/c8fd00143j

We use experimental and computational studies of core–shell metal–semiconductor and metal–molecule systems to investigate the mechanism of energy flow and energetic charge carrier generation in multicomponent plasmonic systems. We demonstrate that the rates of plasmon decay through the formation of energetic charge carriers are governed by two factors: (1) the intensity of the local plasmon induced electric fields at a specific location in the multicomponent nanostructure, and (2) the availability of direct, momentum conserved electronic excitations in the material located in that specific location. We propose a unifying physical framework that describes the flow of energy in all multicomponent plasmonic systems and leads us towards molecular control of the energy flow and excited charge carrier generation in these systems.

1. Introduction

Localized surface plasmon resonance (LSPR) is a collective and coherent oscillation of free electrons in plasmonic nanoparticles stimulated by electromagnetic radiation.[1–3] LSPR is accompanied by large optical extinction cross-sections at the resonant frequencies as well as elevated oscillating electric fields at the surface of the nanostructures. Compared to the electric fields of an incoming photon flux, these fields can be enhanced anywhere from 10^3 times at the surface of single particles to 10^6 times in between two particles separated by ~1 nm.[4–7] LSPR essentially acts to confine incoming electromagnetic energy within small nanovolumes at the surface of the nanoparticles. The energy of these enhanced fields is dissipated either radiatively *via* reemission of photons into the far field (scattering) or non-radiatively *via* the excitation of charge carriers (*i.e.* the formation of energetic electron/hole pairs) within the nanoparticle (absorption). In clean nanoparticles, the energy of these charge carriers is thermalized, *i.e.*, it is

Department of Chemical Engineering, University of Michigan – Ann Arbor, Ann Arbor, Michigan, 48109, USA. E-mail: linic@umich.edu

† Electronic supplementary information (ESI) available. See DOI: 10.1039/c8fd00143j

dissipated as heat directly within the nanostructure *via* electron–electron and subsequently electron–phonon collisions. This thermalization process leads to direct heating of the plasmonic nanostructure under LSPR conditions.[8,9]

Due to their high optical extinction cross-section and their ability to localize electric fields in small volumes, plasmonic nanoparticles are emerging as one of the most promising materials platforms for various applications. Many potential applications of plasmonic nanoparticles rely on spatially controlling the electronic excitation processes in plasmonic nanostructures. Specifically, there is a great deal of interest in creating multifunctional nanostructures where plasmonic nanoparticles that serve as nanoscopic light-energy concentrators are coupled with another material (metal, semiconductor or molecular entities) that utilizes this energy to perform a function.[10–17] In these multifunctional materials, energetic charge carriers would need to transfer from the plasmonic material to the other material before they are thermalized in the plasmonic metal, or preferentially form at high rates (stimulated by LSPR) directly in the other material rather than in the plasmonic metal (Fig. 1). For example, the efficient creation of energetic charge carriers in semiconductors or molecules or the efficient transfer of energetic charge carriers from plasmonic nanoparticles to the attached semiconductors or molecules is crucial in numerous applications including plasmon-enhanced photovoltaics,[18,19] plasmonic sensing,[20] plasmonic photocatalysis[21–25] and photothermal cancer therapy.[26–29] To be able to design such multifunctional nanostructures, a deeper understanding of the energy dissipation pathways in resonantly excited plasmonic multicomponent nanostructures and the factors that govern the flow of energy within these multicomponent plasmonic nanostructures is required.

We have previously reported the design of bimetallic plasmonic nanostructures in the form of Ag–Pt core–shell nanocubes in which relatively large Ag cubes (tens of nanometers) were covered with very thin Pt shells (~1 nm).[30] We demonstrated that in these multimetallic nanostructures, the plasmonic Ag core

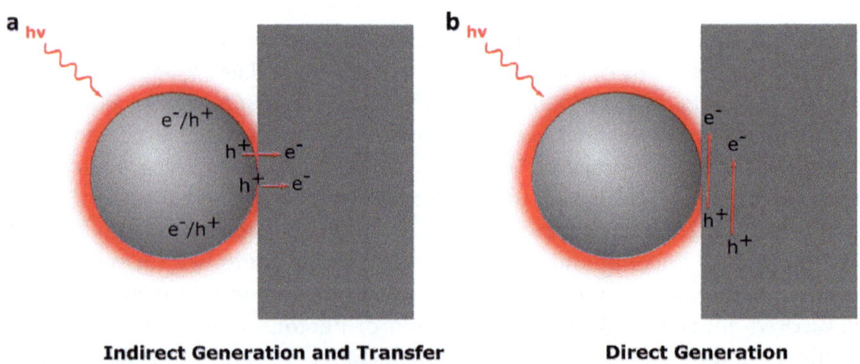

Fig. 1 Mechanisms of energetic charge generation/extraction from plasmonic nanoparticles to surrounding materials. (a) The electron–hole pairs are generated directly in the plasmonic nanostructure. A number of charge carriers with enough energy have an opportunity to transfer over to an attached material. The rest will be thermalized within the bulk of the plasmonic nanostructure. (b) The electron–hole pairs are generated directly in the attached material in regions close to the plasmonic nanostructure. This bypasses thermalization processes in the plasmonic nanostructure.

served to harvest the light energy through the excitation of LSPR and that this LSPR energy was dissipated preferentially through the formation of energetic charge carriers in the thin non-plasmonic Pt shell. We also developed a transparent physical framework that could describe the energy flow in this bimetallic plasmonic system.[31] We postulated that the local rate of formation of energetic charge carriers in the multimetallic nanostructures is to a large degree governed by two factors: (1) the local intensity of the electric field, and (2) the magnitude of the imaginary part of the dielectric function (ε_2) of the non-plasmonic metal relative to the plasmonic metal. In the above-mentioned Ag–Pt core–shell system, the Ag core confined the high field to the surface due to the plasmonic effect, and Pt has a significantly higher ε_2 than Ag at the LSPR frequencies (due to the availability of direct d to s electronic excitations). This led to the dissipation of LSPR energy through the formation of energetic e–h pairs directly in thin Pt shell. This proposed model and the experimental results provided us with a framework for the rational design of multicomponent systems where a plasmonic material guides the light energy to specific, sub-diffraction-limit volumes where this energy is dissipated through the formation of energetic charge carriers at high rates.

To further validate the proposed model for the flow of LSPR energy in multicomponent plasmonic systems, we move away from the above-mentioned bimetallic systems and focus on multicomponent metal–semiconductor and metal–molecule systems. We use experimental and theoretical studies to systematically investigate the LSPR decay mechanisms in core–shell Ag–Cu_2O (metal–semiconductor) and Ag–Methylene Blue (MB) (metal–molecule) multicomponent nanostructures. We show that the high fields generated at the LSPR wavelength can direct energy into the thin Cu_2O and MB shells, respectively, in a manner similar to the one described for the Ag–Pt core–shell structures. This results in the increased absorption in Cu_2O and the MB dye molecules, respectively.

2. Metal–semiconductor systems

To synthesize the Ag–Cu_2O core–shell nanostructures, we first synthesized Ag nanocube seeds using a well-established polyol-based method.[32–34] The resulting nanocubes were characterized using high-angle annular dark field scanning transmission electron microscopy (STEM). The average edge length of the Ag nanocubes was determined to be ~75 nm. The Ag nanocube seeds were then coated with a thin layer of Cu_2O by modifying a recently reported method.[35] Briefly, the Ag–Cu_2O nanoparticles were prepared by reducing a copper(II) nitrate precursor onto the Ag nanocube seeds with hydrazine in the presence of sodium hydroxide and polyvinylpyrrolidone. This resulted in the conformal coating of Cu_2O onto the surface of the Ag nanocubes. Fig. 2a shows a representative bright field STEM image of the Ag–Cu_2O nanocubes investigated in this study. Elemental mapping of the particles (Fig. 2b–e) using energy dispersive X-ray spectroscopy (EDS) shows that the Ag nanocube cores are covered by the Cu_2O shells. From the STEM images, we determined the average thickness of the Cu_2O shell to be ~5 nm.

To investigate the impact of the thin Cu_2O shell on the energy flow in these core–shell nanostructures, we studied the optical properties of the Ag–Cu_2O nanostructures by measuring their extinction, absorption and scattering and comparing these to the optical properties of the pure Ag nanocube seeds of

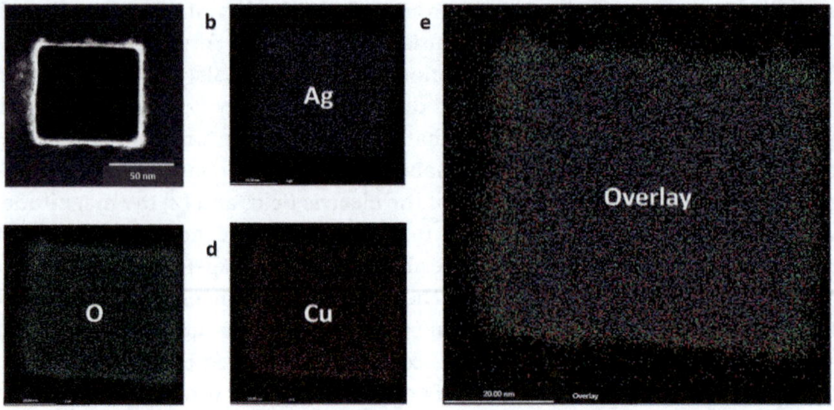

Fig. 2 (a) Bright field STEM image of a representative Ag–Cu_2O nanocube. The average edge length of the nanocubes was ~75 ± 2 nm. EDS elemental maps of Ag (b), O (c), Cu (d) and their overlay (e).

comparable size. The extinction characteristics of the nanocubes were measured using transmission UV-vis spectroscopy. The UV-vis data in Fig. 3a show that the introduction of the thin Cu_2O shell results in a slight red-shift and an increased linewidth of the LSPR extinction peak. This red-shift in the LSPR peak is the result of the higher refractive index of Cu_2O compared to air.[36] Additionally, the extinction linewidth is related to the plasmon lifetime, where larger linewidths correspond to shorter plasmon lifetimes.[37] Furthermore, an optical integrating

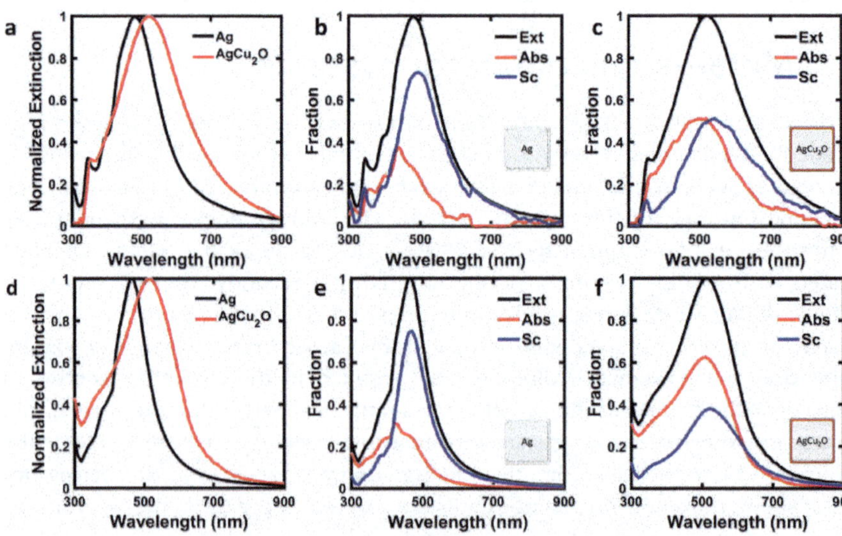

Fig. 3 (a) Measured normalized extinction of the Ag and Ag–Cu_2O nanocubes; (b and c) measured fractional extinction, absorption and scattering of (b) the Ag nanocubes, and (c) the Ag–Cu_2O nanocubes; (d) calculated normalized extinction of the Ag and Ag–Cu_2O nanocubes; (e and f) calculated fractional extinction, absorption and scattering of (e) the Ag nanocubes, and (f) the Ag–Cu_2O nanocubes. The edge dimensions of the Ag cube in the calculations was 75 nm and the thickness of the Cu_2O shell was 5 nm.

sphere was used to measure the partitioning between absorption and scattering for the Ag–Cu_2O nanocubes as well as the pure Ag nanocubes. The data in Fig. 3b and c show that the introduction of the Cu_2O shell modifies the LSPR decay pathway, resulting in an increased ratio of absorption to scattering when compared to the pure Ag nanocubes. Based on these data, we hypothesized that: (1) the Cu_2O shell was providing a faster pathway for the dissipation of the electromagnetic energy than Ag, and that the energy was preferentially dissipated through the absorption process in the thin Cu_2O shell. In other words, the introduction of Cu_2O at the surface increased the rate of formation of energetic e–h pairs at the surface of the nanostructure compared to the case of pure Ag; and (2) the presence of the Ag core increases the absorption rates (the rate of formation of charge carriers) in the Cu_2O shell, compared to the identical Cu_2O shell without the Ag plasmonic core effect.

Since the experimental data in Fig. 3 does not inform us about the location of the energetic e–h pair formation under the LSPR conditions, we performed finite element method (FEM) electrodynamic simulations to spatially resolve the absorption process within the nanostructures. The geometry constructed for the FEM model consisted of a 75 nm Ag nanocube surrounded by a 5 nm shell (corresponding to the dimensions determined from the STEM images) and the dielectric properties for Ag and Cu_2O were obtained from standard databases.[38,39] The data in Fig. 3d–f show that the calculated extinction, absorption and scattering properties for the Ag and Ag–Cu_2O core–shell nanocubes are consistent with the experimental measurements, showing that scattering is the dominant plasmon decay pathway for pure Ag cubes while absorption dominates for the Ag–Cu_2O particles.

We used the FEM model to shed light on the two hypotheses, (1) and (2) discussed above, that emerged from our experimental measurement of the optical properties. The data in Fig. 4a show the calculated power dissipated as a function of wavelength in the Cu_2O shell for the Ag–Cu_2O particle compared to the

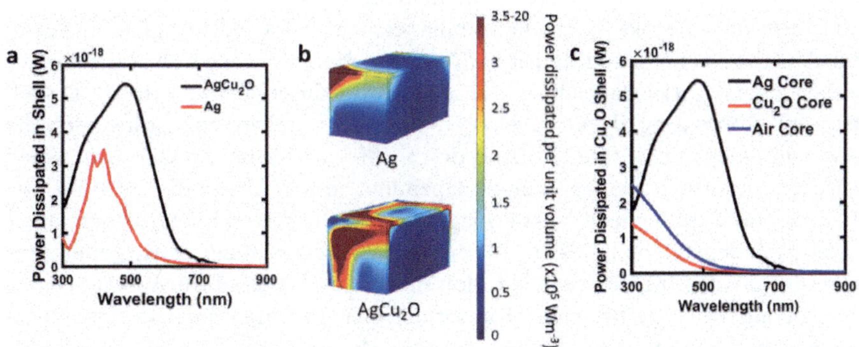

Fig. 4 (a) Total power dissipated as a function of wavelength in the particle shells for the Ag–Cu_2O particle and the Ag–Ag core–shell particle; (b) contour maps of the power dissipation per unit volume at the LSPR frequencies for the pure Ag particle and the Ag–Cu_2O particle. Due to the symmetry of the structures, we show the maps for a quarter of the cubes; (c) total power dissipated in the Cu_2O shell of core–shell particles with varying core materials. The source field for all simulations was 1 V m^{-1} at all frequencies, which for a free space amounts to 2.6×10^{-4} mW cm^{-2}.

calculated energy dissipation rate in a virtual Ag 'shell' of a reference Ag–Ag core–shell nanocube of identical dimensions. The calculated power is essentially a measure of the wavelength-dependent rate of formation of energetic e–h pairs in the shell. The data in Fig. 4a show that, as was hypothesized, the introduction of the thin Cu_2O shell increased the rate of formation of energetic e–h pairs at the surface of the nanostructure (in the thin Cu_2O shell region) compared to the case of pure Ag, *i.e.*, the process of photon absorption is shifted to the surface layers of the multicomponent nanostructure. This is further shown in Fig. 4b, which shows simulated contour maps of the power dissipated throughout the pure Ag nanocube compared to the Ag–Cu_2O nanocube. Upon integrating the power dissipated throughout the shells in these systems (*i.e.* the data in Fig. 4a) with respect to energy, we determined that ~6 times as much energy was being deposited through the shell of the Ag–Cu_2O particle than in the Ag shell of the pure Ag particle of identical size. To test the second hypothesis, we calculated the rate of absorption in the Cu_2O shell of the Ag–Cu_2O nanoparticle and compared it to the absorption of a Cu_2O shell with identical dimensions, supported on either a pure Cu_2O core or in the absence of a core (*i.e.* the core was defined as air). The data in Fig. 4c show that the absorption (*i.e.* generation of charge carriers) in the semiconductor (Cu_2O) shell, at the LSPR wavelengths, is significantly larger for the particle with the Ag plasmonic core. This suggests that in systems where a thin semiconductor shell is surrounding a plasmonic nanoparticle core, the plasmonic core can serve to enhance the rates of absorption in the shell compared to when the shell is supported on non-plasmonic materials.

As discussed above, we postulated that the site-specific rate of formation of energetic charge carriers in the multicomponent plasmonic nanostructures is to a large degree governed by two factors: (1) the intensity of the LSPR-induced electric field at the specific site, and (2) the magnitude of ε_2 of the material that resides at the site. We used the data in Fig. 3 and 4 to test the validity of the model. The first question is related to the impact of the local E-field on the rate of energy dissipation *via* absorption (*i.e.*, the rate of LSPR decay *via* localized absorption). It is well established that plasmonic nanostructures generate high electric fields that are confined to the surface of the nanostructure.[5,40] To investigate how the LSPR decay is affected by the field enhancement, we used the FEM model to calculate the electric field intensity $/E/^2$ as a function of wavelength at the surface of the Ag nanoparticle (specifically, at the Ag–Cu_2O interface) with the Cu_2O shell (Fig. 5a). Comparing these wavelength-dependent E-field intensities with the wavelength-dependent rate of absorption in the surface-bound thin Cu_2O shell (Fig. 4c) shows that the wavelength-dependent rate of absorption in the Cu_2O shell is heavily influenced by the magnitude of the electric field, where larger wavelength-dependent fields induce higher absorption rates. Furthermore, we have also divided the rate of absorption in the Cu_2O shell of the Ag–Cu_2O core–shell nanoparticle by the rate of absorption in the Cu_2O shell with identical dimensions, supported on either a pure Cu_2O core or on an air-filled core (as described above), and compared these ratios to ratios in the E-field intensity of the respective systems. The data in Fig. 5b show there is a quantitative agreement between the ratios in the local field intensities and the ratios in the local rates of absorption in the Cu_2O shells. This analysis shows that for a given multicomponent plasmonic material, it is the electric field induced by the plasmonic component that drives the process of photon absorption, *i.e.*, a higher local field

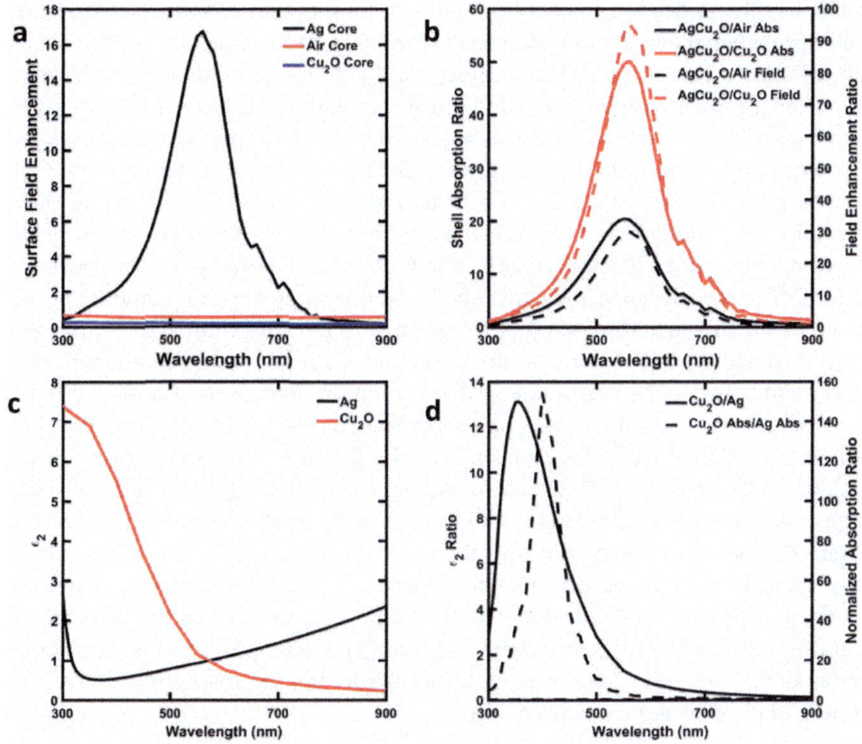

Fig. 5 (a) Calculated electric field enhancements at the core-Cu_2O interface for the varying core materials (Ag, air or Cu_2O); (b) field enhancement ratios (dashed lines) and shell absorption ratios (solid lines) for Ag–Cu_2O with respect to both the Cu_2O–Cu_2O and the Cu_2O–air core–shell systems; (c) the imaginary part of the dielectric function (ε_2) for both Ag and Cu_2O; (d) the ratio of ε_2 of Cu_2O and Ag (solid line) and the E-field-normalized shell absorption ratios of Cu_2O in the Ag–Cu_2O system and Ag in the Ag–Ag system (dashed line).

will induce higher rates of absorption. For a given material (in this case Cu_2O), these local plasmonic E-fields can significantly increase the rates of absorption compared to the situations where these fields are not present.

The second issue we address is related to the impact of the local imaginary part of the dielectric function (ε_2) on the energy dissipation pathway in multicomponent plasmonic nanostructures. The data in Fig. 4a show that the rate of energy dissipation is higher in the Cu_2O shell of the Ag–Cu_2O core–shell nanoparticle than in the Ag shell of the virtual Ag–Ag core–shell structure over a wide range of wavelengths. To explain this result, we plotted the value of ε_2 as a function of wavelength for Cu_2O and Ag in Fig. 5c. The data show that ε_2 is larger for Cu_2O than for Ag at wavelengths below ~590 nm, *i.e.*, at the wavelengths where the LSPR is excited. This difference in the value of ε_2 of the shell material can qualitatively explain the higher rate of energy dissipation through the Cu_2O shell compared to the virtual Ag shell. We note that this discrepancy in the magnitudes of ε_2 for Ag and Cu_2O stems from the difference in electronic structure. The magnitude of ε_2 of a material is directly related to the availability of direct momentum conserving electronic transitions in the material. Since Ag is a d^{10}

metal, its d band lies low in energy and direct, momentum conserved transitions are improbable at low photon energies, resulting in a low value of ε_2 in the visible range.[22] Above 2.1 eV, Cu_2O has a direct band gap, meaning its electronic transitions are momentum-conserved and it can efficiently absorb UV-visible photons with energy higher than 2.1 eV, resulting in a high ε_2 above this threshold energy.

Comparing these values of ε_2 for Cu_2O and Ag (Fig. 5c) with the rate of absorption in the Cu_2O shell in the Ag–Cu_2O particle and the equivalent Ag shell of a pure Ag nanoparticle (Fig. 4a) suggests that the availability of these direct electronic excitations in Cu_2O (that lead to high ε_2) is responsible for the elevated rates of plasmon decay in the Cu_2O shell. We note that the plasmonic E-field is different for the virtual Ag shell of the pure Ag nanocube compared to the Ag–Cu_2O nanocube. This difference in the E-field leads to some deviation in the agreement between the rate of energy dissipation through the Ag and Cu_2O shells and the relative magnitudes of ε_2 of the shell material. To account for this difference in the E-field we have normalized the shell absorption ratio in the Cu_2O and Ag shells on the Ag–Cu_2O and Ag–Ag nanostructures with the ratio of their respective E-field intensities. The data in Fig. 5d shows that the quantity that we obtain this way scales very well with the ratio of ε_2 for Cu_2O and Ag. This essentially means that for the systems where the local E-field is similar, the rate of energy dissipation *via* absorption can be nicely approximated by the value of ε_2. Simply put, by positioning a material with a larger ε_2 (compared to the plasmonic material) on the surface of a plasmonic nanostructure, the absorption channel is preferentially directed to that material.

3. Metal–molecule systems

The analysis presented above describes that the ability of plasmonic nanostructures to guide and dissipate energy in particular locations of a multicomponent plasmonic nanosystem is governed by two critical factors: (i) the local intensity of the electric field, and (ii) the availability of direct electronic transitions (high ε_2) at the wavelengths where the electric fields are high. To further generalize this model, we experimentally and computationally analyzed the optical properties of 75 nm Ag nanocubes interacting with methylene blue (MB) dye molecules. We created our Ag–MB solution by mixing a solution of ~75 nm Ag nanocubes dispersed in water with MB dye molecules. The resulting concentration of the MB in solution was 340 nM. It has been shown that at these low concentrations, the Ag particles are essentially surrounded by thin layers of MB, *i.e.*, solid state MB is adsorbed on the surface of Ag nanoparticles.[41] We used an integrating sphere to measure the optical characteristics (*i.e.* extinction, absorption and scattering) of our Ag–MB samples and compared them to the optical properties of a pure Ag nanocube solution of the same Ag cube concentration. The optical data in Fig. 6a and b show that the introduction of MB into the solution increases the overall absorption in the region around ~665 nm. MB is a dye molecule characterized by strong absorption at ~665 nm. To determine the effect of the Ag nanocubes on the MB absorption, we calculated the absorption in MB on the Ag nanocube surface by subtracting the pure Ag absorption from the Ag–MB absorption (the MB concentration is 340 nM) and comparing it to the absorption of a pure MB solution (340 nM). The data in Fig. 6c show that at the wavelength

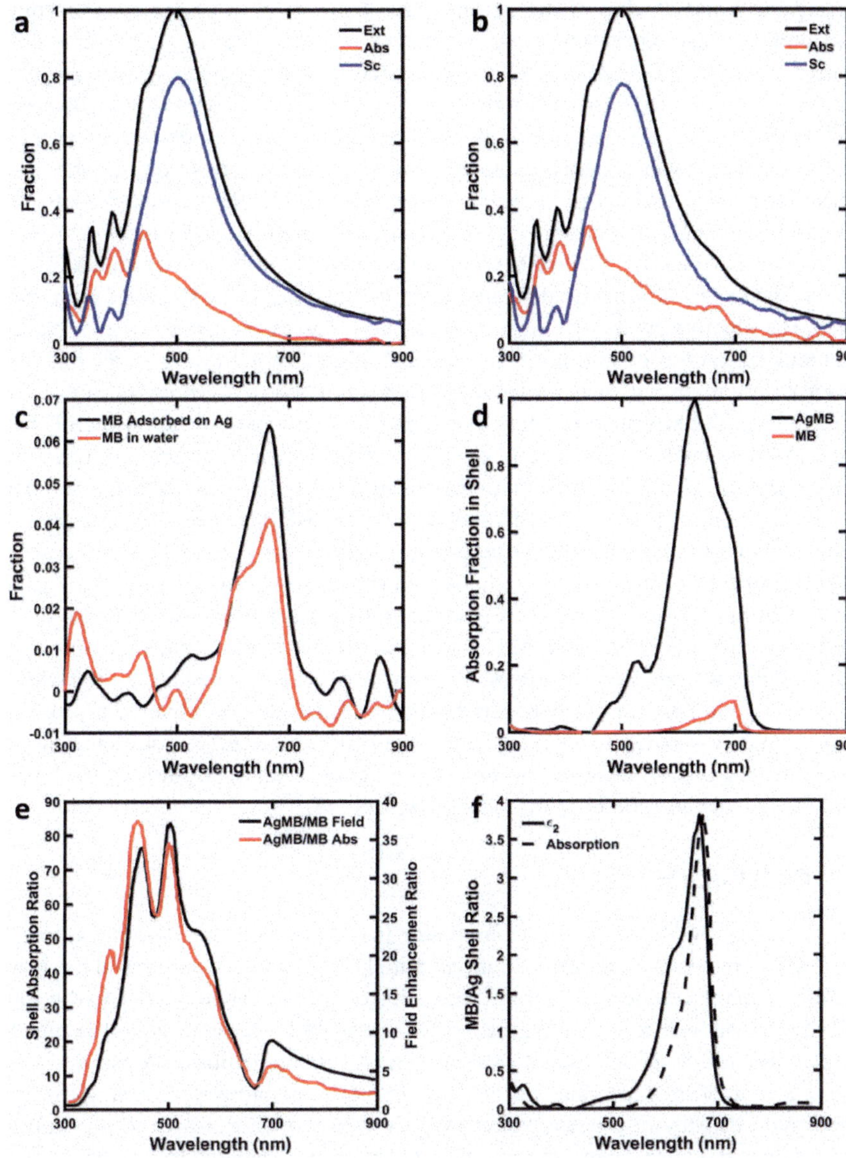

Fig. 6 (a and b) Measured fractional extinction, absorption and scattering of (a) the Ag nanocubes, and (b) the Ag–MB system; (c) absorption of MB in water (red line) and absorption of MB on the Ag nanoparticle surface; (d) calculated absorption in the Ag–MB shell *vs.* the MB shell in a reference MB shell with an MB core system; (e) field enhancement ratios and shell absorption ratios for Ag–MB with respect to the pure MB system; (f) the ratio of ε_2 of MB and Ag *vs.* the field normalized shell absorption ratios of MB in the Ag–MB system and Ag in the Ag–Ag system.

where MB absorbs, the absorption in MB in the presence of the Ag nanocubes is larger than that without the Ag nanocubes.

We used the FEM model to simulate the optical characteristics of these systems. The MB dielectric properties were determined by measuring the absorption cross section of the 340 nM MB solution (with UV-vis spectroscopy) and applying the Kramers–Kronig relations as shown previously.[42] The model system used in the FEM simulations consisted of a Ag nanocube with an edge length of 75 nm covered with a 1 nm MB "shell". To draw conclusions about the factors that govern the energy flow in these systems, we also modeled the behavior of two control model systems: an Ag–Ag core–shell nanocube (similar to one used above) and an MB–MB core–shell nanocube, both of identical dimensions to the Ag–MB nanocube. The data in Fig. 6d show that the simulated absorption spectra for MB on Ag (*i.e.*, the shell of the Ag–MB core shell model system) qualitatively match the experiments. The discrepancy in the spectral peak shape of the simulated MB absorption for the Ag–MB core–shell system compared to the experimentally measured absorption (Fig. 6c, dark line) stems from the fact that the absorption data in Fig. 6c is for solid MB films attached to Ag while the simulated spectra uses the MB dielectric function derived for low concentration mainly isolated MB molecules. We study the impact of the LSPR E-field on the absorption in this system by calculating the ratio of the absorption of MB in the Ag–MB core–shell system to the absorption in the MB shell of a model MB–MB core–shell system and comparing it to the ratio of field intensity at the respective core–shell interfaces (*i.e.* the ratio of the fields with and without a plasmonic Ag core). The data in Fig. 6e show that, like in the Ag–Cu_2O case, the absorption ratio tracks with the field ratio, demonstrating that the enhanced absorption in the shell is plasmon induced. Additionally, we explored the effect of ε_2 in this system by calculating the ratio of shell absorption in the Ag–MB particle to the shell absorption in the pure Ag particle and correcting it by the respective field contributions (in a similar manner to the Ag–Cu_2O case described above). The obtained data quantitatively track very well the calculated ratio of ε_2 for MB and Ag (Fig. 6f).

Conclusions

In summary, by studying multicomponent (*i.e.* core–shell metal–semiconductor and metal–molecule) systems, we demonstrated that both the electric field intensity under resonant LSPR conditions and the value of ε_2 of the shell relative to the core are the critical factors governing the flow of energy in these systems. In the metal–semiconductor case, we studied core–shell Ag–Cu_2O nanocubes. We showed that the presence of the Cu_2O shell altered the LSPR decay pathway, resulting in high absorption directly in the Cu_2O shell. We showed that this enhanced absorption process was field driven and that the field energy was dissipated to a greater extent when the shell material had a higher ε_2 with respect to the core material. Additionally, we showed that this framework also described the flow of energy in metal–molecule systems for the case of MB adsorbed on Ag nanocubes.

Conflicts of interest

There are no conflicts of interest to declare.

Acknowledgements

The work presented in this document was supported by the National Science Foundation (NSF) (CHE-1800197). Secondary support was provided by the

National Science Foundation (NSF) (CHE-17024711) (analysis of molecular interactions with nanostructures) and the Office of Basic Energy Science, Division of Chemical Sciences (FG-02-05ER15686) (materials synthesis). S. L. also acknowledges the partial support of the Technische Universität München – Institute for Advanced Study, funded by the German Excellence Initiative and the European Union Seventh Framework Programme under grant agreement no. 291763.

References

1 M. A. El-Sayed, Some Interesting Properties of Metals Confined in Time and Nanometer Space of Different Shapes, *Acc. Chem. Res.*, 2001, **34**, 257–264.

2 K. L. Kelly, E. Coronado, L. L. Zhao and G. C. Schatz, The Optical Properties of Metal Nanoparticles: The Influence of Size, Shape, and Dielectric Environment, *J. Phys. Chem. B*, 2003, **107**, 668–677.

3 C. F. Bohren and D. R. Huffman, *Absorption and Scattering of Light by Small Particles*, Wiley, 1998, p. 544.

4 L. Gunnarsson, T. Rindzevicius, J. Prikulis, B. Kasemo, M. Käll, S. Zou and G. C. Schatz, Confined Plasmons in Nanofabricated Single Silver Particle Pairs: Experimental Observations of Strong Interparticle Interactions, *J Phys. Chem. B*, 2004, **109**, 1079–1087.

5 E. Hao and G. C. Schatz, Electromagnetic Fields around Silver Nanoparticles and Dimers, *J. Chem. Phys.*, 2004, **120**, 357.

6 J. Jiang, K. Bosnick, M. Maillard and L. Brus, Single Molecule Raman Spectroscopy at the Junctions of Large Ag Nanocrystals, *J. Phys. Chem. B*, 2003, **107**, 9964–9972.

7 S. Linic, P. Christopher and D. B. Ingram, Plasmonic-Metal Nanostructures for Efficient Conversion of Solar to Chemical Energy, *Nat. Mater.*, 2011, **10**, 911–921.

8 C. Frischkorn and M. Wolf, Femtochemistry at Metal Surfaces: Nonadiabatic Reaction Dynamics, *Chem. Rev.*, 2006, **106**, 4207–4233.

9 M. L. Brongersma, N. J. Halas and P. Nordlander, Plasmon-Induced Hot Carrier Science and Technology, *Nat. Nano.*, 2015, **10**, 25–34.

10 A. M. Brown, R. Sundararaman, P. Narang, W. A. Goddard and H. A. Atwater, Nonradiative Plasmon Decay and Hot Carrier Dynamics: Effects of Phonons, Surfaces, and Geometry, *ACS Nano*, 2016, **10**, 957–966.

11 K. Li, N. J. Hogan, M. J. Kale, N. J. Halas, P. Nordlander and P. Christopher, Balancing Near-Field Enhancement, Absorption, and Scattering for Effective Antenna–Reactor Plasmonic Photocatalysis, *Nano Lett.*, 2017, **17**, 3710–3719.

12 M. Bernardi, J. Mustafa, J. B. Neaton, S. G. Louie, W. L. Barnes, A. Dereux, T. W. Ebbesen, M. Tame, J. S. Fakonas, H. Lee, *et al.*, Theory and Computation of Hot Carriers Generated by Surface Plasmon Polaritons in Noble Metals, *Nat. Commun.*, 2015, **6**, 7044.

13 A. Joplin, S. Ali Hosseini Jebeli, E. Sung, N. Diemler, P. J. Straney, M. Yorulmaz, W.-S. Chang, J. E. Millstone and S. Link, Correlated Absorption and Scattering Spectroscopy of Individual Platinum-Decorated Gold Nanorods Reveals Strong Excitation Enhancement in the Nonplasmonic Metal, *ACS Nano*, 2017, **11**, 12346–12357.

14 S. Linic, U. Aslam, C. Boerigter and M. Morabito, Photochemical Transformations on Plasmonic Metal Nanoparticles, *Nat. Mater.*, 2015, **14**, 567–576.

15 K. Wu, J. Chen, J. R. McBride, T. Lian, H. A. Atwater, A. Polman, C. Clavero, S. Linic, P. Christopher, D. B. Ingram, *et al.*, CHARGE TRANSFER. Efficient Hot-Electron Transfer by a Plasmon-Induced Interfacial Charge-Transfer Transition, *Science*, 2015, **349**, 632–635.

16 C. Boerigter, U. Aslam and S. Linic, Mechanism of Charge Transfer from Plasmonic Nanostructures to Chemically Attached Materials, *ACS Nano*, 2016, **10**, 6108–6115.

17 C. Boerigter, R. Campana, M. Morabito and S. Linic, Evidence and Implications of Direct Charge Excitation as the Dominant Mechanism in Plasmon-Mediated Photocatalysis, *Nat. Commun.*, 2016, **7**, 10545.

18 H. a. Atwater and A. Polman, Plasmonics for Improved Photovoltaic Devices, *Nat. Mater.*, 2010, **9**, 205–213.

19 F. J. Beck, A. Polman and K. R. Catchpole, Tunable Light Trapping for Solar Cells Using Localized Surface Plasmons, *J. Appl. Phys.*, 2009, **105**, 114310.

20 M. W. Knight, H. Sobhani, P. Nordlander and N. J. Halas, Photodetection with Active Optical Antennas, *Science*, 2011, 332.

21 P. Christopher, H. Xin and S. Linic, Visible-Light-Enhanced Catalytic Oxidation Reactions on Plasmonic Silver Nanostructures, *Nat. Chem.*, 2011, **3**, 467.

22 U. Aslam, V. G. Rao, S. Chavez and S. Linic, Catalytic Conversion of Solar to Chemical Energy on Plasmonic Metal Nanostructures, *Nat. Catal.*, 2018, **1**, 656–665.

23 S. Mukherjee, F. Libisch, N. Large, O. Neumann, L. V. Brown, J. Cheng, J. B. Lassiter, E. A. Carter, P. Nordlander and N. J. Halas, Hot Electrons Do the Impossible: Plasmon-Induced Dissociation of H_2 on Au, *Nano Lett*, 2013, **13**, 240–247.

24 S. Mubeen, J. Lee, N. Singh, S. Krämer, G. D. Stucky and M. Moskovits, An Autonomous Photosynthetic Device in Which All Charge Carriers Derive from Surface Plasmons, *Nat. Nano.*, 2013, **8**, 247–251.

25 P. Christopher, H. Xin, A. Marimuthu and S. Linic, Singular Characteristics and Unique Chemical Bond Activation Mechanisms of Photocatalytic Reactions on Plasmonic Nanostructures, *Nat. Mater.*, 2012, **11**, 1044.

26 X. Huang, I. H. El-Sayed, W. Qian and M. A. El-Sayed, Cancer Cell Imaging and Photothermal Therapy in the Near-Infrared Region by Using Gold Nanorods, *JACS*, 2006, **128**, 2115–2120.

27 L. B. Carpin, L. R. Bickford, G. Agollah, T.-K. Yu, R. Schiff, Y. Li and R. A. Drezek, Immunoconjugated Gold Nanoshell-Mediated Photothermal Ablation of Trastuzumab-Resistant Breast Cancer Cells, *Breast Cancer Res. Treat.*, 2011, **125**, 27–34.

28 M. R. K. Ali, Y. Wu, Y. Tang, H. Xiao, K. Chen, T. Han, N. Fang, R. Wu and M. A. El-Sayed, Targeting Cancer Cell Integrins Using Gold Nanorods in Photothermal Therapy Inhibits Migration through Affecting Cytoskeletal Proteins, *Proc. Natl. Acad. Sci. U. S. A.*, 2017, **114**, E5655–E5663.

29 I. H. El-Sayed, X. Huang and M. A. El-Sayed, Selective Laser Photo-Thermal Therapy of Epithelial Carcinoma Using Anti-EGFR Antibody Conjugated Gold Nanoparticles, *Cancer Lett.*, 2006, **239**, 129–135.

30 U. Aslam, S. Chavez and S. Linic, Controlling Energy Flow in Multimetallic Nanostructures for Plasmonic Catalysis, *Nat. Nano.*, 2017, **12**, 1000–1005.

31 S. Chavez, U. Aslam and S. Linic, Design Principles for Directing Energy and Energetic Charge Flow in Multicomponent Plasmonic Nanostructures, *ACS Energy Lett.*, 2018, **3**, 1590–1596.

32 Y. Sun and Y. Xia, Shape-Controlled Synthesis of Gold and Silver Nanoparticles, *Science*, 2002, **298**, 2176–2179.

33 S. Linic and P. Christopher, Overcoming Limitation in the Design of Selective Solid Catalysts by Manipulating Shape and Size of Catalytic Particles: Epoxidation Reactions on Silver, *ChemCatChem*, 2010, **2**, 1061–1063.

34 T. Van Cleve, E. Gibara and S. Linic, Electrochemical Oxygen Reduction Reaction on Ag Nanoparticles of Different Shapes, *ChemCatChem*, 2016, **8**, 256–261.

35 H. Jing, N. Large, Q. Zhang and H. Wang, Epitaxial Growth of Cu_2O on Ag Allows for Fine Control Over Particle Geometries and Optical Properties of Ag–Cu_2O Core–Shell Nanoparticles, *J Phys. Chem. C.*, 2014, **118**, 19948–19963.

36 U. Kreibig and M. Vollmer, *Optical Properties of Metal Clusters*, 1995.

37 S. A. Maier, *Plasmonics: Fundamentals and Applications*, Springer, 2007.

38 A. D. Rakić, A. B. Djurišić, J. M. Elazar and M. L. Majewski, Optical Properties of Metallic Films for Vertical-Cavity Optoelectronic Devices, *Appl. Opt.*, 1998, **37**, 5271–5283.

39 C. G. Ribbing and A. Roos, *Handbook of Optical Constants of Solids*, 1997, pp. 875–882.

40 P. L. Stiles, J. A. Dieringer, N. C. Shah and R. P. Van Duyne, Surface-Enhanced Raman Spectroscopy SERS: Surface-Enhanced Raman Spectroscopy Raman Scattering: Inelastic Scattering of a Photon from a Molecule in Which the Frequency Change Precisely Matches the Difference in Vibrational Energy Levels, *Annu. Rev. Anal. Chem*, 2008, **1**, 601–626.

41 B. L. Darby, B. Auguié, M. Meyer, A. E. Pantoja and E. C. Le Ru, Modified Optical Absorption of Molecules on Metallic Nanoparticles at Sub-Monolayer Coverage, *Nat. Photon.*, 2016, **10**, 40–45.

42 A. Djorovic, M. Meyer, B. L. Darby and E. C. Le Ru, Accurate Modeling of the Polarizability of Dyes for Electromagnetic Calculations, *ACS Omega*, 2017, **2**, 1804–1811.

Faraday Discussions

PAPER

Plasmon-induced optical control over dithionite-mediated chemical redox reactions

Junyang Huang, †[a] Bart de Nijs, †[a] Sean Cormier, [a] Kamil Sokolowski, [b] David-Benjamin Grys, [a] Charlie A. Readman, [b] Steven J. Barrow, [b] Oren A. Scherman [b] and Jeremy J. Baumberg *[a]

Received 16th October 2018, Accepted 8th November 2018

DOI: 10.1039/c8fd00155c

External-stimuli controlled reversible formation of radical species is of great interest for synthetic and supramolecular chemistry, molecular machinery, as well as emerging technologies ranging from (photo)catalysis and photovoltaics to nanomedicine. Here we show a novel hybrid colloidal system for light-driven reversible reduction of chemical species that, on their own, do not respond to light. This is achieved by the unique combination of photo-sensitive plasmonic aggregates and temperature-responsive inorganic species generating radicals that can be finally accepted and stabilised by non-photo-responsive organic molecules. In this system Au nanoparticles (NPs) self-assembled *via* sub-nm precise molecular spacers (cucurbit[*n*]urils) interact strongly with visible light to locally accelerate the decomposition of dithionite species ($S_2O_4^{2-}$) close to the NP interfaces. This light-driven process leads to the generation of inorganic radicals whose electrons can then be reversibly picked up by small organic acceptors, such as the methyl viologen molecules (MV^{2+}) used here. During light-triggered plasmon- and heat-assisted generation of radicals, the $S_2O_4^{2-}$ species work as a chemical 'fuel' linking photo-induced processes at the NP interfaces with redox chemistry in the surrounding water environment. By incorporating MV^{2+} as a Raman-active reporter molecule, the resulting optically-controlled redox processes can be followed in real-time.

Introduction

Using light to control chemistry on the nanoscale has gathered much attention over recent years, for instance in light controlled molecular switching,[1]

[a]NanoPhotonics Centre, Cavendish Laboratory, Department of Physics, University of Cambridge, JJ Thompson Avenue, Cambridge, CB3 0HE, UK. E-mail: bd355@cam.ac.uk; jjb12@cam.ac.uk

[b]Melville Laboratory for Polymer Synthesis, Department of Chemistry, University of Cambridge, Lensfield Road, Cambridge CB2 1EW, UK

† These authors contributed equally.

optomechanics,[2] strong coupling,[3] and non-thermal plasmon-enhanced catalytic reactions,[4,5] all shown to be readily accessible using plasmonic nanoconstructs. Moreover, the ability to optically-control energy deposition through light absorption in plasmonic nanoconstructs is of great interest for emerging technologies providing a means to thermally activate localized (electro)chemical processes.[6–15] Such systems are of particular interest for their facile accessibility as they can be readily made using self-assembly, eliminating the need for expensive manufacturing processes, and typically proceed under ambient conditions.

Here, we present direct optical control over dithionite-mediated redox processes in aqueous environments using self-assembled gold nanoparticles (AuNPs). By self-assembling gold nanoparticles using either salt or sub-nm molecular spacers (cucurbit[*n*]urils: CB[*n*]), aggregates with broad spectral absorbance are formed. Both the AuNP aggregates assembled using salt or CB[*n*] can be utilized as efficient SERS substrates.[16–18] A redox reporter molecule such as methyl viologen (MV^{2+}) can be added to the substrates allowing for surface enhanced Raman spectroscopy (SERS) to be used to track *in situ* the dithionite redox processes in real-time.[17] We demonstrate that by switching between high and low laser irradiances, this reduction process can be turned on and off and the change in redox state of the reporter molecules can be directly observed. Such optical control over reduction opens up new avenues for switching (electro-) chemical processes, for example in fuel cells or batteries.

Results and discussion

Our hybrid system comprises (i) light-sensitive SERS-active Au nanoparticles self-assembled *via* sub-nm precise molecular spacers of CB[7] or by increasing the salt concentration, (ii) temperature-responsive inorganic dithionite species, $S_2O_4{}^{2-}$, and (iii) a non-photoresponsive organic electron-accepting molecule, MV^{2+}. The dithionite $S_2O_4{}^{2-}$ species in a water environment undergoes dissociation and complex transformations leading to a variety of free radicals, from which sulphur dioxide radical anions ($SO_2{}^{-\bullet}$) appear as the effective reducing agent in many redox reactions.[18] The dissociation of $S_2O_4{}^{2-}$ is temperature- and pH-dependent. At room temperature and neutral pH the formation of radical species can be thermally accelerated.[21] The redox potential of sulphur oxide radicals allows one-electron reduction of MV^{2+} leading to methyl viologen radical monocations, $MV^{+\bullet}$. Reduction of MV^{2+} is reversible in the presence of dioxygen, and in the case of chemical reduction by dithionite species can only appear once the concentration of sulphur dioxide radical anions is higher than that of dioxygen in a reaction environment ($[SO_2{}^{-\bullet}] > [O_2]$).[20,22] MV^{2+} was selected as the final electron acceptor on account of its characteristic Raman signatures for MV^{2+} and $MV^{+\bullet}$ species which allows one to distinguish between oxidized and reduced states and therefore easily track redox process dynamics by SERS.[19]

By adding cucurbit[*n*]uril to a colloidal suspension of citrate stabilized AuNPs, self-assembly forms aggregates with fractal-like structures,[23] red-shifting the plasmon resonance in extinction spectroscopy (Fig. 1a). As a result the extinction at 785 nm increases from 12% to 83% cm^{-1}. While a fraction of the extinction is plasmonic scattering, the remainder is converted to heat through non-radiative

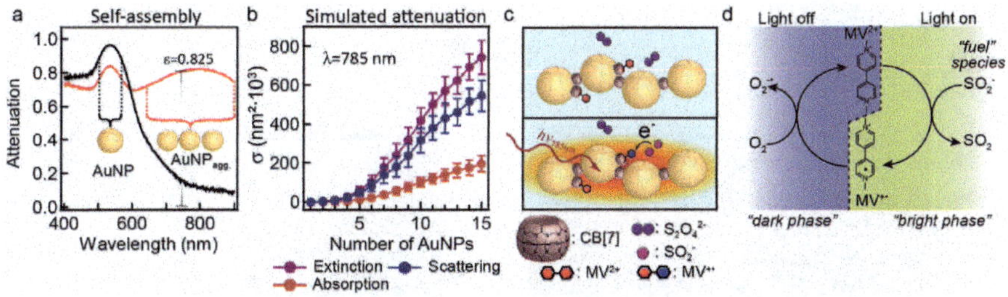

Fig. 1 Plasmonic-mediated chemistry. (a) Extinction spectra of the 60 nm AuNP suspension (black) and CB[7] spaced self-assembled AuNP aggregates (red) showing (b) 83% extinction per cm at 785 nm after 10 minutes of self-assembly. (c) Schematic illustration of thermally-activated dithionite reduction of MV^{2+}. (Top) At room temperature and ambient conditions reactions do not proceed, while (bottom) plasmon-induced heating catalyses dissociation and redox chemistry. (d) Scheme depicting the reduction and oxidation cycle of MV^{2+} by dithionite and oxygen, respectively.

relaxation processes in the Au (Fig. 1b). Such localized heating is used here to catalyse chemical reactions. The elevated temperatures accelerate the dissociation of the chemical 'fuel' $S_2O_4^{2-}$ species (Fig. 1c), leading to the formation of highly reactive $SO_2^{-\bullet}$ radical anions which efficiently reduce MV^{2+} molecules forming $MV^{+\bullet}$ radicals (Fig. 1d; bright phase). The high concentration of $MV^{+\bullet}$ radicals is maintained while the light is kept on and enough $S_2O_4^{2-}$ is present in the system. Once the light is switched off, $MV^{+\bullet}$ radicals are quickly re-oxidized by oxygen molecules, restoring the MV^{2+} (Fig. 1d; dark phase). The system is reversible and this cycle can be repeated until the $S_2O_4^{2-}$ fuel is consumed. Using a 785 nm laser to plasmonically induce the local temperature increase, the activated chemical redox processes can be followed in real time using surface enhanced Raman spectroscopy, giving a much better idea of the complexity involved in such redox systems.

After several seconds of 125 mW laser power irradiation at 785 nm a set of clear changes are observed in the SERS spectra. The appearance of the 1660, 1600, 1357, 1248, and 1025 cm^{-1} bands are characteristic of the cation $MV^{+\bullet}$ species being formed.[24] Upon lowering the laser irradiation intensity to 11 mW (so that SERS spectra can still be collected) the MV^{2+} spectrum returns after several seconds. This redox process is repeatable until the dithionite 'fuel' for the chemical reaction is depleted. The reduction process can also be reproduced through 'bulk' heating of the entire cuvette. The bulk solution temperature at which this reduction is seen in SERS switching is 29.3 °C, 9.3 °C above room temperature. Comparing the different SERS spectra we find that the low power (Fig. 2a) and low temperature (Fig. 2b) spectra closely match the powder Raman spectrum of MV^{2+} (Fig. 2e, black and grey traces), while the reduced state observed at 125 mW (Fig. 2a) and higher temperatures (Fig. 2b) show many similarities to each other but are distinctly different from the MV^{2+} spectrum (Fig. 2e, red and green traces, respectively). Slight differences in the spectra between optically- and temperature-induced reduction suggest variations in the local environment as a result of a potentially different dissociation mechanism of dithionite. Spectra acquired with CB[7]-mediated aggregates and salt induced aggregates for the reduced species have almost identical spectral features.

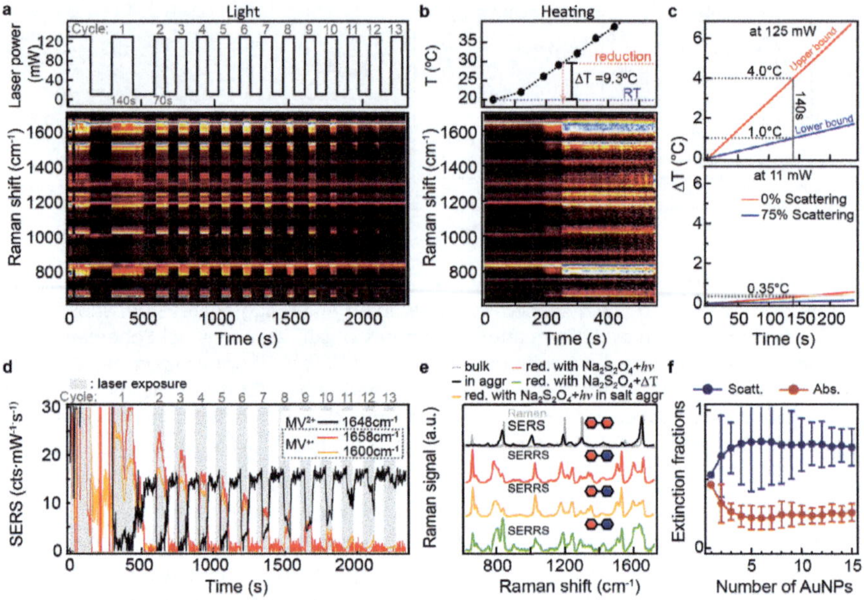

Fig. 2 Optical switching and heat activation of the MV^{2+} redox process. Normalized SERS time scans of methyl viologen in the presence of sodium dithionite (a) under modulated irradiation power, and (b) with temperature ramping of the entire suspension. Both spectra are normalised by the integrated intensity from 1550 to 1700 cm^{-1}. (c) Calculated overall temperature change of the suspension caused by optical absorption with a laser power of 125 mW and 11 mW. Upper bound calculated assuming the incident light is fully absorbed by the AuNP aggregates, while the lower bound is calculated assuming 83% of irradiation at 785 nm is scattered. (d) Dynamics of SERS intensity of MV redox states (MV^{2+}: 1648 cm^{-1}, $MV^{+\bullet}$: 1658 cm^{-1}, 1600 cm^{-1}) under modulated laser power, with high laser exposure (125 mW) marked by shaded areas. (e) Raman spectra of the MV^{2+} solution with (black) and without (grey) the presence of AuNP aggregates, SERRS spectrum of $MV^{+\bullet}$ upon optically (red) and thermally (green) induced reduction with dithionite. (f) Calculated scattering and absorption fraction for different sized aggregates ranging from 1 to 15 AuNPs.

By considering the total laser power on the sample, an upper and lower estimate of the laser-induced bulk temperature increase per second can be estimated from

$$\Delta T = \frac{I_0 t \varepsilon_\lambda A}{C_Q m}, \tag{1}$$

where I_0 is the incident laser intensity in J s^{-1}, t the duration of exposure, ε_λ the extinction of light over the optical path at the laser wavelength, A the fraction of absorbed light, C_Q the specific heat capacity and m the mass of water in the sample. Assuming all of the attenuated light is absorbed ($C_Q = 1$) an upper bound for ΔT is determined (red trace, Fig. 2c). Modelling the scattering properties of individual AuNP aggregates shows 25% absorbance and 75% scattering (Fig. 2f), providing a lower bound for ΔT ($C_Q = 0.25$). From this we find that the average temperature of the bulk suspension is increased by <4 °C after 140 s of laser irradiation at 125 mW.

In this estimate, heat dissipation to the surrounding environment is not considered, however the calculated ΔT_{max} is already insufficient to trigger the reduction by bulk temperature alone. The temperature increase of a single AuNP aggregate, ΔT_{agg}, due to optical absorption in a homogeneous medium can be estimated from

$$\Delta T_{agg} = \frac{\sigma_{abs} E_e}{4\pi R_{eff} \kappa_m}, \tag{2}$$

where σ_{abs} is the absorption cross section, E_e the incident irradiance of light, R_{eff} the effective radius of the aggregate,[25] and κ_m the thermal conductivity of the surrounding medium. For a simulated $\sigma_{abs} = 2 \times 10^5$ nm^2, $\kappa_{water} = 0.6$ W m^{-1} K^{-1}, and $R_{eff} = 160$ nm, we find an estimated average temperature increase of 1.6 °C on a single aggregate. Even if we consider only the gap region, which has a local plasmonic heating efficiency 1.5 times higher compared to the effective heating zone within 10 nm proximity from the aggregate, such a temperature increase is inadequate to reach $\Delta T_{water} > 9$ °C to catalyse the reduction process. This suggests that other local effects in the nanostructure play a key role in the catalysis of the reduction process. For example, the diffusion rate of oxygen molecules can be limited inside the confined gap regions, which in turn can allow the plasmonic heating to locally trigger the redox process at a lower temperature. Additionally, other optical or hot-electron effects on the chemistry process cannot be fully excluded at this stage.

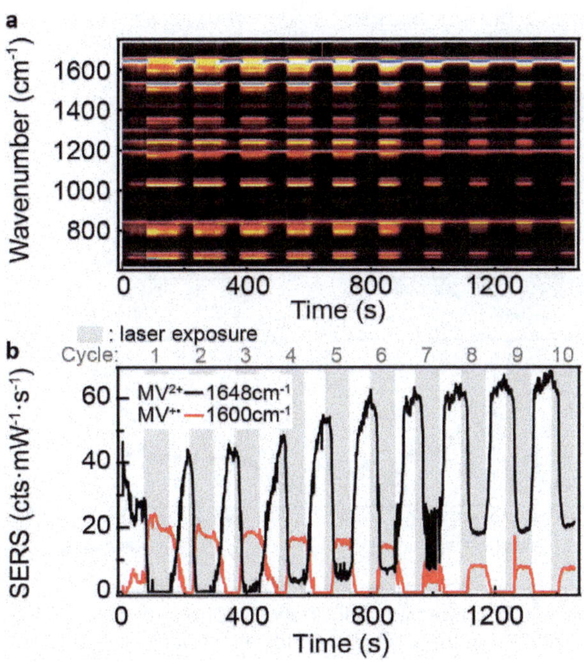

Fig. 3 Optical switching with salt induced AuNP aggregates. (a) Normalized SERS time scans of methyl viologen in the presence of sodium dithionite, under modulated irradiation power. (b) Correspondent SERS dynamics of MV redox states (MV^{2+}: 1648 cm^{-1}, $MV^{+\bullet}$: 1600 cm^{-1}), with high laser exposure (125 mW) marked by shaded areas.

Using salt to induce the AuNP aggregation yields a comparable switching phenomenon, however, two clear differences are observed (see Fig. 3). During the laser power modulation, the MV^{2+} signals continue to rise with each cycle, visible at the low powers . This contrasts with CB[7] in which a consistent peak amplitude is reached through each oxidation cycle. This is also in line with quantitative SERS results observed with CB[n] aggregates in literature.[17,18,26] The second notable difference is the faster (roughly two-fold) and more reliable redox switching observed for the salt induced aggregates. Switching with CB[5] formed aggregates is also investigated here, and comparable redox speeds are observed. However with CB[5], the aggregate-induced redox states are less stable, giving more fluctuating SERS signatures.

To investigate the switching behaviour with CB[7] and salt induced aggregates in more detail SERS signals extracted from the 1600 cm^{-1} ($MV^{+\bullet}$) and 1648 cm^{-1} (MV^{2+}) modes are compared between each reduction and oxidation cycle (Fig. 4a). By taking the first derivative peak this delay time can be plotted as a function of cycle number (Fig. 4b). To reach the triggering temperature of 29.3 °C for each cycle, the switching time is expected to occur faster on successive cycles as the

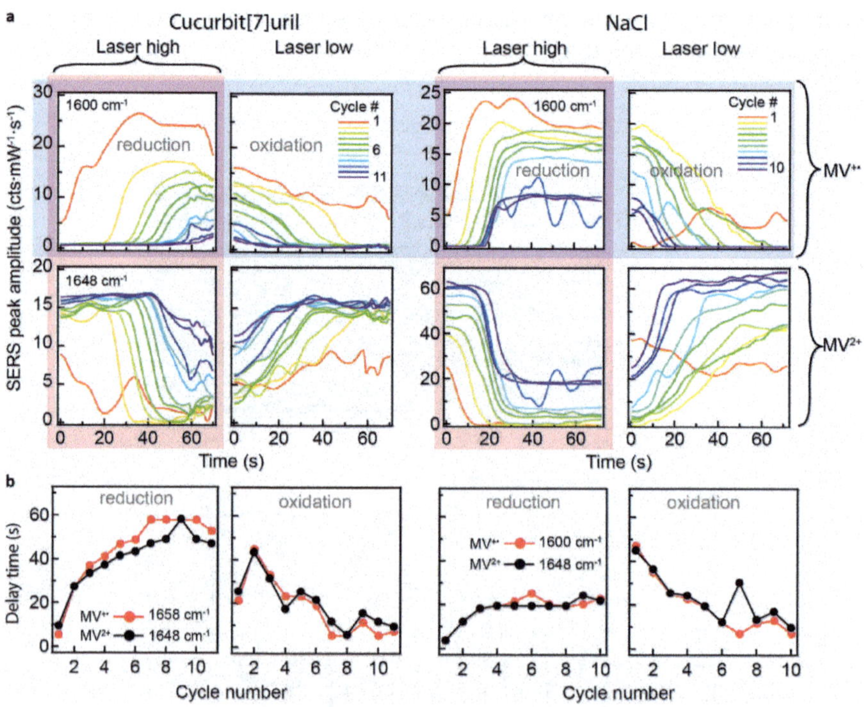

Fig. 4 Kinetics of the plasmon catalysed redox process with CB[7] and salt induced AuNP aggregates. (Left) CB[7] aggregates and (right) NaCl aggregates. (a) Evolution of SERS intensity of MV redox states for successive modulation cycles of laser power, with (top) the increase and decrease of the radical $MV^{+\bullet}$ species and (bottom) the correlated decrease and subsequent increase of the MV^{2+} species. (b) Extracted delay times of reduction and oxidation thresholds *versus* cycle number for high (left) CB[7] aggregates and (right) NaCl aggregates. The delay times are determined by the turning points of the first derivative of the SERS kinetics in (a).

solution accumulates heat. However, an opposite trend is observed in the case with CB[7] mediated aggregates. For these aggregates a clear increase in delay time for the reduction threshold is seen, whereas when salt is used to form aggregates, the delay time is shorter and after 3 cycles stabilises at 20 s. The re-oxidations of both systems show a comparable trend in which the oxidation process seems to consistently speed up with each cycle. These differences between the CB[7] and NaCl show that subtle variations in how the plasmonic substrates are formed can greatly affect the resulting switching behaviour suggesting that there is still much room for improvement. When fully developed, such a system can form a basis for promising radical-on-demand technologies.

Conclusions

We have demonstrated that plasmon-induced effects can be used to catalyse dithionite-mediated chemical reduction with either CB[7] or salt induced AuNP aggregates. By controlling irradiation intensities this reduction process can be switched on and off repeatably. We have shown that the localised effects close to the plasmonic aggregates is what causes redox switching, rather than bulk heating, and that the dithionite fuel controls the system response time.

Materials and methods

Dithionite was purchased from Sigma Aldrich, 60 nm citrate capped AuNPs were purchased from BBI solutions, and both were used as received. Cucurbit[*n*]uril spacer molecules were synthesized as described in ref. 27.

AuNP aggregates were assembled in an aqueous environment by mixing a CB[*n*] solution and a 60 nm citrate-capped colloidal AuNP suspension in a polystyrene cuvette (Fisher Scientific). For CB[*n*]-bridged aggregation, typically, 2 μL of 0.8 g L^{-1} CB[*n*] solution was first prepared in the cuvette and 1 mL of AuNP suspension was added to trigger the self-assembly process. After 1 min of CB[*n*] mediated aggregation, 3.6 μL of 0.1 g L^{-1} methyl viologen solution (Sigma-Aldrich) was mixed in to give $C_{CB[n]}$: $C_{MV^{2+}} = 1$. This was further followed by the addition of a small amount (0.1–0.2 mg) of sodium dithionite powder (Sigma-Aldrich) using a powder spatula. For salt induced aggregation, 10 μL of 5 molar NaCl solution was mixed with 1 mL of AuNP suspension, followed by the addition of 20 μL of 0.1 g L^{-1} methyl viologen solution and 0.2 mg of sodium dithionite powder.

The extinction, absorption, and scattering efficiencies of the AuNP clusters of varying size at a 785 nm wavelength are calculated using Mackowski's Multiple Sphere T-Matrix (MSTM, v3.0) code (available open source). The calculations use 100 iterations of randomly grown 60 nm gold nanoparticle cluster structures with defined inter-particle spacings (0.9 nm). The algorithm for generating the clusters consists of first adding a new AuNP onto the surface of the cluster from a randomized position and direction, and then allowing the new particle to relax to the nearest surface dimer at the specified interparticle distance (thus touching at least two other NPs). The randomly grown AuNP clusters are then surrounded with an external medium of water ($n = 1.33$). The dielectric function of the gold nanoparticles is interpolated from Johnson and Christy.

Raman spectroscopy was performed on a Renishaw inVia Raman microscope, using a 5× objective (NA = 0.12) and a 785 nm excitation diode laser. Such an excitation wavelength is known to resonantly excite the chain plasmons of the AuNP aggregates.[28,29] The laser beam was focused at the liquid–air interface from the top of the open cuvette (1 × 0.4 × 2.5 cm), without having the cuvette wall in the path of either the incident or scattered light. The spot size of the laser at the focal point with the 5× objective was measured to be 12 500 μm^2. Spectra were collected with a 1 s integration time and the irradiation power was alternated between 125 mW and 11 mW, where each power was maintained for 70 s. The optical depth of the suspension in the vertical direction was 2.5 cm. To control the temperature of the mixture the cuvette was placed in a water bath in contact with a hot plate. A thermometer was used to monitor the temperature of the water bath during the SERS measurement.

Conflicts of interest

There are no conflicts to declare.

Acknowledgements

We acknowledge financial support from EPSRC (Grants EP/L027151/1, EP/L015889/1 and EP/N020669/1) and the European Research Council grant BioNet 757850. B. d. N. acknowledges support from the Leverhulme Trust and Newton Trust. K. S. and S. J. B. thank the European Commission for a Marie Skłodowska-Curie Fellowship (ESTIMABLeNANO, 706425, NANOSPHERE, 658360, respectively).

References

1 Y. B. Zheng, *et al.*, Active Molecular Plasmonics: Controlling Plasmon Resonances with Molecular Switches, *Nano Lett.*, 2009, **9**, 819–825.

2 F. Benz, *et al.*, Single-molecule optomechanics in 'picocavities', *Science*, 2016, **354**, 726–729.

3 R. Chikkaraddy, *et al.*, Single-molecule strong coupling at room temperature in plasmonic nanocavities, *Nature*, 2016, **535**, 127–130.

4 X. Zhang, *et al.*, Plasmon-Enhanced Catalysis: Distinguishing Thermal and Nonthermal Effects, *Nano Lett.*, 2018, **18**, 1714–1723.

5 S. Mukherjee, *et al.*, Hot Electrons Do the Impossible: Plasmon-Induced Dissociation of H2 on Au, *Nano Lett.*, 2013, **13**, 240–247.

6 G. L. Liu, J. Kim, Y. Lu and L. P. Lee, Optofluidic control using photothermal nanoparticles, *Nat. Mater.*, 2006, **5**, 27–32.

7 A. M. Gobin, *et al.*, Near-Infrared Resonant Nanoshells for Combined Optical Imaging and Photothermal Cancer Therapy, *Nano Lett.*, 2007, **7**, 1929–1934.

8 P. K. Jain, I. H. El-Sayed and M. A. El-Sayed, Au nanoparticles target cancer, *Nano Today*, 2007, **2**, 18–29.

9 A. G. Skirtach, *et al.*, The Role of Metal Nanoparticles in Remote Release of Encapsulated Materials, *Nano Lett.*, 2005, **5**, 1371–1377.

10 D. Pissuwan, S. M. Valenzuela and M. B. Cortie, Therapeutic possibilities of plasmonically heated gold nanoparticles, *Trends Biotechnol.*, 2006, **24**, 62–67.

11 L. Wang and B. Li, Thermal Memory: A Storage of Phononic Information, *Phys. Rev. Lett.*, 2008, **101**, 267203.

12 L. Cao, D. N. Barsic, A. R. Guichard and M. L. Brongersma, Plasmon-Assisted Local Temperature Control to Pattern Individual Semiconductor Nanowires and Carbon Nanotubes, *Nano Lett.*, 2007, 7, 3523–3527.

13 J. R. Adleman, D. A. Boyd, D. G. Goodwin and D. Psaltis, Heterogenous Catalysis Mediated by Plasmon Heating, *Nano Lett.*, 2009, **9**, 4417–4423.

14 S. Cormier, T. Ding, V. Turek and J. J. Baumberg, Actuating Single Nano-Oscillators with Light, *Adv. Opt. Mater.*, 2018, **6**, 1701281.

15 T. Ding, *et al.*, Light-induced actuating nanotransducers, *Proc. Natl. Acad. Sci. U. S. A.*, 2016, **113**, 5503–5507.

16 C. Andreou, M. R. Hoonejani, M. R. Barmi, M. Moskovits and C. D. Meinhart, Rapid Detection of Drugs of Abuse in Saliva Using Surface Enhanced Raman Spectroscopy and Microfluidics, *ACS Nano*, 2013, 7, 7157–7164.

17 S. Kasera, L. O. Herrmann, J. del Barrio, J. J. Baumberg and O. A. Scherman, Quantitative multiplexing with nano-self-assemblies in SERS, *Sci. Rep.*, 2014, **4**, 6785.

18 B. de Nijs, *et al.*, Smart supramolecular sensing with cucurbit[*n*]urils: probing hydrogen bonding with SERS, *Faraday Discuss.*, 2017, **205**, 505–515.

19 B. Nijs, *et al.*, Plasmonic tunnel junctions for single-molecule redox chemistry, *Nat. Commun.*, 2017, **8**, 994.

20 K. Tsukahara and R. G. Wilkins, Kinetics of reduction of eight viologens by dithionite ion, *J. Am. Chem. Soc.*, 1985, **107**, 2632–2635.

21 R. G. Rinker, S. Lynn, D. M. Mason and W. H. Corcoran, Kinetics and Mechanism of Thermal Decomposition of Sodium Dithionite in Aqueous Solution, *Ind. Eng. Chem. Fundam.*, 1965, **4**, 282–288.

22 S. V. Makarov, E. V. Kudrik, R. van Eldik and E. V. Naidenko, Reactions of methyl viologen and nitrite with thiourea dioxide. New opportunities for an old reductant, *J. Chem. Soc., Dalton Trans.*, 2002, **0**, 4074–4076.

23 R. W. Taylor, *et al.*, Precise Subnanometer Plasmonic Junctions for SERS within Gold Nanoparticle Assemblies Using Cucurbit[*n*]uril 'Glue', *ACS Nano*, 2011, **5**, 3878–3887.

24 H. Feilchenfeld, G. Chumanov and T. M. Cotton, Photoreduction of Methylviologen Adsorbed on Silver, *J. Phys. Chem.*, 1996, **100**, 4937–4943.

25 G. Baffou, R. Quidant and F. J. García de Abajo, Nanoscale Control of Optical Heating in Complex Plasmonic Systems, *ACS Nano*, 2010, **4**, 709–716.

26 S. Kasera, F. Biedermann, J. J. Baumberg, O. A. Scherman and S. Mahajan, Quantitative SERS Using the Sequestration of Small Molecules Inside Precise Plasmonic Nanoconstructs, *Nano Lett.*, 2012, **12**, 5924–5928.

27 S. J. Barrow, S. Kasera, M. J. Rowland, J. del Barrio and O. A. Scherman, Cucurbituril-Based Molecular Recognition, *Chem. Rev.*, 2015, **115**, 12320–12406.

28 R. W. Taylor, R. Esteban, S. Mahajan, J. Aizpurua and J. J. Baumberg, Optimizing SERS from Gold Nanoparticle Clusters: Addressing the Near Field by an Embedded Chain Plasmon Model, *J. Phys. Chem. C*, 2016, **120**, 10512–10522.

29 C. Carnegie, *et al.*, Mapping SERS in CB:Au Plasmonic Nanoaggregates, *ACS Photonics*, 2017, **4**, 2681–2686.

Faraday Discussions

PAPER

The impact of optically rectified fields on plasmonic electrocatalysis

Darby A. Nelson and Zachary D. Schultz *

Received 18th September 2018, Accepted 15th November 2018
DOI: 10.1039/c8fd00135a

Studies have shown that the excitation of plasmon resonances on nanostructured materials can drive catalytic processes. Plasmon resonances can be tuned across the solar spectrum, thus offering intriguing possibilities for their application in plasmonic catalysis. Previous work carried out by our group has indicated that nanostructures with tight junctions can create direct current (DC) electric fields. These fields arise from an optical rectification of the plasmon resonance on the plasmonic surface, and our group has shown that these fields modulate photocatalytic activity. This work looks to shed further light on the impact that optically rectified fields can have on catalytic reactions. Cyclic voltammetry shows that the electrochemical reduction and oxidation potentials of a 2 mM $CuSO_4$ solution occur at ~100 mV lower overpotential on an optically excited Ag nanodendrite electrode. Stark spectroscopy of the nitriles absorbed to these surfaces indicate photo-associated changes in the surface potential across the Ag nanodendrites. Localized areas evince photo-induced changes in the surface potential upwards of 300 mV. These results provide evidence of optically rectified fields altering the electrochemical reactivity on plasmonic surfaces and suggest that optimizing this nonlinear phenomenon may improve plasmonic photocatalysts.

Introduction

Reports have shown that the excitation of electrons in localized surface plasmon resonances (LSPR) can drive photocatalytic reactions on plasmonic materials.[1–3] In fact, the ability to tune the LSPR across the solar spectrum offers tremendous possibilities for photocatalysis. Plasmonic nanostructures promote chemical reactions on their surfaces when illuminated at their plasmon resonance frequency.[1,2,4–7] This behavior is contrary to that traditionally observed on bulk surfaces of the same materials that do not support catalysis. These catalysts are commonly made up of metals that exhibit intra-band electronic transitions. The frequency dependence of the increased reactivity correlates strongly with the plasmon resonance rather than the electronic transition, thus further implicating a plasmonic effect. Interestingly, the

The Ohio State University, Department of Chemistry & Biochemistry, Columbus, OH, 43210, USA. E-mail: Schultz.133@osu.edu

excitation of plasmon resonances has been reported to alter the selectivity of reaction products.[8,9] An improved understanding of the phenomena arising from plasmon resonances would offer tremendous potential for the development of new photocatalysts.

On nanostructured arrays, heterogeneous reactivity has been observed, thus implicating that "hotspots" and step-edges, areas of intense electric fields, are important for catalysis.[10,11] The excitation of localized surface plasmon resonances in metal nanostructures results in confined electric fields that are commonly utilized for chemical transformation and trace detection.[1,12–16] These electric fields are largely responsible for signal enhancements in surface related spectroscopies, where the optical response of molecules that experience the enhanced electromagnetic fields increases. When incident light is resonant with plasmon resonances in nanomaterials, relaxation of the excited electrons can occur both through light scattering and by thermal relaxation to produce hot electrons. These hot electrons are commonly associated with chemical transformations, such as photocatalysis, where exciting plasmon resonances facilitate chemical reactions on the nanostructure surface.[1,2,4–7]

The role of plasmons in catalysis remains an open question. Relaxation of the electrons excited by plasmon resonances can occur through different mechanisms that lead to physical effects such as heating, hot electron formation and optical signal generation.[1,6,14,15] Electron relaxation *via* thermal relaxation can produce sizable temperature increases on the surfaces of nanoparticles depending on the thermal conductivity of the surroundings;[17] however, the role of temperature in catalysis has been recently disputed.[18] The relaxation from the excited state can form hot electrons that promote reactions near nanostructure surfaces. This hot electron mechanism has received the most attention for plasmonic catalysis.[1,2,4,6]

Our lab has been exploring how exciting plasmon resonances can drive static surface potentials, as evidenced by vibrational Stark effects.[19–22] A DC field arising from plasmonic excitation was independently observed to generate a Stark shift in a molecule located in the plasmonic junction.[21,23] Measurement of the change in vibrational frequency of the bond can thus be used to determine the magnitude of the electric field. The vibrational Stark effect has been used to measure electric fields in a variety of chemical systems including the electrochemical double layer,[24–28] proteins,[29–32] biomembranes,[33] and the energy levels of molecules.[34–36]

The vibrational Stark effect is a change in the vibrational frequency that arises from an electric field perturbation of a chemical bond. In plasmonic materials, these perturbations arise from the intense electric fields associated with excited plasmon resonances driving nonlinear optical phenomena.[37] A second order nonlinear phenomenon, optical rectification, appears to increase in the high field regions associated with excited plasmon resonances.[21,23] Specifically, optical rectification results in a DC field at interfaces under the influence of high intensity electric fields, as described by eqn (1):

$$V_{DC} \cong -A\chi^{(2)} E(\omega)E(\omega') \tag{1}$$

In eqn (1), the DC voltage (V_{DC}) observed is related to the second order susceptibility, the electric fields [$E(\omega)$], and a coefficient (A) accounting for the geometry of the interface.[38]

The potential (V_{DC}) induced by exciting a plasmonic junction at the resonant frequency has been determined by measuring the tunneling current through a bow tie gold electrode structure separated by a nm gap when the structure was illuminated.[37] In the bow tie example, a bias was used across the junction to direct the optically induced current. Recent experiments carried out in our lab have shown that when nanostructures have small gap distances, within the quantum regime,[39,40] the tunneling of electrons across the junction also gives rise to a DC potential.[19] In these experiments, the dipole of the molecules in the gap biases the interface and controls the net direction of the tunneling current, which is believed to control the sign of the observed field. The vibrational Stark shift from a nitrile in this electric field can be used to determine the magnitude of the optically rectified field. Interestingly, mapping the CN stretch of a Stark reporter on a plasmonic surface shows evidence of localized regions with both net positive and negative changes in the surface potential.[20]

The magnitude of the rectified field can be quite large. Apkarian and coworkers also reported optical rectification in a nanoparticle dimer giving rise to a vibrational Stark shift from a CO molecule between the two spherical particles that correlated to an electric field enhancement on the order of 10^{12} and single molecule detection.[23] The high fields reported by Apkarian and coworkers are consistent with previous work carried out in our lab, where a 130 cm^{-1} Stark shift in the CN stretch frequency was observed from CN adsorbed onto a roughened gold surface when a nanoparticle TERS tip was brought into contact to form a plasmonic junction.[21] Only the CN in the confined region exhibits a Stark shift, while the CN stretch frequency for molecules outside the plasmonic junction is observed to be unchanged. Previous work investigating Stark shifts on electrochemical surfaces has indicated that a blue shift corresponds to a positive potential on the metal surface, that shifts about 25 cm^{-1} V^{-1}.[24,25,30,41,42] Thus the 130 cm^{-1} Stark shift would correspond to a change in the surface potential of ~3–4 V. Whilst there remains some uncertainty, such as the distance associated with the field for the plasmonic field compared to the applied electrochemical potential, the large fields observed in these examples suggest intriguing possibilities for the photocatalytic applications of nanostructures.

Recently, our lab has been able to show that the V_{DC} observed on a plasmonic Au film was able to modulate the rate of the oxidative and reductive formation of di-mercaptoazobenzene (DMAB) from nitrothiophenol and aminothiophenol upon plasmonic excitation.[22] By using a coadsorbed Stark reporter, the surface potential was shown to correlate with changes in the rate of DMAB formation. Additionally, the photocatalytic changes were shown to correlate with the electrochemically induced oxidation/reduction of the surface species.

The existence of these large DC fields on nanoparticle surfaces suggests an alternative mechanism for photocatalysis, whereby the excitation of the LSPR in nanostructures changes the energy of the Fermi level, as is made evident by the vibrational Stark shift of adsorbed molecules. By providing a substantial overpotential, the DC field resulting from optical rectification thus drives the chemical reaction at accelerated rates. In this report, we present results that show how the overpotential required for electrochemical reactions is modified by the optically rectified field on a plasmonic surface. We combine Stark spectroscopy with electrochemical measurements to assess the impact of plasmon resonances on the energy required for electrochemical reactions. These results illustrate how the previously under-appreciated phenomenon of optical rectification modulates the activity of plasmonic catalysts.

Experimental

Materials and reagents

Silver nitrate (99%), reagent alcohol (95%), copper sulfate pentahydrate (98%) and sodium sulfate (99%) were obtained from Sigma Aldrich (Missouri, USA) and used without modification. Reynolds Wrap aluminum foil was purchased from a local hardware store (Ohio, USA) and used without modification. 4-Mercaptobenzonitrile (MBN) was acquired from Synfine Research, Ltd. (Ontario, CAN) and used without modification.

Electrochemical deposition/measurement

Electrochemical deposition and measurements were performed using a CHI660D potentiostat. Strips of aluminum foil, with one end folded over itself multiple times to provide a sturdy surface for the alligator clip, served as the working electrode. To complete the three-electrode system, a Ag/AgCl reference electrode and platinum wire counter electrode were used. All potentials stated in the manuscript are relative to Ag/AgCl unless otherwise specified.

Photoelectrochemical measurements

All photoelectrochemical experiments were performed in a Zahner PECC-2 cell (Admiral Instruments), with excitation occurring directly on the dendrites through a quartz window at the front of the cell. The impact of optically rectified fields on the oxidation and reduction events at the surface of the Ag dendrites was measured in a solution of 2 mM Cu_2SO_4 and 100 mM Na_2SO_4. Optical excitation was achieved using a 565 nm LED (Thorlabs) at a power density of 2.3 mW mm^{-2}.

Silver dendrite preparation

Silver dendrites were prepared by electrochemical deposition using a 0.05 M silver nitrate solution and by depositing at -1.6 V for 200 seconds onto an aluminum foil working electrode. After deposition, the dendrite/aluminum foil electrode was removed from the electrochemical cell and the dendrites were allowed to dry overnight in order to ensure they adhered to the foil.

Monolayer preparation

An MBN monolayer was adsorbed to the surface of the Ag dendrites to generate a SERS signal. Clean Ag dendrites were soaked in a 0.01 M ethanolic solution of MBN for 24 hours to create the monolayer.

Raman measurements

Raman spectra were acquired using a Renishaw inVia Raman microscope with a 632.8 nm HeNe laser (Thorlabs). For the MBN studies, the scattering between 1912–2374 cm^{-1} was collected. Spectral analysis was performed using MATLAB with a modified peak-fitting script from Mathworks. The spectra were fit with a single Gaussian, with the optimal fit being that with the smallest percentage root mean squared (RMS) error. In general, 15 iterations were sufficient to minimize the RMS error.

For the experiments analyzing the changes in the MBN vibrational frequency from LED exposure, a 455 nm LED (Thorlabs) was used as it provided the highest power density (5.4 mW mm^{-2}) to the sample without exciting at wavelengths in the Raman spectrum collection region from the 632.8 nm laser.

Results and discussion

Silver nanodendrites were prepared by electrochemical reduction of $AgNO_3$ onto aluminum foil. These structures have previously been reported to provide high surface area electrodes[43] and enable surface enhanced Raman scattering.[44] The structure of these electrodes is consistent with heterogeneous surfaces previously shown to evince Stark shifts from nitrile containing molecules adsorbed to the surface.[20] In order to calibrate the surface potential to the vibrational frequency of our Stark reporter molecule adsorbed to the Ag dendrites, we performed spectro-electrochemical measurements. Fig. 1 shows the observed CN stretch frequency as a function of applied electrochemical potential for MBN adsorbed to the Ag nanostructures. The potential was swept from 0 V to −1 V at a scan rate of 10 mV s^{-1} while Raman spectra were recorded every second. As the surface potential becomes increasingly negative, the CN stretch frequency is observed to decrease to lower Raman shifts. Fitting the change in the CN stretch frequency *versus* potential indicates a Stark tuning coefficient of 8 cm^{-1} V^{-1}. This value is consistent with our previous report for MBN on Au nanowire array electrodes.[22]

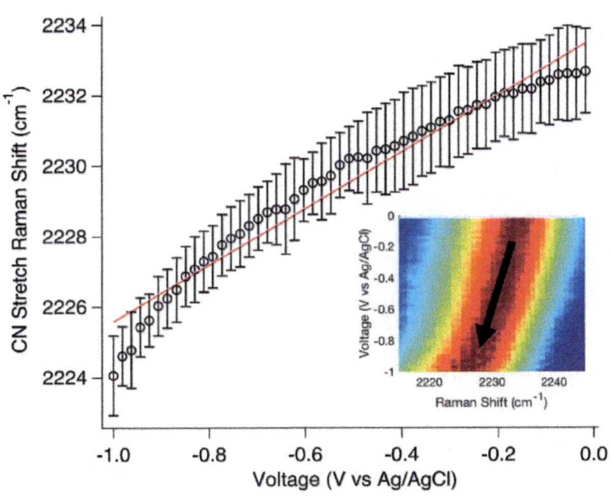

Fig. 1 Change in the CN stretch frequency as a function of applied electrochemical potential, plotted for MBN adsorbed to a Ag nanodendrite electrode. The CN stretch is observed to decrease at a rate of 8 cm^{-1} V^{-1}. The inset shows a heatmap illustrating how the vibrational mode lineshape evolves with potential. Spectra were recorded using a 40× water immersion objective (NA = 0.8), a laser power of 0.20 mW at the sample, and at 1 s acquisition times. The error bars correspond to the standard deviation observed from 6 separate measurements on different spots. A linear fit is plotted to show the expected trend in CN frequency with changing applied potential.

In order to verify the formation of optically rectified fields on the surface of the Ag nanodendrites, the CN stretch mode was recorded at increasing laser power. Fig. 2A shows the change in the CN frequency observed with increasing laser power. The CN stretch frequency was determined from the center of a Gaussian line shape fitted to the observed spectrum. The experiment was performed over 4 measurements from different spots on the same substrate.

From eqn (1), we observe the expected linear change in frequency with increasing power arising from a linear change in V_{DC}. The power range examined is limited by the output of our HeNe laser and the damage threshold of the surface. Over the 4.5 mW range measured, we observe a frequency shift of $-0.32\ cm^{-1}\ mW^{-1}$. Fig. 2B shows the spectra obtained at excitation powers of 0.04

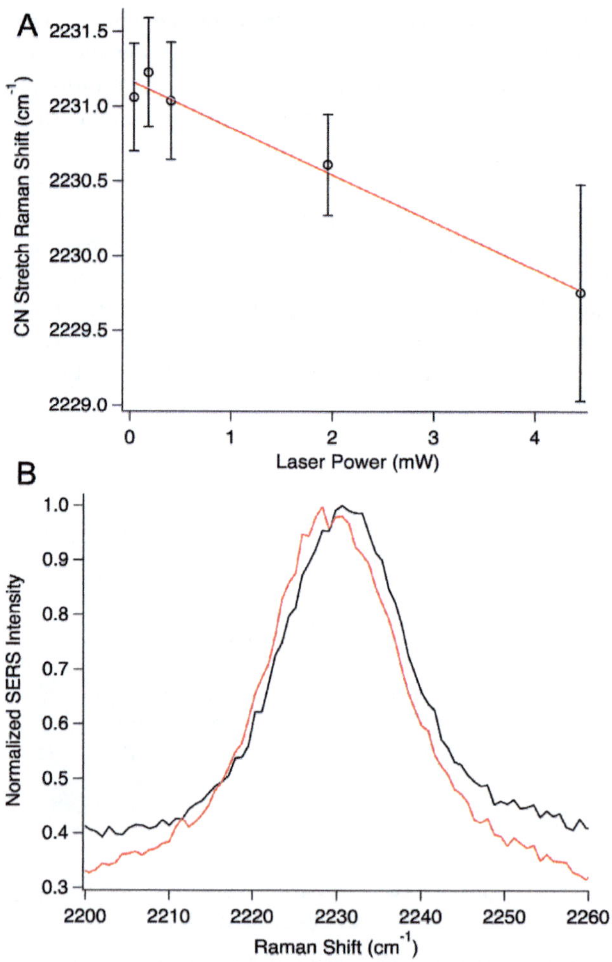

Fig. 2 (A) CN stretch frequency plotted as a function of incident laser power for an MBN functionalized Ag nanodendrite surface. With increasing power, the CN stretch is observed at decreasing Raman shifts. The change in the Raman shift with power indicates the formation of a negatively charged optically rectified field on the electrode surface. (B) Example spectra obtained at 0.04 mW (black) and 4.45 mW (red). The spectra were corrected for acquisition time and laser power and then normalized for comparison. The spectra were obtained using a 50× air objective (NA = 0.75).

mW and 4.45 mW. The peaks are clearly shifted but show negligible broadening. The lack of broadening suggests that temperature is not a dominant factor at these powers, as increasing the temperature should increase the heterogeneous line broadening and give rise to wider peaks. Changing the temperature is not expected to change the vibrational frequency of the CN moiety.

The results in Fig. 1 and 2 combine to suggest that the plasmonic structures give rise to a negatively charged surface with increasing laser power. This negative potential should affect the reduction and oxidation activity on the electrode surface. To assess the impact of the optically rectified electric fields on the electrochemical activity of the surface, spectro-electrochemical measurements were performed consisting of cyclic voltammetry in the presence of 2 mM $CuSO_4$ while the surface was either in the dark or illuminated with a 565 nm LED. Fig. 3 shows the results of these spectro-electrochemical measurements. In the absence of illumination, Fig. 3A shows that as the potential is made more negative, a reduction peak is observed near −0.05 V. Upon switching the direction of the potential sweep, two oxidation peaks are observed at 0.05 V and 0.2 V. There is some minor fluctuation of the peaks observed, likely arising from inhomogeneities on the highly branched electrodes. When the surface is illuminated with the diode, a new reduction peak is observed at 0.1 V during the cathodic sweep, and at 0.25 V upon switching the potential during the anodic sweep. The peak currents associated with these new peaks are observed to diminish and disappear when the light is turned off.

Fig. 3B shows the time dependent behavior observed for the photoinduced peaks on the surface. The reduction and oxidation peak currents are shown to correlate with illumination of the surface. The increases and decreases in peak current occur on a timescale of seconds, thus suggesting that mass transport effects (*e.g.* double-layer charging) impact the formation of the optically rectified fields on the electrode surface. Further studies are required to explore the origins of the observed kinetics. However, the new oxidation and reduction peaks are observed at potentials positive of the events measured in the dark. This is consistent with optical rectification providing an increase in the negative potential to facilitate these electrochemical transitions.

To further examine the origins of the photoinduced electrochemical behavior, we obtained Raman images of the MBN-functionalized surface with and without diode illumination (Fig. 4). To avoid interference from the diode in our Raman measurements, we used a 455 nm LED with an estimated power density of 5.4 mW mm^{-2}. The Raman excitation laser power was minimized to 0.008 mW to avoid inducing an additional Stark shift. From looking at Fig. 2, we observe that this laser power has a negligible impact on the CN stretch frequency. Fig. 4A plots the CN stretch frequency observed for a Ag dendrite SERS surface without LED illumination. Similar to previous results, the magnitude of the CN shift is observed to distribute across different regions of the substrate.

In Fig. 4B, we observe that LED illumination produces a change in the observed CN stretch frequencies. By repeating the Raman map with the LED illuminating this region, the value of the CN stretch is observed to increase in some areas and decrease in others. Using the Stark tuning coefficient (8 cm^{-1} V^{-1}) determined from Fig. 1, we are able to determine the distribution of surface potentials on the nanostructured silver surface. The intensities observed in the

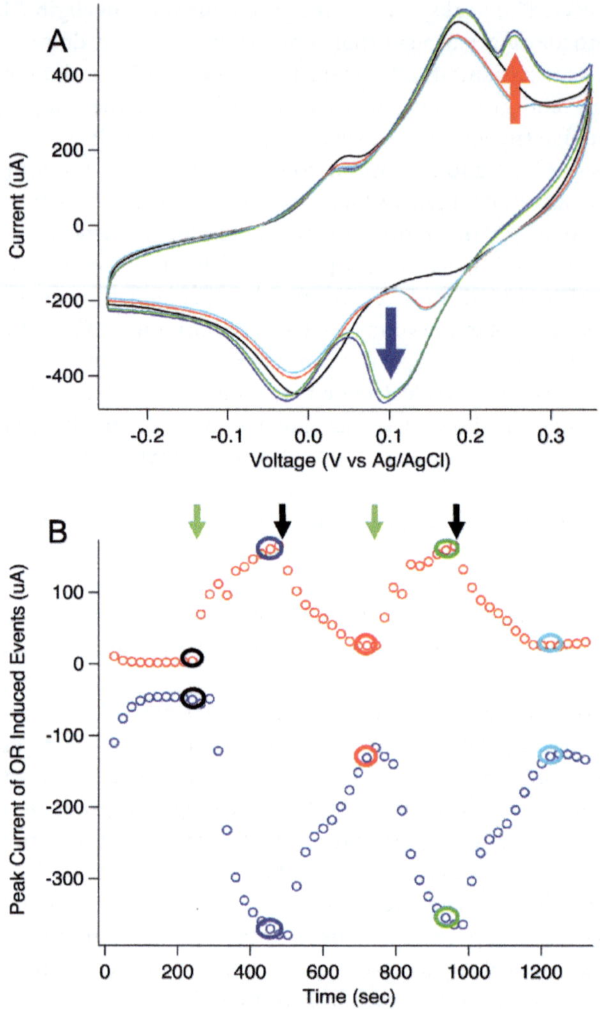

Fig. 3 (A) Cyclic voltammograms taken with and without illumination from a 565 nm LED. When the Ag plasmonic electrode is illuminated, new peaks are observed in the cathodic wave (blue arrow) and anodic wave (red arrow). (B) The sample's peak current of the photoinduced voltammogram peaks (red and blue) plotted *versus* time. The times at which the LED is turned on are indicated by green arrows, whilst the times at which the LED is turned off are indicated by black arrows. The colored circles correspond to the similarly colored voltammograms shown in (A).

repeated Raman maps were consistent, thus indicating that no visible drift or photodamage occurred.

Fig. 4C plots the distribution of surface potentials observed from 3 different experiments. To create the distribution, a linear background was subtracted and the spectra fitted to a single Gaussian lineshape. Only peaks that were 3 standard deviations above the noise were included. The observed frequencies were weighted by the fitted amplitude. The change in Raman shift was then converted to surface potential using the Stark tuning coefficient determined in Fig. 1. The mean calculated surface potential of −0.054 V agrees well with the observed

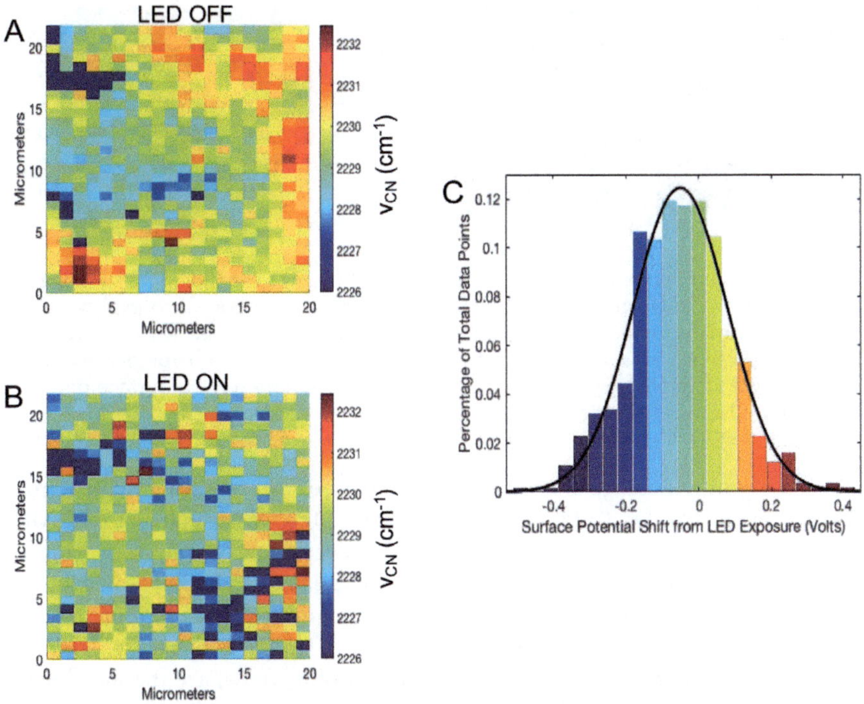

Fig. 4 (A) CN stretch frequency observed for an MBN-functionalized surface without LED illumination. (B) Changes in the CN frequency over the same area plotted in (A) are observed when a 455 nm LED illuminates the surface at a power density of 5.4 mW mm^{-2}. (C) The differences in the CN stretch frequency with and without illumination for each pixel in 3 separate experiments were combined and converted to surface potential using the Stark tuning coefficient (8 cm^{-1} V^{-1}) determined in Fig. 1. The black line is the Gaussian fit of the distribution.

center of the Gaussian distribution of −0.052 V; however, it is evident that there are regions with substantially different charge.

The small net change in surface charge helps explain the new peaks observed in the cyclic voltammetry experiments. If the optically rectified fields altered the surface potential of the entire surface, the applied electrochemical potential relative to Ag/AgCl would shift proportionally. However, with illumination we observe two distinct behaviors: one representative of the electrode in the dark, and the other a photo-induced behavior requiring less overpotential to drive the electrochemical reaction. Based on the distributions shown in Fig. 4C, we assign the photoinduced behavior to those areas that exhibit significantly shifted surface potentials. The ~150 mV shift in peak potential observed in the CV experiment is easily contained in the distribution of potentials shown in Fig. 4C. Previous work has associated defects, step edges and nanoparticle corners with catalytically active sites,[10,11] which is consistent with our observations. However, we believe that the optically rectified field at these sites also plays a role in the observed photo-electrochemical activity.

In order to further evaluate the correspondence between "hotspots" and the observed optically rectified fields, Fig. 5 is included to show how the SERS intensity corresponds to the observed distribution of surface potentials. Fig. 5A

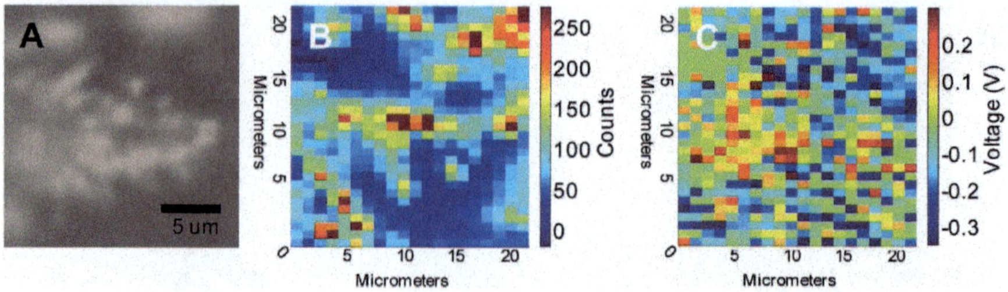

Fig. 5 Brightfield image of the Ag dendrite (A) and the corresponding Raman map (B) plotted from the intensities of the CN stretch. (C) The distribution of surface potentials calculated from the change in CN stretch frequency with and without illumination for the Raman map (B) provides a far field picture of how the change in charge density corresponds to the location of hotspots. In (C), red indicates a more positive surface potential, blue is more negative, and green is unchanged.

shows a brightfield image of the Ag dendrite surface corresponding to the Raman map shown in Fig. 5B. In Fig. 5B, the intensity of the CN stretch from the MBN molecules on the Ag dendrite is plotted. Regions of intense Raman scattering are observed along the dendrites. Fig. 5C shows the calculated charge distribution from the same area determined from maps with and without illumination. The correlation between the hotspots and the charge density is not clear. However, this is reasonable because the hotspots, *i.e.* nanojunctions, are the locations at which the electrons tunnel between the nanostructures. Therefore, one side should be positively charged and the other negatively charged. The Raman microscope used in this experiment does not have the spatial resolution to distinguish spectroscopic changes on this length-scale. What is observed are pockets of altered charge density near to where the hotspots are observed. The physical structures that correlate to these pockets of altered charge density are hypothesized to give rise to the photoinduced features observed in the cyclic voltammetry measurements.

Summary and conclusions

Taken together, our results suggest that the optical rectification of plasmonic fields is relevant to catalytic reactions on these materials. Fig. 6 summarizes our results and presents a schematic of what we believe may occur. In the presence of light to drive the plasmon resonance, electrons can tunnel though nanojunctions and give rise to optically rectified fields (V_{DC}). This optically induced voltage raises the Fermi level of the electrode in localized regions. These localized fields on the plasmonic material enable electrochemical reactions at reduced overpotentials. In Fig. 6, we analyze the cathodic wave of the CV, which appears to demonstrate more straightforward behavior. $E_{p,c}$ corresponds to the reduction of Cu by the electrode without the assistance of plasmons. When the plasmons are excited, regions on the electrode at increased negative potential drive the reaction at a lower applied potential, thus giving rise to $E^*_{p,c}$. The behavior of the anodic wave is also consistent with this explanation; however, the two oxidation peaks observed in the dark require additional analysis.

This model has implications for the mechanisms of plasmonic catalysis. Firstly, no thermal effects are included in this model. Our observations are based

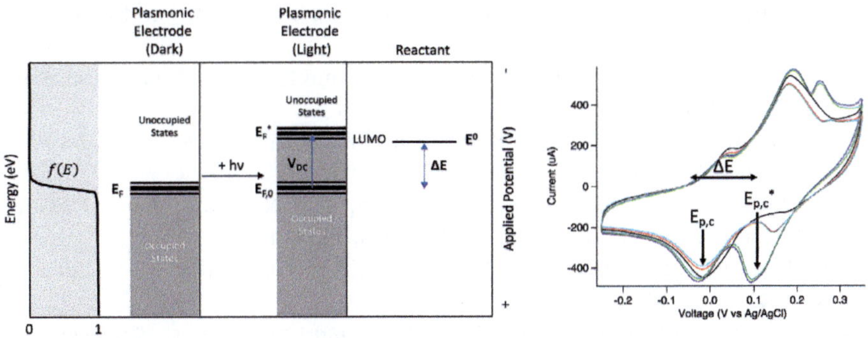

Fig. 6 This schematic illustrates the connection between the optically rectified field (V_{DC}) and the observation of new redox peaks in the CV measurements. $E_{p,c}$ is the reduction peak arising from the regions of the electrode that show no change in potential with optical illumination and which have electrons with a defined Fermi level ($E_{F,0}$). $E^*_{p,c}$ corresponds to the regions of the electrode where the optically rectified field (V_{DC}) raises the energy of the Fermi level, thus enabling reduction with less applied negative potential.

on changes in, and indicators of, potential. Both the frequency of bonds and the redox peak observed in CV are expected to broaden at increased temperature, but not shift. The Fermi–Dirac distribution of electrons into different energy levels, $f(E)$, is sensitive to temperature, but the center of the distribution, the Fermi level, does not change. Whilst temperature clearly has an effect on catalytic rates, this is in addition to the optically rectified fields. Secondly, our previous work indicated that nanojunctions are required for optical rectification.[19] This suggests that reactions on isolated plasmonic structures occur through a different mechanism, such as hot electron generation. In the case of coupled nanostructures, the optically rectified field has been shown to modulate the hot electron associated reaction rates.[22]

In conclusion, this combination of Stark spectroscopy and electrochemical measurements provides evidence that optical rectification on plasmonic surfaces plays an important role in plasmonic catalysis. Future efforts to design catalytic materials will benefit from understanding the impact of this nonlinear process.

Conflicts of interest

There are no conflicts of interest to declare.

Acknowledgements

The authors acknowledge Dr Rafael Masitas for his input on using Ag dendrites for a SERS substrate. This work was supported by the National Science Foundation award CHE-1830994.

References

1 S. Linic, P. Christopher and D. B. Ingram, *Nat. Mater.*, 2011, **10**, 911–921.
2 P. Christopher, H. L. Xin and S. Linic, *Nat. Chem.*, 2011, **3**, 467–472.

3 Y. C. Zhang, S. He, W. X. Guo, Y. Hu, J. W. Huang, J. R. Mulcahy and W. D. Wei, *Chem. Rev.*, 2018, **118**, 2927–2954.

4 S. Navalon, M. de Miguel, R. Martin, M. Alvaro and H. Garcia, *J. Am. Chem. Soc.*, 2011, **133**, 2218–2226.

5 H. Y. Zhu, X. B. Ke, X. Z. Yang, S. Sarina and H. W. Liu, *Angew. Chem., Int. Ed.*, 2010, **49**, 9657–9661.

6 X. Chen, Z. F. Zheng, X. B. Ke, E. Jaatinen, T. F. Xie, D. J. Wang, C. Guo, J. C. Zhao and H. Y. Zhu, *Green Chem.*, 2010, **12**, 414–419.

7 X. Chen, H. Y. Zhu, J. C. Zhao, Z. T. Zheng and X. P. Gao, *Angew. Chem., Int. Ed.*, 2008, **47**, 5353–5356.

8 S. J. Yu, A. J. Wilson, J. Heo and P. K. Jain, *Nano Lett.*, 2018, **18**, 2189–2194.

9 X. Zhang, X. Q. Li, D. Zhang, N. Q. Su, W. T. Yang, H. O. Everitt and J. Liu, *Nat. Commun.*, 2017, **8**, 1–9.

10 L. Kubus, H. Erdogan, S. S. Cetin, E. Biskin and G. Demirel, *ChemCatChem*, 2013, **5**, 2973–2977.

11 J.-H. Zhong, X. Jin, L. Meng, X. Wang, H.-S. Su, Z.-L. Yang, C. T. Williams and B. Ren, *Nat. Nanotechnol.*, 2017, **12**, 132–136.

12 N. J. Halas, S. Lal, W. S. Chang, S. Link and P. Nordlander, *Chem. Rev.*, 2011, **111**, 3913–3961.

13 P. L. Stiles, J. A. Dieringer, N. C. Shah and R. R. Van Duyne, *Annu. Rev. Anal. Chem.*, 2008, **1**, 601–626.

14 P. K. Jain, X. H. Huang, I. H. El-Sayed and M. A. El-Sayed, *Acc. Chem. Res.*, 2008, **41**, 1578–1586.

15 L. Brus, *Acc. Chem. Res.*, 2008, **41**, 1742–1749.

16 J. Jiang, K. Bosnick, M. Maillard and L. Brus, *J. Phys. Chem. B*, 2003, **107**, 9964–9972.

17 Z.-C. Zeng, H. Wang, P. Johns, G. V. Hartland and Z. D. Schultz, *J. Phys. Chem. C*, 2017, **121**, 11623–11631.

18 E. L. Keller and R. R. Frontiera, *ACS Nano*, 2018, **12**, 5848–5855.

19 H. Wang, K. Yao, J. A. Parkhill and Z. D. Schultz, *Phys. Chem. Chem. Phys.*, 2017, **19**, 5786–5796.

20 D. T. Kwasnieski, H. Wang and Z. D. Schultz, *Chem. Sci.*, 2015, **6**, 4484–4494.

21 J. M. Marr and Z. D. Schultz, *J. Phys. Chem. Lett.*, 2013, **4**, 3268–3272.

22 D. A. Nelson and Z. D. Schultz, *J. Phys. Chem. C*, 2018, **122**, 8581–8588.

23 M. Banik, P. Z. El-Khoury, A. Nag, A. Rodriguez-Perez, N. Guarrottxena, G. C. Bazan and V. A. Apkarian, *ACS Nano*, 2012, **6**, 10343–10354.

24 V. Oklejas and J. M. Harris, *Appl. Spectrosc.*, 2004, **58**, 945–951.

25 V. Oklejas, C. Sjostrom and J. M. Harris, *J. Am. Chem. Soc.*, 2002, **124**, 2408–2409.

26 D. K. Lambert, *Electrochim. Acta*, 1996, **41**, 623–630.

27 S. A. Sorenson, J. G. Patrow and J. M. Dawlaty, *J. Am. Chem. Soc.*, 2017, **139**, 2369–2378.

28 J. G. Patrow, S. A. Sorenson and J. M. Dawlaty, *J. Phys. Chem. C*, 2017, **121**, 11585–11592.

29 A. J. Stafford, D. L. Ensign and L. J. Webb, *J. Phys. Chem. B*, 2010, **114**, 15331–15344.

30 S. G. Boxer, *J. Phys. Chem. B*, 2009, **113**, 2972–2983.

31 L. J. Webb and S. G. Boxer, *Biochemistry*, 2008, **47**, 1588–1598.

32 I. T. Suydam and S. G. Boxer, *Biochemistry*, 2003, **42**, 12050–12055.

33 W. H. Hu and L. J. Webb, *J. Phys. Chem. Lett.*, 2011, **2**, 1925–1930.
34 W. A. Bryan, C. R. Calvert, R. B. King, J. B. Greenwood, W. R. Newell and I. D. Williams, *Faraday Discuss.*, 2011, **153**, 343–360.
35 S. Y. Zou, Q. H. Ren, G. G. Balint-Kurti and F. R. Manby, *Phys. Rev. Lett.*, 2006, **96**, 243003.
36 K. Hermansson and H. Tepper, *Mol. Phys.*, 1996, **89**, 1291–1299.
37 D. R. Ward, F. Huser, F. Pauly, J. C. Cuevas and D. Natelson, *Nat. Nanotechnol.*, 2010, **5**, 732–736.
38 Y. R. Shen, *The principles of nonlinear optics*, J. Wiley, New York, 1984.
39 R. Esteban, A. G. Borisov, P. Nordlander and J. Aizpurua, *Nat. Commun.*, 2012, **3**, 825.
40 K. J. Savage, M. M. Hawkeye, R. Esteban, A. G. Borisov, J. Aizpurua and J. J. Baumberg, *Nature*, 2012, **491**, 574–577.
41 V. Oklejas, C. Sjostrom and J. M. Harris, *J. Phys. Chem. B*, 2003, **107**, 7788–7794.
42 G. L. Beltramo, T. E. Shubina, S. J. Mitchell and M. T. M. Koper, *J. Electroanal. Chem.*, 2004, **563**, 111–120.
43 A. T. Sage, J. D. Besant, B. Lam, E. H. Sargent and S. O. Kelley, *Acc. Chem. Res.*, 2014, **47**, 2417–2425.
44 A. Gutés, C. Carraro and R. Maboudian, *J. Am. Chem. Soc.*, 2010, **132**, 1476–1477.

Faraday Discussions

DISCUSSIONS

Applications in catalysis, photochemistry, and photodetection: general discussion

Javier Aizpurua, Jeremy Baumberg, Valérie Caps, Emiliano Cortes, Bart de Nijs, Yuri Diaz Fernandez, Laura Fabris, Simon Freakley, Sylwester Gawinkowski, Daniel Glass, Junyang Huang, Bartlomiej Jankiewicz, Jacob Khurgin, Priyank Vijaya Kumar, Reinhard J. Maurer, Peter McBreen, Niclas Sven Mueller, Jeong Y. Park, Jhon Quiroz, Sebastian Rejman, Rosa Mayela Romero Gómez, Jorge Salmon-Gamboa, Sebastian Schlücker, Zachary Schultz, Amaresh Shukla, Yonatan Sivan, Madasamy Thangamuthu, Laura Torrente-Murciano, Xiaofei Xiao, Hongxing Xu and Chao Zhan

DOI: 10.1039/C9FD90014D

Sylwester Gawinkowski opened a general discussion of the paper by Jorge Salmon-Gamboa: Why you are using SiO_2 nanoparticles? Do they have any function or are they only the substrate to attach other active nanoparticles to? You have shown that only gold nanoparticles attached to silica nanoparticles do not influence the rate of reaction significantly. You have also demonstrated that adding platinum decorations on the gold nanoparticles causes a strong increase in the reaction rate. The significance of the gold nanoparticles would be more clearly shown with a simple experiment in which you have no gold nanoparticles but still have platinum decorations.

Jorge Salmon-Gamboa replied: Silica nanoparticles were chosen as a nano-sized inert substrate for Au nanoparticles. This choice of substrate provides a larger surface area covered with the active Au-Pt nanoparticles, in contrast to the situation when the active particles are placed directly onto a flat substrate. Aiming for applications, the SiO_2–Au–Pt nanoparticles can then in turn be attached onto a flat surface, forming a solid device that can be submerged into water, avoiding the problem of water contamination by nanoparticles. The role of the Au nanoparticles was investigated. Under illumination in the LSP spectral band (556-566 nm), SiO_2–Pt did not enhance the reaction rate. In contrast, under the same conditions, SiO_2–Au–Pt showed a considerably enhanced rate, proving that hot carriers were generated in Au (see Fig. 1 below).

Yuri Diaz Fernandez continued the discussion: I have two questions: (1) Can you comment on the dispersity of the size distribution of metal and SiO_2 particles

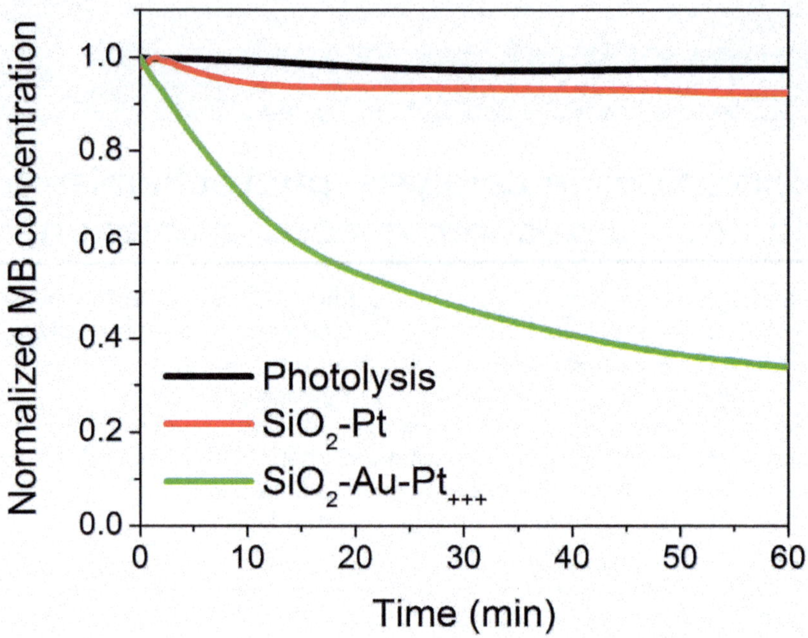

Fig. 1 Degradation of methylene blue over time by catalysts with Au (SiO_2–Au–Pt+++) and without Au (SiO_2–Pt).

and how well controlled are these in your system? (2) In the paper it was not clearly stated which normalisation criteria were used. Was the data normalised considering the extinction efficiency of gold, or considering the metal content in the samples? Did you correct you data for changes in the nanoparticle concentration due to material lost during synthesis/purification? If you consider this dilution effect, your efficiency may be higher.

Jorge Salmon-Gamboa answered: (1) The size distribution of silica nanoparticles had a good uniformity, confirmed by TEM. They were fabricated following the Stöber synthesis method, which is known for its well-controlled growth and size uniformity. On the other hand, the size distribution of Au nanoparticles was broader (9-15 nm in diameter). As the LSP resonances are relatively broad, the size distribution of the nanoparticles does not have a detrimental effect for the studied processes from an optical point of view. (2) The data were normalised considering the maximum optical absorption of the colloidal samples, so that the effect of nanoparticle concentration is taken into account.

Bartlomiej Jankiewicz commented: As we have heard from Professor Halas platinum is considered an endangered element and so it could be reasonable to recover the catalyst you have discussed. Thus, did you consider making core–shell structures with magnetic cores, *i.e.* Fe_3O_4@SiO_2@Au–Pt, so that you can recover them?

Jorge Salmon-Gamboa replied: Using magnetic cores was not considered at the beginning of the project, but it seems to be a very interesting approach for recovery.

Priyank Vijaya Kumar asked: Your schematic shows a nice core–shell structure. However, is this really the case in experiments? Do you see a nice shell of Pt formed on the core or do you see some mixing with the core, like an alloy? Is there any structural characterisation or other experiments you could do to confirm this?

Jorge Salmon-Gamboa responded: Pump–probe experiments were carried out on heterostructures without Pt and with different Pt loadings. It was found that the electron transfer occurred from Au to Pt and it depends on the amount of Pt. This effect would not appear in the case of alloy formation. Also, the localised surface plasmon resonance is sensitive to the material, and optical simulations suggest that alloy formation was not the case. The HAADF-STEM elemental maps presented in the paper also confirm this.

Sebastian Schlücker requested: Could you please comment on the mechanism of this photodegradation reaction on the atomic scale? What analytical techniques did you use or do you plan to use (*e.g.* UV/vis absorption, NMR, MS)?

Jorge Salmon-Gamboa replied: Photogenerated hot carriers interact with the surrounding aqueous medium, producing reactive oxygen species. Hot electrons can ionise adsorbed oxygen molecules forming superoxide anion radicals ($^{\bullet}O_2^{-}$). These highly reactive species can degrade organic molecules.

Optical spectroscopy was used to monitor methylene blue absorbance during degradation experiments. We plan to use complementary analytical techniques in the future to determine the products of the reaction. This was beyond the scope of the present work.

Jhon Jhon Quiroz Torres queried: It has been demonstrated in core–shell multimetallic architectures, comprised of plasmonic and catalytic metals, that the energy concentrated *via* LSPR excitation can be dissipated through absorption processes in the catalytic metal. Have you estimated the contributions from absorption and scattering to the calculated extinction spectra of the different Au@Pt core–shell architecture models that you presented? How does it change?

Jorge Salmon-Gamboa answered: An estimation of the contribution from absorption and scattering to the calculated extinction was determined by numerical simulations. The total absorption, scattering and extinction cross sections of Au and Au–Pt nanoparticles deposited on SiO_2 nanoparticles are presented in the Fig. 2 below. An increase in absorption and scattering cross sections is observed as the quantity of Pt increases on the surface of Au. This effect was also noted experimentally.

Xiaofei Xiao asked: Thanks, I have two questions. Firstly, how did you estimate the concentration of the small particles – did you consider different concentrations, and what is the effect on the final spectrum? Secondly, I am curious whether a method exists to add another layer of silicon or silica on the surface of the finial structure?

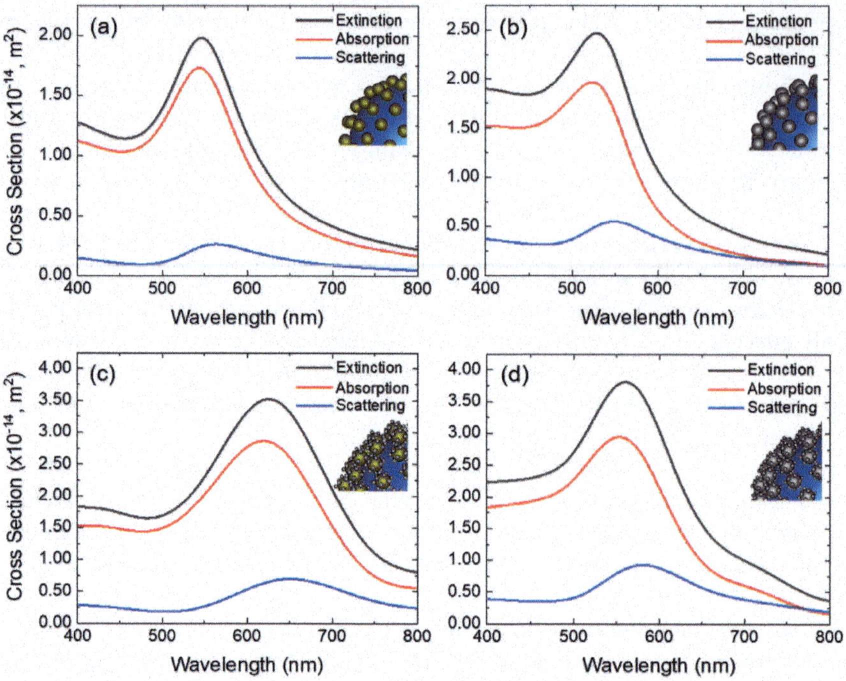

Fig. 2 Numerical simulations of extinction, absorption and scattering cross sections for (a) 15 nm diameter Au nanoparticles on SiO_2 and (b–d) SiO_2–Au–Pt heterostructures with (b) a 0.5 nm thickness of Pt shell, (c) a 2 nm radius of Pt seeds and (d) a 0.5 nm Pt shell and 2 nm radius Pt seeds. Au nanoparticle size is the same for all simulations.

Jorge Salmon-Gamboa replied: The concentration of small Au nanoparticles can be estimated considering the average size of a gold nanoparticle obtained by TEM, the concentration of K-Au used (number of Au^+ ions), and the concentration of the SiO_2 cores. Then it is assumed that all Au^+ ions reduce to form small Au nanoparticles, distributed over all the silica cores. Ultimately, comparison of the absorption spectra allows one to normalise different solutions to the nanoparticle concentrations. Different concentrations of Au were considered. The LSPR spectrum red-shifts as the size of the Au seeds increases. When the Au nanoparticles are big enough they merge together to form a complete Au shell, and finally, for thicker Au layers, the spectrum blue shifts.

As for multilayered coatings, it is possible to add more silica layers as described by P. Wang.[1]

1 P. Wang, A. V. Krasavin, F. N. Viscomi, A. M. Adawi, J.-S. G. Bouillard, L. Zhang, D. J. Roth, L. Tong and A. V. Zayats, *Laser Photonics Rev.*, 2018, **12**, 1800179.

Sebastian Rejman enquired: Have you conducted any kinetic studies of the reaction, and determined the turnover number and frequencies?

Is it possible to do SERS on these nanoparticles? This could provide valuable information about the mechanism of the reaction you studied and possibly other reactions you could run on this system.

Jorge Salmon-Gamboa responded: Turnover number and frequencies were not determined, as it was not within the scope of this work. In principle, they can be calculated. In general, Turnover Frequency is TOF = (number of molecules reacted)/(number of active sites)(time). This measure does not account for the number of incident photons and becomes irrelevant for reporting efficiencies of plasmonic photocatalysts, as pointed out by E. Cortes during this Faraday Discussion.

Performing SERS measurements on these nanoparticles seems to be very interesting, for this and other reactions. We will attempt this in the future.

Simon Freakley remarked: The particle density is very high on the SiO_2 particles – in thermal catalysis a major deactivation pathway would be particle sintering or evolution of the nanostructure to produce alloys or loss of shell material *via* agglomeration. Do you see any signs of plasmon-induced evolution of the materials?

Jorge Salmon-Gamboa replied: We did not see any signs of material evolution and degradation. The absorption spectra were checked before and after the experiments and also, in order to check the behaviour and stability of the heterostructures, under high power laser illumination. The extinction spectra of the samples were monitored and did not change over time. Since the localised surface plasmon resonance is very sensitive to shape, material and agglomeration, we would detect these effects if present. Therefore, we can rule out nanoparticle material evolution.

Javier Aizpurua questioned: In the optical spectroscopy of your structure in Fig. 2 in your paper, there seems to be some response in the long wavelength part of the spectrum, which is not fully captured in the simulations. Have you played with the concentration of metallic particles on the SiO_2 core to reproduce this better? Is there any chance that magnetic modes are induced in these structures? How important is it to faithfully reproduce the concentration and the percolation to properly address the near field?

Rosa Mayela Romero Gómez responded: Indeed, this is very complex system with random particle distributions which may have many collective resonant effects. Potentially, uncontrolled aggregation of nanoparticles in the experiment may result in collective modes with additional long wavelength resonances. This sort of resonance is difficult to reproduce in the simulations due to random realisation of particle configurations. The change of nanoparticle concentration influences the resonance, leading to its broadening.

SiO_2 nanoparticles, with the same dimensions as investigated in this work, do not exhibit resonant modes in the range of 400-800 nm due to their low relative refractive index. Decoration of silica particles with Au and Au–Pt nanoparticles may or may not lead to additional magnetic modes due to a collective interaction[1] but we did not observe these effects.

1 V. Ponsinet, P. Barois, S. M. Gali, P. Richetti, J. B. Salmon, A. Vallecchi, M. Albani, A. Le Beulze, S. Gomez-Grana, E. Duguet, S. Mornet and M. Treguer-Delapierre, *Phys. Rev. B*, 2015, **92**, 220414(R).

Jorge Salmon-Gamboa added: The near-field is of paramount importance for the hot-electron excitation and related processes. Unfortunately, as for the effects discussed above, random realisation of the nanoparticle configurations makes it difficult to simulate and compare the near-field effects. The only conclusion that can be made is that the field is enhanced at the Au nanoparticles, and in the case of their agglomerates, additional near-field features may exist. All these near-field enhancements contribute to hot-electron generation; at the same time, they are different for different random realisations of nanoparticle assemblies. All these effects, combined together, provide the observed averaged effect.

Jeong Y. Park queried: In the paper, it is stated that "SiO_2-core based nanoparticles were chosen instead due to the dielectric properties of SiO_2, so that a Schottky barrier between metal and semiconductor is avoided, and thus more hot-carriers are available for reduction processes." I am not quite sure about this statement. I am wondering if fewer hot electron carriers would be available if the authors use a more reactive support such as TiO_2. In addition, photocatalytic systems using SiO_2 are not effective photocatalysts because SiO_2 doesn't contribute any photocatalytic activity and absorption on metal nanoparticles is quite low. Could the authors comment on this?

Rosa Mayela Romero Gómez replied: The SiO_2 nanoparticles do not contribute catalytically – they were used as a nanosize inert support for Au nanoparticles to increase the surface area. On the other hand, if a semiconductor was used instead (TiO_2 for example), some hot electrons generated from Au would migrate to the semiconductor through the Schottky barrier and fewer electrons would be available on the surface of Pt, thus the overall photocatalytic activity of the heterostructure might be less. This effect was previously noted.[1]

1 S. Mukherjee, L. Zhou, A. M. Goodman, N. Large, C. Ayala-Orozco, Y. Zhang, P. Nordlander and N. J. Halas, *J. Am. Chem. Soc.*, 2014, **136**, 64–67.

Chao Zhan commented: Surface area is very important for catalysts. Are the Pt surface areas of the Pt NPs and Pt-SiO_2–Au the same? You use the term "Pt shell" in the paper, but usually it is small Pt nanoparticles.

Jorge Salmon-Gamboa responded: Surface area is of paramount importance for catalysts. In our experiment, comparison is done between SiO_2–Pt nanoparticles and SiO_2–Au–Pt nanoparticles with different amounts of Pt. So the Pt nanoparticles cover Au differently for different Pt concentrations and evolve from individual Pt nanoparticles to a complete "rough" shell around the Au nanoparticles. Therefore, we did not ensure precisely the same surface areas of the active particles, but, since the Pt nanoparticles on SiO_2–Pt are very small ($\sim$ 3 nm in diameter), they deliver a higher surface area than the Au–Pt particles ($\sim$12 nm). Despite having a higher surface area, the reaction rate was not enhanced in the case of SiO_2–Pt, whereas SiO_2–Au–Pt showed a significant reaction rate enhancement. The goal of the experiment was to show the role of the combination of the plasmonic-Au and catalyst-Pt in the heterostructures. The term "Pt shell" is used for those heterostructures with a higher Pt loading (where Pt shells formed, as confirmed by the HAADF-STEM elemental maps), which showed higher reactions rates.

Peter McBreen opened a general discussion of the paper by Madasamy Thangamuthu: Thank you for your interesting paper. You describe a minor hot electron driven pathway in Fig. 7B in your paper. Have you considered a pathway other than nitrogen splitting through electron injection into the LUMO? For example, a pathway that first involves N–H bond formation. The adsorption energy for nitrogen on TiO_2 or on oxidized aluminum is still quite small. Do you believe that N_2 needs to be strongly coupled to TiO_2 (strongly adsorbed) or to the aluminum triangles to facilitate such electron transfer into the LUMO?

Madasamy Thangamuthu answered: Electron injection into the LUMO of N_2 can potentially break the bonds either directly in one step or in three steps: i) NH, ii) NH_2 and iii) NH_3. We expect that the three step mechanism is more likely to occur in our configuration since the experimental conditions are mild. Currently we are studying the mechanism by analyzing the different by-products of the reaction. A strong adsorption of N_2 on catalytic surfaces is very important to reduce the bond dissociation energy. The TiO_2 surface offers suitable active sites for N_2 adsorption. Moreover, in our configuration aluminium covers less than 10% of the overall electrode surface. It would be very interesting to study the adsorption mechanism using first principles calculations. We are very much interested to collaborate with theoretical chemists to understand and optimize the N_2 adsorption.

Yuri Diaz Fernandez remarked: In your paper the efficiency is calculated considering the total incident photon flux. Is there any particular reason to do this taking into account that with the relatively low extinction values of your samples most of the light will be not be absorbed or scattered?

Madasamy Thangamuthu responded: The interesting point is how much ammonia can be produced per incident flux.

Bart de Nijs asked: Based on the TEM images the structures seem to have a relatively large separation, which means that most of the field enhancement is coming from just the structure. If the structures or different types of structures were closer together a coupling between these structures would occur and provide a coupled plasmon mode. The coupled mode will have the advantage of a much higher field but the optical resonance will be at a longer wavelength (*i.e.* less energy per photon). Could you say something about this tradeoff? Do you think a high field would benefit this reaction or does the change to a longer wavelength make this light catalyzed process less effective?

Madasamy Thangamuthu answered: A coupled mode could definitely improve the efficiency due to the higher field in the gap (hot spot) but smaller gaps also lead to an increased confinement which lowers the volume of action. Energetically it is also possible to induce the reaction at a longer wavelength.

Jeremy Baumberg commented: I am interested in your UV spectroscopy, because the simulations show that 80% extinction is expected, and you measure (in the small peak) about 5%. Understanding this difference is crucial to get a connection to the photocatalysis. Including a thin oxide layer should not make so much difference, and coupling is weak. This seems rather surprising.

Madasamy Thangamuthu replied: Both the simulations and experimental spectra showing the plasmonic resonance wavelength of the aluminium nanotriangles agree well. The simulations have been done by taking the average refractive index of the substrate and air. The variation might be due to the plasmonic interaction with the substrate.

Yonatan Sivan enquired: Can I ask you to elaborate on the thermal control experiments you have done? Did you measure the temperature rise in your sample?

I am asking because the increase in absorption you report can lead either to the generation of non-thermal (*i.e.*, "hot") electrons, or to a purely thermal effect (heating), both of which can lead to faster reactions. In that respect, you claim that you added illumination at 650 nm which did not make a difference. However, your Fig. 1D indicates that the AlNTs absorb rather weakly at that wavelength. In fact, it is not clear what is measured (extinction upon illuminating the whole layer (*i.e.*, extinction = 1 − transmission) or a single AlNT), and whether the simulation is of the exact same configuration? Otherwise, how can you explain the very different spectra of the experiment and simulation? They are similar just in terms of the resonance position, but the latter is pronounced only in the simulation.

Could it be that the absorption due to the 650 nm illumination is just small with respect to the 365 nm illumination? If so, I would not say that this test ruled out the possibility of thermal effects.

Madasamy Thangamuthu responded: No, we didn't measure the temperature changes but we observed that there is no temperature influence on photocurrent measurement by irradiating the electrodes far from the resonance wavelength. The conditions we used in our experiment, 5 mW cm^{-2} CW narrow band light sources, are very mild so we expect that they don't raise the temperature significantly. We measured extinction (1−T) for the whole layer not for a single AlNT. AlNTs absorb about 7% of the light (experimental absorption spectra, Fig. 4*E* in our paper) at 650 nm illumination. A comparison of measurements performed under dark and illuminated conditions revealed no measurable difference in the photocurrent. We conclude that under these conditions there is no thermal contribution.

Bartlomiej Jankiewicz commented: During the presentation of Professor Laura Fabris we have heard that in order to somehow influence TiO_2 reactivity by plasmonics the shell thickness in the case of nanoparticles should be around 4–5 nm to avoid electron quenching. Thus, did you try to reduce the thickness of the TiO_2 layer over the Al nanostructures and how did it affect the efficiency of the process you were investigating?

Madasamy Thangamuthu answered: We didn't reduce the thickness of the TiO_2 layer. All our experiments were performed with 100 nm TiO_2 on top of AlNTs. Thus, we didn't study the electron quenching phenomenon.

Jacob Khurgin remarked: As far as I know there is no shortage of TiO_2 in the earth's crust. So why do we need plasmons to absorb UV light? It will get eventually absorbed in TiO_2 anyway provided we have enough of it. And making the TiO_2 particles small will ensure that absorption takes place near the surface.

Madasamy Thangamuthu replied: UV plasmonics is highly useful for chemical reactions that are require higher energy to break strong chemical bonds. Although TiO_2 is present in sufficiently large amounts in the earth's crust, it fails as a photocatalyst in a number of ways such as by having a fast charge-carrier recombination, poor charge extraction, *etc.* Ammonia is one of the most highly produced chemicals worldwide. To compete against the strongly polluting Haber–Bosch process a large ammonia yield per volume is desirable. Hence, using plasmonic enhancement has the potential to increase the efficiency and, therefore, the yield per volume.

Laura Torrente-Murciano questioned: You are using ethanol as a hole scavenger in these reaction systems. Have you tried to remove it? If no reaction (or very little) takes place in the absence of ethanol, it would confirm that although thermal effects might have a role in the resulting activity, the generation of hot electrons is the main pathway for the reaction to take place.

Madasamy Thangamuthu replied: Yes, we performed experiments in the absence of ethanol and observed that after some time the original photocurrent response as well as ammonia production efficiency is significantly reduced. This confirms that the chemical reaction takes place by electron injection (either by hot electron injection or near-field coupling where TiO_2 generates electrons) into the N_2 molecule rather than from a thermal contribution.

Reinhard J. Maurer said: The Haber-Bosch process is costly in terms of reaction conditions, but existing plants follow highly optimised processes. The main business-critical cost is the cost of methane. There is also the additional aspect of transitioning a heterogeneous catalytic reaction to an electrochemical setup. In this context, I wonder if nitrogen dissociation is the best reaction for which to develop hot-electron-enhancement procedures rather than finding new (potentially plasmonic) routes to deliver hydrogen for this reaction.

Madasamy Thangamuthu answered: Thank you for the comment. I agree that new plasmonic routes to produce hydrogen are as important as the nitrogen reduction reaction.

Amaresh Shukla asked: Regarding the Al nanotriangle structures, usually Al has an oxide layer. You haven't shown this in the band gap or simulations. Does this not affect your predictions?

Madasamy Thangamuthu answered: We agree that the Al nanostructure has a 2-3 nm native oxide layer. We checked with SIE simulation that this 3 nm native oxide does not change the spectral features of the system.

Chao Zhan commented: The experimental conditions are a nitrogen-saturated solution (pH=4) containing 10% ethanol under illumination by a narrowband 365 nm beam with a 5 mW cm^{-2} power. The catalysts are TiO_2 and Al–TiO_2. All of these are as same as the reaction conditions for hydrogen evolution. In the plenty of reported papers of photocatalysis, the product is only hydrogen. And under acidic conditions, hydrogen evolution is more likely to happen. Nitrogen reduction is much more difficult than hydrogen evolution, and for synthetic ammonia isotope experiments are needed.

Madasamy Thangamuthu replied: Thank you for the very nice feedback. Isotopic experiments are definitely needed to support synthetic ammonia production. We agree that the electrons generated in our electrodes can also produce hydrogen gas as it needs less energy than ammonia production. Currently we are investigating it by measuring ammonia as well as hydrogen gas from the reaction chamber.

Sebastian Rejman enquired: What is your opinion about moving your system to the gas phase? In solution, the concentration of nitrogen is limited by its solubility and the proton concentration is limited by pH. Reactions in the gas phase allow for way higher concentrations.

Madasamy Thangamuthu responded: Indeed, nitrogen has a limited solubility in water and also, as mentioned, the proton concentration in solution is an issue. Our long-term goal is to produce ammonia in the gas phase.

Bartlomiej Jankiewicz opened a general discussion of the papers by Valérie Caps: The fabrication process that you describe in your paper seems to be quite random because it involves simply mixing nanoparticles with TiO_2 particles. Did you observe any variation of the CH_4 generation efficiency between various batches? Did you expose your photocatalyst to sunlight to see if you can produce methane?

Valérie Caps answered: The preparation of Au/TiO_2 composites was achieved by colloidal deposition. In this method, a colloidal solution of gold is prepared by chemical reduction (using sodium borohydride as a reducing agent) before deposition on titania. Titania is then introduced, upon vigorous stirring, in this perfectly stable, i.e. homogeneous, solution of $NaBH_4$/$NaBR_4$ (R = O, OH)-protected Au NPs exhibiting a narrow size distribution (3.0 $\pm$ 0.7 nm as evidenced by TEM). In about 30 min, the burgundy colored gold solution becomes colorless and the white titania powder turns pink, highlighting adsorption of Au NPs onto titania. Adsorption is complete, 100% of Au NPs are adsorbed, as determined by ICP. The composite is washed with water to remove the $NaBH_4$-derived protecting agent (Na, B < 100 ppm in the final composite as determined by ICP). The limited increase in Au NP size after deposition (to 3.6 $\pm$ 1.5 nm) indicates that Au NPs are well adsorbed and stabilized by the titania support. So the Au NP size is well controlled within the given standard deviation. Only their distribution over the support is random, adsorption occurring preferentially on titania surface defects. The standard deviation of CH_4 production rates did not exceed 15% between batches. The reaction has not been carried out using direct sunlight as the energy source, only with the lamp mimicking the solar spectrum which contains less UV irradiation (1% *vs.* 4%).

Laura Torrente-Murciano commented: I really enjoyed this systematic study. You claim that there is not a size effect of the gold nanoparticles in the 3 and 15 nm range, however, it is possible that there are competing effects as you increase the particle size that results in overall similar reactivities. Have you tried to normalise the activity per surface area? Such a calculation might reveal the real size effect on the catalytic activity.

Valérie Caps replied: Au/TiO_2 composites containing 3 nm or 15 nm Au NPs on average exhibit similar mass-normalized CH_4 production rates, both under visible irradiation and under solar irradiation. The increase in Au NP average size results in a decrease in metallic surface, i.e. the dispersion or fraction of surface atoms *vs.* total number of atoms in the particle, from about 37% to 10%, based on geometrical calculations for cubo-octahedral NPs, as described in the paper. This means that surface Au atoms are about 4 times more active in Au_{15nm}/TiO_2. Au_{15nm}/TiO_2 also exhibits the highest internal quantum yield of the study (0.012% under simulated solar irradiation). The subsequent decrease in CH_4 production rate observed with Au_{50nm}/TiO_2 of about 80% correlates with the loss of gold dispersion (10% to 3%, i.e. 70% loss). The variation of the solar CH_4 production rates normalized by the gold surface as a function of Au NP size thus consists of an increase from 3 nm to 15 nm and stabilization (plateau) from 15 nm to 50 nm. It seems that the co-catalytic effect may not be limiting and that the effect of metallic dispersion may be counterbalanced by the change in Au NP density (from about 90 NP/μm^2 in Au_{3nm}/TiO_2 to 1 and <<0.1 NP/μm^2 in Au_{15nm}/TiO_2 and Au_{50nm}/TiO_2, respectively (Fig. 8 in our paper)), which affects competitive adsorption of the reactants (CO_2 *vs.* H_2O), as suggested by changes in selectivity. It is noted that the average CH_4 production rate observed over Au_{50nm}/TiO_2 is similar to that observed over bare TiO_2. Hence the primary electron trap function (density effect) of the Au NPs appears inhibited in the solar process while the co-catalyst function is limited by the low Au dispersion (size effect). In the visible process, the variation of the solar CH_4 production rates normalized by the gold surface as a function of Au NP size consists of an increase from 3 nm to 15 nm and an even more drastic increase from 15 nm to 50 nm, which clearly indicates that the plasmonic effect dominates over both electron trap and co-catalytic effects.

Jeremy Baumberg asked: How is the Au NP surface and interface changed after calcination? Does the particle change shape – are there sharp asperities on the nanoparticles? Does it matter if the calcination is above the melting point of the NPs (well below the bulk melting point)?

Valérie Caps responded: In the post-calcined series, the whole Au/TiO_2 composite is submitted to calcination. Au NPs drastically increase in size from 3.6 nm in the uncalcined composite to 5.1 nm, 10.0 nm and 42 nm when calcination is performed at 450°C, 700°C and 900°C, respectively (Table 1 in our paper). This corresponds to a decrease in gold dispersion from 37% to 27%, 15% and 4%, respectively, based on geometrical calculations using cubo-octahedral shapes for the particles. The surface of the Au NPs thus dramatically decreases:

only about 10% of the surface atoms remain on the surface after calcination at 900°C. Interaction of Au NPs with titania also dramatically increases with increasing calcination temperature, as shown by the significant red shift (up to 90 nm, Fig. 5a in our paper) in the position of the surface plasmon resonance (LSPR) maximum, which is much more moderate in the pre-calcined series (Fig. 5b in our paper), and by HRTEM images (Fig. 7 left in our paper), showing inclusion of Au atoms inside the titania lattice at 900°C and the presence of titania layers around the Au NPs. This strong metal–support interaction is key in stabilizing the supported Au NPs and in retaining discrete NPs despite the fact that calcination is performed at a much higher temperature than the melting temperature of the Au NPs (Tammann). The initially pseudo-round-shaped Au NPs become more and more facetted with increasing temperature, but the Au NPs retain a general cubo-octahedral geometry.

In the pre-calcined series, calcination of titania is performed in the absence of Au NPs, before their deposition. The impact of calcination on Au NP size, shape and interaction with titania is thus much more limited. Au NP size varies from 3.6 to 9 nm (Table 1 in our paper), the LSPR maximum is between 530 and 560 nm (Fig. 5b in our paper), and no shape change is detected by HRTEM (cubo-octahedral particles). The more modest, yet effective, variations are due to the change of chemical composition and structure of the titania surface which controls adsorption of the gold sol. In particular, longer adsorption times are needed to adsorb Au NPs onto the more perfect, more crystallized surfaces, which increases the probability for aggregation (*cf.* the larger standard deviations observed at higher calcination temperatures in the pre-calcined vs. post-calcined series, Table 1 in our paper).

Daniel Glass enquired: Your paper discusses the treatment of TiO_2 films by calcination at high temperatures, as a result seeing different properties of the films. Can you comment on how the calcination effects possible phase changes of the TiO_2?

Additionally, heating to high temperatures can introduce or change bulk and surface defects – can you comment on what effect this could have on the reactions described?

Valérie Caps answered: The uncalcined titania semi-conductor (SC) used in our Au/TiO_2 composites exhibits 100% anatase phase. It remains 100% anatase upon calcination up to 700°C. Diffraction peaks related to the rutile phase, namely rutile (101), are observed only after calcination at 900°C (Fig. 4 in our paper). TEM (Fig. 7 right in our paper) and TRMC results (Fig. 12-14 in our paper) suggest that the anatase-to-rutile transition in both pre- and post-calcined Au/TiO_2 composites is initiated from 700°C, leading to the elongation of anatase crystallites, but the crystallised rutile phase appears only at 900°C in the XRD spectra.

Concerning bulk and surface defects, increasing calcination temperature generally induces an increase in the crystallisation degree of TiO_2, as shown by the narrowing of the diffraction peaks related to the anatase phase (Fig. 4 in our paper). So the density of bulk and surface defects is expected to decrease. This should positively impact the photocatalytic performances in the SC-driven solar process, as defects are known to act as recombination centers and to lower charge carrier lifetime. The surface-normalized solar activities of the composites indeed increase with increasing calcination temperature up to 700°C. However, narrowing of XRD peaks with increasing calcination temperature also implies an increase in the crystallite size, which can be estimated from the Debye-Scherrer law. The size of anatase crystallites increases from around 9 nm in the uncalcined sample to 15 nm, 45 nm and 80 nm in composites calcined at 450°C, 700°C and 900°C, respectively. The lower (surface-normalized) solar activities of the composites calcined at 900°C indicates that the beneficial effect expected from the increased crystallinity of the sample and decreased number of defects is counterbalanced by the particularly large crystallite size. The latter indeed significantly lengthens the path for the charge carriers generated within the SC bulk to reach the surface, which increases the probability of charge recombination before charge carriers can react at the surface and carry out any reaction (water oxidation to generate the active reducing agent, CO_2 reduction). After calcination at 900°C, this effect becomes predominant also over the potential increase in charge carrier lifetime expected from the appearance of the anatase-rutile heterojunction.

Hongxing Xu addressed Valérie Caps and Madasamy Thangamuthu: How close is your study to industrialization in terms of both the efficiency and the cost?

Valérie Caps replied: The best internal quantum yield observed in our study under simulated solar irradiation is 0.012%. It is likely to increase to a certain extent under real solar irradiation, due to the higher amount of UV photons (4% of the full solar spectrum *vs.* only 1% in the lamp emission spectrum). This value is however at least two orders of magnitude lower than what would be required for industrialization.

Madasamy Thangamuthu answered: The efficiency of the industrial process is 10-15% and we are at 0.06% and hence our study is far from industrialisation. Furthermore, in terms of cost, it is too early to compare. More optimization is required to move towards industrialisation.

Javier Aizpurua questioned: I am wondering about the spatial resolution of these chemical effects reported in this session. We all assume an improvement of product generation connected to the near field induced at some spot, but we mostly trace the far field. Couldn't we put more emphasis in spatially resolved techniques that tell us something about where particularly designed chemistry is happening in space? Can we identify chemical hotspots using these techniques?

Valérie Caps responded: We can get indirect information about the location of the chemical reaction on *e.g.* Au/TiO_2 composites by using an approach such as the one described in our paper. Indeed, by controlling the size of metal nanoparticles (*via* size-controlled synthesis such as colloidal deposition which leads to narrow size distributions) and surface area of the support (by calcination), the average density of plasmonic objects may be controlled. Distribution of the particles over the support remains random. Nevertheless, the way in which Au NP density impacts the photocatalytic properties (activity + selectivity) of the composite under visible *vs.* solar (UV + visible) irradiation gives us information on the role of Au–TiO_2 perimeter sites (here in the adsorption of CO_2 in the solar process for example) *vs.* the role of longer range chemical processes (such as hydrogen spillover from the titania surface to Au NPs) and titania surface (ensuring water activation). Besides, state of the art developments of EELS-STEM techniques now allow the mapping of a surface plasmon on single metal nanoparticles using sub-nm resolved EELS spectrum-imaging. We intend to perform such a study within the French Microscopy Network METSA in order to complete our experimental study with more direct local information on where reactions take place.

Emiliano Cortes contributed: There are some techniques that could identify the places where the reaction is taking place within a single nanostructure. This has been for a long time an important goal for the whole heterogeneous catalysis community (see for example ref. 1). However, the ultimate resolution may need to reach the atomic scale to disentangle the role of step-edges, surface-defects and other (low coordinated) atoms in the surface, which are usually claimed to be the most reactive places on the surface of the material.[2] We have used a series of different approaches (including chemical tracking of reactive hot-spots, super-resolution microscopy, opto-electrochemistry, *etc.*) to try to resolve in energy and space these transformations.

1 C.-Y. Wu, W. J. Wolf, Y. Levartovsky, H. A. Bechtel, M. C. Martin, F. D. Toste and E. Gross, *Nature*, 2017, **541**, 511–515.
2 J.-H. Zhong, X. Jin, L. Meng, X. Wang, H. S. Su, Z. L. Yang, C. T. Williams and B. Ren, *Nat. Nanotechnol.*, 2017, **12**, 132–136.

Sebastian Schlücker added: I agree that we need experimental techniques probing reactivity with nanoscale resolution. Nanoscale IR spectroscopy, tip-enhanced Raman scattering, and super-resolution fluorescence microscopy are a few examples of experimental approaches which can achieve this.

Jeremy Baumberg noted: This is a very difficult challenge, and electrochemical TERS is perhaps one technique that will be able to access such information, but it is not high throughput and not easy to extract useful chemistry of hot electrons. The use of the Nanoparticle-on-Mirror (NPoM) system allows creation of precise nm-scale hotspots and also may deliver better defined spatial resolution.

Niclas Sven Mueller opened a general discussion of the paper by Junyang Huang: In your paper you estimated the local plasmonic heating efficiency to be 1.5 times higher in the gap region than in the area 10 nm around the aggregate. How did you get to this estimate? Did you consider the effect of plasmonic hot spots between the nanoparticles?

Junyang Huang replied: Thank you. This estimation is based on the steady-state Coulomb temperature profile from the metal surface, assuming the metal has a uniform temperature. The concentrated optical field in the hot-spot is likely to contribute additional heating but is not included in this estimation.

Hongxing Xu questioned: Is this a dimer or an aggregate? How easy is it to get the dimer together?

Junyang Huang responded: The SERS substrates are based on AuNP aggregates bridged by CB[*n*] molecular spacers. The resultant aggregates generate 400-fold near-field enhancement ($|E|/|E_0|$) by trapping incident light in the hot-spot at resonant frequencies and are consequently utilized as SERS substrates for their 10^{10}-fold Raman enhancement.[1] Fabricating dimeric colloidal structures through self-assembly approaches is possible but known to be challenging, especially if an open interface is desired.

1 C. Carnegie, R. Chikkaraddy, F. Benz, B. de Nijs, W. M. Deacon, M. Horton, W. Wang, C. Readman, S. J. Barrow, O. A. Scherman and J. J. Baumberg, *ACS Photonics*, 2017, **4**, 2681–2686.

Sebastian Schlücker queried: In your supramolecular chemistry CB-based approach to make dimers what is the approximate size range (min to max) of molecules that fit into the hot spot and still yield single-molecule SERS?

Junyang Huang answered: In this presented work, the aggregates are used to probe an ensemble of molecules in a large number of hot-spots, however, CB[*n*]s have been shown to facilitate single-molecule SERS.[1] There have been extensive studies carried out on the topic of encapsulation and association of various sized guest molecules in a range of CB[*n*] cavities (*i.e.* $n = 5,6,7,8,10$).[2] However, it has been reported that molecules which cannot fit into CB[*n*] cavities can also be incorporated interstitially in the hotspot between nanoparticles and yield SERS detection.[3]

1 B. de Nijs, F. Benz, S. J. Barrow, D. O. Sigle, R. Chikkaraddy, A. Palma, C. Carnegie, M. Kamp, R. Sundararaman, P. Narang, O. A. Scherman and J. J. Baumberg, *Nat. Commun.*, 2017, **8**, 994.
2 S. J. Barrow, S. Kasera, M. J. Rowland, J. del Barrio and O. A. Scherman, *Chem. Rev.*, 2015, **115**, 12320–12406.
3 B. de Nijs, M. Kamp, I. Szabó, S. J. Barrow, F. Benz, G. Wu, C. Carnegie, R. Chikkaraddy, W. Wang, W. M. Deacon, E. Rosta, J. J. Baumberg and O. A. Scherman, *Faraday Discuss.*, 2017, **205**, 505–515.

Laura Torrente-Murciano commented: It is very impressive the way that you control and manipulate the space between the Au NPs in the agglomerates. Do you think that the molecular spacer might have additional roles beyond keeping the Au nanoparticles at specific distances? Can they accept hot carriers as demonstrated previously in the case of adsorbed molecules?

Junyang Huang replied: Thank you for the comments. We show that the presented optically controlled redox chemistry can be achieved with both CB[*n*]-linked aggregates and salt-induced aggregates. However, there are differences between the two types of aggregates which we take into consideration. It is found that CB[*n*]-induced aggregates exhibit an extended fractal morphology while using salt results in more compact clusters of nanoparticles. This could influence the in-coupling and out-coupling of light into and from the aggregates, which could, in turn, change the heating efficiency and Raman response. Using CB[*n*] also allows us to obtain nanoparticle aggregates at a lower salt concentration. The ionic concentration of the suspension can play an important role as the dithionite ions are negatively charged and how easily they can approach the nanoparticle surfaces depends on the Debye length. In addition, the host–guest chemistry between CB[7] and methyl viologen could also influence the redox dynamics. It has been demonstrated in previous work that using CB[*n*] to capture methyl viologen molecules in plasmonic junctions allows us to observe hot electrons hopping through the junction through the molecule and the subsequent redox chemistry on a single molecule can be tracked by SERS.[1]

1 B. de Nijs, F. Benz, S. J. Barrow, D. O. Sigle, R. Chikkaraddy, A. Palma, C. Carnegie, M. Kamp, R. Sundararaman, P. Narang, O. A. Scherman and J. J. Baumberg, *Nat. Commun.*, 2017, **8**, 994.

Hongxing Xu asked: What is the advantage of using SERS to study hot electrons compared to other technologies?

Junyang Huang responded: Most analytical techniques probe chemical changes in the bulk of the sample or reactor, whereas SERS allows for the probing of chemical properties at or near metal interfaces where charge carrier transfer and heterogeneous catalysis occurs. The immense signal enhancement obtainable using SERS for these interface signals means that they far exceed any Raman signal from the bulk environment. In addition, the large amount of signal allows for real-time observation of chemical reactions,[1,2] and it allows for small sample volumes to be probed enabling rapid screening of substrates.

1 B. de Nijs, F. Benz, S. J. Barrow, D. O. Sigle, R. Chikkaraddy, A. Palma, C. Carnegie, M. Kamp, R. Sundararaman, P. Narang, O. A. Scherman and J. J. Baumberg, *Nat. Commun.*, 2017, **8**, 994.
2 E. A. Sprague-Klein, B. Negru, L. R. Madison, S. C. Coste, B. K. Rugg, A. M. Felts, M. O. McAnally, M. Banik, V. A. Apkarian, M. R. Wasielewski, M. A. Ratner, T. Seideman, G. C. Schatz and R. P. Van Duyne, *J. Am. Chem. Soc.*, 2018, **140**, 10583–10592.

Zachary Schultz said: In your nanoparticle aggregates, the chromophores all undergo redox. Can you comment on the nature of the SERS signal, whether this is resonantly enhanced by either an electronic state or a charge-transfer interaction with the nanoparticles? Also, at what wavelength are your studies performed?

Junyang Huang answered: Thank you for addressing this point. The reduction of the methyl viologen is a result of charge transfer from the chemical fuel dithionite. The Raman intensity of the methyl viologen radical cation ($MV^{+\bullet}$) indeed experiences an electronic resonance enhancement in the Raman scattering. The $MV^{+\bullet}$ radical has a broad absorption from 500 nm to 700 nm which is close to the excitation wavelength used in the experiments (785 nm). However, the methyl viologen dication (MV^{2+}) does not absorb in the visible range and therefore does not experience the same enhancement.

Jeremy Baumberg opened a general discussion of the paper by Zachary Schultz: This is really interesting work, although here the Ag nanostructures are rather complex to understand. My question is how it might be possible to distinguish surface rectification compared to light-induced charging at the interface? I am trying to understand what might be clear evidence for the $\chi^{(2)}$ effect you are discussing. Also, you use a linear fit on your Stark shifts, while it definitely looks much more like a quadratic dependence here. You noted it is different on Au, and it might be nice to include this also, together with a comment about why these are not linear.

Zachary Schultz replied: We have looked at the Stark shift from nitrile (-CN) groups on a number of surfaces. The shift in the frequency corresponds to the electric field experienced by the CN bond. The work on Au was published last year.[1] The behavior on Au is more linear; however, the linearity has been attributed to chemistry that occurs on the surface associated with double layer phenomena. As chemical reactions alter the surface chemistry, for example oxidation of the Ag, the linearity of the shift *vs.* applied potential decreases. Au, a chemically more inert surface, shows better linearity in the double layer region. At potentials where electron transfer reactions are observed on Au, the linearity also goes away. The connection to $\chi^{(2)}$ originates from further back, in our previous work;[2] where plotting SERS intensity against the CN stretch frequency showed a dependence that resembled electric field induced second harmonic generation trends (see ref. 3). We are currently working to correlate the second harmonic intensity to the Stark Spectroscopy results we report. We believe this will address the correlation to $\chi^{(2)}$ more directly.

1 D. A. Nelson and Z. D. Schultz, *J. Phys. Chem. C*, 2018, **122**, 8581–8588.
2 D. T. Kwasnieski , H. Wang and Z. D. Schultz , *Chem. Sci.*, 2015, **6** , 4484–4494.
3 R. M. Corn, M. Romagnoli, M. D. Levenson and M. R. Philpott, *Chem. Phys. Lett.*, 1984, **106**, 30–35.

Hongxing Xu asked: Can you explain what you mean by 'rectify the field'? Is it a general effect? What is the efficiency?

Zachary Schultz responded: 'Rectify the field' means convert the AC field into a DC field. In this case the AC plasmon resonance, and associated field, drives a DC potential on the surface. Optical rectification is a 2nd order nonlinear process and has been previously reported,[1] however, it has not been widely appreciated in the plasmonic community. Our group and a few others have reported this effect in plasmonic materials. As commented on in this discussion by Professor Baumberg, there are still some remaining questions about the origin of the DC potential, but optical rectification is consistent with multiple observations. It is a general effect, though the exact conditions that optimize the process in plasmonic materials is still an open question. In regards to efficiency, this is difficult to address. I think we would have to agree on a definition of efficiency first.

1 Y.R. Shen, *The Principles of Nonlinear Optics*, Wiley, 1984.

Sebastian Schlücker enquired: Have you thought about mapping out the Stark effect at the nanoscale using molecular rulers, *e.g.*, by changing the distance between the nitrile group and the metal surface?

Zachary Schultz replied: This is something we have considered, and others have tried previously. We have used different length thiolate molecules to change the gap distance between surfaces, but the nitrile is always at a fixed distance. Moving the nitrile within the gap may show interesting behavior. The challenge is the availability of molecules, but this is something that could be overcome.

Bart de Nijs queried: In your article you provide a brief description on how the self-assembled monolayers (SAMs) are made. Have you done any characterization on the formed SAMs, could you say something about for example the grafting density of the SAM angle, or are these parameters of less importance for this work?

Zachary Schultz answered: We have not done extensive characterization of the self-assembled monolayers. We use a high concentration of thiols to promote a dense, or complete layer. The roughness of the surface makes detailed characterization complicated. In general, we use replicate measurements for verification as defects should be random and more difficult to reproduce.

Javier Aizpurua commented: Your work involves non-linearities which are very interesting. We have recently reported theoretically the non-linear response in plasmonic gaps, where we have identified that the tunnelling current is a key source of non-linearity of the response, for instance in high-harmonic generation.[1] Since rectification is connected with these non-linearities, of type $\chi^{(2)}$, I think it would be very interesting to understand also the optical rectification originating from the tunneling current. On another issue, could we use some specific molecule that is fluorescent, where this kind of rectifying effect could change the electronic structure, and this change manifested in the fluorescence?

1 G. Aguirregabiria, D. C. Marinica, R. Esteban, A. K. Kazansky, J. Aizpurua and A. G. Borisov, *Phys. Rev. B*, 2018, **97**, 115430.

Zachary Schultz responded: One could look at changes in fluorescence intensity that originate from the rectified fields on the surface. A voltage sensitive dye might provide signal, provided quenching at the metal surface can be avoided. Rather than fluorescence we have reported on changes in the Raman signal that originate from photoreactions on the surface, specifically the conversion of aminothiophenol (or nitrothiophenol) to dimercaptoazobenzene. These reactions are believed to result from hot-electrons, but our work shows that the rectified current alters the rate at which they occur. I would expect the same to be true for the fluorescence indicators of electron transfer. These chemical reactions require electron transfer, which generally implicates hot electrons. The Stark shift from the nitrile is a measure of the surface charge, which provides insight into the rectified field. As we noted in response to the earlier question about $\chi^{(2)}$, we are working to better correlate these effects.

Hongxing Xu asked: Since SERS is a very sensitive technology, is it also good for the study of catalysis? What is your opinion on SERS for general sensing applications so far?

Junyang Huang answered: In order to facilitate further understanding of the fundamental mechanism and catalytic reactions on the active sites during catalysis processes, it is crucial that structural and dynamic insights are acquired on a molecular level. A range of operando methodologies have been developed to track catalysis processes under realistic working conditions, from which SERS is one of the most informative ones, as it provides dynamic chemical information down to a single molecular level and can be integrated with other methods such as electron microscopy and scanning probe techniques to unfold molecular dynamics with a high spatial resolution. Recently a few demonstrations have shown that SERS is highly promising for understanding catalysis processes.[1-3]

1 J.-H. Zhong, X. Jin, L. Meng, X. Wang, H. S. Su, Z. L. Yang, C. T. Williams and B. Ren, *Nat. Nanotechnol.*, 2017, **12**, 132–136.
2 H. Zhang, C. Wang, H.-L. Sun, G. Fu, S. Chen, Y.-J. Zhang, B.-H. Chen, J. R. Anema, Z.-L. Yang, J.-F. Li and Z.-Q. Tian, *Nat. Commun.*, 2017, **8**, 15447.
3 J.-C. Dong, X.-G. Zhang, V. Briega-Martos, X. Jin, J. Yang, S. Chen, Z.-L. Yang, D.-Y. Wu, J. Miguel Feliu, C. T. Williams, Z.-Q. Tian and J.-F. Li, *Nat. Energy*, 2019, **4**, 60–67.

Zachary Schultz answered: SERS is something I am very interested in and believe it is one tool for studying catalysis. In SERS the importance of junctions and chemical enhancements for some molecules may limit the utility. For example, the surface science community has studied the dependence of crystal planes for adsorption and reactivity, but this is difficult to study in nanometer sized junctions. Some molecules have very poor cross-sections and, even with SERS, are hard to detect at low concentrations. For other molecules and reactions, I think SERS is an invaluable tool for monitoring chemical transformations. I also think the work being done with tip-enhanced Raman imaging, though a more complex experiment, has real potential to elucidate the connections between chemical reactions and surface sites.

Chao Zhan added: PEMS which has been widely investigated over the past four decades can enlighten our understanding and development of PMCR. In its earliest incarnations PMCR was often studied using PERS. In such experiments the nanostructure-based plasmonic enhancement was exploited in two ways: enabling a strong photochemical response, and allowing the time-dependent concentration of the products, often produced in meagre quantities, to be reliably measured.

Nevertheless, PEMS and PMCR are different in crucial ways. Usually, there are more challenges in PMCR compared with PEMS because of the molecular transformation. For example, PEMS measurements obtain the better sensitivity when molecules bond to the plasmonic nanostructure, while the over-strong bonding/adsorption causes problems for PMCR, like the blocking of actives sites. Thus, PERS is useful but careful design is needed.

Bartlomiej Jankiewicz remarked: A very interesting example of problems with applications of SERS is the history of Klarite substrate, which I am sure has been used by many researchers gathered at this Faraday Discussion. It has been withdrawn from the market because the main customers of the company producing Klarite were researchers who were just using it to compare the commercially available substrate with their home-made SERS substrates. Because of that production of Klarite was unprofitable and has been stopped. But I would like to mention a very recent initiative, the NATO Exploratory Team SET-ET-109 "Surface-enhanced Raman Scattering for Defense Applications". This team, which was proposed by colleagues from the Edgewood Chemical Biological Center (ECBC) and which includes scientist from several countries, will attempt to answer questions needed for SERS to be applicable to defense applications.

Zachary Schultz replied: The interesting thing about Klarite is that it was originally designed to be a reproducible substrate. I think our understanding of SERS has advanced and a higher signal and reproducible substrate could be produced. I think the outstanding question for SERS applications is understanding why some molecules show enhancement and others do not. For example, rhodamine 6G gives an excellent signal but sugars are difficult to detect. Perhaps a better understanding of chemical effects is what needs to be studied.

Jeremy Baumberg continued: To add to this, as the inventor and initial commercialiser of Klarite, the real challenge was to create a SERS substrate that was completely reproducible, which is why it was of real benefit to the community. In fact it was purchased in large quantity also by bioresearchers working in this area.

One of the main challenges is not just to create a quantitative and reproducible substrate. We note that the Au:CB colloidal aggregate system we now use is a similar low-cost but bottom-up version, see various papers at www.np.phy.cam.ac.uk/publications, including most recently ref. 1. For Klarite a major challenge was to package such a high-fidelity sensor, so that it could be stored and deployed when needed, without the sensor surface showing contamination already. Packaging is to be ignored at peril when developing sensors! We managed to achieve this (first in Mesophotonics, and then transferred to Renishaw), but often people complained about contamination when they opened their sensor, but it merely picked up whatever they had at trace concentrations in their own labs! The advantage of the new colloidal systems is that the sensor is only self-assembled when it is needed (though remaining quantitative and reproducible), and then the component nanoparticles and chemicals can be kept clean much more easily. Our forthcoming publication will detail this aspect.

1 B. de Nijs, M. Kamp, I. Szabó, S. J. Barrow, F. Benz, G. Wu, C. Carnegie, R. Chikkaraddy, W. Wang, W. M. Deacon, E. Rosta, J. J. Baumberg and O. A. Scherman, *Faraday Discuss.*, 2017, **205**, 505–515.

Laura Fabris commented: If one reads the literature on SERS, one realizes that the majority of publications cover the application of SERS to detect some kind of analyte, while fundamental studies have recently become extremely scarce. We all agree that SERS can be quite effective for qualitative detection even by simply adding salts and inducing nanoparticle aggregation around the analyte. However, this approach has led to inconsistencies in the enhancement factors and detection platforms, drastically hindering the applicability of SERS beyond the research lab. It is therefore important to modify our approach so that we can produce results that lead to (at least) some semiquantitative results. I believe that identifying materials design rules to obtain materials that produce homogeneous and reproducible enhancements is mandatory. In addition, designing sensing platforms that are simple to set up would be the way to go to extend the applicability of the technique. If nothing changes, we will be still here in twenty years discussing why SERS has not reached the levels of applicability that it had promised many years ago. We need to start learning from our mistakes and address them systematically with quantifiable metrics that can confirm our success.

Zachary Schultz noted: I think this echoes the earlier comments in the meeting to define standards. We discussed efficiencies for catalysis, but for SERS based sensing, an agreed upon standard for enhancement would also be useful.

Conflicts of interest

There are no conflicts to declare.

Faraday Discussions

PAPER

Hot electron science in plasmonics and catalysis: what we argue about

Jeremy J. Baumberg *

Received 22nd March 2019, Accepted 1st April 2019
DOI: 10.1039/c9fd00027e

Hot electron photochemistry has made strong claims for improved control of chemical reactions. Here I discuss these claims in the light of a plethora of model experiments and theories, asking what are the key issues to solve. I particularly highlight the need to understand nanoscale thermal hot-spots, thermal gradients, and thermal transport, as well as the conventional optical confinement in plasmonics. I note how the 'direct electron transfer' process seems to dominate, and resembles well known 'indirect excitons' in semiconductor quantum wells. I believe a crucial advance still required is a prototype nano-confined geometry which allows reactants and products to access a well-controlled metallic atomic surface.

Introduction

I started this meeting with optimistic views about the possibility of tackling several puzzles, paradoxes, and propaganda that have accreted around the field of hot electrons in plasmonics. While this meeting has approached some of these, and cleared away some of the vegetation to allow a better view, many of the issues remain implacably present, and the aim here is to present these in some context.

This has been a fascinating meeting, with perspectives on the concepts and utilisation of hot electrons for harnessing from a variety of directions. In this summary, I will discuss these in terms of models, of experiments, and of the probes used. I then aim to point to some areas of terminology that the field should harmonise, before discussing the challenges and some possible approaches to tackle them.

Prior experience

I would start by outlining a few recent experiments from my own research group that have suggested the influence of hot electrons, before critiquing them. One approach that we ourselves have used is to explore polymerisation reactions that can be initiated by the hot electrons from plasmonic nanoparticles (NPs). This has

NanoPhotonics Centre, Cavendish Laboratory, Department of Physics, University of Cambridge, JJ Thompson Avenue, Cambridge, CB3 0HE, UK. E-mail: jjb12@cam.ac.uk

utilised the nanoparticle-on-mirror or NPoM (also sometimes called particle-over-film or other descriptions, see ref. 1 and below) immersed in pure monomers for radical polymerisation such as divinylbenzene (DVB), styrene (St), or methyl methacrylate (MMA). Our aim in this work is to avoid complications from oxides by using only Au, and prevent Schottky barrier effects by eliminating semiconductor layers such as ZnO or TiO_2, to avoid any competing photochemistries. We mix quenchers into the monomer, and omit the radical initiators normally used to start the reaction. When illuminated (which is most efficient for wavelengths tuned to the trapped plasmon in the nanometre gap between the spherical nanoparticle and the planar mirror), hot electrons create radicals only at the metal surface, and these initiate the polymerisation which terminates within 20–50 nm of the surface.[2,3] Upon washing away unreacted monomer, this leaves a nanostructured polymer surface-coating on the NP around the gap (Fig. 1). The growing polymer forces the NP above the mirror surface, thus tuning the gap plasmon, and allowing the process to be controlled in real time by monitoring the dark-field scattering of the plasmonic construct. The hot electrons thus allow control of where monomers cross-link, in a similar way to conventional UV photoresists.

A second series of experiments (Fig. 2) uses the same NPoM geometry but inserts a few redox-active molecules into the gap whose oxidation state can be measured in real-time by the surface-enhanced Raman scattering (SERS) of their vibrations. When the electronic state changes, the modified electron density in the bonds changes their vibrational frequencies allowing remote tracking of their charge state, even at the single molecule level.[4] Essentially this allows us to watch chemistry in real time, with hot electrons jumping from the Au driving the process stochastically, depending on the conductivity of the molecule–metal interfaces. More recent experiments use bis-phthalocyanine nanogap spacers which are only 0.4 nm wide but contain single lanthanide ion cores that can tune the electronic and magnetic properties of the nanogap, thus changing the plasmonic mode.[5] The difficulty of full theoretical simulations that combine both the quantum mechanics of the molecular electronic states with the classical electromagnetism of the plasmonic confinement, means that such experiments cannot yet be accounted for by theory.

A final class of experiments uses light or voltage to change the redox state around such nano-constructs, for instance reducing copper oxide nanoparticles back to plasmonic copper[6] or creating tuneable wallpapers for building-scale display applications.[7] In this case electrons are involved in the redox processes but it is not clear if they can be termed hot or not.

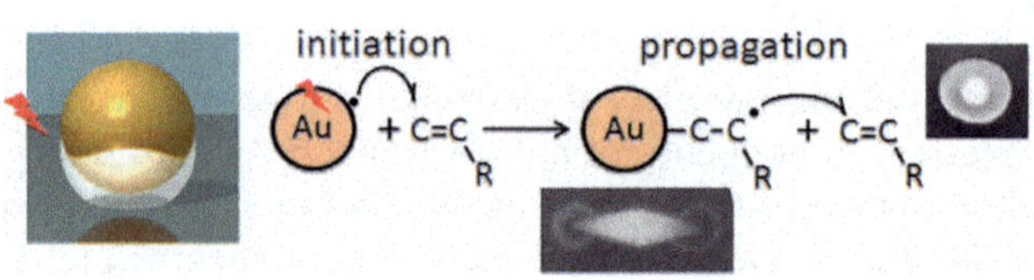

Fig. 1 Plasmon-induced radical polymerisation, illuminating a single NPoM with μW CW light immersed in monomers. Adapted from ref. 2 and 3 (http://creativecommons.org/licenses/by/4.0/).

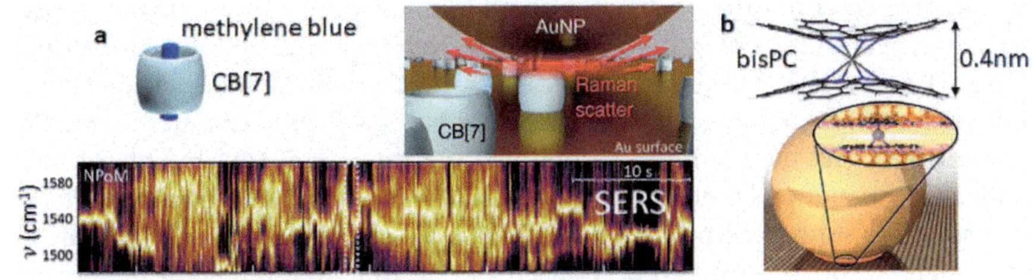

Fig. 2 (a) Single molecule redox, using methylene blue in CB[7] cage spacers within the NPoM nanogap, whose Raman emission (below) shows discrete jumps on single electron changes (adapted from ref. 4, http://creativecommons.org/licenses/by/4.0/). (b) Bis-phthalocyanine spacer with a lanthanide ion in the centre. Reprinted (adapted) with permission from ref. 5. Copyright 2019 American Chemical Society.

Such experiments are perhaps typical examples of the enthusiasm for exploring hot electron effects within plasmonics. They can be critiqued more generally in that all hot electron effects can be inferred: it is extremely hard to understand the local phonon temperatures in the metals and molecules at the same time, even though they are well defined precise constructs. Most work in the field as we see at this meeting involves much less well defined geometries making the task of separating the causes of the observations much harder. A theme of this summary will be that we need to find new constructs that allow full measurements and control of hot electrons in precise nano-geometries.

One particular issue is understanding the surface electronic states. Electrons at the surface of a metal such as gold experience their own image charges, thus forming an attractive potential above the surface which contains quantised states (Fig. 3a). Outside a vacuum environment, these states are screened. Even in air, nanoscale morphologies instantly collect monolayer-thick water films with ions which heavily screen away such confinement (Fig. 3b), however this is considerably altered by physisorbed and chemisorbed molecules. These combined surface electronic states are rarely characterised.

A second discussion is what is meant by 'hot electrons': this can mean (a) non-thermal electrons, (b) thermalized electron temperatures greater than the lattice temperature, or (c) locally hot but still lattice-thermalized regions of electrons above the global temperature (Fig. 3c). We will return to this discussion later.

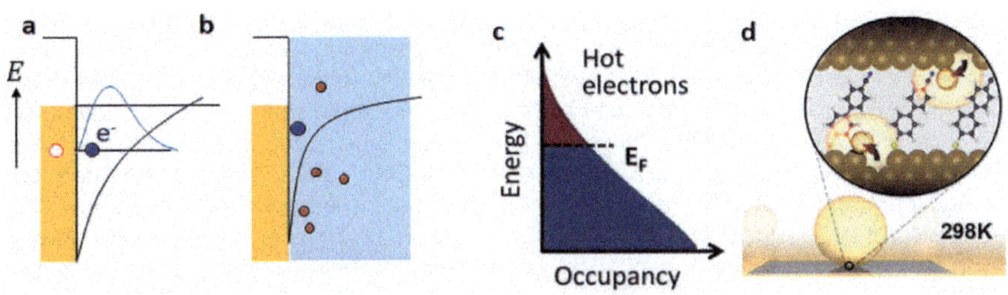

Fig. 3 (a) Electron surface states outside the metal, which (b) are screened with aqueous ions. (c) Hot Boltzmann distributions allow hot electrons to interact with surface states. (d) Gold atom movement tracked in real time in NPoMs. Reprinted (adapted) with permission from ref. 8. Copyright 2018 American Chemical Society.

Thirdly, it is important to realise that metal nanoparticles below their volume melting temperature (which is already suppressed below the bulk metal melting point) all possess surface atomic layers with much higher mobilities, so their liquid-like surface atoms scurry around at room temperature. We have recently demonstrated that both at low temperature[9] and room temperature,[8] plasmonic nanoparticle structures even at μW illumination powers exhibit dynamic surface reconstruction. The extremely tight plasmonic volumes allow the movement of even single Au or Ag atoms to be tracked in real time, reinforcing a picture in which binding of molecules to the surface is mostly transient and constantly reforming (Fig. 3d).

Models

At this meeting we have seen a number of models for hot electron production at plasmonic surfaces (Fig. 4), including a beautiful review of the field from Naomi Halas (DOI: 10.1039/C9FD00001A). Basic analytic understanding (DOI: 10.1039/C8FD00200B) allows us to see that the 'surface damping' (Fig. 4a) dominates hot electron production. This is despite much calculation in the literature picking out the quantum absorption of plasmons to estimate hot electron production, and

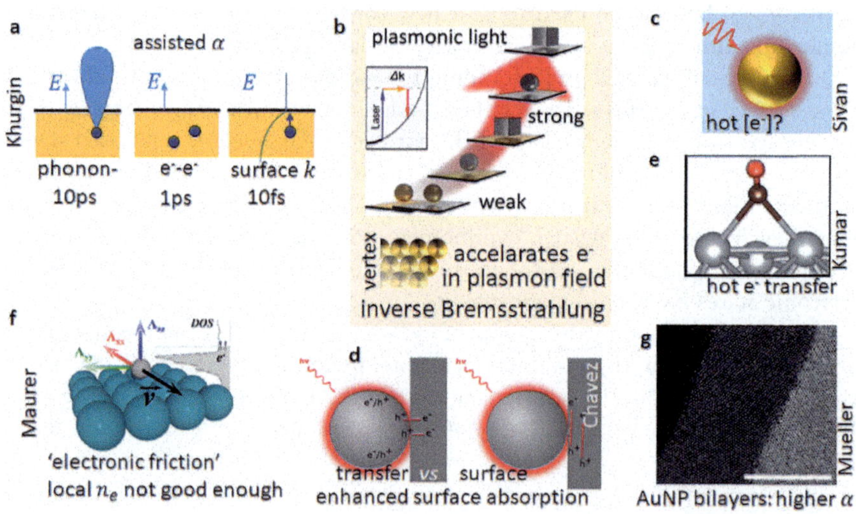

Fig. 4 (a) Hot-e⁻ timescales, needing different scattering processes. Only surface generated e⁻ are significant (DOI: 10.1039/C8FD00200B). (b) Model of electronic-Raman scattering (ERS) giving light emission at surfaces from the rapid spatial decay of plasmons, in a type of inverse Bremsstrahlung,[11] which will be enhanced at tip vertices around nanostructures. Adapted with permission from ref. 11. Copyright 2017 American Chemical Society. (c) Model of thermal heating on nanostructures in embedded thermal environments (DOI: 10.1039/C8FD00147B). (d) Comparison of transfer of electrons across a surface, and the indirect excitation of carriers in the near-field (DOI: 10.1039/C8FD00143J). (e) Full calculation of indirect excitation of e⁻ in CO on a Ag nanocluster (DOI: 10.1039/C8FD00154E). (f) Beyond using local surface electron density for molecular damping (DOI: 10.1039/C8FD00140E). (g) Model system of a bilayer of Au nanoparticles (DOI: 10.1039/C8FD00149A). (a) and (c)–(g) reproduced with permission from the Royal Society of Chemistry.

which generally has not been tested quantitatively against experiments. This process is similar to 'plasmon Bremsstrahlung' (Fig. 4b) which gives plasmon-enhanced light emission from the electronic-Raman scattering (ERS) process that depends on the plasmon field gradient at the surface of any nanostructure (see also ref. 10).

A crucial discussion opened by Yonatan Sivan *et al.* (DOI: 10.1039/C8FD00147B) is the need for a proper analysis of the local temperature at the nanostructure surface, when illuminated (Fig. 4c). We will return to this below, but it generated heated and interesting dialogues at the meeting. The proposal is that this analysis explains all purported plasmonic hot electron catalysis through highly localised thermal heating at hot spots. Since catalysis has energy barriers, reaction rates are exponentially sensitive to this localised heating. One crucial point is that comparing the temperature measured by a sample thermometer when heating the entire sample thermally *vs.* heating optically, is not enough. This is because complicated plasmonic nanostructures will absorb light in different locations and the temperature rise will depend on thermal Kapitza resistances at all interfaces that are not well quantified.

Besides the surface hot electrons (the weak ballistic electrons we now ignore as noted above), we should also consider enhanced near-field absorption in the surface-bound molecules (Fig. 4d), as well as 'direct transfer' where an electron from the metal is excited into the LUMO level of a surface molecule, or *vice versa* for holes. Full DFT can be used to calculate electron yields in the molecule (Fig. 4e), though these are not 'hot' since they are directly injected into the molecule. We will discuss later the (confusing) different terminologies for this type of process in different research fields. A further difficulty highlighted was of calculating damping effects on surface-bound molecules from metal underlayers (Fig. 4f and DOI: 10.1039/C8FD00140E), while finally a nice model system was shown to form from bilayers of Au nanoparticles (Fig. 4g).

Experimental progress

Many presentations at this meeting show intriguing materials nanostructures for enhancing catalytic interaction with plasmonic components, as well as different ways to characterise the hot electrons involved (Fig. 5).

These involve different plasmonic metal/semiconductor interfaces in a variety of geometries, with predominantly Schottky barriers to harvest electrons from the mechanisms discussed above. In these cases the nanostructures are intricate but complex, and hence the precise quantitative understanding is rather difficult since the control is limited. Even for thin semiconductors inside arrays of metal-insulator-metal patches (Fig. 5d) (or NPoMs), it is hard for molecules to access the tightly confined gaps under the patches, so likely most photo-redox activity takes place at the edges of these patches. While optimisation for specific reactions is clearly of interest, without the understanding of electron yields at different facets and vertices, material development is not well focussed.

Photocatalytic probes

A third focus of this meeting has been the development of probes to be able to study the photocatalytic processes in more detail (Fig. 6). One possibility is to use SERS gaps in the vicinity of the photocatalytic sites to track the reactants and

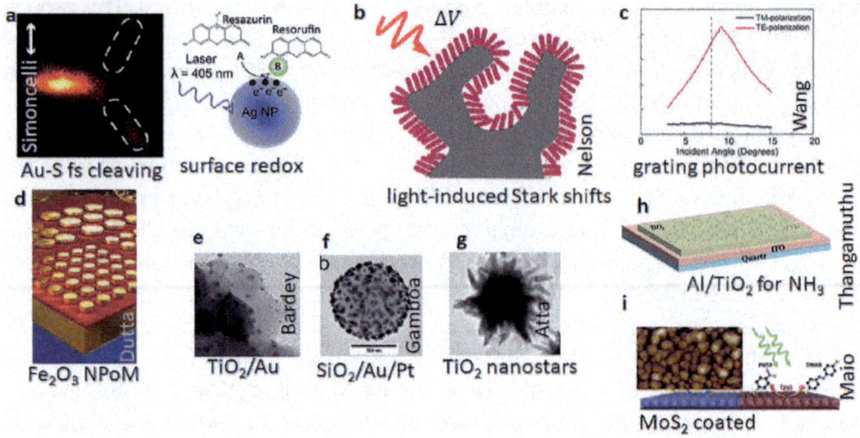

Fig. 5 (a) Polarisation-dependent plasmons select Au–S bonds to cleave with fs pulses (DOI: 10.1039/C8FD00138C), as well as studies of surface redox. (b) Measuring light-induced charging through Stark-shifts in SERS (DOI: 10.1039/C8FD00135A). (c) Photocurrents produced at angled gratings (DOI: 10.1039/C8FD00141C). (d) Enhancing photocatalytic efficiency in metal-insulator-metal geometries (DOI: 10.1039/C8FD00148K). (e–i) Varieties of nanostructures for improving and enhancing photocatalysis (DOI: 10.1039/C8FD00139A, DOI: 10.1039/C8FD00152A, DOI: 10.1039/C8FD00150B, DOI: 10.1039/C8FD00146D and DOI: 10.1039/C8FD00144H). Reproduced with permission from the Royal Society of Chemistry.

products in real time (Fig. 6a). A completely different approach is to break bonds of an adsorbed molecule on a single-crystal surface watching where the reactants end up using STM (Fig. 6b). Evaluating NPoM systems with different materials (Fig. 6c) showed that absorption from lower quality plasmonic materials is not necessarily as much a problem in other plasmonic applications.

A further range of ideas were also presented at the meeting, using TERS (Fig. 6d) and theory (Fig. 6f) to consider TMD hot electrons, or controlling Schottky nanointerfaces (Fig. 6e). Several theories also explored the nonlinear response of electrons, from the high intensity pulses in nanogaps (Fig. 6g) and from fluid models of plasma transport (Fig. 6h).

Terminologies

There are three among many overlapping terminological discussions that I will open up here. These arise because of the highly interdisciplinary nature of our community, and are a healthy sign of cross-fertilisation of ideas, but also remain a confusing block to both younger and more experienced researchers as they are presented as separate or unknown processes, while in fact they can be slotted into much more established paradigms.

The first concerns the name of the process in which electrons from the metal are optically excited directly into the molecule, so far referred to as '*direct transfer*' (Fig. 7a). This can also cover the excitation of an electron from the molecule being excited above the Fermi energy in the metal. This process has also been titled '*chemical-induced damping*', presumably as near-surface electrons have a larger range of states to now be excited into, and may indeed by related to the surface

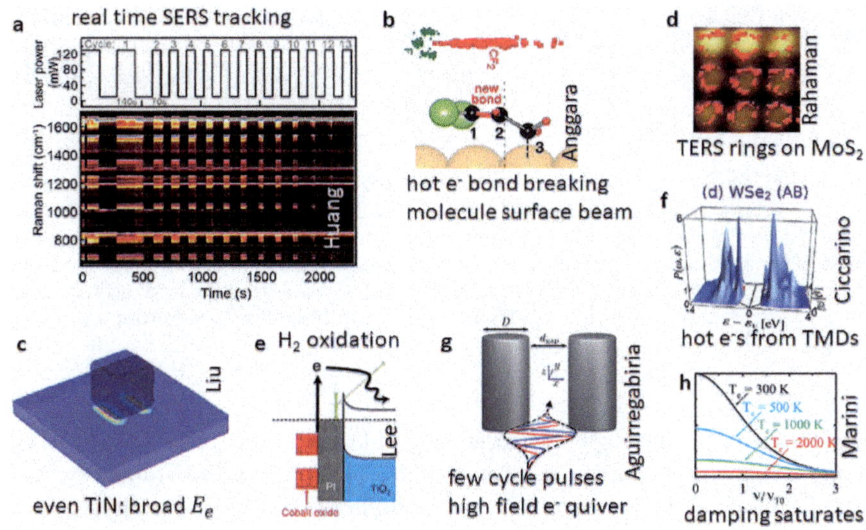

Fig. 6 (a) SERS spectra as a function of time of reactants, intermediates, and products (DOI: 10.1039/C8FD00155C). (b) STM tracking of molecular fragments (DOI: 10.1039/C8FD00137E). (c) Cube on mirror (*viz* NPoM) geometries evaluated for photocatalytic activity with different material systems (DOI: 10.1039/C8FD00145F). (d) MoS_2 clad pillars show TERS enhanced in spatial halos (DOI: 10.1039/C8FD00142A). (e) Cobalt oxide nanocubes on Pt/TiO_2 to quantify interface currents (DOI: 10.1039/C8FD00136G). (f) Optimising ballistic hot electrons from TMDs (DOI: 10.1039/C8FD00159F). (g) Quantifying nonlinear optical field driven electron currents (DOI: 10.1039/C8FD00158H). (h) Nonlinear damping in metals (DOI: 10.1039/C8FD00153G). Reproduced with permission from the Royal Society of Chemistry.

damping concept added to the Drude model accounting for metal interfaces (but is not yet treated consistently within that framework). Another term is '*hybridisation*' which describes how the spatial leakage of molecular wavefunctions into the metal, and spillout of electron wavefunctions from the metal allow the wavefunctions to mix together. This produces regions near the surface where both the leaking wavefunctions are co-located, thus allowing such optical transitions to occur at a single spatial site. Another term is '*plasmon resonant energy transfer*' or PRET which describes dipole–dipole induced transfer of energy from donor to absorber mediated by (virtual) plasmons or metal-dressed photons, and I would argue that it is not clear how such a process where a plasmon is dipole-coupled to the molecular transition would be experimentally distinguished from the '*direct transfer*' we have been discussing, apart from semantically. DFT calculations may be useful in this respect, as shown in the meeting, but only if they can span systems large enough to hold the key physics. Finally, it is important to make a connection to a well-established (but apparently unknown to this community) similar scientific idea, which is unfortunately termed '*indirect absorption*' known from semiconductor physics for >30 years. This forms exquisitely controllable potential landscapes for electrons using planar layers of different semiconductor materials typically grown by MBE and composed of III–V materials such as GaAs and AlGaAs alloys. Double quantum wells of width 5–20 nm can be grown with very thin barriers (<5 nm) in between which act as analogues for the metal and molecule systems here, since the

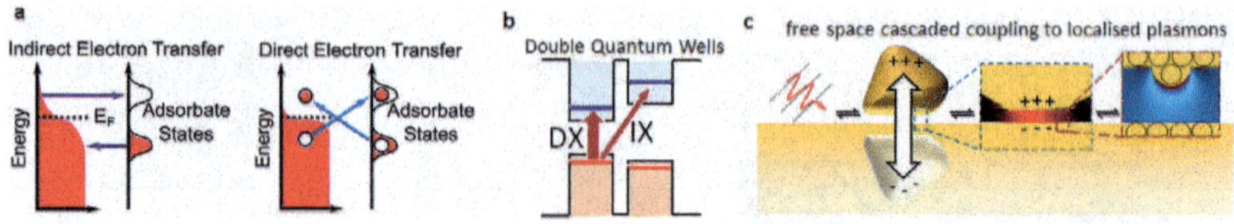

Fig. 7 (a) Concepts of indirect (hot electrons in metal and ballistic transport) and direct electron transfer. Reproduced from DOI: 10.1039/C8FD00151K with permission from the Royal Society of Chemistry. (b) Indirect exciton optical resonances in double quantum wells of semiconductor layers, from ref. 12. (c) Cascaded coupling of free space light into tightly localised plasmons, through plasmonic antennas, nanogaps, and even atomic protrusions. Reproduced from ref. 1 with permission from Springer Nature, copyright 2019.

wavefunctions leak through the barriers (Fig. 7b). One optical transition is between a valence band to conduction band state in the same quantum well (direct exciton, or DX), and this is independent of a voltage applied between the quantum wells. A second transition is between the valence band in one quantum well and the conduction band in the neighbouring quantum well whose electronic wavefunction is exponentially leaking through the barrier, but provides some smaller oscillator strength. This is known in the field as a '*spatially-indirect exciton*', but is identical to the '*direct electron transfer*' here, and can be spectrally tuned by applying a voltage, as seen in many experiments[12] (and references therein). Stark shifts are also sometimes seen for molecules on surfaces, but it is not easy to support a potential across this interface due to screening and conduction, but the spatially-indirect nature of the transition is well established and quantified exactly. I note that exactly the same spatially-indirect transition occurs for metal/semiconductor Schottky barriers which thus do not need to invoke ballistic transport.

A second confusion is more trivial, and it is the profusion of names that have grown up around '*nanogap plasmons*' which provide the most effective concentration of optical field. These range over the terms '*metal-insulator-metal*' (MIM) waveguides, nanogap '*patch antennas*', '*metastructures*', and my own favoured '*nanoparticle-on-mirror*' (NPoM). These all support similar plasmons, with a number of modes which can radiate more or less efficiently (see ref. 1 for a review and Fig. 7c). The names reflect fabrication approaches from top-down to bottom-up, but remain somewhat confusing for those outside the field. The real trade-off here is between sharply spiked nanoparticles (from nanorods to nanostars) in which the asperities are hard to control and maintain in time but are highly accessible to reactants and allow products to escape, compared to nanogaps which have reliable and deep confinement of light but present steric barriers to reactants and products.

A third confusion in the meeting has been the discussion about what is a '*hot electron*'. In the community, it is generally agreed that the vision of interest is to trigger photocatalytic reactions that are in some way different from those that are simply thermally driven. Just heating a catalytic nanoparticle sample creates a uniform temperature distribution, with minimal thermal gradients so that most surface reconstructions are static. Conversely irradiating with light typically causes highly localised heating enhanced by the plasmonic optical properties, and thus very inhomogeneous thermal distributions (literally 'thermal hot-spots') and high thermal gradients with significant dynamic surface atomic reconstruction. At the same time, it may also be possible for non-thermal electrons to be harvested by molecules adsorbed onto the plasmonic surfaces for chemical reactions that are otherwise not feasible, but these add onto the thermal hot-spot processes. Since catalysis is thermally activated, reaction rates are extremely sensitive to a small fraction of localised hot spots. I also emphasise that local lattice temperatures can dynamically exceed melting temperatures (one might term it '*superheating*'), and certainly surface reconstruction takes place at much lower irradiations, and is optically catalysed.[9]

In this context, it is not clear that 'hot electrons' should only be those which are non-thermal (a very tiny fraction) or whose temperature exceeds the local lattice temperature. It is also not clear that we can call the direct electron transfer (which dominates over ballistic electron transport) 'hot' since it is a driven excitation of a two level system. However it is certainly plausible and observed that optical irradiation can enhance catalysis at a given substrate temperature, and change reaction pathways.

Prospects

In conclusion then, I would emphasise that while a huge amount of research has focussed on optical field enhancements in plasmonics, rather less effort has looked at the phonon transport in such nanostructures. The profusion of interfaces between dissimilar materials and hence the large Kapitza thermal resistances, the effects of morphology and nano-contacts on thermal transport, the effects of gas and solvent diffusion and convection in nanoscale pore geometries, and the effects of strong thermal gradients on atomic reconstructions that are crucial to provide step edges and facets that are active in catalysis, are all vital pieces of science to now study. The role of shape is subtle, since optical fields are localised around metal apexes, and penetration into the metal is enhanced at these points creating both thermal hot-spots and excited electrons in the metal and direct transfer to adsorbed molecules. Tight confinement at such points also gives large wavevector components that enhance absorption. It is thus likely that points and gaps are important in their photocatalytic effects, but they have not so far been well controlled.

There is thus now a need for individual nanogeometries in which all these electromagnetic and thermal configurations are fully controlled. This would provide a testbed to investigate many of the issues raised here. These include influence of inhomogeneous temperatures, thermal transport, thermal gradient forces and potentials, direct and indirect electrons, shape, diffusion of species, reconstruction and degradation of activity, as well as their influence on a range of reactions. The latter is also a significant issue for the field, with a need to move beyond degradation reactions to bond formation. On the other hand, it is clear that surface reconstruction at the nanoscale depends on the molecular species being adsorbed, hence there is another subtle feedback where the hot electron/spot effects depend on the reaction being presented at the metal surface. The precise type of adsorption will also be important, with the orientation, proximity, atomic configuration of binding site, solvation, and screening which all affect hybridisation and wavefunction leakage that control electron/hole transfer to molecules.

Investigating such single nanostructures (or arrays of identical nanostructures) gives problems in detecting the small quantities of molecular product. While SERS techniques such as DOI: 10.1039/C8FD00138C and DOI: 10.1039/C8FD00155C may be help in some cases, more developments in this area are also needed, in solution phase as well as gas phase catalysis. The challenge of fabricating nanostructures remains significant, since both atomic scale and nanoscale to wavelength-scale features are all important. While I favour nanogap schemes such as the NPoM to provide some of this optical and structural control, the barrier remaining to solve is enabling molecular diffusion at this scale and control of facetting. Hence this research field has plenty of challenges for the next years that promise development of a real understanding of what is going on inside the photocatalytic (nano)box.

Conflicts of interest

The authors declare no competing financial interest.

Acknowledgements

We acknowledge funding from the EPSRC Cambridge NanoDTC, EP/L015978/1 and EP/N016920/1, and EP/L027151/1. SM, BM, BU each acknowledge support through an Australian Government Research Training Program Scholarship. AT also acknowledges key support from UOW SMAH research funding.

References

1 J. J. Baumberg, J. Aizpurua, M. H. Mikkelsen and D. R. Smith, Extreme nanophotonics from ultrathin metallic gaps, *Nat. Mater.*, 2019, DOI: 10.1038/s41563-019-0290-y.

2 T. Ding, J. Mertens, A. Lombardi, O. A. Scherman and J. J. Baumberg, Light-Directed Tuning of Plasmon Resonances *via* Plasmon-Induced Polymerization Using Hot Electrons, *ACS Photonics*, 2017, **4**, 1453–1458.

3 Y. Wang, *et al.*, Plasmon-directed polymerization: Regulating polymer growth with light, *Nano Res.*, 2018, **11**, 6384–6390.

4 B. De Nijs, *et al.*, Plasmonic tunnel junctions for single-molecule redox chemistry, *Nat. Commun.*, 2017, **8**, 1–7.

5 C. Readman, *et al.*, Anomalously Large Spectral Shifts near the Quantum Tunnelling Limit in Plasmonic Rulers with Subatomic Resolution, *Nano Lett.*, 2019, **19**, 2051–2058.

6 G. Di Martino, V. A. Turek, P. Braeuninger-Weimer, S. Hofmann and J. J. Baumberg, Laser-induced reduction and *in situ* optical spectroscopy of individual plasmonic copper nanoparticles for catalytic reactions, *Appl. Phys. Lett.*, 2017, **110**, 071111.

7 J. Peng, *et al.*, Scalable electrochromic nano-pixels using plasmonics, *Sci. Adv.*, 2019, **5**, eaaw2205.

8 C. Carnegie, *et al.*, Room-Temperature Optical Picocavities below 1 nm 3 Accessing Single-Atom Geometries, *J. Phys. Chem. Lett.*, 2018, **9**, 7146–7151.

9 F. Benz, *et al.*, Single-molecule optomechanics in "picocavities", *Science*, 2016, **354**, 726–729.

10 F. J. García de Abajo, Optical excitations in electron microscopy, *Rev. Mod. Phys.*, 2010, **82**, 209–275.

11 J. Mertens, M.-E. Kleemann, R. Chikkaraddy, P. Narang and J. J. Baumberg, How Light Is Emitted by Plasmonic Metals, *Nano Lett.*, 2017, **17**, 2568–2574.

12 P. Cristofolini, *et al.*, Coupling Quantum Tunneling with Cavity Photons, *Science*, 2012, **336**, 704–707.

Poster titles

Hot-electron generation in plasmonic nanocones, **R. M. Córdova-Castro, A. V. Krasavin, P. Wang, C. McPolin, W. Dickson, V. A. Podolskiy, A. V. Zayats**, *King's College London, UK*

Time dependent density functional theory for nano-plasmonics, **R. D'Agosta, W. Zhao, F. Baletto**, *Universidad del Pais Vasco EHU/UPV, Spain*

Colliding molecules at selected impact parameters, **K. Anggara, L. Leung, M. Timm, Z. Hu, J. Polanyi**, *University of Toronto, Canada*

Controlling reaction selectivity over hybrid plasmonic nanocatalysts, **J. Quiroz, E. Barbosa, T. Araujo, J. Fliorio, Y.-C. Wang Wang, Y.-C. Zou, T. Mou, T. Alves, D. Oliveira, B. Wang, S. Haigh, L. Rossi, P. Camargo**, *University of São Paulo (IQ/USP), Brazil*

Cu(I) adlayer impact on optically rectified fields and plasmonic electrocatalysis, **D. Nelson, E. Landaeta, Z. Schultz**, *Ohio State University, USA*

Spectral screening of the energy of hot holes over a particle plasmon resonance, **E. Pensa, J. Gargiulo, A. Lauri, S. Schlücker, E. Cortés, S. A. Maier**, *Imperial College London, UK*

Photothermal vs hot charge carrier effects in plasmon-driven syntheses of nanoparticles, **R. Kamarudheen, G. Castellanos, L. Kamp, H. Clercx, A. Baldi**, *DIFFER, Netherlands*

Size-dependent solar catalysis by hollow plasmonic silver nanocubes, **B. K. Dadhich, B. Bhushan, A. Priyam**, *Kalinga Institute of Industrial Technology, Bhubaneswar, India*

Ultrafast intraband spectroscopy of hot-carrier cooling in lead-halide perovskites, **T. Hopper, A. Gorodetsky, J. Frost, C. Müller, R. Lovrincic, A. Bakulin**, *Imperial College London, UK*

How to detect molecular photochemistry products on nanoparticles with chemical specificity?, **R. Grzeschik, A. Letzel, F. Uteschil, S. Barcikowski, O. Schmitz, S. Schlücker**, *Universität Duisburg-Essen, Germany*

Coulomb effects on the photo-excited dynamics of electrons in a plasmonic nanosphere, **A. Crai, A. Pusch, D. E. Reiter, T. Kuhn, O. Hess**, *Imperial College London, UK*

Kinetics of hot electron induced reactions on gold nanoparticle arrays, **A. Dutta, R. Schürmann, A. Nagel, C. Pacholski, I. Bald**, *University Potsdam, Germany*

Real-time dynamics of plasmons in nanoparticles: First-principles atomic-scale analysis of plasmon formation and decay into hot electrons and holes, **T. Rossi, P. Erhart, M. Kuisma**, *Chalmers University of Technology, Sweden*

Hot carrier generation from single plasmons, **L. Roman Castellanos, J. Lischner, O. Hess**, *Imperial College London, UK*

Platinum doped carbon dots and its hybridization with TiO_2 for enhanced visible light photocatalytic hydrogen evolution, **H. Luo, A. Sobrido, M. Titirici**, *Queen Mary University of London, UK*

Wafer scale plasmonic dimer array with sub nanometer gap for hot electron generation: application in phase transformation of MoS_2, **P. Deshpande, H. J. Singh, A. Ghosh**, *Indian Institute of Science, India*

Thermal effects - an alternative mechanism for plasmonic photo-catalysis, **Y. Dubai, Y. Sivan**, *Ben-Gurion University, Israel*

An insight into the dynamics of photo-induced oxygen vacancy defects in photocatalytic metal-oxide semiconductors, **D. Glass, E. Cortés, S. Ben-Jaber, T. Brick, W. Peveler, Y. Zhu, C. Blackman, C. Howle, R. Quesada-Cabrera, I. Parkin, S. A. Maier**, *Imperial College London, UK*

Plasmonic sensing via chemical interface damping in core@shell metal@semiconductor nanoparticles, **M. Parente, S. Sheikholeslami, G. Naik, J. Dionne, A. Baldi**, *DIFFER, Netherlands*

Spatial distribution of hot electron nonlinear photoluminescence in Au nanowires, **A. Agreda, D. Sharma, S. Viarbitskaya, R. Hernandez, B. Cluzel, O. Demichel, G. C. des Francs, G. V. Pavan Kumar, A. Bouhelier**, *Universite de Bourgogne Franche Comte, France*

Plasmonic broadband perfect absorbers for enhanced hot carrier extraction, **C. Ng, J. Cadusch, S. Dligatch, A. Roberts, T. J. Davis, D. Gomez**, *Leibniz Institute of Polymer Research Dresden, Germany*

Super-resolution mapping of plasmonic hot electrons, **R. Hamans, M. Parente, A. Baldi**, *DIFFER, Netherlands*

Ultrafast dynamics of optical excitations in plasmonic nanocone arrays, **V. Podolskiy, R. M. Córdova-Castro, A. V. Zayats**, *University of Massachusetts Lowell, USA*

Theoretical study on synthesis of hydrogenated graphene via hot electron and birch reduction, **T. K. Lee, S. K. Kwak**, *Ulsan National Institute of Science and Technology (UNIST), South Korea*

Plasmonic hybrid gap waveguides - a platform for sensing, detection and electro-optical modulation, **N. Güsken, M. Nielsen, N. Nguyen, L. Lafone, R. Oulton, S. A. Maier**, *Imperial College London, UK*

Modelling the properties of plasmon-induced hot carriers in metallic nanoparticles for photocatalysis, **S. Dal Forno, L. Ranno, J. Lischner**, *Imperial College London, UK*

Isolation of photothermal and non-thermal effects in plasmonic catalysis, **X. Li, X. Zhang, S. He, H. O. Everitt, J. Liu,** *Duke University, USA*

Electro-photo-chemistry in plasmonic tunnel junctions, **P. Wang, A. Krasavin, M. Nasir, W. Dickson, A. V. Zayats,** *King's College London, UK*

Plasmons in suspended graphene, **X. Xiao, J. D. Caldwell, S. A. Maier, V. Giannini,** *Imperial College London, UK*

Towards hot electron nano sensing with quantum dots, **S. Morozov, E. L. Pensa, S. Mignuzzi, E. Cortés, S. Vezzoli, S. A. Maier, I. Moreels, R. Sapienza,** *Imperial College London, UK*

Ab initio plasmon quasiparticle and self-energy description of hot-electron generation, **T. Rossi, P. Erhart, M. Kuisma,** *Chalmers University of Technology, Sweden*

Effect of plasmonic enhancement and valance band states in the hot electron induced dimerization of nitrothiophenol, **R. Schürmann, S. Kogikoski Jr, A. Dutta, K. Ebel, C. Nicolas, A. R. Milosavljević, I. Bald,** *University Potsdam, Germany*

Nonlinear optics with metallic optical antennas, **S. Gennaro, Y. Li, S. A. Maier, R. Oulton,** *Imperial College London, UK*

Plasmonic nanorods and nanotubes for photo-electrochemical catalysis, **A. Zaleska, W. P. Wardley, F. Lotti, A. V. Zayats, W. Dickson,** *King's College London, UK*

Development of Ag@CeO_2 nanocubes with photocatalytic application, **D. O'Neill, D. Prezgot, A. Ianoul, G. Mul, A. Huijser,** *University of Twente, Netherlands*

Dark interlayer plasmons in gold nanoparticle metafilms for efficient hot electron generation, **N. S. Mueller, B. G. M. Vieira, F. Schulz, D. Höing, E. B. Barros, H. Lange, S. Reich,** *Freie Universität Berlin, Germany*

Simulations on plasmonic properties of gold nanoparticle few-layers for hot electron generation, **B. G. M. Vieira, N. S. Mueller, E. B. Barros, S. Reich,** *Universidade Federal do Ceará, Brazil*

Oxide and nitride thin films as alternative plasmonic materials, **R. Bower, M. P. Wells, B. G. Doiron, A. P. Mihai, B. Zou, G. Mallia, L. F. Cohen, S. A. Maier, N. M. Alford, P. K. Petrov,** *Imperial College London, UK*

Spatio-temporal separation of plasmon-induced hot carriers by time- and energy-resolved photoemission electron microscopy, **M. Hartelt, A.-K. Mahro, T. Eul, B. Frisch, E. Prinz, D. Kilbane, B. Stadtmüller, M. Cinchetti, M. Aeschlimann,** *TU Kaiserslautern, Germany*

Perovskite nanoantenna for plasmonic photocatalysis, **S. Saris, A. Loiudice, R. Buonsanti,** *EPFL, Switzerland*

Nanoparticles for plasmonic carbon dioxide reduction, **V. Niemann, S. Saris, A. Loiudice, R. Buonsanti,** *École Polytechnique Fédérale de Lausanne, Switzerland*

Light amplification by plasmon induced hot carriers, **K. Braun, F. Laible, X. Wang, D. Zhang, M. Fleischer, A. J. Meixner,** *University of Tuebingen, Germany*

Silver diffusion in metal-titania core-shell nanostructures, **B. Jankiewicz, B. Bartosewicz, M. Liszewska, B. Budner, M. Michalska-Domańska**, *Military University of Technology, Poland*

Investigating hot electrons with plasmonic waveguides and metal-semiconductor gratings, **R. Hernandez, O. Demichel, R. Juliano Martins, M. Lodari, M. Celebrano, M. Finazzi, L. Duò, G. Isella, M. Bollani, A. Agreda, M. Petit, J.-C. Weeber, A. Bouhelier, P. Biagioni, B. Cluzel**, *Laboratoire Interdisciplinaire Carnot de Bourgogne, France*

Surface generation and emission of hot electrons into electrolyte from nanoporous gold, **M. Graf, D. Jalas, J. Weissmüller, A. Petrov, M. Eich**, *Helmholtz Center Geesthacht, Germany*

CuS nanodisks: synthesis, thin films and optical characterization, **A. Shukla, A. Bykov, M. V. Schilfgaarde, M. Green, A. V. Zayats**, *King's College, London, UK*

Aluminium metamaterial structures for UV-plasmonic induced hot carrier generation, **R. M. Romero Gómez, W. P. Wardley, F. J. Rodríguez-Fortuño, A. V. Zayats, W. Dickson**, *King's College London, UK*

Ultrafast studies of hot hole dynamics in Au/p-GaN heterostructures, **G. Tagliabue, J. S. DuChene, M. Abdellah, A. Habib, R. Sundararaman, J. Sá, H. A. Atwater**, *EPFL, Switzerland*

Tailoring dynamic optical properties of free-electron gases with metamaterial design, **L. H. Nicholls, F. J. Rodríguez-Fortuño, T. Stefaniuk, M. E. Nasir, G. A. Wartz, A. V. Zayats**, *King's College London, UK*

The affiliation of the presenting author is listed above.

The Faraday Discussions Poster Prize for the best poster was awarded to Mr Michael Hartelt of TU Kaiserslautern, Germany, for his poster on spatio-temporal separation of plasmon-induced hot carriers by time-and energy-resolved photoemission electron microscopy.

List of participants

Miss Katie Ackermann, *Royal Society of Chemistry*, *United Kingdom*
Mr Adrian Agreda, *Laboratoire ICB CNRS Universite de Bourgogne Franche Comte*, *France*
Professor Javier Aizpurua, *Center for Materials Physics in San Sebastián (CSIC-UPV/EHU)*, *Spain*
Dr Kelvin Anggara, *Department of Chemistry and Institute of Optical Sciences, University of Toronto*, *Canada*
Dr Helen Armes, *Royal Society of Chemistry*, *United Kingdom*
Professor Michael Ashfold, *University of Bristol*, *United Kingdom*
Dr Jeremie Asselin, *University of Cambridge*, *United Kingdom*
Professor Dr Ilko Bald, *University of Potsdam*, *Germany*
Dr Francesca Baletto, *King's College London*, *United Kingdom*
Mr Steven Bardey, *ICPEES, Strasbourg / IFP Energies Nouvelles, Lyon.*, *France*
Professor Jeremy Baumberg, *University of Cambridge*, *United Kingdom*
Miss Meltiani Belekoukia, *Heriot-Watt University*, *United Kingdom*
Professor Alexandra Boltasseva, *Purdue University*, *USA*
Mr Ryan Bower, *Imperial College London*, *United Kingdom*
Dr Kai Braun, *University of Tuebingen*, *Germany*
Professor Eleanor Campbell, *University of Edinburgh*, *United Kingdom*
Dr Valérie Caps, *CNRS (ICPEES, UMR 7515)*, *France*
Professor Phillip Christopher, *UC Santa Barbara*, *USA*
Dr Ilaria Cianchetta, *Cell Press*, *USA*
Professor Dr Emiliano Cortes, *Ludwig-Maximilians-Universität München*, *Germany*
Miss Alexandra Crai, *Imperial College London*, *United Kingdom*
Professor Steve Cronin, *University of Southern California*, *USA*
Miss R. Margoth Córdova-Castro, *King's College London*, *United Kingdom*
Dr Roberto D'Agosta, *Universidad del Pais Vasco EHU/UPV*, *Spain*
Mr Bhavesh Kumar Dadhich, *Kalinga Institute of Industrial Technology, Bhubaneswar, Odisha 751024*, *India*
Dr Bart de Nijs, *University of Cambridge*, *United Kingdom*
Dr Angela Demetriadou, *University of Birmingham*, *United Kingdom*
Mrs Preeti Deshpande, *Indian Institute of Science*, *India*
Mr Yonatan Dubi, *Ben-Gurion University*, *Israel*
Dr Anushree Dutta, *Universität of Potsdam*, *Germany*
Professor Laura Fabris, *Rutgers University*, *USA*
Dr Yuri Fernandez, *University of Liverpool*, *United Kingdom*
Dr Simon Freakley, *University of Bath*, *United Kingdom*
Dr Julian Gargiulo, *Imperial College London*, *United Kingdom*
Dr Sylwester Gawinkowski, *Institute of Physical Chemistry PAS*, *Poland*
Dr Sylvain Gennaro, *Imperial College London*, *United Kingdom*
Mr Joseph Gibson, *Royal Society of Chemistry*, *United Kingdom*
Mr Josh Gill, *EPSRC (UKRI)*, *United Kingdom*
Mr Daniel Glass, *Imperial College London*, *United Kingdom*

Mr Bruno Gondim de Melo Vieira, *Universidade Federal do Ceará, Freie Universität Berlin, Brazil*
Professor Alexander Govorov, *UESTC (China) and Ohio Univ. (USA), China*
Dr Matthias Graf, *Helmholtz Center Geesthacht, Germany*
Mr Roland Grzeschik, *Universität Duisburg-Essen, Germany*
Mr Nicholas Güsken, *Imperial College London, United Kingdom*
Professor Naomi Halas, *Rice University, USA*
Miss Kirsten Hall, *Royal Society of Chemistry, United Kingdom*
Ms Eleanor Hall, *Royal Society of Chemistry - Press team, United Kingdom*
Mr Ruben Hamans, *DIFFER, Netherlands*
Mr Michael Hartelt, *TU Kaiserslautern, Germany*
Professor Dwayne Heard, *University of Leeds, United Kingdom*
Mr Romain Hernandez, *Laboratoire Interdisciplinaire Carnot de Bourgogne, France*
Mr Benedict Hewitt, *Royal Society of Chemistry, United Kingdom*
Mr Tom Hopper, *Imperial College London, United Kingdom*
Mr Junyang Huang, *University of Cambridge, United Kingdom*
Dr Andrea Jacassi, *Imperial College, United Kingdom*
Dr Bartłomiej Jankiewicz, *Military University of Technology, Institute of Optoelectronics, Poland*
Ms Yunlu Jiang, *King's College London, United Kingdom*
Mr Rifat Kamarudheen, *DIFFER, Netherlands*
Professor Jacob Khurgin, *Johns Hopkins University, USA*
Dr Sergio Kogikoski Junior, *University of Potsdam, Institute of Chemistry, Physical Chemistry, Germany*
Dr Mikael Kuisma, *University of Jyväskylä, Finland*
Dr Priyank Vijaya Kumar, *ETH Zurich, Switzerland*
Mr Tae Kyung Lee, *Ulsan National Institute of Science and Technology (UNIST), South Korea*
Dr Johannes Lischner, *Imperial College London, United Kingdom*
Professor Jie Liu, *Duke University, USA*
Miss Hui Luo, *Queen Mary University of London, United Kingdom*
Dr Sandra Macaskill, *Royal Society of Chemistry, United Kingdom*
Mr Julien Mahin, *University of Cambridge, United Kingdom*
Professor Stefan Maier, *Imperial College London, United Kingdom*
Dr Maria Maragkou, *Nature Materials, United Kingdom*
Dr Andrea Marini, *University of L'Aquila, Italy*
Dr Reinhard Maurer, *University of Warwick, United Kingdom*
Dr Peter McBreen, *Laval University, Canada*
Dr Cillian McPolin, *King's College London, United Kingdom*
Mr Paul Minett, *Royal Society of Chemistry, United Kingdom*
Mr Sergii Morozov, *Imperial College London, United Kingdom*
Professor Andrew Mount, *The University of Edinburgh, School of Chemistry, United Kingdom*
Mr Niclas Sven Mueller, *Freie Universität Berlin, Department of Physics, Germany*
Miss Fiona Nalden, *Royal Society of Chemistry, United Kingdom*
Mr Darby Nelson, *Ohio State University, USA*
Dr Charlene Ng, *Leibniz Institute of Polymer Research Dresden, Germany*
Mr Luke Nicholls, *King's College London, United Kingdom*
Miss Valerie Niemann, *École Polytechnique Fédérale de Lausanne, Switzerland*
Dr Wendy Niu, *Royal Society of Chemistry, United Kingdom*
Mr Devin O'Neill, *University of Twente, Netherlands*
Dr Rupert Oulton, *Imperial College London, United Kingdom*
Mr Matteo Parente, *DIFFER, Netherlands*

Professor Jeong Y. Park, *IBS/KAIST*, *South Korea*
Professor Viktor Podolskiy, *University of Massachusetts Lowell*, *USA*
Professor John Polanyi, *University of Toronto*, *Canada*
Dr Jhon Jhon Quiroz Torres, *Chemistry Institute of the University of São Paulo (IQ/USP)*, *Brazil*
Mr Mahfujur Rahaman, *Chemnitz University of Technology*, *Germany*
Professor Paul Raithby, *University of Bath*, *United Kingdom*
Professor Stephanie Reich, *Freie Universitaet Berlin*, *Germany*
Mr Sebastian Rejman, *University of Cambridge*, *United Kingdom*
Professor David Richards, *King's College London*, *United Kingdom*
Miss Lara Roman Castellanos, *Imperial College London*, *United Kingdom*
Miss Rosa Mayela Romero Gómez, *King's College London*, *United Kingdom*
Ms Philippa Ross, *Royal Society of Chemistry*, *United Kingdom*
Dr Tuomas Rossi, *Chalmers University of Technology*, *Sweden*
Mr Jorge U. Salmon-Gamboa, *King's College London*, *United Kingdom*
Dr Riccardo Sapienza, *Imperial college london*, *United Kingdom*
Mr Seryio Saris, *EPFL*, *Switzerland*
Professor Sebastian Schlücker, *Universität Duisburg-Essen, Department of Chemistry and CENIDE*, *Germany*
Dr Zachary Schultz, *The Ohio State University*, *USA*
Dr Robin Schürmann, *University Potsdam, Institute of Chemistry, Physical Chemistry*, *Germany*
Professor John Seddon, *Imperial College London*, *United Kingdom*
Mr Amaresh Shukla, *King's College, London*, *United Kingdom*
Dr Yonatan Sivan, *Ben-Gurion University*, *Israel*
Miss Giulia Tagliabue, *EPFL*, *Switzerland*
Dr Madasamy Thangamuthu, *Swiss Federal Institute of Technology Lausanne (EPFL)*, *Switzerland*
Mr Christopher Tiburski, *Chalmers University of Technology*, *Sweden*
Dr Laura Torrente Murciano, *University of Cambridge*, *United Kingdom*
Professor Claire Vallance, *University of Oxford*, *United Kingdom*
Dr Pan Wang, *King's College London*, *United Kingdom*
Dr Susan Weatherby, *Royal Society of Chemistry*, *United Kingdom*
Mr Xiaofei Xiao, *Imperial College London*, *United Kingdom*
Professor Dr Hongxing Xu, *School of Physics and Technology, Wuhan University*, *China*
Miss Anastasiia Zaleska, *King's College London*, *United Kingdom*
Professor Anatoly Zayats, *King's College London*, *United Kingdom*
Dr Chao Zhan, *Xiamen University, iChem*, *China*
Mr Wei Zhao, *King's College London*, *United Kingdom*